The Role of Heat in the Development of Energy and Mineral Resources in the Northern Basin and Range Province

SPECIAL REPORT NO. 13

Second Edition

Published By
GEOTHERMAL RESOURCES COUNCIL
P.O. Box 1350 • Davis, California 95617-1350 U.S.A. • (916) 758-2360

ISSN 0193-5933
ISBN 0-934412-13-8

TABLE OF CONTENTS

Thermogenics and Hydrocarbon Resources 177

Fossil Hydrothermal Systems 243

Regional Geophysics of the Northern Great Basin 305

OVERVIEW OF THE

NORTHERN BASIN AND RANGE PROVINCE

ISOTOPE GEOCHEMISTRY OF PLUTONS IN THE NORTHERN GREAT BASIN

Ronald W. Kistler

U. S. Geological Survey, Menlo Park, California 94025

ABSTRACT

Exposed plutons in the northern Great Basin range in age from Permian (270 m.y.) to Miocene (14 m.y.). Initial $^{87}Sr/^{86}Sr$ values of these plutons range from 0.7035 to 0.7372. For a few of these same plutons, Sm, Nd and $^{143}Nd/^{144}Nd$ have been measured with the resultant $\in Nd(T)$ (initial $^{143}Nd/^{144}Nd$) ranging from +5.1 to -17.3. Oxygen isotopes have a range of $\delta18O$ from +0.6 to +13.0. The type II lead province of Zartman encompasses most of the region, but type Ib and type III leads occur within eastern and western boundaries, respectively. All isotopic systems have regular geographic variations that are independent of the ages of the plutons investigated.

Oxygen isotopes correlate well with lead isotopes, in that plutons with $\delta18O$ greater than +9.0 occur in the type II lead isotopic province. Both isotopic characteristics are independently interpreted to indicate a significant sedimentary rock component in the source of the lead and the oxygen in the plutons.

Neodymium isotopes correlate well with strontium isotopes, with positive $\in Nd(T)$ values occuring in plutons with low initial $^{87}Sr/^{86}Sr$ values and negative $\in Nd(T)$ values occuring in plutons with high initial $^{87}Sr/^{86}Sr$ values. The boundary between plutons with initial $^{87}Sr/^{86}Sr$ greater than 0.7060 and less than 0.7060, interpreted to approximate the western limit of Precambrian sialic crust, is now well constrained in the region. The isotopic boundary coincides with the eastern border of the Antler orogenic highland and the western exposures of Mississippian foreland basin clastic sedimentary rocks.

Combined strontium, neodymium and oxygen isotopic characteristics of plutons with initial $^{87}Sr/^{86}Sr$ less than 0.7060 are compatible with these granitoid rocks having source materials of basalts like those from ophiolite sequences. These same isotopic characteristics in plutons with initial $^{87}Sr/^{86}Sr$ greater than 0.7060 are compatible with these granitoid rocks having source materials of mafic lower continental crust about 1700 m.y. old, altered oceanic basalts, or mixtures of mantle derived basalts and sediments with a cratonal provenance. Scattered muscovite bearing granitoids with initial $^{87}Sr/^{86}Sr$ greater than 0.7100, $\delta18O$ greater than +10.0, and $\in Nd(T)$ of -10 to -17 could have source materials composed entirely of continentally derived sediments.

The western boundary of Precambrian sialic crust, defined by the strontium isotopic boundary, is offset in a right-lateral sense in western Nevada and eastern California. Major displacement apparently occurred during the middle Jurassic, about 150 m.y. ago, with cumulative northwestward displacement of sialic crust for about 300 km.

INTRODUCTION

This report summarizes the results of published and ongoing studies of age, strontium, neodymium, lead, and oxygen isotopic characteristics of plutons in the western United States. Emphasis of the discussion is on the northern Great Basin. All isotopic systems show regular geographic variations that are independent of the ages of the plutons investigated, and each isotopic system yields information about possible source materials for the magmas that formed the plutons.

ISOTOPIC SIGNATURES OF POSSIBLE GRANITIC ROCK SOURCE MATERIALS

During a partial melting of some source rock, Rb/Sr will increase and Sm/Nd will decrease in the melt relative to the residual material. Therefore, since the time of formation of a chondritic earth, successive melting events have resulted in an increase in Rb/Sr and therefore $^{87}Sr/^{86}Sr$, and in a decrease in Sm/Nd, and therefore $^{143}Nd/^{144}Nd$, in crustal materials, relative to undisturbed chondritic material, by partitioning melts into the earth's crust. The residual mantle, conversely, has had a decrease in Rb/Sr, and therefore $^{87}Sr/^{86}Sr$, and an increase in Sm/Nd, and therefore $^{143}Nd/^{144}Nd$, relative to an undisturbed chondritic reservoir (fig.1).

When comparing initial $^{87}Sr/^{86}Sr$ to initial $^{143}Nd/^{144}Nd$ of a pluton, the ratios are reported in the $\in$-notation (DePaolo and Wasserburg, 1976) as deviations in parts in 10^4 from the reference mantle reservoir CHUR. Present-day reference values are $(^{143}Nd/^{144}Nd)_{CHUR}^0 = 0.511836$ and $(^{147}Sm/^{144}Nd)_{CHUR}^0 = 0.1967$ (Jacobsen and Wasserburg, 1980); $\lambda_{Sm} = 6.54 \times 10^{-12} yr^{-1}$. $(^{143}Nd/^{144}Nd)INIT$ is the

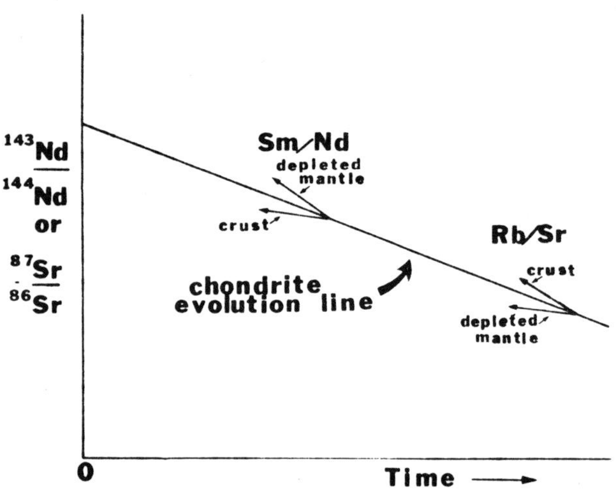

Figure 1. Diagram illustrating the relative
change in Rb/Sr and Sm/Nd and subsequent rate
of change of 143Nd/144Nd and 87Sr/86Sr after
partial melting of a chondritic earth.

measured ratio in the rock, corrected for the
decay since the time of crystallization T. A
similar notation is used for Sr with $(87Sr/86Sr)^0_{UR}$=
0.7045, $(87Rb/86Sr)^0_{UR}$=0.0827 (Jacobsen and
Wasserburg, 1980), and λ_{Rb}=1.42x10^{-11}yr^{-1}

A generalized isotopic model for present-day
ϵ_{Nd} and ϵ_{Sr} values in the mantle and the crust as a
function of crustal ages (fig. 2) is given by
DePaolo (1981). This diagram indicates the Nd and
Sr isotopic signature that would characterize a
partial melt of a crustal or mantle source material.

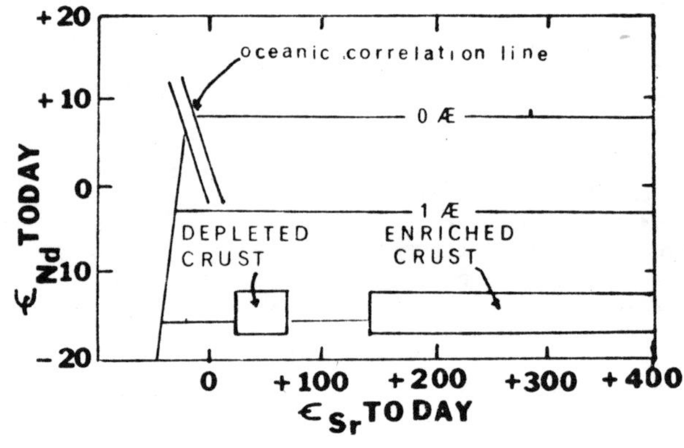

Figure 2. Generalized isotopic model for present-
day ϵNd and ϵSr values in the mantle (oceanic
correlation line) and the crust, as a function
of crustal age (modified from DePaolo, 1981).
One AE is 10^9 yrs.

Zartman (1974) has shown that lead isotope
compositions of plutons in the western United
States can be separated geographically into three
groups. The type 1 leads are comparatively un-
radiogenic with a fairly wide range in isotopic

composition. Rocks with this lead isotopic characteristic are interpreted to have a source of Precambrian crystalline rocks of the lower crust, or
possibly the mantle. The type 11 leads are comparatively radiogenic with a small range in isotopic
composition. The source for plutons with this type
of lead signature is interpreted to be isotopically
homogenized miogeosynclinal sedimentary rocks
eroded from the adjacent Precambrian sialic upper
crust. The type 111 leads are isotopically intermediate between leads of types 1 and 11 and have
a small range in isotopic composition. The source
for plutons with this type of lead is interpreted
to be eugeosynclinal plutonic, volcanic, and sedimentary rocks (fig. 3).

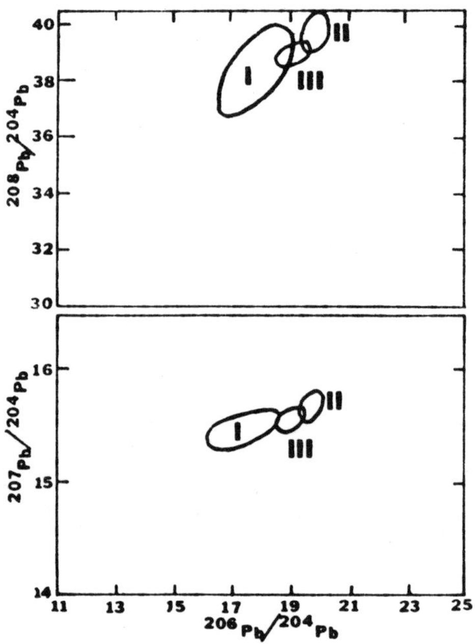

Figure 3. Diagram showing the range of lead
isotopic ratios that characterize the three
lead provinces in the western United States
(from Zartman, 1974).

Oxygen isotopic compositions of rocks are reported as $\delta 18O$, which is defined as the relative
difference in $18O/16O$ between the sample and SMOW
(Standard Mean Ocean Water) standard in per mil.
Primary $\delta 18O$ values of most of the major plutonic
granites of the world lie between +7.0 and +10.0
(Taylor, 1968). Plutons with these oxygen isotopic
values have been interpreted to represent partial
melts of material that was probably never weathered
on the surface of the earth (O'Neil and Chappell,
1977). Plutons with $\delta 18O$ greater than about +9.0
or +10.0 have been interpreted to have source
materials that include weathered sedimentary rocks.

ISOTOPES IN PLUTONS OF THE NORTHERN GREAT BASIN

Figure 4 is a location map for plutons in the
western United States, analysed for Rb, Sr, and

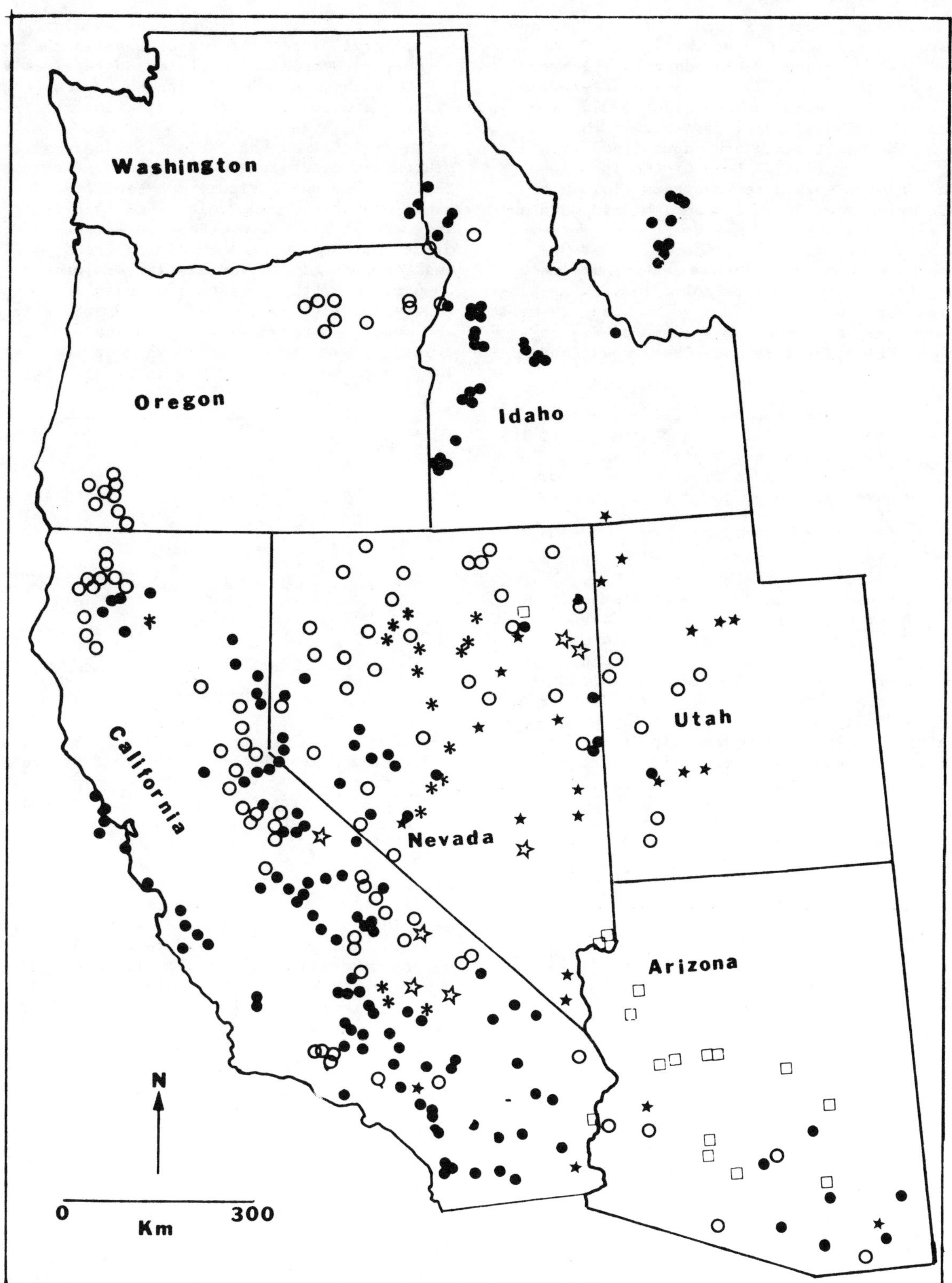

Figure 4. Location map for plutons that have been analysed for Rb, Sr, and initial $^{87}Sr/^{86}Sr$. References are given in the text. Symbols: open square, Precambrian; asterisk, Permian; open star, Triassic; open circle, Jurassic; solid circle, Cretaceous; solid star, Tertiary.

initial 87Sr/86Sr, that are used to locate the edge
of Precambrian sialic crust on the basis of their
initial 87Sr/86 values. Published data are from
Armstrong and others (1977), Kistler and Peterman
(1973,1978), Kistler and others (1973,1981), and
Lee and others (1981). Data for most of the locations in the Great Basin are unpublished (Kistler
and Lee, in preparation). Each of the locations
shown in the Great Basin represents a minimum of
two Rb-Sr determinations. Oxygen isotopic data are
available for all locations shown (Lee and others,
1981; Masi and others, 1981), whereas a few
locations have neodymium isotope determinations
(DePaolo, 1981; Farmer and DePaolo, 1983). Ages
of the plutons range from 270 m.y. to about 14 m.
y. old, and they are shown by symbols on Figure 4
as Permian, Triassic, Jurassic, Cretaceous, and
Tertiary.

If Rb/Sr is plotted against initial 87Sr/86Sr
for these plutons (fig. 5), the data scatter to the
right of two limiting lines. Apparent ages of 1750
m.y. and 450 m.y. calculated from the slopes of
these lines have been interpreted to represent
minimum ages for the source materials of the
granitic rocks (Kistler and Peterman, 1973; 1978).
Because the 1750 m.y. line is defined by specimens
with initial 87Sr/86Sr greater than 0.7060, these
plutons are considered to be emplaced into areas
underlain by Precambrian sialic crust. Plutons
with initial 87Sr/86Sr less than 0.7060 are considered to be emplaced into areas underlain by
mafic crust of oceanic or transitional character
that is early Paleozoic or younger age (Kistler and
Peterman, 1973; 1978). Major element compositions
of the plutons seem to correlate with the strontium
isotopic groups, whereas plutons with initial 87Sr/
86Sr greater than 0.7060 are calc-alkalic, but
those with initial 87Sr/86Sr less than 0.7060 are
calcic (Kistler, 1974). Both the major element as
well as the isotopic characteristics of these
plutons are interpreted to be primarily source
related.

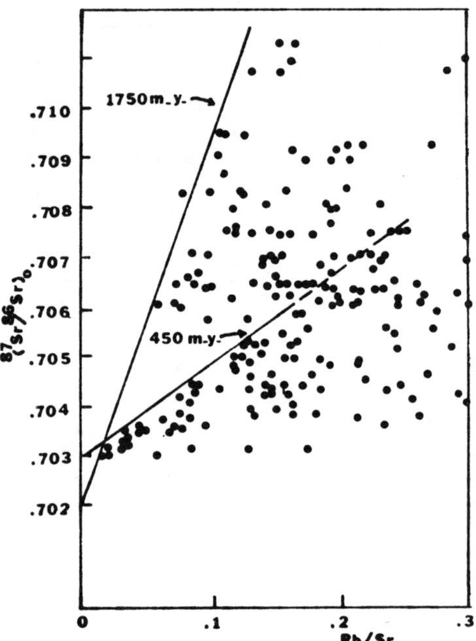

Figure 5 (bottom left). Diagram showing Rb/Sr vs.
initial 87Sr/86 Sr for plutons in the western
United States south of the Snake River Plain.
Only those specimens with Rb/Sr less than
0.3 are included on the diagram.

The initial Nd and Sr isotopic compositions of
plutons in a traverse across the northern Great
Basin are shown in Figure 6. Neodymium determinations are from DePaolo (1981) and from Farmer
and DePaolo (1983), whereas the strontium determinations for the same specimens are from Kistler
and Peterman (1973) and Kistler and others (1981).
Comparison of the data on this diagram to the
generalized isotopic model for Nd and Sr in crust
and mantle sources (fig. 2), indicates an old
crustal source component in those plutons with
large positive ∈ Sr and large negative ∈ Nd values.

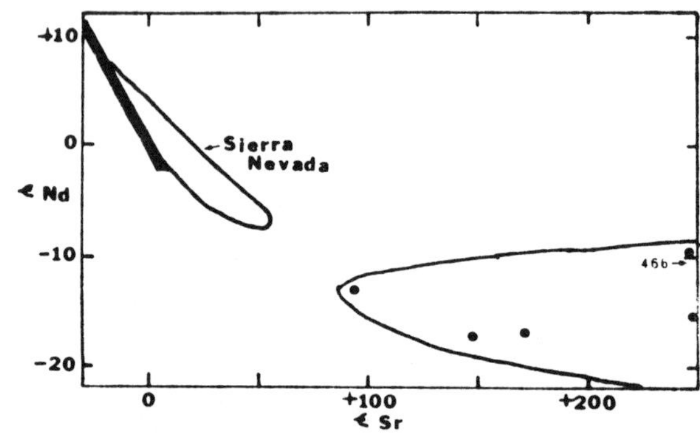

Figure 6. Diagram showing ∈ Nd vs. ∈ Sr for plutons
in the northern Great Basin and Sierra Nevada
(modified from Farmer and DePaolo, 1983). Solid
circles represent values for plutons in the
Ruby and Snake Ranges.

Initial 87Sr/86Sr and δ18O of plutons from the
Great Basin is plotted in Figure 7. The isotopic
signatures of various possible granitic rock source
materials are also shown on the diagram. Plutons
with initial 87Sr/86Sr less than 0.7060 could have
source materials like basalts from ophiolites.
Plutons with initial 87Sr/86Sr greater than 0.7060
but less than 0.7100 could be derived from altered
oceanic basalt or Precambrian crystalline lower
continental crust. Plutons from the Ruby
Mountains and Snake Range, Nevada with initial
87Sr/86Sr greater than 0.7120 are probably derived
entirely from continental sedimentary rocks of
Precambrian age (Kistler and others, 1981; Lee and
others, 1981; Farmer and DePaolo, 1983).

Kistler and Peterman (1973, 1978) and Armstrong
and others (1977) have interpreted the area in the
western Unitied States where plutons have initial
87Sr/86Sr greater than 0.7060 to be underlain by
Precambrian sialic crust. The line that encompasses this area, the initial 87Sr/86Sr= 0.706
line, is taken to approximate the western margin
of Precambrian sial in the region. Existing
published and unpublished strontium isotopic data
closely constrain this line (fig.8). The
boundaries between the three lead isotopic

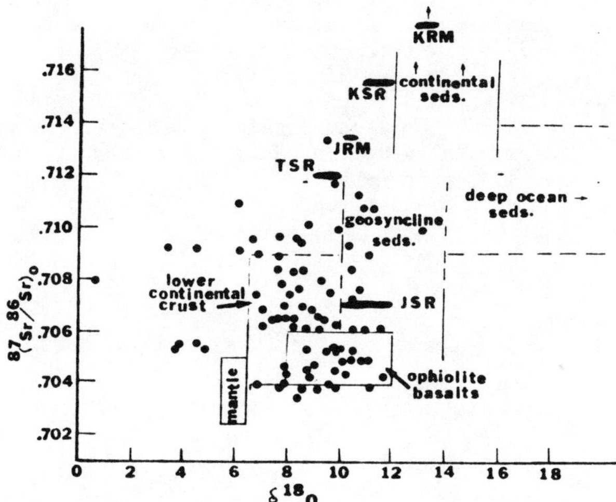

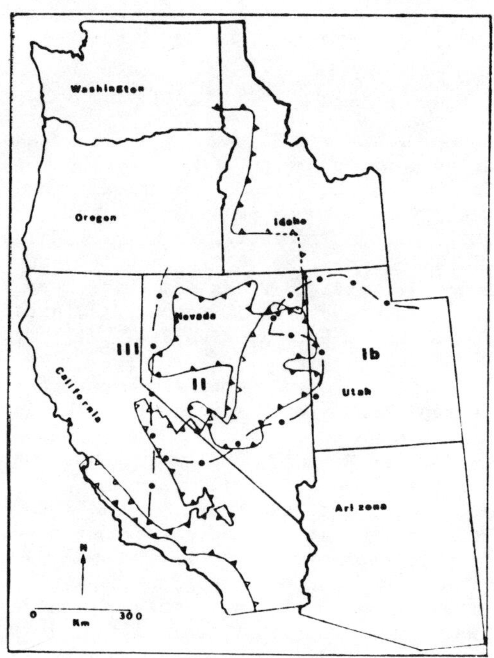

Figure 7. Diagram showing initial 87Sr/86Sr vs.
δ180 in plutons from the Great Basin and fields
for these isotope ratios in some possible
granitic rock source materials. JRM and KRM
represent fields of values for Jurassic and
Cretaceous plutons, respectively, in the Ruby
Mountains, Nevada. JSR, KSR, and TSR represent fields of values for Jurassic, Cretaceous,
and Tertiary plutons, respectively, in the
Snake Range, Nevada.

Figure 8. Outline map of the western United
States comparing Sr, Pb, and oxygen isotopic
compositions of plutons in the northern Great
Basin. Line with open triangles is the initial
87Sr/86Sr=0.706 line. Triangles point toward
the area intruded by plutons with initial
87Sr/86Sr greater than 0.706. Line with solid
triangles encloses an area where plutons have
δ180 values greater than +9.0. The triangles
point toward the area of 180 enriched plutons.
The dashed lines with solid circles are the
boundaries between the lead isotopic provinces
(Ib,II,III) of Zartman (1974).

provinces of Zartman (1974), along with a line that
encloses exposures of plutons with δ180 greater
than +9.0 in the northern Great Basin (Lee and
others, 1981) are also shown on Figure 8. The area
with plutons that have δ180 greater than +9.0
corresponds quite closely with the area of type II
lead of Zartman (1974). Both of these isotopic
characteristics are interpreted to indicate a large
sedimentary rock component in the source materials
of the plutons (Lee and others, 1981; Zartman,
1974). This is surprising, especially in the area
where initial 87Sr/86Sr is less than 0.7060. If
the plutons in this area had sedimentary rocks as
a significant component of their source materials,
as the oxygen and lead data suggest, the sediment
provenance would have to have been mafic rocks with
low 87Sr/86Sr values. However, basalts from ophiolites (fig. 7) could have given these granitoid
plutons their strontium and oxygen isotopic signatures without the necessity of a large sedimentary
component in the source region. The area underlain by Precambrian sialic crust indicated by both
strontium and neodymium isotopes (Kistler and
Peterman, 1973, 1978; Farmer and DePaolo, 1983)
does not correspond exactly with the area underlain by Precambrian crystalline crust indicated by
the type Ib lead (fig. 8). This could be due to a
much smaller sampling density for the lead data.

The western limit of Precambrian sialic crust,
defined by the strontium isotopic boundary, is
offset in a right-lateral sense in western Nevada

and eastern California (fig. 8). Major displacement apparently occurred during the middle
Jurassic, about 150 m.y. ago, with cumulative
northwestward displacement of sialic crust for
about 300 km (Kistler, 1983).

SUMMARY

In the northern Great Basin, combined strontium,
neodymium, and oxygen isotopic characteristics of
plutons with initial 87Sr/86Sr less than 0.7060 are
compatible with these granitoid rocks having source
materials of basalts like those found in ophiolite
sequences. These same isotopes in plutons with
initial 87Sr/86Sr greater than 0.7060 indicate
these granitoid rocks could have source materials
of mafic lower continental crust about 1700 m.y.
old, of altered oceanic basalts, or of mixtures of
mantle derived basalts and sediments with a

cratonal provenance. Scattered muscovite bearing granitoids with initial $^{87}Sr/^{86}Sr$ greater than 0.7100, $\delta 180$ greater than +10.0, and $\epsilon Nd(T)$ of -10 to -17 could have source materials composed entirely of continentally derived sediments.

REFERENCES CITED

Armstrong, R. L., and Suppe, J., 1973, Potassium-argon geochronometry of Mesozoic igneous rocks in Nevada, Utah, and southern California: Geol. Soc. America Bull., v. 84, p. 1375-1392.

Armstrong, R. L., Taubeneck, W. H., and Hales, P. O., 1977, Rb-Sr and K-Ar geochronometry of Mesozoic granitic rocks and their strontium isotopic composition, Oregon, Washington, and Idaho: Geol. Soc. America Bull., v. 88, p. 397-411.

DePaolo, D. J., 1981, A neodymium and strontium isotopic study of Mesozoic calc-alkaline granitic batholiths of the Sierra Nevada and Peninsular Ranges, California: Jour. Geophys. Res., v.86, p. 10470-10488.

DePaolo, D. J., and Wasserburg, G. J., 1976, Nd isotopic variations and petrogenetic models: Geophys. Res. Lett., v. 3, p. 249-252.

Farmer, G. Lang, and DePaolo, Donald J., 1983, Origin of Mesozoic and Tertiary granite in the western United States and implications for pre-Mesozoic crustal structure 1. Nd and Sr isotopic studies in the geocline of the northern Great Basin: Jour. Geophys. Res., v. 88, no. B4, p. 3379-3401.

Jacobsen, S. B., and Wasserburg, G. J., 1980, Sm-Nd isotopic evolution of chondrites: Earth and Planet. Sci. Lett., v. 50, p. 139-155.

Kistler, R. W., 1974, Phanerozoic batholiths in western North America: A summary of some recent work on variations in time, space, chemistry, and isotopic compositions: Ann. Rev. Earth and Planet. Sci., v.2, p. 403-418.

Kistler, R. W., 1983, Configuration of the pre-middle Jurassic continental margin in the western United States: Geol. Soc. America abs. with prog., v. 15, no. 5, p. 272.

Kistler, R. W., Ghent, E. D., and O'Neil, J. R., 1981, Petrogenesis of garnet two-mica granites in the Ruby Mountains, Nevada: Jour. Geophys. Res., v. 86, p. 10591-10606.

Kistler, R. W., and Peterman, Z. E., 1973, Variations in Sr, Rb, K, Na, and initial $^{87}Sr/^{86}Sr$ in Mesozoic granitic rocks and intruded wall rocks in central California: Geol. Soc. America Bull., v. 84, p. 3489-3512.

Kistler, R. W., and Peterman, Z. E., 1978, Reconstruction of crustal blocks of California on the basis of initial strontium isotopic compositions of Mesozoic granitic rocks: U. S. Geol. Survey Prof. Paper 1071, 17p.

Kistler, R. W., Peterman, Z. E., Ross D. C., and Gottfried, D., 1973, Strontium isotopes and the San Andreas fault, in Kovach, R. L., and Nur, Amos, eds., Proceedings of the conference on tectonic problems of the San Andreas fault system: Stanford Univ. Pubs. Geol. Sci., v. XIII, p. 339-347.

Lee, D. E., Kistler, R. W., Friedman, I., and Van Loenen, R. E., 1981, Two-mica granites of northeastern Nevada: Jour. Geophys. Res., v. 86, p. 10607-10616.

Masi, Umberto, O'Neil, J. R., and Kistler, R. W., 1981, Stable isotope systematics in Mesozoic granites of central and northern California and southwestern Oregon: Contrib. Mineral. Petrol., v. 76, p. 116-126.

O'Neil, J. R., and Chappell, B. C., 1977, Oxygen and hydrogen isotope relations in the Berridale batholith: Jour. Geol. Soc. London, v. 133, p. 559-571.

Zartman, R. E., 1974, Lead isotopic provinces of the western United States and their geologic significance: Econ. Geol., v. 69, p. 792-805.

Regional Hydrology in the Northern Basin and Range Province

M.D. Mifflin

Water Resources Center
Desert Research Institute
1500 E. Tropicana Blvd., Ste 201
Las Vegas, NV 89109

ABSTRACT

The theory of groundwater flow provides a useful approach for understanding groundwater hydrology of the Basin and Range Province. Theoretical and conceptual frameworks are useful for interpreting sparse data, and the approach may help to place into context geothermal systems which occur within the province. Recharge distribution, fluid potential theory and data, terrain capacity to transmit flow, geologic structure, distribution of discharge, hydrogeochemistry, and groundwater budgets combine to permit delineation and characterization of groundwater flow systems. Regional groundwater hydrology of the province is reviewed within the context of flow system delineation, and implications with respect to geothermal systems are pointed out. Regions vary with respect to terrain capacity to transmit flow, moisture availability for recharge in the mountains and basins, and structure; the character of groundwater flow systems vary accordingly. The distribution and frequency of the occurrence of thermal water at and near land surface also can be generally related to the flow system characteristics. The majority of thermal groundwater, when viewed from this perspective, seems related to depth of circulation; local shallow heat sources may be less common local anomalies.

PRECENOZOIC TECTONIC EVOLUTION OF NORTHEASTERN NEVADA

R. C. Speed

Department of Geological Sciences
Northwestern University
Evanston, Illinois 60201

ABSTRACT

Nevada has undergone three major sequential tectonic regimes since late in Precambrian time: development of a passive continental margin formation in late Proterozoic and Early Cambrian time, maintenance of a passive continental margin together with collisions of displaced terranes through mid-Triassic time, and active margin phenomena ever since. The record of pre-Jurassic continental margin tectonics is well preserved in Nevada but is fragmentary elsewhere in western North America due to widespread tectonic erosion of the sialic margin. The edge of sialic Precambrian North America lies in Nevada. It formed mainly in the late Proterozoic west of an unusually wide zone of attenuated sialic crust. The sialic edge may have persisted in Nevada as the general ocean-continent transition until early in the Triassic. In that duration, collisions of the edge with migrating arc terranes (Sonomia, Antleria) caused the obduction of the oceanic Golconda and Roberts Mountains allochthons above the slope and outer shelf in Triassic and Mississippian times, respectively. After each collision, an oceanic configuration was regained at the sialic edge, perhaps due to thermal contraction of the arc lithosphere. Mesozoic active margin features in Nevada relate to subduction near the western margin of the Sierra Nevada in California. They include an extensive continental arc that persisted in western Nevada from late in the Triassic to the Late Cretaceous and a deforming foreland that spanned the rest of the state. The foreland tectonics include heterogeneous zones of thin-skinned contraction among cover rocks and heating and ductile deformation of deep seated rocks. Contraction of the foreland may have occurred across Nevada throughout Jurassic and Cretaceous times. Certain ancient structures in Nevada have been important in controlling the position and form of younger structures.

Introduction

The Phanerozoic* evolution of the North American continent and adjacent oceans in what is now Nevada was governed by three sequential tectonic regimes. The first, beginning late in Proterozoic time and continuing to Middle Cambrian time, created a passive margin to western sialic North America (Stewart, 1976). It caused the rifting and drifting away of a part of the continent of unknown size and the growth of oceanic lithosphere against the new sialic edge. The second regime maintained a passive continental margin of western North America from Middle Cambrian to Middle or Late Triassic time but permitted collisions of outboard terranes with the sialic margin in Mississippian and Early Triassic times (Speed, 1983). Since late in Triassic time, western North America has existed in a regime of active margin tectonics (Hamilton, 1969). This third and currently active regime has included diverse phenomena that have varied with time and position: subduction of oceanic lithosphere below the continent, phases of highly oblique convergence and suture-zone or intra-arc spreading, ridge-trench collision, growth of a continental arc, major foreland contraction and extension, and the accretion of displaced terranes to the sialic edge.

Although similarly eventful histories probably occurred along the entire western margin of North America, a record of pre-Jurassic events is best preserved in Nevada, and in fact, many elements of the record are known only in Nevada. The pre-Jurassic margin of North America in Nevada evidently escaped strong tectonic erosion which elsewhere removed and rafted away sizeable fragments of the sialic continent and early accreted terranes (Speed, 1983). Thus, Nevada provides an almost unique glimpse into the past of marginal western North America.

Pretertiary Tectonostratigraphy
of Northwestern Nevada

Figure 1 shows the tectonic affiliations of Pretertiary rocks of northwestern Nevada and vicinity except in regions of nearly continuous Cenozoic cover (unit 1). Unit 7 is Precambrian sialic North America and its Paleozoic and Mesozoic cover. The unit has probably maintained near-coherency in Phanerozoic time but has undergone Mesozoic contraction and Cenozoic extension. The western 150 km or so of sialic North America (unit 7) are overlain by major allochthons of greatly displaced Paleozoic oceanic rocks. Such rocks compose unit 6a and include the Roberts Mountains allochthon which was obducted across the

*began at about 700 mybp.

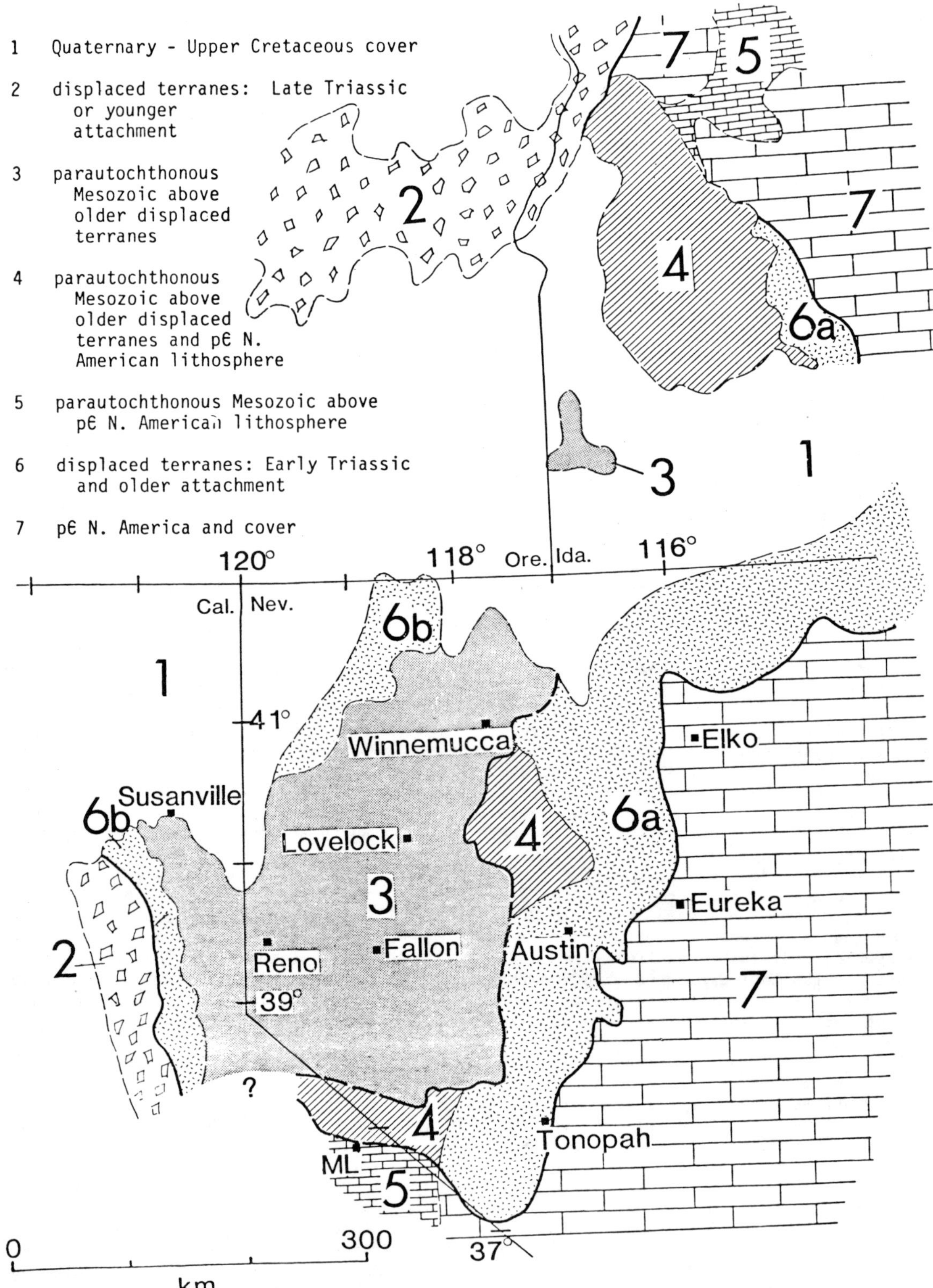

Figure 1. Map of northwestern Nevada and vicinity showing tectonic affiliations of principal outcrop units, emphasizing Pretertiary rocks where possible and excluding basin-range structure. Heavy lines are tectonic boundries with moderate to major displacement. ML is Mono Lake.

12

slope and outer shelf of North America in Mississippian time and the Golconda allochthon, which was emplaced early in Triassic time.

Deformed Paleozoic rocks also exist in the Black Rock Desert region and northern Sierra Nevada (unit 6b, Fig. 1). These are also thought to be greatly displaced with respect to their sites of attachment to North America. However, they are probably structurally discrete from the obducted allochthons of unit 6a and probably occupy the lower structural levels of the Sonomia microplate (Speed, 1979) that collided with the passive margin of North America in Triassic time.

Units 3-5 (Fig. 1) are composed of Mesozoic sedimentary and igneous rocks that were deposited or intruded at or near (<300 km) their present sites after the collision of Sonomia and the onset of active margin tectonism. The threefold division of Mesozoic rocks in Figure 1 is based on the different tectonostratigraphy of their basements. Unit 3 includes mainly basinal strata continental arc rocks that were deposited on and intruded into the Sonomia microplate (Speed, 1978). Sonomia's cryptic suture is approximately coincident with the eastern boundary of unit 3. Unit 4 consists mainly of shelfal and carbonate platform deposits that overlie the Roberts Mountains and Golconda allochthon of unit 6b. Unit 5 is constituted by scarce remnants of Mesozoic rocks that appear to be depositional on or intrusive into ensialic North America. Unit 2 consists of displaced and suture-related terranes (Saleeby and Sharp, 1980; Schweickert, 1981) that attached to the edge of North America and early accretants during active margin tectonism in Late Triassic to Late Cretaceous time.

Figure 8 provides a sectional view of the tectonostratigraphic stacking of units 3-7 from the Jackson to the Shoshone Mountains in northwestern Nevada.

Edge of Sialic North America

Lines that approximate the present edge of contiguous sialic Precambrian North America in Nevada, Idaho, and California and probable ages at which segments of the margin were generated are shown in Figure 2. The sialic edge is a basement feature and is generally cryptic due to burial by younger rocks and nappes or obscuration by magmatism and metamorphism. The margin's surface trace is estimated by outermost outcrops of autochthonous continental platform or shelf facies (Roberts and others, 1958; Kay and Crawford, 1964; Matti and McKee, 1977), by ratios of initial Sr and Pb isotopes and mineralogy of autochthonous Phanerozoic magmatic rocks (Kistler and Peterman, 1973; Doe, 1973; Armstrong and others, 1977; Zartman, 1974; Miller and Bradfish, 1980), and locally (south of 41°N), by maxima in gravity gradients that indicate rapid lateral changes in crustal thickness and/or composition (Speed, 1983).

The present edge of Precambrian sialic crust evolved from a late Proterozoic passive margin that developed along entire western North America.

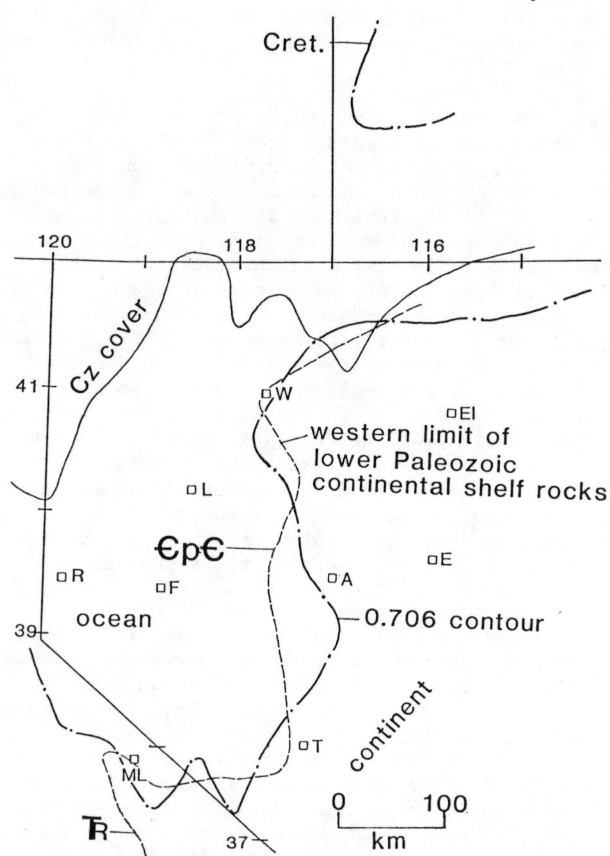

Figure 2. Approximate position of present edge of Precambrian sialic continental North America defined by 1) initial Sr isotopic ratio in Mesozoic igneous rocks with value of 0.706 and 2) present limit if lower Paleozoic shelf strata. Age coding gives time of formation of sialic edge. Letters abbreviate town names as given in Figure 1.

This ancient sialic edge probably existed at varied distances outboard of the present edge. The present north-trending edge of sialic crust in Nevada (Fig. 2) may be the only preserved segment of the passive margin, its absence elsewhere due to tectonic removal during Mesozoic and Cenozoic active margin tectonics (Speed, 1983).

The existence of an ancient passive margin along the Pacific reach of North America was interpreted from autochthonous and parautochthonous upper Precambrian and lower Paleozoic shelf strata that crop out between the platform-shelf hinge and the present sialic edge (Fig. 2) by Stewart (1972, 1976) and Stewart and Suczek (1977). Lower strata of this sequence compose a terrigenous marginward-thickening clastic prism that reflects arching, rifting, and early drifting phases of passive margin development between about 550 and 700 myBP. Such strata lie above crystalline rocks and older Precambrian beds with highly discordant structural trends suggesting major reconfiguration of the continental margin in late Precambrian time. The clastic prism is succeeded by rocks of shallow marine subsiding shelf environments (Kay and Crawford, 1964; Matti and McKee, 1977).

The sialic crust in the region affected by passive margin tectonics (west of the platform-shelf hinge in central Utah) is between 30 and 35 km thick except where the Mesozoic continental arc (Fig. 6) is superposed on it, there producing crustal thicknesses as great as 50 km. The prevalent thicknesses are less than those in the craton of central Utah (Smith, 1978) east of the Paleozoic subsiding shelf. The region of thinner sialic crust possibly reflects an unusually broad zone of rifting and attenuation created during passive margin formation. Although the existence of upper Precambrian diamictite and mafic igneous rocks in the clastic wedge section just west of the shelf-craton hingeline in Utah support the idea of a broad zone of rifting, the magnitude of attenuation cannot be calculated because of probable strong Mesozoic thickening and Cenozoic thinning of the sialic crust west of the hingeline (Armstrong, 1972).

Evidence that the late Precambrian - Early Cambrian passive margin may be preserved in Nevada (Fig. 2) comes from differences among autochthonous and parautochthonous lower Paleozoic strata that can be interpreted as inner and outer shelf facies (Kay, 1960; Rowell and others, 1979; Matti and McKee, 1977). Moreover, shoalwater carbonate rocks that locally form outermost outcrops of the Paleozoic shelf succession are considered deposits on a ridged outer shelf edge. The tectonic removal of an outer zone of continental rock including the Precambrian sialic edge at other sites in western North America (Idaho and central California in Figure 2) is inferred from the absence of the pre-Jurassic tectonostratigraphic units that are sutured against or obducted on the sialic continent in Nevada (Speed, 1983) and the truncation of the Paleozoic craton-shelf hingeline and projected Paleozoic structures by the present sialic edge (Hamilton and Myers, 1966).

Passive Margin Accretionary Tectonics

The Foothills suture in the Sierra Nevada (contact of units 2 and 6a, Fig. 1) is a major accretionary surface that separates terranes that attached to the passive margin of North America before Middle Triassic time from those that arrived during the later active margin phase (Saleeby and Sharp, 1980; Schweickert, 1981; Saleeby, 1982).

East of the Foothills suture, three accretionary terranes are exposed, the Roberts Mountains and Golconda allochthons and Sonomia (Figs. 3 and 4). A cryptic fourth terrane, Antleria, is postulated. The Roberts Mountains allochthon consists of a tectonic assemblage of pelagic, hemipelagic, turbiditic, and volcanic rocks of early Paleozoic age and probable oceanic derivation. It laps over lower Paleozoic strata of the North American continental shelf at least 130 km from the sialic edge and was almost certainly emplaced from the west early in Mississippian time (Roberts

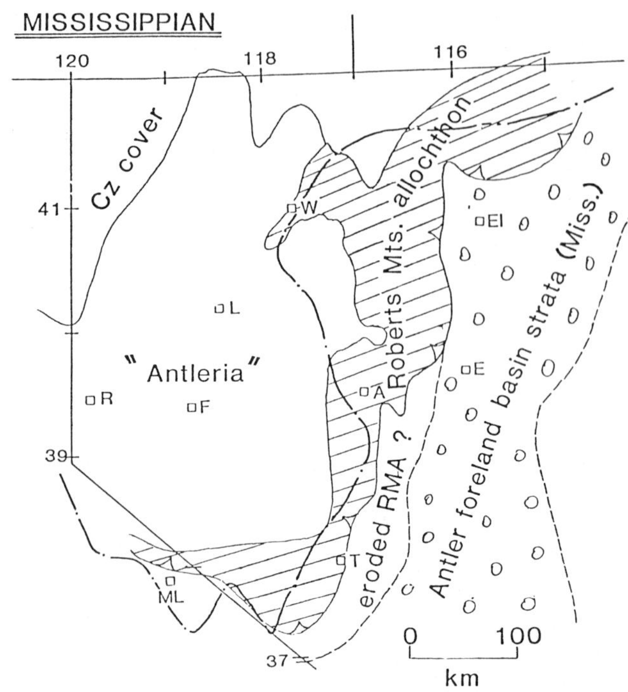

Figure 3. Major tectonic features resulting from arc-passive margin collision in Mississippian time. Dash-dot line is isotopic 0.706 line.

and others, 1958; Smith and Ketner, 1968; Stewart and Poole, 1974; Poole, 1974; Speed and Sleep, 1982; Dickinson and others, 1983). The Golconda allochthon possesses similar rocks and architecture to the Roberts Mountains except that the rocks are of Mississippian to Permian age (Silberling, 1973; Stewart and others, 1977; Speed, 1977, 1979; Miller and others, 1981, 1983; Snyder and Brueckner, 1983). It was emplaced in Early Triassic time at least 100 km inboard of the sialic edge and above the earlier Roberts Mountains allochthon and the late Paleozoic and Early Triassic cover to the Roberts (Speed, 1979).

Sonomia (Fig. 4) is thought to be a lithospheric fragment of Paleozoic arc-related tectonostratigraphic units, surmounted by a Permian magmatic arc (Speed, 1979). It collided with the edge of sialic North America early in the Triassic. Its main exposures are at the microplate margins where late Mesozoic deformation has transported Sonomia's rocks to the surface, either by imbricate thrusting as in Nevada or by rotation and flattening as in the northern Sierra Nevada. The central regions of Sonomia are deeply buried below thick Triassic flysch and continental arc volcanics that succeeded Sonomia (Speed, 1978). Sonomia is the earliest of exposed displaced lithospheric terranes to have attached to western North America. It is a question whether Sonomia was a contiguous mass upon initial collision or whether it includes sequentially accreted fragments. Paleomagnetic data show no anomalies in the area of Sonomia with respect to North America

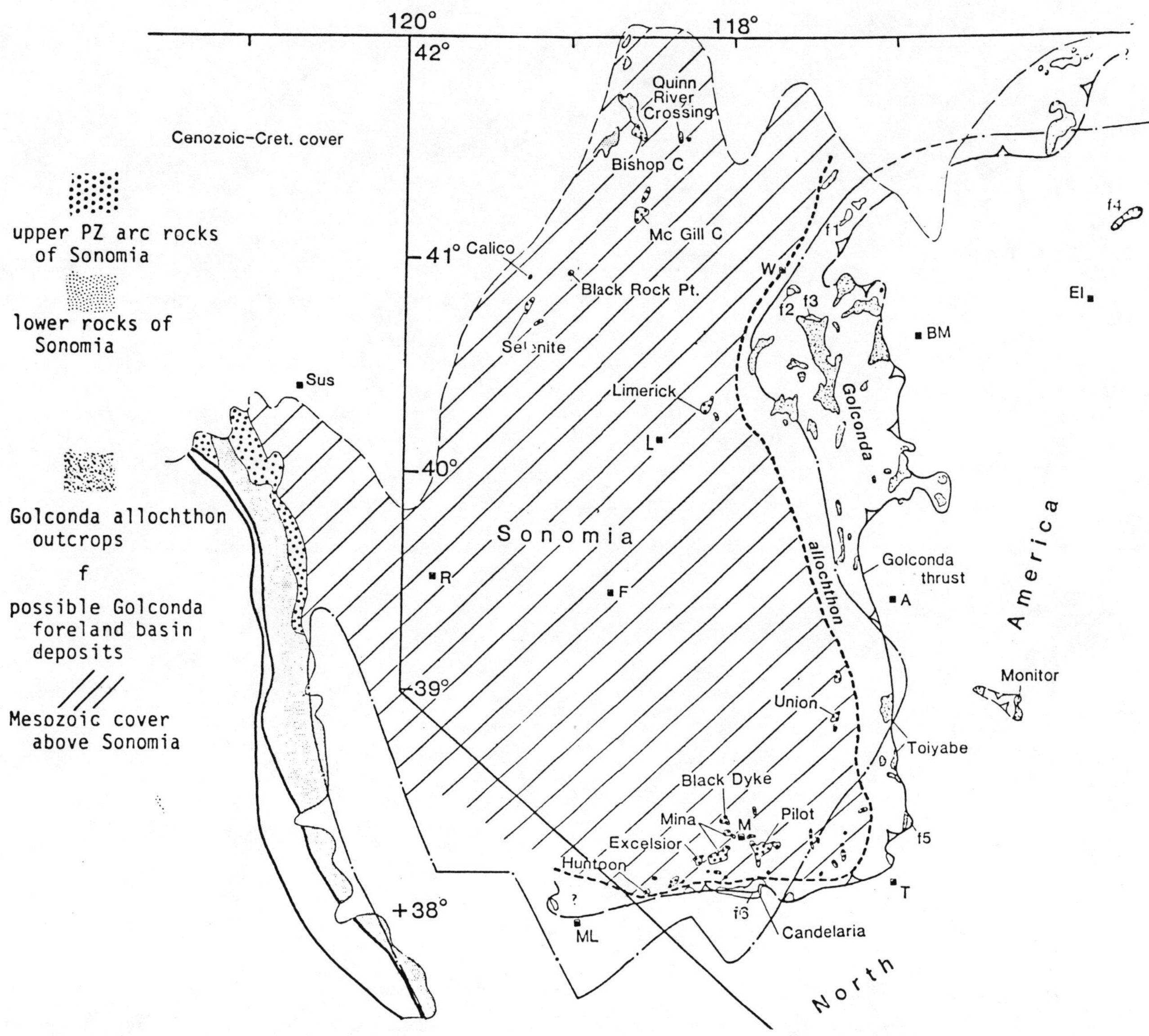

Figure 4. Major tectonic features resulting from arc–passive margin collision in Early Triassic time. Place names indicate outcrop areas of rocks of Sonomia. Heavy dash line is approximate locus of suture between Sonomia nad North America. Dash–dot line is isotopic 0.706 line.

in Cretaceous and possibly Late Jurassic times (Gromme and others, 1967, 1982; Hannah and Verosub, 1980; Russell and others, 1982; Frei and others, 1982), thus permitting western Sonomia to have been autochthonous since at least those times. Moreover, the absence of recognized major sutures within the perimeter of Sonomia suggests that it may have been initially coherent.

Because of the closeness in minimum ages of rocks in the Golconda allochthon and Sonomia, their long reach of probable contact, and the occurrence of copious Early Triassic magmatic arc-derived debris in a foreland basin of the Golconda allochthon at Candelaria (Fig. 4), the emplacement of the two terranes was probably paired and constituted the Sonoma orogeny of Silberling and Roberts (1962).

The emplacements of both the Golconda and Roberts Mountains allochthons had similar manifestations: transport from an oceanic region as a predeformed tectonic mass, absence of related magmatism and metamorphism within the continent, and lack of pervasive crustal shortening or mountain-building within the continental crust. Deformation within the overridden continental shelf strata consists only of local shear strain and(or) thrust duplexing in a thin zone below the obducted allochthon (Gilluly and Gates, 1965; Kay and Crawford, 1964).

15

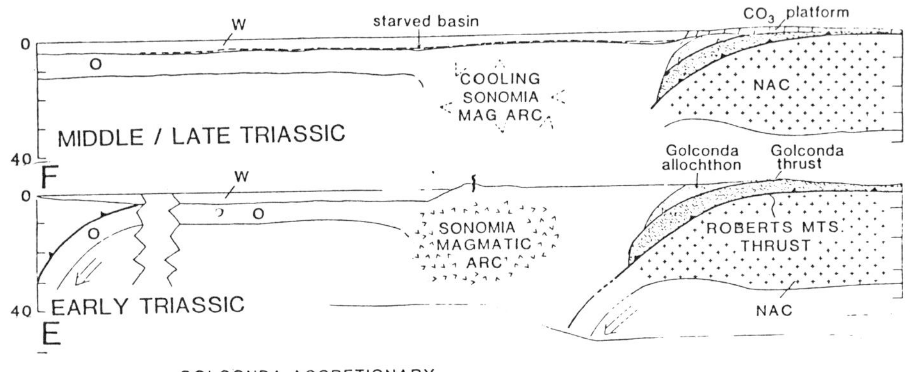

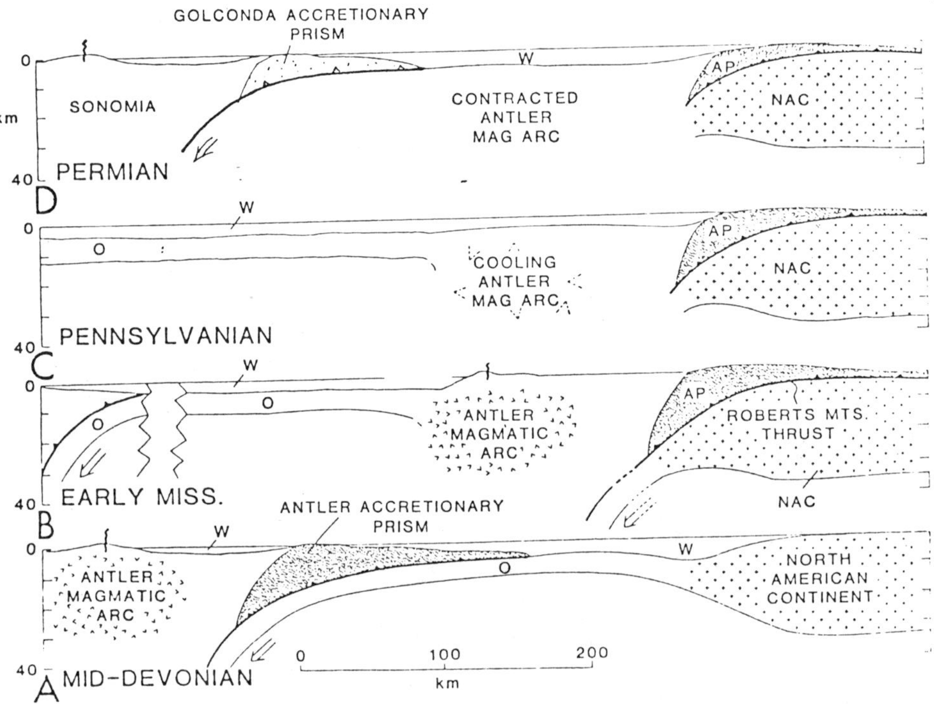

Figure 5. Model of Antler and Sonomia orogenies as collisions of arc systems with the passive margin of western North America, modified after Speed (1979) and Speed and Sleep (1982).

Such phenomena indicate that the margin of North America remained passive during the two obductions. Because Sonomia was active as a magmatic island arc until at least Late Permian time, the Golconda allochthon can be regarded as an accretionary prism in Sonomia's forearc that was driven across the slope and outer shelf of North America during its collision with the arc system (Speed, 1977, 1979). Figures 5d to 5f illustrate a model of such events. In Figure 5d, Sonomia is shown as an island arc system migrating with closing and lateral components along coastal North America (Speed, 1977, 1979). The arc surmounted a subduction zone in which the downgoing slab was oceanic lithosphere attached to the passive margin of North America. Closure ceased when continental lithosphere started down below Sonomia by which time the forearc, the Golconda allochthon, was almost fully emplaced on the outer continental shelf. If the oceanic slab attached to Paleozoic North America broke off and sank upon cessation of closure, a finite but probably minor width of

sialic continental edge may have been taken into the mantle with it. Upon welding to North America, a new ocean-facing convergent zone probably developed somewhere west of Sonomia, and Sonomia underwent thermal contraction and subsidence due to loss of subduction-related heating. The origin of the deepwater Triassic successor basin that is approximately coextensive with Sonomia is thus explained.

Because of the tectonic similarity of the emplacements of the Roberts Mountains and Golconda allochthons, a similar collisional process is envisioned for the Roberts. The Roberts is also regarded as an accretionary prism (Dickinson, 1977; Speed and Sleep, 1982) that was underrun by the continental shelf before final collision between the continent and a migrating island arc system (Figs. 5a-5c). The Antler magmatic arc, however, is a postulate because it is nowhere exposed. The Antler arc is assumed to have thermally contracted, like Sonomia, because upon collision, the subduction zone jumped west and the arc lost its heat source (Speed and Sleep, 1982). The contracted Antler arc may have been completely subducted below Sonomia.

There is no evidence for the direction or magnitude of transport of Sonomia and Antleria and their respective forearcs relative to their position of attachment in Nevada. Magnetizations in Sonomian rocks measured at sites in California (Hannah and Verosub, 1980) and in Nevada (Russell and others, 1982) were reset during late Mesozoic magmatism such that original latitude anomalies are unknown. The turbidites within fault packets in the Golconda and Roberts Mountains allochthons are certainly or probably of North American provenance (Speed, 1977; Miller and others, 1983) but the dimensions of the region over which they were swept up before obduction onto the shelf is unknown. The lack of correlation of beds in many adjacent packets and the tectonic intercalation of packets of mainly pelagic and of mainly turbiditic facies suggest a large lateral distance along the passive margin.

A major effect of the Mississippian collision was the generation of an asymmetric foreland basin (Fig. 3) with amplitude ≤ 3.5 km that rimmed the continentward edge of the Roberts Mountains allochthon (Poole, 1974). The basin probably reflects elastic or elasticoviscous downflexing of strong continental lithosphere during vertical loading by the allochthon and broadening during progressive sedimentation which widened the region of loading (Speed and Sleep, 1982). Certain stratigraphic features below the foreland basin fill suggest that an upbulge migrated eastward ahead of the downwarp, as predicted by the theory of flexure.

In contrast to the extensive foreland basin developed during the Mississippian continental margin event, foreland basin deposits associated with the Triassic Golconda allochthon are recognized only adjacent to the southern third of that allochthon (Fig. 4). This may be because the northern Golconda allochthon was too small to cause significant flexure, because it remained submarine and provided no orogenic sediment to such a basin, or because it was later thrust over related foreland basin strata.

The principal effects of the accretion of Sonomia to the passive sialic margin were the addition of large girth to the morphologic continent and the generation of a deepwater basin west of the Sonomian suture and the edge of sialic North America. The Triassic basin was successor to and approximately coextensive with Sonomia and accumulated early pelagic and hemipelagic sediments followed by great thicknesses of Late Triassic terrigenous turbidite (Speed, 1978). The subsidence of the successor basin is hypothesized to have been due to thermal contraction of the lithosphere of Sonomia (Speed, 1977, 1979).

Extensive Cenozoic rocks north of the Pretertiary outcrops of northwestern Nevada and California presumably cover one or more sutures between late Mesozoic displaced terranes (unit 2 of Fig. 1) and Sonomia and North America. Such terranes are partly represented in the Blue Mountains of Oregon and are there attached to Precambrian late Mesozoic sialic North America. This suture was created by rafting away from southeastern Oregon and western Idaho a fragment of sialic North America, terranes that were earlier accreted to the passive margin, and perhaps, other terranes attached early in the Mesozoic during the active margin phase (Speed, 1983).

<u>Active Margin Tectonics</u>

An active margin developed on western North America (Hamilton, 1969; Burchfiel and Davis, 1972) in Middle or Late Triassic time. The subduction trace between an east-dipping oceanic slab and the morphologic continent then existed at or west of the Foothills suture (Saleeby and Sharp, 1980; Schweickert, 1981) in the Sierra Nevada (Fig. 1) such that at least part of Sonomia and perhaps other early accreted terranes were incorporated into Triassic North America.

During the passive-to-active margin transition in the Triassic, the outer 100-200 km of Precambrian sialic North America and obducted allochthons subsided as carbonate-deltaic shelf (Silberling and Wallace, 1969; Nichols and Silberling, 1977; Speed, 1978). The present distribution of such strata (Fig. 6) probably is a good approximation of the original shelf configuration except for their southern outcrops which may be oroclinal (Albers, 1967), although this is disputed by Oldow (1983). The successor basin also accumulated basinal sediment during the Triassic (Fig. 6), but the distribution of such sediment has been more perturbed by Mesozoic deformation than that of shelfal strata. A set of thrust sheets of Mesozoic volcanic rocks partly overlies the basinal strata south of Lovelock (Fig. 6). The third major Mesozoic rock facies is Triassic to Cretaceous volcanic and intrusive rocks that made up a long-standing continental magmatic arc (Fig. 6).

Sections below discuss several of the main effects of active margin tectonism on Nevada and Idaho: the Mesozoic continental arc, thin-skinned foreland contraction of Jurassic-Paleogene age, and tectonic erosion.

<u>Mesozoic Continental Arc</u>: A continental magmatic arc (Hamilton, 1969) emerged in eastern California and western Nevada (Figs. 6 and 7) late in Triassic time, probably between 215 and 235 myBP according to oldest dates of silicic volcanic and plutonic rocks obtained so far (Kistler, 1966; Brook, 1980; R. W. Kistler, 1982, oral commun.; Wright and others, 1983). The arc crossed the suture between earlier accreted terranes and sialic North America at about $38°N$ (Speed, 1978). Manifestations of the arc are silicic and intermediate volcanogenic (breccia, lava, sediments) and plutonic (granodiorite, quartz diorite, quartz monzonite) rocks that range as young as about 80 my. The arc region is defined by the existence of Mesozoic volcanic rocks and by a high proportion of plutonic rocks (at least half) among Preterti-

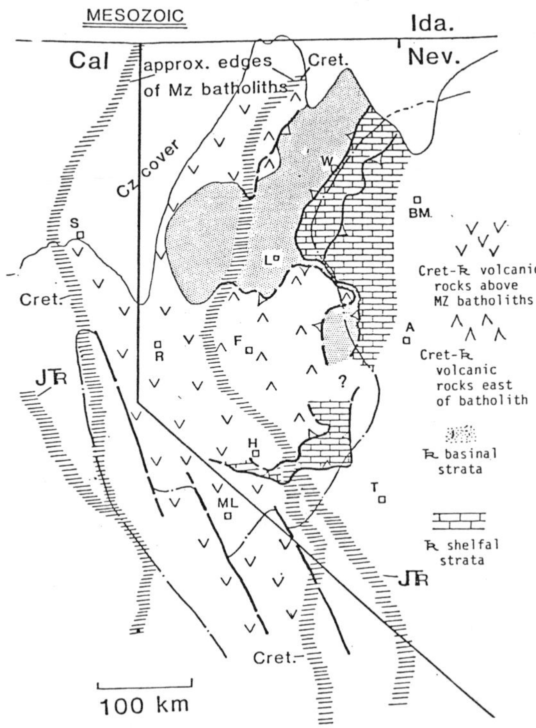

Figure 6. Map showing paleogeographic terranes of early Mesozoic marine and younger volcanic and intrusive rocks of the western Great Basin. Dash-dot line is isotopic 0.706 line.

ary exposures (Speed, 1983).

The locus of Mesozoic continental arc magmatism varied within eastern California and western Nevada (Fig. 6), although its track is imprecisely known because crystallization ages of both volcanic and plutonic rocks older than Late Cretaceous age are not sufficiently well known. However, the arc may have existed nearly in place between 38° and 39° with only minor east-west migration and with almost continuous magmatism (Evernden and Kistler, 1970; Stern and others, 1981; Chen and Moore, 1982). An exception is a possible magmatic lull in Early Cretaceous time. North of that latitude band, the existence of an autochthonous Cretaceous plutonic arc is well established, but the existence of Triassic and Jurassic precursors can be inferred only from scattered volcanic remnants (Speed, 1983; Russell, 1983). South of 38°N, the locus of Cretaceous magmatism seems to diverge from the belt of earlier Mesozoic plutonic and volcanic rocks (Fig. 6). Cretaceous and possibly latest Jurassic magnetization directions in the region of the continental arc show no significant anomalies with respect to North America (Gromme and others, 1967; Hannah and Verosub, 1980; Russell and others, 1982; Frei and others, 1982; Geissman and others, 1984).

The main effects of arc development on the continent were crustal thickening and intra-arc deformation. Where continental arc rocks invade the sialic continent (Fig. 6), crustal thickness is 5-15 km greater than that of normal craton, such as the presumably undeformed crust of the Colorado Plateau (Eaton, 1963; Johnson, 1968; Pakiser and Jones, 1975; Prohdehl, 1979; Smith, 1979). Moreover the crust below Sonomia thickens substantially below the Mesozoic continental arc (Eaton, 1963; Prohdehl, 1979; Stauber, 1980). Chemical and isotopic studies imply that major magma additions to the crust were mantle-derived but reflective of the different lithospheres in which they were generated and/or contaminated (Hamilton, 1969; Kistler and Peterman, 1973, 1978; Miller and Bradfish, 1980).

Deformation related to the continental arc consists of protracted shortening normal to the arc, arc-parallel shear, and at least local arc-vergent thrusting at its backside. Progressive ENE-coaxial shortening of Jurassic and mid-Cretaceous age in arc-related strata, maximum at about 80% in older rocks, was demonstrated by Tobisch and others (1982). Their arc-parallel flattening plane parallels the most pervasive Mesozoic axial planar fabric orientations of the arc region between 37° and 40°N (Kistler and Bateman, 1966; Schweickert, 1981; Saleeby and Sharp, 1980; Speed, 1978; Nokelberg and Kistler, 1982). Shear on mainly discrete NNW-striking planes led to Mesozoic right-lateral displacements of decreasing magnitude east toward the arc-foreland boundary (Speed, 1978, 1983; Kistler and others, 1971, 1980). Such motions may be the primary cause of the clockwise Mina deflection of Pretertiary features in western Nevada (Albers, 1967; Stewart and others, 1968) and of the northerly salient of sialic continental crust within the central Sierra Nevada (Kistler and Peterman, 1978; R. W. Kistler, 1982, written commun.). The partitioning of shear and normal displacements among discrete structures evidently occurred in this ancient arc as in some modern arcs during convergence with moderate obliquity (Fitch, 1972; Walcott, 1978; Beck, 1983). Arcward thrusting of Cretaceous age at the backside of the continental arc has been recognized in northwestern Nevada (Russell, 1983) and could be widespread to the south.

Magmatism and metamorphism at deep crustal levels also occurred during Mesozoic time within the continental foreland from the arc nearby as far east as the Paleozoic craton-shelf hingeline. The timing of such phenomena is not precisely known but seems to have spanned 70 to 165 myBP with possible peaks of plutonism at 155-165 and 70-80 myBP (Armstrong and Suppe, 1973; Allmendinger and Jordan, 1981; Lee and others, 1983). Thermal phenomena of the foreland differ from those of the arc by 1) the restriction to relatively deep levels and absence of evident volcanism (except for one site, the Late Jurassic Pony Trail Group, Muffler, 1964), 2) the smallness and discreteness of individual plutons, and 3) the compositions of foreland plutons indicating cru-

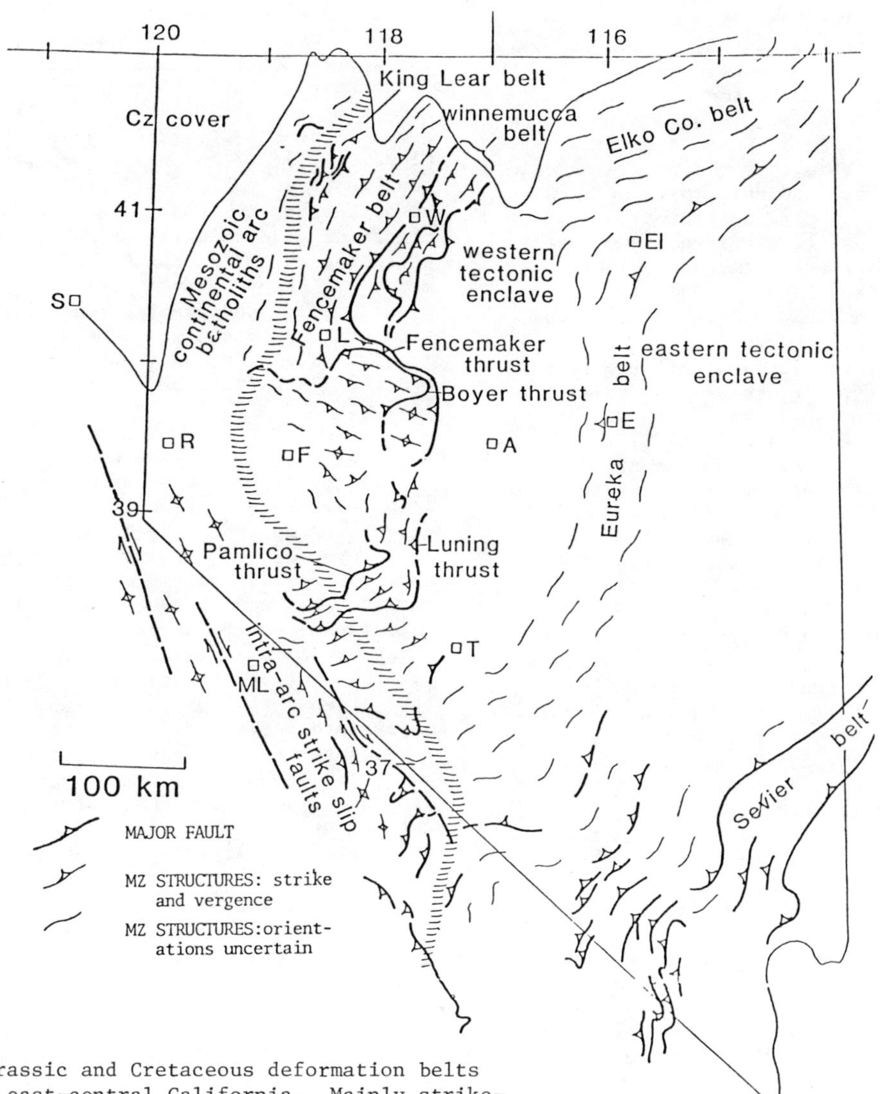

Figure 7. Jurassic and Cretaceous deformation belts in Nevada and east-central California. Mainly strike-slip faults and(or) shear zones within continental arc and thin-skinned thrust and fold belts in foreland to east.

stal or sedimentary precursors (mainly S-type vs. I-type of arc). Mesozoic metamorphism and deformation of Precambrian and lower Paleozoic strata occurred deep within the continental shelf succession but such effects are apparently absent in the upper few kilometers of Paleozoic strata (Howard, 1971, 1979; Snoke, 1979). Such relationships are recognized in Cenozoic uplifts (core complexes). Mesozoic magma generation and metamorphism in the foreland are likely to have been related processes. Metamorphic minerals indicate depths as great as about 15 km, far deeper than the undisturbed, expected stratigraphic depth of the metamorphosed strata. Thus, either metamorphism occurred in strata of anomalous depositional thickness or where strata were thickened tectonically. In the latter case, thickening may reflect contraction at deep levels of foreland during the Mesozoic (Armstrong, 1972).

It is a fundamental question whether the elevation of the geotherms at depth in the Mesozoic foreland was directly related to subduction or whether it was a product of motions within the continental lithosphere and(or) subjacent asthenosphere. An appeal to foreland heating by the same slab that caused magmatism in the continental arc would require the unlikely condition that melting occurred across a zone about 600 km wide. Present dating indicates that foreland plutons are not contemporaneous with the possible Early Cretaceous lull in continental arc magmatism, a time when a shallowly-dipping slab could have existed. Thus, Armstrong's (1972) hypothesis of heating due to crustal thickening seems to account best for the broad reach of foreland thermal phenomena and their restriction to deeper crustal levels in the Mesozoic.

Thin-Skinned Foreland Deformation: The region between the continental arc and the Paleozoic platform-shelf hinge in the western Colorado Plateau was the foreland of North America (Fig. 7) that underwent mainly contractile deformation from Jurassic to Paleocene times. Post-Early Triassic Mesozoic structures within this region are almost entirely within Mesozoic and Paleozoic cover rocks. The structures record nonthermal, thin-skinned tectonics with little evident involvement of the crystalline basement (Armstrong, 1968; Burchfiel and others, 1980; Royce and others, 1975; Speed, 1978, 1983; Dunne and others, 1978; Oldow, 1981, 1982; Allmendinger and Jordan, 1981). Thin-skinned contractile displacements are heterogeneous and are concentrated in several belts (Fig. 7), some of which splay about two regional enclaves in which there was apparently little Mesozoic shortening of cover rocks (Armstrong, 1972; Speed, 1983). The enclaves do contain low angle faults with younger-on-older juxtaposition which may indicate extensional deformation. It is not generally clear, however, whether such faults are Mesozoic or Cenozoic.

The foreland thrust and fold belts (Fig. 7) are mainly north-striking and underwent east-west contraction between 39° and 41°N. The deflection of the Luning belt (Fig. 7) (Oldow, 1981, 1982) is probably related to motions in the intra-arc shear zone (Speed, 1978). The origin of the northeasterly trend of the thrust belt of northeastern Nevada is unclear. Vergence varies among the thrust and fold belts, indicating both easterly and westerly overriding between 39° and 41°N. Local vergence can be related to facing of ramps at deep structural levels (Speed, 1983).

Figure 8 illustrates possible structural relationships among three thrust belts in northwestern Nevada. The Jackson Mountains at the northwestern end are underlain by Mesozoic volcanic and plutonic rocks of the continental arc, including the Lower Cretaceous intra-arc basin deposits of the King Lear Formation (Willden, 1958; Russell, 1983). West-vergent thrust slices of Triassic arc and basinal facies (Russell, 1983) are piled against the arc edifice. The Triassic successor basin fill between the Jackson Mountains and Sonoma Range is greatly thickened by deformation and has ridden ESE at its eastern margin over the Triassic basin-shelf margin in the Fencemaker belt. Below the east-verging Fencemaker allochthon lies a series of west-verging thrust slices of the Jurassic Winnemucca thrust belt (Speed, 1983). The Winnemucca belt cuts mainly Triassic shelfal strata, subjacent Golconda and Roberts allochthons, and beds of the Paleozoic continental shelf. East of the Winnemucca belt is the western tectonic enclave (Fig. 7) with nearly undisturbed Triassic beds. The vergence of the belts is thought to be due to the facing of proximal basement ramps, except in the case of the Fencemaker belt for which the earlier Winnemucca belt provided a west-facing ramp.

The timing of displacements in the Sevier belt (Fig. 8), dated by a nearly continuous series of foreland basin deposits, is Late Jurassic to Paleocene (Royce and others, 1975; Jordan, 1981) at least near the Utah-Wyoming border. Motions in the western belts are more difficult to date but indicate as a whole that deformational events occurred from Early Jurassic to Late Cretaceous times (Speed, 1978, 1984; Speed and Kistler, 1980; Oldow, 1981). The data are suggestive of either continuous or periodically reactivated movements at each site, but not a systematically east-migrating wave of deformation. Shortening is estimated at 50% in the Sevier belt at the eastern margin of the foreland (Armstrong, 1968; Royce and others, 1975) and greater than 50% in the Fencemaker belt (Wiens and others, 1982). Assuming similar values for each belt, the foreland contracted EW at least 300 km or 33% relative to initial width.

The question arises whether the displacements within the cover were continuous vertically down into the foreland basement, or on the other hand, were taken up at a detachment within the cover or at greater depth. The first case predicts heterogeneous crustal thickening with maxima below the thrust belts and is supported by the approximate coincidence of the western edge of Paleozoic shelf strata and sialic basement. The second alternative implies the existence of a detachment fault that spanned the entire foreland (Speed, 1983) to a presumed point of arrested propagation on the continentward side. Assuming rigid media, the total displacement by thrusting within the foreland cover would equal the offset across the detachment at its western limit. In this case, there would be little or no expression of foreland deformation in basement thicknesses, nor in isostatic responses; the absence of both phenomena are suggested by the lack of related effects in gravity, mountain building, and orogenic sedimentation.

Tectonic Erosion of the Sialic Edge During Oblique Convergence: Local truncation of the outer sialic continent and rafting away of continental fragments occurred at various times and places during and possibly associated with the onset of active margin tectonism.

A probably large raft of sialic North America and early accreted terranes (Sonomia and possibly others) is thought to have been extracted from the region of southeastern Oregon, southwestern Idaho, and possibly northern Nevada and California (Hamilton, 1976; Speed, 1983). Evidence is that 1) the probably preserved late Precambrian sialic edge, pre-Jurassic accreted terranes, and parautochthonous Mesozoic stratal units of central and western Nevada, all mainly north-striking, are apparently missing from the region to the north, 2) the sialic edge in Idaho intersects the Paleozoic craton-shelf hinge, an unlikely primary arrangement, and 3) displaced terranes with large late-Mesozoic rotations (Hillhouse and others, 1982), which are absent from rocks of Nevada, sutured in the Cretaceous against the sialic edge in Idaho. The withdrawal of a continental raft in Oregon and Idaho may have resulted from the propagation of a rift-transform system inboard of the

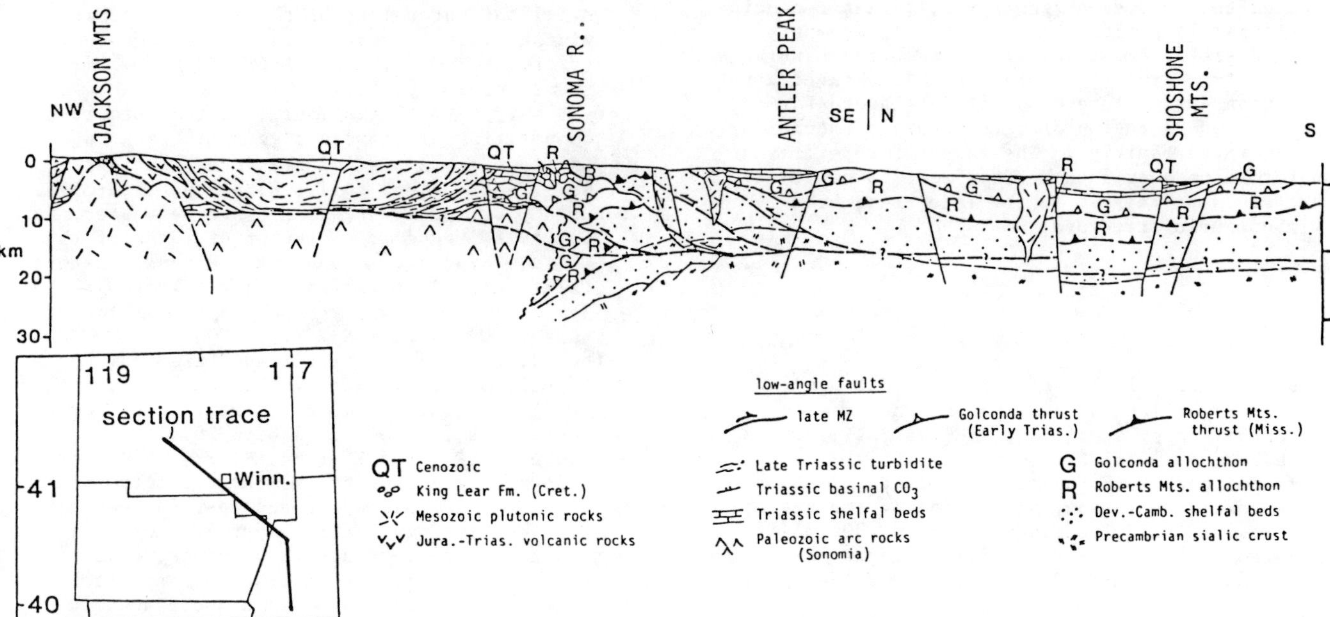

Figure 8. Cross section illustrating Mesozoic thrust fold belts and tectonostratigraphy of northwestern Nevada.

sialic margin during highly oblique convergence in mid-Mesozoic time. The formation of the modern Gulf of California and north-westward rafting of continental Baja California may be an analog. The fragment removed from Oregon and Idaho was presumably replaced by oceanic lithosphere whose spreading centers migrated north with respect to a southern boundary.

Structural Inheritance

East of the Sierra Nevada, there is abundant evidence for the influence of older features on the form and position of younger structures, but reactivation of sutures as new major displacement zones or continental boundaries seems not to have occurred.

The ancient sialic edge of possible Precambrian age in Nevada (Fig. 2) has been maintained through Phanerozoic time as a distinct but covered boundary. It influenced later ocean-continent configurations by forming a buttress to which island arcs sutured in Mississippian and Early Triassic times. It approximately underlies a major Triassic transition between a carbonate shelf to the east and deepwater successor basin to the west. This transition is thought to reflect large differences in temperature gradients and degree of subsequent thermal contraction across the suture between the sialic continent and the arc terrane of Sonomia which became inactive upon collision. The ancient sialic edge also controls zones of surface breakthrough of displacements during Jurassic and Cretaceous foreland contraction. The imbrication may reflect a ramp on the basement surface across the sutured edge of the continent (Speed, 1983).

Other evident influences of older on younger structures occur in the foreland deformation belts (Figs. 7, 8). The Sevier belt is coincident over 1000 km with the hingeline between craton and subsiding shelf. The hingeline thus defines a west-facing ramp on the top of the crystalline basement and also in horizons in suprajacent layered rocks. Another ramp effect is the localization of the Eureka thrust belt (Fig. 7) at the eastern edge of the Roberts Mountains allochthon, a locus of maximum rate of change of downbuckling of horizons in the autochthon due to loading by the allochthon (Speed and Sleep, 1982).

The zonation of Neogene faulting in the Basin and Range province between a westernmost belt (Walker Lane) containing strike-slip together with dipslip faults and the classic mainly dipslip faults to the east may be influenced by older structure. The westernmost belt includes at least part of the region in which intra-arc strike-slip faulting occurred in Mesozoic time. The most conspicuous set of Neogene strike-slip faults parallels the Mesozoic faults and may represent reactivations of the older faults. It is not clear whether the NNW-trending right slip faults of the western Basin-Range take up some of Pacific-North American relative plate motion or whether they are simply faults of a conjugate set that take up pure shear with east-west extension. If the first hypothesis is correct, the very wide zone of Neogene plate boundary deformation may be attributable to inherited structures of appropriate orientation.

Finally, there is an evident spatial correspondence in the extent of deformation in western sialic North America during three discrete

pulses that represent greatly different tectonic
settings: 1) region of attenuated, presumably
rifted sialic crust in late Precambrian passive
margin formation, 2) region of thin-skinned contraction during late Mesozoic foreland deformations, and 3) region of late Cenozoic extensional
basin-range faulting. The cause of repetitive
deformation over such a broad zone is not clear
but could be related to pre-late Precambrian
lithospheric structure.

References

Albers, J. P., 1967, Belt of sigmoidal bending
   and right-lateral faulting in the western
   Great Basin: Geol. Soc. America Bull., v. 78,
   no. 2, p. 143-156.

Allmendinger, R. W. and Jordan, T. E., 1981,
   Mesozoic evolution, hinterland of the Sevier
   orogenic belt: Geology, v. 9, p. 309-314.

Armstrong, R. L., 1968, Sevier orogenic belt in
   Nevada and Utah: Geol. Soc. Amer. Bull., v.
   79, p. 429-458.

Armstrong, R. L., 1972, Low-angle (denudation)
   faults, hinterland of the Sevier orogenic
   belt, eastern Nevada and western Utah: Geol.
   Soc. Amer. Bull., v. 83, p. 1729-1754.

Armstrong, R. L., and Suppe, J., 1973,
   Potassium-argon geochronometry of Mesozoic
   igneous rocks in Nevada, Utah, and southern
   California: Geol. Soc. Amer. Bull., v. 84,
   p. 1375-1392.

Armstrong, R. L., Taubeneck, W. H., and Hales, P.
   O., 1977, Rb-Sr and K-Ar geochronometry of
   Mesozoic granitic rocks and their Sr isotopic
   composition, Oregon, Washington and Idaho:
   Geol. Soc. Amer. Bull., v. 88, p. 397-441.

Beck, M. E. Jr., 1983, On the mechanism of tectonic transport in zones of oblique subduction: Tectonophysics, in press.

Brook, C. A., 1977, Stratigraphy and structure of
   the Saddlebag Lake roof pendant, Sierra
   Nevada, CA: Geol. Soc. Amer. Bull., v. 88, p.
   321-331.

Burchfiel, B. C., Pelton, P. J., and Sutter, J.,
   1970, An early Mesozoic deformation belt in
   south-central Nevada-southeastern California:
   Geol. Soc. Amer. Bull., v. 81, no. 1, p. 211-
   216.

Burchfiel, B. C. and Davis, G. A., 1972, Structural framework and evolution of the southern
   part of the Cordilleran orogen, Western United
   States: Amer. Jour. Sci., v. 272, p. 97-118.

Chen, J. H., and Moore, J. G., 1982, U-Pb isotopic ages from the Sierra Nevada batholith, CA:
   Jour. Geophys. Res., v. 87, p. 4761-4785.

Dickinson, W. R., 1977, Paleozoic plate tectonics
   and the evolution of the Cordilleran continental margin: SEPM, Pacific Coast Paleogeography Symposium I, Paleozoic Paleogeography of
   the Western U.S., p. 137-157.

Dickinson, W. D., Harbaugh, D. W., Antler, A. H.,
   Heller, P. L., and Snyder, W. S., 1983, Detrital modes of upper Paleozoic sandstones
   derived from Antler orogen in Nevada: Amer.
   J. Science, in press.

Dilles, J. H., Wright, J. E., and Proffett, J.
   M., 1983, Chronology of the early Mesozoic
   plutonism and volcanism in the Yerington District, NV: Geol. Soc. Amer. Absts. w. Programs, v. 15, p. 383.

Doe, B. R., 1973, Variations in lead-isotopic
   compositions in Mesozoic granitic rocks of
   California - A preliminary investigation:
   Geol. Soc. Amer. Bull., v. 84, p. 3513-3526.

Dunne, G. C., Gulliver, R. M., and Sylvester, A.
   G., 1978, Mesozoic evolution of rocks of the
   White, Inyo, Argus, and State Ranges, eastern
   California: SEPM Pacific Coast Paleogeography
   Symposium 2, Mesozoic Paleogeography of the
   Western U.S., p. 189-209.

Eaton, J. P., 1963, Crustal structure from San
   Francisco, California, to Eureka, Nevada, from
   seismic refraction measurements: Jour. Geophys. Res., v. 68, p. 5789-5806.

Evernden, J. F., and Kistler, R. W., 1970, Chronology of emplacement of Mesozoic batholithic
   complexes in California and western Nevada:
   U.S. Geol. Surv. Prof. Paper 623, 67 p.

Fitch, T. J., 1972, Plate convergence, transcurrent faults, and internal deformation
   adjacent to southeast Asia and the western
   Pacific: Jour. Geophys. Res., v. 77, p.
   4432-4461.

Frei, L. S., Magill, J. R., and Cox, A., 1982,
   Paleomagnetic results from the central Sierra
   Nevada: constraints on reconstruction of the
   western U.S.: EOS, v. 63, p. 916.

Geissman, J. W., Callian, J. T., Oldow, J. S.,
   and Humphries, S. E., 1984, Paleomagnetic
   assessment of oroflexural deformation in
   west-central Nevada and relations to Paleozoic
   and Mesozoic implacement of allochthonous
   assemblages: Tectonics, in press.

Gilluly, J., and Gates, O., 1965, Tectonic and
   igneous geology of the northern Shoshone
   Range, Nevada: U.S. Geol. Survey Prof. Paper
   465, 153 p.

Gromme, C. S. and Merrill, R. T., 1965,
   Paleomagnetism of Late Cretaceous granitic
   plutons in the Sierra Nevada, CA: further
   results: Jour. Geophys. Res., v. 70, p.
   3407-3420.

Gromme, C. S., Merrill, R. T., and Verhoogen, J.,
   1967, Paleomagnetism of Jurassic and Cretaceous plutonic rocks in the Sierra Nevada,
   California, and its significance for polar
   wandering and continental drift: Jour. Geophys. Res., v. 72, p. 5661-5684.

Hannah, J. L., and Verosub, K. L., 1980, Tectonic
   implications of remagnetized upper Paleozoic
   strata of the northern Sierra Nevada: Geology, v. 8, p. 520- 524.

Hamilton, W., 1969, Mesozoic California and the
   underflow of the Pacific mantle: Geol. Soc.
   Amer. Bull., v. 80, p. 2409-2430.

Hamilton, W., 1978, Mesozoic tectonics of the
   western U.S.; Mesozoic Paleogeog. of western
   U.S., Pacific Coast Paleogeography Symposium
   2, Soc. Econ. Pal. and Min.

Hamilton, W., and Myers, W. B., 1966, Cenozoic
   tectonics of the western United States: Rev.
   Geophys., v. 4, p. 509-549.

Hillhouse, J. W., Gromme, C. S., and Vallier, T.
   L., 1982, Paleomagnetism and Mesozoic tectonics of the 7 Devils arc, Oregon: Jour. Geo-

phys. Res., v. 87, p. 3777-3794.

Howard, K. A., 1979, Metamorphic intrastructure in the northern Ruby Mts., Nev: Geol. Soc. Amer. Mem. 153, p. 335-349.

Jordan, T. E., 1981, Thrust loads and foreland basin evolution, Cretaceous, western United States: Amer. Assoc. Petr. Geol. Bull., v. 65, p. 2506-2520.

Kay, M., 1960, Paleozoic continental margin in central Nevada, western United States: Int'l. Geol. Cong., 21st, Copenhagen 1960, Rept., pt 12, p. 94-103.

Kay, M., and Crawford, J. P., 1964, Paleozoic facies from the miogeosynclinal to the eugeosynclinal belt in thrust slices, central Nevada: Geol. Soc. Amer. Bull., v. 75, no. 5, p. 425-454.

Kistler, R. W., 1966, Structure and metamorphism in the Mono Craters quadrangle, Sierra Nevada, CA: U.S. Geol. Surv. Bull. 1221E, 53 p.

Kistler, R. W., and Peterman, Z. E., 1973, Variations in Sr, Rb, K, Na and initial 87Sr/86Sr in Mesozoic granitic rocks and intruded wall rocks in Central California: Geol. Soc. Amer. Bull., v. 84, p. 3489-3512.

Kistler, R. W., Robinson, A. C., and Fleck, R. J., 1980, Mesozoic right lateral faults in eastern California: Geol. Soc. Amer. Absts. w. Programs, v. 12, p. 115.

Lee, D., Kistler, R. W., and Robinson, A. C., 1983, Strontium isotopic composition of granitoid rocks of the southern Snake Range, NV: U.S. Geol. Survey Prof. Paper, in press.

Matti, J. C., and McKee, E. H., 1977, Silurian and Lower Devonian paleogeo- graphy of the outer continental shelf of the cordilleran miogeocline, central Nevada: SEPM Pacific Coast Paleogeography Symposium 1, p. 181-217.

Miller, C. F., and Bradfish, L. J., 1980, An inner Cordilleran belt of muscovite-bearing plutons: Geology, v. 8, p. 412-416.

Miller, E. L., Bateson, J., Dinter, D., Dyer, J., Harbaugh, D., and Jones, D., 1981, Thrust emplacement of the Schoonover sequence, northern Independence Range, NV: Geol. Soc. Amer. Bull. I, v. 92, p. 730-737.

Miller, E. L., Holdsworth, B. K., Whiteford, W. B., and Rodgers, D., 1983, Stratigraphy and structure of the Schoonover complex, northeastern Nevada: Geol. Soc. Amer. Bull., in press.

Muffler, L. J. P., 1964, Geology of the Frenchie Creek quadrangle, north- central Nevada: U.S. Geol. Survey Bull. 1179, 99 p.

Nichols, K. M., and Silberling, N. J., 1977, Stratigraphy and depositional history of the Star Peak Group (Triassic), northwestern Nevada: Geol. Soc. Amer. Spec. Paper 178, 73 p.

Nokelberg, W. J. and Kistler, R. W., 1982, Paleozoic and Mesozoic deformations in the central Sierra Nevada, CA: U.S. Geol. Survey Prof. Paper 1145, 24 p.

Oldow, J. S., 1981, Kinematics of late Mesozoic thrusting, Pilot Mountains, Nevada: Jour. Struct. Geol., v. 3, p. 39-51.

Oldow, J. S., 1983, Spatial variability in the structure of the Roberts Mountains allochthon: Geol. Soc. Amer. Bull., in press.

Poole, F. G., 1974, Flysch deposits of the Antler foreland basin: SEPM Spec. Publ. 22, p. 58-82.

Prohdehl, C., 1979, Crustal structure of the western United States: U.S. Geol. Surv. Prof. Paper 1034, 74 p.

Roberts, R. J., Hotz, P. E., Gilluly, J., and Purgusan, H. G., 1958, Paleozoic rocks of north-central Nevada: Amer. Assn. Petrol. Geol. Bull., v. 42, p. 2813-2857.

Rowell, A. J., Rees, M. N., and Suczek, C. A., 1979, Margin of the North American continent in Nevada during late Cambrian time: Am. Jour. Sci., v. 279, p. 1-18.

Royce, F., Warner, M. A., and Resse, D. L., 1975, Thrust belt geometry and related stratigraphic problems Wy-Ida-Ut: Rocky Mt. Assn. Geols. Symp., p. 41-54.

Russell, B. R., 1983, Mesozoic geology of the Jackson Mts., NV: Geol. Soc. Amer. Bull., in press.

Russell, B. J., Beck, M. E., Jr., Burmester, R. F., and Speed, R. C., 1982, Cretaceous magnetizations in northwestern Nevada and tectonic implications: Geology, in press.

Saleeby, J. B., 1982, Polygenetic ophiolite belt of the California Sierra Nevada - Geochronological and tectonostratigraphic development: Jour. Geophys. Res., in press.

Saleeby, J. B. and Sharp, W., 1980, Chronology of structural and petrologic development of the SW Sierra Nevada foothills: Geol. Soc. Amer. Bull. I, p. 317-320.

Schweickert, R. A., 1981, Tectonic evolution of the Sierra Nevada Range: in Geotectonic Development of California, ed. Ernst, W: New Jersey, Prentice Hall, p. 87-132.

Silberling, N. J., 1973, Geologic events during Permian-Triassic time along the Pacific margin of the U.S.: Alberta Soc. Petrol. Geologists Mem. 2, p. 345-362.

Silberling, N. J., and Roberts, R. J., 1962, Pretertiary stratigraphy and structure of northwestern Nevada: Geol. Soc. Amer. Spec. Paper 72, 58 p.

Silberling, N. J., and Wallace, R. E., 1969, Stratigraphy of the Star Peak Group and overlying Mesozoic rocks in the Humboldt Range, Nevada: U.S. Geol. Surv. Prof. Paper 592, 50 p.

Smith, J. F., and Ketner, K. B., 1968, Devonian and Mississippian rocks and the date of the Roberts Mountains thrust in the Carlin-Pinon Range area, Nevada: U.S. Geol. Surv. Bull. 1251-I, p. 11-118.

Snoke, A. W., 1979, Transitions from infrastructure to suprastructure in the northern Ruby Mts., Nev: Geol. Soc. Amer. Mem. 153, p. 287-334.

Snyder, W. S., and Brueckner, H. K., 1983, Tectonic evolution of the Golconda allochthon, NV: problems and perspectives: in Pre-Jurassic Rocks in Western North American Suspect Terranes, ed. C. H. Stevens: Pacific Section SEPM.

Speed, R. C., 1977, Island arc and other paleogeographic terranes of late Paleozoic age in the western Great Basin: SEPM, Pacific Coast Paleogeo- graphy Symposium 1, Paleozoic Paleogeography of the western U.S., p. 349-362.

Speed, R. C., 1978, Paleogeographic and plate tectonic evolution of the early Mesozoic marine providence of the western Great Basin: SEPM Pacific Coast Paleogeography Symposium 2, Mesozoic Paleogeography of the western U.S., p. 253-270.

Speed, R. C., 1979, Collided Paleozoic microplate in the western United States: Jour. Geol., v. 87, p. 279-292.

Speed, R. C., 1983, Evolution of the sialic margin in the central-western United States: Amer. Assoc. Petrol. Geol. Mem., Hedberg Volume (in press).

Speed, R. C., 1984, Mesozoic foreland thrusting in the Great Basin: Geology, in press.

Speed, R. C., and Kistler, K. W., 1980, Cretaceous volcanism, Excelsior Mountains, Nevada: Geol. Soc. Amer. Bull., v. 91, p. 392-298.

Speed, R. C. and Sleep, N. H., 1982, Antler orogeny and foreland basin: A model: Geol. Soc. Amer. Bull., v. 93, p. 815-828.

Stern, T. W. and others, 1981, Isotopic U-Pb ages of zircon from granitoids of the central Sierra Nevada, CA: U.S. Geol. Surv. Prof. Paper 1185.

Stewart, J. H., 1972, Initial deposits of the Cordilleran geosyncline: evidence of a Late Precambrian (850 m.y.) continental separation: Geol. Soc. Amer. Bull., v. 83, p. 1345-1360.

Stewart, J. H., 1976, Late Precambrian evolution of North America: plate tectonics implications: Geology, v. 4, p. 11-15.

Stewart, J. H., and Poole, F. G., 1974, Lower Paleozoic and uppermost Precambrian Cordilleran miogeocline, Great Basin: SEPM Spec. Publ. 22, p. 27-57.

Stewart, J. H., and Suczek, C. A., 1977, Cambrian and latest Precambrian paleogeography and tectonics in the western United States: SEPM Pacific Coast Paleogeography Symposium 1, p. 1-19.

Stewart, J. H., Albers, J. P., and Poole, F. G., 1968, Summary of regional evidence for right-lateral displacement in the western Great Basin: Geol. Soc. Amer. Bull., v. 79, no. 10, p. 1407-1414.

Stewart, J. H., MacMillan, J. R., Nichols, K. M., and Stevens, C. H., 1977, Deep-water upper Paleozoic rocks in north-central Nevada - a study of the type area of the Havallah Formation, in Stewart, J. H., Stevens, C. H., and Fritsche, A. E., eds., Paleozoic paleogeography of the western United States: Soc. Econ. Pal. and Min., Pacific Sec., Pacific Coast Paleogeography Symposium 1, p. 337-347.

Tobisch, O. T., and Fiske, R. S., 1982, Repeated parallel deformation in part of the eastern Sierra Nevada, CA, and its implications for dating structural events: Jour. Struct. Geol., v. 4, p. 177-195.

Walcott, R. I., 1978, Present tectonics and late Cenozoic evolution of New Zealand: Geophys. Jour. Roy. Astro. Soc., v. 52, p. 137-164.

Wiens, D. A., Engeln, J. F., and Speed, R. C., 1982, Deformation of Triassic turbidites, Santa Rosa Mtns., Nevada: Geol. Soc. Amer. Absts. w. Progr., Cordilleran Section, v. 14, p. 244.

Willden, C. R., 1958, Cretaceous and Tertiary orogeny in Jackson Mountains Humboldt County, Nevada: Am. Assoc. Petrol. Geol. Bull., v. 42, no. 10, p. 2378-2398.

Zartman, R. E., 1974, Lead isotopic provinces in the Cordillera of the western United States and their geologic significance: Econ. Geology., v. 69, p. 792- 805.

CENOZOIC STRUCTURE AND TECTONICS OF THE NORTHERN BASIN
AND RANGE PROVINCE, CALIFORNIA, NEVADA, AND UTAH

By John H. Stewart

U.S. Geological Survey
345 Middlefield Road
Menlo Park, CA 94025

ABSTRACT

During the early Cenozoic, the northern
Basin and Range province was largely an upland
from which detritus was shed eastward into basins
in the western part of the Colorado Plateaus
province. Within this upland area, broad
sedimentary basins locally formed that may in
part have been created by extensional faulting.

During the middle and part of the late
Cenozoic, the major tectonic features in the
northern Basin and Range province were related to
widespread volcanic activity that began in the
northern part of the province about 43 m.y. ago
and spread southward along an arcuate east-west-
trending front and ended about 6 m.y. ago in
southern Nevada and Utah. East-west-trending
volcano-tectonic and graben structures, igneous
and mineral belts, and structural lineaments are
related to this southward sweep of igneous
activity.

During part of the middle and throughout the
late Cenozoic, local and regional extension and
local strike-slip faulting dominated the
tectonics of the northern Basin and Range
province. Low-angle extensional faults that
juxtapose younger over older rocks developed
primarily from 20 to 10 m.y. ago. They occur
within a diffuse 200- to 300-km-wide zone
extending north-northeastward in eastern Nevada
and westernmost Utah, and within a 100-km-wide
northwest-trending zone along the Walker Lane,
astride the California-Nevada state line. Block
faulting that has created the characteristic
physiography of the present-day northern Basin
and Range province is controlled by north-south
or north-northeast-trending normal faults that
have broken the province into blocks, generally
about 30 km across. Such block faulting is
commonly superimposed across areas of low-angle
faulting and, in addition, extends into areas
previously unaffected by extension. Block
faulting has occurred primarily during the past
10 m.y.

INTRODUCTION

The northern Basin and Range province in
Nevada, western Utah, and parts of adjacent
states is a high desert area of largely interior
drainage characterized by north or
north-northeast-trending mountains and valleys.
The elevations of the valleys are generally 1,300
to 1,600 m; mountain crests are commonly 2,000 to
3,000 m and locally about 3,600 m. The entire
Basin and Range province (fig. 1) extends from
southern Oregon and Idaho, through most of Nevada
and parts of California, Utah, Arizona, and New
Mexico, to northern Mexico--a total distance of
more than 2,500 km.

This article summarizes information on the
character and development of Cenozoic structures
in the northern Basin and Range province and
emphasizes the regional distribution of
structures. Concepts of the Cenozoic history
have developed greatly in the past 15 years as a
result of plate tectonic theories and studies of
metamorphic core complexes and low angle
detachment faults. Ideas continue to evolve
rapidly with many different viewpoints (Stewart,
1978; Zoback and others, 1981; Eaton, 1982).

EARLY CENOZOIC STRUCTURES

The Paleocene to middle Eocene history of
the northern Basin and Range region is obscure
because rocks of this age are sparse. During
most of this time, the region was probably an
upland from which debris was shed mainly eastward
into basins in the western part of the Colorado
Plateau province (Hintze, 1973). Within the
broad upland area, large sedimentary basins
locally formed. Strata in some of these basins
are as thick as 1,000 m and consist of alluvial
fans deposits adjacent to mountains fronts and
lake deposits in the central parts of basins
(Fouch, 1979). Igneous activity was slight or
may not have occurred at all. Only a few
uncertainly dated Paleocene to middle Eocene
igneous rocks are known from the region (Carlson
and others, 1975).

Evidence of tectonic activity in the early
Cenozoic is seen in the Sheep Pass Formation of
east-central Nevada, which represents fluvial and
lacustrine deposition in a broad internal
drainage system. The Sheep Pass Formation rests
unconformably on faulted and broadly folded
Paleozoic strata (Kellogg, 1964; Moores and
others, 1968), presumably deformed in the middle
and late Mesozoic, and contains boulder

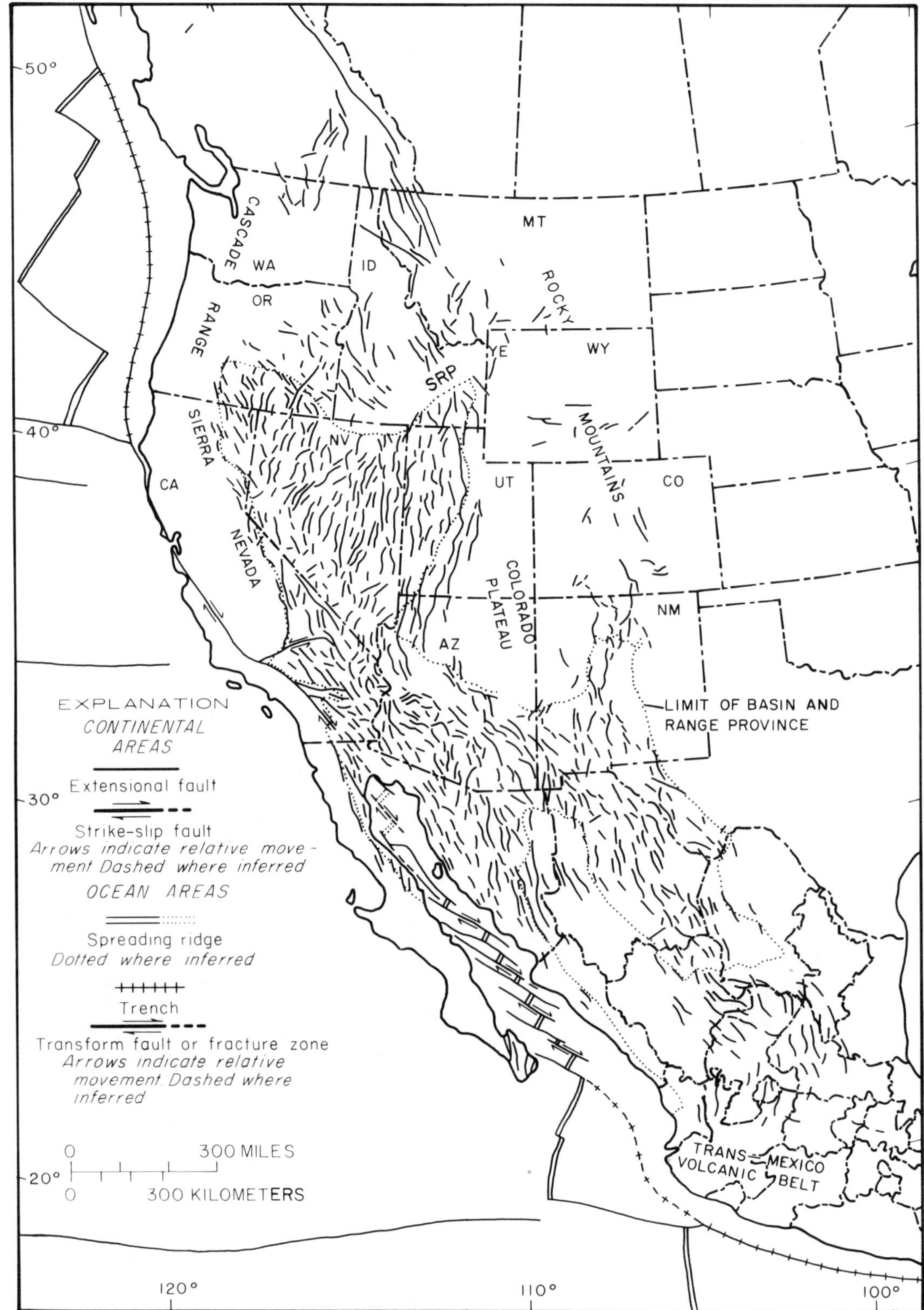

Figure 1. Distribution of late Cenozoic extensional faults, major strike-slip faults, and physiographic provinces in western North America and present-day lithospheric plate boundaries. From Stewart (1978). States: WA, Washington; OR, Oregon; CA, California; CO, Colorado; ID, Idaho; MT, Montana; WY, Wyoming; NV, Nevada; UT, Utah; AZ, Arizona; NM, New Mexico. Localities SRP, Snake River Plain: YE, Yellowstone.

conglomerate (Kellogg, 1964). Although much of this pre-Sheep Pass structure may be related to Mesozoic deformation, some may be related to deformation during deposition of the Sheep Pass Formation, as small faults cut conglomerate in the lower part of the Sheep Pass but do not affect overlying limestone and mudstone (Kellogg, 1964). Large slide blocks of Paleozoic rocks in the Sheep Pass (Newman, 1979) indicate considerable relief in the areas adjacent to the depositional basin. Newman (1979) suggests that the basin in which the Sheep Pass was deposited resulted from normal faulting.

MIDDLE AND LATE CENOZOIC TECTONICS RELATED TO IGNEOUS ACTIVITY

During the middle and part of the late Cenozoic, many major tectonic and structural features in the northern Basin and Range province are related to widespread igneous activity. The pattern of igneous activity is complex (fig. 2), but one element of this pattern is a southward migration of igneous activity starting in the northern part of the province about 43 m.y. ago, spreading southward along an arcuate east-west-trending front, and ending 6 m.y. ago in southern Nevada and Utah (Snyder and others, 1976; Stewart and others, 1977; Cross and Pilger, 1978). In

Utah and easternmost Nevada, middle and late Cenozoic igneous activity is confined to four distinct igneous belts (fig. 3), progressively younger to the south and each with a concentration of Cenozoic mineral deposits (Hilpert and Roberts, 1964; Stewart and others, 1977). These belts merge and become indistinct features to the west.

Many major aeromagnetic anomalies in Nevada and Utah trend east-west, east-southeast, or east-northeast (fig. 3). In Utah, these anomalies closely follow the trends of the Cenozoic igneous and mineral belts and clearly are related to magnetic Cenozoic intrusive rocks (Stewart and others, 1977; Shawe and Stewart, 1976). In Nevada, the cause of the generally easterly trending anomalies is less clear, although some appear to be related to Cenozoic and others to Mesozoic igneous rocks (Stewart and others, 1977). The anomalies in many places may mark the position of the slightly arcuate leading edge of the southward-moving front of Cenozoic igneous activity.

East-west-trending volcano-tectonic troughs locally developed in Nevada and eastern California as a consequence, or possible consequence, of middle and late Cenozoic igneous

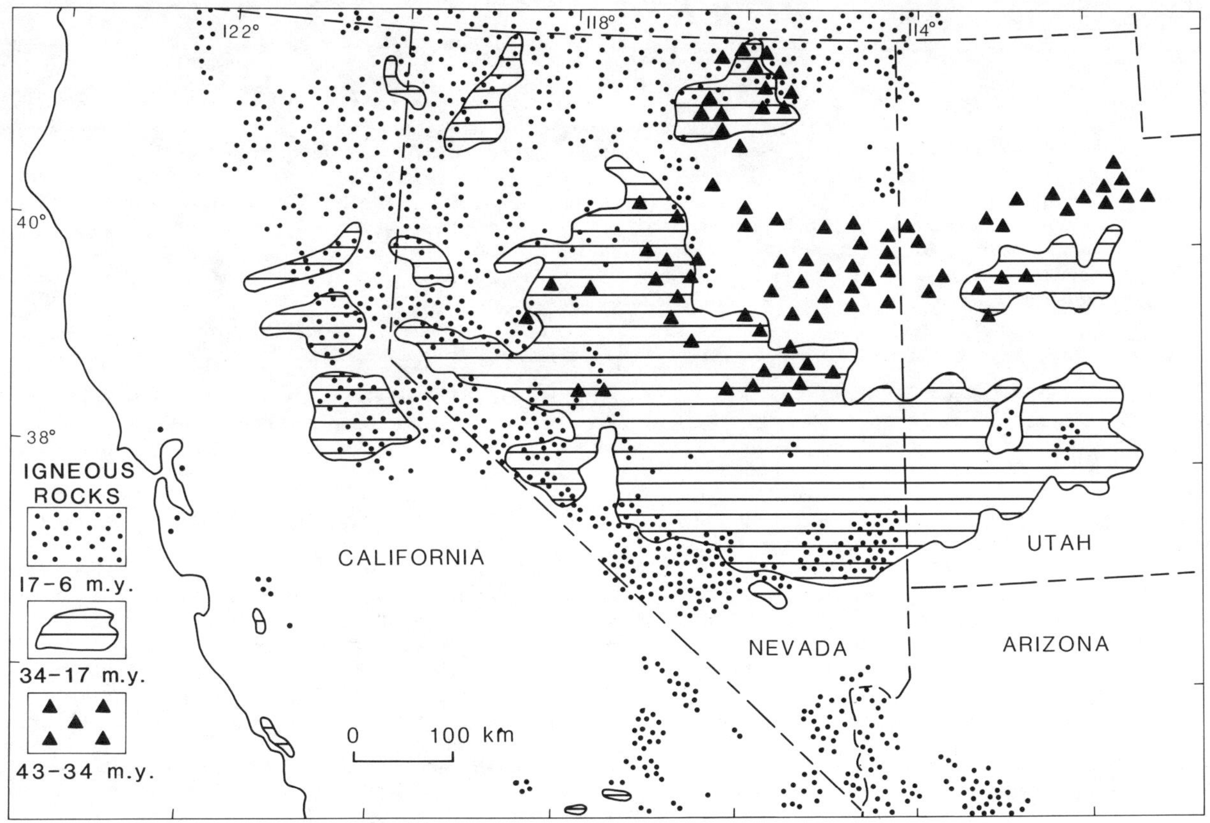

Figure 2.   Distribution of 43- to 6-m.y.-old igneous rocks in Nevada, Utah, and parts of adjacent states.   From Stewart and others (1977).

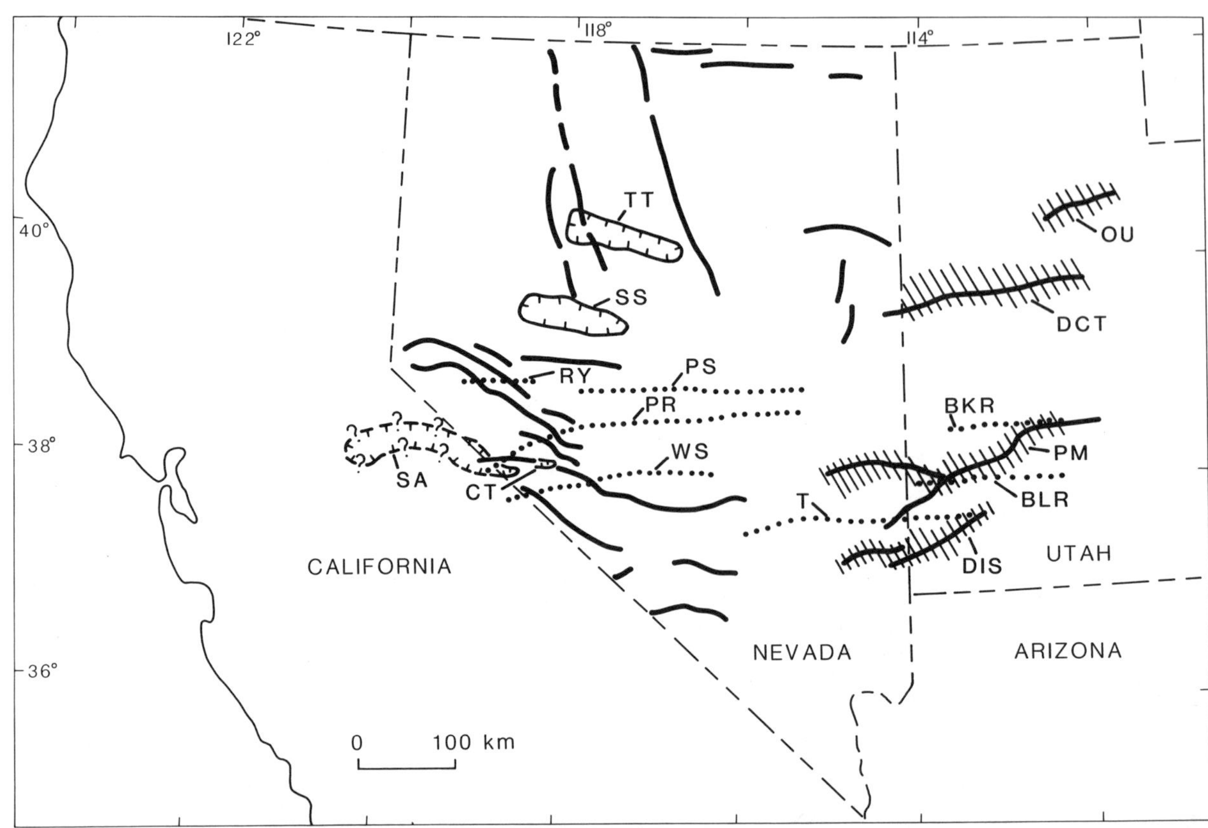

Figure 3. Distribution of positive aeromagnetic anomalies (solid lines), igneous and minerals belts (diagonal line pattern), volcano-tectonic troughs (hachured pattern), and lineaments (dotted pattern) in Nevada, Utah, and part of California. Igneous and mineral belts: OU, Oquirrh-Uinta; DCT, Deep Creek-Tintic; PM, Pioche-Marysvale; DIS, Delamar-Iron Springs. Volcano-tectonic troughs: TT, Tobin-Toiyabe; SS, Stillwater-Shoshone; CT, Candelaria; SA, Sierra Nevada-Adobe Hills. Lineaments: RY, Rawhide-Yerington; PS, Pritchard Station; PR, Pancake Range; WS, Warm Spring; BKR, Black Rock; BLR, Blue Ribbon; T, Timpiute. Based on Ekren and others (1976), Shawe and Stewart (1976), Rowley and others (1978, 1979), Speed and Cogbill (1979), and Burke and McKee (1979).

activity. In central Nevada, Burke and McKee (1979) have described two large east-west-elongate fault-bounded troughs (TT and SS, fig. 3) which were the sites of intense Oligocene and early Miocene volcanic activity (fig. 3). In western Nevada, Speed and Cogbill (1979) have described a deep east-northeast-trending fault trough (CT, fig. 3) filled with a thick accumulation of Oligocene and Miocene volcanic rocks. In eastern California and western Nevada, 9- to 11-m.y.-old ash-flow tuffs and lava are concentrated in an arcuate east-west trending belt (SA, fig. 3) (Gilbert and others, 1968, fig. 5; Noble and others, 1974; Stewart and others, 1982). The Little Walker caldera (Noble and others, 1974) lies within this belt and is the source of part of the tuff. The east-west belt may be a volcano-tectonic trough, although G. F. Brem (oral commun., 1982) has suggested that the Little Walker caldera lay at the crest of the ancestral Sierra Nevada and that the east-west pattern is in part the consequence of flows

dispersed downslope both east and west from the crest. The east-west trend of these volcano-tectonic troughs suggests general north-south extension during their development.

Generally east-west-trending lineaments (fig. 3), some well defined and other highly speculative, have been described in several parts of Nevada and Utah (Crosby, 1973; Ekren and others, 1976; Rowley and others, 1978; Rowley and others, 1979). These lineaments are defined on the basis of topographic and structural features that coincide with lithologic boundaries, range and valley termini, east-west-striking high-angle faults, caldera boundaries, alinement of eruptive centers and hydrothermally altered rock, alinement of mineral deposits, and east-west-trending magnetic highs and interruptions of magnetic highs. The possible alinement of Oligocene and Miocene volcanic centers locally along these trends suggests that these lineaments may have developed in part during the middle

Cenozoic, although the extent to which these alinements follow early Cenozoic or Mesozoic structures is unclear (Ekren and others, 1976; Rowley and others, 1978). Parts of these lineaments have been interpreted to be the result of strike-slip faulting, whereas others have been interpreted as due to high-angle normal faulting.

The east-west trends of igneous activity, aeromagnetic anomalies, volcano-tectonic troughs, and lineaments are difficult to explain in terms of a subduction system, or arc, extending generally north-south along the western margin of North America. Such a subduction system might more logically be expected to produce igneous activity parallel to and inland of the subduction system and extension approximately perpendicular to it. Possibly the east-west trends are related to old structures, perhaps Precambrian in part. For example, the Oquirrh-Uinta mineral belt in Utah (fig. 3) is alined along the Cortez-Uinta axis in Utah (Hilpert and Roberts, 1964; Tooker, 1971). This axis is a major tectonic zone in which Paleozoic sedimentary trends are disrupted (Roberts and others, 1965; Stewart and Poole, 1974) and is online with a major Precambrian epicratonal trough (or aulocogen) in which the Uinta Mountain Group was deposited in the Uinta Mountains of northern Utah. Perhaps the concentration of middle Cenozoic igneous activity in the Oquirrh-Uinta mineral belt is in a zone of structural weakness first developed in the Precambrian. Other east-west trends in Nevada and Utah locally follow east-west elongate Mesozoic plutons, again suggesting that some of these features were first developed in pre-Cenozoic time.

The systematic southward migration of igneous activity during the middle and late Cenozoic argues against a purely pre-Cenozoic control for the east-west trends in Nevada and Utah. If the east-west trends are controlled by pre-Cenozoic structures, why would the Cenozoic activity be progressively younger to the south? More likely pre-Cenozoic structural zones have in places localized igneous activity, but other factors have controlled the regional pattern. Rowley and others (1978) have suggested that the east-west lineaments are related to major transcurrent faults, but here again, why the ages of the volcanic activity should decrease to the south is not known. Stewart and others (1977) following a suggestion by P. W. Lipman have hypothesized that the southward migration of igneous activity along an arcurate east-west front is related to a southward migration of an east-west warp in the subducting plate. None of these ideas seem very satisfying to explain the peculiar east-west orientation of Cenozoic igneous activity in Nevada and Utah. Nevertheless, the east-west trends are important tectonically because the development of east-west volcano-tectonic troughs during the igneous activity suggests local north-south extension in part synchronous in part with east-west extension (discussed next) during detachment faulting in metamorphic core complexes and elsewhere.

MIDDLE AND LATE CENOZOIC EXTENSIONAL FAULTING

During part of the middle Cenzoic and throughout the late Cenozoic, local and regional extension and local strike-slip faults dominated the tectonics of the northern Basin and Range. Extensional structures vary significantly in style, age, and distribution and are here described under two categories: low-angle extensional faults and Basin-Range faults. Strike-slip faults are considered separately.

Low-angle extensional faults

A variety of low-angle extensional faults that juxtapose younger over older strata are recognized in the northern Basin and Range province. Such faults include low-angle detachment faults that probably were originally horizontal, gently dipping, or gently warped, as well as faults that were originally steep and were subsequently tilted to a low-angle position. The nomenclature of low-angle faults is not clearly established, and the term "detachment fault" has been used to describe a variety of structures, from low-angle thrust faults to widespread extensional faults that presumably were originally nearly horizontal (Davis and others, 1980), as well as widespread low-angle faults that may have evolved from tilting of originally high-angle faults into low-angle positions (Howard and others, 1982). In some places, distinguishing between these categories of low-angle faults is difficult. Current usage favors use of the term "detachment" fault for widespread fault surfaces that probably were originally horizontal, gently inclined, or gently folded. The Whipple Mountains detachment fault (Davis and others, 1980) in the Mohave Desert in southeastern California has perhaps become the type example of such a fault (Davis and others, 1980); it forms the decollement, or dislocation surface, that in this case separates a metamorphic core complex from its relatively unmetamorphosed carapace.

Low-angle extensional faults are exposed (fig. 4) within a diffuse 200- to 300-km-wide zone extending north-northeastward in eastern Nevada and westernmost Utah and within a 100-km-wide northwest-trending zone astride the California-Nevada state line. In addition, seismic reflection profiles indicate a widespread, gently west-dipping detachment surface, not shown on figure 4, in the subsurface in the Sevier Desert (Sevier Desert detachment fault) in western Utah (McDonald, 1976; Allmendinger and others, 1983; Zoback, 1983). The zone astride the state line is coextensive with the Walker Lane, a belt of disrupted topography characterized by discontinuous strike-slip faults (Gianella and Callaghan, 1934; Locke and others, 1940; Albers, 1967). The northeast-trending belt in eastern Nevada and western Utah, however, does not lie in a region of conspicuous strike-slip faulting. The zones containing low-angle extensional faults are characterized by generally high dip of middle

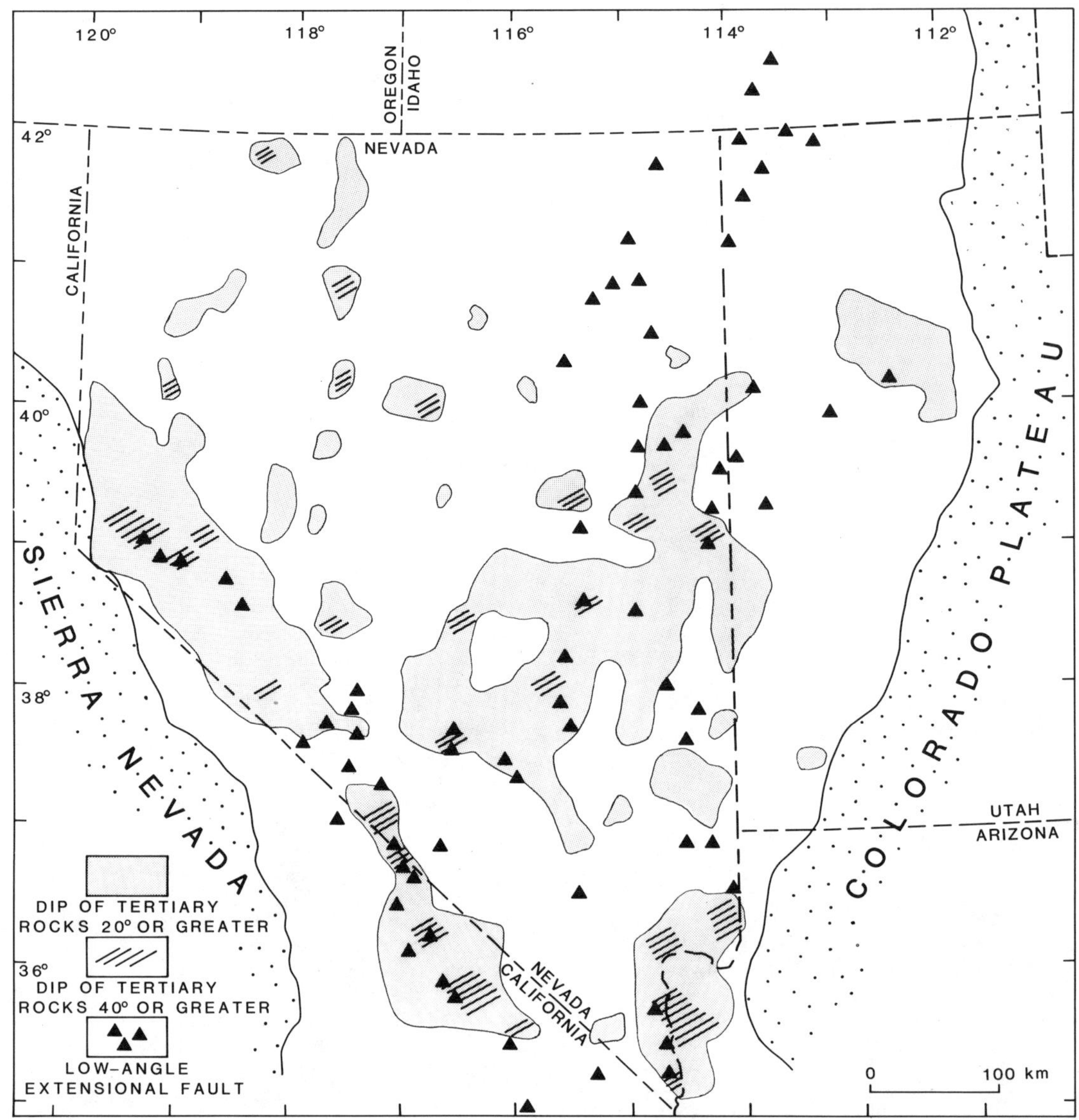

Figure 4.   Distribution of low-angle faults and generalized dip of Tertiary rocks.
Based in part on Hose and Danes (1973) and Stewart and Johannesen (1979).

Cenozoic rocks (fig. 4) and by the presence of metamorphic core complexes (fig. 5) and muscovite-bearing granitic rocks (Miller and Bradfish, 1980).

Many different models of low-angle extensional faults have been proposed (fig. 6). Most show a widespread low angle fault (detachment fault) with either relatively untilted or with highly tilted overlying strata (Davis and Coney, 1979; Davis and others, 1980; Wernicke, 1981; Wernicke and Burchfiel, 1982). Tilting may occur either along a system of listric faults (fig. 6) or as a consequence of a system of planar faults that bound blocks tilted domino-style (fig. 6).   Study of exposed fault surfaces in the upper plate of the Whipple

Mountain detachment fault (Gross and Hillemeyer, 1982) suggests that most of the faults are planar, that they join and divide in complex ways, and that they generally intersect the detachment surface at a high angle.

Other models of low-angle faulting (fig. 6) indicate listric or planar faults that do not bottom downward against a detachment fault, but rather die out downward into an intact, but presumably plastically extending, substratum (Wright and Troxel, 1973; Morton and Black, 1975; Proffett, 1977).   In some of these models, faults are originally steeply dipping but are rotated into low angle attitudes as the blocks are progressively tilted.   In other models, high-angle faults are rotated into low-angle

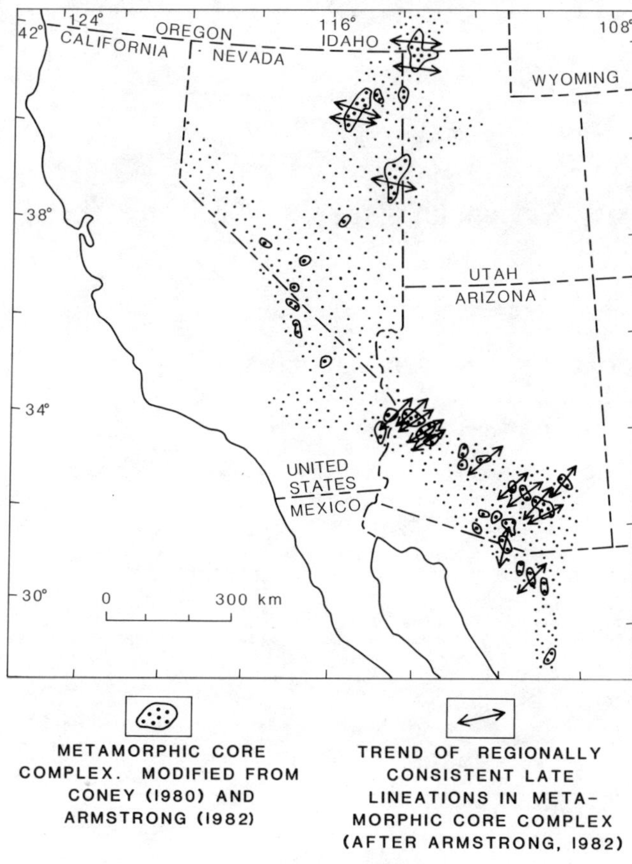

METAMORPHIC CORE
COMPLEX. MODIFIED FROM
CONEY (1980) AND
ARMSTRONG (1982)

TREND OF REGIONALLY
CONSISTENT LATE
LINEATIONS IN META-
MORPHIC CORE COMPLEX
(AFTER ARMSTRONG, 1982)

GENERALIZIED AREA OF
LARGE-SCALE EXTENSION

Figure 5. Distribution of metamorphic core
complexes and areas of large-scale extension.
In part from Coney (1980) and Armstrong (1982).

positions and are then cut by new high angle
faults which are in turn rotated.

Still other models (fig. 6) show extension
as thin skinned (Osmond, 1960; Wallace, 1965;
Anderson, 1971) in which surface blocks moved
over a substratum that was not extending or was
extending at a lower rate than the surface
blocks. In some examples, blocks of this type
may virtually be surface landslides.

Finally, Hardyman and others (1975) and
Hardyman (1978) have proposed that in a part of
the Walker Lane area of west-central Nevada,
detachment surfaces that separate Cenozoic
volcanic rocks from underlying Mesozoic igneous
and metamorphic rocks, and that also occur within
Cenozoic sequences, are due to strike-slip motion
that breaks the Mesozoic rocks along a few major
faults, but causes diffuse strain and the
development of many small-scale strike-slip and
normal faults in the overlying Tertiary rocks.

In this scheme, a detachment fault develops at
the boundary between the lower block cut by a few
major faults and the upper block of diffuse
strain.

Many of the low angle extensional faults
appear to be deep structures that extend, or
originally extended, deep into the crust.
Wernicke (1981) has proposed that low-angle
extensional faults cut through the entire crust
(fig. 6). Recent results from a COCORP seismic
reflection profile in the Sevier Desert of
western Utah shows that the subsurface Sevier
Desert detachment is shallow on the east, dips on
the average 12° to the west, and reaches a
maximum depth of 12-15 km in westernmost Utah
(Allmendinger and others, 1983; Zoback, 1983).
Elsewhere steeply dipping Cenozoic and pre-
Cenozoic rocks in stratigraphic sequences from 6
to 13 km thick are exposed in blocks bounded by
low-angle extensional faults (Proffett, 1977;
Gans, 1982; Howard and others, 1982). The steep
dip of these rocks and the low-angle of the
bounding faults is considered to result from
rotation during extension of originally gently
dipping strata and originally steeply dipping
normal faults. If so, the originally steeply
dipping faults must have extended to depths of
6-13 km in order to account for the thickness of
rocks now exposed at the surface.

The amount of extension in areas of
low-angle extensional faulting is believed to be
great, from 50 to greater than 200 percent of the
original width of the area. Proffett (1977)
estimated 100 percent extension in the Yerington
district of western Nevada, where geometry of
faulting and the tilt of strata is well
documented. Gans (1982) estimated more than 200
percent extension in the Egan Range in east-
central Nevada. Although impossible to state in
terms of percentage, I have speculated (Stewart,
1983) that the Panamint Range block in the Death
Valley area has moved 80 km to the northwest
along a low-angle detachment surface.
Large-scale extension appears to have affected
only that part of the Basin and Range province
where metamorphic core complexes and low-angle
detachment faults are abundant (figs. 4 and 5).

Most low-angle faulting in the northern
Basin and Range province probably occurred during
the middle Miocene, approximately 20 to 10 m.y.
ago. The middle Miocene age of faulting
corresponds to pre-basin-range extension
described by Zoback and others (1981) that
preceded the break-up of the region into ranges
resembling the modern basins and ranges. A pre-
basin-range age of most of the low-angle faults
is indicated in many parts of the Basin and Range
province where low-angle extentional faults are
cut by, and are thus older than, high-angle
basin-range faults.

A middle Miocene age for low-angle faulting
is clearly indicated in the Yerington area of
western Nevada, where faults that are now nearly
horizontal are interpreted to have been

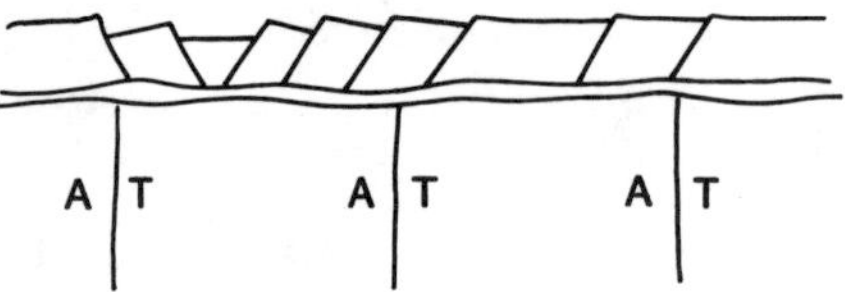

Figure 6. Models of low-angle extensional faults.

originally high-angle faults during eruption of andesitic rocks about 17 to 14 m.y. ago and to have been rotated into their present horizontal position largely before eruption of basaltic rocks 8 m.y. ago (Proffett, 1977). Directly west of the Yerington area, nearly flat faults similar to those near Yerington are cut by mafic dikes dated as 15.7 m.y. old (Hudson and Oriel, 1979), indicating that movement on some of these low-angle faults had ceased by then.

Detachment faults in metamorphic core complexes of the northern Basin and Range may also be mostly middle Miocene in age. In the Grouse Creek metamorphic core complex in northwestern Utah, a 24.9-m.y.-old stock contains foliation and late-stage lineation parallel to that in surrounding rocks, indicating that metamorphism and associated east-west extension is at least as young as this stock (Compton and others, 1977). In the Grouse Creek Range (Compton, 1983) and in the Matlin Mountains (Todd, 1983), large allochthonous sheets override 11 m.y.-old Miocene sedimentary rocks. In the Ruby Range metamorphic core complex, a low-angle fault system cuts 17-m.y.-old volcanic rocks (Snoke, 1980).

A Miocene age of low-angle faulting is also suggested by potassium-argon dating in metamorphic core complexes. In the Snake Range metamorphic complex, potassium-argon ages on mica decrease toward the Snake Range decollement or detachment surface; cataclastic rocks near the detachment are about 17 m.y. old (Lee and others, 1970). This relationship is interpreted (Lee and others, 1970) to indicate reduction in ages due to thermal stresses related to Tertiary activity along the detachment surface. In the Ruby Mountains metamorphic core complex, potassium-argon ages decrease progressively across the range and reach a minimum age of 21 m.y. in cataclastic rocks along a detachment surface (Kistler and Willden, 1969). The 21- m.y. age may be the approximate age of detachment faulting (Kistler and Willden, 1969), although Kistler and O'Neil (1975) have suggested this is a cooling age related to uplift and erosion of the range. Potassium-argon ages (E. H. McKee, written commun., 1983) of about 11 m.y. and 14 m.y., respectively, have been obtained from metamorphic minerals in the Bullfrog Hills in western Nevada and from the Trappman Hills in south-central Nevada. The metamorphic rocks in these two areas have been considered to be part of the Precambrian crystalline basement (Cornwall and Kleinhampl, 1964; Ekren and others, 1971), although they lie far from outcrops of known Precambrian basement rocks. More likely, they represent terranes metamorphosed during the Mesozoic or Cenozoic (Stewart, 1980) and are similar to the metamorphic core complexes of the Snake Range and Ruby Mountains. If so, the Miocene potassium-argon ages may indicate the time of detachment faulting or uplift related to detachement faulting. A potassium-argon age of about 17 m.y. (E. H. McKee, written commun., 1983) for metamorphic minerals on Mineral Ridge in western Nevada may also indicate a time of detachment faulting. A detachment surface on Mineral Ridge separates a core of metamorphic rocks from a relatively unmetamorphic carapace. Mineral Ridge is considered a turtleback dome by Kirsch (1971) and a metamorphic core complex by G. A. Davis and Dayton Marcout (oral commun., 1982).

Some low-angle faulting, on the other hand, may be older than middle Miocene and some is clearly younger. Gans (1982) reported that 36-m.y.-old rhyolite dikes intrude and are cut by a system of faults that rotated domino-style to low-angle attitudes. These relations suggest to Gans (1982) that initial faulting occurred during the intrusion of the rhyolite. R. E. Anderson (written commun., 1983) described features in western Utah that may indicate large-scale low-angle faulting during Oligocene time. Evidence of young detachment faulting is found in the Death Valley area, where rocks that are probably about 6 to 8 m.y. old or younger are incorporated in the Amargosa Chaos of Noble (1941) above a widespread detachment surface (Wernicke, 1981; Stewart, 1983). In the Sevier Desert, high-angle normal faults that offset Quaternary surface deposits terminate at the Sevier Desert detachment fault, implying that Quaternary movement on the detachment fault triggered movement on the high-angle faults (M. L. Zoback, oral commun., 1982).

Present data do not give a clear picture of the direction of extension during the development of low-angle faults, particularly during the Miocene, when most of these structures apparently developed. Extension directions during development of late-stage lineation in the metamorphic core complexes in northwestern Nevada and northwest Utah were nearly east-west (fig. 5). In the Grouse Creek Range, this lineation is developed in a 24.9-m.y.-old pluton, indicating that east-west extension is locally at least this young. If potassium-argon dates on cataclastic rocks in the Ruby Mountains metamorphic core complex (21 m.y.), and in the Snake Range (17 m.y.) indicate the time of detachment faulting, east-west extension may be at least as young. Zoback and others (1981), on the other hand, have outlined evidence of the preferentially oriented dike swarms and fault-slip vectors for 20 to 10 m.y. ago that indicates a uniform west southwest-east northeast extension direction, based mostly on evidence from the Arizona segment of the Basin and Range province. Zoback and Thompson (1978) indicated this same extension direction during the emplacement of a 13.8- to 16.3-m.y.-old north-northwest-trending system of dikes and lava flows in northern Nevada. Perhaps the extension direction changed from largely east-west during development of the core complexes to west southwest-east northeast at a slightly younger time, but the exact timing of these events, if they are indeed related to different times and directions of extension, is not clear. A further complexity is indicated by the southward sweep of Cenozoic igneous activity from 43 to 6 m.y. ago. This southward sweep, as described above, was accompanied by the development of east-west-trending igneous belts and east-west-trending volcano-tectonic toughs or grabens suggestive of north-south extension. This period of north-south extension overlaps in time both the east-west extension in the core complexes and the presumably younger west southwest-east northeast extension indicated by dike emplacement.

Basin-range faulting

The term "basin-range faulting" is restricted (Zoback and others, 1981) to faulting that produced the elongate mountains and valleys that characterize the present-day Basin and Range province (fig. 1). The characteristic crest to crest spacing of the mountain blocks is generally 25 to 35 km. The mountains are usually about 10 to 20 km across and are separated by alluvial valleys of comparable width. In detail, the patterns of mountains and valleys are highly complex and the ranges, although generally elongate, are locally equidimensional or elliptical in plan view.

Basin-range structure is related to block faulting, in which ranges are formed by vertical movements along major faults on one or both sides of the mountain block. The concept of block faulting may imply that the structure is simple, whereas in reality, faulting is not merely confined to the sides of mountains but is distributed throughout the mountain areas and in suballuvial rocks of the valleys as well. Major structural blocks, therefore, should not be viewed as rigid coherent masses but rather as aggregate structural units that generally move in a more or less uniform manner relative to adjacent structural units.

Three general models (fig. 7) of basin-range structure have been proposed. One model relates basin-range structure to a system of horsts and grabens in which individual horsts form mountains and individual grabens form valleys. A second model relates basin-range structure to a system of buoyant blocks that float on a substratum. The third relates basin-range structure to a system of structural blocks rotated along curving, downward-flattening (listric) faults. Tilting of major blocks, a characteristic of many basin-range blocks, is most easily visualized in the buoyant block and listric fault model, although moderate tilting also can be accommodated in the horst and graben model.

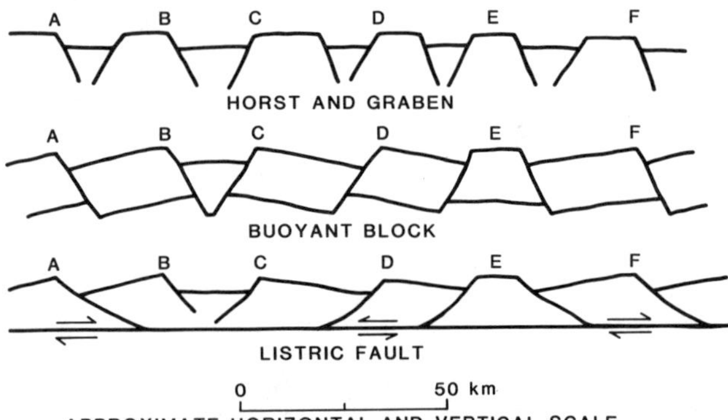

Figure 7. Models of basin-range structure. After Stewart (1979).

In a detailed review of selected subsurface data (mostly seismic reflection data) Anderson and others (1983) concluded that no single structural model explains basin-range structure but that the structure is in places due to major steep planar normal faults, in other places to moderately to deeply penetrating listric normal faults, and in still other places to sharply curving shallow listric faults that merge downward with a detachment surface.

Estimates of the amount of extension necessary to produce basin-range structure have varied widely, and at present no unanimity of opinion exists on the amount. Difficulties arise in part from uncertainties of the subsurface structure. If the major faults that bound basin-range blocks are largely planar, estimates of 10 to 20 percent are reasonable (Stewart, 1971, 1978; Thompson and Burke, 1974), whereas if the faults flatten with depth, estimates of 30 to perhaps several hundred percent are reasonable (Wright and Troxel, 1973; Proffett, 1977). An additional problem is the uncertainty in places as to how much extension is due to pre-basin-range low-angle faulting and how much is related to the basin-range block faulting. In places the

distinction between what is here considered to be largely two distinct events (pre-basin-range and basin-range) is unclear and the two may overlap in time.

The most definitive estimate of extension related to basin-range block faulting is in north-central Nevada, where Zoback (1979) estimated about 20 percent extension. Her estimate is based on an analysis of faulting on four major fault sets where the range-bounding faults have considerable oblique (left-lateral) components of motion. The total offset along each major fault set was calculated by using gravity data to constrain the vertical offset along the faults and magnetic data to contain lateral offsets of a zone of dikes.

The grain of present-day topography, and thus the development of basin-range fault block mountains, probably develop in the northern Basin and Range province during the past 10 m.y. (Stewart, 1978; Zoback and others, 1981). A young age for present-day topography is indicated in southern Nevada (Ekren and others, 1968) where an 11-m.y.-old tuff does not vary significantly in thickness between present-day valleys and mountains, a situation that would be impossible if present-day topography had developed by then. By the time another tuff had erupted about 7 m.y. ago, however, the topographic grain was much as it is today. This younger tuff lapped against some of the ranges and in places flowed in valleys that are the sites of present-day streams. In west-central Nevada, large sedimentary basins, presumably formed by extensional faulting, were well defined about 13 to 9 m.y. ago (Robinson and others, 1968; Gilbert and Reynolds, 1973). These basins were much more extensive (Gilbert and Reynolds, 1973) than present-day basins in the region, and studies of sedimentary transport directions in fluvial units in these basins indicate an integrated drainage system over large areas (fig. 8). Clearly, the present-day topography was not developed until after the close of deposition in these basins, about 9 m.y. ago. This interpretation is supported by widespread faulting of rocks younger than 9 m.y. in west-central Nevada (Stewart and others, 1982). In north-central Nevada, widespread basaltic andesite flows 17 to 14 m.y. old and an olivine basalt flow 10 m.y. old (Stewart and others, 1975; Stewart and McKee, 1977; Zoback and Thompson, 1978) are cut by major basin-range faults related to present-day ranges, whereas a 6-m.y.-old basalt (Zoback and Thompson, 1978) flowed across alluvium in what are now valleys. Here again, a young age (less than 10 m.y.) is indicated for the development of basin-range structure.

The general north to north-northeast orientation of major basin-range normal faults (fig. 1) in the northern Basin and Range province indicates extension in an east-west or east southeast-west southwest direction. Present-day extension also has this direction, based on earthquake focal mechanisms and _in situ_ stress

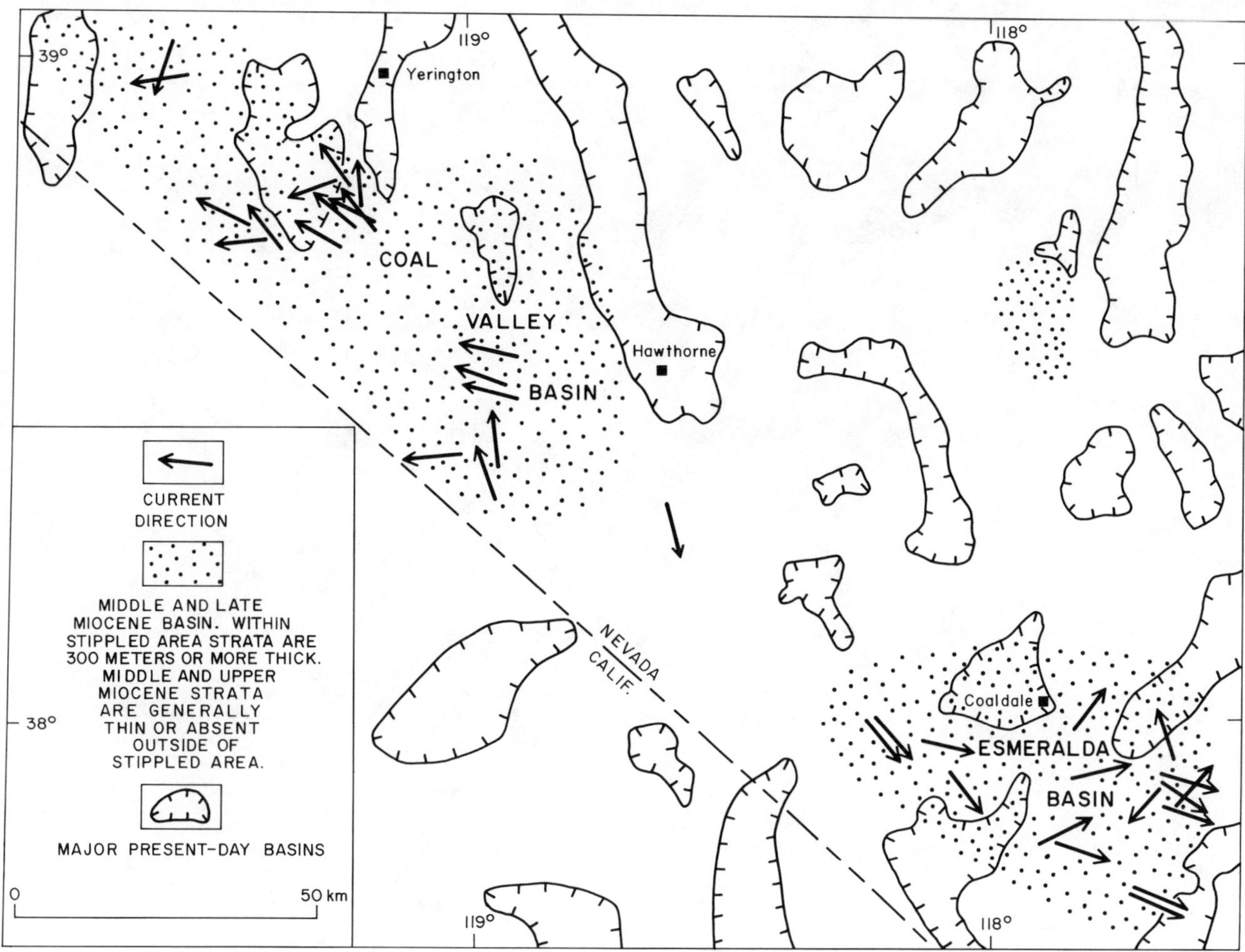

Figure 8. Distribution of middle and late Miocene sedimentary basins, paleocurrent directions in middle and upper Miocene fluvial units, and distribution of major present-day basins, eastern California and western Nevada.

measurements (Zoback and Zoback, 1980; Zoback and others, 1981).

Strike-slip faults

Major strike-slip faults, along which most movement was probably late Cenozoic in age, occur in the Walker Lane and in a diffuse east-northeast-trending zone (Garlock fault to Lake Mead fault system) in eastern California and southern Nevada. Both northwest-trending right-lateral faults and northeast-trending left-lateral faults occur. Estimates of offsets on individual faults are as great as 80 km (Smith, 1962; Stewart and others, 1968; Bohannon, 1979).

The restriction of strike slip faults to the Walker Lane and to the diffuse east-northeast zone in eastern California and southern Nevada may indicate that these zones allowed for adjustment between terranes of different tectonic character. The Walker Lane may mark the boundary between a northwest-moving Sierra Nevada block and the west-northwest-extending Basin and Range province (Wright, 1976). The east-northeast zone of strike-slip faults (Garlock fault to Lake Mead fault system) may mark a diffuse boundary between areas to the south that extended in a southwest direction and those to the north that extended in a west-northwest direction (fig. 9). Such a difference in extension direction north and south of the east-northeast zone of strike-slip faulting is compatible with evidence of extension directions during development of late-stage lineation in the metamorphic core complexes (fig. 5). In Arizona and southeastern California this extension direction is southwest-northeast, whereas in Nevada and western Utah it is east-west.

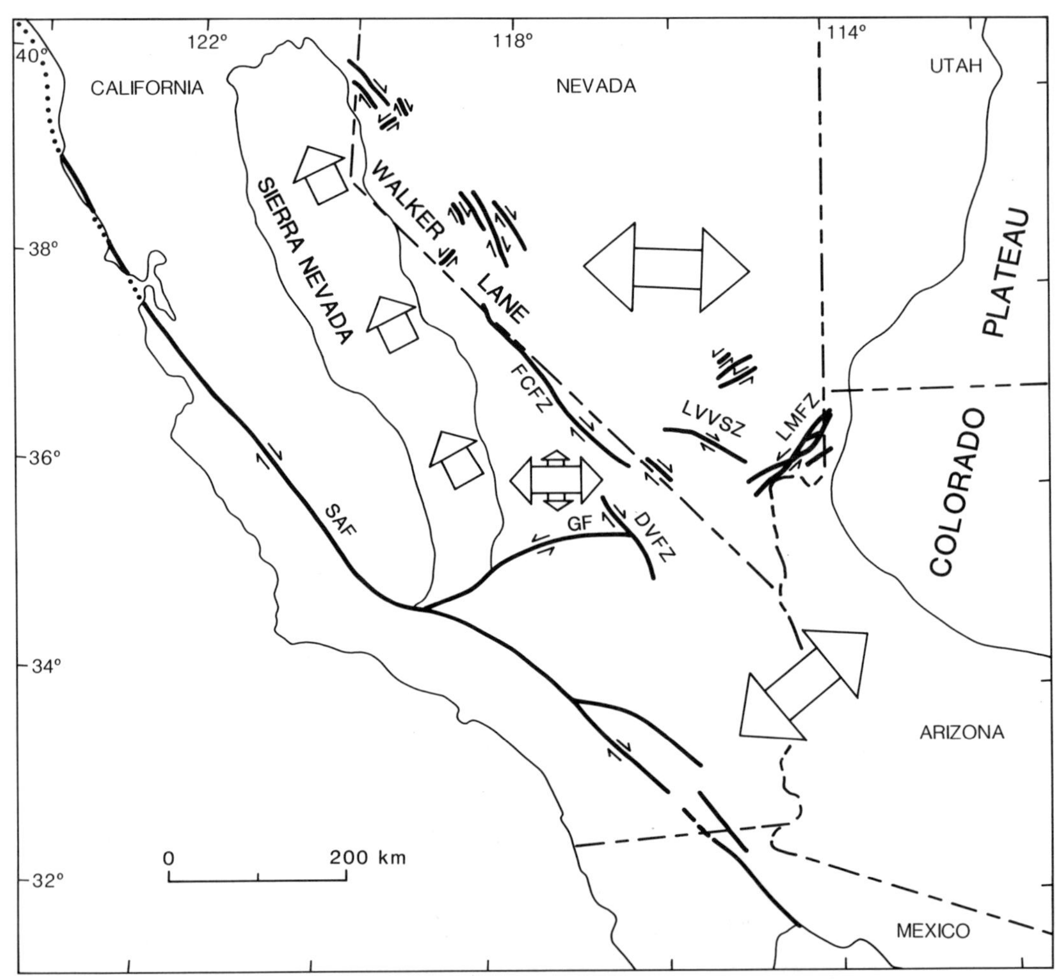

Figure 9. Strike-slip faults and an interpretation of regional extensional
directions during their development. Faults: FCFZ, Furnace Creek fault zone;
LVVSZ, Las Vegas Valley shear zone; LMFZ, Lake Mead fault zone; GF, Garlock
fault; DVFZ, Death Valley fault zone.

REFERENCES

Albers, J. P., 1967, Belt of sigmoidal bending
and right-lateral faulting in the western
Great Basin: Geological Society of America
Bulletin, v. 78, no. 2, p. 143-156.

Allmendinger, R. W., Sharp, J. W., Von Tish, D.,
Serpa, L., Brown, L., Oliver, J. E., and
Kaufman, S., 1983, COCORP seismic reflection
data from the eastern Basin and Range, west-
central Utah: Geological Society of America
Abstracts with Programs, v. 15, no. 5, p.
287

Anderson, R. E., 1971, Thin skin distension in
Tertiary rocks of southeastern Nevada:
Geological Society of America Bulletin, v.
82, no. 1, p. 43-58.

Anderson, R. E., Zoback, M. L., and Thompson, G.
A., 1983, Implications of selected
subsurface data on the structural form and
evolution of some basins in the northern
Basin and Range province: Geological
Society of America Bulletin (in press).

Armstrong, R. L., 1982, Cordilleran metamorphic
core complexes--from Arizona to southern
Canada: Annual Review Earth and Planetary
Science, v. 10, p. 129-154.

Bohannon, R. G., 1979, Strike-slip faults of the
Lake Mead region of southern Nevada, in
Armentrout, J. M., Cole, M. R., and Ter
Best, Harry, Jr., eds., Cenozoic
paleogeography of the western United
States: Society Economic Paleontologist and
Mineralogist, Pacific Section, Pacific Coast
Paleogeography Symposium 3, p. 129-139.

Burke, D. B., and McKee, E. H., 1979, Mid-
Cenozoic volcano-tectonic troughs in central
Nevada: Geological Society of America
Bulletin, v. 90, no. 2, part 1, p. 181-184.

Carlson, J. E., Laird, D. W., Peterson, J. A.,
Schilling, J. H., Silberman, M. L., and
Stewart, J. H., 1975, Preliminary map

showing distribution and isotopic ages of Mesozoic and Cenozoic intrusive rocks in Nevada: U.S. Geological Survey Open-File Report 75-499, scale 1:1,000,000.

Colleta, Bernard, and Angelier, Jacques, 1982, Sur les systemes de blocs failles bascules associes aux fortes extensions: etude preliminaire d'exemples ouest-americains (Nevada, U.S.A. et Basse-California, Mexique): C. R. Academy of Sciences Paris, t. 294, Serie II, p. 467-469.

Compton, R. R., 1983, Displaced Miocene rocks of the west flank of the Raft River-Grouse Creek core complex, Utah, in Miller, D. M. ed., Geological Society of America Memoir 157 (in press).

Compton, R. R., Todd, V. R., Zartman, R. E., and Naeser, C. W., 1977, Oligocene and Miocene metamorphism, folding, and low-angle faulting in northwestern Utah: Geological Society of America Bulletin, v. 88, no. 9, p. 1237-1250.

Coney, P. T., 1980, Cordilleran metamorphic core complexes: An overview, in Crittenden, M. D., Jr., Coney, P. J., and Davis, G. H., 1980, Cordilleran metamorphic core complexes: Geological Society of America Memoir 153, p. 7-31.

Cornwall, H. R., and Kleinhampl, F. J., 1964, Geology of the Bullfrog quadrangle and ore deposits related to the Bullfrog Hills caldera, Nye County, Nevada, and Inyo County, California: U.S. Geological Survey Professional Paper 454-J, 25 p.

Crosby, G. W., 1973, Regional structure in southwestern Utah, in Hintze, L. F., and Whelan, J. A., eds., Geology of the Milford area, 1973: Utah Geological Association Publication 3, p. 27-32.

Cross, T. A., and Pilger, R. H., Jr., 1978, Constraints on absolute motion and plate interaction from Cenozoic igneous activity in the western United States: American Journal of Science, v. 278, no. 7, p. 865-902.

Davis, G. A., Anderson, J. L., Frost, E. G., and Shackelford, T. J., 1980, Mylonization and detachment faulting in the Whipple-Buckskin-Rawhide Mountains terrane, southeastern California and western Arizona, in Crittenden, M. D., Jr., Coney, P. J., and Davis, G. H., eds, Cordilleran metamorphic core complexes: Geological Society of America Memoir 153, p. 79-129.

Davis, G. H., and Coney, P. J., 1979, Geologic development of the Cordilleran metamorphic core complexes: Geology, v. 7, no. 3, p. 120-124.

Eaton, G. P., 1982, The Basin and Range province: Origin and Tectonic significance: Annual Review Earth and Planetary Science, v. 10, p. 409-440.

Ekren, E. B., Anderson, R. E., Rogers, C. L., and Noble, D. C., 1971, Geology of the northern Nellis Air Force Base Bombing and Gunnery Range, Nye County, Nevada: U.S. Geological Survey Professional Paper 651, 91 p.

Ekren, E. B., Bucknam, R. C., Carr, W. J., Dixon, G. L., and Quinlivan, W. D., 1976, East-trending structural lineaments in central Nevada: U.S. Geological Survey Professional Paper 986, 16 p.

Ekren, E. B., Rogers, C. L., Anderson, R. E., and Orkild, P. P., 1968, Age of basin and range normal faults in Nevada Test Site and Nellis Air Force Range, Nevada, in Nevada Test Site: Geological Society of America Memoir 110, p. 247-250.

Fouch, T. D., 1979, Character and paleogeographic distribution of Upper Cretaceous(?) and Paleocene nonmarine sedimentary rocks in east-central Nevada, in Armentrout, J. M., Cole, M. R., and Ter Best, Harry, Jr., eds., Cenozoic paleogeography of the western United States: Society of Economic Paleontologists and Mineralogists, Pacific Coast Paleogeography Symposium 3, p. 97-111.

Gans, P. B., 1982, Geometry of mid Tertiary extensional faulting, northern Egan Range, east-central Nevada: Geological Society America Abstracts with Programs, v. 14, no. 4, p. 165.

Gianella, V. P., and Callaghan, Eugene, 1934, The earthquake of December 20, 1932, at Cedar Mountain, Nevada, and its bearing on the genesis of Basin-Range structure: Journal of Geology, v. 42, p. 1-22.

Gilbert, C. M., Christensen, M. N., Al-Rawi, Yehya, and Lajoie, K. R., 1968, Structural and volcanic history of Mono Basin, California-Nevada, in Studies in volcanology--A memoir in honor of Howel Williams: Geological Society of America Memoir 116, p. 275-329.

Gilbert, C. M., and Reynolds, M. W., 1973, Character and chronology of basin development western margin of the Basin and Range province: Geological Society of America Bulletin, v. 84, no. 8, p. 2489-2509.

Gross, W. W., and Hillemeyer, F. L., 1982, Geometric analysis of upper-plate fault patterns in the Whipple-Buckskin detachment terrane, California and Arizona, in Frost, E. G., and Martin, D. L., eds., Mesozoic-Cenozoic tectonic evolution of the Colorado River region, California, Arizona, and Nevada: San Diego, California, Cordilleran Publishers, p. 257-265.

Hardyman, R. F., 1978, Volcanic stratigraphy and structural geology of Gillis Canyon quadrangle, northern Gillis Range, Mineral County, Nevada: University of Nevada, Reno, Ph.D. thesis, 248 p.

Hardyman, R. F., Ekren E. B., and Byers, F. M., Jr., 1975, Cenozoic strike-slip, normal and detachment faults in northern part of the Walker Lane, west-central Nevada: Geological Society of America Abstracts with Programs, v. 7, no. 7, p. 1100.

Hilpert, L. S., and Roberts, R. J., 1964, Economic geology, in Minerals and water resources of Utah: U.S. Geological Survey Report, p. 28-37.

Hintze, L. F., 1973, Geologic history of Utah: Brigham Young University Geological Studies, v. 20, pt. 3, Studies for Students No. 8, 181 p.

Hose, R. K., and Danes, Z. F., 1973, Development of late Mesozoic to early Cenozoic structures in the eastern Great Basin, in De Jong, K. A., and Scholten, Robert, eds., Gravity and tectonics: New York, John Wiley, p. 429-442.

Howard, K. A., Goodge, J. W., and John, B. E., 1982, Detached crystalline rocks of the Mahare, Buck and Bill Williams Mountains, western Arizona, in Frost, E. G., and Martin, D. L., eds., Mesozoic-Cenozoic tectonic evolution of the Colorado River region, California, Arizona, and Nevada: San Diego, California, Cordilleran Publishers, p. 377-390.

Hudson, D. M., and Oriel, W. M., 1979, Geologic map of the Buckskin Range, Nevada: Nevada Bureau of Mines and Geology Map 64, scale 1:18,000.

Kellogg, H. E., 1964, Cenozoic stratigraphy and structure of the southern Egan Range, Nevada: Geological Society of America Bulletin, v. 75, no. 10, p. 949-968.

Kirsch, S. A., 1971, Chaos structure and turtleback dome, Mineral Ridge, Esmeralda County, Nevada: Geological Society of America Bulletin, v. 82, no. 11, p. 3169-3176.

Kistler, R. W., and O'Neil, J. R., 1975, Fossil thermal gradients in crystalline rocks of the Ruby Mountains, Nevada, as indicated by radiogenic and stable isotopes: Geological Society of America Abstracts with Programs, v. 7, no. 3, p. 334-335.

Kistler, R. W., and Willden, Ronald, 1969, Age of thrusting in the Ruby Mountains, Nevada: Geological Society of America Abstracts with Programs, v. 1, pt. 5, p. 40-41.

Lee, D. E., Marvin, R. F., Stern, T. W., and Peterman, Z. E., 1970, Modification of potassium-argon ages by Tertiary thrusting in the Snake Range, White Pine County, Nevada, in Geological Survey research 1970: U.S. Geological Survey Professional Paper 700-D, p. D92-D102.

Locke, Augustus, Billingsley, R. R, and Mayo, E. B., 1940, Sierra Nevada tectonic patterns: Geological Society of America Bulletin, v. 51, no. 4, p. 513-540.

McDonald, R. E., 1976, Tertiary tectonics and sedimentary rocks along the transition, Basin and Range province to plateau of thrust belt province, Utah, in Hill, J. G., eds., Symposium on Geology of the Cordilleran Hingeline: Rocky Mountain Association of Geologists, Denver, p. 281-317.

Miller, C. F., and Bradfish, L. J., 1980, An inner Cordilleran belt of muscovite-bearing plutons: Geology, v. 8, no. 9, p. 412-416.

Moores, E. M., Scott, R. B., and Lumsden, W. W., 1968, Tertiary tectonics of the White Pine-Grant Range region, east-central Nevada, and some regional implications: Geological Society of America Bulletin, v. 79, no. 12, p. 1703-1726.

Morton, W. H., and Black, R., 1975, Crustal alteration in Afar, in Pilger, A., and Rosler, A., eds., Afar depression of Ethopia, Inter-Union Commission on Geodynamics: International Symposium on the Afar Region and Related Rift Problems, E. Schweizerbartische Verlagsbuchhandlung, Stuttgard, Germany, Proceedings, Scientific Report No. 14, p. 55-65.

Newman, G. W., 1979, Late Cretaceous(?)-Eocene faulting in the east-central Basin and Range, in Newman, G. W., and Goode, H. P., eds., 1979, Basin and Range symposium and Great Basin field conference: Rocky Mountain Association of Geologists and Utah Geological Association, p. 167-173.

Noble, D. C., Slemmons, D. B., Korringa, M. K., Dickinson, W. R., Al-Rawi, Yehya, and McKee, E. H., 1974, Eureka Valley Tuff, east-central California and adjacent Nevada: Geology, v. 2, no. 3, p. 139-142.

Noble, L. F., 1941, Structural features of the Virgin Spring area, Death Valley, California: Geological Society of America Bulletin, v. 52, p. 921-1000.

Osmond, T. C., 1960, Tectonic history of the Basin and Range province in Utah and Nevada: Mining Engineering, v. 12, no. 3, p. 251-265.

Proffett, J. M., Jr., 1977, Cenozoic geology of the Yerington district, Nevada, and implications for the nature and origin of Basin and Range faulting: Geological Society of America Bulletin, v. 88, no. 2, p. 247-266.

Roberts, R. J., Crittenden, M. D., Jr., Tooker, E. W., Morris, H. T., Hose, R. K., and Cheney, T. M., 1965, Pennsylvanian and Permian basins in northwestern Utah, northeastern Nevada, and south-central Idaho: American Association of Petroleum Geologists Bulletin, v. 49, no. 11, p. 1926-1956.

Robinson, P. T., McKee, E. H., and Moiola, R. J., 1968, Cenozoic volcanism and sedimentation, Silver Peak region, western Nevada and adjacent California, in Coats, R. R., Hag, R. L., and Anderson, C. A., eds., Studies in volcanology--A memoir in honor of Howel Williams: Geological Society of America Memoir 116, p. 577-611.

Rowley, P. D., Lipman, P. W., Mehnert, H. H., Lindsey, D. A., and Anderson, J. J., 1978, Blue Ribbon lineament, an east-trending structural zone within the Pioche mineral belt of southwestern Utah and eastern Nevada: U.S. Geological Survey, Journal of Research, v. 6, no. 2, p. 175-192.

Rowley, P. D., Steven, T. A., Anderson, J. J., and Cunningham, C. G., 1979, Cenozoic stratigraphic and structural framework of southwestern Utah: U.S. Geological Survey Professional Paper 1149, 22 p.

Shawe, D. R., and Stewart, J. H., 1976, Ore deposits as related to tectonics and magmatism, Nevada and Utah: American

Institute of Mining, Metallurgical and Petroleum Engineers Transactions v. 260, p. 225-232.

Smith, G. I., 1962, Large lateral displacement on Garlock fault, California, as measured form offset dike swarms: American Association of Petroleum Geologists Bulletin, v. 46, no. 1, p. 85-104.

Snoke, A. W., 1980, Transition from infrastructure to suprastructure in the northern Ruby Mountains, Nevada, in Crittenden, M. D., Jr., Coney, P. J., and Davis, G. H., Cordilleran metamorphic core complexes: Geological Society of America Memoir 153, p. 287-333.

Snyder, W. S., Dickinson, W. R., and Silberman, M. L., 1976, Tectonic implications of space-time patterns of Cenozoic magmatism in the western United States: Earth and Planetary Science Letters, v. 32, p. 91-106.

Speed, R. C., and Cogbill, A. H., 1979, Deep fault trough of Oligocene age, Candelaria Hills, Nevada: Summary: Geological Society of America Bulletin, pt. 1, v. 90, no. 2, p. 145-148.

Stewart, J. H., 1971, Basin and Range structure--a system of horsts and grabens produced by deep seated extension: Geological Society of America Bulletin, v. 82, no. 4, p. 1019-1044.

Stewart, J. H., 1978, Basin-range structure in western North America: A review, in Smith, R. B., and Eaton, G. P., eds., Cenozoic tectonics and regional geophysics of the western Cordillera: Geological Society of America Memoir 152, p. 1-31.

Stewart, J. H., 1979, Regional tilt patterns of late Cenozoic basin-range fault blocks in the Great Basin: Geological Society of America Bulletin, pt. 1, v. 91, no. 8, p. 460-464.

Stewart, J. H., 1980, Geology of Nevada: Nevada Bureau of Mines and Geology Special Publications 4, 136 p.

Stewart, J. H., 1983, Extensional tectonics in the Death Valley area, California--transport of the Panamint Range structural block 80 km northwestward: Geology, v. 11, p. 153-157.

Stewart, J. H., Albers, J. P., and Poole, F. G., 1968, Summary of regional evidence for right-lateral displacement in the western Great Basin: Geological Society of America Bulletin, v. 79, no. 10, p. 1407-1413.

Stewart, J. H., Carlson, J. E., and Johannesen, D. C., 1982, Geologic map of the Walker Lake 1° by 2° quadrangle, California and Nevada: U.S. Geological Survey Miscellaneous Field Studies Map MF-1382-A, scale 1:250,000.

Stewart, J. H., and Johannesen, D. C., 1979, Map showing regional tilt patterns of late Cenozoic basin-range fault blocks in western United States: U.S. Geological Survey Open-File Report 79-1134.

Stewart, J. H., and McKee, E. H., 1977, Geology, Pt. I, in Geology and minerals deposits of Lander County, Nevada: Nevada Bureau of Mines and Geology Bulletin 88, p. 1-59.

Stewart, J. H., Moore, W. J., and Zietz, Ididore, 1977, East-west patterns of Cenozoic igneous rocks, aeromagnetic anomalies, and mineral deposits, Nevada and Utah: Geological Society of America Bulletin, v. 88, no. 1, p. 67-77.

Stewart, J. H., and Poole, F. G., 1974, Lower Paleozoic and uppermost Precambrian Cordilleran miogeocline, Great Basin, Western United States, in Dickinson, W. R., eds., Tectonics and Sedimentation: Society of Economic Paleontologists and Mineralogists Special Publication 22, p. 28-57.

Stewart, J. H., Walker, G. W., and Kleinhampl, F. J., 1975, Oregon-Nevada lineament: Geology, v. 3, p. 265-268.

Thompson, G. A., and Burke, D. B., 1974, Regional geophysics of the Basin and Range province: Annual Review Earth and Planetary Science, v. 2, p. 213-238.

Todd, V. R., 1983, Late Miocene tectonic displacement of pre-Tertiary and Tertiary rocks in the Matlin Mountains, northwestern Utah, in Miller, D. M., ed., Geological Society of America Memoir 157 (in press).

Tooker, E. W., 1971, Regional structure in southwestern Utah, in Hintze, L. F., and Whelan, J. A., eds., Geology of the Milford area, 1973: Utah Geological Association Publication 3, p. 27-32.

Wallace, R. E., 1965, Possible origin of homoclinal structure in West Humboldt Range by landslide-like mechanism, an illustration, in Basin and Range Geologic Field Conference, 1st, Mackay School of Mines, Reno, Nevada, 1965, Guidebook.

Wernicke, Brian, 1981, Low-angle normal faults in the Basin and Range province: nappe tectonics in an extending orogen: Nature, v. 291, no. 5817, p. 645-648.

Wernicke, Brian, and Burchfiel, B. C., 1982, Modes of extensional tectonics, Journal of Structural Geology, v. 4, no. 2, p. 105-115.

Wright, L A., 1976, Late Cenozoic fault patterns and stress fields in the Great Basin and western displacement of the Sierra Nevada block: Geology, v. 4, no. 8, p. 489-494.

Wright, L A., and Troxel, B. W., 1973, Shallow-fault interpretation of Basin and Range structure, southwestern Great Basin, in de Jong, K. A., and Scholten, Robert, eds., Gravity and tectonics: New York, John Wiley, p. 397-407.

Zoback, M. L., 1979, Direction and amount of late Cenozoic extension in north-central Nevada: Geological Society of America, Abstracts with Programs, v. 11, no. 3, p. 137.

Zoback, M. L., 1983, Shallow structure of the Sevier Desert detachment: Geological Society of America Abstracts with Programs, v. 15, no. 5, p. 287.

Zoback, M. L., Anderson, R. E., and Thompson, G. A., 1981, Cenozoic evolution of the state of stress and style of tectonism of the Basin and Range province of the western United States, in Vine, F. J., and Smith, A. D., eds., Extensional tectonics associated with convergent plate boundaries: The Royal Society, London, p. 189-216.

Zoback, M. L., and Thompson, G. 1978, Basin and Range rifting in northern Nevada: Clues from a mid-Miocene rift and its subsequent offset: Geology, v. 6, no. 2, p. 111-116.

Zoback, M. L., and Zoback, Mark, 1980, State of stress in the conterminous United States: Journal of Geophysical Research, v. 85, no. B11, p. 6113-6156.

Active and fossil hydrothermal-convection systems of the Great Basin

By Donald E. White and Chris Heropoulos

U.S. Geological Survey, Menlo Park, CA 94025

ABSTRACT

The Great Basin contains hundreds of active thermal spring systems, most discharge at temperatures much below boiling and generally lack characteristics likely to be preserved. Twenty-eight systems have measured or geochemically indicated subsurface temperatures from 150°C to 265°C.

The epithermal group of Au-Ag vein deposits of the Great Basin are the "fossil" equivalents of some active systems, including Steamboat Springs, Nev., and Broadlands, New Zealand, but other active systems of nearly equal temperature are not known to be depositing Au and Ag.

This paper compares the characteristics of the active and fossil epithermal vein systems, including their present distribution, physical and chemical characteristics, and evolution through space and time. The fossil systems ranged from ~0.2 to 7% salinity and averaged at least 1% (about twice the present systems).

Many epithermal Au-Ag deposits are also enriched in As, Sb, and Hg, and some contain Tl, Bi, Se, Te, and B. We are analyzing hot-spring sinters for these "mobile" elements as well as Au and Ag. Steamboat Springs is the only presently active Great Basin system whose sinters commonly exceed our detection limits of 0.1 ppm Au and Ag and also are highly enriched in most other mobile elements. Long Valley, Calif., is another possible example, but Roosevelt hot springs, a volcanic-centered system in the eastern Great Basin, is high in most of the mobile elements, but not in Ag and Au.

The "disseminated" or "sediment-hosted" gold deposits, from available data, are older than most epithermal veins. Carlin and Getchell(?) are enriched in the "mobile" element group, but our data are inadequate for a clear understanding of interrelations.

INTRODUCTION

Most known active hydrothermal convection systems discharge at the surface as hot springs. Some active systems are not discharging now but have discharged in the recent past, and others may never have discharged directly to the surface.

Fossil hydrothermal convection systems have probably been abundant through much of the Tertiary period, presumably varying greatly in number and intensity. Fossil low-temperature systems would leave little evidence of their former activity. High-temperature systems that altered the rocks along their channels, deposited metalliferous ores near these channels, or formed siliceous spring deposits where they discharged at the ground surface of the time are the fossil equivalents of some present-day high-temperature systems. This paper focusses on former systems that deposited ores generally known as epithermal gold-silver deposits. We compare the distributions, salinities, and temperatures of these two groups through study of the active systems and any available evidence on temperatures, salinities, and rare chemical elements of the fossil systems.

ACTIVE SYSTEMS

In a recent assessment of geothermal resources of the United States, 28 convection systems within or immediately adjacent to the Great Basin were identified as indicating reservoir temperatures of at least 150°C (Brook and others, 1979). In part, temperatures were physically measured in drilled wells, but in others, minimum temperatures were estimated from chemical geothermometers, with an uncertainty generally less than 10°C. Other high-temperature systems surely exist, either not discharging at the surface or mixing with normal ground water below the surface and not yet identified.

Table 1 lists the 28 identified systems, and figure 1 shows their locations. Most of these 28 high-temperature systems are clustered along or near the west boundary of the Great Basin and in the "Battle Mountain High" of above-normal conductive heat flow of northern Nevada (Muffler, 1979, pl. 1). The eight systems in table 1 that have indicated reservoir temperatures >200°C are identified by name in figure 1.

Three systems along the western margin of the Great Basin are associated with silicic volcanism younger than ~1 m.y., including Steamboat Springs, Nev., Long Valley, Calif., and Coso, Calif. Roosevelt Springs, Utah, is the only system near the east basin margin with similarly young volcanism; its reservoir temperature, near 265°C, is the highest yet identified in the Great Basin.

Four systems, Mickey, Oreg., and Humboldt House, Beowawe, and Desert Peak, Nev., have no

Table 1. Active geothermal systems of the Great Basin with measured or geochemically indicated temperatures >150°C[1]

State	Name	N. Lat		W. Long		No.[1]	Temp., °C[2]
CA	Surprise Valley	41	40.0	120	12.0	35	152
CA	Long Valley	37	40.0	118	52.0	56	<u>227</u>
CA	Coso	36	3.0	117	47.0	57	<u>220</u>
[3]CA	Randsburg	35	23.0	117	32.0	58	172
NV	Baltazor	41	55.3	118	42.6	130	158
NV	Pinto	41	21.0	118	47.0	132	173
NV	Gerlach	40	39.7	119	21.7	137	178
NV	San Emedio Desert	40	24.0	119	25.0	138	166
NV	Steamboat Springs	39	23.0	119	45.0	141	<u>228</u>
NV	Lee	39	12.6	118	43.4	143	166
[3]NV	Soda Lake	39	34.0	118	51.0	144	157
[3]NV	Stillwater	39	31.0	118	33.1	145	159
[3]NV	Fernley	39	35.9	119	6.4	146	182
NV	Brady	39	47.2	119	0.0	147	155
[3]NV	Desert Peak	39	45.0	118	57.0	148	<u>221</u>
NV	Humboldt House	40	32.1	118	16.1	151	<u>217</u>
NV	Kyle	40	24.4	117	52.9	152	159
NV	Leach	40	36.2	117	38.7	154	162
NV	Beowawe	40	34.2	116	34.8	162	<u>229</u>
NV	Hot Sulfur (Tusc.)	41	28.2	116	9.0	164	165
NV	Sulphur Hot Springs	40	35.2	115	17.1	169	178
OR	Crumps	42	13.8	119	53.0	190	167
OR	Mickey	42	40.5	118	20.7	196	<u>205</u>
OR	Alvord	42	32.6	118	31.6	197	181
OR	Hot Borax Lake	42	20.0	118	36.0	198	191
OR	Trout Creek	42	11.0	118	23.0	199	154
[3]UT	Cove Fort	38	36.0	112	33.0	208	167
UT	Roosevelt	38	30.0	112	50.9	209	<u>265</u>

[1]Identification number used by Brook and others, 1979.

[2]Estimated reservoir temperature; underlined systems exceed 200°C.

[3]No present surface discharge of water.

clear relations to young volcanism. Instead, they seem more closely related to high regional conductive heat flow.

Southern Nevada is notably lacking in high-temperature spring systems for reasons that are not yet clear. Their apparent absence may be related to the dry climate, low annual rainfall of the region, and no young silicic volcanism. In addition, many of the intermontane basins have deep water tables, with subsurface drainage through carbonate aquifers.

Low-temperature systems (<150°C) are not considered here because they seldom form sinter deposits, veins, or hydrothermal mineral associations likely to be preserved or recognized as fossil geothermal systems.

FOSSIL PRECIOUS METAL SYSTEMS OF THE GREAT BASIN

Most of the ore deposits called "epithermal" by Lindgren (1933 and earlier studies) consist of Au-Ag (or Ag-Au) veins in or closely associated with Tertiary volcanic rocks of the Great Basin, although he also included deposits mined primarily for Hg, Sb, and As. The first worker to concentrate on epithermal precious metal deposits was Nolan (1933), who focused on their physical characteristics, mineralogy, and great differences in Au/Ag ratios.

The most thorough recent review of precious-metal veins, largely or almost entirely in volcanic rocks of western U.S. and Mexico, was by Buchanan (1981), who objected to the term "epithermal" as a misnomer with respect to temperature. He also excluded similar veins entirely in non-volcanic rocks, as well as other precious metal deposits of recent interest, variously called "disseminated," "carbonate-hosted," "invisible gold," "bulk minable," and "Carlin type." The relations of most of these precious metal deposits to the classic volcanic-hosted veins are still being debated and are reviewed briefly here. Table 2 lists the available data on locations, ages, temperatures, and salinities of these systems, and figure 2 shows their distributions.

COMPARISON OF ACTIVE HIGH-TEMPERATURE CONVECTION SYSTEMS WITH FOSSIL Au-Ag ORE DEPOSITS

Table 3 compares the active convection systems with the fossil Au-Ag ore deposits of the Great

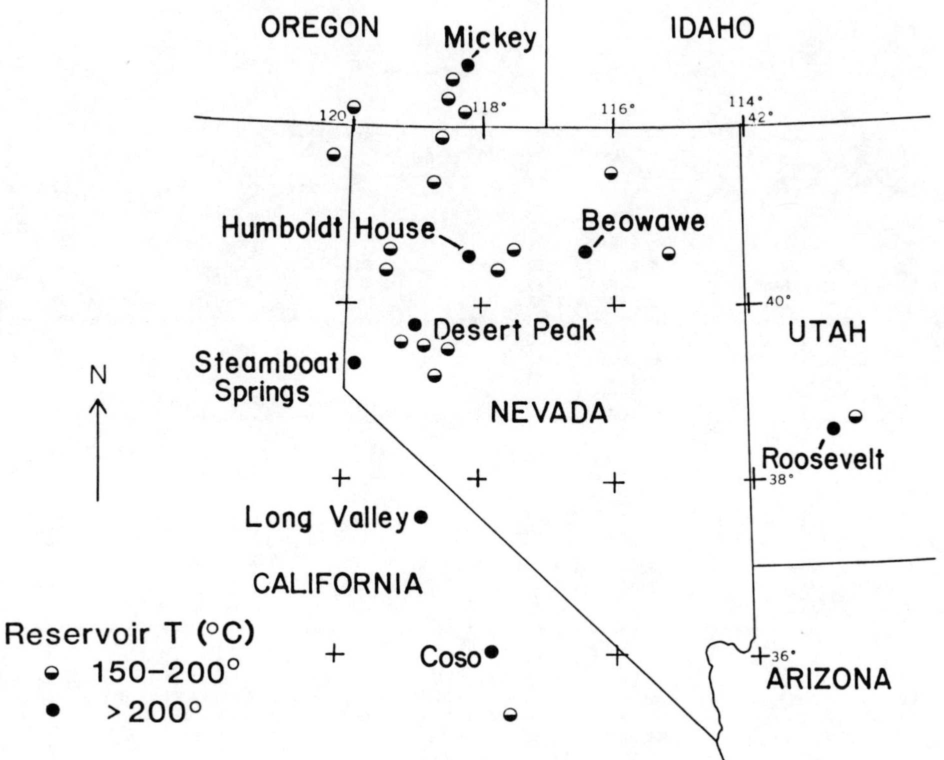

Figure 1. Active geothermal systems of the Great Basin, >150°C

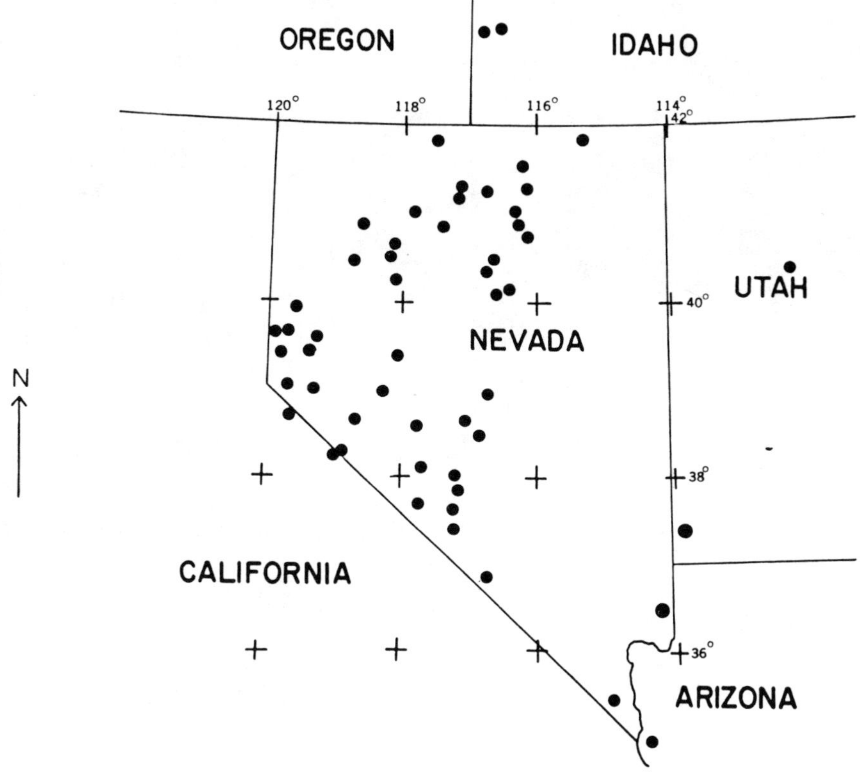

Figure 2. Gold-silver deposits of Nevada and nearby areas

43

Table 2. Gold-Silver deposits of Nevada and nearby areas, indicating ages, salinities, and temperatures of ore solutions.

State	No.	District	Lat.	Long.	Age, m.y.[1]	Salinity[2]	Temps, °C[3]	Refs[4]
NV	1	Adelaide	40 51	117 30	14			S
NV	2	Aurora	38 17	118 54	~11	0.2 – 1.7	227 – 255	S, N, Bu
[5]NV	3	Blue Star	40 54	116 19	37.5			B
[5]NV	4	Borealis	38 40	118 45	5			B
NV	5	Buckhorn	40 10	116 25	14.6			S, Bu
NV	6	Buckskin	38 59	119 20				B
NV	7	Bullfrog	36 55	116 48	9			S, Bu
[5]NV	8	Bullion	40 23	116 43	35			S
NV	9	Camp Douglas	38 21	118 12	15			S, G
[5]NV	10	Carlin	40 45	116 18	90		~150 – 200	N
NV	11	Cedar Mt. (Bell)	38 35	117 47		0.2 – 1.7	250	Bu
NV	12	Comstock Lode	39 03	119 37	~13		250 – 300	S, Bu
NV	13	Cornucopia	41 32	116 16	15			Bu
[5]NV	14	Cortez	40 08	116 37	35		~150 – 200	B, N, S
NV	15	Cuprite	37 31	117 13				A
NV	16	Divide	37 59	117 15	~16			S, Bu
[5]NV	17	Getchell	41 12	117 16	90			B
NV	18	Gilbert	38 09	117 40	8		275	Bu, S
[5]NV	19	Gold Acres	40 15	116 45	94	5.4 – 7.3	160 – 185	B, N
NV	20	Gold Circle	41 15	116 47	15		175 – 200	S, Bu
[5]NV	21	Gold Strike	40 59	116 22	78.4			B
NV	22	Goldfield	37 43	117 13	20		200 – 300	A, Bu, S
[5]NV	23	Hasbrouch	37 59	117 16	16			B
NV	24	Humboldt	40 36	118 11	73			Bu
NV	25	Jarbidge	41 52	115 26	14			S, Bu
NV	26	Manhattan	38 33	117 02	16	0.4 – 1.9	220	S, N
NV	27	National	41 50	117 35	15.5			Bu
[5]NV	28	Northumberland	38 57	116 47	84.6			B
NV	29	Peavine	39 36	119 55				A
[5]NV	30	Pinson	41 10	117 17	90?			B, S
NV	31	Pyramid	39 52	119 37	21			A
NV	32	Ramsey (Lyon)	39 27	119 19	10		221	S, Bu
NV	33	Rawhide	39 01	118 20	16			S, Bu
[5]NV	34	Rochester	40 17	118 09	58 – 79	~6	270 – 310	V, N
[5]NV	35	Round Mountain	38 42	117 04	25	0.2 – 1.4	250 – 260	B, N, S
NV	36	Searchlight	35 27	114 55				Bu
NV	37	Seven Troughs	40 29	118 46	14		240 – 318	S
NV	38	Silver Dike	38 19	118 12	17.3		300	S, G
NV	39	Silver Peak	37 47	117 43	5			S, Bu
[5]NV	40	Standard	40 31	118 12	73			B
NV	41	Steamboat Springs	39 23	119 45	<3	0.2	(90) – 230	S
NV	42	Sulphur	40 53	118 40	1.9			A
[5]NV	43	Talapoosa	39 27	119 19	10			S
NV	44	Tenmile	41 02	117 53	16	0.4 – 7.3	(135) – 330	N, S
NV	45	Tonopah	38 04	117 14	19	<1	240 – 265	S, Bu
NV	46	Tuscarora	41 17	116 14	38			S, Bu
NV	47	Widekind	39 35	119 45				A
NV	48	Wonder	39 24	118 06	22			S, Bu
CA	1	Bodie	38 12	119 00	7		215 – 245	S
CA	2	Monitor	38 42	119 40	5			S
AZ	1	Oatman	35 02	114 23			220	Bu
ID	1	DeLamar	43 02	116 50				Bu
ID	2	Silver City	43 02	116 44	~15			Bu
[5]UT	1	Gold Strike	37 23	113 53	78.4			Bu
[5]UT	2	Mercur	40 19	112 12				Bu

[1]Age, in millions of years; ~ indicates approximate average of range. Adularia most common mineral dated; see Silberman et al., 1976, and Buchanan, 1981, for summaries of minerals and dates.

[2]Salinity, from freezing-point depression of fluid inclusions, in NaCl equivalent.

[3]Temperatures of homogenization of fluid inclusions, with no pressure correction; probably valid for most of these deposits (Buchanan, 1981).

[4]References: S (Silberman and others, 1976, 1979); B (Bonham, 1982); Bu (Buchanan, 1981); A (Roger Ashley, personal communication, 1983); N (Nash, 1972); V (Vikre, 1981); G (Garside, 1979); and W (Andrew Wallace, personal communication, 1983).

[5]Deposits variously called disseminated, carbonate-hosted, Carlin-type, bulk mining, or invisible gold, and differing from classic epithermal veins; many other similar deposits recently discovered but no age or fluid-inclusion data available. The Rochester district has characteristics of both types.

Table 3. Active high-temperature convection systems compared with "fossil" Au-Ag ore deposits

	Active systems[1]	"Fossil" Au-Ag ore systems[2]
Temperature range, °C	150-265	~130-330
Surface Discharge	Most but not all	Some; evidence generally lacking
Sinter deposited from surface springs	Many; can be intermittent	A few; evidence generally lacking
Typical dissolved matter in fluids, %	0.1 to 0.8	0.2 to > 3
Water composition	$Na-K-Cl-HCO_3$	? - Mainly Na-Cl
Origin of water	Dominantly meteoric[3]	Dominantly meteoric[3]
Water pressure	Near hydrostatic for depth & T's	Near hydrostatic?
Evidence for boiling	Upper parts of most discharging systems[4]	Upper parts of most most systems[5]
Post-sinter erosion	0 to few 10's of m	Generally unknown
Source of energy	Volcanic-centered + high regional heat flow	Volcanic-centered?
Host rocks	Mainly alluvium near surface; variety of igneous, sedimentary, and metamorphic rocks at depth	Mainly volcanic rocks, in part underlain by sedimentary and metamorphic rocks
Open-space depositional banding	Some, but rarely studied (generally no core	Prevalent in most
Ore minerals Au Ag	 Unknown; Au low in most? Unknown (some pyrargyrite)[6]	 Au, electrum, selenides, tellurides[5] Argentite, electrum, various sulfosalts[5]
Au-Ag distribution with depth	Au and Au/Ag decrease downward[7]	Au and Au/Ag decrease downward, generally with increasing base metals[5]
Maximum depth range of ore	Not well defined	Au-Ag ore generally < 350 m; related to initial boiling?[5]
Hydrothermal gangue SiO_2 near surface SiO_2 at depth Feldspars Carbonates Sulfides	 Opal, crystobalite Chalcedony, quartz Nearly pure K-spar[6] in upper part; some deeper albite Calcite, in part lamellar, overgrown by quartz, partly dissolved Pyrite, stibnite, cinnabar, rarely marcasite; arsenopyrite, base metals at depth	 Chalcedony Chalcedony, quartz Nearly pure K-spar[6], some deeper albite Calcite commonly overgrown by quartz, then partly dissolved Pyrite, other sulfides (Buchanan, 1981)
Alteration minerals	Fn.gr. micas, clays; propylitic zoning; kaolinite, alunite, opal near and above water table	Similar to active systems (Buchanan, 1981)
Depth zoning of rare elements	Au, Hg, B, Tl, As, Sb enriched upwards; Ag/Au, Cu, Pb, Zn may increase downward	Au, Hg, Sb, commonly enriched upward; Ag/Au and base metals increase downward (Buchanan, 1981)
Ratio, vein length/depth	?	Commonly 2/1 or greater[5]

[1]Restricted to systems of at least 150°C, either by direct measurements or chemical geothermometry.

[2]Dominantly the classic "bonanza" veins in volcanic rocks, excluding the disseminated or carbonate-hosted deposits of Table 2.

[3]Most active and fossil systems of Great Basin are dominantly meteoric water (O'Neal and Silberman, 1974; White, 1974).

[4]Direct measurement; monoclinic, nearly pure K-feldspar (adularia) and tabular vein calcite crystals, commonly with quartz overgrowths.

[5](Buchanan, 1981); the best evidence for boiling is associated vapor-rich and liquid-rich fluid inclusions, vein calcite, and adularia.

[6]At Steamboat Springs, uncertain elsewhere.

[7]Au and Ag decrease downward as base metals increase at Steamboat Springs, NV, Broadlands, N.Z., and Waiotapu, N.Z., but data not available for most systems.

Basin. These fossil systems are dominantly the classic "bonanza" veins in volcanic rocks, excluding the disseminated or carbonate-hosted deposits of table 2, discussed below.

TEMPERATURES

The active systems (excluding all systems below 150°C) range as high as ~265°C (Roosevelt, Utah). In contrast, most ore deposits with temperatures of deposition determined from fluid inclusions (Table 2) range from 200° to 300°C, with an extreme range of 150°C (excluding Steamboat Springs and post-gold calcite inclusions from Tenmile) to 330°C in two of the tabulated deposits, with no data available from the others. Average temperatures of the studied deposits are close to 250°C. For comparison, temperatures of deposition of only 3 of the disseminated deposits are available: Carlin, 150° to 200°C (Nash, 1972); Gold Acres, 160° to 185°C (Nash, 1972); and Rochester, 270° to 310°C (Vikre, 1981, which has characteristics of both ore types).

SURFACE DISCHARGE

Most active geothermal systems discovered to date in the Great Basin discharge at the surface. Six systems of Table 1 have had no recent discharge, including Coso, Calif. (formerly discharge but now vents only gas); Lee and Brady, Nev. (discharge at very low rates as compared to former rate); and Roosevelt, Utah, which decreased to a seeping discharge of high-silica water by 1957 but self-sealed its vent with amorphous SiO_2 about 1960. Many of these spring systems discharged at much higher rates and temperatures in the past, probably during the late Pleistocene when water supply was abundant and water tables intersected the ground surface. The fossil precious-metal systems generally provide no direct evidence for discharging on the ground surface of the time. True sinter provides the most conclusive evidence but, if present over these fossil systems, it has generally been eroded. At least 17 of the active systems of Table 1 deposited sinter at some time in their past, thereby providing evidence for discharge temperatures close to boiling, and subsurface reservoir temperatures near or above 180°C (White, 1973).

SALINITIES ANE WATER COMPOSITIONS

Most active systems of the Great Basin have low contents of dissolved matter, generally ranging from <0.1 to 0.25 weight percent. However, a few systems listed in Table 1 range as high as 0.8 percent (Desert Peak, Nev., and Roosevelt, Utah), and Gerlach and Stillwater have intermediate salinities. Dissolved matter of most high temperature springs is dominated by NaCl and bicarbonate. Beowawe, however, is unusual for its low content of dissolved solids (~0.1 percent), containing only ~50 ppm Cl but considerably higher total carbonate and bicarbonate (generally near 400 ppm; Garside and Schilling, 1979).

In comparison, the epithermal vein fluids ranged from ~0.2 to 7.3 percent in salinity (NaCl equivalent from freezing-point depression of fluid inclusions), with most ranging from 0.2 to 2 percent. No data exist on actual compositions of these fluids. Ore-fluid salinities are seldom given the attention that Nash (1972) gave to the Tenmile district, where most inclusions indicate 0.9 to 2.1 weight percent, but late quartz fluids of two mines indicate ~7 percent, and a late (later?) calcite only 0.4 percent. The available data indicate that the fluids of epithermal veins are nearly always more saline than active high-temperature hot springs. Ore fluid salinities of the disseminated deposits are even less well known because of the scarcity of satisfactory fluid inclusions. Only the data from Gold Acres are "fairly satisfactory," according to Nash (1972), as judged from fluid proportions in tiny inclusions from Carlin and Cortez that seemed to homogenize at 175 ± 25°. No inclusions were suitable for freezing-stage salinities.

ORIGIN OF WATER

Water of the active systems is entirely or dominantly meteoric (Craig and others, 1956), as indicated mainly by the hydrogen isotopes of water. Volcanic-centered systems are suspected of having some magmatic water (in or from molten magma), but much of this water may ultimately have been meteoric, evolved connate, or metamorphic rather than derived from the mantle. A few thermal springs in California (White and others, 1973) are now known to contain a significant proportion of evolved connate or metamorphic water, based on contrasts in their δD values relative to present meteoric waters.

Fluid inclusion waters of fossil epithermal systems are also dominantly meteoric (O'Neil and Silberman, 1974), although climatic and elevation differences since formation of the ore deposits account for some of the contrast with present meteoric waters. The only epithermal district these authors suspected of having a significant direct magmatic content was the Comstock Lode, where one high-grade ore sample differed by ~50°/oo in δD from present meteoric waters of the area. More extensive systematic study of other epithermal districts will be required to determine the possible relations of the δD values of individual districts with depositional age and fluid inclusions. Multiple analyses from each district are necessary to establish whether other types of water than meteoric are present, but the considerably higher salinities of most inclusion fluids, relative to the dilute active systems, favor the possibility of detecting other origins.

WATER PRESSURES AND EVIDENCE FOR BOILING

Temperatures and depths in the active high-temperature systems indicate that their upper parts are generally adjusted to the hydrostatic boiling-point curve (White, 1968). A few systems need a small correction for high salinity or high gas content. At greater depths in most systems, pressure is sufficiently great to prevent boiling or the separation of a vapor phase if gas contents are high. Steamboat Springs (White, 1968) initially had pressures slightly above hydrostatic in several drill holes, but no special attention was given to the phenomenon. Later research drilling in Yellowstone Park (White and others, 1975) established the existence of overpressures locally as much as ~30% above hydrostatic because of high silica contents and impeded flow (self-sealing) of the upflowing water. Careful measurements at Beowawe, Nev., and Roosevelt, Utah, before their disturbance by geothermal exploration, probably would have identified overpressures in these systems.

Fluid-inclusion studies of some epithermal veins indicate two populations of coexisting inclusions: One set homogenizes to liquid with increasing temperature while the second set homogenizes to vapor (or largely vapor). This result is interpreted as due to boiling, with individual inclusions trapping different proportions of vapor and liquid in a boiling system. Boiling can alternate with non-boiling through time as a system becomes self-sealed, overpressure develops as the cold, more dense recharge part of the system becomes effective in response to decreasing flow rates; hydrofracturing may eventually relieve the overpressure.

POST-SINTER EROSION

Erosion of the active systems is generally slight, ranging from essentially 0 to at least a few tens of meters at Steamboat Springs (White and others, 1964) and probably lesser amounts in younger systems. Most epithermal systems have been eroded so extensively that no direct evidence of sinter deposition is preserved, and precise estimates of the extent of erosion generally cannot be made. Critical evidence can be provided from preserved sinter and evidence of fossil water-table relations. Both kinds of evidence require careful identification and correlation with the ore deposit. Of the deposits listed in Table 2, sinter is stated to exist near the Aurora, Nev., and Bodie, Calif., systems (M. L. Silberman and F. J. Kleinhampl, oral commun.); Borealis, Nev. (Don Strachan, oral commun.); DeLamar, Idaho (Pansze, 1975); Silver City, Idaho (Dan Shawe, oral commun., 1981); Divide and Sulphur, Nev. (Andrew Wallace, oral commun., 1981); and Steamboat Springs, Nev. (White and others, 1964). We have examined some of these deposits and consider them to be chalcedonic sinter, initially deposited as opaline sinter but, after burial, later chalcedonized with time and higher temperatures. The different characteristics of various types of sinter were described by White and others (1968). Previously existing rocks are also commonly converted into sinterlike rocks through acid leaching and chalcedonization related to boiling from a subsurface water table; some mercury deposits of the "opalite" type in Nevada (Bailey and Phoenix, 1944) are of this origin. All claimed sinters need to be restudied in the light of petrographic, chemical, and field characteristics; we are reluctant to identify several doubtful sinters until such studies are made.

SOURCES OF ENERGY

Of the active convection systems of Table 1, Long Valley, Coso, Steamboat Springs, Soda Lake, and Roosevelt are closely associated with volcanic systems about 1 m.y. old or less. Volcanic heat is viewed as the major source of their energy with the possible exception of Soda Lake (near a young basaltic cinder cone in a region of high conductive heat flow).

However, the energy source for most of the other active systems listed in table 1 is less certain. Most of these are in or near the "Battle Mountain heat-flow high," where conductive heat flow is at least 50 percent above the continental average, but silicic volcanic rocks younger than 10 m.y. are absent (Lachenbruch and Sass, 1977).

Much recent interest has focused on the abundance of the helium isotopes ^{3}He and ^{4}He. ^{3}He significantly above atmospheric is now generally viewed as indicating a mantle or primitive constituent, in contrast to ^{4}He, which forms mainly in the earth's crust from radiogenic reactions of U and Th. ^{3}He/^{4}He ratios 4 to 16 times above atmospheric have been identified along oceanic spreading ridges, in Iceland, Yellowstone Park, Lassen Volcanic National Park, Steamboat Springs, Nev., and in many other volcanic areas. In contrast, Beowawe, Nev., and other systems driven by high conductive heat flow have only ~10 percent of the atmospheric ratio, a value indicating a dominantly crustal heat source (which could also be old degassed magma bodies, as favored by Lachenbruch and Sass, 1977). Systematic study of the helium ratios of all high temperature spring systems of the Great Basin is urgently needed for a better understanding of present-day heat sources.

Many of the fossil epithermal systems listed in table 2 are closely related in space and time to contemporaneous active volcanism (Silberman, and others, 1976). Were all of these convection systems as closely related to volcanism as present-day Yellowstone and Steamboat Springs? Were there any epithermal ore-generating systems like present-day Beowawe, where subsurface temperatures are within the epithermal precious-metal range but the energy is supplied by crustal conductive heat?

HOST ROCKS

The host rocks of the active systems include varieties of igneous and sedimentary rocks, some of which are old igneous and metamorphic rocks unrelated to the present convection systems. The fossil ore systems also include a wide variety of associated rocks but are mainly near-surface Tertiary volcanic rocks not differing greatly in age from that of the mineralization (Buchanan, 1981); many of these volcanic rocks are underlain by sedimentary and metamorphic rocks. The main heat sources must normally be at least several thousand meters below the surface to account for the high hydrostatic pressures and temperatures required by the data of Table 2. Thus, the composition of the host rocks immediately adjacent to ore bodies can be largely coincidental.

OPEN-SPACE DEPOSITIONAL BANDING OF VEINS

One of the most characteristic features of epithermal veins is banded open space fillings from the vein walls inward; one wall is commonly almost a mirror image of the other; paired zones of inward-facing euhedral crystals are also very common. Characteristic shattering and brecciation indicate intermittent growth, perhaps with alternate filling or "self sealing," followed by renewed reopening and further separation of the walls, probably commonly by hydrofracturing but with tectonic fracturing also involved. Lindgren (1933, and his earlier reports on individual districts) was one of the first to recognize these features as indicative of low pressure and shallow depth. This concept is still generally supported, although most present-day economic geologists no longer agree that the concept of low temperatures is valid. The tectonic style is also consistent with crustal extension.

Banded or crustified veins also characterize the few active geothermal systems that have been core-drilled and studied systematically. Many examples of banding or crustification were observed in core from Steamboat Springs, Nev. (White, 1968), and in Yellowstone Park (Keith and others, 1978). Selected intervals of drill holes cored in New Zealand yielded some similar growth-banded veins. In commercial geothermal drilling, however, rock bits are normally used, with consequent loss of detailed vein characteristics.

ORE MINERALS

No gold mineral has yet been identified in the active systems, although black siliceous muds at Steamboat Springs contain as much as 15 ppm Au, and orange amorphous metastibnite as much as 60 ppm Au (White, 1981). Au and some other elements may be adsorbed on amorphous silica and fine-grained sulfides. In contrast, Au occurs in a various minerals in epithermal vein deposits but is "invisible" (adsorbed?) on carbon and other minerals of disseminated deposits.

The silver minerals deposited in active systems are also generally unknown, except that pyrargyrite has been positively identified in core from Steamboat Springs (White, 1981). A silver telluride was identified in tiny inclusions in sphalerite from Broadland, New Zealand, and both Au and Ag were much enriched in surface-deposited metastibnite. Both Au and the ratio Au/Ag decrease markedly downward in the three active systems most thoroughly studied (Steamboat Springs, Nev., and Broadlands and Waiotapu, New Zealand), both Au as Ag and the base metals increase.

A similar zonation of decreasing Au and Au/Ag with depth as base metals increase clearly occurs in most epithermal ore deposits (Buchanan, 1981). The few exceptions are generally Au-rich and acidic, in contrast to most vein deposits, where Ag generally dominates greatly over Au (by weight).

MAXIMUM DEPTH RANGE OF ORE

A significant aspect of most epithermal ore deposits is their abrupt termination of commercial ore with depth ("flat bottoming"), especially as indicated by the ore stopes. This observation is generally viewed as due to decreasing Au content, a less abrupt but effective decrease in Ag, and increasing base metals. The depth of bottoming is seldom >350 m below the surface, although mining in a few districts extends below 1000 m or more (Buchanan, 1981), commonly because Ag continues downward as base metals increase sufficiently to constitute minable grades. In a few epithermal districts, ore shoots are described as also terminating abruptly upward. Veins and vein structures, in contrast to ore grade, commonly extend both upward (where preserved) and downward beyond commercial grades.

Much less is known about the distribution of Au and Ag in the few active systems enriched in precious metals. These systems are never mined for their precious metals but have been explored principally for geothermal energy, so continuity or grade of precious metal "ores" has received little attention.

RATIOS OF VEIN LENGTH TO DEPTH

Buchanan (1981) tabulated the ratios of vein length to depth, based on stope dimensions rather than ore grade. In contrast to many other kinds of hydrothermal ore deposits, vein length commonly exceeds depth by a factor of at least 2. Erosion of the upper parts of some veins accounts in part for these ratios, but rather abrupt downward termination of the mining grade of ore (even though the veins continue downward) is viewed as the dominant factor.

Buchanan (1981) and others have suspected a chemical control over downward terminations, probably because of changes in the ore fluids, especially resulting from boiling. With upward flow, hydrostatic pressure decreases, the fluid start to boil, and CO_2 and H_2S are selectively lost to the vapor phase, thereby causing instability in the ore metal complexes. Buchanan (1981) concluded that evidence of boiling, especially from coexisting vapor-rich and liquid-rich fluid inclusions, commonly starts near the bottom of ore stopes. For further support, more systematic observations of the distribution of vein adularia, and calcite (especially the lamellar variety, overgrown with quartz) are needed for correlation, as Browne and Ellis (1970) did for the Broadlands system in New Zealand. These minerals are likely to correlate with boiling, loss of CO_2, and consequent increase in pH of the liquid.

Geothermal veins and vein structures in active systems have not yet been studied systematically with respect to the distribution of Au, and so the ratios of individual vein lengths to depths are not yet known. The permeability of vein structures is significant to the geothermal industry, indicating permeable aquifers likely to yield high flows of thermal fluids. Vein adularia has long been considered as a favorable indicator of permeable fluid channels (Browne and Ellis, 1970), unless these channels are later filled by self-sealing.

OTHER COMPARISONS OF ACTIVE AND FOSSIL SYSTEMS

Table 3 also compares other important characteristics of active and fossil convection systems that are adequately summarized by Buchanan (1981) for the epithermal volcanic-hosted vein systems, and summarized in Table 3 for the active systems.

MINOR ELEMENTS AS POSSIBLE INDICATORS OF INTERRELATED PRECIOUS-METAL ORE DEPOSITS

For many years we have been interested in a group of minor elements that may be significant in interrelating high-temperature hot springs, epithermal Au-Ag vein deposits, and, more recently, the disseminated or Carlin-type gold deposits. The association of Ag, Au, As, Sb, and Hg at Steamboat Springs has been known for many years (Lindgren, 1933; White and others, 1964; White, 1981). As, Sb, and Hg are associated with many epithermal veins (emphasized by Nolan, 1933), and Radtke and others (1972) demonstrated that As, Sb, Hg and Tl are characteristic of the disseminated gold deposit of Carlin, Nev., where the gold was invisible. As, Sb, and Hg were already known as sulfide minerals in the somewhat similar Getchell deposit, but little attention was given to the rare element Tl until its abundance and new thallium minerals were identified at Carlin (Radtke and others, 1972). Detailed study of the distribution of these trace elements was limited in part because their detection limits by ordinary emission spectrographic methods were appreciably higher than crustal abundances levels. Chris Heropoulos, in collaboration with Radtke and others, 1972; Heropoulos and others, in press), had initialed the development of new sensitive spectrographic methods, concentrating on the short-wave length radiation area of the spectrum, which is not normally utilized. As a consequence, detection limits for most volatile elements of this group in hot spring systems are now near or below the crustal abundances (Table 4). Elements not yet detected by these methods at average crustal abundance levels are Ag, Au, and Hg (close for all except Hg, but Hg is generally much enriched in near-surface materials). Bi, Se, and Te are also members of this group of volatile elements; however, their sensitivity is not high enough to make them detectable at the crustal abundance levels.

Hot spring sinters were selected as appropriate material for comparison because all true sinters were deposited at the ground surface from near-boiling springs discharging from reservoirs near or above 180°C. Uncertainties for each system are the actual discharge rate and reservoir temperature at the time of deposition of each sample, and post-depositional changes (diagenetic) after each sinter was first deposited. Chalcedonization of amorphous opaline sinter, generally through an intermediate cristobalitic stage (White and others, 1964), probably decreases the metal contents by remobilizing elements adsorbed on amorphous opal.

Table 4 summarizes our first reconnaissance effort to compare the active high-temperature systems of the Great Basin. None of the sinter samples was collected specifically for this purpose, and so the selected samples are not necessarily representative; only 7 of the 15 systems are represented by two or more samples, and only those systems with four or more analyses seem likely to approximate a representative spread of elemental contents.

Steamboat Springs is the only system of the group that is generally enriched in all elements, commonly by a factor of 10 or more. Au is strongly enriched above its crustal abundance to depths of ~70 m, but is below the limit of detection at greater depths (White, 1981, Table 2). Near-surface enrichments of As, Sb, and Hg are especially notable; the abundances of most of these elements decrease significantly with depth except Ag, which is 0.2 to 2 ppm in sinters to 25-m depth but is more strongly enriched in chalcedony-quartz-calcite veins to the greatest depth (136 m).

Most samples from Steamboat Springs are strikingly enriched above crustal abundances in Sb and As and generally much enriched in Hg, B, and Tl. In contrast, Zn, representative of near-surface base-metal concentrations, is always within detection limits but much below its crustal abundance.

Of all other active systems listed in Table 4, none shows much evidence for generating an epithermal ore deposit. Long Valley caldera, Calif., probably has the best possibilities: two of four samples contain detectable Ag (0.3 and 10 ppm), but Au was detected in only one (1.0 ppm); Sb, As, Hg, and B are strongly enriched, but Tl is low. Long Valley waters have undergone much near-surface dilution, so ores could be forming below the surface in the dilution zone.

Beowawe sinters are generally low in Ag, and only one of the four samples contains detected Au (0.2 ppm). As, Sb, Hg, and B are modestly enriched in some of these samples, but the general contents are modest in comparison with Long Valley and, especially, Steamboat Springs.

The major surprise of Table 4, at least for us, is the Roosevelt system, Utah, which has the highest salinity and temperature of those systems closely associated with young silicic volcanism. We had suspected that the minor elements would show close ties to epithermal Au-Ag ore deposits. Of seven Roosevelt sinter samples, most are approximately equivalent to Steamboat Springs in As, Sb, and B; Tl is higher, but Hg is lower. However, the sinters are almost lacking in precious metals: only one of the seven samples contained detectable Ag and none contained detected Au. Therefore, Roosevelt is a "bust" as far as Au and Ag contents of sinters are concerned! However, this is the only system containing notable Zn, with two of the seven samples (70 and 300 ppm) equaling or exceeding the crustal average.

TIME-SPACE RELATIONS OF THE EPITHERMAL VEINS

Au-Ag deposits of Nevada and adjacent areas occur mainly in a broad north-northeast striking belt through west-central Nevada (fig. 2).

Table 4. "Volatile" elements and their concentration ranges in ppm in Hot Spring Sinters[1] as compared to crustal abundances

Locality	Temp.[2]	No. of Samples	Ag 0.07[3]	Au <0.05[3]	As 1.8	Sb 0.2	Tl 0.45	Hg 0.08	B 10	Zn 70
Nevada										
Steamboat Springs	228	10	2(<0.1) 6(0.1–1.0) 5.0,15	6(<0.1) 4(0.2–1.5)	6(2.0–70) 4(100–700)	3(70–300) 7(700–>2000)	3(<1.0) 5(1.5–7.0) 2(70)	1(<1.0) 6(1.0–30) 3(1000–>2000)	5(20–70) 5(200–1000)	5(0.5–2.0) 5(5.0–10)
Beowawe Geysers	229	4	1(<0.1) 3(0.5–1.0)	3(<0.1) 0.2	4(5.0–10)	3(20) 5.0	2(<1.0) 1.0,2.0	4(2.0–10)	4(70–200)	2.0,3.0 10,30
Pinto	173	3	2(<0.1) 0.3	3(<0.1)	10,30,100	20,30,30	2(<1.0) 7.0	20,300,1000	30,150,300	1.0,2.0,10
Brady	155	2	2(0.5)	2(<0.1)	20,30	30,50	2(<1.0)	20,20	300,300	10,15
Leach	162	1	7.0	<0.1	20	20	<1.0	1.5	150	15
Gerlach	178	1	1.5	<0.1	30	10	1.0	5.0	1500	50
Baltazor	158	1	<0.1	<0.1	20	15	<1.0	<1.0	300	20
Howard	?[4]	1	<0.1	<0.1	30	30	<1.0	50	30	<1.0
Jersey Valley	?[4]	1	<0.1	<0.1	15	10	<1.0	5.0	70	3
Lee	166	1	0.3	<0.1	20	30	1.5	5.0	200	15
California										
Long Valley	227	4	2(<0.1) 0.3,10	3(<0.1) 1.0	3(50)	3(200–300) 100	3(<1.0) 30	2.0,5.0 2.0	30,100 50,700	4(10–20) 700,1000
Coso	220	2	<0.1	<0.1	7.0,20	100,200	<1.0	20,50	200	<1.0,7.0
Surprise Valley	152	1	<0.1	<0.1	20	20	<1.0	50	100	3.0
Oregon										
Mickey	205	1	<0.1	<0.1	70	50	<1.0	50	1500	5.0
Utah										
Roosevelt	265	7	6(<0.1) 0.5	7(<0.1)	5(5.0–20) 100,300	3(30–100) 3(200) 2000	5(1.0–5.0) 50,150	4(1.0) 3(2.0–20)	6(200–1000) 50	1(<1.0) 4(2.0–20) 70,300

[1] The short wavelength radiation (SWR) method (Heropoulos, et al, in press) was used to determine concentrations and to establish limits of detection.

[2] Temperatures measured in drill holes or indicated by chemical geothermometry (Brook, Mariner, and others, 1979)

[3] Crustal abundance of each element in ppm from Krauskopf (1967).

[4] Discharge too low for chemical geothermometry

Northwestern and southeastern Nevada are almost barren of these deposits. The most favorable terrane, as presently known, terminates southwestward near the Nevada-California stateline (close to the Walker Lane), and abruptly southeastward; the northwest boundary is equally abrupt.

In a broad way, mid-Tertiary volcanism started in northeastern Nevada, expanded in a crudely arcuate form (Stewart and others, 1977) to the south and west through time, finally leaving most of Nevada barren of volcanism during the last 6 m.y. (except for local basalt).

Epithermal vein mineralization, as evaluated from the limited data in table 2, broadly followed the same migrating arcuate system except for greater irregularity and shorter, sharper ore-depositing pulses (Silberman and McKee, 1974). All four dated ore systems of the oldest mid-Tertiary volcanic group (43 to 34 m.y.; Stewart and others, 1977) range in age from 38 to 35 m.y. The second mid-Tertiary volcanic group (34 to 17 m.y.) has five associated dated ore systems, all of which, if representative, formed within a short period of 25 to 19 m.y. B.P. These deposits, for unknown reasons, are concentrated only in the southwestern part of the corresponding volcanic belt, approximately parallel to the Walker Lane. A total of 15 districts indicate a sharp climax in epithermal mineralization within the brief timespan of 3 m.y. from 17 to 14 m.y. B.P. (the older portion of Stewart's 17 to 6 m.y. volcanic group), and are distributed rather uniformly over most of the broad northeast-striking belt shown in Figure 2. During the later part of this volcanic period (<14 to 6 m.y. B.P.), mineralization waned; the six dated districts range in age from 7 to 13 m.y.; their locus shifted southward, being restricted to a northwest-striking belt adjacent to the Walker Lane (near the California-Nevada stateline).

The youngest ore group, 0 to 6 m.y. B.P., includes three districts ~5 m.y. old that continue the Walker Lane trend, followed by two young hot-spring systems. The Sulphur system, adjacent to the Black Rock Desert (1.9 m.y. old from dating of alunite, according to Andrew Wallace (oral commun., 1981), has a superb fossil sinter system, reportedly underlain by ore just east of Sulphur Station, where warm gases, H_2S, and elemental S indicate continuing activity after the dated alunite. Steamboat Springs has been discussed above; its activity started prior to a well-dated basaltic andesite flow (2.5 m.y.) and still continues (Silberman and others, 1979). The high metal contents of present-day Steamboat fluids prove that metal transport and ore generation was not restricted to early activity but is still continuing. Together, the activity at Steamboat Springs and Sulphur suggest either a new northeastward trend or the northeastern limb of a westward-advancing arc, depending on whether Long Valley is indeed a young active epithermal Au-Ag system.

DISSEMINATED DEPOSITS

Interrelations between the middle- and late-

Cenozoic epithermal vein systems and the pre-Tertiary sediment-hosted "disseminated" deposits are still not clearly understood. Meager data on enrichment of the minor elements As, Sb, Tl, Hg, and B at Carlin and probably also at Getchell suggest a close similarity to the active epithermal vein systems of Steamboat Springs and Broadlands, New Zealand. We had suspected that these minor elements might establish a common tie between all epithermal veins, the active high-temperature spring systems, and the disseminated group, and that the principal difference would be the nature of the host rocks. However, the data on these systems listed in table 2, if representative, suggest that the disseminated group is generally older. Highly inadequate fluid inclusion data also suggest that temperatures may have been slightly lower and the salinity of ore fluids higher than the mid-Tertiary and younger epithermal veins. Rochester may be a special exception in being older than most epithermal veins but having the same general temperature range and Ag enrichment appropriate for most of these veins (Vikre, 1981). However, Rochester fluid inclusions also have unusually high salinity, perhaps more appropriate for the disseminated group.

Systematic study of the distribution of As, Sb, Tl, Hg, and B is needed for all active and fossil ore systems. These minor elements are strongly zoned with depth in the active Steamboat and Broadlands systems, which are "open" and may be dispersing most of these minor elements near the surface by means of discharging springs. These elements decrease rapidly downward in abundance at Steamboat Springs (White, 1981), where cinnabar was identified to a maximum depth of only 15 m below the present surface in eight drill holes, and stibnite only to 45 m. Although distinct As, Tl, and B minerals have not been identified at Steamboat, the abundances of these elements decrease rapidly to ~25 m and are lower (close to the system's background) at greater depths (White, 1981), wherever temperatures exceed ~140°C. In contrast, these elements were concentrated in ores yielding fluid-inclusion temperatures of 150° to 200°C in the Carlin deposit (Nash, 1972) and presumably at similar temperatures at Getchell. These relations suggest that Carlin, Getchell, and, possibly, the other pre-Tertiary disseminated gold deposits differ from epithermal veins not only in age but also in structural, chemical, and other depositional controls. The fluids may have been entrapped below a permeability barrier where they could not discharge directly to the surface but were cooled by conduction, and eventually mixed with other pore fluids below the surface.

CONCLUSIONS

Among the active high temperature convection systems of the Great Basin, Steamboat Springs may be the only one that is forming an epithermal ore deposit now. Long Valley, Calif., may also have good possibilities, as may some others if Au is almost totally deposited below the surface. Coso, Calif., is difficult to evaluate from its sinter compositions; total sinter is meager, and the system has discharged at the surface only briefly in the past when the water table was higher than at present. Most of the thermal discharge from this system must be subsurface into ground water under the basin to the east or, possibly, in other directions.

Present data suggest that the disseminated deposits differ in several important respects from epithermal veins, but further detailed systematic study is necessary.

ACKNOWLEDGEMENTS

We are deeply indebted to our many close associates of the U.S. Geological Survey for assistance and stimulation in this latest effort to interrelate active hydrothermal convection systems and their fossil equivalents that have generated the epithermal Au-Ag vein deposits of the Great Basin. We especially thank Roger Ashley, William Bagby, R. O. Fournier, Manuel Nathenson, M. L. Silberman, J. H. Stewart, and E. W. Tooker. H. F. Bonham, Jr., of the Nevada Bureau of Mines and Geology, was of major help in providing maps, locations, and other data for the first critical task--identifying the gold and silver districts and prospects to be included as epithermal and disseminated types in Nevada from other mines and districts mainly of interest for other metals. He and others may not agree with all deposits listed in table 2; many more of these types no doubt exist but either were too small or lacked sufficient data on age, salinity, and temperature.

REFERENCES

Bailey, E. H., and Phoenix, D. A., 1944, Quicksilver deposits in Nevada: Univ. Nev. Bull., v. 38, (Geol. Min. Ser. no. 41), 206 p.

Bonham, H. F., Jr., 1982, Reserves, host rocks, and ages of bulk-minable precious metal deposits in Nevada: Nev. Bur. Mines Geol. Open-File Rep. 81-1 and Prelimin. Rep. 82-9.

Brook, C. A., Mariner, R. H., Mabey, D. R., Swanson, Jr., Guffanti, Marianne, and Muffler, L. J. P., 1979, Hydrothermal convection systems with reservoir temperatures >90°C in Muffler, L. J. P., ed., Assessment of Geothermal Resources of the United States--1978: U.S. Geol. Survey Circular 790, p. 18-85.

Browne, P. R. L., and Ellis, A. J., 1970, The Ohaki-Broadlands hydrothermal area, New Zealand: Mineralogy and related geochemistry: Am. Jour. Sci., v. 269, p. 97-131.

Buchanan, L. J., 1981, Precious metal deposits associated with volcanic environments in the Southwest: in Dickinson, W. R., and Payne, W. D., eds., Relations of tectonics to ore deposits in the Southern Cordillera: Ariz. Geol. Soc. Digest, v. 14, p. 237-262.

Craig, Harmon, Boato, Giovanni, and White, D. E., 1956, Isotopic geochemistry of thermal waters: U.S. Natl. Acad. Sci.-Nat. Res. Counc. Publ. 400, p. 29-38.

Garside, L. J., and Schilling, J. H., 1979, Thermal waters of Nevada: Nev. Bur. Mines Geol. Bull. 91, 163 p.

Heropoulos, Chris, Seeley, J. L., and Radtke, A. S., in press, Spectrographic determination of selenium in stibnite: Appl. Spectrosc. J.

Keith, T. E. C., White, D. E., and Beeson, M. H., 1978, Hydrothermal alteration and self-sealing in Y-7 and Y-8 drillholes, Yellowstone Park, Wyo.: U.S. Geol. Surv. Prof. Pap. 1054-A, 26 p.

Krauskopf, K. B., 1967, Introduction to geochemistry: New York, McGraw Hill, 721 p.

Lachenbruch, A. H., and Sass, J. H., 1977, Heat flow in the United States and the thermal regime of the crust in Heacock, J. G., ed., The Earth's crust: Its nature and physical properties: Am. Geophys. Union Monograph 20, p. 626-675.

Lindgren, Waldemar, 1933, Mineral deposits, 4th ed.: New York, McGraw-Hill, 930 p.

Muffler, L. J. P., ed., 1979, Assessment of geothermal resources of the United States--1978: U.S. Geol. Surv. Circular 790, 163 p.

Nash, J. T., 1972, Fluid inclusion studies of some gold deposits in Nevada: Geological Survey research, 1972: U.S. Geol. Surv. Prof. Pap. 800-C, p. C15-C19.

Nolan, T. B., 1933, Epithermal precious-metal deposits, in Ore deposits of the western states: New York, Am. Inst. Mining Metall. Eng., p. 623-640.

O'Neil, J. R., and Silberman, M. L., 1974, Stable isotope relations in epithermal Au-Ag deposits: Econ. Geology, v. 69, p. 902-909.

Pansze, Arthur J., Jr., 1975, Geology and ore deposits of the Silver City-DeLamar-Flint region, Owyhee County, Idaho: Idaho Bur. Mines Geol. Pamph. 161, p. 1-79.

Radtke, A. S., Heropoulos, Chris, Fabbi, B. P., Scheiner, B. J., and Essington, M., 1972, Data on major and minor elements in host rocks and ores, Carlin gold deposit, Nevada: Econ. Geology, v. 67, p. 975-978.

Silberman, Miles H., and McKee, Edwin, H., 1974, Ages of Tertiary volcanic rocks and hydrothermal precious metal deposits in central and western Nevada: Nev. Bur. Mines Report 19, p. 67-72.

Silberman, Miles L., Stewart, John H., and McKee, Edwin H., 1976, Igneous activity, tectonics, and hydrothermal precious-metal mineralization in the Great Basin during Cenozoic time: Soc. Mining Engineers Trans., v. 260, p. 253-263.

Silberman, M. L., White, D. E., Keith, T. E. C., and Dockter, R. D., 1979, Duration of hydrothermal activity at Steamboat Springs, Nevada, from ages of spatially associated volcanic rocks: U.S. Geol. Surv. Prof. Pap. 458-D, 14 p.

Stewart, J. H., Moore, W. J., and Zietz, I., 1977, East-west patterns of Cenozoic rocks, aeromagnetic anomalies, and mineral deposits, Nevada and Utah: Geol. Soc. Am. Bull., v. 88, p. 67-77.

Taylor, Hugh P., Jr., 1973, Oxygen-18/oxygen-16 evidence for meteoric- hydrothermal alteration

and ore deposition in the Tonopah, Comstock
Lode, and Goldfield mining districts, Nevada:
Econ. Geology, v. 68, p. 747-764.

Taylor, Hugh P., Jr., 1974, The application of
oxygen and hydrogen isotope studies to
problems of hydrothermal alteration and ore
deposition: Econ. Geology, v. 69, p. 843-883.

Vikre, P. G., 1981, Silver mineralization of the
Rochester district, Pershing Co., Nev.: Econ.
Geology, v. 76, p. 580-609.

White, D. E., 1968, Hydrology, activity, and heat
flow of the Steamboat Springs thermal system,
Washoe County, Nevada: U.S. Geol. Surv. Prof.
Pap. 458-C, 109 p.

White, D. E., 1967, Mercury and base-metal
deposits with associated thermal and mineral
waters, in Barnes, H. L., ed., Geochemistry of
hydrothermal ore deposits: New York, Holt,
Rinehart, and Winston, Inc., p. 575-631.

White, D. E., 1973, Characteristics of geothermal
resources, in Paul Kruger, and Carel Otto,
eds., Geothermal Energy: Resources,
production, stimulation: Stanford Univ.
Press, Stanford, California, p. 69-94.

White, D. E., 1981, Active geothermal systems and
hydrothermal ore deposits, in Econ. Geology,
75th Anniv. Vol., p. 392-423.

White, D. E., Barnes, Ivan, and O'Neil, J. R.,
1973, Thermal and mineral waters of
non-meteoric origin, California Coast Ranges:
Geol. Soc. America Bull., v. 84, p. 547-560.

White, D. E., Fournier, R. O., Muffler, L. J. P.,
and Truesdell, A. H., 1975, Physical results
of research drilling in thermal areas of
Yellowstone National Park, Wyoming: U.S.
Geol. Surv. Prof. Pap. 892, 70 p.

White, D. E., Thompson, G. A., and Sandberg, C.
A., 1964, Rocks, structure, and geologic
history of Steamboat Springs thermal area,
Washoe County, Nevada: U.S. Geol. Surv. Prof.
Pap. 458-B, 63 p.

ACTIVE HYROTHERMAL SYSTEMS

A REVIEW OF HIGH-TEMPERATURE GEOTHERMAL DEVELOPMENTS
IN THE NORTHERN BASIN AND RANGE PROVINCE

Walter R. Benoit

Robert W. Butler

Phillips Petroleum Company

Chevron Resources Company

ABSTRACT

Intensive geothermal exploration in the northern Basin and Range province has resulted in the the discovery of nine high-temperature (>200°C) geothermal reservoirs:

1) Roosevelt Hot Springs, Utah
2) Beowawe, Nevada
3) Humboldt House, Nevada
4) Brady's Hot Springs, Nevada
5) Desert Peak, Nevada
6) Northern Dixie Valley, Nevada
7) Soda Lake, Nevada
8) Steamboat Springs, Nevada
9) Coso, California

In addition, there is geological, geophysical, and geochemical evidence to indicate an undiscovered reservoir in the Long Valley caldera, California. Delays in Federal leasing are the main reason this reservoir has not yet been located.

Four of these areas occur along the east or west margins of the province and are spatially associated with Quaternary or Recent siliceous volcanic centers. Five are in or near the Carson Basin in northwestern Nevada and lack evidence for magmatic heating. The Beowawe reservoir has a unique occurrence near the east-west center of the province. Most reservoirs are closely associated with known or suspected Basin and Range normal faults.

With the exceptions of the localized shallow steam production at Coso and bicarbonate-rich water at Beowawe, the known reservoir waters have a dilute sodium chloride composition. Reservoir temperatures typically range from 200 to 220°C. The maximum reported temperature in the northern Basin and Range province is 271°C at Roosevelt Hot Springs.

The most thoroughly evaluated reservoirs are Roosevelt Hot Springs and northern Dixie Valley with 13 and 10 deep wells respectively. The most limited data are from the Soda Lake, Steamboat Springs, Humboldt House, and Long Valley prospects where only two or three deep wells per prospect have been drilled. Depths of the producing intervals vary from about 300 to 3000 m, but production is often from less than 1200 m.

Only one high-temperature, geothermal power plant at Roosevelt Hot Springs is under construction in the province. Extensive negotiations between developers and utilities have taken place regarding the Beowawe, Dixie Valley, and Desert Peak reservoirs.

INTRODUCTION

During the past ten years there has been a major effort by private industry, government agencies, research organizations, and universities to locate and study geothermal resources capable of generating electrical power. This has resulted in hundreds of published papers covering many geothermal areas in the northern Basin and Range province in a wide variety of scientific journals, plus a great amount of unpublished data generated by private industry. It is impossible to know how many potential high-temperature areas have been considered as prospects but the number must exceed 100. A complete exploration history of the northern Basin and Range province would be incomplete and sporadic at best. Therefore, this paper will briefly review the history and geology of ten areas where exploration for high-temperature (>200°C) geothermal reservoirs has been successful.

Exploration for high-temperature geothermal reservoirs in the northern Basin and Range province started in 1950 with the drilling of the Rodeo well at Steamboat Springs, Nevada specifically searching for steam to generate electricity (White, 1983). In the 33 years since this little-noticed beginning, several hundred million dollars and untold man years have been spent in exploration. At least 171 wells intended to produce high-temperature geothermal fluids have been drilled by many different entities to depths from 28 to 3854 m. The net result has been the discovery of nine high-temperature reservoirs in eight widely separated areas (Fig. 1). The total industry cost per discovery in the northern Basin and Range province is estimated at $20 million (Edmiston, 1982). Of these nine discoveries, one power plant is presently under construction. Four additional power plants have been seriously discussed.

There was little exploration activity after the unsuccessful Rodeo well until 1959, when Magma Power Company began a major drilling program in search of dry-steam reservoirs. By late 1962, Magma ceased this initial exploration program after drilling 48

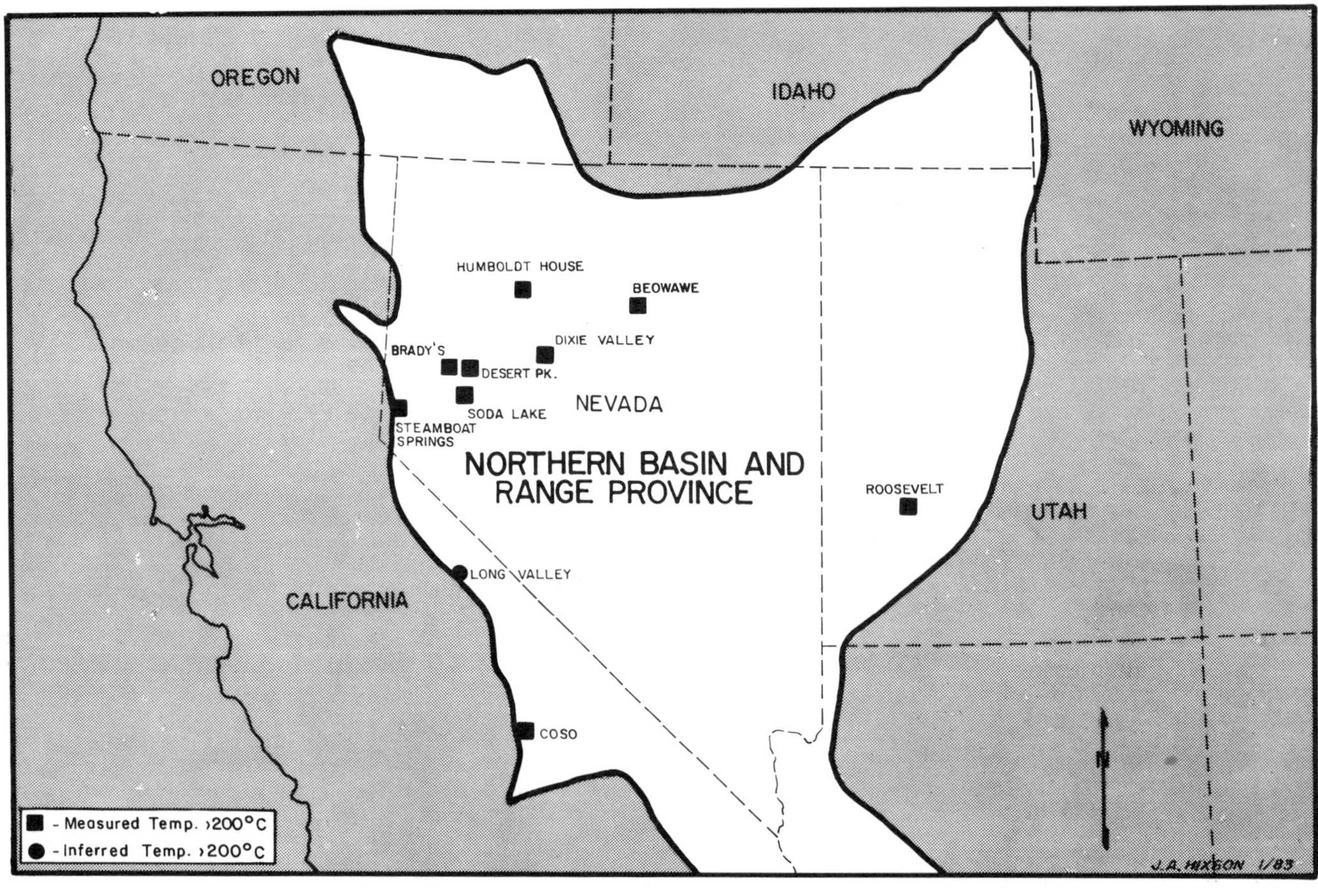

Figure 1.   Locations of High-Temperature Geothermal Reservoirs in the Northern Basin and Range Province (modified from Edmiston, 1982).

wells in the immediate vicinity of 15 thermal springs. This activity proved to Magma that the high-temperature reservoirs, if present in the Basin and Range province, would not be dry-steam systems like the Geysers in California. In retrospect, Magma was at a further disadvantage in their exploration because they were limited to privately owned, hot spring properties. Until 1974, Federal lands were unavailable for geothermal leasing.

Between 1962 and 1973 geothermal exploration proceeded slowly. The main exploration thrust was deeper drilling, generally in the more promising areas drilled by Magma. Additional companies joined in the search during this time, including some not cited in this report, but no major new successes resulted. At least 76 wells were drilled prior to 1974 resulting in measured temperatures greater than 200°C at Beowawe and Brady's Hot Springs. In the early 1970's geothermal exploration in the Basin and Range province began to revive for several reasons. The Geothermal Steam Act of 1970 permitted future leasing and development on Federal lands. The true potential of the Geysers in California was becoming evident, proving that geothermal power could be generated at competitive prices and in sufficient quantity to interest large corporations. The early 1970's were dominated by OPEC policies--dramatic oil-price increases, embargoes,and energy shortages--that shook America's complacency regarding energy supplies. Finally, geothermal had enhanced environmental benefits when compared to coal, nuclear, oil, and new large scale hydroelectric power. These factors encouraged large energy companies to commit substantial sums of money, derived primarily from oil production, to geothermal exploration. In addition, the Federal government committed tens of millions of dollars to geothermal exploration, research and development, and managing leasing activities in the province. The commitment of funds reached an initial high point in 1975 (Edmiston, 1982) when private industry first obtained enough Federal leases and completed enough preliminary exploration to warrant drilling 12 large-diameter production wells. Drilling peaked in 1979 at 19 wells, many partially funded by the U. S. Department of Energy's Industry-Coupled Geothermal Reservoir Assessment Program (Fiore, 1980). A large share of the Basin and Range geothermal literature resulted from this program. Between the enactment of the Geothermal Steam Act in 1974 and May, 1983, 95 large-diameter exploratory or production wells were drilled.

ROOSEVELT HOT SPRINGS

The Roosevelt Hot Springs geothermal reservoir is located in southwestern Utah near the east margin of the province (Fig. 1) along the western side of the Mineral Mountains about 19 km northeast of Milford, Utah (Fig. 2).

Roosevelt Hot Springs is the only known high-temperature geothermal reservoir in the eastern half of the province. It is also the hottest known reservoir in the province with a maximum publicly reported temperature of 271°C (Rudisill, 1976). The discovery well, 3-1, was the second drilled by Phillips Petroleum Company in 1975, following a comprehensive three-year exploration program (Lenzer et al., 1977). The high reservoir temperature and relatively early discovery date, which coincided with abundant grant money made available by the U. S. Department of Energy, has made Roosevelt Hot Springs the site of extensive geological, geophysical, and geochemical research performed primarily by the University of Utah Research Institute. This extensive data base has been integrated into a comprehensive case study by Ross et al. (1982). An older, more extensive bibliography was prepared by McKinney (1978).

The Roosevelt Hot Springs geothermal system was classified as a Known Geothermal Resource Area (KGRA) by the U. S. Geological Survey in 1972, mainly because of its encouraging surficial features which include 190°F thermal springs, siliceous sinter, mercury deposits, nearby young obsidian flows, and an 82-m deep "steam" well which was drilled in 1967 and 1968. The well flowed for six weeks before being plugged. The Federal lands were leased in July, 1974 at one of the earliest KGRA sales. Since that sale, 13 exploration and production wells (Appendix 1) and eight deep temperature-gradient holes have been drilled (Fig. 2.), making it the most thoroughly drilled reservoir in the province and outlining a productive area of almost 14 km^2.

The Roosevelt Hot Springs geothermal reservoir is a fractured complex of competent Tertiary granitic and Precambrian metamorphic rocks. It underlies an area 2.4 km wide by 3.7 km long between the Dome Fault on the west and the irregular front of the Mineral Range on the east (Peterson, 1975; Ross et al., 1982). The reservoir is elongate in a north-northeast direction and coincides closely with a series of range-bounding normal faults as suggested by seismic-refraction data (Ross et al., 1982).

A line of rhyolite domes dated at .5 to .8 my occur along the crest of the Mineral Range both north and south of the reservoir. These domes indicate the possibility of magma in the vicinity of the reservoir and may explain the substantially higher than normal temperature of this Basin and

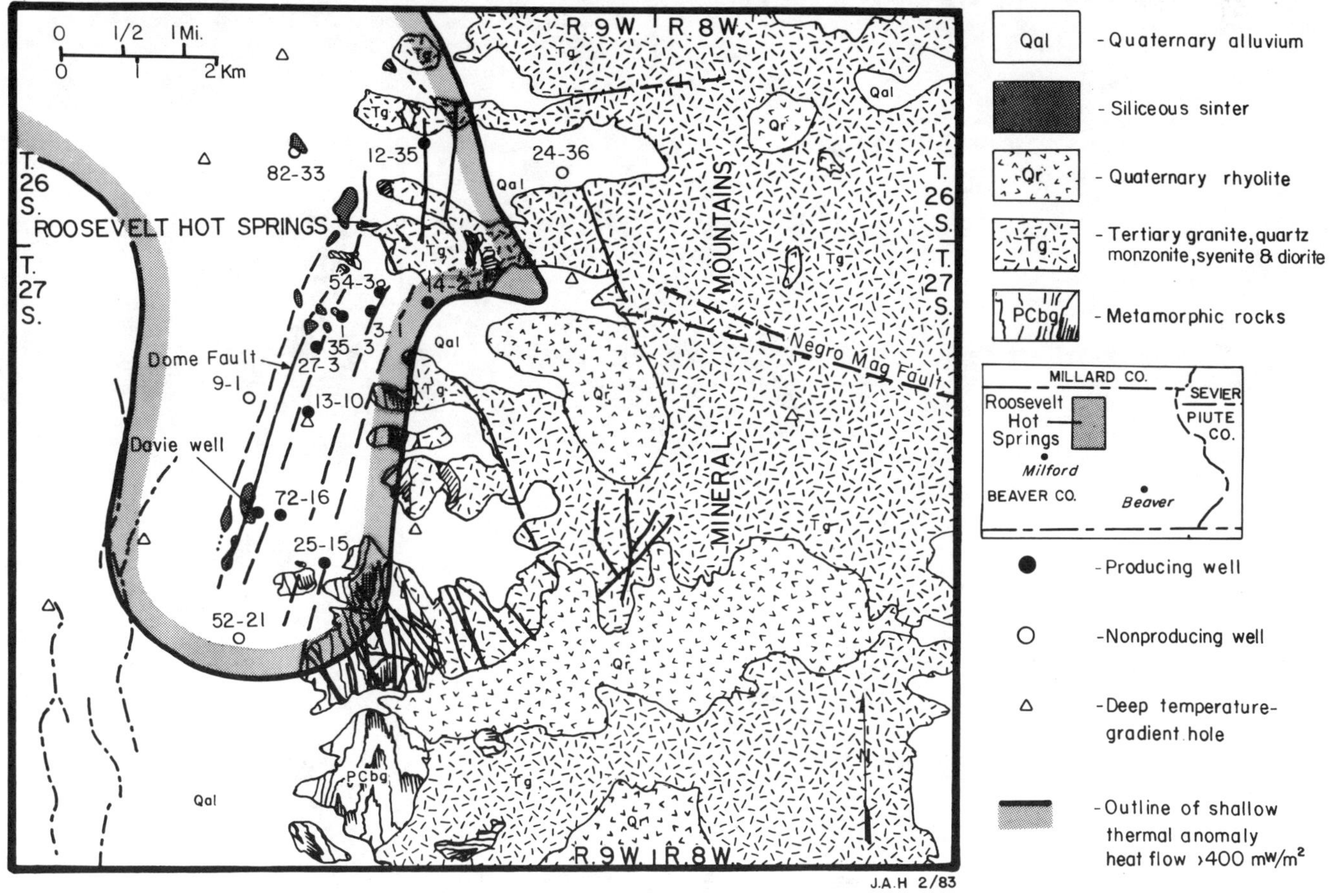

Figure 2. Map of the Roosevelt Hot Springs, Utah Area (modified from Ross et al., 1982).

Range reservoir.

Production wells at Roosevelt Hot Springs produce a sodium chloride water with a total dissolved solids content of about 7000 mg/l. These wells are somewhat unique in the Basin and Range province because the abnormally high temperature has created a reservoir pressure greater than the hydrostatic pressure, requiring no stimulation to begin flow. Mass-flow rates for the Roosevelt wells are variable but can be very high. Wells 27-3 and 35-3 have respective flow rates of 454,000 and 635,000 kg/hr, making them among the most prolific hot-water geothermal producers in North America (National Geothermal Service, 1983).

Negotiations for a heat-sales agreement with Utah Power and Light began in 1977. In September, 1980, a contract was signed for the construction of a 20-mw, single-flash demonstration plant to be followed by two additional 50-mw units contingent upon the success of the first plant. The 20-mw power plant is currently under construction and is expected to be operational early in 1984.

The Roosevelt Hot Springs area was the site of a one-mw helical-screw expander that generated the first electricity produced from a geothermal resource in the province during a long-term flow test of well 54-3 in March, 1978. In April, 1983 a 1.6-mw Biphase rotary-separator turbine completed a six-month endurance test, again powered by fluid from well 54-3.

BEOWAWE

The Beowawe geothermal area is located in north-central Nevada, 32 km southeast of the town of Battle Mountain near the center of the province (Fig. 3). It has some of the most spectacular surface manifestations of any area within the province, including a 75-m high sinter terrace made up almost exclusively of opal. In addition, there are boiling springs, fumaroles, steaming ground, geysers, mud pots, and abundant hydrothermally altered rock and ground. Thus it is no surprise that in 1960, Beowawe was the first high-temperature geothermal reservoir to be discovered in the Basin and Range province.

The geothermal literature on Beowawe is extensive. The local geology is discussed by Struhsacker (1980) who also included a very thorough bibliography. The regional setting of Beowawe is emphasized in papers by Stewart et al. (1975), and Zoback and Thompson (1978). Geophysical studies have been published by Smith (1979), Smith et al. (1979), Swift (1979), and Zoback (1979). Reservoir-engineering studies have been prepared by Middleton (1961) and Epperson (1982).

The basement geology at Beowawe is intensely deformed, thrust-faulted Paleozoic carbonate and clastic sedimentary rocks. Overlying the Paleozoic section is a gently dipping cap of basaltic-andesite and dacite flows with minor tuffs, related to a north-northwest-striking, mid-Miocene rift just a few km west of the sinter terrace. The rift at Beowawe is part of a 700-km long belt of extensional faulting and volcanic centers with associated dike swarms and graben-filling volcanic rocks, extending from central Oregon to central Nevada (Stewart et al., 1975; Zoback and Thompson, 1978). The Beowawe reservoir is located along an east-northeast-striking, Basin and Range normal fault, the Malpais fault, in the Paleozoic sedimentary and Tertiary volcanic rocks.

Geothermal exploration at Beowawe started in 1959 with shallow well drilling by Magma and Sierra Pacific Power Companies (Appendix 1). By 1965 they had completed 12 wells and discovered a resource

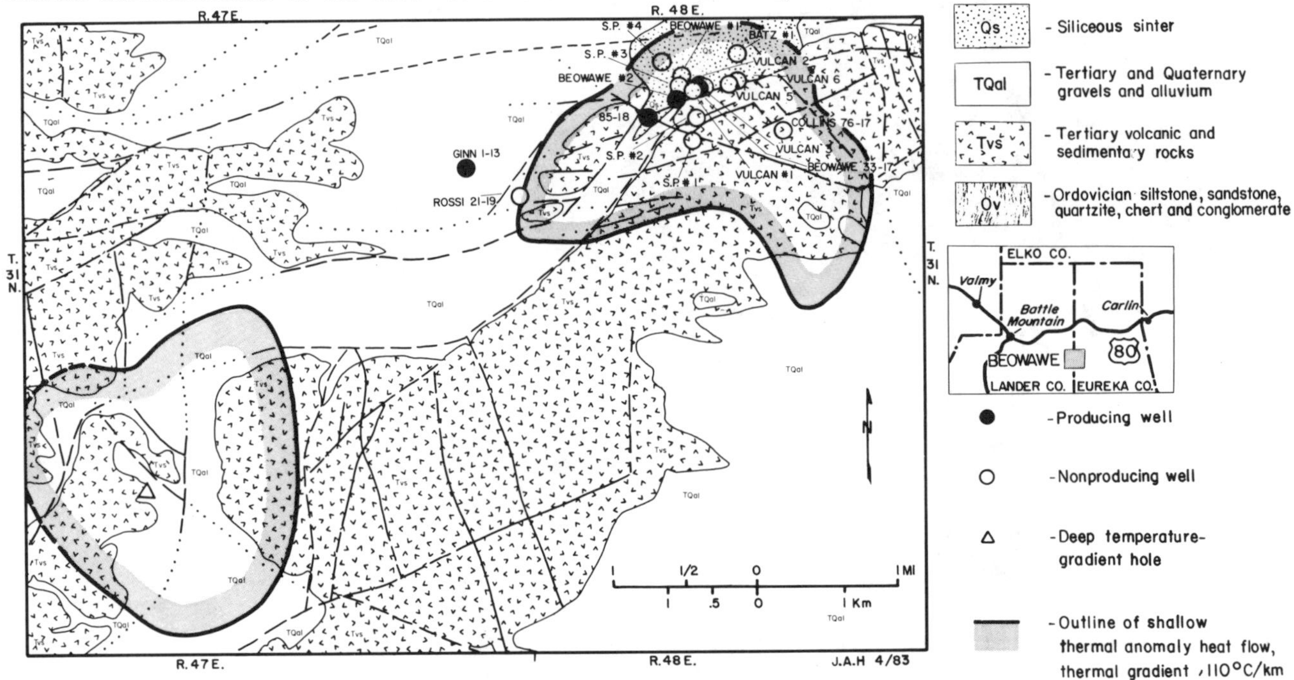

Figure 3. Map of the Beowawe, Nevada Area (modified from Struhsacker, 1980)

> 200°C at depths of 215 to 245 m (Garside, 1974; Garside and Schilling, 1979). Vulcan wells 1, 2, and 3 had flow rates of 680,000 kg/hr at wellhead pressures of 7.5 bars $\pm$ (Middleton, 1961). However, temperature and flow-rate decreases led to the conclusion that the reservoir was depleting and being invaded by cold water (Epperson, 1982). Activity then ceased until the early 1970's.

Exploration resumed in the early 1970's when Chevron drilled the Ginn 1-13. This 2911-m test, the deepest in the province at the time (1974), encountered a 216°C reservoir within fractured Paleozoic quartzites along the Malpais fault. The Rossi 21-19 was next drilled closer to the Malpais fault, but lacks sufficient permeability. In 1975 exploration drilling returned to the sinter terrace. However, the Batz #1 well at the east end of the terrace was the final disappointment for Magma. Chevron later drilled two producing wells on the terrace, 85-18 and 33-17. The most recent well, the Collins No. 1, was drilled by Getty Oil Company in 1981 on the back side of the terrace, and was unsuccessful.

The current interpretation of the Beowawe reservoir is that there is a shallow 185°C $\pm$ producing interval in the terrace area within the Tertiary volcanic section. There is also deeper and hotter fracture production between 2500 and 3000 m with a temperature of 216°C $\pm$ in Paleozoic rocks along the Malpais fault. Interference tests indicate a high degree of continuity between all wells. Pressure responses are observed in less than one hour for wells up to 2 km apart, even for those completed in different geologic units (Epperson, 1982).

The productivity of the recently tested wells is around 185,000 kg/hr total mass flow. Epperson (1982) believes the lower flow rates compared to those of Middleton (1961), are due primarily to mechanical completion restrictions. A more conventional casing program would allow the Ginn 1-13 to flow two-phase brine at 450,000 kg/hr with a pressure of 9.3 kg/cm^2 $\pm$.

The water produced at Beowawe is a dilute sodium bicarbonate water with a dissolved solids content of 1200 mg/l. This is the only high-temperature resource in the province which is not a typical sodium chloride water, possibly indicating a deeper reservoir in Paleozoic carbonate rocks.

Between 1979 and 1982, Chevron and a consortium of utilities known as NORNEV (Keilman, 1982) negotiated to build a 13-mw (gross) binary-power plant at Beowawe. Currently Chevron is negotiating with a field-development partner for a joint-venture, 10 to 20-mw power plant to be built by a third-party manufacturer. A power-sales contract would be negotiated with Sierra Pacific Power Company.

BRADY'S HOT SPRINGS - DESERT PEAK

The Brady's Hot Springs-Desert Peak area is located in the northern part of the Hot Springs Mountains, 32 km northeast of Fernley, Nevada (Fig. 4). Brady's Hot Springs and Desert Peak are apparently separate, high-temperature geothermal systems located about six km apart.

A lengthy case history and bibliography of these two areas has recently been published by Benoit et al

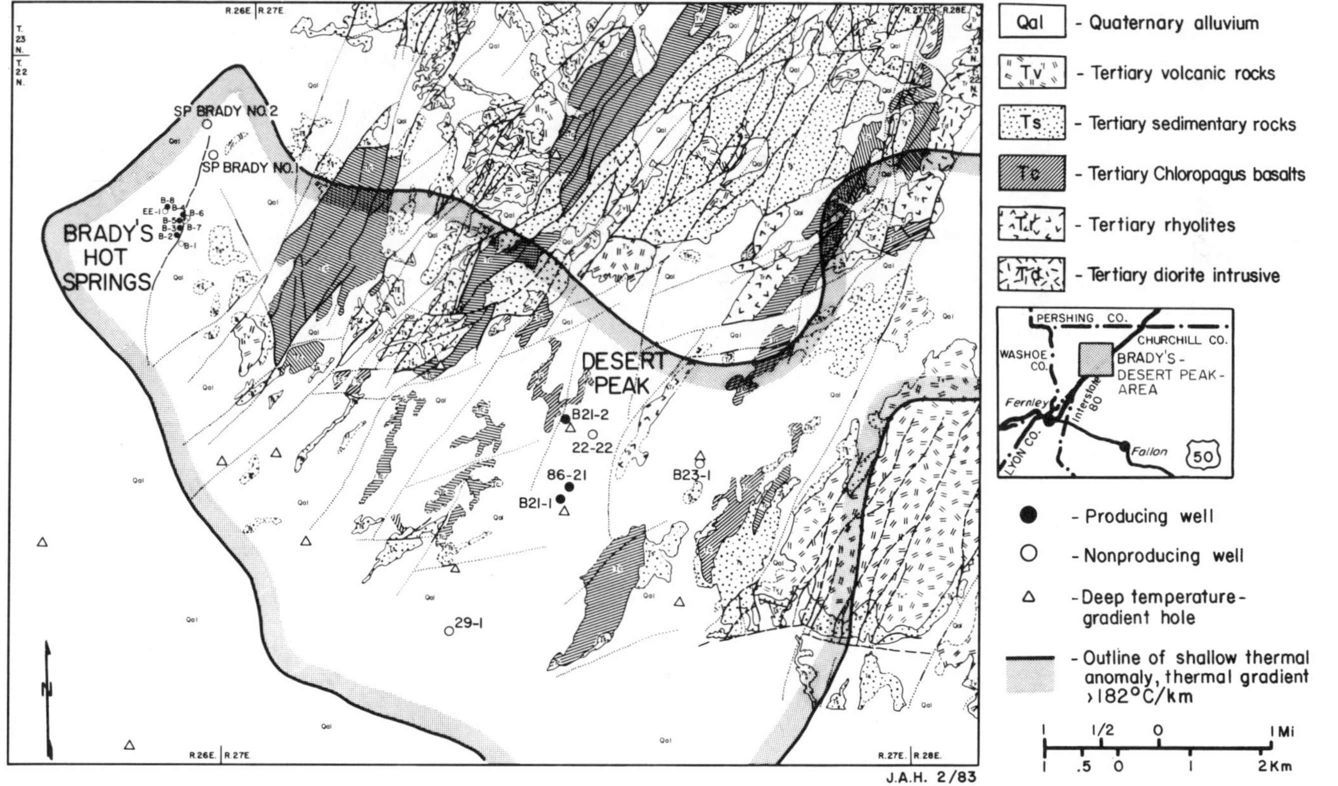

Figure 4. Map of the Brady's Hot Springs-Desert Peak, Nevada Area (after Benoit et al., 1982)

(1982). A more recently published work on Desert
Peak is a temperature study of well B23-1 by Urban
and Diment (1982).

The Desert Peak geothermal system is significant
in two aspects. It was the first blind geothermal
discovery in the province with minor surface manifestations to indicate the presence of this reservoir. Second, the near-surface thermal anomaly associated with the Desert Peak reservoir is the largest
and most intense in the province. An area of approximately 195 km^2 has temperature gradients in excess
of 110°C/km. The local surficial geology consists
of Tertiary volcanic and sedimentary rocks that are
drape folded over numerous, seldom-exposed, north-
northeast and east-northeast-striking normal faults.

Geothermal exploration began in 1959 at Brady's
Hot Springs solely because of the once-impressive
surficial thermal features. Magma Power Company initially drilled six shallow wells, several of which
produced large volumes of sodium chloride water with
temperatures over 160°C (Appendix 1) and dissolved
solids contents near 3500 mg/l after steam separation. In 1964 Earth Energy drilled the first deep
well, encountered 212°C temperatures but failed to
produce significant amounts of fluid. The last
major exploration at Brady's Hot Springs occurred
in 1974 and 1975 when two unsuccessful deep wells
confirmed the small size of this reservoir. This
does not mean that Brady's Hot Springs is an insignificant resource. The world's only commercial
geothermal food-processing plant has been in operation since 1978 at Brady's Hot Springs. This
plant is the largest commercial geothermal operation
in the Basin and Range province and will remain so
until the 20-mw power plant at Roosevelt Hot Springs
comes on line in 1984.

Geothermal exploration at Desert Peak began in
1973 as a result of a shallow temperature-gradient
hole program centered on Brady's Hot Springs. The
Desert Peak reservoir was discovered solely by
shallow and deep temperature-gradient drilling, a
technique which was successful primarily because
the Desert Peak thermal anomaly is so large. This
large size is a result of subsurface thermal-water
discharge into subhorizontal aquifers at shallow
depths. Phillips Petroleum Company's second deep
exploratory well, B21-1, discovered the Desert Peak
reservoir in November, 1976. To date, six production
wells and 12 deep temperature-gradient holes have
been drilled. The first three producing wells at
Desert Peak suggested the reservoir was areally
extensive and confined to pre-Tertiary metasedimentary, metavolcanic and granitic rocks. In
1982, wells 86-21 and 22-22 confirmed that discrete north-northeast-and east-northeast-striking
faults provide the shallow permeability and proved
production from the Tertiary volcanic rocks. Well
86-21 is currently the largest producer at Desert
Peak with a maximum flow rate between 340,000 and
410,000 kg/hr. The sodium chloride reservoir water
at Desert Peak is about twice as saline as the
Brady's water with a dissolved solids content of
6700 mg/l.

Calcium carbonate scaling in the wellbore during
production is a problem at Desert Peak as it is in
most, or possibly all, of the high-tempererature
Basin and Range reservoirs. A 30-day, scale-
inhibition test by EFP Systems Inc. on well B21-2
was successful in alleviating this problem. They
used recycled carbon dioxide as a gas-lift pump to
increase the wellhead pressure and lower the pH of
the geothermal fluid from 7.0 to 5.6, thus preventing scaling during flashing in the wellbore (Kuwada,
1982).

Negotiations are underway between Sierra Pacific
Power Company and Phillips Petroleum which will
hopefully result in the construction of a 10-mw
demonstration power plant. The ultimate potential
of this large and promising prospect has yet to be
determined.

HUMBOLDT HOUSE

The Humboldt House geothermal reservoir is located in northwestern Nevada midway between the
towns of Lovelock and Winnemucca (Fig. 5) and underlies a series of coalescing alluvial fans descending
from the west flank of the Humboldt Range. This
prospect has been characterized by initial success
followed by increasing frustration in exploration.
Geothermal literature on this prospect is quite
limited. The regional geology is described by
Johnson (1977) and the local geology has been mapped
by Silberling and Wallace (1967). Some geothermal
history and information is briefly presented by
Desormier (1979). Data and interpretations from the
most recent exploratory well, Campbell E-2, have been
published by Phillips Petroleum Company (1979), and
Sibbitt and Glenn (1981).

The geothermal potential of the Humboldt House
area was recognized during a regional shallow
temperature-gradient-hole drilling program. As the
near-surface thermal anomaly was being outlined,
recently extinct, siliceous and calcareous spring
deposits were noted (Garside and Schilling, 1979)
and a small volume of 75°C water was discovered
leaking from an old shallow mineral-exploration
hole. The Na-K-Ca geothermometer predicts a subsurface temperature of 260°C for this water. This
may be the highest predicted subsurface temperature
in Nevada, but no drill hole at Humboldt House has
yet come close to this temperature (Appendix 1).

The first production well, Campbell E-1, was
drilled by Phillips Petroleum Company in November,
1977 and is capable of producing about 363,000 kg/hr
of fluid with a maximum subsurface temperature of
183°C. The produced fluid is a dilute sodium chloride
water with a total dissolved solids content of about
5000 mg/l, and is chemically very similar to the thermal water from the old shallow mineral-exploration
hole well 6.6 km to the north-northwest. The Campbell
E-1 well is believed to produce from an unconsolidate
zone of alluvial limestone boulders at depths between
546 and 559 m (Desormier, 1982). This is the only
known geothermal well in Nevada with a high shut-in
pressure, 10.5 kg/cm^2.

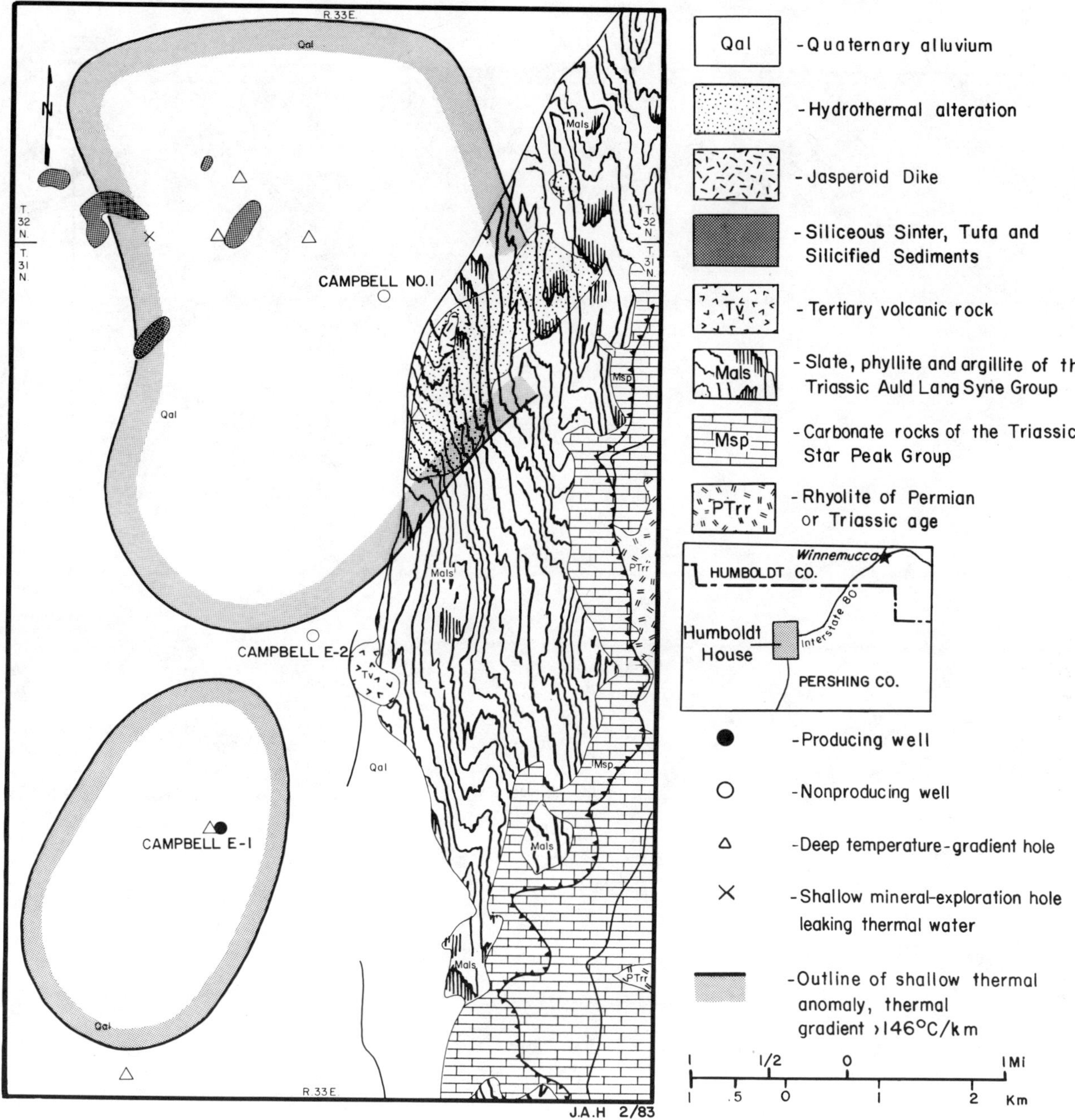

Figure 5. Map of the Humboldt House, Nevada Area (geology after Silberling and Wallace, 1967).

Two other dry exploratory wells were drilled in 1978 and 1979 (Appendix 1). Union Oil Company's Campbell No. 1 well has the maximum measured temperature at Humboldt House (205°C) but produced only about 10 Lpm of brine. These two wells encountered thick sections of the Triassic Auld Lang Syne Group--shales, slates and phyllites which appear too incompetent to maintain fracture permeability. The lack of permeability in the Auld Lang Syne Group has also been a problem at other geothermal prospects in northwestern Nevada. Thick sequences of siliceous sinter have been found interbedded within the Quaternary alluvium and underlying Tertiary lacustrine sedimentary rocks in both wells and deep temperature-gradient holes, indicating a long history of geothermal activity at Humboldt House.

No drilling has occurred on this prospect since 1979. Additional drilling is needed to confirm the reservoir, possibly one of the hottest in Nevada, but the major problem remains an apparent inability to locate successful wells.

NORTHERN DIXIE VALLEY

The Northern Dixie Valley geothermal area in west-central Nevada is about 95 km northeast of Fallon (Fig. 6). This is the second most developed area in the province, with 10 deep wells. Actually, there are several prospective, high-temperature areas in Dixie Valley, but only the area of the potentially commercial SUNEDCO development near Senator fumaroles will be discussed.

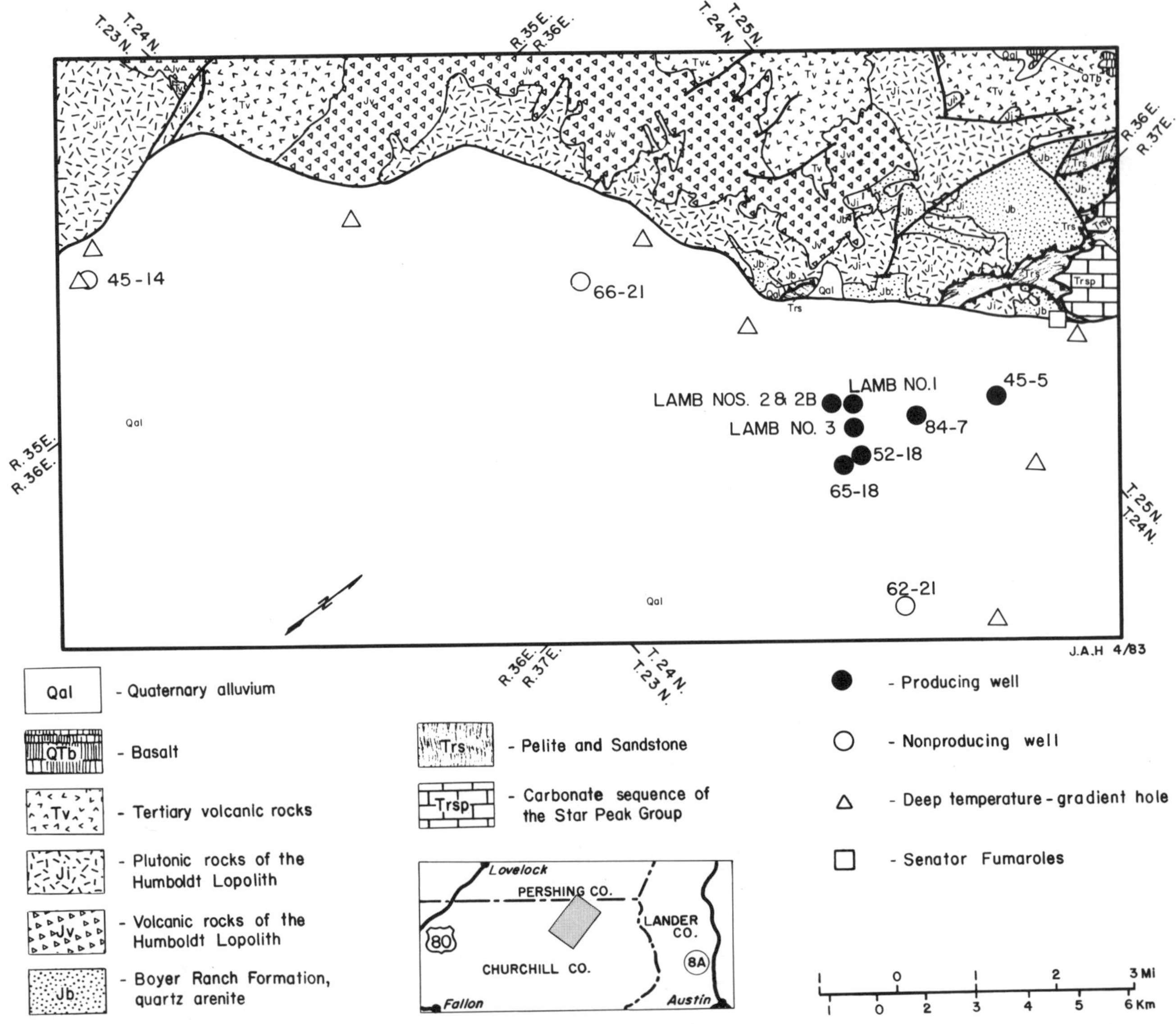

Figure 6. Map of the Dixie Valley, Nevada Area (modified from Speed, 1976).

SUNEDCO has released few results of their exploration. Therefore, the geothermal literature on Dixie Valley is mostly regional studies or discussions of data from the two wells drilled by Thermal Power Company and Southland Royalty Company as part of the Industry-Coupled Program (Fiore, 1980; Denton et al., 1980). A compilation of published literature and data can be found in the June, 1981 Geothermal Resources Council Bulletin. The most recent report summarizing SUNEDCO's exploration activities is by Parchman and Knox (1981). Geothermal developments in the southern part of Dixie Valley are reported by Waibel (1983).

Dixie Valley is a typical Basin and Range graben with a northeasterly strike and interior drainage into the Humboldt Salt Marsh. The geothermal reservoir is associated with the major normal fault(s) separating Dixie Valley from the Stillwater Range to the west. The Stillwater Range is a structurally complex block of Mesozoic sedimentary and igneous rocks overlain by a thick and variable Tertiary volcanic sequence (Willden and Speed, 1974; Speed, 1976; Waibel, 1983). Similar geology has been penetrated by the wells in Dixie Valley (Bard, 1980). Dixie Valley is known primarily for the 6.8-magnitude, 1954 Dixie Valley-Fairview Peak earthquake and associated swarms that produced fault scarps as high as 6 m (Slemmons, 1957). A spreading rate of 1 mm/yr for the past 12,000 yrs has been estimated by Thompson and Burke (1974).

Surface thermal manifestations in northern Dixie Valley are obvious and abundant at the base of the east scarp of the Stillwater Range. They include Dixie and Sou Hot Springs, Senator and other unnamed fumaroles, and several hydrothermally altered and mineralized areas. Although the predicted subsurface temperatures based on the standard chemical geothermometers from the thermal springs were low, SUNEDCO continued exploration because silica—mixing models at Dixie Hot Springs and temperature—gradient data indicated an estimated 180-210°C reservoir at depth (Parchman and Knox, 1981).

SUNEDCO's first well, Lamb #1, was the discovery well (Appendix 1). Completed at 2211 m in 1978 about 4-1/2 km southeast of the Senator Fumaroles, this well produces from fractured Tertiary volcanic rocks and possibly the underlying Mesozoic intrusive and volcanic rocks (Bard, 1980). SUNEDCO has subsequently drilled seven stepout or delineation wells, six of which are producers. The only non-producer, Federal 62-21, is the deepest test to date and the furthest from the range-front fault. As all the producing wells are between 2211 and 3005 m deep, Dixie Valley is the deepest geothermal reservoir yet discovered in the province. SUNEDCO has not released temperature data but a bottomhole temperature of the SUNEDCO development in excess of 238°C has been published (Keilman, 1982).

Two additional deep wells were drilled along the range front southwest of SUNEDCO's wells in 1979 by Thermal Power Company. Although both of these wells are hot, neither is productive. Bard (1980) reports that the Thermal Power Company wells encountered the same general lithology as the Lamb #1. However, he believes the reason the permeability is reduced is because of a "missing" red clay-layer cap overlying the volcanic sequence, and the absence of a sizeable intrusive body.

The limited fluids produced from the Thermal Power wells are sodium chloride in composition and have a total dissolved solids content ranging from 1600 to 5400 mg/ml (Bohm et al., 1980). The SUNEDCO wells produce a similar low-salinity water.

The SUNEDCO flow rates have not been released, but they are considered favorable for potential electric-power development. SUNEDCO has talked to a number utilities about developing Dixie Valley and at the present time, the next likely step is a small demonstration plant.

SODA LAKE

The Soda Lake geothermal area is located in the southwestern part of the Carson Sink, some 10 km northwest of Fallon, Nevada (Fig. 7). This prospect is relatively unknown as industry exploration results to date have not been widely publicized, and the surficial geology largely conceals the active geothermal manifestations.

The geothermal potential of the Soda Lake area was first indicated in 1903, when water well drilling at an extinct hot spring hit boiling water at 18 m (Garside and Schilling, 1979). This well furnished steam for a bathhouse as late as 1964. Morrison (1964) mapped the area as part of a larger study of Lake Lahontan and the southern Carson Desert. Olmsted et al. (1975) studied the hydrology plus outlined and interpreted the large near-surface thermal anomaly. Industry began exploratory work in 1973. Hill et al. (1979) presented a brief

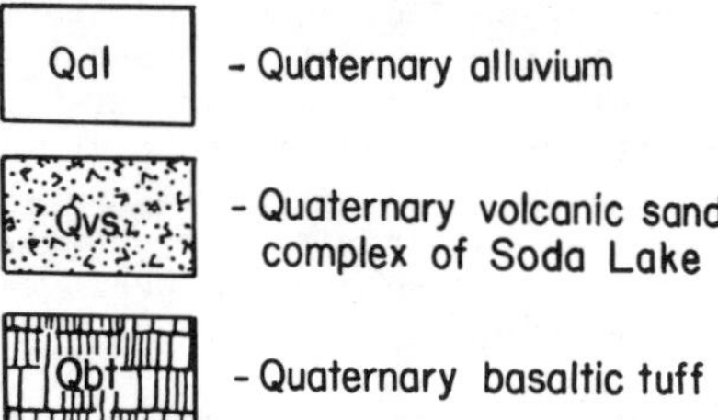

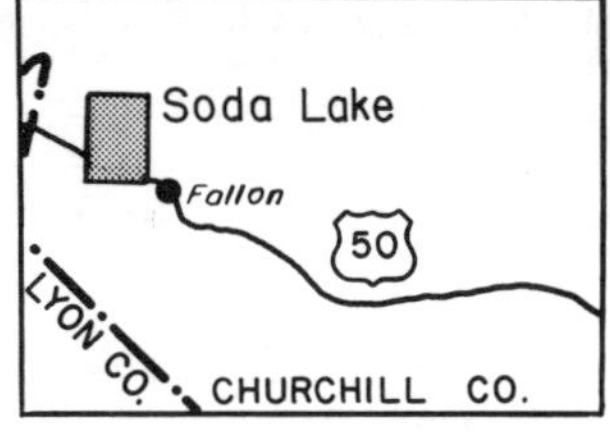

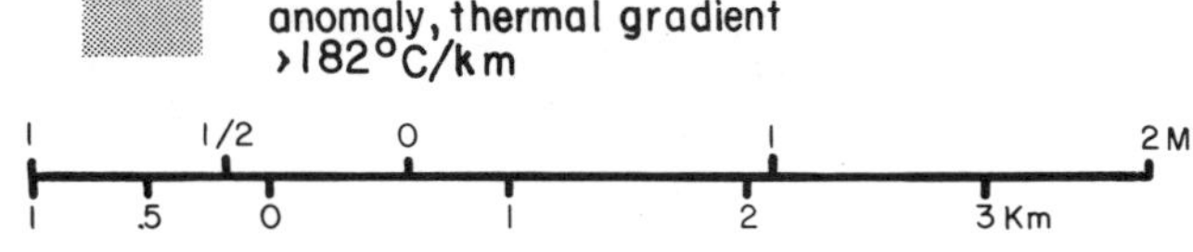

Figure 7. Map of the Soda Lake, Nevada Area (geology modified from Morrison, 1964).

exploration history and geothermal interpretation. Detailed lithologic logs from several of the inter-mediate-depth and deep wells in the area are presented by Sibbett (1979).

The Soda Lake geothermal area is the only proven high-temperature Basin and Range geothermal prospect not adjacent to range-bounding, frontal-fault systems or located within an exposed small-relief horst block. It is located in the southern Carson Sink, about 18 km from pre-Quaternary bedrock exposures. The Carson Sink is the major drainage sump of the northwestern Great Basin and is one of the largest and deepest basins in northern Nevada. The surficial geology is almost entirely late Pleistocene lacustrine and aeolian sediments. The flat, monotonous sedi-mentary surficial cover is broken by two volcanic-related deposits. Soda Lake and Little Soda Lake occupy phreatic explosion craters within cones of Quaternary sand and basaltic lapilli ejecta about 30 m high. These explosive craters are believed to have been active as recently as 10,000 yrs ago and hot-spring activity apparently is present near the center of Soda Lake (Breese, 1968). Quaternary magmatic activity at Upsal Hogback, about 9 km north-east of Soda Lake, has produced from four to seven overlapping cones of subaerially deposited basaltic tuff. The alignment of these volcanic and hydro-thermal features provides evidence for a major northeast-striking structure controlling the geo-thermal system. Hill et al. (1979) have inter-preted this structure from seismic data as a narrow graben.

There is gravity and magnetic evidence for other interesting, and as yet poorly understood, features beneath the surficial cover. A 6-mgal, arcuate Bouguer gravity high with a diameter of about 10 km is approximately centered on Soda Lake. A magnetic low correlates with the central gravity low. Directly beneath Soda Lake a small positive gravity residual may indicate an intrusive plug.

The Soda Lake resource is largely defined by temperature-gradient holes. Three large-diameter wells and six intermediate-depth temperature-gradient holes have helped define the thermal regime at depth. However, stratigraphic correla-tions between these holes, especially in the pre-Quaternary rocks, have met with limited success to date.

The first deep exploratory well, 1-29, (Appen-dix 1) was drilled in 1974 just west of the old steam well. Production was found at 238 m in un-consolidated alluvium with a maximum temperature of 172°C. The 44-5 well was unsuccessfully drilled in 1978 on a resistivity anomaly near the south margin of the shallow thermal anomaly (Hill et al., 1979). Although 188°C has been measured as shallow as 610 m, the hottest measured temperature to date is 204°C in the 84-33 well. A short-term flow test of well 84-33 has indicated potential for commercial production. It produced 115,000 kg/hr total-mass flow through 75 m of perforated casing completed in the Tertiary volcanic section. The reservoir fluid is a low-salinity, 5000 mg/l, sodium chloride water.

Soda Lake is not yet a commercial success and no power plants have been proposed, but the results to date are encouraging. Additional drilling, test-ing and evaluation are required before the potential of this reservoir can be determined.

STEAMBOAT SPRINGS

Steamboat Springs is located about 16 km south of downtown Reno, Nevada (Fig. 8). Systematic research on the Steamboat Springs geothermal system began in 1945, making it the first such system to be extensively studied in the Basin and Range pro-vince (Thompson and White, 1964; White et al., 1964; White, 1968). This early work, along with the impressive thermal features, has made Steamboat Springs known to geothermal and economic geologists worldwide. The water geochemistry has been studied by Bateman and Scheibach (1975), and Nehring (1979, 1980). White et al. (1974) have studied the geo-chemical processes which created the extensive areas of acid-leached and hydrothermally altered rock at Steamboat Springs. Geophysical data have been presented by Hoover et al. (1975a, 1975b), Long and Brigham (1975), and Peterson (1975). The most recent geological paper has been on the duration of hydrothermal activity (Silberman et al., 1979). The only published information on the recent geo-thermal exploration is by Desormier (1983).

Steamboat Springs has a wide variety of geo-thermal features covering an area of about 10 km^2, making it one of the most obvious and interesting geothermal exploration targets in the province. The thermal springs are located at the northeast end of Steamboat Hills, a small northeast-striking range transverse to the dominant regional trends. The Steamboat Hills consist of granodiorite and metamorphosed sedimentary and volcanic rocks part-ially buried by Tertiary volcanic rocks and Quater-nary volcanic and sedimentary rocks. A north-east-striking line of four Quaternary rhyolite domes (Thompson and White, 1964) indicates a possible magmatic heat source. The area is highly faulted and these faults appear to control the location of the known reservoir (Desormier, 1983). Steamboat Springs has a long documented history of geothermal activity.It has been intermittently active for at least 2.5 million years (Silberman et al., 1979).

Geothermal exploration at Steamboat Springs began about 1920 when a local resort owner drilled shallow wells to supply a spa. In 1950 the Rodeo well (Appendix 1) was the first well to be drilled at Steamboat Springs, and probably the first in the Basin and Range province, specifically search-ing for steam for generating electricity (White, 1983). The initial intermediate-depth exploratory well was drilled in 1959 by Nevada Thermal Power Company (Magma) to 558 m (White, 1968). Five other wells soon followed with the maximum measured temp-erature of 186°C in Nevada Thermal Power well 4. No production or deep exploratory wells were drilled at Steamboat Springs between 1962 and 1979.

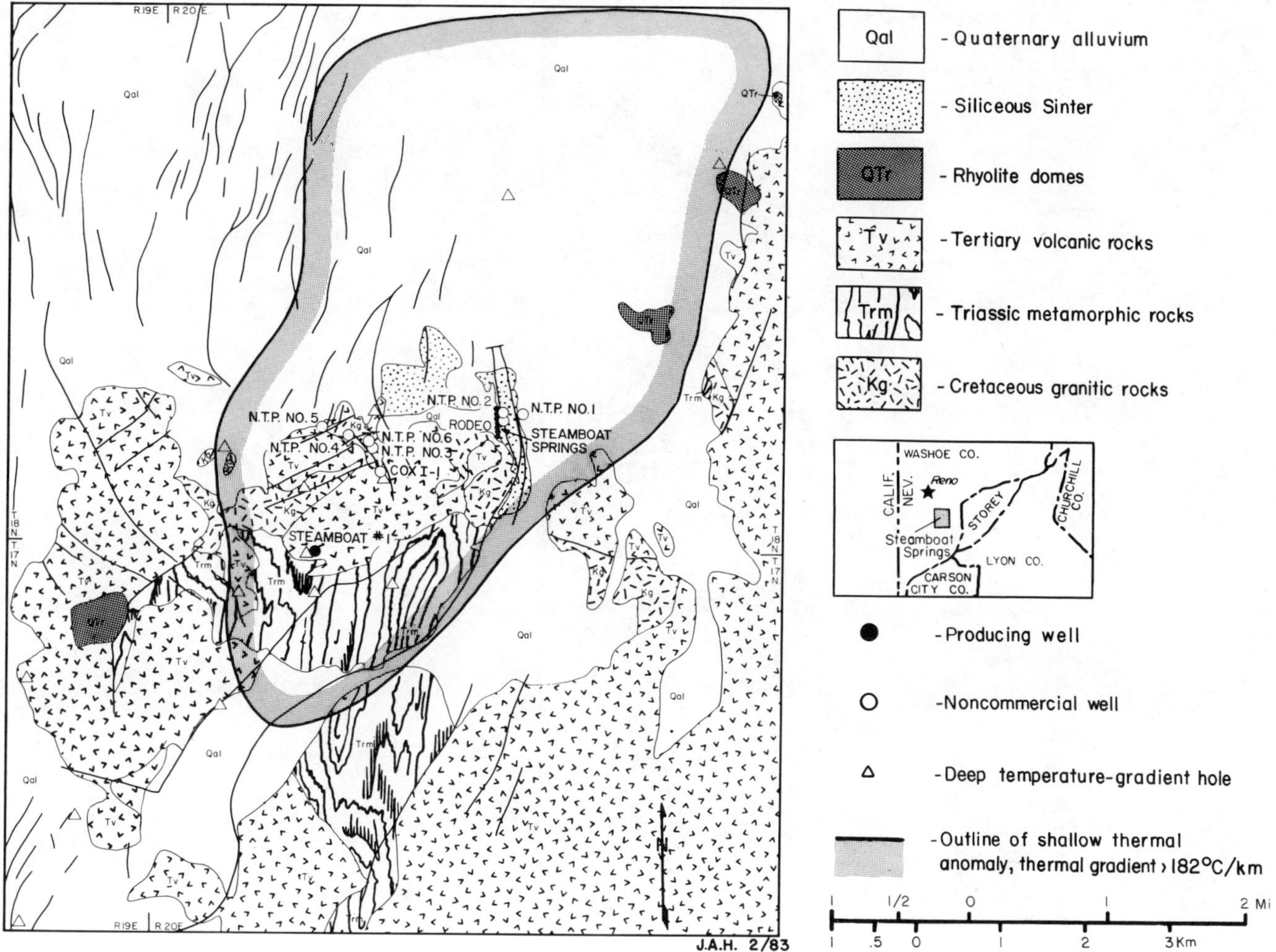

Figure 8. Map of the Steamboat Springs, Nevada Area (modified from Silberman et al., 1979)

In the mid 1970's private industry resumed exploration at Steamboat Springs in anticipation of the September, 1975 KGRA sale. Phillips Petroleum conducted an integrated exploration program between 1975 and 1979 which resulted in drilling five intermediate-depth, temperature-gradient holes. Data from these holes were used to locate Steamboat #1, the discovery well (Desormier, 1983).

Steamboat #1 was drilled to a depth of 930 m in the summer of 1979 and is capable of producing 272,000 kg/hr of fluid from fractured granodiorite and metamorphic rocks. The maximum measured subsurface temperature is 228°C. The chemical geothermometers indicate subsurface temperatures near 220°C (Nehring, 1979). The produced sodium chloride water is chemically similar to the hot springs water with a salinity of 2200 mg/l.

Steamboat #1 appears to have an unusual location, on top of the Steamboat Hills, almost three km southwest of the main thermal springs. However, Steamboat #1 has demonstrated that this area is a deeper source for the Steamboat Springs thermal water which flows laterally to the northeast from beneath Steamboat Hills.

Nine intermediate-depth temperature-gradient holes and one other non-commercial production well, the Cox I-1, have been drilled since 1979. These have confirmed the Steamboat Springs geothermal reservoir underlies the higher parts of the Steamboat Hills. Between 1981 and 1983 there was no drilling activity because one of the major lease holders was liquidating their position. This was accomplished early in 1983 so a second attempt at a confirmation well is now feasable.

COSO

The Coso geothermal field is located in the Coso Mountains of southern California about 55 km north of the town of Ridgecrest and mostly within the borders of the China Lake Naval Weapons Center (Fig. 9).

Coso lies a short distance east of the scenic eastern scarp of the Sierra Nevada Range amid a spectacular cluster of Quaternary rhyolite domes.

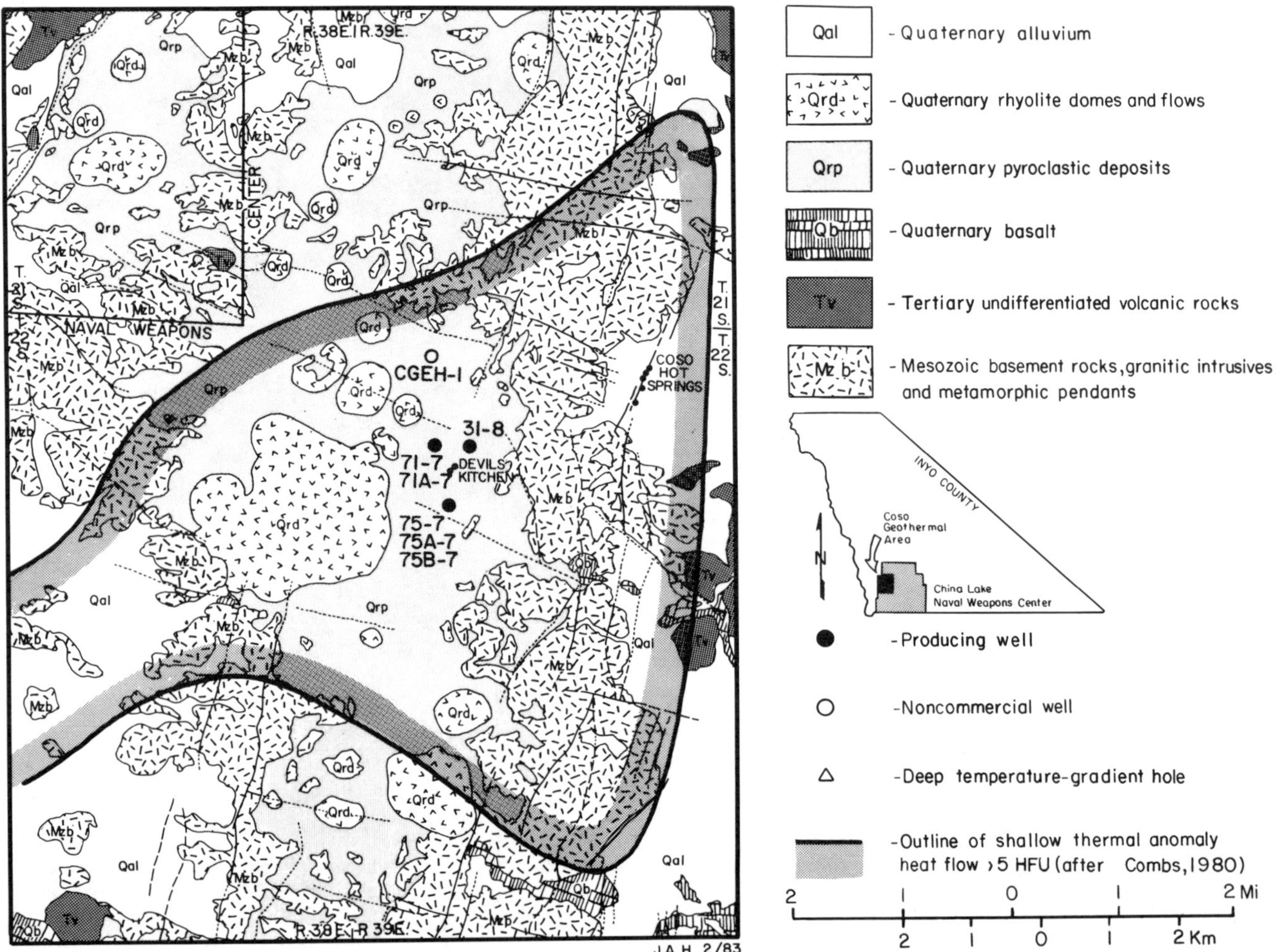

Figure 9. Map of the Coso, California Area (modified from Duffield and Bacon, 1981).

These domes, together with the Coso thermal springs and the Devil's Kitchen fumaroles, clearly indicate high geothermal potential. Coso has been intensively studied by the U. S. Geological Survey, University of Utah Research Insitute, U. S. Navy, the U. S. Department of Energy, and private industry.

The literature on Coso is extensive. The collection of papers in the Journal of Geophysical Research (1980) examine many facets of Coso. More recent contributions are a geologic map (Duffield and Bacon, 1981) and recent production-drilling results near the Devil's Kitchen (Moore et al., 1982).

Numerous geophysical studies were conducted at Coso prior to drilling the first large-diameter exploratory well, CGEH-1 late in 1977 (Galbraith, 1978). CGEH-1 was drilled by the U. S. Department of Energy in search of hot dry rock. CGEH-1 has a maximum temperature of 195°C (Appendix 1) and during drilling produced 27,000 kg/hr of sodium chloride water with a total dissolved solids content near 4,500 mg/l. The chloride water has been interpreted to indicate a hot-water reservoir (Fournier et al., 1980).

After the drilling and testing of CGEH-1 there was no additional drilling until late 1981. During this time, legal and political problems requiring congressional and U. S. Navy action to permit private development within the Naval Weapons Center were resolved. In December, 1979, almost six years after the Geothermal Steam Act was enacted, California Energy Company contracted with the U. S. Navy to explore for and develop geothermal resources on an initial 3000-ac tract. This arrangement is unique because the U. S. Navy has retained title to the geothermal resource. In September, 1981 the remaining Federal KGRA lands outside the Naval Weapons Center were offered for lease. The Los Angeles Dept. of Water and Power bid $1262 and $1012 per acre for two parcels which are by far the highest KGRA bonus bids in the province.

In December, 1981, California Energy drilled well 75-7 (Appendix 1) near the Devil's Kitchen to a depth of 405 m and encountered dry steam with a temperature of 213°C and a pressure of 17.9 kg/cm^2. The 75-7 well is capable of a sustained steam-flow rate well in excess of 45,000 kg/hr (Moore et al., 1982; The Oil and Gas Journal, 1982). The dry steam was largely a surprise because of the chloride water present in CGEH-1. Moore et al. (1982) believe the steam results from a local zone of vigorous flashing rather than from an extensive steam cap. Fournier (pers comm.) interprets the dry steam to result from the partial obstruction of hot water flow along a fault which was intersected by the 75-7 wellbore. He believes that there also is no steam cap, but that the steam only forms as the pressure is decreased above the obstruction. The reservoir at Coso apparently consists of north-south striking fractures in Mesozoic granitic rocks and variable high-grade metamorphic rocks of uncertain age.

California Energy has drilled five additional wells, all of which are reported to be capable of commercial flow rates and produce two-phase fluids (National Geothermal Service, 1982). The drilling strategy to date at Coso has varied from most other Basin and Range prospects. Temperature-gradient holes deeper than 150 m are conspicuously absent.

The maximum reported temperature at Coso is 213°C in well 75-7. Temperatures as high as 245°C in the deeper chloride-water part of the reservoir are possible based on geochemical evidence (Fournier et al.,

1980). As all of the wells to date at Coso have been relatively shallow, deeper drilling may encounter these higher indicated temperatures.

California Energy Company has announced that they intend to be generating a substantial amount of electrical power by the end of 1984. If this schedule is met, the elapsed time between the first commercial well and power production will be only 3 years. This would be the most rapid commercial geothermal-power development in the United States.

LONG VALLEY

The Long Valley caldera is located at the base of the eastern scarp of the Sierra Nevada, 50 km north of the town of Bishop. The scenic resort town of Mammoth Lakes is nestled within its southwest quadrant (Fig. 10). The caldera is located directly along the western margin of the Basin and Range province and may be more closely related to the province boundary than the province proper. However, the 450 km^2 Quaternary caldera has all the geological pre-requisites to contain the largest

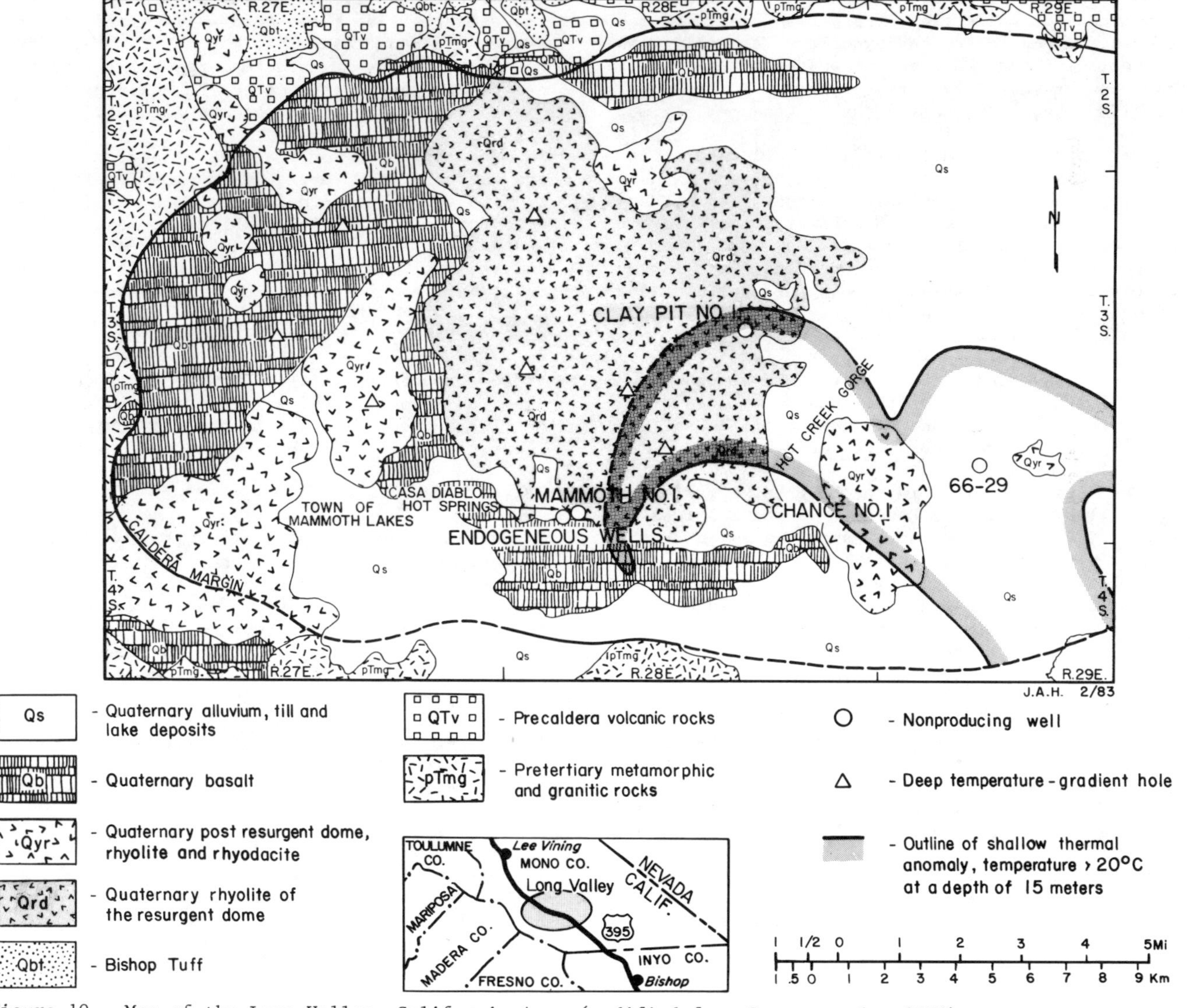

Figure 10. Map of the Long Valley, California Area (modified from Sorey et al., 1978).

geothermal reservoir in the Basin and Range province. Recent obsidian flows, along with gravity, seismic refraction, and P-wave delay data indicate a magmatic heat source underlies the western portion of the caldera. The hot springs in Hot Creek Gorge are the largest volume boiling springs in the province and the chemical geothermometers when applied to these waters indicate a reservoir temperature between 200 and 282°C. Large areas of intense hydrothermal alteration are present at Casa Diablo, Hot Creek, and the Clay Pit. Fairly intense seismic activity, probably related to magma movement, is presumably maintaining or enhancing reservoir permeability. On the other hand, this activity has initiated a volcano-watch designation by the U. S. Geological Survey. In spite of these favorable geological characteristics, no large reservoir has yet been discovered. The main reason for this is U. S. Forest Service delays in leasing Federal land within this most scenic of the high-temperature geothermal prospects.

The literature on Long Valley is very extensive. The U. S. Geological Survey's Long Valley Symposium (Journal of Geophysical Research, 1976) is a very comprehensive collection of papers on geology, geochemistry, geophysics and hydrology. A complete integration of the hydrothermal system of Long Valley has been presented by Sorey et al (1978). Diment et al (1980) have studied the shallow thermal regime.

Geothermal exploration at Long Valley began in 1959 at Casa Diablo Hot Springs on private lands. Nine shallow production holes were drilled by Magma Power Company and Endogeneous Power Company to a maximum depth of 324 m (Appendix 1). A maximum temperature of 178°C was measured and the maximum-reported flow rate was 246,000 kg/hr (McNitt, 1963). After this initial burst of activity which lasted through 1962, little happened until the U. S. Geological Survey began a comprehensive program in 1971 to study Long Valley as its type hot-water geothermal system. Some of the interpretations from this study, made without any deep drill holes, have since proven to be almost prophetic.

The first deep well in the Long Valley caldera was drilled in 1976 by Republic Geothermal to a depth of 2109 m in the southeastern quadrant (Smith and Rex, 1977). Well 66-29 is located on land leased at the first KGRA sale in January 1974. This apparently had much to do with the location. Well 66-29 turned out to be surprisingly cold with a maximum unstabilized temperature of 72°C, obtained 90 hrs after last circulation.

Between 1976 and 1979 no deep wells were drilled in the caldera. This was not because of the discouraging results of the Republic well, but because the U. S. Forest Service was not making Federal lands in the central and western parts of the caldera available for lease. In 1979, the U. S. Forest Service, together with the Bureau of Land Management and U. S. Geological Survey, proposed a lease sale in the central part of the caldera. However, the special stipulations attached to the leases included such items as forced unitization, a commitment to drill three wells at least four km apart within two years of unitization, and phased leasing. Industry felt it could not effectively operate under these requirements and filed an appeal. The Chief of the U. S. Forest Service mitigated these problems in March 1981.

In preparation for this sale Union Oil Company drilled two unsuccessful deep production wells in the summer of 1979. Union verbally presented data from these wells to the geothermal industry prior to the rescheduled KGRA sale at a Bay Area section meeting of the Geothermal Resources Council. Some of this material has later appeared in print (Gambill, 1981).

The Union Mammoth No. 1 well has a double temperature reversal with a 60°C temperature decrease below the reversal. The Clay Pit well was also quite discouraging with a bottomhole temperature of 147°C at 1846 m. More importantly, these two wells demonstrated that a high-temperature geothermal reservoir is not present beneath two of the most impressive surficial manifestations in the caldera. The three deep wells within the caldera have demonstrated that much of the caldera is not underlain by a geothermal reservoir. Consequently the exploration focus has shifted to the unexplored western third of the caldera.

The lease sale for the central part of the caldera was held in October, 1981, after the appeal had run its course. A final lease sale was later scheduled for September 1982, to include most of the western third of the caldera. This lease sale was delayed due to appeals by environmental factions to July, 1983.

Long Valley has been the site of the most friction between the geothermal industry, government, and environmental factions in the province. However, additional deep exploratory wells will be drilled in the caldera and hopefully this elusive reservoir will be discovered in the near future.

THE FUTURE

Currently geothermal exploration for high-temperature reservoirs is stagnant (Edmiston, 1982), a result of the modest decline in the price of the principal forms of energy. Most known reservoirs in the Basin and Range province have marginal temperatures under present economic conditions. Any significant additional decline in energy prices could render most and possibly all of these reservoirs noncompetitive. A collapse of OPEC could create such a decline. Until energy prices stabilize or increase, geothermal exploration for undiscovered reservoirs will continue at a low level with fewer than four wildcat wells being drilled in the province each year.

There could still be significant development drilling activity on the discovered reservoirs. It is crucial that the geothermal industry prove as soon as possible, that at least one of the discovered reservoirs in the temperature range of 200-220°C is capable of power generation. In all likelihood the first power plants on each reservoir will be demonstration facilities of 10 or 20 mw. At this time it is not possible to predict when or how many of the reservoirs will be developed.

Long-term exploration for new reservoirs will continue, although at a much slower pace than in the past decade. Most future discoveries will either have to be blind, like Desert Peak; deep, like Dixie Valley; or reinterpretations of some already drilled areas. Thermal systems with boiling springs and large exposed siliceous-sinter deposits such as Beowawe, Roosevelt Hot Springs, Steamboat Springs, and Brady's Hot Springs have been drilled. Temperatures of new discoveries generally will not exceed 220°C as there is no evidence to expect any new discoveries, other than Long Valley and possibly the Mono Craters, to be closely associated with shallow silicic-magma chambers.

In spite of all the exploration to date in the northern Basin and Range province, many areas in excess of 200 km^2 do not contain even a single shallow temperature-gradient hole. There has not been a deep well drilled in any of the major, high-relief mountain ranges. The same may generally be said for deep temperature-gradient holes. Similarly, the areas of Recent mafic volcanism such as Lunar Crater or the Owens Valley have also been virtually ignored. It is not likely that all the high-temperature reservoirs have been discovered. Future discoveries will probably be concentrated along the east or west margins of the province, or in the vicinity of the Carson Sink.

<u>ACKNOWLEDGEMENTS</u>

The authors are indebted to the following companies and individuals who have allowed previously confidential data to be published or have reviewed and corrected portions of this manuscript: Chevron Resources Company, Phillips Petroleum Company, Union Oil Company, Carl Austin, Bill Desormier, Ron Forrest, John Knox, Erik Layman, Dick Lenzer, Jim Moore, and Don White. The drafting was done by Alice Hixson and Dianne Feist did the typing.

<u>BIBLIOGRAPHY</u>

Bard, T. R., 1980, Petrologic alteration studies in Geothermal Reservoir Assessment Case Study-- Northern Dixie Valley, Nevada: Southland Royalty Company, U. S. Dept. of Energy Contract No. DE-ACO8-79 ET27006, p. 88-158.

Bateman, R. L., and Scheibach, R. B., 1975, Evaluation of geothermal activity in the Truckee Meadows, Washoe County, Nevada: Nevada Bur. of Mines and Geol. Report 25, 38 p.

Benoit, W. R., Hiner, J. E., and Forest, R. T., 1982, Discovery and geology of the Desert Peak geothermal field: A case history: Nevada Bur. of Mines and Geol. Bull. 97, 82 p.

Bohn, B. W., Jacobson, R. L., Campana, M. E., and Ingraham, N. L., 1980, Hydrology and hydro-chemistry in Geothermal Reservoir Assessment Case Study-- Northern Dixie Valley, Nevada: Southland Royalty Company, U. S. Dept. of Energy Contract No. DE-ACO8-79ET27006, p. 159-186.

Breese, C. R. Jr., 1968, A general limnological study of Big Soda Lake: M. S. thesis, Univ. of Nevada, Reno, 83 p.

Combs, J., 1980, Heat flow in the Coso geothermal area, Inyo County, California: Jour. of Geophys. Res. Vol. 85, No. B5, p. 2411-2424.

Denton, J. M., Bell, E. J., and Jodry, R. L., 1980, Geothermal reservoir assessment case study-- Northern Dixie Valley, Nevada: Southland Royalty Company, U. S. Dept. of Energy Contract No. DE-ACO8-79ET27006, 223 p.

Desormier, W. L., 1979, Desert Peak to Humboldt House and Winnemucca, in Lane, M. A., (ed.) Nevada Geothermal Areas: Desert Peak, Humboldt House, Beowawe: Guidebook for Field Trip #6 Geothermal Resources Council 1979 Annual Meeting, p. 9-18.

_____, 1982, Humboldt House lease evaluation and preliminary development plan Pershing County, Nevada: unpublished report for Phillips Petroleum Company, 15 p.

_____, 1983, Steamboat Springs Geothermal Project: Geothermal Resources Council Field Trip Guidebook for the Role of Heat in the Development of Energy and Mineral Resources in the Northern Basin and Range Province, 7 p.

Diment, W. H., Urban, T. C., and Nathenson, M., 1980, Notes on the shallow thermal regime of the Long Valley caldera, Mono County, Calif-Vol. 4, p. 37-40.

Duffield, W. A., and Bacon, C. R., 1981, Geologic map of the Coso volcanic field and adjacent areas, Inyo County, California: U. S. Geol. Surv. Misc. Invest. Map I-1200.

Edmiston, R. C., 1982, A review and analysis of geothermal exploratory drilling results in the Northern Basin and Range geologic province of the USA from 1974 through 1981: Geothermal Resources Council Trans. Vol. 6, p. 11-14.

Edmiston, R. C., 1982, Lets face the facts about geothermal power development in the U. S.: Geothermal Resources Council Bull. Vol. 11, No. 11, p. 19-20.

Epperson, I. J., 1982, Beowawe, Nevada, well testing: history and results: Geothermal Resources Council Trans., vol. 6, p. 257-260.

Fiore, J. N., 1980, Overview and status of the U. S. Department of Energy's Industry-Coupled Geothermal Reservoir Assessment Program: Geothermal Resources Council Trans. Vol. 4, p. 201-204.

Fournier, R. O., Thompson, J. M., and Austin, C. F., 1980, Interpretation of chemical analyses of waters collected from two geothermal wells at Coso, California: Jour. of Geophys. Res., Vol. 85, p. 2405-2410.

Galbraith, R. M., 1978, Geological and geophysical analysis of Coso geothermal exploration hole no. 1 (CGEG-1), Coso Hot Springs KGRA, California: Univ. of Utah Research Inst., Earth Science Lab. Report, 39 p.

Gambill, D. T., 1981, Preliminary hot dry rock geothermal evaluation of Long Valley caldera, California: Los Alamos National Laboratory, LA-8710-HDR, 22 p.

Garside, L. J., 1974, Geothermal exploration and development in Nevada through 1973: Nevada Bureau of Mines and Geology Report 21, 12 p.

Garside, L. J., and Schilling, J. H., 1979, Thermal Waters of Nevada: Nevada Bur. of Mines and Geology Bull. 91, 163 p.

Hill, D. G., Layman, E. B., Swift, C. M., and Yungul, S. H., 1979, Soda Lake, Nevada, thermal anomaly: Geothermal Resources Council Trans. Vol. 3, p. 305-308.

Hoover, D. B., O'Donnell, J. O., Batzle, M., and Rodriguez, R., 1975, Map of telluric profiles, Steamboat Hills, Nevada: U. S. Geol. Survey Open-File Report 75-445.

Hoover, D. B., Batzle, M., and Rodriguez, R., 1975b, Self-potential map, Steamboat Hills, Nevada: U. S. Geol. Survey Open-File Report 75-446.

Johnson, M. G., 1977, Geology and mineral deposits of Pershing County, Nevada: Nevada Bur. Mines and Geol. Bull. 89, 115 p.

Journal of Geophysical Research, 1976, Vol. 81, No. 5, p. 721-860.

_____, 1980, Vol. 85, p. 2379-2516.

Keilman, L., 1982, Beowawe #1 - a 10 mw geothermal unit in northern Nevada: Geothermal Resources Council Trans. Vol. 6, p. 351-353.

Kuwada, J. T., 1982, Field demonstration of the EFP system for carbonate scale control: Geothermal Resources Council Bull. Vol. 11, No. 9, p. 3-9.

Lenzer, R. C., Crosby, C. W., and Berge, C. W., 1977, Recent developments at the Roosevelt Hot Springs KGRA: Trans. Am. Nuclear Soc. Topical Mtg., Golden, Colo., April 12-14, p. 60-67.

Long, C. L., and Brigham, R. H., 1975, Audio-magnetotelluric data log for Steamboat Hills, Nevada: U. S. Geol. Survey Open-File Report 75-447, 7 p.

McKinney, D. B., 1978, Annotated bibliography of the geology of the Roosevelt Hot Springs known geothermal resource area and the adjacent Mineral Mountains, March 1978: Univ. of Utah Research Inst., Earth Science Lab Report, 15 p.

McNitt, J. R., 1963, Exploration and development of geothermal power in California: California Div. of Mines and Geol: Spec. Report 75, 45 p.

Middleton, W. M., 1961, Report on Beowawe, Nevada Geothermal Steam Wells: prepared for Magma - Vulcan Thermal Power Project (unpublished).

Moore, J. L., Austin, C. F., and Prostka, H. J., in press, Geology and geothermal energy development at the Coso KGRA: in Energy Resources of the Pacific Region, Proceedings 1982 Circum-pacific Conference, Amer. Assoc. of Petrol. Geol.

Morrison, R. B., 1964, Lake Lahontan: Geology of southern Carson Desert, Nevada: U. S. Geol. Survey Prof. Paper 401, 156 p.

National Geothermal Service, 1982, Vol. 4, No. 13.

_____, 1983, Vol. 5, No. 21.

Nehring, N. L., 1979, Reservoir temperature, flow, and recharge at Steamboat Springs, Nevada: Geothermal Resources Council Trans. Vol. 3, p. 481-483.

_____, 1980, Geochemistry of Steamboat Springs: U. S. Geol. Survey Open-File Report 80-887, 66 p.

Oil and Gas Journal, 1982, Geothermal strike drilled on Navy's China Lake land: Vol. 80, No. 6, p. 66.

Olmsted, F. H., Glancy, P. A., Harrill, J. R., Rush, F. E., and VanDenburgh, A. S., 1975, preliminary hydrologic appraisal of selected geothermal systems in northern and central Nevada: U. S. Geol. Survey Open-File Report 76-76, 274 p.

Parchman, W. L., and Knox, J. W., 1981, Exploration for geothermal resources in Dixie Valley, Nevada a case history: Geothermal Resources Council Bull. Vol. 10, No. 5, p. 3-6.

Peterson, C. A., 1975, Geology of the Roosevelt Hot Springs area, Beaver County, Utah: Utah Geology, Vol. 2, No. 2, p. 109-116.

Peterson, D. L., 1975, Principal facts for gravity stations in Steamboat Hills and Wabuska, Nevada: U. S. Geol. Survey Open-File Report 75-443, 8 p.

Phillips Petroleum Company, 1979, Geothermal reservoir assessment case study, northern Basin and Range province: U. S. Dept. of Energy, Division of Geothermal Energy Report DOE/ET/ 27099-1, 38 p.

Ross, H. P., Nielson, D. L., Moore, J. N., 1982, Roosevelt Hot Springs Geothermal System, Utah-Case study: Amer. Assoc. of Petrol. Geol. Bull. Vol. 66, No. 7, p. 879-902.

Rudisell, J. M., 1976, Geothermal well Utah State 14-2, 48-hour flow test, Nov. 16, 1976 to Nov. 18, 1976: released by U. S. Dept. of Energy under contract No. EG-77-C-08-1525, 15 p.

Schoen, R., White, D. E., and Hemley, J. J., 1974, Argillization by descending acid at Steamboat Springs, Nevada: Clays and Clay Minerals, Vol. 22, p. 1-22.

Sibbett, B. S., 1979, Geology of the Soda lake geothermal area: Univ. of Utah Research Inst., Earth Science Lab. Report, 14 p.

Sibbett, B. S., and Glenn, W. E., 1981, Lithology and well log study of Campbell "E-2" geothermal test well, Humboldt House geothermal prospect, Pershing County, Nevada: Univ. Utah Research Inst., Earth Science Lab Report 53, 17 p.

Silberling, N. J., and Wallace, R. E., 1967, Geologic map of the Imlay quadrangle, Pershing County, Nevada: U. S. Geol. Survey map GQ-666.

Silberman, M. L., White, D. E., Keith, T. E. C., and Dockter, R. D., 1979, Duration of hydrothermal activity at Steamboat Springs, Nevada, from ages of spatially associated volcanic rocks: U. S. Geological Survey Prof. Paper 458-D, 14 p.

Slemmons, D. B., 1957, Geological effects of the Dixie Valley - Fairview Peak, Nevada, earthquakes of December 16, 1954: Seis Soc. America Bull. 47, No. 4, p. 353-375.

Smith, J. L. and Rex, R. W., 1977, Drilling results from the eastern Long Valley caldera: Am. Nuclear Soc. Mtg. on Energy and Mineral Recovery Research, Golden,Colo., April 12-14, p. 529-540.

Smith, C., 1979, Interpretation of electrical resistivity and shallow seismic reflection profiles, Whirlwind Valley and Horse Heaven areas, Beowawe KGRA, Nevada: Earth Science Lab Report 25, 43 p.

Smith, C., Struhsacker, E. M., and Struhsacker, D. W., 1979, Structural inferences from geologic and geophysical data at Beowawe KGRA, north-central Nevada: Geothermal Resources Council Trans. Vol. 3, p. 659-662.

Sorey, M. L., Lewis, R. E., and Olmsted, F. H., 1978, The hydrothermal system of Long Valley caldera, California: U. S. Geol. Surv. Prof. Paper 1044-A, 60 p.

Speed, R. C., 1976, Geologic map of the Humboldt Lopolith and surrounding terrain, Nevada: Geol. Soc. America Map MC-14, 4 p.

Stewart, J. H., Walker, G. W., and Kleinhampl, F. J., 1975, Oregon-Nevada lineament: Geology, Vol. 3, p. 265-268.

Struhsacker, E. M., 1980, The geology of the Beowawe geothermal system, Eureka and Lander Counties, Nevada: Earth Science Lab Report 37, 78 p.

Swift, C. M., 1979, Geophysical data, Beowawe geothermal area, Nevada: Geothermal Resources Council Trans. Vol. ·13, p. 701-703.

Thompson, G. A., and White, D. E., 1964, Regional geology of the Steamboat Springs area, Washoe County, Nevada: U. S. Geological Survey Prof. Paper 458-A, 52 p.

Thompson, G. A., and Burke, D. B., 1974, Regional Geophysics of the Basin and Range Province: Ann. Rev. Earth and Planetary Sci., Vol. 2, p. 213-238.

Urban, T. C., and Diment, W. H., 1982, An interpretation of precision temperature logs in a deep geothermal well near Desert Peak, Churchill County, Nevada: Geothermal Resources Council Trans. Vol. 6, p. 317-320.

Waibel, A. F., 1983, Field Trip #1 Reno, NV. to Dixie Valley, NV. 15 May 1983: Geothermal Resources Council Field Trip Guidebook for the Role of Heat in the Development of Energy and Mineral Resources in the Northern Basin and Range Province, 24 p.

White, D. E., 1968, Hydrology, activity and heat flow of the Steamboat Springs thermal system, Washoe County, Nevada: U. S. Geological Survey Prof. Paper 458-C, 109 p.

______, 1983, Summary of Steamboat Springs Geothermal Area, Nevada, with attached road-log commentary: Geothermal Resources Council Field Trip Guidebook for the Role of Heat in the Development of Energy and Mineral Resources in the Northern Basin and Range Province, 20 p.

White, D. E., Thompson, G. A., and Sandberg, C.H.,
1964, Rocks, structure, and geologic history of
Steamboat Springs thermal area, Washoe County,
Nevada: U. S. Geol. Survey Prof. Paper 458-B,
63p.

Willden, R., and Speed, R. C., 1974, Geology and
Mineral Deposits of Churchill County, Nevada:
Nevada Bureau of Mines and Geology Bull. 83,
95 p.

Zoback, M. L., 1979, A geological and geophysical
investigation of the Beowawe geothermal area,
north-central Nevada: Stanford University
Publication, Geological Sciences, Vol. 16, 79 p.

Zoback, M. L., and Thompson, G. A., 1978, Basin
and Range rifting in northern Nevada: Clues from
a mid-Miocene rift and its subsequent offsets:
Geology, Vol. 6, p. 111-116.

APPENDIX I

<u>WELL SUMMARY FOR ROOSEVELT HOT SPRINGS, UTAH</u>

Well	Date	Location	Depth (m)	Status	T Max (°C)	Operator
Unnamed	1967	NW NE Sec. 16 T27S, R9W	82	Plugged	132	Dr. Eugene Davie
Roosevelt KGRA 9-1	1975	NE NW Sec. 9 T27S, R9W	2098	Dry	227	Phillips
Roosevelt KGRA 3-1	1975	SW NE Sec. 3 T27S, R9W	830	Producer	NA	Phillips
Roosevelt KGRA 54-3	1975	SW NE Sec. 3 T27S, R9W	878	Producer	NA	Phillips
Roosevelt KGRA 12-35	1975	NW NW Sec. 35 T26S, R9W	2232	Non-Commercial Producer	NA	Phillips
Roosevelt KGRA 13-10	1975	SW NW Sec. 10 T27S, R9W	1631	Producer	NA	Phillips
Roosevelt KGRA 82-33	1975	NE NE Sec. 33 T26S, R9W	1837	Injector	NA	Phillips
Utah State 14-2	1976	SW NW Sec. 2 T27S, R9W	1862	Producer	271	Thermal Power
Roosevelt HSU 25-15	1976	NW SW Sec. 15 T27S, R9W	2286	Producer	NA	Phillips
Utah State 72-16	1976	SE NE Sec. 16 T27S, R9W	382	Producer	243	Thermal Power/ O'Brien
Utah State 24-36	1977	SW NW Sec. 36 T27S, R9W	1865	Dry	NA	Thermal Power
Roosevelt HS KGRA 52-21	1978	NW NE Sec. 21 T27S, R9W	2286	Dry	206	Getty
Roosevelt HSU 27-3	1982	SW SW Sec. 3 T27S, R9W	1042	Producer	NA	Phillips
Roosevelt HSU 35-3	1982	NE SW Sec. 3 T27S, R9W	795	Producer	NA	Phillips

<u>WELL SUMMARY FOR BEOWAWE, NEVADA</u>

Well	Date	Location	Depth (m)	Status	T Max (°C)	Operator
Beowawe 1	1959	SW NW Sec. 17 T31N, R48E	164	Plugged	NA	Magma
Beowawe 2	1960	SW NW Sec. 17 T31N, R48E	218	Producer	212	Magma
Vulcan 1	1961	SE NW Sec. 17 T31N, R48E	200	Plugged	208	Vulcan Thermal* Power Co.
Vulcan 2 redrill	1962	SE NW Sec. 17 T31N, R48E	220	Producer	208	Vulcan Thermal* Power Co.

Well	Date	Location	Depth (m)	Status	T Max (°C)	Operator
Vulcan 3 redrill	1962	SE NW Sec. 17 T31N, R48E	244	Plugged	210	Vulcan Thermal* Power Co.
Vulcan 4	1961	Sec. 17 T32N, R48E	234	Plugged	NA	Vulcan Thermal* Power Co.
Vulcan 5	1963	SW NE Sec. 17 T31N, R48E	72	Plugged	NA	Vulcan Thermal* Power Co.
Vulcan 6	1963	SW NE Sec. 17 T31N, R48E	145	Plugged	139	Vulcan Thermal* Power Co.
Sierra Pacific 1	1964	NE SW Sec. 17 T31N, R48E	283	Idle	81	Sierra Pacific
Sierra Pacific 2	1964	NE SW Sec. 17 T31N, R48E	127	Idle	NA	Sierra Pacific
Sierra Pacific 3	1965	SW NW Sec. 17 T31N, R48E	625	Abandoned	196	Sierra Pacific
Sierra Pacific 4	1965	NW NW Sec. 17 T31N, R48E	306	Idle	133	Sierra Pacific
Ginn 1-13	1973	SE SE Sec. 13 T31N, R47E	2911	Producer	216	Chevron/American Thermal Resources
Batz No. 1	1975	NW NE Sec. 17 T31N, R48E	1829	Idle/Inj	124	Magma/Dow
Rossi 21-19	1975	NW NW Sec. 19 T31N, R48E	1732	Non-Producer	NA	Chevron
Rossi 21-19 deepened	1979	NW NW Sec. 19 T31N, R48E	2199	Non-Producer	204	Chevron
Beowawe 33-17	1979	SE NW Sec. 17 T31N, R48E	360	Producer	185	Chevron
Beowawe 85-18	1979	NE NE Sec. 18 T31N, R48E	1807	Producer	180	Chevron
Collins No. 1	1981	NE SE Sec. 17 T31N, R48E	2745	Abandoned	134	Getty

* Associated/affiliated with Magma

WELL SUMMARY FOR BRADY'S HOT SPRINGS--DESERT PEAK, NEVADA

Well	Date	Location	Depth (m)	Status	T Max (°C)	Operator
Brady No. 1	1959	NE SW Sec. 12 T22N, R26E	546	Dry	179	Magma
Brady No. 2	1959/ 1960	NE SW Sec. 12 T22N, R26E	73	Producer	165	Magma
Brady No. 3	1960	SE NW Sec. 12 T22N, R26E	610	Producer	157	Magma

Well	Date	Location	Depth (m)	Status	T Max (°C)	Operator
Brady No. 4	1961	SE NW Sec. 12 T22N, R26E	220	Producer	171	Magma
Brady No. 5	1961	SE NW Sec. 12 T22N, R26E	549	Producer	171	Magma
Brady No. 6	1961/ 1962	SW NE Sec. 12 T22N, R26E	234	Producer	162	Magma
R. Brady EE No. 1	1964	SE NW Sec. 12 T22N, R26E	1543	Unknown	212	Earth Energy
Desert Peak 29-1	1974	SE SE Sec. 29 T22N, R27E	2335	Dry	165	Phillips
S.P. Brady No. 1	1974	SW SE Sec. 1 T22N, R26E	2217	Non-Commercial	188	Union
S.P. Brady No. 2	1975	NE SE Sec. 1 T22N, R26E	1355	Non-Commercial	148	Magma
Brady No. 8	1975	SE NW Sec. 12 T22N, R26E	1057	Producer	171	Magma
Desert Peak B21-1	1976	SW SE Sec. 21 T22N, R27E	1264	Producer	208	Phillips
Desert Peak B21-2	1976	NE NE Sec. 21 T22N, R27E	973	Producer	200	Phillips
Desert Peak B23-1	1979	SW NW Sec. 23 T22N, R27E	2938	Idle	213	Phillips
Desert Peak 86-21	1982	NE SE Sec. 21 T22N, R27E	966	Producer	208	Phillips
Desert Peak 22-22	1982	NW NW Sec. 22 T22N, R27E	2051	Idle	209	Phillips

WELL SUMMARY FOR HUMBOLDT HOUSE, NEVADA

Well	Date	Location	Depth (m)	Status	T Max (°C)	Operator
Campbell E-1	1977	SE SE Sec. 21 T31N, R33E	559	Producer	183	Phillips
Campbell No. 1	1978	SE NE Sec. 3 T31N, R33E	2080	Dry	205	Union
Campbell "E" No. 2	1979	NE SW Sec. 15 T31N, R33E	2457	Dry	193	Phillips

WELL SUMMARY FOR DIXIE VALLEY, NEVADA

Well	Date	Location	Depth (m)	Status	T Max (°C)	Operator
S.W. Lamb No. 1	1978	NW NW Sec. 18 T24N, R37E	2211	Producer	NA	SUNEDCO
Dixie Federal 45-14	1979	NE SW Sec. 14 T23N, R35E	2750	Non-Producer	197	Thermal Power/ Southland Royalty

Well	Date	Location	Depth (m)	Status	T Max (°C)	Operator
Dixie Federal 66-21	1979	NW SE Sec. 21 T24N, R36E	2981	Non-Producer	205	Thermal Power/ Southland Royalty
S. W. Lamb No. 2	1979	NW NW Sec. 18 T24N, R37E	2713	Producer	NA	SUNEDCO
S. W. Lamb No. 2b	1979	NW NW Sec. 18 T24N, R37E	2591	Suspended	NA	SUNEDCO
S. W. Lamb No. 3	1979	NW NW Sec. 18 T24N, R37E	2809	Producer	NA	SUNEDCO
Federal 62-21	1980	NE NE Sec. 21 T24N, R37E	3810	Non-Producer	NA	SUNEDCO
Federal 84-7	1980	SE NE Sec. 7 T24N, R37E	2481	Producer	NA	SUNEDCO
Dixie Federal 52-18	1980	SE NE Sec. 18 T24N, R37E	3005	Producer	NA	SUNEDCO
Federal 45-5	1981	SE NW Sec. 5 T24N, R37E	2518	Producer	NA	SUNEDCO
Federal 65-18	1981	NE SE Sec. 18 T24N, R37E	2836	Producer	NA	SUNEDCO

WELL SUMMARY FOR SODA LAKE, NEVADA

Well	Date	Location	Depth (m)	Status	T Max (°C)	Operator
1-29	1974	SE SE Sec. 29 T20N, R28E	1312	Producer	172	Chevron/ Phillips
44-5	1978	SE NW Sec. 5 T19N, R28E	1545	Dry	121	Chevron/ Phillips
84-33	1981	SE NE Sec. 33 T20N, R28E	2587	Producer	204	Chevron/ Phillips

WELL SUMMARY FOR STEAMBOAT SPRINGS, NEVADA *

Well	Date	Location	Depth (m)	Status	T Max (°C)	Operator
Rodeo Well	1950	NE NW Sec. 33 T18N, R20E	86	Unknown	169	Unknown
Nevada Thermal Power 1	1959	NW NE Sec. 33 T18N, R20E	558	Non-Commercial	NA	Nevada Thermal** Power Co.
Nevada Thermal Power 2	1959	NE NW Sec. 33 T18N, R20E	294	Unknown	NA	Nevada Thermal** Power Co.
Nevada Thermal Power 3	1960 (?)	NW NE Sec. 32 T18N, R20E	385	Unknown	NA	Nevada Thermal** Power Co.

Well	Date	Location	Depth (m)	Status	T Max (°C)	Operator
Nevada Thermal Power 4	1960	NE NW Sec. 24 T18N, R20E	221	Non-Commercial	186	Nevada Thermal** Power Co.
Nevada Thermal Power 5	1961	NW NW Sec. 32 T18N, R20E	252	Unknown	175	Nevada Thermal** Power Co.
Nevada Thermal Power 6	1961	NW NE Sec. 32 T18N, R20E	218	Unknown	179	Nevada Thermal** Power Co.
Steamboat # 1	1979	NW NW Sec. 5 T17N, R20E	929	Producer	228	Phillips
Cox I-1	1981	SW NE Sec. 32 T18N, R20E	1058	Non-Commercial Producer	177	Phillips/Gulf

* Other shallow wells have been drilled in the vicinity of the hot springs but they are poorly documented and it is not known which, if any, were intended for electrical generation.

** Associated/affiliated with Magma

WELL SUMMARY FOR COSO, CALIFORNIA

Well	Date	Location	Depth (m)	Status	T Max (°C)	Operator
CGEH-1	1977	SW NE Sec. 6 T22S, R39E	1477	Plugged	195	Dept.of Energy
75-7 Coso	1981	NE SE Sec. 7 T22S, R39E	405	Producer	213	California Energy
31-8 Coso	1982	NE NW Sec. 8 T22S, R39E	1226	Producer	NA	California Energy
71-7 Coso	1982	NE NE Sec. 7 T22S, R39E	623	Producer	NA	California Energy
75A-7 Coso	1982	NE SE Sec. 7 T22S, R39E	465	Producer	NA	California Energy
75B-7 Coso	1982	NE SE Sec. 7 T22S, R39E	523	Producer	NA	California Energy
71A-7 Coso	1982	NE NE Sec. 7 T22S, R39E	640	Producer	NA	California Energy

WELL SUMMARY FOR LONG VALLEY, CALIFORNIA

Well	Date	Location	Depth (m)	Status	T Max (°C)	Operator
Mammoth No.1	1959	NW NE Sec. 32 T3S, R28E	324	Producer	148	Magma
Endogeneous No.1	1960	SW NW Sec. 32 T3S, R28E	192	Producer	178	Magma/ Endogeneous Power Co.
Endogeneous No.2	1960	SW NW Sec. 32 T3S, R28E	247	Producer	174	Magma/ Endogeneous Power Co.

Well	Date	Location	Depth (m)	Status	T Max (°C)	Operator
Endogeneous No.3	1960	SW NW Sec. 32 T3S, R28E	174	Producer	172	Magma/ Endogeneous Power Co.
Endogeneous No.4	1961	SW NW Sec. 32 T3S, R28E	NA	NA	NA	Magma/ Endogeneous Power Co.
Chance No.1	1961	SW NW Sec. 35 T3S, R28E	245	NA	135	Magma/ Endogeneous Power Co.
Endogeneous No.5	1962	SW NW Sec. 32 T3S, R28E	123	NA	NA	Magma/ Endogeneous Power Co.
Endogeneous No.6	1962	SE NW Sec. 32 T3S, R28E	230	NA	NA	Magma/ Endogeneous Power Co.
Endogeneous No.7	1962	SW NW Sec. 32 T3S, R28E	204	NA	NA	Magma/ Endogeneous Power Co.
Long Valley 66-29	1976	NW SE Sec. 29 T3S, R29E	2109	Dry	72	Republic
Clay Pit No. 1	1979	NE NE Sec. 15 T3S, R28E	1846	Dry	147	Union
Mammoth No. 1	1979	SE NW Sec. 32 T3S, R29E	1604	Dry	157	Union

HEAT FLOW IN THE NORTHERN BASIN AND RANGE PROVINCE

David D. Blackwell

Department of Geological Sciences
Southern Methodist University
Dallas, Texas 75275

ABSTRACT

The heat flow in the Basin and Range province of northern Nevada is extremely complex. It is a product of superposition of the regional effects of extension and volcanism/intrusion modified by the local conductive effects of thermal refraction (complicated structural settings), variations in radioactive heat production, erosion and sedimentation. In addition to these conductive effects, groundwater flow, both on a local and a regional basis, affects heat-flow measurements. Typical heat-flow values for the Basin and Range province average 85 ± 10 mWm^{-2}. The higher estimates are probably based on biased sets of heat-flow measurements, and actual averages are on the order of 100 ± 10 mWm^{-2}. Geothermal systems appear to be related to deep fluid circulation in an active tectonic setting rather than to young silicic volcanic rocks. Young volcanoes occur along the borders of the Basin and Range province, but not in the part of northern Nevada discussed in this paper.

INTRODUCTION

The Basin and Range province of the western Cordillera of the United States is an exceedingly complex area resulting from the superposition of 600 million years of recurring tectonic and volcanic activity. In the late Cenozoic, the already highly fragmented geology of the Basin and Range was further disrupted by the subsidence and covering of approximately half the province by alluvial deposits in the valleys and by the exposure of different stratigraphic and structural levels in adjacent ranges. The heat flow in the Basin and Range province reflects this complicated tectonic development in both regional and detailed ways.

The object of this paper is to discuss the distribution of, and controls on, heat flow in the part of the Basin and Range province in northwestern Nevada from an observational point of view. Thermal variations may not have as distinctive an imprint as structural and stratigraphic evolution; however, certainly the varied thermal events during the mid and late Cenozoic have left their marks. Furthermore, the interaction between the present-day thermal background and the detailed structural and hydrologic settings is complex. Consequently, even with a relatively high density of heat flow data, compared to many areas of the earth, there are still many uncertainties and unknowns in the actual magnitude of the background, the detailed geographic and vertical distribution, and the way in which variations relate to structural geology, hydrology and volcanic history.

The history of heat flow studies in the Basin and Range province dates back to the late 1960s. Reconnaissance data in the Basin and Range were discussed by Roy et al. (1968b) and Sass et al. (1971). The most recent detailed heat flow map of northwestern Nevada was presented by Sass et al. (1981, Fig. 1). Sass et al. (1971) divided the Basin and Range province in Nevada into three heat flow regimes: a region of heat flow typical of the province average (surface heat flow values of about 85 ± 10 mWm^{-2}*); a region of above average heat flow (surface heat flow values of $100 +$ mWm^{-2}) which was designated the Battle Mountain Heat Flow High; an area of below average heat flow values which was named the Eureka Heat Flow Low (surface heat flow values of less than 60 mWm^{-2}). These heat flow subdivisions were maintained in later discussions by Lachenbruch and Sass (1977, 1978). In contrast, Blackwell (1978) argued that the highest overall energy loss within the Basin and Range province was along the eastern and western boundaries near the Wasatch and the Sierra Nevada Mountains, and not within the Battle Mountain Heat Flow High.

Extensive heat flow or subsurface temperature data have become available for many geothermal systems subsequent to the major discussions in 1978, especially in northern and northwestern Nevada. These data have yet to be integrated into the pre-existing regional data set. Neither space nor time permit the integration of these data in this paper, but some salient aspects of these studies which have a bearing on the current understanding of heat transfer in the Basin and Range province will be discussed. The order of the discussion in the paper proceeds more or less according to Table 1, a list of major effects on the heat flow pattern in the Basin and Range province. Following a brief summary of the

---

TABLE 1. THERMAL EFFECTS

Volcanism and Intrusion
Extension
Thermal Refraction-Structure
Radioactive Heat Production
Erosion and Deposition
Local Groundwater Flow
Geothermal Systems

---

* 41.84 mWm^{-2} = 1 x 10^{-6} cal/cm^2 sec = 1 HFU

observations, the regional effects on heat flow will be discussed. These regional effects are grouped as thermal effects related to extension and to volcanism. Following this discussion, the more local effects which may cause heat flow variations on the scale of a few kilometers will be discussed. The effects specifically to be discussed include the conductive effects of variation of basement radioactivity, effects of structure (i.e., lateral variations of thermal conductivity), erosion and deposition, and the effects of regional and local convective geothermal fluid-flow systems.

OBSERVED HEAT FLOW DISTRIBUTION

The regional heat flow data described by Sass et al. (1981) are shown in Figure 1. A total of 93 heat-flow measurements for the Nevada portion of the Basin and Range province are shown. The spacing of heat-flow stations is quite dense for a continental region, but as will subsequently become obvious, it is not dense enough to determine many of the characteristics of the heat-flow pattern. Superimposed on the heat-flow map are the contours of the Eureka Heat Flow Low and Battle Mountain Heat Flow High. Heat-flow values generally exceed 100 mWm^{-2} within the area identified as the Battle Mountain Heat Flow High. Elsewhere within the Basin and Range province, heat-flow values are typically 85 ± 10 mWm^{-2}. The thermal boundaries of the Basin and Range province are very sharp against the Wasatch Mountains-Colorado Plateau region on the east (Bodell and Chapman, 1982; Keller et al., 1979; Reiter et al.,

1979) and the Sierra Mountains on the west (Roy et al., 1968b, 1972; Sass et al., 1971). To the north there is no distinct thermal boundary with the Columbia Plateau region and, in fact, contours of the Battle Mountain Heat Flow High include part of the High Lava Plains in Oregon, and the Snake River Plain region in Idaho (see Brott et al., 1978, 1981).

In addition to the "regional heat flow" data set, an extensive data set including 10 to 100 times as many holes as shown on Figure 1 is available from exploration activities associated with individual geothermal systems. In general, these data are not included in Sass et al. (1981), nor in earlier descriptions of heat flow in the Basin and Range province. The existence of this new data set gives an additional complexity to the heat-flow character, poses many interesting questions, and also opens the possibility of investigating in more detail local conditions affecting observation.

Most of the geothermal exploration drilling has been along the boundaries between a range and valley, or within a valley. On Figure 1, no "regional" heat flow data from the valleys are shown, so the two data sets do not overlap geographically or geologically, and it cannot necessarily be anticipated that the geothermal regimes will be the same in the ranges and in the valleys. Another important aspect of the geothermal systems is the maximum temperature of the system, which is controlled by the depth and rate of circulation. No systematic summary of such temperatures based on drilling has yet been described. Edmiston (1982) has presented a map showing the location of deep geothermal tests and the geothermal tests which have been successful (i.e., have found temperatures in excess of 200°C). Analysis and integration of the new data set with regional data will take effort, but has the potential to greatly refine our understanding of the heat flow distribution in the Basin and Range province.

REGIONAL CONTROLS ON HEAT FLOW

Introduction. The two dominant regional thermal effects are mechanical-thermal effects associated with the late Cenozoic extension, and the thermal effects of mid and late Cenozoic intrusion and volcanism. Although these two effects will be discussed separately, in fact they are interrelated, and individual components are difficult to identify separately.

Thermal Effects of Extension. There has been an emphasis on extension effects on the thermal pattern because the extensional activity is so prominent in the most recent history of the Basin and Range province. Lachenbruch and Sass (1977, 1978) have developed a model for the thermal regime in the Basin and Range province emphasizing extension. The primary thermal effect of extension is to move deeper, hotter material to shallower depths as the lithosphere is stretched. However, there are definite limits to the enhancement of surface heat flow by this mechanism if the spreading does not lead rapidly to ocean basin formation, so the favored models discussed by Lachenbruch and Sass (1978) actually depend on basalt intrusion to supply a large part of the high heat flow observed at the surface. Indeed, the Lachenbruch and Sass (1978) models have a one-to-one correlation between extension rate and intrusive emplacement. These models were also applied to areas of large-scale silicic volcanism such as Yellowstone and Long Valley. In these areas, much more of the anomalous

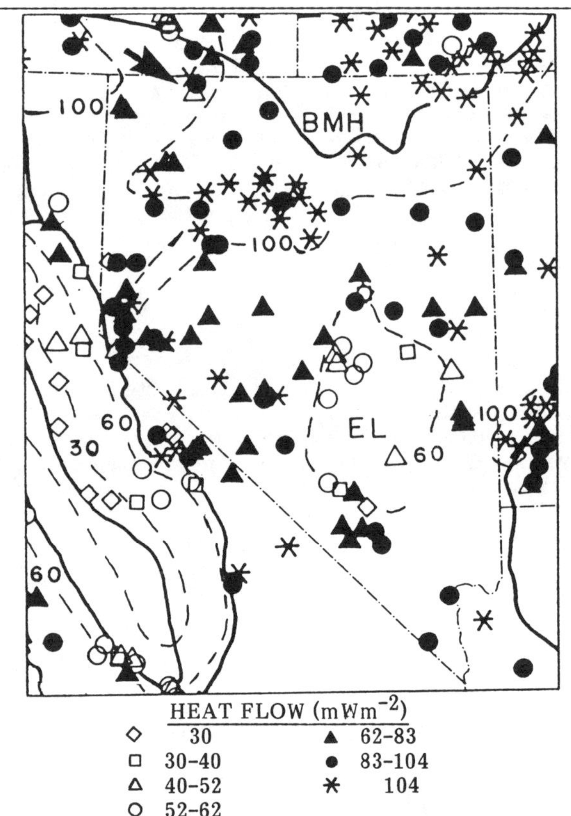

HEAT FLOW (mWm^{-2})

◇	30	▲	62–83
□	30–40	●	83–104
△	40–52	✳	104
○	52–62		

FIGURE 1. Heat flow in Nevada and surrounding areas, after Sass et al. (1978). Contours are from Lachenbruch and Sass (1978). Arrow in northwestern Nevada indicates three holes in granite discussed in text.

surface heat flow is associated with intrusion than with extension!

Limits on the amount of extension are related to isostatic effects. As stretching occurs, continental crustal material is replaced by mantle material with a large increase in density and, consequently, subsidence. The subsidence is offset to some extent by thermal effects of extension. A comparison of these two quantities is shown in Table 2. The heat-flow model includes one-dimensional time-dependent stretching (Jarvis and McKenzie, 1980). Example extension rates, total extension associated with a 17 M.Y. period of extension in the form of ratios (βs), the amount of mechanical subsidence which would be associated with extension of an originally 40 km thick continental crust, the amount of thermal uplift, and the difference between the mechanical subsidence and the thermal uplift (a net subsidence) are shown in Table 2. The associated heat flow anomaly for each case is also shown. This model includes a time-dependency, so the heat flow values are slightly lower than would be the case if extension had occurred long enough for thermal equilibrium to be attained (see Figure 2). The 17 M.Y. period was chosen as it is the maximum period of time of extension in the northern Basin and Range province during the late Cenozoic, and was the period of time of extension assumed in the analysis of Lachenbruch and Sass (1978).

Typical extension proposed (Lachenbruch and Sass, 1978) ranges from 50 to over 100%, corresponding to β values of 1.5 to 2. Taking isostatic effects into account, a net subsidence of 1.5 to 2.4 km is associated with the required heat-flow anomaly of 40 to 60 mWm^{-2}. These observations can be compared to the mean topographic height of the Basin and Range province in Nevada at the present time (from 1.5 km in the Lahontan and Bonneville Basins to over 2 km in the center of the province), and the regional elevations of approximately 2 to 2.5 km in the Wasatch Range and the Sierra Nevada Mountains. Unless the Basin and Range province started at an extraordinarily great elevation, it seems unlikely that the total subsidence could exceed 0.5 km, as the Basin and Range province still stands at a high elevation. The results of this analysis suggest that extension is not the main mechanism responsible for high heat flow in the Basin and Range province.

Figure 2 shows more detail of the heat-flow contribution associated with a transient thermal event as summarized in Table 2. In this calculation, an initial lithospheric thickness of 94 km and a background heat flow of 40 mWm^{-2} were assumed. Extension rates in percent per million years are shown for each curve. The dashed line on the plot is the locus of points with a β value of 1.5.

These results may be compared to the various heat flow subregions in the Basin and Range province. A typical background heat flow in the Basin and Range is approximately 85 mWm^{-2}. The reduced or mantle heat flow (the heat flow from below the upper crustal radioactivity layer) is 59 mWm^{-2}, approximately 50% of

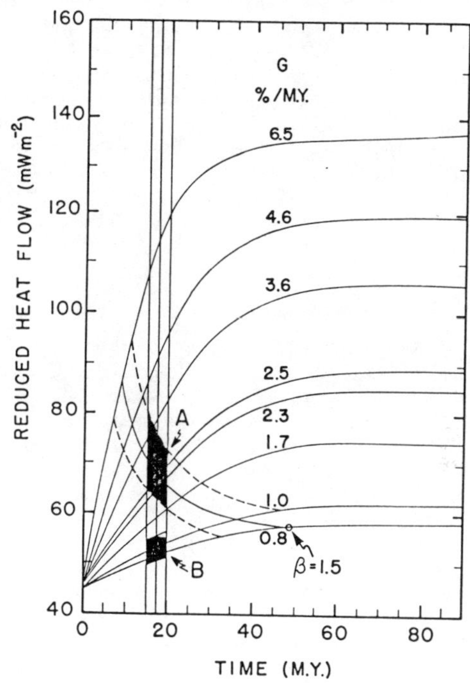

FIGURE 2. Thermal effect of extension calculated using model of Jarvis and McKenzie, 1980. Age range of extension in the Basin and Range province is shown by the vertical lines (17 ± 2 M.Y.). G is extensional strain rate in %/M.Y. Shaded area A is the region of the graph corresponding to reduced heat flow values typical of the Battle Mountain Heat Flow High. Shaded area B is the region of the graph consistent with present crustal thickness.

TABLE 2. COMPARISON OF MECHANICAL SUBSIDENCE AND THERMAL UPLIFT IN BASIN AND RANGE PROVINCE FOR 17 M.Y.

Extensional Strain Rate (%/M.Y.)	β (17 M.Y.)	Mechanical Subsidence (km)	Thermal Uplift (km)	Net Subsidence (km)	Surface Heat Flow (mWm^{-2})
0.8	1.14	0.67	0.24	0.43	78
1.7	1.29	1.33	0.54	0.79	84
2.5	1.43	1.83	0.73	1.10	93
3.6	1.61	2.43	0.91	1.52	104
4.6	1.78	2.87	1.02	1.85	117
5.6	1.95	3.25	1.11	1.11	128
6.5	2.11	3.54	1.17	2.37	140

which is anomalous with respect to a thermally "normal" continental lithosphere (Roy et al., 1968a, 1972). Heat flow values in the Battle Mountain Heat Flow High range from 100 to 150 mWm^{-2}, and the reduced heat flow in the Battle Mountain Heat Flow High according to Lachenbruch and Sass (1977, 1978) ranges from 85-100 mWm^{-2}, or approximately 30-45 mWm^{-2} in excess of the normal Basin and Range reduced heat flow, and 60-75 mWm^{-2} in excess of a "normal" continental reduced heat flow.

It is clear from the consideration of Table 2 and Figure 2 that the magnitude of heat flow observed in the Battle Mountain Heat Flow High cannot be associated with the thermal effects of extension alone. To add emphasis to this conclusion, it is clear that the present-day distribution of active earthquake zones in the Basin and Range province is not coincident with the highest heat flow. One zone of earthquake activity extends north through the Battle Mountain Heat Flow High, but much of the heat-flow high is essentially aseismic at the present time (Smith, 1978). Other zones of active seismicity along the Sierra Nevada-Basin and Range transition and along the Colorado Plateau-Basin and Range transition are not apparently characterized by quite as high surface heat flow (see Lachenbruch and Sass, 1978; Blackwell, 1978).

Thermal Effects of Intrusive and Volcanic Activity. There is no difficulty in generating high heat flow values in association with volcanic and intrusive activity, although many of the highest values are obviously associated with geothermal systems rather than with conductive heat flow from magma chambers. A correlation between the age of volcanic activity in a particular area and the heat flow, especially in the case of the continental crust characterized by rhyolitic volcanic activity, might be expected. A generalized volcanic age map for the area shown in Figure 1 is shown in Figure 3. The pattern has been discussed by many people, recently including Stewart and Carlson (1978) and Snyder et al. (1976). The areas of oldest, or lack of, Cenozoic volcanism are in southern Nevada and in east-central Nevada. Most of the rest of Nevada is characterized by volcanism in the age range 6-17 M.Y. and only along the margins of the province are younger, extensive silicic volcanic features found.

Various types of volcano/thermal models have been discussed, including the extension/intrusion models of Lachenbruch and Sass (1978). If volcanism is not associated one-to-one with extension, then the event will lead to some sort of heating of the crust followed by cooling after the end of the volcanic/intrusive event. The peak heat flow will depend on the actual distribution of magma within the crust and lithosphere. However, after 1-5 M.Y. or so of cooling, all models approach the same sort of behavior. The long time asymptote of the cooling depends on the assumption of the background heat flow. The general sort of behavior is shown in Figure 4. The curves in Figure 4 are shown to cool off to three different backgrounds, depending on the assumed steady-state mantle heat flow. These models are discussed in more detail by Blackwell (1978). On Figure 4, the typical heat flow in the Basin and Range province and the heat-flow range for the Battle Mountain Heat Flow High are shown. Based on these results, it would appear that the higher estimates of heat flow in the Battle Mountain Heat Flow High are too high to be explained by the volcanic model if no silicic volcanism in the area is

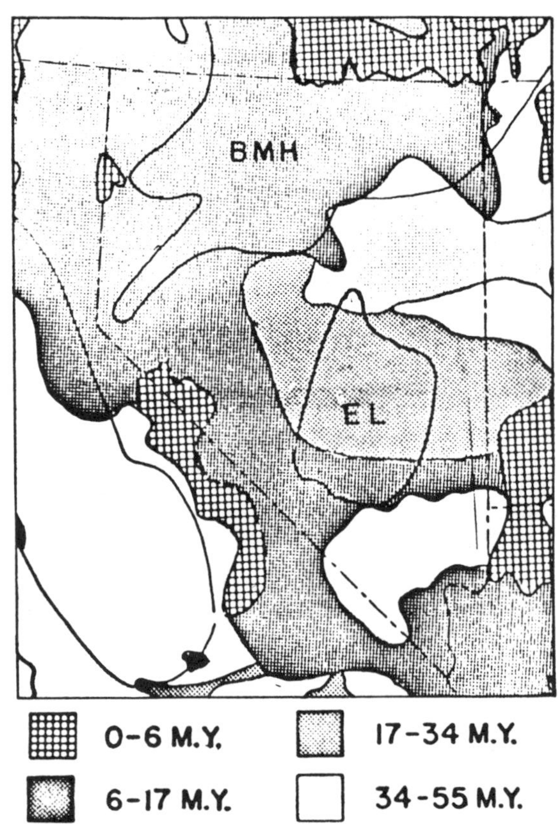

FIGURE 3. Age of silicic volcanism (generalized from Stewart and Carlson, 1978). Light lines are 60 mWm^{-2} heat-flow contours.

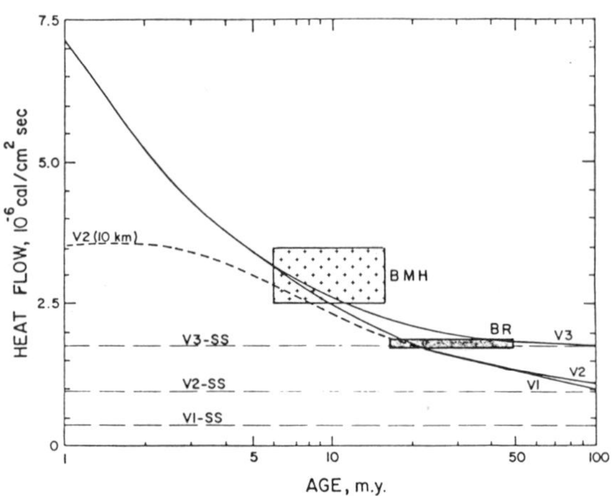

FIGURE 4. Thermal models for regional volcanic and intrusive events in the Basin and Range province. Observed heat-flow ranges for the models are explained by Blackwell (1978). SS indicates the steady-state asymptotic heat-flow values.

younger than 6 M.Y. Because the heat flow is also too high to be explained by extension, the apparent magnitude of the regional heat flow pattern is somewhat difficult to explain.

LOCAL CONDUCTIVE HEAT FLOW EFFECTS

Radioactive Heat Production.   If the geology of the Basin and Range province were similar to the Sierra Nevada Mountains, and consisted of a large, homogeneous, relatively unfractured, granitic terrain, then the heat flow evaluation would be relatively straightforward.   The surface heat flow within the granite would show a linear correlation with the local heat production of the rocks from uranium, thorium and potassium.   In this simple case, local lateral heat-flow variations due to sub- and intracrustal sources, groundwater flow, thermal refraction and so forth, would not have significant effects on the heat flow distribution.   Such linear arrays have been observed in many places throughout the world (Roy et al., 1968a; 1972).   Early studies of heat flow versus heat production for the Basin and Range province indicated a regional linear relationship with an intercept heat flow of approximately 59 mWm^{-2} and a slope of approximately 9.4 km (Roy et al., 1968a).   Subsequent studies of the relationship between heat flow and heat production in granitic rocks in the Basin and Range have led to a considerably more complicated pattern.

A summary of the available data is shown in Figure 5.   This figure was presented by Lachenbruch and Sass (1978), but the detailed data on which this figure is based have not been published.   It is difficult to identify a linear relationship between heat flow and heat production

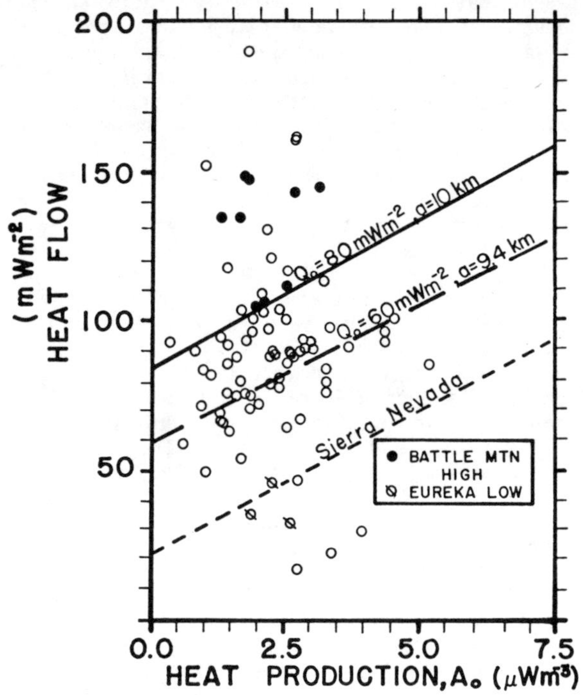

FIGURE 5.   Surface heat flow as a function of surface heat production (after Lachenbruch and Sass, 1977).

in this data set, although most of the data lie between two lines with intercept values of 50 mWm^{-2} and 90 mWm^{-2} (with slopes of 9.4 to 10 km).

In the Basin and Range province, unlike the Sierra Nevada Mountains, it is not possible to avoid systematic

effects on the heat flow data, even by drilling heat-flow holes only in granitic rocks, in large part because granitic rocks represent only a small fraction of the exposed bedrock of Nevada (2.8%, Archbold, 1972).   Therefore, the data shown in Figure 5 have all sorts of extraneous (from the regional point of view) effects present in the measurements.   The difficulties of attempting to use these data to determine regional quantities are illustrated by three points in granite in close proximity in north-central Nevada (see arrow in Figure 1).   The radioactive heat production values are approximately the same for the three points, and yet the heat-flow values are respectively 45, 85 and 150 mWm^{-2}.   These points are within a few km of one another, and all are in granite.   If only one of the three holes were available, it might be inferred that this area had heat flow typical of (1) the Eureka Heat Flow Low,   (2) normal Basin and Range, or (3) the Battle Mountain Heat Flow High.

The sites are close to the Baltazor geothermal system (Earth Power Prod. Co., 1980), and it is possible that there is some interaction between the fluid convection represented by the geothemal system and the extreme variation in heat-flow values shown by these three data points.   Thus it seems quite clear that both high and low heat flow values are associated with convection systems, as concluded by Lachenbruch and Sass (1977).

Because of the very noisy data set, the heat flow-heat production relationship for various parts of Nevada and its relationship to other known continental patterns is not presently resolvable in an accurate way.   It is clear that heat flow determinations in granitic rocks in Nevada, without considerable attention to each individual measurement, are insufficient for the determination of typical regional heat-flow values and heat-flow distribution in contrast to standard practice in some geologic terrains.

Thermal Effects of Structure: Refraction.   In an ideal case, to avoid anomalies due to the spatial distribution of thermal conductivity, the thermal conductivity of the rocks should either be uniform or should vary only in the vertical direction.   If there are variations only in the vertical direction, then a hole penetrating various units will show an inverse correlation between the thermal conductivity of the rock and the geothermal gradient, resulting in constant heat flow with depth.   However, the geologic structure of the Basin and Range province is anything but layer cake.   In order to investigate possible systematic effects of thermal conductivity on heat flow, determinations from the Battle Mountain Heat Flow High are plotted as a function of thermal conductivity in Figure 6.   Most of the early heat-flow determinations in the Battle Mountain Heat Flow High were made in very high thermal conductivity sedimentary rocks.   Average thermal conductivities at these sites are over 4 Wm^{-1}K^{-1}.   The thermal conductivity of the granites where subsequent heat-flow values were obtained is typically about 3 Wm^{-1}K^{-1} (Sass, personal communication, 1980).   Typical thermal conductivities from some of the basins are 1-1.5 Wm^{-1}K^{-1}.   The basin data are primarily from the Black Rock Desert, which has been extensively studied (Sass et al., 1979; Mase and Sass, 1980), and the Grass Valley area (Sass et al., 1976; Welch et al., 1981).   A strong positive correlation between the heat flow and thermal conductivity is demonstrated in Figure 6.   This correlation suggests that thermal refraction is important in the results; consequently, heat-flow values from any one geologic terrain may not represent true regional values.

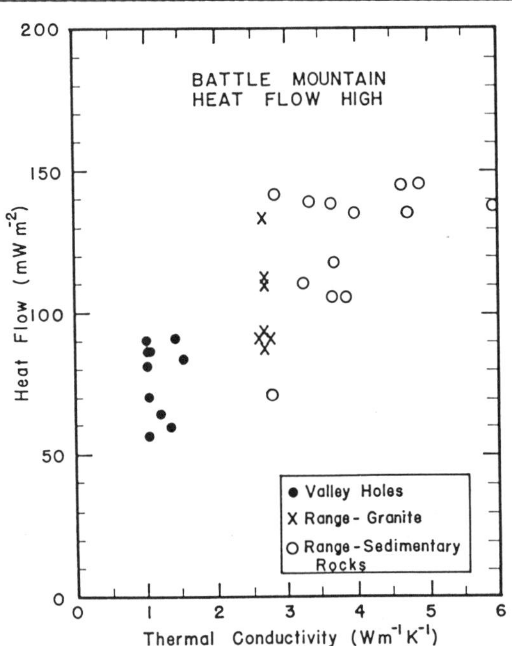

FIGURE 6.   Comparison of heat flow and thermal conductivity from heat-flow sites in the Battle Mountain Heat Flow High. Data from reference in text and from J.H. Sass (personal communication, 1980).

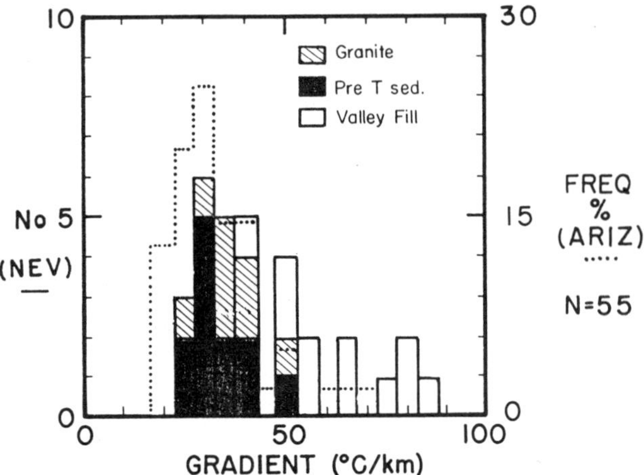

FIGURE 7.   Histograms of gradient in Battle Mountain Heat Flow High and Arizona Basin and Range province (dotted lines). Arizona data from Shearer and Reiter (1981) and Sass et al. (1982).

Another way to investigate systematic effects between heat flow and thermal conductivity is to look at variations in gradient. Measurements in granite, in pre-Tertiary sedimentary rocks, and in valley fill are shown by separate patterns in Figure 7. The gradient distribution from both the granite and the sedimentary rock lithologies overlap, with the range of values being 25-40°C/km. The fact that there is a correlation between heat flow and thermal conductivity, but not between heat flow and gradient, also suggests that structural effects are controlling the variation of the heat-flow values. The reasoning proceeds as follows: if the measurements had been made in horizontal units of varying thermal conductivity (the ideal case), then the gradients would be inversely proportional to the thermal conductivity, there would be no correlation between heat flow and thermal conductivity, and there would be good correlation between gradient and conductivity. In fact, the reverse situation is observed in each case.

Also shown in Figure 7 is a histogram of gradient values from the Basin and Range province of Arizona. This data set includes values recently published by Shearer and Reiter (1981). The Basin and Range province in Arizona has a considerably larger percent of granitic bedrock, generally lower relief, fewer known geothermal systems, and older volcanic rocks. Thus some of the geologic complexities of the northern Basin and Range in Nevada are more subdued in the part of the Basin and Range province in Arizona. A histogram of gradients from the ranges in Arizona shows almost identical distribution to that in the Battle Mountain Heat Flow High, whereas an average of the heat flow values in the

two provinces differs by 25 to 50% (lower in Arizona). Figure 8 shows a plot of thermal conductivity versus heat flow for the Arizona data. Included in this plot are the data from several holes in the valleys in Arizona which are, as in the case of Nevada, undersampled in the present data set. The correlation between thermal conductivity and heat flow is weak, and there is a large overlap between the data from the Battle Mountain Heat Flow High and from central Arizona, excluding the very high heat flow values in the high thermal conductivity sedimentary rocks in the northern Basin and Range province. The conclusion of this discussion is that there may be a bias in heat flow toward too-high values if data only from the ranges (including the high thermal conductivity sedimentary rocks) are used to calculate province average values.

Of course, detailed evaluation of the structural effects on heat flow has to be considered individually for each hole. However, there is one large-scale effect which needs to be considered in this discussion. This thermal effect is the large-scale distortion of heat flow by Basin and Range structure. In general, the valleys are 10-20 km wide and 1-2 km deep. They are generally filled with low thermal conductivity Cenozoic sedimentary rocks. In the ranges, older sedimentary, igneous and metamorphic rocks generally have thermal conductivity values two to three times greater. As a result of this contrast, heat flows preferentially into the ranges, resulting in systematically high values in the ranges, and systematically low values in the valleys with respect to the regional mean value. This large-scale effect of a valley is sometimes modeled as a single, semi-elliptical or semi-circular cylinder embedded in an otherwise uniform media (see Jaeger, 1965). In this case, for typical geometries observed in the Basin and Range province, there would be no appreciable expected effect of thermal refraction except in the immediate proximity of a range-bounding fault, where higher-than-normal values would be observed in the range side, and lower-than-normal values would be observed on the basin side. Two effects are not considered in this simple model. The first of these is that the basins and ranges repeat, so that it is not appropriate to consider only a single valley

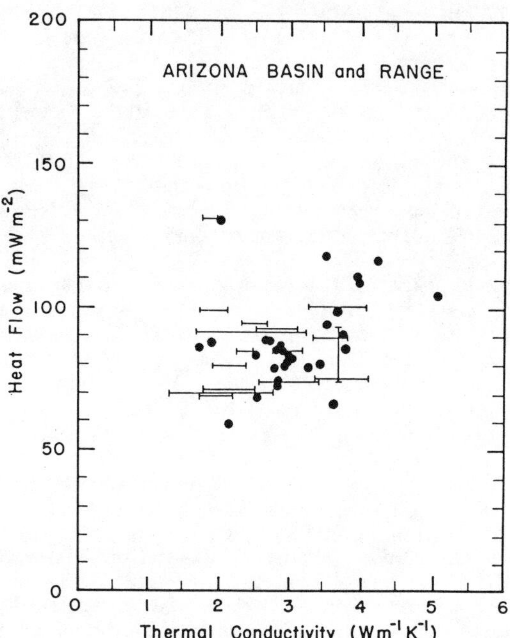

FIGURE 8.   Comparison of heat flow and thermal conductivity at heat-flow sites in Arizona. The bars represent ranges of values as observed in specific holes or areas.

to consider only a single valley embedded in an otherwise semi-infinite media with conductivities typical of the ranges. The effect of repeated ranges and valleys is interaction (Lee and Henyey, 1974), so that larger refraction effects than calculated from the single ellipse model of Jaeger (1965) are observed.

Another complexity not considered in either of these models is the fact that the heat source is within the crust. The boundary condition for the models is that there is constant heat flow from "great depth" and the heat flow is allowed to adjust at great depths below the inhomogeneity. In the Basin and Range province the heat sources (extension effects and magma) may be in the mid- to upper levels of the crust, and there may be inter-action between the heat source and the variations in thermal conductivity, with even more heat being forced through the ranges (and less heat forced through the valleys) than calculated assuming a constant heat flow at great depth. In the extreme limit, the model would be characterized more by constant temperature than by constant heat flow, in which case (except again in the immediate vicinity of the boundaries of the inhomogen-eities) the mean heat flow would be simply proportional to the integrated thermal resistance from the constant temperature plane to the surface.

A typical model is shown in Figure 9. A Basin and Range valley filled with rocks having a thermal conductivity of 1.7 $Wm^{-1}K^{-1}$ is embedded in material with a conductivity of 3.35 $Wm^{-1}K^{-1}$. The geometry of the model is constant into and out of the paper, and repeats that shown side-to-side. In a steady-state case, it is assumed that the heat flow is uniform at great depth. If the average regional heat flow is assumed to be 72 mWm^{-2}, then the heat flow observed in the valley

away from the bounding fault would be 60 mWm^{-2}, while the heat flow observed in the range away from the bounding fault would be 80 mWm^{-2}; this represents a 25% variation between the high and low heat-flow values and an error in regional heat-flow determination (if the range value is used) of about +12%. If heat-flow determinations were made in the ranges without avoiding areas close to the bounding fault, however, the mean observed heat flow would be 16% higher than the true mean. Therefore, based on this simple model, a random set of measure-ments in the ranges would be 16% higher than the regional average, and would be greater than 25% higher than the values measured in the adjoining valleys.

These results are consistent with the distribution of gradients shown in Figure 7, and to some extent with the distribution of heat-flow values shown in Figure 6, if the most thermally conductive sedimentary rocks are not considered typical and the values in the granites are used for comparison instead. It is iteresting to note that peak heat-flow values associated with the boundary of the range would be 96 mWm^{-2}, or 25% high with respect to the average in a steady-state case. Since the amount of the granitic exposure in the northern Basin and Range

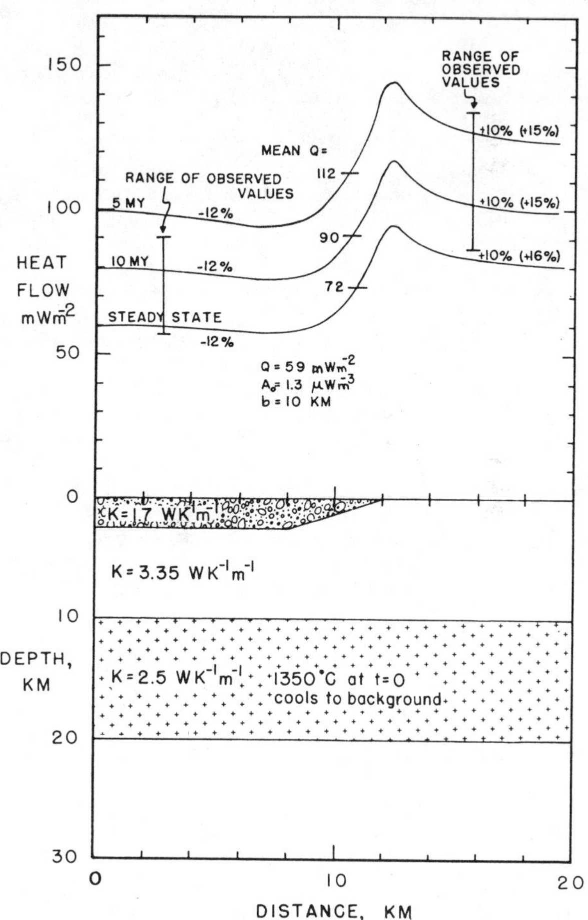

FIGURE 9.   Surface heat flow effects of Basin and Range structure and a crustal intrusive.   Range of observed values refers to Battle Mountain Heat Flow High.

province is so limited, it is argued that use of these sites for the determination of regional heat-flow values may have a large bias.

Effects of Erosion and Deposition.   The thermal effects associated with erosion of mountain ranges and the deposition of sediments in basins have been discussed by numerous authors (including Jaeger, 1965; England and Richardson, 1980).   The general effect is to decrease observed heat-flow values in the sedimenting basins as the colder sediments deposited in the basin are heated up, and to increase heat-flow values in the ranges as hotter rocks are exposed by erosion.   In view of the great tectonic relief developed in the last few million years in the Basin and Range province, significant perturbations from these effects may exist in the heat-flow data set. Unfortunately, there are not sufficient data to estimate these corrections in general.   However, it is possible that systematic effects of up to 10 or 20% of the observed values could be related to the systematic differences in the erosional and depositional environments of the ranges and valleys.   These effects might or might not be superimposed on the refraction effects in an additive way, depending on the detailed timing of the development of the basins and ranges, and of the thermal conductivity contrast associated with various structural settings. The sense of these effects is in the same direction as the structural effects, i.e. values would appear to be higher than the true regional value in the ranges and lower than the true regional value in the valleys.

CONVECTIVE HEAT-FLOW EFFECTS

Shallow Groundwater Aquifers.   The effects of subsurface water movement on heat flow can be many, complex, and on different scales.   The most common effect is that of water table fluid flow, which is probably present in varying degrees at almost all sites.   Water movement in very large aquifers may affect the heat flow so severely that it is virtually impossible to use holes of any reasonable depth to do classical heat-flow studies.   An example of this situation is the Snake River Plain aquifer in Idaho, where heat flow values are much subnormal over an area 50 km wide by 200 km long (Brott et al., 1981).   Because of the rapid flow rates in the aquifer (up to 1000 m/year), about 75% of the total amount of heat conducted into the bottom of the Snake Plain aquifer is advected out the end of the aquifer, leaving only about 25% of the heat to be measured by conductive heat-flow studies.   On the other end of the scale, if the aquifer moves slowly enough that no heat is actually advected at the discharge zone of the aquifer, then the overall heat budget would remain the same as in the conductive heat-flow case, but the distribution would reflect the downflow and upflow parts of the system. This situation has been modeled by Domenico and Palciauskas (1973) and applied to basins in the Rio Grande rift by Morgan et al. (1981).

Mifflan (1968, 1983) has discussed in detail the hydrology of the Basin and Range province.   The dominant hydrologic system in the Basin and Range province is a range-to-valley flow system.   Aquifer heads are typically highest in the ranges and lowest in the valleys.   As drill holes are deepened in the ranges, each successive aquifer usually has a lower head, and as drill holes are deepened in the valleys, each successive aquifer usually has a higher head.   This observation implies recharge in the ranges and discharge in the valleys of the groundwater flow systems.   Thus low heat-flow values

would be observed in the ranges and high heat-flow values would be observed in the valleys, if this effect dominates the shallow heat transfer.

In the Battle Mountain Heat Flow High, no low heat-flow values which might be characteristic of recharge areas in the ranges have been described.   Even if the regional heat flow is high, it seems likely that some evidence of local circulation would have been discovered, particularly in the sedimentary rocks.

Superimposed on the local hydrologic pattern are (in some situations) complex interbasin groundwater flow systems.   These interbasin groundwater flow systems are particularly characteristic of the carbonate terrain in eastern and south-central Nevada.   This phenomenon may be related to the existence of the Eureka Heat Flow Low (Sass et al., 1971).

Geothermal Systems.   There is also abundant evidence, in the form of many geothermal anomalies, for large-scale water flow in northern Nevada.   If water flow goes deep enough, high temperatures and a commercially attractive geothermal system may result.   The depth of circulation required is a function of the permeability of the rocks (rate of fluid flow) and the background geothermal gradient.   These geothermal convective systems can be local or they can be regional.   Near many active volcanic and intrusive centers, the water flow systems may have lateral dimensions of only a few kilometers.   On the other hand, some geothermal systems may involve fluid which moves across ranges and valleys (for example, the Desert Peak geothermal anomaly; Yeamans, 1983).   The characteristics of some of the geothermal systems in the Basin and Range province have been discussed by Blackwell and Chapman (1977), Sass et al. (1976), Benoit et al. (1983), and Benoit and Butler (1983).   Extensive thermal data are now available from many areas, in too great a number to be discussed in detail here.   Many of these data were collected through DOE-Industry coupled geothermal programs, and are available from the Earth Science Laboratory of the University of Utah.   Brief discussions of many of these areas are given in papers of the Geothermal Resources Council Transactions.   References to discussions of thermal data from areas in northern Nevada are listed in Table 3.

Mase and Sass (1980) have discussed an extensive heat-flow study in the Black Rock Desert area of northern Nevada (Figure 10).   This area includes several geothermal systems, one of which is Gerlach Hot Springs, but in addition areally extensive data outside the geothermal systems were also calibrated.   Typical heat-flow values in the Black Rock Desert are 40-60 mWm^{-2} along the axis of the valley, rising to 80-100 mWm^{-2} at the margins of the valley.   Several large high heat-flow anomalies are identified, particularly north of Gerlach along the east side of the Granite Range, along the southeast margin of the Black Rock Desert against Pahisupp Mountain, and along the west side of the Black Rock Range.   In addition, MacFarlane's Hot Spring (Swanberg and Bowers, 1982) is just off the map to the east.   The major gap in the data is that few values are available from the ranges to compare to the basin.   A heat-flow value measured in the granite of Pahisupp Mountain was 188 mWm^{-2}, clearly much in excess of any possible regional value, and affected in some way by a geothermal system.

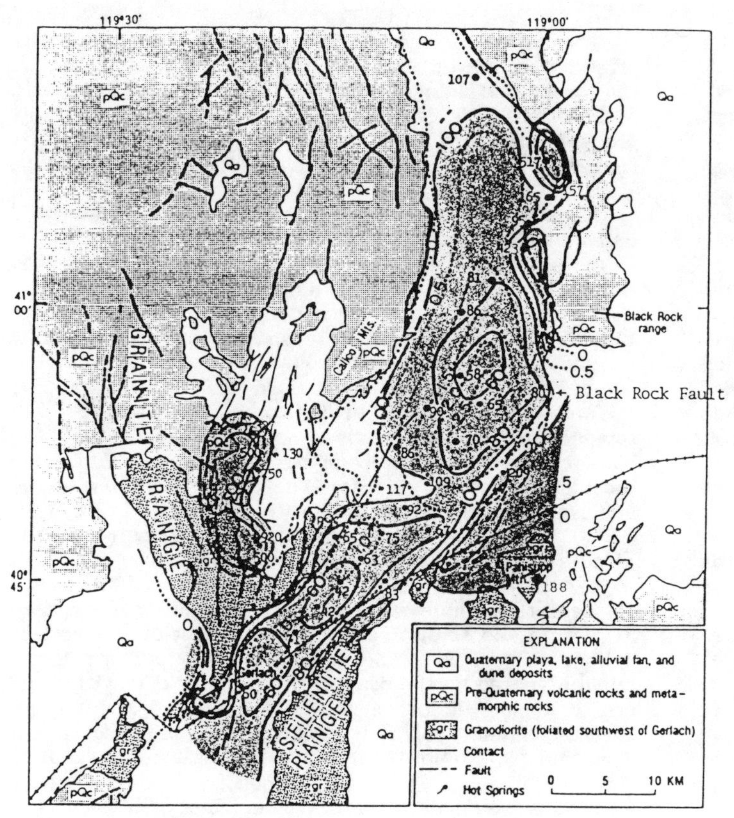

FIGURE 10.   Heat-flow contours (solid line: 60, 80, 100, 200, 300, 500 mWm^{-2}) and depth-to-basement contours (dotted line: 0, 0.5, 1.0, 1.5, 2.0, 2.5 km) in the western Black Rock Desert. Location of heat-flow points (solid circles) for reference. Major normal faults are also indicated. After Mase and Sass (1980).

TABLE 3.   Geothermal systems and published geothermal gradient/heat flow studies or data reports.

Location	Geothermal System	Reference
Black Rock Desert	McFarland H.S.	Swanberg and Bowers, 1982
	Gerlach H.S.	Sass et al., 1979
		Sass et al., 1976
San Emedio Desert	San Emedio West	Mackelprang et al., 1980
Humbolt Sink	Colado	Mackelprang, 1982
	Humbolt House	
Carson Sink	Desert Peak	Benoit et al., 1982
	Brady H.S.	
	Stillwater	
	Soda Lake	Hill et al., 1979
	Fallon-NAS	
Clan Alpine Mts.	McCoy	Olson et al., 1979
Pueblo Mountains	Baltazor H.S.	Earth Power Prod. Co., 1980
Steamboat Hills	Steamboat H.S.	White, 1968
Buena Vista Valley	Kyle H.S.	McMannes et al., 1981
Crescent Valley	Beowawe H.S.	Smith, 1983
Independence Valley	Tuscarora	Pilkington et al., 1980
Grass Valley	Leach H.S.	Sass et al., 1977
	Parker Canyon	Welch et al., 1981
		Mase and Sass, 1980
Buffalo Valley		Sass et al., 1976

Superimposed on the heat-flow values are the depth-to-basement contours, also from Mase and Sass (1980). There is a good correlation between the heat-flow values and the depth to basement, with the lowest heat-flow values being associated with the deepest part of the basin. Mase and Sass (1980) suggested that regional downflow is occurring in the valleys with upflow along the margins of the valleys. The proposed hydrologic circulation would be counter to the "normal" hydrologic circulation pattern of downflow in the ranges and upflow in the valleys (Mifflan, 1968). Unfortunately, the lack of data in the ranges does not allow a complete test of the Mase and Sass (1980) hypothesis. Furthermore, geothermal systems are often associated with the contact between the valleys and the ranges and may be controlled by hydrologic barriers between the ranges and the valleys. In this case, if water flow were down in the range, it might be expected to come back up again along the bounding fault so that the whole system would be contained within the range. This particular circulation pattern seems to be characteristic of the Roosevelt area in southeastern Utah (Ward et al., 1978). The data of Mase and Sass (1980) not immediately adjacent to geothermal systems are included in Figure 6, and it is clear that the gradients and heat flow values in the Black Rock Desert are consistent with those observed in the Battle Mountain Heat Flow High, if an allowance is made for the refractive effect of the low thermal conductivity sedimentary basins, and no absorption of heat by fluid downflow in the basins is necessary to make the heat flow consistent with the inclusion of this area as part of the Battle Mountain Heat Flow High. Mase and Sass (1980) included this area in the Battle Mountain Heat Flow High based on addition of the total amount of energy lost in the geothermal systems to the actual observed heat-flow values in the valleys. If refractive effects are significant, as is implied by Figure 6, then the observed values are consistent with such an association, without hypothesizing any large-scale water flow effects on the heat-flow data. An additional complexity associated with the geothermal systems is that flow is usually transient. Temperature gradient reversals with depth in exploration holes are almost the rule rather than an exception (see Benoit et al., 1982). Ziagos and Blackwell (1980, 1983) have discussed modeling of temperature-depth curves where this phenomenon is observed, to determine aquifer characteristics and flow regimes.

There is no doubt that the fluid flow patterns are extremely complicated and have a major effect on heat flow at some locations in the Basin and Range province. However, the nature of these patterns remains to be sorted out, as in no place are there sufficient heat-flow data to thoroughly investigate the total geographic extent of a water flow system. There is evidence for many different kinds of flow patterns from the geothermal systems themselves. Many of the geothermal systems are associated with the range-valley contacts; however, some systems appear to be confined to the valley side, some to the range side, some to the faults, and some involve intrabasin/range fluid flow. Consequently, convective systems can be imagined which might involve flow in ranges only, valleys only, ranges and valleys, and perhaps even only along the normal faults. At the present time, neither the data nor the interpretations are sufficiently constrained to allow description of the nature of a "typical" geothermal flow system in the Basin and Range province.

DISCUSSION

The object of this paper has been to investigate the effects and interactions that make the thermal pattern in the Basin and Range province so complex. The basic observations of geothermal gradient (Figure 7) are probably the most reliable and useful information. The heat-flow pattern remains to be explained.

A very simple conceptual model of heat flow in a Basin and Range setting is shown in Figure 9. The model includes the effects of a crustal thermal disturbance and refraction in the Basin and Range setting. The thermal source is approximated by a body at a temperature of $1350^\circ C$ emplaced between 10 and 20 km. The $1350^\circ C$ includes the regional background, which was calculated for a Basin and Range heat flow of 80 mWm^{-2} at the surface, and a mantle heat flow of 59 mWm^{-2}. The temperature corresponds approximately to the emplacement temperature of basalt with some allowance for the latent heat effects. The model is similar to the one used for the western Snake River Plain by Brott et al. (1978). The model was designed to investigate the effects of refraction and a thermal anomaly.

Heat-flow values as a function of position are shown at 5 M.Y. and 10 M.Y. after emplacement of this sort of a body (Figure 9). The basic heat-flow pattern is not much affected by the source, even though it is within the crust. If such a source had been emplaced 5 M.Y. ago, then the mean heat flow would be 112 mWm^{-2}, with a variation from below 100 to over 120 mWm^{-2}, except in the vicinity of the range-bounding faults, where variations would be more extreme. If the intrusive has cooled for a period of 10 M.Y., then the mean heat flow drops to 90 mWm^{-2} while the heat flow varies from about 80 mWm^{-2} in the valleys to 100 mWm^{-2} in the ranges. The variation of observed heat-flow values for the valleys and the ranges in the Battle Mountain Heat Flow High, ignoring the extremely high heat-flow values in the sedimentary rocks and those that are certainly associated with geothermal systems, is shown on the plot. The heat-flow pattern associated with a cooling period of 10 ± 5 M.Y. would approximate the midpoint of the range of observed data. There is slightly larger variation in heat flow than is predicted by this model between the ranges and the valleys, which might suggest some superimposed small effects of fluid circulation. The calculated age has little significance because the model is very approximate, and if extension effects are superimposed on a magmatic effect, the heat flow might be maintained at high values for some period of time after it would ordinarily have dropped. At the same time less extension would be required to keep the heat flow at a high level.

The conclusion is that a reasonable heat-flow model for the northern Basin and Range province involves a background heat flow which in superposition of a late Cenozoic thermal (intrusive and extrusive) event followed by regional extension active to within the last 1 M.Y. In the northern Nevada region, intrusive or extrusive silicic rocks of young age are not involved with the geothermal systems. The actual distribution of values is profoundly affected by the thermal conductivity contrasts between the ranges and the valleys. A combination of a very high permeability crust (due to the extension) with relatively high gradients due to high regional heat flow results in the observed pattern of heat flow and geothermal systems. Geothermal manifestations are associated in many cases with range-bounding faults, although there is no simple relationship between the faults and the geothermal systems. Some of the geothermal systems are on the range side, some are on the valley side, some involve ranges and valleys, and some may be within the fault system itself. Although there are a lot of heat-flow data available, much analysis remains before we thoroughly understand the heat-flow pattern. Drilling in the geothermal systems has added a new set of data which gives information on areas which have not been previously included in the regional data set, and need to be included for complete understanding of the heat flow pattern. Future studies will need to thoroughly assimilate these data, so that a more realistic and accurate pattern of heat-flow distribution and the controls on that distribution in the Basin and Range province can be developed.

ACKNOWLEDGMENTS

The calculations illustrated in Table 2 and Figure 2 were carried out by S. Chockalingam. The calculations in Figure 9 were made by Charles A. Brott. Heat-flow values from some U.S. Geological Survey studies in Nevada, published only as points on maps, were made available by J.H. Sass.

REFERENCES

Archbold, N.L., 1972, Modified geologic map of Nevada, Nevada Bur. Mines and Geol. Map 44.

Benoit, W.R. and D. Butler, 1983 (this volume).

Benoit, W.R., J.E. Hiner and R.T. Forest, 1982, Discovery and geology of the Desert Peak geothermal field: a case history, Nevada Bur. Mines and Geol. Bull. 97, 81 pp.

Blackwell, D.D., 1978, Heat flow and energy loss in the western United States, pp. 175-208 in Cenozoic Tectonics and Regional Geophysics of the Western Cordillera, Memoir 152, Smith, R.B. and G.P. Eaton (eds.), Geol. Soc. Amer., Boulder, Colorado.

Blackwell, D.D. and D.S. Chapman, 1977, Interpretation of geothermal gradient and heat flow data for Basin and Range geothermal systems, Trans. Geothermal Resources Council, 1, 19-20.

Bodell, J.M. and D.S. Chapman, 1982, Heat flow in the Northern Central Colorado Plateau, Jour. Geophys. Res., 87, 2869-2884.

Brott, C.A., 1978, Tectonic implications of the heat flow of the western Snake River Plain, Idaho, Geol. Soc. Amer. Bull., 89, 1697-1707.

Brott, C.A., D.D. Blackwell and J.P. Ziagos, 1981, Thermal and tectonic implications of heat flow in the eastern Snake River Plain, Jour. Geophys. Res., 86, 11709-11734.

Domenico, P.A. and V.V. Palciauskas, 1973, Theoretical analysis of forced convective heat transfer in regional ground-water flow, Geol. Soc. Amer. Bull., 84, 3803-3814.

Earth Power Prod. Co., 1980, Geochemical map, geologic cross section, sulfate map, microearthquake survey map, DOE Industry Coupled Program, Earth Science Laboratory of University of Utah Research Institute Open-File Rept. NV/BAL/EPP-7.

Edmiston, R.C., 1982, A review and analysis of geothermal exploratory drilling results in the northern Basin and Range geologic province of the USA from 1974 through 1981, Trans. Geothermal Resources Council, 6, 11-14.

England, P.C. and S.W. Richardson, 1980, Erosion and the age dependence of continental heat flow, Geophys. Jour. Royal Astr. Soc., 62, 421-437.

Hill, D.G., E.B. Layman, C.M. Swift and S.H. Yungul, 1979, Soda Lake, Nevada thermal anomaly, Trans. Geothermal Resources Council, 3, 305-308.

Jaeger, J.C., 1965, Application of the theory of heat conduction to geothermal measurements, pp. 7-23 in Terrestrial Heat Flow, Geophys. Mono. Ser., 8, Lee, W.H.K. (ed.), Amer. Geophys. Union.

Jarvis, G.T. and D.P. McKenzie, 1980, Sedimentary basin formation with finite extension rates, Earth Planet. Sci. Lett., 48, 42-52.

Keller, G.P., L.W. Braile and P. Morgan, 1979, Crustal structure, geophysical models, and contemporary tectonism of the Colorado Plateau, Tectonophysics, 61, 131-148.

Lachenbruch, A.H., and J.H. Sass, 1977, Heat flow in the United States and the thermal regime of the crust, pp. 626-675 in The Earth's Crust, Geophys. Mono. Ser., 20, Amer. Geophys. Union, Heacock, J.G. (ed.).

Lachenbruch A.H. and J.H. Sass, 1978, Models of extending lithosphere and heat flow in the Basin and Range province, pp. 209-250 in Cenozoic Tectonics and Regional Geophysics of the Western Cordillera, Memoir 152, Smith, R.B. and G.P. Eaton (eds.), Geol. Soc. Amer., Boulder, Colorado.

Lee, T.C. and T.L. Henyey, 1974, Heat-flow retraction across dissimilar media, Geophys. Jour. Royal Astr. Soc., 39, 319-333.

Mackelprang, C.E., 1982, Interpretation of geophysical data from the Colado KGRA, Pershing County, Nevada, Univ. Utah Res. Inst. Rept. DOE/ID/12079-58, ESL-71, 27 pp.

Mackelprang, C.E., J.N. Moore and H.P. Ross, 1980, A summary of the geology and geophysics of the San Emedio KGRA, Washoe County, Nevada, Trans. Geothermal Resources Council, 4, 221-224.

Mase, C.W. and J.H. Sass, 1980, Heat flow from the western arm of the Black Rock Desert, Nevada, U.S. Geol. Surv. Open-File Rept. 80-1238, 38 pp.

McMannes, D., B. Quillin and D. Butler, 1981, Granite Mountain, Nevada geothermal prospect: a case history, Trans. Geothermal Resources Council, 5, 107-110.

Mifflan, M.D., 1968, Delineation of groundwater flow systems in Nevada, Desert Research Institute, Center for Water Resources Research Tech. Rept. Ser., Pub. 4.

Mifflan, M.D., 1983 (this volume).

Morgan, P., V. Harder, C.A. Swanberg and P.H. Daggett, 1981, A groundwater convection model for Rio Grande Rift geothermal resources, Trans. Geothermal Resources Council, 5, 193-196.

Olson, H.J., F. Dellechaie, H.D. Pilkington and A.L. Lange, 1979, The McCoy geothermal prospect, status report of a possible new discovery in Churchill and Louder Counties, Nevada, Trans. Geothermal Resources Council, 3, 515-518.

Pilkington, H.D., A.L. Lange and F.E. Barkman, 1980, Geothermal exploration of the Tuscarora prospect in Elko County, Nevada, Trans. Geothermal Resources Council, 4, 233-236.

Reiter, M., A.J. Mansure and C. Shearer, 1979, Geothermal characteristics of the Colorado Plateau, Tectonophysics, 61, 183-195.

Roy, R.F., D.D. Blackwell and F. Birch, 1968a, Heat generation of plutonic rocks and continental heat flow provinces, Earth Planet. Sci. Lett., 5, 1-12.

Roy, R.F., E.R. Decker, D.D. Blackwell and F. Birch, 1968b, Heat flow in the United States, Jour. Geophys. Res., 73, 5207-5221.

BLACKWELL

Sass, J.H., A.H. Lachenbruch, R.J. Munroe, G.W. Green and T.H. Moses, Jr., 1971, Heat flow in the western United States, Jour. Geophys. Res., 76, 6356-6431.

Sass, J.H., F.H. Olmsted, M.L. Sorey, H.A. Wollenberg, A.H. Lachenbruch, R.J. Munroe and S.P. Galanis, Jr., 1976, Geothermal data from test wells drilled in Grass Valley and Buffalo Valley, Nevada, U.S. Geol. Surv. Open-File Rept. 76-85.

Sass, J.H., J.P. Ziagos, H.A. Wollenberg, R.J. Munroe, D.E. diSomma and A.H. Lachenbruch, 1977, Application of heat-flow techniques to geothermal energy exploration, Leach Hot Springs area, Grass Valley, Nevada, U.S. Geol. Surv. Open-File Rept. 77-762, 125 pp.

Sass, J.H., M.L. Zoback and S.P. Galanis, Jr., 1979, Heat flow in relation to hydrothermal activity in the southern Black Rock Desert, Nevada, U.S. Geol. Surv. Open-File Rept. 79-1467, 39 pp.

Sass, J.H., R.J. Munroe and C. Stone, 1981, Heat flow from five uranium test wells in west-central Arizona, U.S. Geol. Surv. Open-File Rept. 81-1089, 42 pp.

Sass, J.H., D.D. Blackwell, D.S. Chapman, J.K. Costain, E.R. Decker, L.A. Lawver and C.A. Swanberg, 1981, Heat flow from the crust of the United States, pp. 503-548 in Physical Properties of Rocks and Minerals, McGraw-Hill/CINDAS Data Series on Material Properties, Vol. II-2, Touloukian, Y.S., W.R. Judd and R.F. Roy (eds.), McGraw-Hill.

Shearer, C. and M. Reiter, 1981, Terrestrial heat flow in Arizona, Jour. Geophys. Res., 87, 6249-6260.

Smith, C., 1983, Thermal hydrology and heat flow of Beowawe geothermal area, Nevada, Geophysics, 48, 618-626.

Smith, R.B., 1978, Seismicity, crustal structure, and intraplate tectonics of the interior of the western Cordillera, pp. 111-144 in Cenozoic Tectonics and Regional Geophysics of the Western Cordillera, Memoir 152, Smith, R.B. and G.P. Eaton (eds.), Geol. Soc. Amer., Boulder, Colorado.

Snyder, W.S., W.R. Dickinson and M.L. Silberman, 1976, Tectonic implications of space-time patterns of Cenozoic magmatism in the Western United States, Earth Planet. Sci. Lett., 32, 91-106.

Stewart, J.H. and J.E. Carlson, 1978, Generalized maps showing distribution, lithology, and age of Cenozoic igneous rocks in the western United States, pp. 263-264 in Cenozoic Tectonics and Regional Geophysics of the Western Cordillera, Memoir 152, Smith, R.B. and G.P. Eaton (eds.), Geol. Soc. Amer., Boulder, Colorado.

Swanberg, C.A. and R.L. Bowers, 1982, Downward continuation of temperature gradients at MacFarlane's Hot Spring, northern Nevada, Trans. Geothermal Resources Council, 6, 177-180.

Ward, S.H., W.T. Parry, W.P. Nash, W.R. Sill, K.L. Cook, R.B. Smith, D.S. Chapman, F.H. Brown, J.A. Whelan and J.R. Bowman, 1978, A summary of the geology, geochemistry and geophysics of the Roosevelt Hot Springs thermal area, Utah, Geophys., 43, 1515-1542.

Welch, A.H., M.L. Sorey and F.H. Olmsted, 1981, The hydrothermal system in southern Grass Valley, Pershing County, Nevada, U.S. Geol. Surv. Open-File Rept. 81-915, 193 pp.

White, D.E., 1968, Hydrology, activity and heat flow of the Steamboat Springs thermal system, Washoe County, Nevada, U.S. Geol. Surv. Prof. Paper 458-C, 109 pp.

Yeamans, 1983 (this volume).

Ziagos, J.P. and D.D. Blackwell, 1981, A model for the effect of horizontal fluid flow in a thin aquifer on temperature-depth profiles, Trans. Geothermal Resources Council, 5, 221-224.

Ziagos, J.P. and D.D. Blackwell, 1983, A model for the transient temperature effects of horizontal fluid flow in geothermal systems, Jour. Geophys. Res., in press.

Controls on the Location and Intensity of Magmatic and Non-Magmatic Geothermal Systems in the Basin and Range Province

James B. Koenig, President
and
James R. McNitt, Vice President

GeothermEx, Inc.
5221 Central Avenue, Suite 201
Richmond, CA 94804

ABSTRACT

Geothermal systems of known or probable magmatic control are located along the eastern and western margins of the Basin and Range province, at or very near its contact with the Colorado Plateau and Sierra Nevada provinces. These include Roosevelt Hot Springs and possibly Cove Fort, Utah, along the eastern boundary; and Coso Hot Springs and Long Valley, California, and Steamboat Springs, Nevada, along the western boundary. No geothermal system located away from the margin of the Basin and Range province is believed to have active magmatic source.

Magmatic systems are characterized by very youthful silicic volcanism, with emplacement of major plutons occurring at depths of perhaps 5 to 10 km. There is limited evidence that these occur at the intersection of northeast-trending fracture zones or lineaments with the province boundaries. Where depth to groundwater is relatively great, thermal emissions are weak to absent. Shallow groundwater is associated with stronger thermal emissions, as at Steamboat Springs or Long Valley. Temperatures in magmatic systems are higher at comparable depth than in non-magmatic systems. Therefore, it is not surprising that magmatic systems are commercially productive at shallower depth. At least 3 of the magmatic systems (Coso, Steamboat and Roosevelt) are present within fractured granitic basement.

High-temperature, non-magmatic systems (over 200^{0}C) discovered to date are located in the western half of the northern Basin and Range province. Several of these fall along important northeast-trending zones of topographic flexures, linear river valleys and left-lateral strike-slip faults that have been active during the Quaternary. From this, it is assumed that a deep-seated set of northeast-trending fractures leaks heated fluids more intensely than do fractures of other orientation.

Although most non-magmatic systems have surface expression along range-front faults, several (Soda Lake, Desert Peak, Humboldt House and Salt Wells) have little or no leakage manifestation. In these cases, either the water table is deep, or controlling faults are not well-defined. The ultimate source of heat for these systems is the very thinness of the crust in the northern Basin and Range (22-25 km). However, it appears that the temperature in the crystalline basement is, in many cases, related to or controlled by the thickness of low-conductivity sedimentary cover. Undoubtedly, there is convective transfer of heat along faults and other fractures, to depth of perhaps 3 to 4 km. This would account for the linear or arcuate pattern of both the surface emissions and the underlying high-temperature system.

In the final analysis, the distribution of high-temperature versus moderate-temperature non-magmatic systems may be controlled by the thinness of the crust and the thickness of low-conductivity cover.

GEOCHEMISTRY OF ACTIVE GEOTHERMAL SYSTEMS IN THE NORTHERN BASIN AND RANGE PROVINCE

R. H. Mariner, T. S. Presser, and W. C. Evans

U.S. Geological Survey, Menlo Park, California

ABSTRACT

Numerous thermal springs occur in the northern Basin and Range Province due primarily to the structure and high regional heat flow. Dilute to slightly saline (200 to 3,000 mg/L TDS) Ca and/or Na-HCO$_3$ type waters, many associated with travertine, are dominant in eastern and northeastern Nevada. CO$_2$-charged Na-HCO$_3$ waters are particularly common along the eastern side of the Sierra Nevada from Long Valley north to Bridgeport. Moderately to very saline (3,000 to 35,000 mg/L TDS) Na-Cl type waters predominate near major topographic lows. Na-SO$_4$ type waters are common in western Nevada and in northeastern California. Na-mixed anion waters are common along the north side of the Black Rock Desert in northwestern Nevada and the Alvord Desert in southeastern Oregon.

Measured temperatures in deep geothermal wells in the northern Basin and Range Province are about 14°C cooler than the average temperature calculated using two chemical and one isotope geothermometer on waters discharged by nearby thermal springs and shallow wells (<100m). Isotopic data (δD) for thermal springs in the northern Basin and Range have the same general pattern as modern precipitation. However, the few detailed studies of recharge areas for specific systems have generally found local cold waters to be slightly more enriched in deuterium than water currently discharged by the thermal springs. This probably indicates that the water currently being discharged by the hot springs was recharged during colder time periods, perhaps the Pleistocene.

Introduction

Although chemical analyses of a few hot springs in the northern Basin and Range were published in the late 1800's (Peale 1886, Church 1878), the first detailed studies were begun by Don White at Steamboat Springs in the late 40's and continued into the early 60's (White and Brannock, 1950, White and others, 1964). Research activities exploded in the 70's as interest in the geothermal resource increased. Most thermal springs in the northern Basin and Range Province have now been sampled and one or more analyses are available in the literature.

Compilations of chemical data for thermal springs in the various states which include parts of the northern Basin and Range Province have been presented by Mundorff (1970), and Goode (1978) for Utah, Garside and Schilling (1979) for Nevada, Majmundar (in press) for California, and U.S. Geological Survey and Oregon Department of Mineral Industries (1979) for Oregon. Locations and temperatures of thermal springs and wells have also been compiled by Waring (1965) and Berry and others (1980). Figure 1 shows the outline of the northern part of the Basin and Range Province (Great Basin) along with physical features mentioned in the text.

Chemical Composition of Water

Chemically, the thermal waters of the Basin and Range Province range from dilute (<1,000 mg/L TDS) Na-HCO$_3$ or Ca-HCO$_3$ type waters to very saline (10,000-35,000 mg/L TDS) Na-Cl type waters (Table 1). The chemical constituents in thermal waters are derived primarily from the country rock by dissolution of, or exchange with, the rock-forming minerals (Ellis and Mahon, 1964, 1967). Ideally, concentrations of silica, sodium, potassium, calcium and magnesium are controlled by mineral-water equilibria. However, other constituents such as chloride don't attain concentrations sufficient to reach saturation with respect to any chloride bearing minerals.

Na-Cl type thermal waters of moderate salinity occur in northwestern Nevada and west-southwest of Tonopah (Fig. 2). Na-Cl type thermal waters also discharge along the eastern edge of the Basin and Range Province in Utah and southeastern Idaho. These waters result from the extensive reaction of thermal fluids with sedimentary rocks deposited in a marine environment. Sulfate in the Na-Cl type water of Utah is primarily of marine origin based on both the oxygen and sulfur isotopic compositions of dissolved sulfate (Nehring and Mariner, 1979, Cole, 1982). Locally, chemical and isotopic data indicate some admixture with chloride-rich saline-lake or playa waters may have occurred. High chloride concentrations can also be attained by circulation through some granites (Moore and others, 1983), however, the extent of

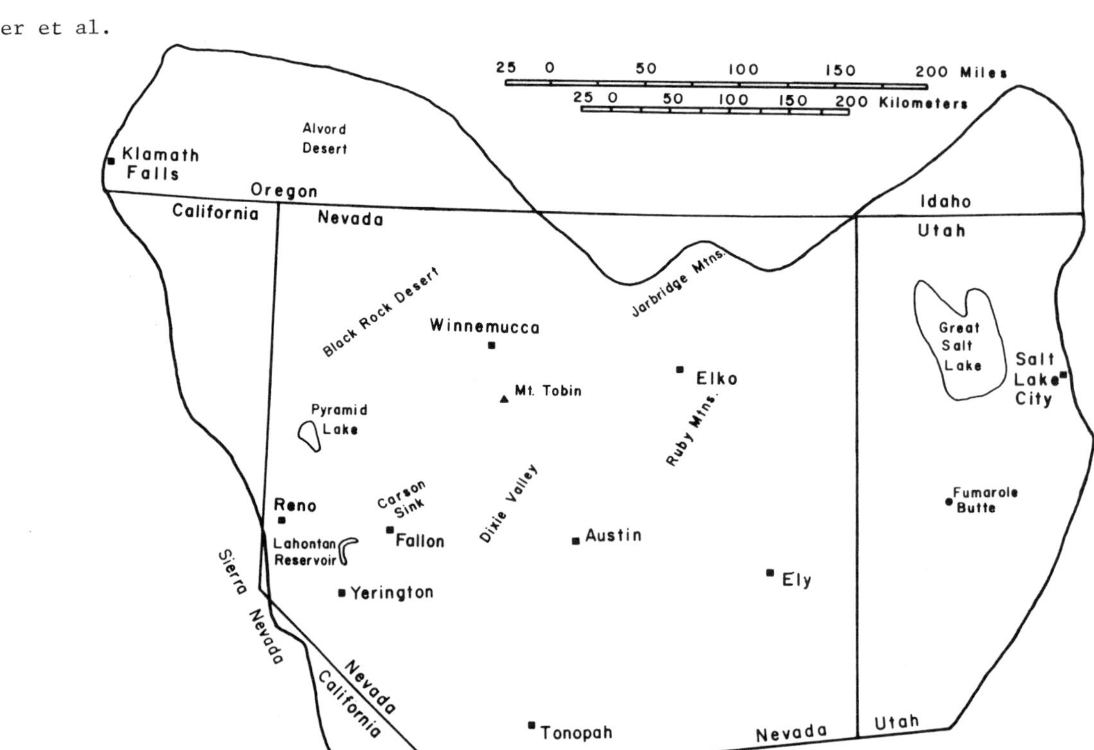

Figure 1.   Location map of the major population centers and features mentioned in the text.

this process in the Basin and Range is unknown. High chloride concentrations are an indication that more extensive water-rock reaction has taken place, and this may infer longer flow paths and generally deeper circulation. Most high temperature geothermal wells in the Great

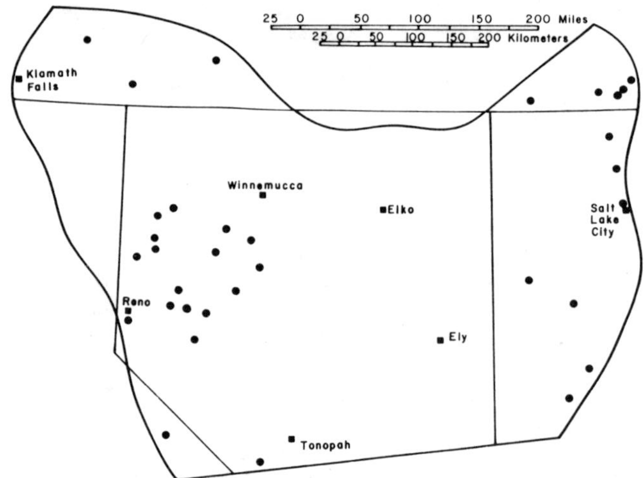

Figure 2. Distribution of Na-Cl waters in the northern Basin and Range Province.

Basin have encountered Na-Cl waters.

Bicarbonate waters range from dilute to slightly saline depending largely on the source of the dissolved CO_2. If CO_2 is generated at depth and dissolved under pressure in the thermal water, then a high TDS CO_2-charged water will be produced. High TDS CO_2-charged waters are particularly common adjacent to the southern Sierra Nevada. The more saline CO_2-charged waters are slightly radioactive due to their high radium and radon content (Femlee and Cadigan, 1982, and Wollenberg, 1974). Enough calcium is present in most of these waters to form travertine mounds and terraces. Fales and Travertine hot springs, near Bridgeport, California discharge this type of water as does Hyder Hot Springs, in Dixie Valley, Nevada. However, if the CO_2 is dissolved from the atmosphere, or soil gas, then the dissolved carbon concentrations and related total dissolved solids will be very low. Darrough and Soldier Meadows hot springs in central and northwestern Nevada are good samples of dilute $Na-HCO_3$ waters. Dilute $Na-HCO_3$ and $Ca-HCO_3$ type waters are particularly common in central and eastern Nevada (Fig. 3). A dilute $Ca-HCO_3$ water will be produced if the principal bedrock is a limestone and these waters are common in eastern Nevada where Paleozoic limestones crop out. The dilute Na–HCO$_3$ waters develop in areas where the rock contains appreciable sodium silicate or sodium aluminosilicate minerals and only small amounts of CO$_2$ are available.

Dilute to slightly saline Na–SO$_4$ ($\pm$ Cl) waters occur principally in northeastern California and the adjacent part of Nevada (Fig. 4). High sulfate concentrations can result from dissolution of gypsum (anhydrite) in sedimentry rocks or dissolution of minerals such as alunite, jarosite, anhydrite, or pyrite from mineralized zones (Hem, 1970). Oxidation of sulfide to sulfate can also produce large sulfate concentrations and acid sulfate waters

(White and others, 1971) but these are rare in the Basin and Range Province. The oxygen isotopic compositions of marine and hydrothermal sulfate are quite different (Longinelli and Craig, 1967, and McKenzie and Truesdell, 1977) thus it is relatively easy to differentiate between them. Sulfates in the waters of the western part of the Great Basin generally range from 0 to −10 o/oo in oxygen-18 (Nehring and Mariner, 1979) and have a possible source in the sulfate minerals associated with ore deposits and gossans. Gypsum and anhydrite deposits in Mesozoic marine rocks are also fairly common in western Nevada, however, only one of the hot springs for which isotopic data is available,

Table 1. Chemical Composition of Selected Thermal Waters of the Northern Basin and Range--continued.
[Concentrations are in mg/L, temperatures are in °C; sodium (Na) values followed by K represent sodium plus potassium; dashes (−) indicate no data.]

Name/Location	Temp	pH	SiO$_2$	Ca	Mg	Na	K	HCO$_3$	Cl	SO$_4$	F	Reference
CALIFORNIA												
Alpine County												
Unn. sprs. on the Carson River												
SENESE, sec. 34, T. 11 N., R. 20 E.	84	6.63	178	76	6.5	510	40	388	355	550	6.3	Unpub. data, Mariner and others
Unn. sprs. on the Carson River												
NWSW, sec. 26, T. 11 N., R. 20 E.	65	6.52	110	125	4.0	480	21	333	295	700	2.4	Unpub. data, Mariner and others
Lassen County												
Amadee Hot Springs												
NESW, sec. 08, T. 28 N., R. 16 E.	96	8.36	98	15	<.1	235	5.7	57	155	280	4.6	Mariner and others, 1976a
Bassett Hot Springs												
NWSE, sec. 12, T. 38 N., R. 7 E.	79	8.53	65	30	<.1	220	3.2	32	93	370	2.0	Reed, 1975
Kellog Hot Springs												
SWSE, sec. 15, T. 38 N., R. 8 E.	78.4	8.63	85	30	<.1	240	5.9	35	110	370	2.6	Reed, 1975
Wendell Hot Springs												
NESW, sec. 23, T. 29 N., R. 15 E.	95.5	8.26	125	20	<.1	280	8.0	53	185	340	4.2	Mariner and others, 1976a
Zamboni Hot Spring												
NWNW, sec. 24, T. 24 N., R. 17 E.	41	9.37	36	2.2	.02	66	.57	74	14	57	2.0	Unpub. data, Mariner and others
Modoc County												
Hot Springs Motel												
NESW, sec. 06, T. 42 N., R. 17 E.	98	8.40	100	16	<.01	280	5.5	61	200	320	5.1	Reed, 1975
Kelly Hot Springs												
NENW, sec. 29, T. 42 N., R. 10 E.	91.5	8.08	110	20	<.1	250	6.5	47	160	300	2.1	Reed, 1975
Leonards Hot Springs												
NENE, sec. 13, T. 43 N., R. 16 E.	61.8	7.82	110	26	.6	330	8.5	84	220	390	5.2	Reed, 1975
Little Hot Springs												
NWSW, sec. 9, T. 39 N., R. 5 E.	75.7	7.59	87	44	.2	230	5.2	49	120	400	1.9	Reed, 1975
Seyforth Hot Springs												
NWNW, sec. 12, T. 39 N., R. 5 E.	85	7.66	110	28	<.1	300	9.0	63	220	370	5.4	Reed, 1975
West Valley Reservoir (spring)												
NWNE, sec. 29, T. 39 N., R. 14 E.	77.3	7.79	130	19	<.1	330	11.	63	150	510	4.0	Reed, 1975
Mono County												
Benton Hot Springs												
SW, sec. 2, T. 2 S., R. 31 E.	56.5	9.32	63	1.4	<.1	80	1.0	96	22	50	3.8	Mariner and others, 1977
Fales Hot Springs												
SE, sec. 24, T. 6 N., R. 23 E.	61	6.55	114	41	10	560	37	1130	160	260	4.7	Mariner and others, 1977
Long Valley-Hot Creek Gorge												
NE, sec. 25, T. 3 S., R. 28 E.	90	6.6	150	1.6	.1	400	24	549	225	100	9.6	Mariner and Wiley, 1976
Mono Lake – North Shore												
sec. 11, T. 2 N., R. 26 E.	66	7.68	76	13	2.9	430	8.8	454	350	100	4.8	Mariner and others, 1977
– South Shore												
sec. 18, T. 1 N., R. 27 E.	33	6.38	130	120	61	410	34	1560	105	28	.4	Mariner and others, 1977
Travertine Hot Spring												
SW, sec. 34, T. 5 N., R. 25 E.	69	6.73	100	64	18	1100	55	1800	200	920	4.5	Mariner and others, 1977
IDAHO												
Bear Lake County												
Bear Lake Hot Spring												
SW, sec. 13, T. 15 S., R. 44 E.	47.5	6.6	35	210	55	180	61	256	79	800	7.1	Young and Mitchell, 1973
Cassia County												
RRGE-1												
sec. 23, T. 15 S., R. 26 E.	95	8.9	144	58	0.3	505	35	46	896	59	6.2	Nathenson and others, 1982
Franklin County												
Maple Grove Hot Springs												
NE, sec. 7, T. 13 S., R. 41 E.	76	7.3	55	89	24	490	40	491	630	260	1.1	Young and Mitchell, 1973
Wayland Hot Springs												
NE, sec. 8, T. 5 S., R. 39 E.	77	7.0	80	160	16	3100	660	699	5400	50	12	Young and Mitchell, 1973
Oneida County												
Woodruff Hot Springs												
NE, sec. 10, T. 16 S., R. 36 E.	27	7.3	29	130	45	910	87	454	1600	58	.6	Young and Mitchell, 1973

Table 1. Chemical Composition of Selected Thermal Waters of the Northern Basin and Range--continued.
[Concentrations are in mg/L, temperatures are in °C; sodium (Na) values followed by K represent
sodium plus potassium; dashes (-) indicate no data.]

Name/Location	Temp	pH	SiO_2	Ca	Mg	Na	K	HCO_3	Cl	SO_4	F	Reference
NEVADA												
Carson City												
Carson Hot Springs, SENE, sec. 5, T. 15 N., R. 20 E.	49	9.3	-	2.6	.4	96	-	92	29	96	-	Worts & Malmberg, 1966
Pinyon Hills Well sec. 23, T. 15 N., R. 20 E.	46	8.6	-	275	3	254	-	41	33	135	4.3	CWRR, 1973
Churchill County												
Dixie Valley Hot Springs SE. sec. 5, T. 22 N., R. 35 E.	72	8.6	115	3.2	.02	190	6.5	133	126	111	16.3	Mariner and others, 1974
Dixie Federal 52-18 NE, sec. 18, T. 24 N., R. 37 E.	boiling	8.27 @32°C	383	4.4	.03	385	36	379	307	127	7.0	Unpub. data, Mariner and others
Brady's Hot Spring (well) SW, sec. 12, T. 22 N., R. 26 E.	-	6.78 @24°C	164	45	.32	850	36	111	1140	320	5.8	Unpub. data, Mariner and others
Eagle Salt Works Spring sec. 34, T. 22 N., R. 26 E.			259	32	2	839	-	99	955	334		Adams, 1944
Soda Lake-Upsal Hogback SW, sec. 28, T. 20 N., R. 28 E.	boiling	7.86 @42°C	160	82	2.1	1000	48	144	1500	360	.6	Mariner and others, 1975
Stillwater Area SW, sec. 07, T. 19 N., R. 31 E.	96	7.57	170	108	1.7	1480	42	90	2200	190	5.0	Mariner and others, 1974
Lee Hot Springs Unsurveyed (39°12'N b 118°43'W)	88	7.4	180	44	.6	450	26	114	380	470	7.9	Mariner and others, 1974
Douglas County												
Hobo Hot Spring SESE, sec. 23, T. 14 N., R. 19 E.	46	8.9	47	6	.7	125	1.7	85	74	109	7.1	Glancy and Katzner, 1975
Saratoga Hot Springs SESE, sec. 23, T. 14 N., R. 20 E.	50	9.0	20	172	-	160k	-	18	39	678	9.0	Glancy and Katzner, 1975
Walley's Hot Spring NE, sec. 22, T. 13 N.,R 19 E.	62	8.8	58	10	.01	145	3.6	68	44	235	4.9	Mariner and others, 1974
Elko County												
Nile Spring SW, sec. 30, T. 47 N., R. 70 E.	43	7.2	31	40	11.5	10	5.6	149	8.7	37	.4	Mariner and others, 1974
Trout Creek Ranch Well NWNW, sec. 23, T. 46 N., R. 69 E.	43	8.3	21	16	5.7	24	5.6	120	2	22	.6	Moore and Eakin, 1968
San Jacinto Ranch Spring NWNW, sec. 23, T. 46 N., R. 64 E.	26	8.1	18	25	8.6	13	3.9	132	3.9	11	.5	Moore and Eakin, 1968
Rizzi Ranch Hot Spring sec. 29, T. 45 N., R. 54 E.	41	7.4	23	29	7.7	110	8.3	380	4.4	36	3.4	Moore and Eakin, 1968
Mineral (Contact) Hot Springs sec. 16, T. 45 N., R. 64 E.	60	9.1	83	1.6	<.01	75	2.2	108	15	45	8.9	Mariner and others, 1974
Wild Horse Hot Spring SESE, sec. 4, T. 43 N., R. 55 E.	54	7.2	40	48	12	130	22	482	14	40	5.2	Mariner and others, 1974
Hot Creek Springs NW, sec. 12, T. 28 N., R. 52 E.	26	7.30	20	46	23.5	10	2.1	228	4.6	27	<.1	Mariner and others, 1974
Hot Creek Springs NW, sec. 34, T. 43 N., R. 60 E.	37.5	6.76	139	62	20	23	17	307	3	51	.78	Mariner and others, 1975
Hot Sulphur Spring NE, sec. 8, T. 41 N. , R. 52 E.	92	7.32	165	9.4	.22	160	15	367	16	54	9.6	Mariner and others, 1975
Wine Cup Ranch Well NWNW, sec. 25, T. 41 N., R. 64 E.	59	8.4	-	49	17	139k	-	426	30	69	-	Rush, 1968b
Hot Lake NNW, sec. 25, T. 38 N., R. 46 E.	18	7.2	57	29	5.8	33	7.0	132	22	34	.4	Unpub. data, Mariner and others
Unnamed spring on Rock Creek SWSW, sec. 1,T. 39 N., R. 47 E.	35	6.87	23	41	11	86	14	330	13	62	2.4	Unpub. data, Mariner and others
Humboldt Wells Area SE, sec. 20, T. 38 N., R. 62 E.	60	6.58	110	78	36	300	30	1210	26	24	6.1	Mariner and others, 1975
Hot Hole NE, sec. 21, T. 34 N., R. 55 E.	56	7.21	65	60	15.5	120	39	490	16	72	1.9	Mariner and others, 1974
Hot Spring near Carlin sec. 33, T. 33 N., R. 52 E.	79	7.6	70	60	15	45	16	335	12	52	-	Mariner and others, 1974
Sulphur Hot Spring NW, sec. 11, T. 31 N., R. 59 E.	93	8.53	210	1.0	.03	135	8.9	274	23	40	17.7	Mariner and others, 1974
Smith Ranch (Unn. spr. - Ruby Marsh) NW, sec. 2., T. 27 N., R. 58 E.	65	8.0	50	45	12	58	14	377	6.5	24	-	Mariner and others, 1974
Esmeralda County												
Alkali Springs NW, sec. 26, T. 01 S., R. 41 E.	60	8.1	-	46	4.6	349k	-	348	68	492	-	Rush, 1968a
Silver Peak (Waterworks) Hot Springs SE,sec. 15, T 02 S., R. 39 E.	40	7.18 @19°C	105	540	71	10000	750	559	17000	410	4.1	Unpub. data, Mariner and others
Eureka County												
Beowawe SW, sec. 08, T. 31 N., R. 48 E.	98	8.98	320	1.0	<.1	230	16	383	69	130	17	Mariner and others, 1974
Hot Springs Point NW, sec. 11, T. 29 N., R. 48 E.	54	6.63	67	53	35	230	58	913	1	7	6.6	Mariner and others, 1974
Bruffey's (Mineral Hill) Hot Spring sec. 14, T. 27 N., R. 52 E.	66	7.0	58	52	16	39	8.7	287	14	27	.7	Roberts, Montgomery, and Lehner, 1967
Walti Hot Springs SW,sec. 33, T. 24 N., R. 48 E.	72	6.47	68	56	12	44	14	264	12	64	2.5	Mariner and others, 1974
Shipley Hot Springs NESE, sec. 23, T. 24 N., R. 52 E.	39	7.2	40	57	21	29	5.9	279	21	35	.2	Eakin, 1962a
Klobe Hot Springs Bartholomae) SE, sec.28, T. 18 N., R. 50 E.	54	9.25	85	1	<.1	64	.7	144	6.3	18	-	Mariner and others, 1974

Table 1. Chemical Composition of Selected Thermal Waters of the Northern Basin and Range--continued.
[Concentrations are in mg/L, temperatures are in ^{o}C; sodium (Na) values followed by K represent sodium plus potassium; dashes (-) indicate no data.]

Name/Location	Temp	pH	SiO_2	Ca	Mg	Na	K	HCO_3	Cl	SO_4	F	Reference
Humboldt County												
Cordero Mercury Mine well SE, sec. 28, T. 47 N., R. 37 E.	60	-	57	36	10	62k	-	195	34	56	-	Visher, 1957
Bog Hot Springs SWNW, sec. 07, T. 46 N., R. 28 E.	54	9.05	57	.2	<.1	81	1.0	138	15	45	1.7	Mariner and others, 1974
Baltazor Hot Springs (well) NW, sec. 13, T. 46 N., R. 28 E.	90	7.50	150	10	.1	180	8.2	156	47	230	6.8	Mariner and others, 1974
Howard Hot Springs NE, sec. 04, T. 44 N., R. 31 E.	56	9.2	85	3	<.1	88	1.7	127	10	62	-	Mariner and others, 1974
Dyke Hot Springs SE, sec. 25, T. 43 N., R. 30 E.	66	8.86	85	1.8	<.1	150	4.3	277	21	82	8.0	Mariner and others, 1974
The Hot Springs NE, sec. 20, T. 41 N., R. 41 E.	58	8.0	55	10	8	296	36	881	26	36	-	Mariner and others, 1974
Soldier Meadows Hot Springs sec. 23, T. 40 N., R. 24 E.	54	8.53	63	3.1	<.1	74	1.1	98	18	41	12	Mariner and others, 1974
Pinto Hot Springs ESE,sec. 17, T. 40 N., R. 28 E.	93	7.14	150	14	.4	330	23	497	160	120	12	Mariner and others, 1974
Double Hot Springs sec. 4, T. 36 N., R. 26 E.	80	7.93	105	4.8	.1	180	4.5	265	59	120	10	Mariner and others, 1974
Macfarlane's Bath House Spring NW, sec. 27, T. 37 N., R. 29 E.,	75	6.61	82	43	11	1350	28	2050	870	210	2.6	Unpub. data, Mariner and others
Well SWSE, sec. 03, T. 37 N., R. 39 E.	70	7.4	-	30	7.1	450	26	1240	14	52	-	Cohen, 1962
Golconda Area SE, sec. 29, T. 36 N., R. 40 E.	74	6.53	66	33	6.8	130	22	429	18	56	1.8	Mariner and others, 1974
Hot Pot SW, sec. 11, T. 35 N., R. 43 E.	57	6.95	39	18	5.3	660	28	1625	61	155	9.4	Unpub. data, Mariner and others
Hot Spring Ranch (Tipton) SE, sec. 05, T. 33 N., R. 40 E.	85	8.36	125	16	.9	200	18	385	41	140	-	Mariner and others, 1974
Lander County												
Buffalo Valley Hot Springs SE, sec. 23, T. 28 N., R. 41 E.	49	6.53	80	45	4.9	250	34	813	29	110	4.8	Mariner and others, 1974
" " " " " " " "	73	-	71	43	5.2	295	33	-	29	104	4.6	Unpub. data, Mariner and others
Hot Springs Ranch (Valley of the Moon) NE, sec. 23, T. 27 N., R. 43 E.	53	8.0	40	20	9	118	21	333	21	64	-	Mariner and others, 1974
South Smith Creek Valley NE, sec. 25, T. 17 N., R. 39 E.	86	7.72	110	4.8	.06	170	8.4	256	22	102	8.9	Mariner and others, 1974
Spencer Hot Springs SE, sec. 13, T. 17 N., R. 45 E.	72	6.49	77	43	9.4	200	36	672	22	51	4.7	Mariner and others, 1974
Unnamed spring near Walti Hot Springs Unsurveyed 39°56.6'N by 116°40.8'W	64	6.51	83	66	11	135	42	559	32	58	4.1	Unpub. data, Mariner and others
Lyon County												
Hazen Area SW, sec. 18, T. 20 N., R. 26 E.	86	7.05	150	70	1.5	620	38	100	820	400	4.2	Mariner and others, 1975
Wabuska Hot Springs SE, sec. 16, T. 15 N., R. 25 E.	94	8.06	110	39	.1	300	14	74	55	620	8.2	Mariner and others, 1975
Hind's (Nevada) Hot Springs SE, sec. 16, T. 12 N., R. 23 E.	61	8.65	52	4.5	.01	102	2.5	69	17	169	3.1	Mariner and others, 1974
Wedell Springs SW, sec. 07, T. 12 N., R. 34 E.	60	7.83 @50°C	153	13	.2	270	9.2	214	78	315	12	Unpub. data, Mariner and others
Nye County												
Diana's Punch Bowl SE, sec. 22, T. 14 N., R. 47 E.	59	7.14	46	50	11	55	15	277	8	59	2.8	Mariner and others, 1974
Darrough's Hot Spring sec. 08, T. 11 N., R. 43 E.	95	8.29	98	1.3	.1	110	2.6	152	12	53	14	Mariner and others, 1974
Hot Creek Ranch Springs sec. 29, T. 08 N., R. 50 E.	63	8.0	135	51	15.1	197	13.4	545	42	86.4	8	Sanders and Miles, 1974
Upper Hot Creek Ranch Springs NESE, sec. 29, T. 08 N., R. 50 E.	67	8.1	161	33	9.5	193	1.4	501	37	64	8.3	USGS, 1977
Warm (Nanny Goat) Spring NWSW, sec. 20, T. 04 N., R. 50 E.	61	8.1	60	43	24	175	24	714	32	120	-	Mariner and others, 1974
Pershing County												
Colado (wells) SE, sec. 33, T. 28 N., R. 32 E.	60	7.56 @38°C	85	110	6.5	1450	120	199	2400	120	4.6	Mariner, Brook, Swanson and Mabey, 1978
Kyle Hot Springs SW, sec. 01, T. 29 N., R. 36 E.	77	6.50	150	95	25.5	540	80	544	770	51	5.7	Mariner and others, 1974
Trego Area 40°46'N by 119°7'W	84.5	7.93	79	11	.4	430	8.6	163	500	180	4.1	Mariner and others, 1976
Leach Hot Springs SE, sec. 36, T. 32 N., R. 38 E.	92	7.4	135	8.8	.5	160	13	368	29	53	7.8	Unpub. data, Mariner and others
Humboldt House (Rye Patch) SE, sec. 21, T. 31 N., R. 33 E.	-	-	340	43	3.6	1350	240	202	2250	18	6.2	Unpub. data, Mariner and others
art. well	77	-	162	120	7.2	1400	240	380	2300	39	5.3	Unpub. data, Mariner and others
	-	-	440	90	1.0	2000	240	43	3600	90	-	Benoit, 1978
Sou Hot Springs SE, sec. 29, T. 26 N., R. 38 E.	70	7.3	64	106	19.8	167	26	324	77	354	5.5	Sanders and Miles, 1974
Hyder Hot Springs SW, sec. 28, T. 25 N., R. 38 E.	78	6.77	63	41	10	390	20	926	45	120	8.6	Mariner and others, 1976

Table 1. Chemical Composition of Selected Thermal Waters of the Northern Basin and Range--continued.
[Concentrations are in mg/L, temperatures are in ^{o}C; sodium (Na) values followed by K represent
sodium plus potassium; dashes (-) indicate no data.]

Name/Location	Temp	pH	SiO_2	Ca	Mg	Na	K	HCO_3	Cl	SO_4	F	Reference
Washoe County												
Fly Ranch (Ward's)												
sec. 02, T. 34 N., R. 23 E.	80	7.91	82	31	4.2	340	17	466	240	46	7.0	Mariner and others, 1974
Great Boiling Spring												
NW, sec. 15, T. 32 N., R. 23 E.	86	7.15	165	68	1.2	1400	130	83	2200	400	4.5	Mariner and others, 1974
San Emidio Desert												
Unsurveyed - 40°23'N 119°24'W	89	6.57	205	160	2.2	1400	110	102	2300	220	5.1	Mariner and others, 1976
The Needle Rocks												
NWSWSW, sec. 06, T. 26 N., R. 21 E.	56	8.43	110	260	.1	1100	160	24	1900	340	3.0	Mariner and others, 1974
Lawton Hot Springs												
SWNE, sec. 13, T. 19 N., R. 18 E.	49	9.0	46	6.2	.1	117	5.4	52	57	144	2.5	Cohen and Loeltz, 1964
Moana Springs Area (Biglin well)												
NESE, sec. 26, T. 19 N., R. 19 E.	85	8.0	106	23	.08	236	8.0	86	48	455	5.1	Bateman and Scheibach, 1975
Steamboat Springs												
NE, sec. 33, T. 18 N., R. 20 E.	94	7.19	270	16	.7	680	66	368	837	73	2.1	Mariner and others, 1974
Bowers Mansion Hot Springs												
NW, sec. 03, T. 16 N., R. 19 E.	46	9.36	47	2.9	.2	50	0.6	76	6.3	38	2.8	Mariner and others, 1975
White Pine County												
Cherry Creek Hot Springs												
sec. 6, T. 23 N., R. 63 E.	61	7.77	105	12	.3	150	4.8	380	16	1	1.2	Mariner and others, 1975
Monte Neva Hot Springs												
sec. 24, T. 21 N., R. 63 E.	79	6.35	52	63	21	16	5.6	303	5.0	26	1.0	Mariner and others, 1975
OREGON												
Harney County												
Alvord (Indian) Hot Springs												
SE, sec. 33, T. 34 S., R. 34 E.	78.5	6.90	128	12	2.2	1000	63	1230	770	180	11	Mariner and others, 1974
Crane Hot Springs												
SW, sec. 34, T.24 S., R. 33 E.	78	8.1	83	3.7	.1	170	3.9	208	79	86	9.0	Mariner and others, 1974
Mickey Hot Springs												
sec. 13, T.33 S., R. 35 E.	86	8.31	214	1.0	.1	550	30	814	240	220	17	Mariner and others, 1974
Trout Creek Hot Springs												
sec. 16,T. 39 S., R. 37 E.	52	6.77	105	18	.8	270	10.8	441	24	204	12.8	Mariner and others, 1974
Unn. Sprs. near Hot Lake												
SE, sec. 36, T. 27 S., R. 29 E.	96	7.30	160	14	.3	450	28	382	250	434	7.2	Mariner and others, 1974
Unn. Sprs. near Harney Lake	68	7.26	92	12	1.8	630	13	568	590	140	3.3	Mariner and others, 1974
Lake County												
Barry Ranch Hot Springs												
SE, sec. 27, T. 39 S., R. 20 E.	88	7.76	130	8.8	.1	280	9.0	236	170	240	5.4	Mariner and others, 1974
Crump												
SW, sec. 34,T. 38 S., R. 24 E.	78	7.26	180	16	.2	280	11	155	240	200	4.9	Mariner and others, 1974
Fisher Hot Springs												
NW, sec. 10, T. 38 S., R. 25 E.	68	7.93	77	8.4	1.0	92	7.9	107	56	59	3.5	Mariner and others, 1974
Hunters Hot Springs												
NW, sec. 04, T.39 S., R. 20 E.	96	7.77	140	13	<.1	210	8.5	81	120	260	4.4	Mariner and others, 1974
Summer Lake Hot Springs												
NE, sec. 12, T. 33 S., R. 17 E.	43	8.43	94	2.1	.1	390	4.6	426	280	120	2.2	Mariner and others, 1974
UTAH												
Beaver County												
Roosevelt Seep												
NWSWSW, sec. 34, T. 26 S., R. 9 W	25	5.60	165	110	22	1800	260	298	3150	110	3.5	Unpub. data, Mariner and others
Roosevelt Steam Well												
NWSWNE, sec. 03, T. 27 S., R. 9 W.	-	5.80 @24°C	590	7.0	0.1	1950	400	200	3400	61	5.7	Unpub. data, Mariner and others
Thermo Hot Springs												
sec. 28, T. 30 S., R. 12 W.	89.5	7.98	113	71	10	380	52	360	255	480	6.6	Mariner and others, 1977
Box Elder County												
Crystal (Madsen's) Hot Springs												
SE, sec. 29, T. 11 N., R. 2 W.	56	7.3	26	784	186	13600	654	371	22600	444	19	Mundorff, 1970
Udy Hot Springs												
NW, sec. 23, T. 13 N., R. 3 W.	43	7.9	26	212	55	2690	118	366	4470	90	1.5	Mundorff, 1970
Juab County												
Baker Hot Springs												
SE, sec. 10, T. 14 S., R. 8 W.	85	7.36 @61°C	66	360	54	860	58	150	1550	720	2.9	Mariner and others, 1977
Millard County												
Meadow Hot Springs												
SW, sec. 26, T. 2 S., R. 6 W.	41	7.5	47	433	144	1040	13.8	408	1800	1130	-	Mundorff, 1970
Salt Lake County												
Becks Hot Spring												
SE, sec. 14, T. 1 N., R. 1 W.	56	7.4	32	746	131	4250	156	221	7470	985	3.3	Mundorff, 1970
Toole County												
Wilson Springs												
sec. 33, T. 10 S., R. 14 W.	61	7.4	33	741	224	7090	18	178	11900	1560	4.0	Mundorff, 1970
Weber County												
Ogden Hot Springs												
SW, sec. 23, T. 6 N., R. 1 W.	56	6.87	52	330	7.2	2700	350	196	4950	93	3.3	Unpub. data, Mariner and others
Utah Hot Springs												
SE, sec. 14, T. 7 N., R. 2 W.	57	6.28	38	890	24	5900	790	206	11500	180	3.7	Unpub. data, Mariner and others

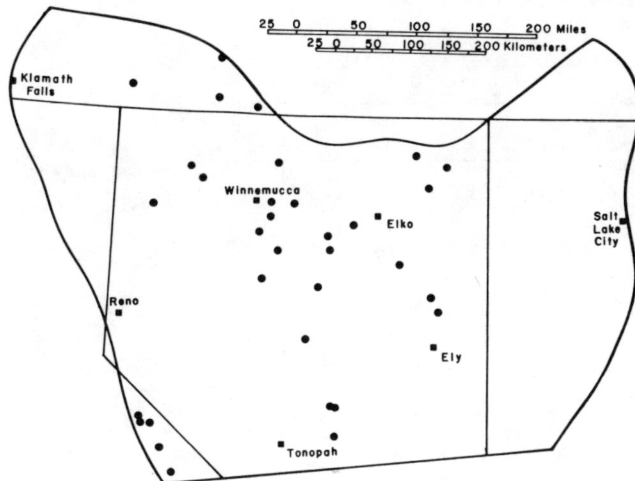

Figure 3. Distribution of Na-HCO₃ waters in the northern Basin and Range Province.

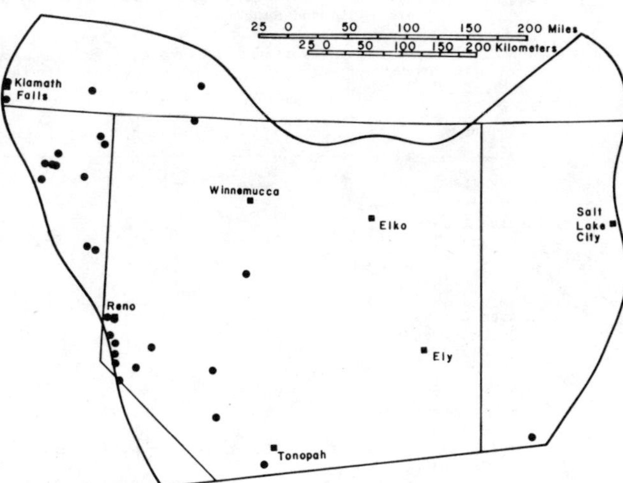

Figure 4. Distribution of Na-SO₄ waters in the northern Basin and Range Province.

discharges sulfate as enriched in oxygen-18 as gypsum of marine origin (+15 o/oo). Either marine gypsum is not available in areas where thermal springs develop or, more likely, the oxygen isotopic composition of the sulfate is being or has been reset in high temperature thermal-environments such as the intrusion of the Sierra Nevada Batholith or the development of mineralized zones. Sulfate-rich waters in some areas, such as northeastern California, are very constant in chemical composition (Table 1). The dissolved sulfate concentration in the water is apparently controlled by the solubility of a common mineral, probably hydrothermal anhydrite.

Mixed anion waters commonly develop in areas where the predominant bedrock or depth of circulation is changing. For instance, the Na-mixed anion waters common in northwestern Nevada west and northwest of the Black Rock Desert, represent a transition zone where rock type and depth of circulation of the thermal waters is changing. Instead of deep circulation in Mesozoic marine strata, which is common to the east and southeast, shallower circulation in predominantly unaltered volcanic rocks of Cenozoic age is more prevalent. The supposition of shallower circulation is evidenced by cooler spring temperatures and lower geothermometer temperatures. Sulfate is a major anion only locally where faults intersect mineralized areas, probably in the Mesozoic rock.

Many of the springs in the Basin and Range Province have been analyzed several times over the last 100 years. These analyses are always slightly different and it is not possible to determine how much of the variation is due to improved analytical techniques and how much to actual changes in the chemical composition with

time. Cole (1982) demonstrated that the major constituents of Becks Hot Springs and Wasatch Hot Springs in Utah varied by as much as 15 % during a single year. This variation was caused by dilution on six and three month cycles. Long-term variations certainly occur in many hot springs but these have not been documented.

Chemical Compositions of Gases

Most thermal springs discharge gas along with the water at rates which range from low to high. Chemically these discharges include nitrogen and/or carbon dioxide with lesser amounts of argon, methane, hydrogen, helium, and hydrogen sulfide (Table 2). These gases probably originate from the atmosphere (N_2 and and Ar_2), soil (CO_2), radiogenic processes (He and Ar), and metamorphic or volcanic processes (CO_2). Ratios of nitrogen to argon range from 1377/1 to 33/1 although most have N_2/Ar ratios are near the 50/1 expected from an atmospheric source. Organic decay product nitrogen is present in at least one sample (Wedell Spring – N_2/Ar = 131/1). Methane concentrations are low, indicating that breakdown of organic material is contributing relatively little at most springs. Hydrogen concentations are occasionally well above that expected from an atmospheric source and indicate that hydrogen is being generated at depth. Detectable(>0.005 %) hydrogen concentrations occur only where high temperature systems are indicated by geothermometry. Helium concentrations are more than an order of magnitude higher than expected from an atmospheric source in most springs and are almost certainly due to radiogenic decay of uranium, thorium, and/or their daughter products. Carbon dioxide makes up 99% or more of the gas phase in the CO_2-charged slightly to moderately saline Na-HCO₃ (± Cl) waters.

Table 2. Compositions of gas discharging from thermal springs, fumaroles and wells in the
Northern Basin and Range.

[Compositions reported in volume %]

Name	O_2	Ar	N_2	CO_2	CH_4	C_2H_6	He	H_2	H_2S	Total
CALIFORNIA										
Alpine County										
Unnamed Spr. E. Fk. of the Carson River (9/3/81)	.31	.33	30.51	68.59	.52	<.01	.03	<.005	–	100.29
Lassen County										
Zamboni Hot Springs (9/3/81)	.93	1.26	97.45	.007	.34	<.01	.025	<.005	–	100.01
Mono County										
Hot Creek Gorge (5/29/80)	<.02	.02	1.00	99.83	.02	<.01	.005	.005	<.04	100.88
Fales Hot Springs (11/5/77)	.23	.08	4.27	94.86	<.005	<.05	<.02	<.01	–	99.44
Mammoth Mountain fumarole (7/28/82)	.01	.11	8.27	91.22	.008	<.01	<.005	.010	<.02	99.63
Travertine Hot Springs (11/5/77)	.02	<.02	.15	99.17	<.005	<.05	<0.02	<.01	–	99.34
Unn. Spring S. Mono Lake (11/5/77)	.18	.02	1.05	99.27	<.005	<.05	<0.02	<.01	–	100.52
NEVADA										
Churchill County										
Brady #8 (7/6/79)	.1	1.31	90.14	2.48	2.63	.03	.01	2.94*	<.01	99.54
Dixie Valley well Lamb 4 (5/29/80)	<.02	.04	2.15	97.14	.99	.03	.005	.05	.13	100.80
Elko County										
Hot Sulphur Springs (Tuscarora) (5/30/80)	<.02	.12	3.95	95.54	.34	<.01	.005	.02	–	100.71
Sulphur Hot Springs (Elko) (5/31/80)	<.44	.46	25.31	74.58	.60	<.01	.02	<.005	<.04	100.41
Unnamed Spring on Mary's River (8/11/82)	.03	.91	60.88	37.84	.014	<.01	.170	<.005	–	99.84
Eureka County										
Beowawe (7/31/82)	<.02	1.69	85.40	7.67	4.72	.11	.130	.290	<.02	100.01
Hot Spring near Walti (7/31/82)	.03	.54	27.85	70.71	.47	<.01	.020	<.005	–	99.62
Humboldt County										
Golconda Hot Spring (8/4/82)	.15	.84	39.58	58.09	.97	<.01	.030	<.005	–	99.66
Hot Pot (8/3/82)	.07	.74	48.74	50.06	.16	<.01	.11	<.005	–	99.88
Macfarlanes Hot Spring (8/3/82)	.02	<.02	.09	99.26	.034	<.01	.005	.005	–	99.40
Lander County										
Smith Creek Valley (7/31/82)	.20	1.95	90.31	6.13	1.32	<.01	.050	.045	–	100.00
Mineral County										
Wedell Springs @ (7/30/82)	<.5	.7	92	8.1	.95	<.2	.08	<.05	–	101.83
Pershing County										
Leach Hot Springs (5/--/77)	.03	1.57	73.49	24.53	.90	<.05	.05	<.01	–	100.57

@ Very low pressure sample
* Hydrogen probably from high temperature steam reaction with steel pipe

''Fumaroles'' occur along the west side of Dixie Valley, at Fumarole Butte near Baker Hot Springs in Utah, and at Mammoth Mountain and Casa Diablo in Long Valley, California (Berry and others, 1980). The fumaroles in Dixie Valley in Nevada discharge water vapor mixed with air (Mariner and Evans, unpublished data). The fumarole on Mammoth Mountain discharges mostly CO_2 with minor amounts of nitrogen and traces of hydrogen (Table 2). Chemically the gas discharged by the ''fumarole'' is more like the gases discharged from hot springs in the Hot Creek Gorge part of Long Valley than a fumarole associated with a volcano (analysis of a fumarole on Mt. Hood is included in Table 2 for comparison).

Chemical Geothermometers

The primary use of chemical data in geothermal exploration has been to estimate the temperature of the deep thermal-aquifer associated with a hot spring. The most useful geothermometers for estimating aquifer-temperatures are either the quartz geothermometer of Fournier and Rowe (1966), the Na-K-Ca geothermometer of Fournier and Truesdell (1973), or the Mg-corrected Na-K-Ca geothermometer of Fournier and Potter (1979). Other means of estimating aquifer-temperatures include the sulfate-water isotope geothermometer (McKenzie and Truesdell, 1977), the gas geothermometer (D'Amore and Panichi, 1980), the Na/K geothermometer (Fournier, 1979), and in a very few systems, the solubilities of minerals such as anhydrite (Sakai and Matsubaya, 1974).

Geothermometer temperatures based on the chemical composition of water discharged from hot springs and shallow wells are shown on Table 3. The high temperature areas in the northern Basin and Range Province ($>150°C$ in Table 3) include: Long Valley, Seyferth Hot Springs, unnamed springs on the East Fork of the Carson River in California; Raft River, Idaho; Baltazor Hot Springs, Beowawe, Great Boiling Springs, Hazen, Hot Springs Ranch (Tipton), Hot Sulphur Spings (Tuscarora), Humboldt House, Leach Hot Springs, Lee Hot Springs, Pinto Hot Springs, San Emidio Desert, Soda Lake-Upsal Hogback, Steamboat Springs, Stillwater and Sulphur Hot Springs in Nevada; Alvord Hot Springs, Crump, Hot Lake, Hunters Hot Springs, and Mickey Hot Springs in Oregon; and Roosevelt Hot Springs in Utah. By types of waters, 10 are Na-Cl, 8 are Na-HCO_3, 4 are Na-mixed anion waters, and 3 are Na-SO_4 waters. '' Successful'' geothermal wells have been drilled at roughly half of the Na-Cl discharging springs but only at one of the Na-HCO_3 discharging springs (Table 4). ''Successful'' high-temperature geothermal wells have not been drilled near any of the Na-mixed anion or Na-SO_4 springs.

Plots of t_{silica} and $t_{Na-K-Ca}$ (Fig. 5) show generally good agreement, with relatively few large disparities. Arbitrarily, the quartz

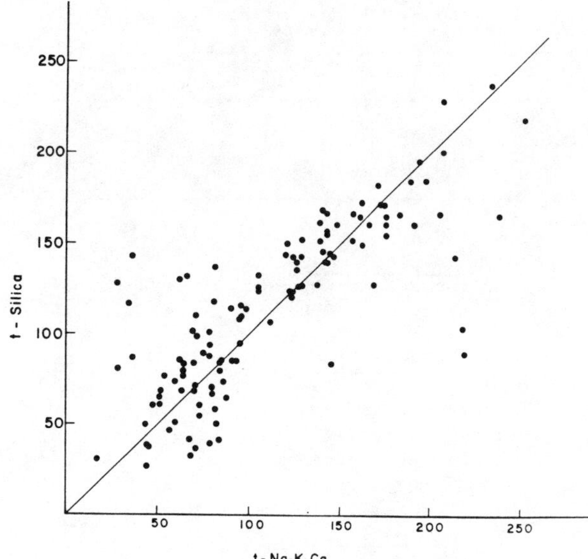

Figure 5. Comparison of temperatures estimated from t_{silica} and $t_{Na-K-Ca}$.

geothermometer was used when the Na-K-Ca geothermometer indicated a temperature of $100°C$ or more, the chalcedony geothermometer was used when the Na-K-Ca geothermometer indicated a temperature of less than $100°C$. The few large disparities occur where waters discharge from silicic tuffs, CO_2-charged waters, waters contaminated with high-chloride saline lake or playa waters, and dilute high-pH waters. Springs issuing from silicic tuffs such as Hot Creek Ranch Springs in Nye County, Nevada and CO_2-charged water such as the water discharged by Travertine and Fales hot springs in California have higher temperatures estimated from the silica geothermometer than from the Na-K-Ca geothermometer. Silica is apparently being taken into solution faster than quartz or chalcedony can be precipitated, supersaturation with respect to quartz or chalcedony is maintained and the quartz (or chalcedony) geothermometer give excessively high subsurface temperature estimates. The magnesium corrected Na-K-Ca geothermometer in these waters generally gives estimated aquifer-temperatures within $25°C$ of the measured spring temperature. High CO_2 concentrations generally require high temperatures for generation, but these conditions may be very deep (Barnes and others, 1978). Some of the Na-Cl waters issue near saline lakes or playas and may contain some admixed saline lake waters (Utah Hot Springs adjacent to Great Salt Lake is an example). Since the saline water contains almost no calcium or magnesium, abnormally high Na-K-Ca geothermometer temperatures are calculated. The Na/K geothermometer is no better since the proportion of Na to K in the saline water was controlled initially by reactions which included calcium. Finally, the dilute high-pH waters

Table 3. Geothermometer temperatures for springs of the Northern Basin and Range.

[All temperatures in $^{o}C.$]

Name	Silica	Geothermometers		Measured	
		Na-K-Ca	SO_4-H_2O	Surface	Depth
CALIFORNIA					
Alpine County					
Unn. sprs. on the Carson River	172	175		84	
Unn. sprs. on the Carson River	146	140		65	
Lassen County					
Amadee Hot Springs	109	95		96	
Bassett Hot Springs	86	62		79	
Kellog Hot Springs	101	78		78	
Wendell Hot Springs	150	121		96	122
Zamboni Hot Springs	(65)	51		41	
Modoc County					
Hot Springs Motel	110	96	200	98	
Kelly Hot Springs	116	95	198	92	
Leonards Hot Springs	143	124		62	
Little Hot Springs	102	69		79	
Seyforth Hot Springs	143	129	205	85	
West Valley Reservoir	152	129	247	77	
Mono County					
Benton Hot Springs	40	79		56	
Fales Hot Springs	118	81	684	61	
Long Valley-Hot Creek Gorge	161	191	224	90	
Mono Lake - North Shore	94	79		66	
- South Shore	126	28		33	
Travertine Hot Springs	110	71	173	69	
IDAHO					
Bear Lake County					
Bear Lake Hot Springs	55	73		48	
Unn. spring	39	44		42	
Cassia County					
RRGE-1	159	171		145	142
Franklin County					
Maple Grove Hot Springs	77	64		76	
Wayland Hot Springs	125	105		77	
Oneida County					
Woodruff Hot Springs	47	57		27	
NEVADA					
Carson City					
Carson Hot Springs	insufficient data			49	
Pinyon Hills Well	insufficient data			46	
Churchill County					
Brady's Hot Spring (well)	167	157	165	–	188
Dixie Valley Hot Springs	145	144	127	72	
Dixie Federal 52-18	229	207	268	191 gas geot.	
Eagle Salt Works Spring	198	–		–	
Lee Hot Springs	173	162	282	88	
Soda Lake-Upsal Hogback	165	161	127	100	188
Stillwater Area	169	140	177	96	178

Table 3. Geothermometer temperatures for springs of the Northern Basin and Range –
continued.

[All temperatures in ^{o}C.]

| Name | Silica | Geothermometers | | Measured | |
		Na-K-Ca	SO_4-H_2O	Surface	Depth
Douglas County					
Hobo Hot Spring	69	70		46	
Saratoga Hot Springs	31	–		50	
Walley's Hot Spring	80	84		62	
Elko County					
Hot Creek Springs	31	17		26	
Hot Creek Springs	132	66		37	
Hot Hole	86	85		56	
Hot Lake	79	67		18	
Hot Spring near Carlin	90	75		79	
Hot Sulphur Spring	167	183	175	92	117
Humboldt Wells Area	117	34		60	
Mineral (Contact) Hot Springs	127	129		60	
Nile Spring	50	43		43	
Rizzi Ranch Hot Spring	37	71		41	
San Jacinto Ranch Spring	27	44		–	
Sulphur Hot Spring (Ruby Valley)	183	181	161	93	
Smith Ranch	72	71		65	
Trout Creek Ranch Well	33	69		43	
Unnamed spring (Rock Creek)	37	69		35	
Wild Horse Hot Spring	61	73		54	
Wine Cup Ranch Well	insufficient data			59	
Esmeralda County					
Alkali Springs	insufficient data			60	
Silver Peak (Waterworks) Hot Springs	140	142		40	
Eureka County					
Beowawe	196	194	251	98	201
Bruffey's (Mineral Hill) Hot Spring	80	64		66	
Hot Springs Point	87	36		54	
Klobe Hot Springs	69	72		54	
Shipley Hot Springs	61	48		39	
Walti Hot Springs	88	78		72	
Humboldt County					
Baltazor Hot Springs (well)	161	148	158	90	
Bog Hot Springs	65	88		54	
Cordero Mercury Mine well	108	–		60	
Double Hot Springs	140	126		80	
Dyke Hot Springs	128	137		66	
Golconda Area	86	92		74	
Hot Pot	59	81		57	
Hot Spring Ranch (Tipton)	150	162		85	
Howard Hot Springs	71	80		56	
Macfarlane's Bath House Spring	99	71		75	
Pinto Hot Springs	161	176	207	93	
Soldier Meadows Hot Springs	84	64		54	
The Hot Springs	77	54		58	
Well	–	81		70	
Lander County					
Buffalo Valley Hot springs	119	126	140	73	
Hot Springs Ranch(Valley of the Moon)	61	51		53	
South Smith Creek Valley	143	156	143	86	
Spencer Hot Springs	95	95		72	
Unn. spring near Walti Hot Spring	127	129		64	

Table 3. Geothermometer temperatures for springs of the Northern Basin and Range – continued.

[All temperatures in ^{o}C.]

Name	Silica	Geothermometers Na-K-Ca	SO_4-H_2O	Measured Surface	Depth
Lyon County					
Hazen Area	161	166	220	86	
Hind's (Nevada) Hot Springs	74	86		61	
Wabuska Hot Springs	143	146	140	94	108
Wedell Springs	162	139		60	
Nye County					
Darrough's Hot Spring	136	126		95	129
Diana's Punch Bowl	67	80		59	
Hot Creek Ranch Springs	130	62		63	
Upper Hot Creek Ranch Springs	143	36		67	
Warm (Nanny Goat) Spring	81	29		61	
Pershing County					
Colado	128	169		61	155
Humboldt House (Rye Patch Reserv.)	219	252	176	–	156
– artesian well	166	238		77	
Hyder Hot Springs	84	70		78	
Kyle Hot Springs	137	81	154	77	
Leach Hot Springs	155	176	176	92	126
Sou Hot Springs	85	84		70	
Trego Area	124	124		84	
Washoe County					
Bowers Mansion Hot Springs	38	45		46	
Fly Ranch (Ward's)	126	105		80	
Great Boiling Spring	167	205	93	86	
Lawton Hot Springs	84	145		49	
Moana Springs Area (Biglin well)	114	98		85	
San Emidio Desert	185	189		89	115
Steamboat Springs	201	207	207–220	94	208
The Needle Rocks	143	214		56	
White Pine County					
Cherry Creek Hot Springs	114	90		61	
Monte Neva Hot Springs	74	60		79	
OREGON					
Harney County					
Alvord (Indian) Hot Springs	152	157	231	78	
Crane Hot Springs	124	124		78	
Mickey Hot Springs	185	197	273	86	
Trout Creek Hot Springs	140	143	235	52	
Unn. spr. near Hot Lake	165	176	231	96	
Unn. spr. near Harney Lake	133	105		68	
Lake County					
Barry Ranch Hot Springs	152	139		88	
Crump	172	173	202	78	
Fisher Hot Springs	123	123		68	
Hunters Hot Springs	157	143	158	96	
Summer Lake Hot Springs	107	112	189	43	
UTAH					
Beaver County					
Roosevelt Seep	167	142	216	25	208
Roosevelt Steam Well	238	234Na-K	278	208	
Thermo Hot Springs	144	120	142	90	

Table 3. Geothermometer temperatures for springs of the Northern Basin and Range –
continued.

[All temperatures in ^{o}C.]

| Name | Silica | Geothermometers | | Measured | |
		Na–K–Ca	SO_4–H_2O	Surface	Depth
Box Elder County					
Crystal (Madsen's) Hot Springs	42	84		56	
Udy Hot Springs	42	68		43	
Juab County					
Baker Hot Springs	86	91	22	85	
Millard County					
Meadow Hot Springs	69	63		41	
Salt Lake County					
Beck's Hot Springs	51	82		56	
Toole County					
Wilson Hot Springs	52	60		61	
Weber County					
Ogden Hot Springs	104	218		56	
Utah Hot Springs	90	219		57	

Table 4. Geothermal Systems with Estimated Reservoir-Temperatures >150°C

Na–HCO$_3$ Waters	Na–Mixed Anion Waters	Na–Cl Waters
*Beowawe	Alvord Hot Springs	Crump
Hot Springs Ranch	Hot Lake (Alvord Desert)	Great Boiling Spring
Hot Sulphur Springs	Lee Hot Springs	Hazen
(Tuscarora)	Unn. Springs-Carson River	*Humboldt House
Leach Hot Springs		*Raft River
Long Valley	**Na–SO$_4$ Waters**	*Roosevelt
Mickey Hot Springs		San Emidio Desert
Pinto Hot Springs	Baltazar Hot Springs	*Soda Lake-Upsal Hogback
Sulphur Hot Springs	Hunters Hot Springs	*Steamboat Springs
(Ruby Valley)	Seyferth Hot Springs	*Stillwater

*Locations of "successful" geothermal wells.

such as Zamboni or Benton hot springs which
discharge from granites near the contact of the
Basin and Range with the Sierra Nevada contain
unusually large silica concentrations due to
dissociation of silicic acid (H_4SiO_4 to $H_3SiO_4^-$
and $H_2SiO_4^=$). With one of the computer codes
such as SOLMINEQ (Kharaka and Mariner, 1977) the
temperature at which the thermal water is in
equilibrium with chalcedony or quartz, as
appropriate, can be determined. These values
are enclosed in parentheses in Table 3.

The apparent agreement between the
temperatures estimated from the silica and
cation geothermometers in most waters of the
Great Basin could be fortuitous. A more
important question is, how do the estimated
temperatures compare with measured temperatures
in geothermal wells? Deep-well temperature data
are available for only 15 systems (Table 5).
Surprisingly, when measured and estimated
temperatures are compared for all 15, the
measured temperatures are, on the average,
only 14°C cooler than the estimated temperatures. The standard deviation is however,

Table 5. Expected and Measured Temperatures of Geothermal Systems
in the Northern Basin and Range.

Name	Temperatures	
	Expected*	Measured
High Discharge Springs		
Beowawe	214	201
Hot Sulfur Springs (Tuscarora)	175	117
Leach Hot Springs	167	126
Long Valley	196	170
Steamboat Springs	205	204
Wendel Hot Springs	136	122
Low Discharge Springs		
Humboldt House	184	156
Roosevelt Hot Springs	175	208
San Emidio Desert	187	115
Wells		
Brady's Hot Springs	162	155
Colado	148	155
Raft River	161	150
Soda Lake	151	188
Stillwater	162	178
Wabuska	143	108

*Average of t_{silica}, t_{cation}, and $t_{SO_4-H_2O}$ when available.

rather large ($\pm$ 28 C). The estimated temperature for each system is an average which includes values from the quartz or chalcedony geothermometer, as appropriate, the Na–K–Ca geothermometer, and when available, the SO_4-H_2O isotope geothermometer. Surprisingly, the average difference between expected temperatures calculated from geothermometry and the measured temperatures for low discharge rate ($<$100 L/min) and high discharge rate hot springs are nearly the same (22°C vs 26°C). Normally, high discharge rate springs should give better estimates of subsurface temperature than low discharge rate springs (Fournier and others, 1974). This apparent anomaly may be due, in part, to the small number of systems for which data is available, and, in part, may indicate that the accessible part of the thermal reservoirs may be smaller than previously thought. Long Valley is the best example of an area where recent drilling has shown that the accessible part of the thermal reservoir is smaller than predicted in the last assessment of high–temperature of geothermal resources of the United States (Brook and others, 1978). If geothermal reservoirs are appreciably smaller than anticipated by Brook and others (1978), geothermal development in the northern Great Basin may be seriously curtailed. The six areas where thermal waters were collected from shallow wells (50 – 1000 feet deep) gave remarkably good agreement between estimated and measured temperatures (less than 2°C average difference).

The disparity between estimated and observed temperatures is considerably larger for the areas where data is available only from hot springs, measured temperatures in geothermal wells average 24°C cooler than the estimated temperatures.

Isotopic Data

Craig (1963) demonstrated that for high-temperature chloride-rich waters or steam discharges, the deuterium content is approximately equal to that of the local meteoric water. Nehring (1979) reached a similar conclusion in a restudy of the high-temperature chloride-rich system at Steamboat Springs as did Mariner and Wiley (1976) in a study of the Long Valley area. However, other detailed studies of recharge areas for specific systems in the northern Basin and Range Province have generally had difficulty locating cold spring waters in the adjacent highlands which were as depleted isotopically as water discharged by the thermal springs. Welch and others (1981) concluded that the water discharged at Leach Hot Springs was paleo-water which recharged during a colder time-period since it was more depleted in terms of deuterium than any of the water discharged by cold springs in the adjacent mountain ranges. Summit Spring (Table 6; $\delta D = -126.8$ o/oo), a perenial spring on the north side of Mt. Tobin near Leach Hot Springs, is as depleted in deuterium as the thermal water discharging at Leach if you assume that the small oxygen shift (+0.3 o/oo δ^{18}O) observed in the cold spring water is due to nonequilibrium evaporation (Fig. 6). This evaporation could have taken place prior to recharge, or in an unconfined aquifer. Although Mt. Tobin is the highest mountain in the region, and could be the recharge area for Leach Hot Springs, this interpretation forces the data to the limits of credibility, only a small summit area on Mt. Tobin exists as a catchment basin and, most damaging, two travertine depositing springs, Buffalo Valley Hot Springs east of Mt. Tobin and Hyder Hot Springs in Dixie Valley south of Mt. Tobin are even more depleted in deuterium (-132 per mil and -133 per mil, respectively).

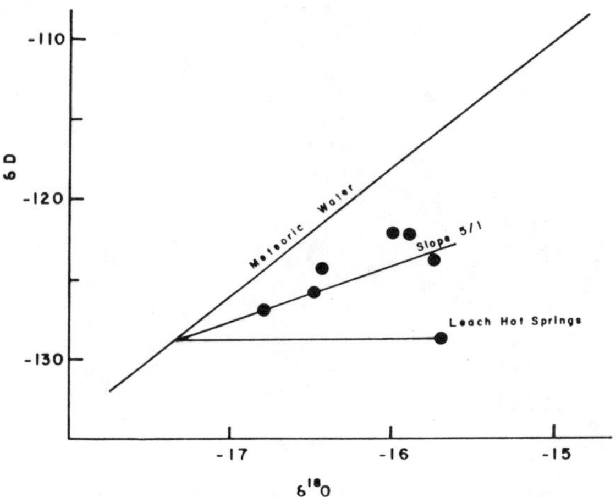

Figure 6. Isotopic composition of thermal and cold waters in the Leach Hot Springs Area.

A detailed study of the Tuscarora Area (Hot Sulphur Springs) at the west end of the Jarbridge Mountains (Bowman and Cole, 1982) also did not find any cold springs as depleted in deuterium as the hot springs. Most of the cold spring water however, appear to have undergone some nonequilibrium evaporation. A study of Ogden and Utah hot springs near Great Salt Lake by Cole (1982) also showed that the thermal waters were more depleted in deuterium than nearby cold springs.

Are these cases the exception or the rule? When the deuterium contents of hot springs in the northern Basin and Range Province (Table 6) are plotted and contoured, the map produced (Fig. 7) is similar to the map for meteoric waters of North America as presented in Taylor (1974). However, based on the map of Taylor (1974), most precipitation in the northern Great Basin should range in deuterium composition from about -110 to -130 o/oo. Most of the hot springs however are in the -120 to -140 o/oo range (figure 7), and a few near Elko are even more depleted (-145 o/oo δD). Appealing to the presence of regional aquifers is not valid since

the nearest area with precipitation as depleted as -145 o/oo is in western Montana. Since the isotopic composition of precipitation is a function of temperature, δD = 5.56t - 98.5 (Dansgaard, 1964), the water being discharged by most of the hot springs in the northern Basin and Range Province apparently recharged during times of colder climate (2 to 3°C colder than at present). Although there have been recent fluctuations in mean annual air temperature with a maximum about 1930, and several recent periods of colder climates (1900-1800, 1700-1575, and 1525-1400) the temperatures were not cold enough to account for the 15 o/oo difference in deuterium observed in most of the Basin and Range. The work of Dansgaard and others (1969) and Johnson and others (1970) has shown that the isotopic composition of precipitation changed from more depleted values to roughly modern values abruptly about 10,000 years ago at the ''end'' of the Pleistocene. Thus most of the water currently discharged by hot springs in the northern Great Basin, apparently recharged at least 10,000 years ago.

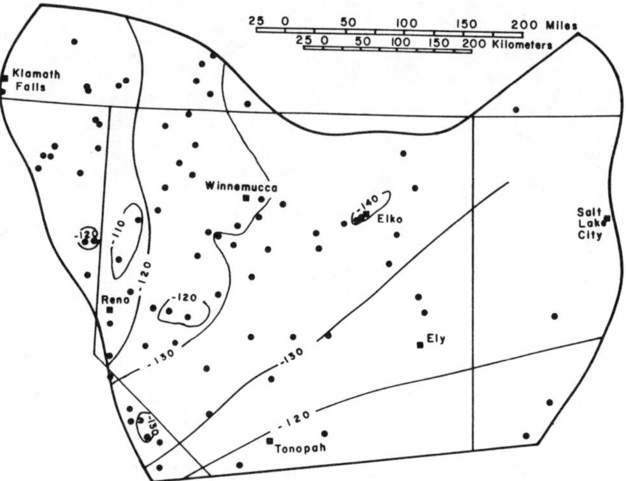

Figure 7. Contour map of the deuterium composition of thermal springs in the northern Basin and Range Province.

Unfortunately, times of circulation for geothermal waters are generally unknown, due in part to the difficulty and uncertainty in interpreting tritium or carbon-14 values. Tritium has a half-life of only 12.3 years and generally concentrations of tritium in thermal waters are less than 0.1 TU. This merely shows that the waters are more than 60 years old (Barnwell, 1963; Wilson, 1963). Carbon-14 has a half-life of 5,670 years and should be much more useful than tritium. However, carbon-14 readily exchanges with reservoir carbon at high temperatures (Craig, 1963).

In contrast to deuterium, which generally does not change in concentration during passage through a geothermal system (boiling excepted), the oxygen isotopic composition of the water

Table 6. Isotopic data for thermal springs of the northern Basin and Range.

	del D	del 18O
CALIFORNIA		
Alpine County		
Unn. sprs. E.Fk. Carson River		
(84°C spring)	-126.5	-15.56
(65°C spring)	-125.3	-15.54
Lassen County		
Amadee Hot Springs	-120.0	-
Bassett Hot Springs	-115.1	-13.54
Kellog Hot Springs	-115.5	-14.09
Wendell Hot Springs	-120.8	-14.04
Zamboni Hot Springs	-118.1	-15.34
Modoc County		
Lake City Mud Eruption	-113.0	-14.79
Kelly Hot Springs	-115.1	-13.54
Little Valley Hot Springs	-116.9	-14.20
Hot Springs Motel (well)	-117.0	-13.81
Menlo Hot Springs	-112.3	-15.30
Seyforth	-121.2	-14.05
Mono County		
Benton Hot Springs	-135.5	-17.46
Fales Hot Springs	-132.8	-17.46
Long Valley (Hot Creek Gorge)	-120.3	-14.83
- LDMW	-115 to -130	
Mono Lake - North Shore	-126.6	-16.91
- South Shore	-126.9	-15.69
The Hot Springs	-137.3	-16.29
Travertine Hot Springs	-139.3	-16.64
NEVADA		
Churchill County		
Brady's Hot Spring (well)	-121.2	-14.22
Dixie Federal 52-18	-133.9	-14.72
Dixie Valley Hot Springs	-126.1	-15.89
- LDMW	-120.0	-15.22
Lee Hot Springs	-125.8	-13.34
Stillwater Area (well)	-110.0	-12.36
Douglas County		
Walley's Hot Springs	-119.5	-15.55
Elko County		
Hot Creek	-126.7	-16.28
- LDMW	-121.4	-15.69
Hot Creek Springs	-135.7	-17.40
- LDMW	-128.9	-16.20
Hot Hole	-144.7	-15.31
Hot spring near Carlin	-132.7	-16.64
Hot Sulfur Spring (Tuscarora)	-138.6	-16.65
Humboldt Wells	-134.7	-
- LDMW	-122.1	-15.81
Mineral (Contact) Hot Spring	-139.0	-17.61
Nile Spring	-139.1	-18.24
Smith Ranch Hot Spring	-132.8	-16.24
Sulphur Hot Spring (Ruby V.)	-130.1	-16.09
- LDMW	-124.6	-16.87
Sulphur Hot Spring (Elko)	-145.9	-17.67

Table 6. Isotopic data for thermal springs of the northern Basin and
Range--continued.

	del D	del 18O
Esmeralda County		
Silver Peak (Water Works spr.)	-118.2	-13.50
Eureka County		
Beowawe	-130.0	-14.76
Hot Springs Point	-136.1	-15.97
Klobe Hot Springs	-127.9	-16.28
Walti Hot Springs	-129.8	-16.87
Humboldt County		
Baltazor Hot Springs	-125.3	-15.26
Bog Hot Springs	-124.3	-15.30
Double Hot Springs	-128.8	-15.93
Dyke Hot Springs	-128.0	-16.29
Hot Pot	-136.7	-16.70
Hot Springs Ranch (Tipton)	-131.4	-15.74
Howard Hot Spring	-127.1	-16.17
Macfarlanes Hot Springs	-127.2	-12.54
Pinto Hot Springs	-129.2	-14.48
Soldier Meadows Hot Springs	-129.2	-16.56
The Hot Springs	-134.6	-16.44
Lander County		
Buffalo Valley Hot Springs	-131.6	-15.85
	-135.2	-13.61
- LDMW	-117.3	-14.95
Hot Springs Ranch		
(Valley of the Moon)	-127.8	-16.28
Smith Creek Valley	-130.4	-16.68
Spencer Hot Spring	-135.8	-16.01
Unn. Spr. (Grass V. nr Walti)	-134.8	-16.73
Lyon County		
Hazen Area	-121.5	-13.30
Hinds (Nevada) Hot Springs	-123.2	-16.01
Wabuska Hot Springs	-129.7	-15.38
Wedell	-131.9	-15.90
Mineral County		
Soda Springs	-130.3	-16.13
Nye County		
Diana's Punchbowl	-124.9	-16.24
Darroughs Hot Springs	-122.5	-15.50
Pershing County		
Colado	-125.5	-14.01
Hyder Hot Springs	-133.2	-15.66
Humboldt House - deep well	-130.6	-14.64
- shallow well	-127.2	-14.09
- LDMW	-119.9	-15.25
Kyle Hot Springs	-130.0	-15.50
- LDMW	-121.1	-14.71
Leach Hot Springs	-128.6	-15.70
Summit Spring - LDMW	-126.8	-16.80
Trego Hot Springs	-124.5	-14.40

Table 6. Isotopic data for thermal springs of the northern Basin and Range--continued.

	del D	del 18O
Washoe County		
Bowers Mansion Hot Springs	-102.3	-14.79
Fly Ranch	-120.7	-14.72
Great Boiling Springs	-100.5	-10.83
San Emidio Desert	-108.3	-12.05
Steamboat Springs	-116.7	-12.16
The Needle Rocks	-106.5	- 6.33
White Pine County		
Cherry Creek Hot Springs	-127.8	-16.20
Monte Neva Hot Springs	-127.8	-16.68
OREGON		
Harney County		
Alvord (Indian) Hot Springs	-123.6	-13.23
Crane Hot Springs	-133.3	-16.17
Mickey Hot Springs	-124.3	-13.42
Pike Creek - LDMW	-108.4	-14.05
Trout Creek Hot Spring	-127.4	-16.17
Trout Creek- LDMW	-115.3	-15.50
Unn. Sprs. near Hot Lake	-125.4	-14.36
Unn. H. Sprs near Harney Lake	-128.5	-
Lake County		
Barry Ranch Hot Springs	-119.4	-13.72
Crane Creek- LDMW	-101.2	-13.40
Crump	-115.5	-13.28
Deep Creek - LDMW	-106.6	-13.46
Fisher Hot Springs	-117.0	-
Hunters Hot Springs	-119.0	-14.32
Summer Lake Hot Springs	-115.0	-13.32
Malheur County		
Unn. Spr. nr. McDermitt	-134.6	-16.95
UTAH		
Beaver County		
Roosevelt Seep	-113.0	-12.95
Thermo Hot Springs	-118.3	-14.32
Steam Well at Roosevelt H.S.	-115.9	-12.99
Juab County		
Crater (Baker,Abraham) Hot S.	-126.3	-16.0

changes due to exchange with minerals in the confining rock (Craig, 1963). In the northern Great Basin, dilute $Na-HCO_3$ waters and $Na-SO_4$ waters generally have less oxygen enrichment than Na-Cl waters (Figs. 8, 9, and 10). Bicarbonate-rich waters, however, occasionally have very large oxygen shifts (up to 4 o/oo; Fig. 8). Carbonates in limetsones are usually +20 to +30 per mil in $\delta^{18}O$ while silicates in most igneous rocks range from +5 to +15 o/oo. Larger oxygen shifts are observed in water associated with limestones than in waters associated with silicate rocks because of the concentration difference and the faster exchange rates between carbonates and waters. The lack of correlation between oxygen shift and amount of dissolved solids in the Na-Cl waters is an indication that although these waters generally occur where most water-rock reaction has taken place, the chloride concentration is at least, in part, a function of chloride availability.

The oxygen isotopic compositions of a few sulfates in thermal waters of the Basin and

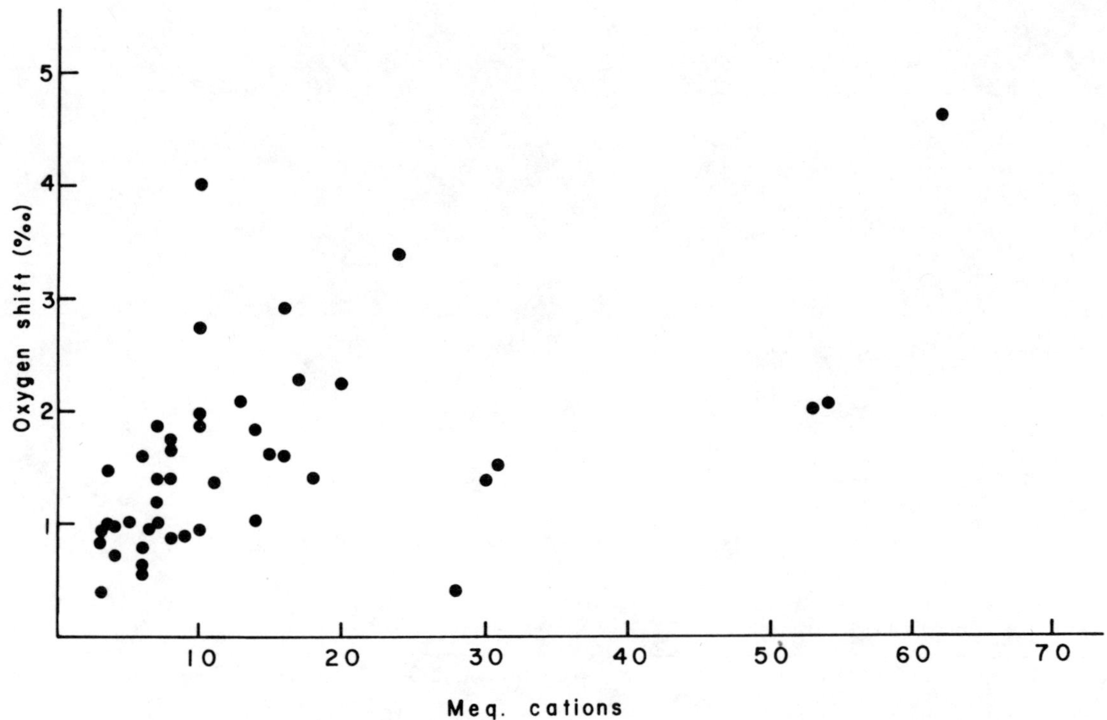

Figure 8. Oxygen shift of HCO_3-rich thermal water as a function of milliequivalents cations (specific conductivity).

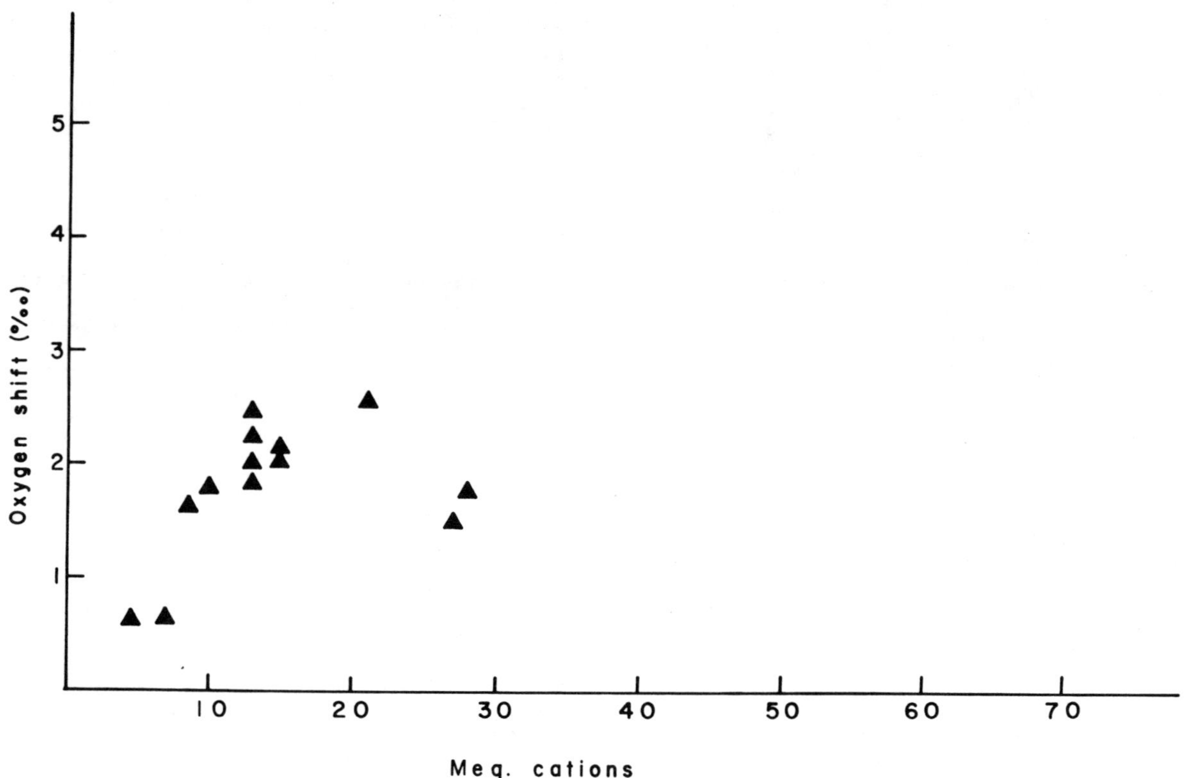

Figure 9. Oxygen shift of SO_4-rich thermal waters as a function of milliequivalents cations (specific conductivity).

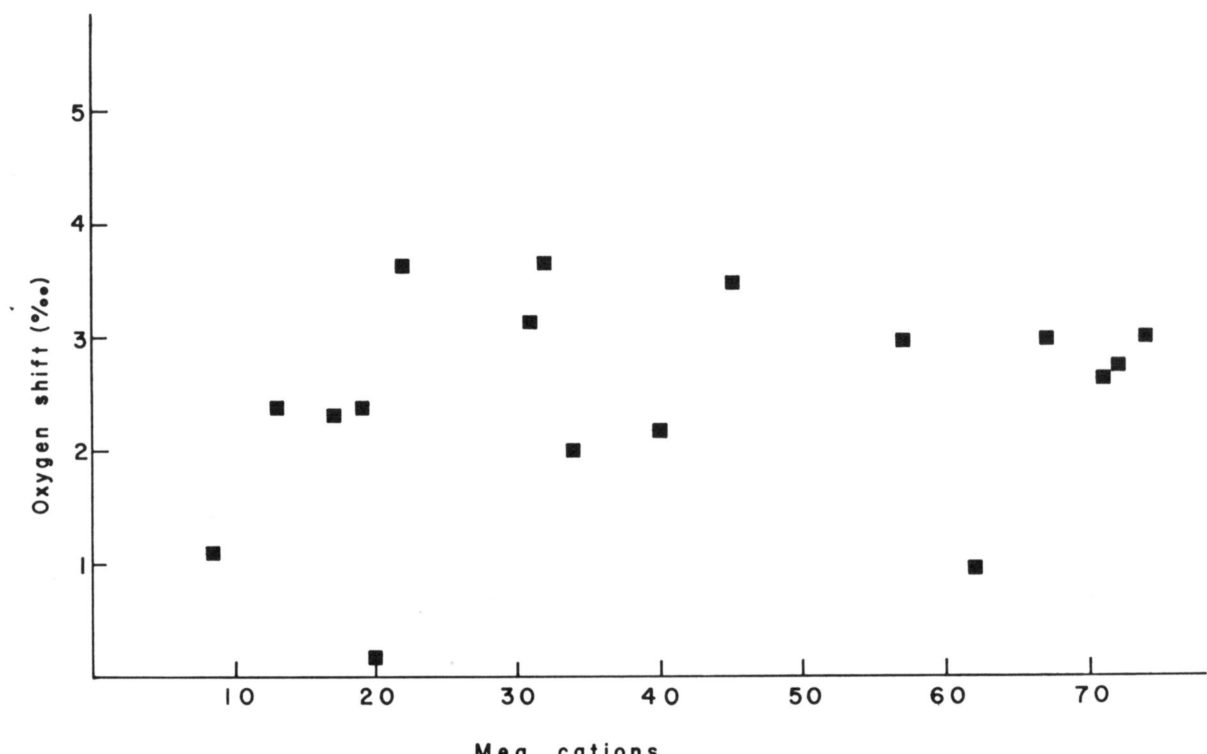

Figure 10. Oxygen shift of Cl-rich thermal waters as a function of milliequivalents cations (specific conductivity).

Range Province have been reported by Nehring and Mariner (1979). Our intent was to estimate the thermal-aquifer temperature using the sulfate-water isotope geothermometer of McKenzie and Truesdell (1977). The sulfate-water isotope geothermometer generally gives calculated temperatures that are slightly hotter than those obtained from the quartz or Na-K-Ca geothermometers. However, in some areas, the sulfate-water isotope geothermometer gives temperatures considerably lower than the measured surface temperatures or considerably higher than the temperatures estimated from the other geothermometers (Table 3). It is possible for sulfate to be dissolved from the country rock after the thermal fluid leaves the thermal aquifer or the thermal fluid may mix with sulfate-rich nonthermal water before it discharges at the surface. This added sulfate probably will have a different original isotopic composition and could significantly change the isotopic composition of the total sulfate in the discharge.

In the northern Basin and Range Province, the sulfate-water isotope geothermometer indicates aquifer-temperatures which are generally 50 to 100°C higher than those estimated from the silica or cation geothermometers. A possible explanation for these higher apparent temperatures is that the depleted sulfate is from dissolution of minerals formed during previous high temperature

hydrothermal or metamorphic events. However, these minerals must be situated along the flow path from the reservoir to the surface or the residence times of fluids in the thermal reservoir must be very short. The latter possibility does not appear likely due to the old apparent age of most of the waters. However, in mineralized areas, there is the possibility of dissolving isotopically depleted sulfate which does not have enough time to attain equilibrium with the dissolving fluid. For example, pickeringite (ideal formula $MgAl_2(SO_4)_4 \cdot 22H_2O$) from a site near Lahontan Reservoir west of Fallon was -6.53 o/oo $\delta^{18}O$. Dissolution of such a mineral without concomitant reequilibration would result in excessively high apparent SO_4-H_2O isotopic equilibrium temperatures. Sulfate minerals are often associated with ore deposits in western Nevada and so isotopically depleted sulfate is readily available.

Alternatively, although the sulfate-water isotope temperatures are no closer to the measured temperatures in the deep wells then the temperatures calculated from the cation or silica geothermometers, the apparent sulfate-water equilibrium temperatures could be correct. Calculations with the computer code SOLMNEQ (Kharaka and Barnes, 1973, as modified by Kharaka and Mariner, 1977) indicate saturation with respect to anhydrite ($CaSO_4$) at temperatures near those estimated from the sulfate–water isotope geothermometer in several of the systems in northeastern California (Table 7). This may be an indication that the temperatures calculated from the sulfate–water isotope geothermometer are accurate in this area. The cooler temperatures estimated from the quartz and Na–K–Ca geothermometer may indicate that chemical equilibrium was approached in a shallow aquifer at temperatures near the spring temperature but the isotopic composition remained unchanged. Anhydrite saturation temperatures (Table 7) for Travertine Hot Springs, Hot Lake (Oregon), Stillwater, Soda Lake–Upsal Hogback, Hazen and Alvord Hot Springs are also reasonably near the sulfate–water isotope equilibrium temperatures. The differences in temperatures estimated from theoretical anhydrite saturation (175°C) and sulfate–water isotopic data (127°C) at the Soda Lake–Upsal Hogback area may indicate that reequilibration or mixing has taken place in a shallow aquifer since deep wells have encountered temperatures more than 50°C hotter. Lee Hot Springs, located south of Fallon, has an apparent anhydrite saturation temperature of only 173°C, almost 100°C cooler than the sulfate–water isotopic equilibrium temperature. The sulfate at Lee must be from a near surface hydrothermal mineral source. At the other extreme, Abraham Hot Springs in Utah had a sulfate–water isotopic equilibrium temperature less than the measured spring temperature. Apparently, the sulfate discharged at Abraham Hot Springs is of marine origin (initially about +15 o/oo in δ^{18}O) and it never attained, isotopic equilibrium with the thermal water.

Summary

Thermal waters in the Basin and Range Province range from dilute $Na–HCO_3$ and $Ca–HCO_3$ waters to very saline Na–Cl waters. The most saline Na–Cl waters occur near Great Salt Lake or near the sinks and playas of northwestern Nevada. Slightly saline CO_2–charged $Na–HCO_3$ waters are common near the Sierra Nevada. $Na–SO_4$ ($\pm$ Cl) waters occur in northeastern California and western Nevada. Sulfate in these waters may be from sulfate minerals, initially deposited during previous hydrothermal events.

Meteoric waters in the probable recharge areas for most hot springs in the Northern Basin and Range Province are generally not as depleted in deuterium as the waters currently discharged by the hot springs. This difference is largest for waters associated with travertine and is almost certainly an indication that the thermal waters recharged during times of colder climate, probably the Pleistocene.

Measured temperatures in deep wells are, on the average, 14°C cooler than expected from the chemical geothermometry on waters from nearby

Table 7. Anhydrite saturation temperatures and sulfate–water isotope equilibrium temperatures for thermal waters of the northern Basin and Range.

Name of Sample	T – Anhydrite Saturation	T – SO_4–H_2O
California		
Fales Hot Springs	310	184
Hot Springs Motel (Surprise Valley)	211	200
Kelly Hot Springs	193	198
Seyferth Hot Springs	189	205
Travertine Hot Springs	190	173
West Valley Reservoir (Hot Spring)	220	247
Nevada		
Hazen (Hot Springs)	185	220
Lee Hot Springs	173	282
Soda Lake – Upsal Hogback (well)	125	127
Stillwater (well)	180	177
Wabuska (shallow well)	176	140
Oregon		
Alvord Hot Springs	275	231
Spring near Hot Lake (Alvord Desert)	215	230
Utah		
Abraham (Baker) Hot Springs	159	22
Thermo Hot Springs	172	142

hot springs and shallow wells. The measured
temperatures in the geothermal reservoirs are,
on the average, $22^{\circ}C$ cooler than expected when
spring waters are used to estimate the deep
aquifer-temperature, however measured
temperature are only $2^{\circ}C$ lower than expected
when waters from shallow wells are used to
estimate the temperature of the deep aquifer.

Anhydrite saturation temperatures are
similar to sulfate-water isotope equilibrium
temperatures for the more saline thermal waters
of the northern Basin and Range Province.
However, some systems in northeastern California
have aquifer-temperatures of 200 to 220 C based
on anhydrite saturation and SO_4-H_2O isotopic
equilibrium temperatures. These temperatures
are roughly $100^{\circ}C$ above the temperatures
estimated from silica or Na-K-Ca
geothermometers.

The oxygen-18 enrichment of $Na-HCO_3$ thermal
waters generally increases as the total
dissolved solids increase. Concentrations of
dissolved solids in Na-Cl waters generally are
higher than either $Na-HCO_3$ or $Na-SO_4$ waters.
Although Na-Cl waters are generally more
enriched in oxygen-18, no correlation seems to
exist between their oxygen-18 enrichment and
amount of dissolved solids.

REFERENCES

Adams, W. B., 1944, Chemical analysis of
municipal water supplies, bottled mineral
waters and hot springs, Nevada: Nevada
University, Reno, Department of Food and
Drugs, Public Services Division, 16 p.

Banwell, C. J., 1963, Oxygen and hydrogen
isotopes in New Zealand thermal
areas, in Tongiorgi, E., ed., Nuclear
Geology on Geothermal Areas: National
Research Council, Laboratory of Nuclear
Geology, Pisa, p. 95-138.

Barnes, Ivan, Irwin, W. P., and White, D. E.,
1978, Global distribution of carbon
dioxide and major zones of seismicity:
U.S. Geological Survey, Water-
Resources Investigations 78-38, 12 p.

Bateman, R. L. and Scheibach, R. B., 1975,
Evaluation of geothermal activity
in the Truckee Meadows, Washoe County,
Nevada: Nevada Bureau of Mines
and Geology, Report 25, 38 p.

Benoit, W. R., 1978, The discovery and geology
of the Desert Peak, Nevada, geothermal
field, in Energy exploration and politics:
California Division of Oil and Gas, 11 p.

Berry, G. W., Grim, P. J. and Ikelman, J. A.,
compilers, 1980. Thermal springs list
for the United States: National Oceanic
and Atmospheric Administration key to
Geophysical Records Documentation 12,
59 p.

Bowman, J. R., and Cole, D. R., 1982, Hydrogen
and oxygen isotope geochemistry of cold
and warm springs from Tuscarora, Nevada
thermal area: Geothermal Resources
Council Annual Meeting, Transactions, San
Diego, California, v. 6, p. 77-80.

Brook, C. A., Mariner, R. H., Mabey, D. R.,
Swanson, J. R., Guffanti, Marianne, and
Muffler, L. J. P., 1979, Hydrothermal
convection systems with reservoir temperatures $\geq 90^{\circ}C$, in Muffler, L. J. P.,
ed., Assessment of geothermal resources of
the United States - 1978: U.S. Geological
Survey Circular 790, p. 18-85.

Center for Water Resources Research, 1973,
Nevada University, Reno, Desert
Research Institute, Center for Water
Resources Research: Computer data
bank.

Church, J. A., 1878, Heat of the Comstock mines:
American Institute of Mining, metallurgic
and Petroleum Engineers Transactions,
v. 7, p. 45-76.

Cohen, Phillip, 1962, Preliminary results of
hydrogeochemical studies in the Humboldt
River Valley near Winnemucca, Nevada:
Nevada Department of Conservation and
Natural Resources, Water Resources
Bulletin 19, 27 p.

Cohen, Phillip and Loeltz, O. J., 1964,
Evaluation of hydrogeology and hydro-
geochemistry of Truckee Meadows area,
Washoe County, Nevada: U.S. Geological
Survey Water-Supply Paper 1779-S, 63 p.

Cole, D. R., 1982, Chemical and sulfur isotope
variations in a thermal spring system
sampled through time: Transactions Geothermal Resources Council Annual Meeting,
San Diego, California, v. 6, p. 81-84.

Craig, Harmon, 1963, The isotopic geochemistry
of water and carbon in geothermal areas,
in Tongiorgi, E., ed., Nuclear Geology
on Geothermal Areas: National Research
Council, Laboratory of Nuclear Geology,
Pisa, p. 17-53.

Craig, H., 1961, Isotopic variations in meteoric
waters: Science, v. 133, p. 1702-1703.

D'Amore, Franco and Panichi, Costanzo, 1980,
Evaluation of deep temperatures of hydrothermal systems by a new gas geothermometer: Geochimica et Cosmochimica Acta,
v. 44, p. 549-556.

Dansgaard, W., 1964, Stable isotopes in precipitation: Tellus, v. 16, no. 4, p. 436-468.

Dansgaard, W., Johnson, S. J., Moller, J., and Langway, C. C., Jr., 1969, One thousand centuries of climatic record from Camp Century on the Greenland ice sheet: Science, v. 166, p. 377–381.

Eakin, T. E., 1962, Ground-water appraisal of Diamond Valley, Eureka and Elko Counties, Nevada: Nevada Department of Conservation and Natural Resources-Reconnaissance Series Report no. 6., 60 p.

Ellis, A. J., and Mahon, W. A. J., 1967, Natural hydrothermal systems and experimental hot water-rock interactions (Part II): Geochimica et Cosmochimica Acta, v. 31, p. 519–538.

Ellis, A. J., and Mahon, W. A. J., 1964, Natural hydrothermal systems and experimental hot water-rock interactions: Geochimica et Cosmoschimica Acta, v. 28, p. 1323–1357.

Femlee, J. K. and Cadigan, R. A., 1982, Radioactivity of selected mineral-spring waters in the Western United States--Basic data and multivariate statistical analysis: U.S. Geological Survey Open-File Report 82-324, 102 p.

Fournier, R. O., 1979, A revised equation for the Na/K geothermometer: Geothermal Resources Council, Transactions, v. 3, p. 221–224.

Fournier, R. O., and Potter, R. W., 1979, A magnesium correction for the Na-K-Ca geothermometer: Geochimica et Cosmochimica Acta, v. 43, p. 1543–1550.

Fournier, R. O., and Rowe, J. J., 1966, Estimation of underground temperatures from the silica content of water from hot springs and wet-stream wells: American Journal of Science, v. 264, p. 685–697.

Fournier, R. O., and Truesdell, A. H., 1973, An empirical Na-K-Ca geothermometer for natural waters: Geochemica et Cosmochimica Acta, v. 37, p. 1255–1275.

Fournier, R. O., White, D. E., and Truesdell, A. H., 1974, Geochemical indications of subsurface temperature-Part I, Basic assumptions: U.S. Geological Survey, Journal of Research, v. 2, no. 3, p. 259–262.

Garside, L. J., and Schilling, J. H., 1979, Thermal waters of Nevada: Nevada Bureau of Mines and Geology, Bulletin 91, 163 p.

Glancy, P. A. and Katzner, T. L., 1975, Water-resources appraisal of the Carson River Basin, western Nevada: Nevada Department of Conservation and Natural Resources-Reconnaissance Series Report 59, 126 p.

Goode, H. D., 1978, Thermal waters of Utah: Utah Geological and Mineral Survey, Report of Investigations No. 129, 183 p.

Hem, J. D., 1970, Study and interpretation of the chemical characteristics of natural water: U.S. Geological Survey Water-Supply Paper 1473, 363 p.

Johnson, S. J., Dansgaard, W., and Clausen, H. B., 1970, Climatic oscillations 1200-2000 A.D.: Nature, v. 277, p. 482–483.

Kharaka, Y. K. and Mariner, R. H., 1977, Solution-mineral equilibrium in natural water-rock systems, in Paquet, H. and Tardy, Y., eds., Proceedings of the Second International Symposium on Water-Rock Interaction, Strasbourg, 1977, p. 66–75.

Kharaka, Y. K., and Barnes, I., 1973, SOLMNEQ: Solution-mineral equilibrium computations: National Technical Information Service, Report P.B. 215899, Springfield, Virginia

Longinelli, A., and Craig, H., 1967, Oxygen isotope composition of sulfate ions in seawater and saline lakes: Science, v. 156, p. 56–59.

Majmundar, H., 1983?, Thermal springs of California, in press.

Mariner, R. H., Swanson, J. R., Orris, G. J., Presser, T. S., and Evans, W. C., 1980, Chemical and isotopic data for water from thermal springs and wells of Oregon: U.S. Geological Survey, Open-File Report 80-737, 50 p.

Mariner, R. H., Brook, C. A., Swanson, J. R. and Mabey, D. R., 1978, Selected data for hydrothermal convection systems in the United States with estimated temperatures $>90^{\circ}C$: Back-up data for U.S. Geological Survey Circular 790: U.S. Geological Survey Open-File Report 78-858, 493 p.

Mariner, R. H., Presser, T. S. and Evans, W. C., 1977, Hot springs of the Central Sierra Nevada, California: U.S. Geological Survey Open-File Report, 27 p.

Mariner, R. H., and Willey, L. M., 1976, Geochemistry of thermal waters in Long Valley, California: Journal of Geophysical Research, v. 81, p. 792–800.

Mariner, R. H., Presser, T. S. and Evans, W. C., 1976a, Chemical composition data and calculated aquifer temperature for selected wells and springs of Honey Lake Valley, California: U.S. Geological Survey Open-File Report 76-783, 10 p.

Mariner, R. H., Presser, T. S. and Evans, W. C., 1976b, Chemical data for eight springs in northwestern Nevada: U.S. Geological Survey Open-File Report, 13 p.

Mariner, R. H., Presser, T. S., Rapp, J. B. and Willey, L. M., 1975, The minor and trace elements, gas and isotope compositions of the of the principal hot springs of Nevada and Oregon: U.S. Geological Survey Open-File Report, 27 p.

Mariner, R. H., Rapp, J. B., Willey, L. M. and Presser, T. S., 1974a, Chemical composition and estimated minimum thermal reservoir temperatures of the principal hot springs of northern and central Nevada: U.S. Geological Survey Open-File Report, 32 p.

Mariner, R. H., Rapp, J. B., Willey, L. M. and Presser, T. S., 1974b, Chemical composition and estimated minimum thermal reservoir temperatures of selected hot springs in Oregon: U.S. Geological Survey Open-File Report, 32 p.

McKenzie, W. F., and Truesdell, A. H., 1977, Geothermal reservoir temperatures estimated from the oxygen isotope compositions of dissolved sulfate in water from hot springs and shallow drillholes: Geothermics, v. 5, p. 51-61.

Mitchell, J. C., 1976, Geochemistry and geologic setting of thermal water of the northern Cache Valley area, Franklin County, Idaho, part 5 of Geothermal Investigations of Idaho: Idaho Department of Water Resources Water Information Bulletin 30, 47 p.

Mitchell, J. C., 1976, Geochemistry and geologic setting of the Blackfoot Reservoir area, Caribou County, Idaho, part 6, of Geothermal Investigations in Idaho: Idaho Department of Water Resources Water Information Bulletin 30, 44 p.

Moore, D. E., Morrow, C. A., and Byerlee, J. D., 1983, Chemical reactions accompanying fluid flow through granite held in a temperature gradient: Geochimica et Cosmochimica Acta, v. 47, p. 445-453.

Moore, D. O. and Eakin, T. E., 1968, Water-resources appraisal of Snake River Basin in Nevada: Nevada Department of Conservation and Natural Resources, Water Resources-Reconnaissance Series Report no. 48, 103 p.

Mundorff, J. C., 1970, Major thermal springs of Utah: Utah Geological and Mineral Survey Water-Resource Bulletin, v. 13, 60 p.

Nathenson, M., Nehring, N. L., Crosthwaite, E. G., Harmon, R. S., Janik, C., and Borthwick, J., 1982, Chemical and light-stable isotope characteristics of water from the Raft River Geothermal Area and environs, Cassia County, Idaho, Box Elder County, Utah: Geothermics, v. 11, p. 215-237.

Nehring, N. L., 1979, Reservoir temperature, flow, and recharge at Steamboat Springs, Nevada: Transactions Geothermal Resources Council Annual Meeting, Reno, Nevada, v. 3, p. 481-484.

Nehring, N. L., and Mariner, R. H., 1979, Sulfate-water isotopic equilibrium temperatures for thermal springs and wells of the Great Basin: Transactions, Geothermal Resources Council Annual Meeting, Reno, Nevada, v. 3, p 485-488.

Peal, A. C., 1886, Lists and analyses of the mineral springs of the United States (a preliminary study): U.S. Geological Survey Bulletin 32, 235 p.

Reed, M. J., 1975, Chemistry of thermal waters in selected geothermal areas of California: California Division of Oil and Gas Technical Report TR15, 37 p.

Roberts, R. J., Montgomery, K. M. and Lehner, R. E., 1967, Geology and mineral resources of Eureka County, Nevada: Nevada Bureau of Mines Bulletin 64, 152 p.

Rush, F. E., 1968a, Ground-water appraisal of Clayton Valley-Stonewall Flat area, Nye County, Nevada: Nevada Department of Conservation and Natural Resources, Water Resources-Reconnaissance Series Report no. 45, 54 p.

Rush, F. E., 1968b, Ground-water appraisal of Thousand Springs Creek Valley, Elko County, Nevada: Nevada Department of Conservation and Natural Resources, Water Resources-Reconnaissance Series Report no. 47, 61 p.

Sakai, H., and Matsubaya, O., 1974, Isotopic geochemistry of the thermal waters of Japan and its bearing on the Kuroko ore solutions: Economic Geology, v. 69, p. 974-991.

Sanders, J. W. and Miles, M. J., 1974, Mineral content of selected geothermal waters: Nevada University, Reno, Desert Research Institute, Center for Water Resources Research, Project Report 26, 37 p.

Taylor, H. P., 1974, The application of oxygen and hydrogen isotope studies to problems of hydrothermal alteration and ore deposition: Economic Geology, v. 69, p. 843-883.

U.S. Geological Survey and Oregon Department of
Geology and Mineral Industries, compilers,
1979, Chemical analyses of thermal springs
and wells in Oregon: Oregon Department of
Geology and Mineral Industries Open-File
Report 0-79-3, 169 p.

U.S. Geological Survey, 1977, WATSTORE water
quality file (computer data bank).

Visher, F. N., 1957, Geology and ground-water
resources of Quinn River Valley, Humboldt
County, Nevada: State of Nevada, Office
of the State Engineer Water Resources
Bulletin 14, 55 p.

Waring, G. A., 1965, Thermal springs of the
United States and other countries of
the world - A summary, revised by
Blankenship, R. R. and Bartall, R.,
U.S. Geological Survey Professional
Paper 492, 383 p.

Welch, A. H., Sorey, M. L. and Olmstead, F. H.,
1981, The hydrothermal system in southern
Grass Valley, Pershing County, Nevada:
U.S. Geological Survey Open-File Report
81-915, 193 p.

White, D. E., Muffler, L. J. P., and Truesdell,
A. H., 1971, Vapor-dominated hydrothermal
systems compared with hot-water systems:
Economic Geology, v. 66, p. 75-97.

White, D. E., Thompson, G. A., and Sandberg,
C. H., 1964, Rocks, structure, and geologic
history of Steamboat Springs thermal area,
Washoe County, Nevada: U.S. Geological
Survey Professional Paper 458-B.

White, D. E., and Brannock, W. W., 1950, Sources
of heat, water supply, and mineral content
of Steamboat Springs, Nevada: American
Geophysical Union Transactions, v. 51,
p. 566-574.

Wilson, S. H., 1963, Tritium determinations on
bore waters in the light of chloride-
enthalpy relations, in Tongiorgi, E., ed.,
Nuclear Geology on Geothermal Areas:
National Research Council, Laboratory
of Nuclear Geology, Pisa, p. 173-184.

Wollenberg, H. A., 1974, Radioactivity of Nevada
hot-spring systems: Geophysical Research
Letters, v. 1, no. 8, p. 359-362.

Worts, F. G., Jr. and Malmberg, G. T., 1966,
Hydrologic appraisal of Eagle Valley,
Ormsby County, Nevada: Nevada Department
of Conservation and Natural Resources,
Water Resources-Reconnaissance Series
Report 39, 55 p.

Young, H. W., and Mitchell, J. C., 1973, Geochemistry of geologic setting of selected
thermal waters, part 1 of Geothermal
Investigations in Idaho: Idaho Department
of Water Resources Water Informtion
Bulletin 30, 43 p.

GEOPHYSICAL STUDIES OF ACTIVE GEOTHERMAL SYSTEMS IN
THE NORTHERN BASIN AND RANGE

Stanley H. Ward

Earth Science Laboratory/University of Utah Research Institute
420 Chipeta Way, Suite 120
Salt Lake City, Utah 84108

ABSTRACT

Most of the geophysical data in the public domain, acquired in exploration for high temperature geothermal systems in the Northern Basin and Range Province, have been reviewed. Sufficient data are available to compare 14 methods at 13 sites, but only 110 entries occur in the 14 by 13 matrix whereas 182 entries would have been optimum. Only four of the sites studied are believed to be capable of production of commercial electricity while three others probably will be placed in this category in the next few years. For three additional systems believed to be capable of commercial production, insufficient geophysical data are available, in the public domain, to permit review.

On a rating scale of 1 equals good through 4 equals poor, no geophysical method has a mean ranking of 1. Five methods rank about 2, six rank about 2.5, and three rank about 3, while none ranks 4. This ranking system is subjective, but uniformly applied to the question, "what contribution has the method made to understanding the reservoir or the presumed reservoir, at site X". The averages of the evaluations at each site are the rankings given above. No combination of any four methods has been successful at more than one site where "successful" means a ranking of 1 or 2. The most useful of the methods, judging by their average rankings, are heat flow, microearthquakes, gravity, resistivity, and self-potential. The least effective methods are earth noise, reflection seismology, magnetics, magnetotellurics and tellurics. The radiometric and induced polarization methods were excluded from the comparative study due to a paucity of data. The CSAMT method has been included with CSEM for purposes of this comparative study.

Twenty-three applications of geophysical techniques were judged to be good, 29 were judged to be fair, 48 were judged to be questionable, while 10 were judged to be poor. Only 52 of 110 entries in the 14 by 13 matrix were judged to be fair or good; i.e. 53 percent of the geophysical applications gave questionable or poor results. Of the 58 applications where questionable or poor ratings were assigned, 41 applications were judged to be simple failure of the geophysical method to solve the problem at hand. However, 17 of the applications could have been better if improved technology or interpretation were available; some poor technology was applied in the middle of the 1970-80 decade.

Overall, this review provides a somewhat discouraging picture. This may be due in part to inadequate subsurface control and to inadequate survey design, execution and interpretation as a result of lack of experience. I would tend to use heat flow, microearthquakes, gravity, resistivity (or CSAMT), and self-potential methods at all prospects. Once these data were interpreted and correlated, I would then decide whether or not additional geophysical surveying was required and/or justified. All geophysical surveys should be designed and executed only after the geology has been mapped carefully, an integrated interpretation has been made of any and all other available earth science data, and one or more specific questions have been formulated for the survey to answer. The surveys should be designed with one or more conceptual geological models in mind and the density and extent of the geophysical coverage should be designed to provide adequate cover of the area dictated by such conceptual geologic models.

Within this manuscript, I have discussed the advantages and limitations of all of the geophysical methods considered; such a discussion is essential to their evaluation. Examples of overlapping geophysical data sets are given for three igneous-related geothermal systems and for two systems without obvious igneous relationships. No systematic difference in the past application to geophysical exploration for igneous-related systems versus those with no obvious igneous relationships is evident.

.1.0 OBJECTIVES

Heat is the essence of a geothermal resource, and its role in the development of such resources, the theme of this symposium, is fundamental. However, this leads to the question of how much surface manifestation of heat is necessary to define a geothermal prospect worthy of exploration. Clearly, heat flow specialists at this conference will attempt to address this question directly. I have been asked to address it indirectly and to concentrate on the other geophysical methods used in geothermal exploration and to provide a "common thread" among geophysical results for prospect-sized areas of 10 to 1000 square miles.

The objective of this paper, accordingly, is to focus on a comparative study of the problems and successes encountered with the gravity, magnetic, passive seismic, active seismic, self-potential, resistivity, passive electromagnetic, and active electromagnetic methods. Such a comparative study would be incomplete if I failed to draw conclusions as to the cost-effectiveness and preferred role of each of the methods listed above. At the risk of criticism from special interest groups, I shall draw such conclusions. The extent to which I am able to justify my conclusions must be judged by the individual reader in relation to his own experience. With the passing of time and further drilling, my conclusions undoubtedly will be modified. At least, however, I shall have provided a datum to which further analyses may be referenced.

For the purposes of this paper, I use Edmiston's (1982) definition of the Northern Basin and Range Province (Fig. 1). The number and evaluation of geothermal prospects in Figure 1 differs, however, from that given by Edmiston (1982). It is my intent to convey an overview of the contributions made by each geophysical method to understanding the geothermal reservoir of most of the prospects of Figure 1, independent of whether the prospect may now be classified as a discovery or otherwise. In this process, I shall use examples of geophysical data from only a few of the prospects. This permits the paper to be reasonably concise and yet illustrative. The zone of enhanced extension, shown in Figure 1, is my own interpretation.

2.0 DISTRIBUTION OF KNOWN HIGH TEMPERATURE RESOURCES

As noted by Edmiston (1982), "The northern Basin and Range Geologic Province of the western U.S.A. has been widely recognized as a highly prospective area for high-temperature geothermal reservoirs. Yet, only six apparent discoveries resulted from the drilling of 53 geothermal wildcat wells in this area from 1974 through 1981. This relatively lower success rate can be partly attributed to the difficulty of developing accurate geological and geophysical models in this area prior to drilling. However, it may also indicate that large, high-temperature geothermal reservoirs may be less common in this area than thought previously." Mansure and Brown (1982) support Edmiston's observations by forecasting (Figure 2) that the rate of drilling of geothermal wells through the year 2000, will be about the same for northern Nevada as for Roosevelt Hot Springs or Valles Caldera. If this forecast is correct, then the electric power generated by geothermal energy in northern Nevada will indeed be modest, i.e. of the order of 500 MWe. One might conclude, from the works of Edmiston (1982), Mansure and Brown (1982), and Benoit and Butler (1983), that for the whole of the northern Basin and Range Province there will be electrical production, by the year 2000, only at Beowawe, Coso, Desert Peak, Dixie Valley, Humboldt House, Roosevelt Hot Springs and Steamboat Springs. Soda Lake and Long Valley should possibly be added to this list. All of these potential resources lie in the eastern and western margins of the northern Basin and Range Province as Figure 1 shows; these are regions of enhanced crustal extension. Of course, moderate- to low-temperature geothermal resources are much more widespread.

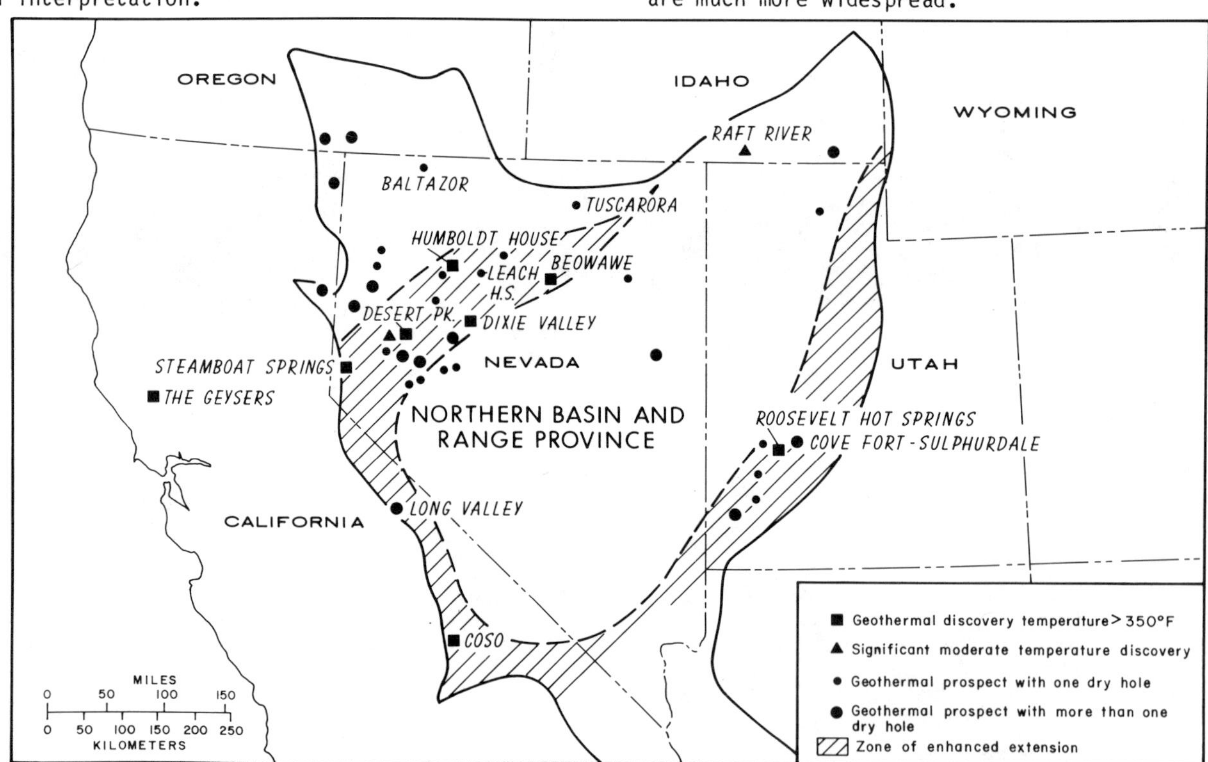

Figure 1. Map showing the location of geothermal discoveries and unsuccessful geothermal wildcat wells in the northern Basin and Range Province (after Edmiston, 1982).

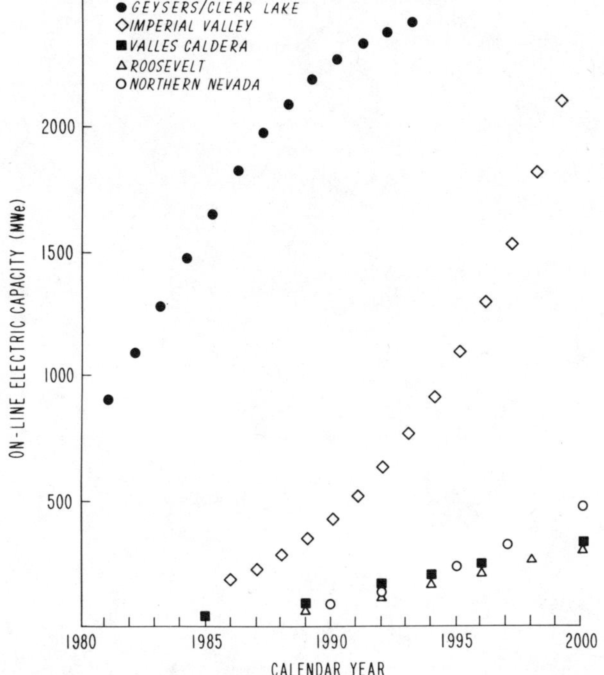

Figure 2.   Forecast of geothermal electric power capacity (after Mansure and Brown, 1982).

earth noise

Earth noise is associated with hot spring activity, fluid circulation, and fluid phase changes in geothermal systems, i.e. active hydrothermal processes, and can be a direct indicator of the presence of a geothermal system. This noise, and noise generated in other ways, may also be used to map the three dimensional distribution of velocity and attenuation (Iyer and Hitchcock, 1975; Liaw and McEvilly, 1979; Liaw and Suyenaga, 1982).

microearthquakes

Microearthquakes are associated with active faulting. They can occur more or less continuously, but more frequently they are episodic. Hypocenter locations give information on regions where faults are active and can help determine strike and dip. Fault-plane solutions can yield information on direction of movement. The three-dimensional distribution of seismic wave attenuation may be mapped. By measuring both compressional and shear wave velocities, Poisson's ratio can be computed; it is expected to be low over a vapor-dominated system and high over a fractured, liquid-dominated system. The source parameters seismic moment, stress drop, fault slip, and source radius may be estimated, but their value in reservoir definition and delineation is unknown at present. Minimum hypocentral depth of microearthquakes is useful in estimating the thermal regime surrounding geothermal reservoirs (Ward, R.W. et al., 1979, Majer and McEvilly, 1979).

teleseisms

If a sufficiently distant earthquake is observed with a closely spaced array of seismographs, changes in P-wave travel-time from station to station can be taken to be due to velocity variations near the array. Travel-time residuals are computed as the observed arrival time minus that calculated for a standard earth. A magma chamber beneath the geothermal system would give rise to low P-wave velocities and hence to late observed travel times (Iyer et al., 1979; Reasenberg et al., 1980; Robinson and Iyer, 1981).

refraction and reflection

The seismic refraction and reflection methods can be used to map the depth to the water table, stratigraphy, faulting, intrusions, and geologic structure in general. They may also yield the subsurface distribution of seismic P-wave and S-wave velocities, attenuations and Poisson's ratio. Detection of a characteristic attenuation or a "bright" spot, as found over reservoirs in petroleum exploration, would be a useful feature (Ward, R.W. et al., 1979; Applegate et al., 1981), but this has not been reported clearly for any prospect in the northern Basin and Range province.

Is this distribution of known, high temperature resources due to the geologic environment created by enhanced crustal extension, or is it because, in part, that we have not developed geophysical techniques capable of detecting and delineating high temperature resources when these resources are hidden? One would certainly have expected more than one exploitable high temperature resource along the eastern margin of the Basin and Range, if high heat flow, thin crust, and Quaternary volcanism are the key indicators of prime prospecting ground.

gravity

Density contrasts among rock units permit

3.0   GEOPHYSICAL METHODS FOR GEOTHERMAL EXPLORATION

In the Northern Basin and Range Province, the basic target for geophysical surveys in most cases is a fracture and fault system filled with thermal fluids. The reservoir itself may be relatively near surface as in parts of Roosevelt Hot Springs, or it may be 5,000 to 10,000 feet deep as at Beowawe or Desert Peak. Many geophysical techniques are only capable of delineating the top of the plumbing system and this may or may not contribute to understanding the reservoir (C.M. Swift Jr., pers. com.). With this thought in mind, let us turn to the geologic feature or features each method is usually directed toward, as listed in Table 1. A brief discussion of each target, in relation to the appropriate method, follows.

Table 1.

GEOPHYSICAL TARGETS IN GEOTHERMAL EXPLORATION

METHOD	TARGETS
SEISMIC	
EARTH NOISE	Active hydrothermal processes, distribution of velocity and attenuation.
MICROEARTHQUAKES	Active faulting, fluid filled fracturing.
TELESEISMS	Deep magma chamber.
REFRACTION	Structure, distibution of velocity and attenuation.
REFLECTION	Structure, distibution of velocity and attenuation.
GRAVITY	Structure, alteration, densification, intrusions, distribution of density.
MAGNETICS	Structure, hydrothermal alteration, intrusives, extrusives.
ELECTRICAL	
RESISTIVITY	Faulting, brines, hydrothermal alteration.
INDUCED POLARIZATION	Hydrothermal alteration.
CSEM & SCALAR AMT	Faulting, brines, hydrothermal alteration.
MT / AMT	Structure, deep reservoir, magma chamber ?, partial melt in deep crust or upper mantle.
SELF POTENTIAL	Fluid and heat flow.
TELLURICS	Faulting, brines, hydrothermal alteration.
RADIOMETRIC	Alteration, ^{226}Radon, ^{222}Radium.
HEAT FLOW	Reservoir temperature.

use of the gravity method to map intrusions, faulting, deep valley fill, and geologic structure in general. A subsurface distribution of densities can be obtained in the interpretation process. Densification of porous sediments gives rise to gravity highs over some geothermal systems (Isherwood and Mabey, 1978).

magnetics

Magnetic susceptibility contrasts in the subsurface permit use of the aeromagnetic and ground magnetic methods to map the distribution of magnetite. In some instances, this distribution may be related to rock type, and hence rock units can be mapped. This can be useful in structural studies. In some convective hydrothermal systems, alteration will lead to destruction of magnetite and hence the magnetic method can be used to map zones of active hydrothermal alteration. It can also used to determine the depth to the Curie isotherm (Isherwood and Mabey, 1978).

resistivity, controlled source electromagnetics (CSEM), controlled source audiomagnetotellurics (CSAMT), tellurics and scalar audiofrequency magnetotellurics

These electrical methods are capable of delineating low resistivity associated with brine-saturated and hydrothermally altered rocks in geothermal systems. Usually, the brine and alteration occur predominantly along faults, so these methods may map faults controlling a fractured reservoir. Alternatively, they may map a stratigraphic unit that contains thermal brines and/or alteration. By virtue of resistivity contrasts among rock units, each of these methods can map faults, stratigraphy, intrusions, and geologic structure in general, independent of the presence of brine or alteration (Hoover et al., 1978; Ward, S.H. and Sill, 1982).

induced polarization

The induced polarization method is theoretically capable of mapping the distribution of pyrite and clays, alteration products in convective hydrothermal systems (Zohdy et al., 1973; Risk, 1975a; Chu et al., 1983).

MT/AMT

The tensor magnetotelluric/audiofrequency magnetotelluric method is usually too expensive to be used for mapping the resistivity distribution in the shallow parts of a geothermal system. Hence, it is more logically used to map regional structure, to map the deeper parts of convective hydrothermal systems, to attempt to map magma chambers, and to detect and delineate zones of partial melt in the deep crust and upper mantle (Ward, S.H. and Wannamaker, 1983).

self-potential

Self-potential anomalies over convective hydrothermal systems arise from electrokinetic and thermoelectric effects. Accordingly, the self-potential method can give information on fluid flow and on the distribution of heat sources in the subsurface. This measurement can be made in the static state or dynamically when an injection experiment is under way (Anderson and Johnson, 1976; Corwin and Hoover, 1979; Sill, 1982a,b,c; Sill, 1983a,b).

radiometric

Gamma-ray spectrometry may be used to map the areal distributions of ^{40}K, ^{238}U, and ^{232}Th. The gamma-ray peaks used are at 1.46 MeV for ^{40}K, 1.76 MeV (^{214}Bi) for ^{238}U, and 2.62 MeV (^{208}Tl) for ^{232}Th. If ^{226}Rn or ^{222}Ra are present in a geothermal system, they will be detected in the ^{214}Bi peak, since they also are daughter products of ^{238}U decay (Ward, S.H., 1981). "An examination of hot-spring waters in Nevada indicates the presence of [^{226}Rn and ^{222}Ra], in varying abundances, in spring systems where $CaCO_3$ is the predominant material being deposited. Systems where silica predominates are relatively low in radioactivity" (Wollenburg, 1975). The use of alpha-cup detectors for radon emanating from geothermal systems has been reported by Wollenburg (1975) and Nielson (1978).

heat flow

Measurement of thermal gradients in holes drilled to 100 m provide shallow temperature data which may be direct evidence of a major source of heat at depth. However, flattening or reversal of thermal gradients frequently occurs so that the data from shallow holes may be misleading. For this reason, temperature gradient holes of 1000 m or greater are frequently employed (Blackwell, this volume).

4.0 THE PRINCIPAL PROBLEMS WITH GEOPHYSICAL METHODS IN GEOTHERMAL APPLICATIONS

The following discussion covers the principal problems encountered in applying geophysical methods in geothermal exploration. To facilitate understanding the problems, the elementary basis for each method is described where considered necessary. The amount of space devoted to each method largely reflects the author's experience. For example, someone who has been directly involved in applying reflection seismology in geothermal exploration would undoubtedly expand upon the section I have written. However, it is expected that most of the significant problems encountered in applying geophysical methods are at least mentioned here.

earth noise

The velocity V, frequency f, wavelength λ, and wavenumber k, of seismic waves are related through the equations

$$V = f\lambda = \frac{2\pi f}{k} \ . \qquad (1)$$

It has been demonstrated by several workers (e.g. Liaw and Suyenaga, 1982) that hydrothermal processes deep in the reservoir radiate seismic body waves in the frequency band 1 to 100 Hz. Contours of noise power on the surface should delineate these reservoir-associated sources of noise, i.e. the noise radiating from a deep reservoir ought to be evident as body waves of high phase velocity (Liaw and McEvilly, 1979).

Noise in the 1 to 100 Hz band also arises in nearby sources such as freeway traffic, trains, rivers, canals, waterfalls, pipelines, wind, cattle, interfering seismic wavetrains, etc. There is also a distant source of noise of unknown origin. Thus, there is always an ambient noise background upon which any seismic radiation due to hydrothermal processes is superimposed. This ambient noise exhibits a diurnal variation, being lowest in the early morning hours. It may be, in part, related to the diurnal solar heating cycle.

It is well known that seismic noise amplitudes are usually higher over alluvium and soft sedimentary basins than over hard rock. (Iyer and Hitchcock, 1975). Thus, noise power anomalies may merely reflect a local increase in sediment cover. The noise in valley alluvium is dominantly of high-wavenumber, i.e. low velocity, (Liaw and McEvilly, 1979) and is propagated as fundamental-mode Rayleigh waves with the alluvium serving as a waveguide.

An array of geophones spaced on 100 m centers, as commonly used in ground noise studies in the past, would give spurious results because spatial aliasing folds high-wavenumber noise into low-wavenumber noise. The spatial aliasing results in the appearance of noise of erroneously high velocity, such noise being interpreted as body waves, whereas in reality they are surface waves of low velocity. From equation (1), one can see that aliasing of k implies aliasing of V.

If a geophone array is sufficiently dense that wavenumber aliasing is avoided, then the fundamental mode Rayleigh waves can be used to obtain the three-dimensional distributions of velocity and attenuation. These distributions may be used with the horizontal component of the vector wavenumber, via ray tracing, to locate a source region of radiating microseisms (Liaw and McEvilly, 1979). Random directions of propagation (i.e. random vector wavenumbers) are characteristic of low velocity waves (Iyer and Hitchcock, 1975).

Liaw and Suyenaga (1982) detected high-velocity body waves at Beowawe, but did not detect any body waves at Roosevelt Hot Springs. Liaw and McEvilly (1979) failed to find any body waves at Grass Valley. Thus, even when full f-k analysis is performed, high-velocity body waves may not be detected because they are not generated in all geothermal systems.

microearthquake (MEQ)

Microearthquakes, those having magnitudes between -2 and +3, frequently are closely related spatially to major geothermal systems. Accurate locations of these earthquakes can provide data on the locations of active faults that may channel hot water toward the surface (Ward, P.L. et al., 1969; Lange and Westphal, 1969; Ward, P.L. and Bjornsson 1971; Ward, P.L., 1972; Hamilton and Muffler, 1972).

From the arrival times of P and S waves at 8 to 12 geophones, spaced several kilometers apart, the source location, i.e. hypocenter of a seismic event, is calculated. The locus of many hypocenters may define a fault zone. Using first motion polarities, a fault-plane solution may indicate the type of motion along the fault, i.e. dip-slip vs. strike-slip, etc.

P- and S-wave velocities are retrievable from microearthquake data. Majer and McEvilly (1979) report locally high P-wave velocities in the production zone at The Geysers as determined from refraction surveys. Gupta et al. (1982) used microearthquake data to obtain regional P- and S-wave velocities for The Geysers. Usually, detailed velocity models, obtained from refraction surveys, are used to control the hypocenter determinations of the microearthquakes.

Measurement of either the absorption coefficient or a differential attenuation number called the "Q" may reveal the presence of exceptionally lossy materials in a reservoir due to fluid-filled fractures, or it may reveal the presence of low-loss materials due to steam-filled fractures or to silica- or carbonate-filled fractures. Majer and McEvilly (1979) found a shallow, high Q in the production zone at The Geysers from refraction and microearthquake surveys while they found a deeper, lower Q from the refraction survey. Majer (1978) reported that a refraction survey yielded high Q at Leach Hot Spring due to silica densification of sediments. Gertson and Smith (1979) found high Q over the geothermal system at Roosevelt Hot Springs, using refraction data.

The ratio, K, of P-wave to S-wave velocity may be estimated using a Wadati diagram in which S-P arrival times are plotted versus the P-wave arrival time at many different stations for a single event. From such a plot, a value for Poisson's ratio may be found. Nur and Simmons (1969) observed, experimentally, that fluid saturation in rocks leads to high values of Poisson's ratio ($\sigma \geq 0.25$) while dry rocks exhibit low values of Poisson's ratio ($\sigma < 0.20$). Thus determination of Poisson's ratio in MEQ surveys can conceivably result in determining whether a geothermal reservoir is vapor or water dominated (Combs and Rotstein, 1975; Majer, 1978; Majer and McEvilly, 1979; Gupta et al., 1982).

Table 2 lists values of Poisson's ratio for several geothermal systems. At Baltazor,

TABLE 2
POISSON'S RATIO

GEOTHERMAL SYSTEM	AUTHOR	POISSON'S RATIO
Baltazor	Senturion Sciences	0.22 ?
Coso	Combs and Rotstein	0.16 Steam
Grass Valley	Beyer et al.	0.25 to 0.30 No Anomalies
McCoy	Lange	0.15 Dry or Siliceous Volcanic Fill
		>0.35 Saturated Alluvium
Roosevelt Hot Springs	Ward	0.25 ?
The Geysers	Majer	0.15 to 0.24 Prod. Zone
	Gupta et al.	0.13 to 0.16 Prod. Zone
Tuscarora	Nicholl and Lange	to 0.35 Saturated Allvuvium

Senturion Sciences, Inc. (1977) reported a Poisson's ratio of 0.22. Combs and Rotstein (1975) reported a Poisson's ratio of 0.16 for Coso, a system dominated by vapor, at least in its shallower parts. In Grass Valley, Poisson's ratios of 0.25 to 0.30 were found over the entire region with no apparent anomalies (Beyer et al., 1976). Lange (1980) presented a plan map at McCoy showing contours of Poisson's ratios ranging from 0.15 over the central volcanic fill area, indicating dry or siliceous competent material, to greater than 0.35 over saturated alluvium. P-wave advances accompanied the higher values of Poisson's ratio while P-wave delays accompanied the lower values of Poisson's ratio at McCoy.

I have analyzed microearthquake swarm data apparently related to an east-west fault at Roosevelt Hot Springs and found a Poisson's ratio of 0.25. For the few events directly beneath the geothermal reservoir, a Poisson's ratio of 0.24 was found. Majer and McEvilly (1979) found Poisson's ratios ranging from 0.15 to 0.24 over the production zone at The Geysers, with higher values outside of it. The low Poisson's ratio in part corresponds to a decrease in P-wave velocity. Gupta et al. (1982), in a more extensive study at The Geysers, noted Poisson's ratios of 0.13 to 0.16 over the production zone and values 0.25 and higher outside of it. Nicholl and Lange (1981) reported Poisson's ratios as high as 0.35 due to saturated, fractured, basin fill at Tuscarora. While low values of Poisson's ratio seem to be associated with vapor-dominated systems at Coso and The Geysers, low values at McCoy would appear to be due to silification. A lack of an anomaly in Poisson's ratio at Roosevelt Hot Springs indicates that this parameter may not always contribute useful information about the reservoir.

Amplitude calibration of the microearthquake recording system is essential for meaningful microearthquake surveys. If microearthquake data are recorded on magnetic tape, then amplitude spectra are readily obtained, from which the source parameters seismic moment, stress drop, fault slip, and source radius may be estimated (Majer and McEvilly, 1979). The usefulness of these parameters in reservoir assessment requires extensive study.

For any of the above analyses of microearthquake data, a good model of the subsurface velocity distribution is required. Lack of good velocity control is a principal problem in analysis of MEQ data. Some geothermal systems, such as Roosevelt Hot Springs, have a generally low, episodic level of occurrence of microearthquakes. Swarms of earthquakes occur, but in the intervals between them, insufficient activity may preclude any of the foregoing analyses. Indeed, one can record passive seismic data for a two- or three-week period or longer and come to the conclusion that the geothermal system is unimportant since it is not seismically active.

In some geothermal systems, a fault-plane solution may not be meaningful because the microearthquakes occur not on a single fault, but on several intersecting faults. Caution must be exercised in accepting fault plane solutions in such cases. Further, the active fault(s) might not be related to the fault zone(s) serving as the reservoir.

teleseisms

Steeples and Iyer (1976a,b) found relative P-wave delays of 0.3 sec at stations in the west central part of the Long Valley caldera. Reasenberg et al. (1980) recorded relative P-wave delays of 0.2 sec at Coso. Iyer et al. (1979) found relative P-wave delays as large as 0.9 sec at The Geysers. Robinson and Iyer (1981) reported relative P-wave delays up to 0.3 sec at Roosevelt Hot Springs. Lange (1980) deduced P-wave delays as large as 0.25 sec at McCoy. Berkman and Lange (1980) recorded relative P-wave delays and advances at Tuscarora; the delays were attributed in part to alluvium and in part to zones of hydrothermal alteration, while the advances were attributed to silicification along fracture zones.

While one can speculate that relative P-wave delays are caused by partial melts or magmas, as may be the case at Coso, Long Valley, and The Geysers, they can also be caused by alluvium, alteration, compositional differences, lateral variations in temperature or locally fractured rock (Iyer and Stewart, 1977). Wechsler and Smith (1979) suggest that the P-wave delays found by Robinson and Iyer (1981) at Roosevelt Hot Springs may well be due to fluid-filled fractures or to a compositional change. Thus, teleseismic P-wave delay studies will not always produce conclusive results.

refraction seismology

The seismic refraction method has been used mainly as a geophysical reconnaissance method for mapping velocity distributions and, hence, faults, fracture zones, stratigraphy, and intrusions (Williams et al., 1975; Hill, 1976; Combs and Jarzabek, 1977; Majer, 1978; Ackerman, 1979; Gertson and Smith, 1979). The seismic refraction method does not give resolution of structure as well as does the seismic reflection method. Sentiment today calls for performing seismic refraction at the same time as seismic reflection, with little added cost. Some attempts have been made to map velocity and amplitude attenuation anomalies, of both P- and S-waves, coinciding with a geothermal system (Goldstein et al., 1978b). Beyer et al. (1976), Combs and Jarzabek (1977), Majer (1978), and Gertson and Smith (1979) found anomalous velocities and amplitudes of refracted waves passing through the reservoir region. "The limited results at hand are not easy to explain and are contradictory" (Goldstein et al., 1978b). On the other hand, the potential contribution to the understanding and mapping of the reservoir seems large.

reflection seismology

The seismic reflection method provides better resolution of horizontal or shallow-dipping layered structures than any other method and, hence, is invaluable in mapping stratigraphic geothermal reservoirs of the Imperial Valley type. However, where the structure becomes highly faulted or folded, diffraction of seismic waves occurs at sharp corners and makes the task of interpreting structure difficult.

"Conventional [reflection] seismic surveys appear to give good definition of Basin and Range border faulting and depths to the base of alluvial fill at Roosevelt Hot Springs, UT, Soda Lake, NV, San Emidio, NV and Grass Valley, NV". "One seismic line which crosses the Mineral Mountains at Roosevelt Hot Springs shows little obvious lithologic or structural information within the range itself, or within the reservoir, but substantial structural information along the range front. At Beowawe, extensive and varied digital processing was ineffective in eliminating the ringing due to a complex near-surface intercalated volcanic-sediment section. Majer (1978) found reflection data extremely useful in delineating structure in Grass Valley, NV" (Ward S.H., et al., 1981). At Soda Lake, "in 1977, Chevron obtained modern, 1200% CDP seismic reflection coverage (12 line miles, 48 channel, explosion). The seismic data yielded a complex NE-SW trending graben from the shore of Soda Lake passing south of Upsal Hogback. The reflectors dip to the southwest, consistent with a small basin over the gravity low. The maximum depths of reliable seismic data are governed by a thin basalt unit and vary from 2,400-4,000 ft" (Swift, 1979).

Zoback (1983) has nicely demonstrated the use of seismic reflection data in mapping the style of initial faulting, infill and subsequent slumping and faulting in some basins in the province.

It is clear from the above remarks that high-resolution reflection seismology is capable of mapping faults, fractures, and stratigraphy in the vicinity of a geothermal system. In the Northern Basin and Range province, it will be useful in mapping range front faults and sedimentary stratigraphy. However, reflection seismology will seldom be of value in outlining reservoirs developed in volcanic, metamorphic, or igneous rocks. It may become almost useless when a volcanic cover or an intercalated volcanic cover is present as at Beowawe. The expense of modern multifold high resolution seismic surveys has deterred their use in geothermal exploration.

gravity

Gravity surveys are used in geothermal exploration as a relatively inexpensive means of obtaining structure and thickness of alluvium. "In the Basin and Range Province, large gravity lows are associated with the low-density basin fill. The gravity anomalies are used to estimate the thickness of the basin fill and the gross structure of the basin, including the location of the major normal faults that are important in understanding a geothermal system. Geothermal-related anomalies in the basins are most commonly residual gravity highs that are interpreted to reflect densification of porous sediments, structural highs, or anomalous geometry of fault zones" (Isherwood and Mabey, 1978). Gravity lows are sometimes found over siliceous magma bodies (Isherwood, 1976). At other times, gravity highs are expected due to rhyolite domes and hydrothermal alteration (Hochstein and Hunt, 1970; Macdonald and Muffler, 1972). Goldstein and Paulsson (1979), Berkman and Lange (1980), and Edquist (1981) found gravity particularly useful in mapping range-front normal faults in the Basin and Range province.

In the above, as in all other geophysical applications, the gravity method is hampered by ambiguity in interpretation. This ambiguity can be reduced by using drill hole, seismic refraction, or seismic reflection data for control. If these additional data are not available, then lateral and vertical resolution of geological features can be quite uncertain.

magnetics

Magnetic surveys, either airborne or ground, have been conducted at many geothermal prospects. Their use can be either for structural mapping or for mapping of changes in the magnetization of rocks caused by hydrothermal fluids. Magnetic anomalies in New Zealand geothermal fields have been interpreted as being due to a conversion of magnetite to pyrite (Studt, 1964). Such an effect would, of course, remain in extinct hydrothermal systems.

"Opinion has been divided on the usefulness of magnetic surveys in geothermal exploration (Cheng, 1970; Banwell, 1970). The magnetization at different rock units may be quite variable, especially in volcanic areas" (Palmason, 1975). "We examine the data from broad magnetic surveys as part of the effort to determine the regional setting ----. Some geothermal anomalies appear to relate to regional magnetic lineaments and zones suggestive of structure within the basement. Regional magnetic data can also be used to estimate the depth to the Curie isotherm by analysis of the spatial wavelengths of the field" (Isherwood and Mabey, 1978).

My own analysis indicates that magnetics have seldom been of major importance in assessing a geothermal reservoir in the Northern Basin and Range province. An outstanding exception appears to occur at Coso, where a magnetic low caused by destruction of magnetite from hydrothermal alteration draws attention, when correlated with other methods, to the heart of the reservoir. A similar magnetic low occurs over a part of the hot spring area at Long Valley (Plouff and Isherwood, 1980), and is interpreted by Kane et al. (1976) as due to magnetite destruction. At the Cove Fort-Sulphurdale KGRA, aeromagnetic data were important for an entirely different reason; structural controls on the potential reservoir could be deduced rather inexpensively from aeromagnetic data.

resistivity

Geothermal reservoirs frequently exhibit low resistivities due to high temperature, enhanced porosity, salinity of the interstitial fluid, and alteration of feldspars and other silicate minerals to clay minerals.

The geothermal use of the Schlumberger and Wenner arrays have been referenced in Banwell and Macdonald (1965), Hatherton et al. (1966), Macdonald and Muffler (1972), Meidav and Furgerson (1972), Zohdy et al. (1973), Arnorsson et al. (1975), Gupta et al. (1975), Stanley et al. (1976), Tripp et al. (1978) and Razo et al. (1980). Dipole-dipole arrays were used in surveys reported by Klein and Kauahikaua (1975), Jiracek et al. (1975), McNitt (1975), Garcia (1975), Beyer (1977), Fox (1978b), Ward, S.H. et al. (1978), Patella et al. (1979, 1980), Baudu et al. (1980), Smith (1980), Wilt et al. (1980a,b), Edquist (1981), and Mackelprang (1982). The bipole-dipole array was first used in geothermal exploration by Risk et al. (1970) and subsequently studied by Bibby and Risk (1973), Keller et al. (1975), Risk (1975a,b), Williams et al. (1975), Beyer et al. (1975), Stanley et al. (1976), Jiracek and Smith (1976), and Souto (1978). The bipole-dipole array achieved early success over broad areas of resistivity lows caused by hydrothermal alteration (Risk et al., 1970; Hohmann and Jiracek, 1979), but it has subsequently fallen into disfavor because of its failure to produce distinctive anomalies over geothermal systems lacking a broad surface manifestation (Dey and Morrison, 1977; Frangos and Ward, 1980).

The Schlumberger array is the most convenient one for depth sounding, i.e. estimation of the thicknesses and resistivities of the layers of a horizontally layered earth. The dipole-dipole array is used for continuous sounding-profiling, i.e. determination of both lateral and vertical variations in resistivity.

Several problems arise with the Schlumberger and dipole-dipole arrays (Ward and Sill, 1982):

1) Natural electric fields constitute noise for resistivity surveys and these result in slower productivity, higher costs, data of poorer quality, or in the extreme, no data at all. Modern data processing reduces but does not eliminate this problem.

2) Cultural features such as fences, powerlines, and pipelines redistribute current from the transmitter electrodes of the resistivity array; spurious resistivity anomalies result.

3) Strong noise voltages are present in the vicinity of powerlines. While notch filtering in the receiver will reduce this noise, it does not eliminate it.

4) Conductive overburden, generally in the form of porous alluvium or weathered bedrock, tends to prevent current from penetrating to the bedrock. Hence detection of bedrock features is less certain when overburden is present than when it is absent. When the overburden is of irregular resistivity, the geologic noise produced by the near-surface features may readily obscure the anomaly due to the target in the bedrock. Of course, this is common to all geophysical methods. Anomalies due to geological heterogeneities of no geothermal significance can also obscure, or partly obscure, the anomaly due to a geothermal system. In the Northern Basin and Range province these usually arise in Quaternary alluvial valley fill, and in salt playas.

5) Much geothermal exploration is done in mountainous terrain where topography can produce spurious resistivity anomalies. In a recent study, Fox et al. (1980) showed that a valley can produce a large, spurious resistivity low which could easily be misinterpreted as evidence for a buried conductor. Similarly, they showed that a hill can produce an apparent resistivity high.

In general, topographic effects are important where slope angles are 10° or more for slope lengths of one dipole or more. The solution to the problem is to include the topographic surface in numerical models used for interpretation.

6) The practical limitation on the depth of exploration of the Schlumberger array is the large separation needed between current electrodes. A large current electrode separation usually means that lateral resistivity variations outside the array will affect the measurements, thus rendering interpretation difficult. The dipole-dipole array minimizes this difficulty, but introduces a new one; a dipole-dipole array is sensitive to shallow lateral resistivity variations beneath the array whereas a Schlumberger array is not so sensitive (Palmason, 1975).

The depth of exploration of resistivity arrays is difficult to assess, but values in the range 0.5 km to 1.5 km are common.

controlled source electromagnetic (CSEM) and controlled source audiomagnetotellurics (CSAMT)

The transmitter for a CSEM method consists of either a loop of wire or a grounded bipole. Either source is energized by one or more frequencies in the audio frequency range or by a step current. One or more orthogonal components of magnetic and electric fields are recorded by a receiver located at a distance from the transmitter. The resistivity of the earth is measured by noting the phase and amplitude relationship of the voltage in the receiver to the current in the transmitter. This relationship is termed the impedance and is defined as

$$Z = \frac{V(t + \phi)}{I(t)} \qquad , (2)$$

in which both the voltage V and the current I are functions of time. The received voltage is shifted in phase by ϕ relative to the current in the transmitter.

Keller (1970) reviewed the applications of active and passive electromagnetic methods in geothermal exploration. His article constituted a baseline for reference to controlled-source electromagnetic methods (CSEM) in geothermal environments. Subsequent to Keller's review, a number of articles have appeared which illustrate the success and failure of these methods in geothermal exploration. Included are the articles by Lumb and MacDonald (1970), Jackson and Keller (1972), Keller and Rapolla (1974), Jacobson and Pritchard (1975), Morrison et al. (1978), Tripp et al. (1978), Kauahikaua (1981), Wilt et al. (1980c,d), Wilt et al. (1981a,b), Goldstein et al. (1982), and Keller et al. (1982).

The method will suffer all of the problems listed under the *resistivity* section above.

CSAMT is a subset of CSEM, or is a subset of AMT in which the transmitter is a grounded bipole. It is the only CSEM method that does not utilize a loop source. Two orthogonal, horizontal components of electric and magnetic field are measured (as in magnetotellurics). It offers advantages over resistivity methods in that it is faster and suffers less from the effects of lateral resistivity variations when providing sounding information (Ward, S.H., 1983). The method is used for combined sounding-profiling.

scalar audiomagnetotellurics (AMT)

The AMT method utilizes natural electromagnetic fields in the 10 Hz to 20 kHz band. These natural fields arise in atmospheric lightning discharges worldwide. Hoover et al. (1976), Hoover and Long (1976), Hoover et al. (1978), and Long and Kaufman (1980) described reconnaissance AMT surveys. Keller (1970), Whiteford (1975), Williams et al. (1975), Isherwood and Mabey (1978), and Jackson and O'Donnell (1980) have also reported on its use in geothermal exploration.

Two orthogonal magnetic-field components and two orthogonal electric-field components are measured and the scalar apparent resistivities are computed from the formulae

$$\rho_{a_1} = 0.2T \left|\frac{E_x}{H_y}\right|^2 \quad , \qquad (3)$$

and

$$\rho_{a_2} = 0.2T \left|\frac{E_y}{H_x}\right|^2 \quad , \qquad (4)$$

in which the electric field is measured in mv/km and the magnetic field in gammas; T is the period in seconds. The apparent resistivities are then given in ohm-meters.

The method suffers from all of the problems listed under resistivity, except natural field noise. It exhibits two other problems, however. The first is that the natural fields occasionally are too weak to obtain useful information. The second, and far more important, is that the simple formulae of equations (9) and (10) are totally inadequate for interpretation in two- and three-dimensional terrains, in which the tensor AMT method should be used. However, the scalar AMT method has proven useful for reconnaissance surveys of the type performed by USGS personnel who have fostered its use in this country. Very little application of the method has been made by industry, probably because equipment for the method has not been available commercially.

The CSAMT method is a substantial improvement over scalar AMT insofar as the direction of the inducing fields can be controlled, thus simplifying interpretation in two- and three-dimensional environments. The uncertainty over the strength of the fields used in AMT disappears when CSAMT is used.

MT/AMT

The tensor MT/AMT method utilizes natural electromagnetic fields in the 10^{-4} Hz to 10^4 Hz band. Below about 1 Hz these fields arise in geomagnetic perturbations brought about by interaction of the solar wind with the main geomagnetic field. Above 1 Hz the natural fields arise in atmospheric discharges worldwide. Three components of magnetic field and two components of electric field are measured. Data acquisition, processing and interpretation must be treated in highly sophisticated ways whose theory is beyond the scope of this paper.

Papers describing application of the tensor MT/AMT method in geothermal areas include Hermance et al. (1975), Hermance and Pedersen (1977), Stanley et al. (1977), Goldstein et al. (1978a), Morrison et al. (1979), Dupis et al. (1980), Gamble et al. (1980), Musmann et al. (1980), Ngoc (1980), Wannamaker et al. (1980), Aiken and Ander (1981), Berktold (1982), Berktold and Kemmerle (1982), Goldstein et al. (1982), Hutton et al. (1982), Martinez et al. (1982), Stanley (1982) and Wannamaker et al. (1983). A comprehensive review of data acquisition, processing, and interpretation for the method, plus a full discussion of the problems it encounters in geothermal exploration, has been prepared by Ward, S.H. and Wannamaker (1983). In the following I seek to summarize their review of these problems.

1) Source dimensions

All of the formulation for interpretation of MT/AMT data over one-, two-, or three-dimensional earths assumes that the MT fields are propagated as plane waves. This assumption was the source of much controversy in the early days of MT, but Madden and Nelson (1964) showed that the field is usually plane wave at frequencies greater than 10^{-3} Hz in mid latitudes.

At frequencies below 1 Hz, the primary concern appears to be whether or not the fields due to equatorial and auroral electrojet ring currents in the E-layer of the ionosphere can be treated as planar. Hermance and Peltier (1970) and Peltier and Hermance (1971) have studied the effects of such ring currents. They conclude that in conductive environments, the plane-wave assumption is valid in the frequency range 10^{-4} Hz to 1 Hz. However, significant errors can occur at frequencies less than 10^{-1} Hz in areas where high resistivities are encountered if measurements are made within 500 km of the position vertically beneath the electrojet.

At frequencies above 1 Hz, the proximity of lightning discharges becomes important. Bannister (1969) studied the fields radiated from a vertical electric dipole over a homogeneous earth and concluded that the plane-wave assumption is valid for distances greater than seven skin depths from the source.

If however, the plane wave assumption is not valid, then the extra field components associated with non-planar waves will be processed so as to produce bias in MT estimates.

2) Random noise

Random noise may arise in a) the electrodes
for E-field measurement via chemical
disequilibrium, b) movement of the E-field wires
in the earth's magnetic field when wind agitates
them, c) movement of the H-field sensors in the
earth's magnetic field due to wind or seismic
activity, d) microphonics in the H-field sensors
due to any motion, e) thermal noise in the E- and
H-field preamplifiers, f) quantization noise in
A/D converters, g) non-linear behavior of the
total recording system, h) sporadic departure from
plane wave propagation, and i) sporadic cultural
noise due to power lines, telephone lines, rail
electrification, pipeline corrosion protection,
radio interference, and the power sources in the
recording instrumentation. The processing system
must be designed to minimize, evaluate, and place
statistical limits on errors introduced into MT
transfer functions by random noise.

3) Systematic noise

Most of the noise sources described in 2)
above are also capable of introducing systematic
noise into estimates of the MT transfer
functions. As Stodt (1983) points out, the
systematic noise must be treated independently of
the random noise in any statistical evaluation of
noise in MT data. As a result of systematic
noise, biased estimates of the MT transfer
functions result. To attempt to eliminate this
problem, the use of a remote reference has become
common practice (Gamble et al., 1979a,b).

4) Geological noise due to overburden

In areas where there is irregular conductive
overburden, current channeling into a patch of
deeper or more conductive overburden will produce
anomalies even to the lowest frequencies. Unless
these anomalies are interpreted via 2D or 3D
modeling, they can be mistaken for deep-seated
features. Wannamaker (1983a) illustrates these
effects.

5) Topography

The effect of topography on the results of an
MT survey may be significant. Anomalous secondary
electric and magnetic fields result. Topography
must be included in the numerical modeling used in
interpretation to prevent topographic effects from
being interpreted as subsurface effects.

6) Depth of exploration and detectability

Depth of exploration is often stated to be one
skin depth, δ , where

$$\delta = 500 \sqrt{\frac{2}{\omega \mu \sigma}} \qquad (5)$$

and σ = the conductivity of the earth, ω is the
angular frequency of the signal under consideration, while μ is magnetic permeability. This
simplification is misleading, because noisy data

or surface geological noise can obscure the
responses of deep bodies. However, with care in
both data acquisition and interpretation, depths
of exploration well in excess of 100 km can be
achieved for infinite interfaces.

For 2D or 3D bodies, depth of exploration can
be considerably less. Newman et al. (1983) have
explored the possibility of detecting deep magma
chambers with MT. If the magma chamber is
electrically connected to a highly conducting
layer below it, the magma chamber probably will
not be detected. On the other hand, if the basal
half-space is resistive or if the earth is not
layered, the magma chamber is more readily
detected.

self-potential

In principle, self-potential surveys are very
simple: two non-polarizing electrodes, a length of
wire, and a D.C. voltmeter are all the equipment
needed to perform a survey. However, much
attention must be paid to details if the desired
reproduceability of $\pm$ 5 mV is to be achieved. Two
methods of moving the electrodes along the
traverse line are used; they are *leapfrog* and
long wire methods. In the former, the back
electrode is leapfrogged past the forward
electrode for each move. Only a short wire is
required. In the long wire method, the back
electrode is left fixed and the forward electrode
is moved farther and farther away. A long length
of wire is then required.

Corwin and Hoover (1979) reviewed the self-
potential method in geothermal exploration. Other
pertinent references to the use of the self-
potential method in geothermal exploration include
Zohdy et al. (1973), Corwin (1975), Anderson and
Johnson (1976), DeMoully and Corwin (1980), Hoover
(1981), Sill (1982a,b,c), and Sill (1983a,b).
Noise in self-potential surveys arises in telluric
currents, electrode drift, topographic effects,
variations in soil moisture, cultural noise,
vegetation potentials, and electrokinetic
potentials due to running surface water (Ward,
S.H. and Sill, 1982). The cultural noise sources
include: radiated fields from power lines,
telephone lines, and electrified rails; corrosion
potentials from pipelines, fences, and well
casings; and spurious potentials from corrosion
protection systems associated with pipelines. The
following discussions of these problems follow
from Ward, S.H. and Sill (1982).

1) Telluric currents

Time-varying voltages induced in the earth by
the geomagnetic field, of frequencies within the
passband of the voltmeter, may reach several
hundred mv/km over resistive terrain (Keller and
Frischknecht, 1966). These time-varying voltages
constitute noise which inhibits repeatability of a
measurement of the steady-state self-potentials.
The magnitude of this noise is proportional to the
separation between the two electrodes and
accordingly is largest for the *long wire* method.

If telluric noise is dominant, then the *leapfrog* method is preferred.

2) Electrode drift

A voltage will be measured across an electrode pair if either or both of the electrodes are not in equilibrium. Departure from zero electrode potential will occur if the electrolyte in the non-polarizing electrode is diluted or contaminated by groundwater or if there is a temperature differential between the two. These effects will vary with time, moisture content of the soil, and ambient temperature. Repeated checks of drift must be made periodically by placing both measuring electrodes in a bath of electrolyte solution and connecting them in order to establish equilibrium. Leapfrog surveys usually result in irregular electrode drift so that long wire surveys are preferred where electrode drift is the dominant source of noise.

3) Topography

Topographic relief will distort self-potentials and this effect must be taken into account in interpretation of field data. Another topographic effect is due to the movement of shallow groundwater. More negative potentials are sometimes correlated with an increase in elevation with observed gradients as large as -6 mv/m (Hoover, 1981).

4) Variations in soil moisture

As noted by Corwin and Hoover (1979), variations in soil moisture often give rise to self-potential variations, with the electrode in the wetter soil usually becoming more positive. Watering of electrodes to improve electrical contact can produce the same effect, but even worse, electrokinetic potentials are generated as the water moves through the soil. Electrode watering should be avoided for self-potential surveys in geothermal areas because the geothermal anomalies frequently are small, so that reducing noise to a minimum becomes essential.

5) Cultural noise

We have noted above some sources of cultural noise encountered when performing S.P. surveys. All of them can lead to potentials, which vary with time, while corrosion potentials and potentials from corrosion protection systems additionally will produce spurious anomalies. Results from self-potential surveys in a developing or developed geothermal field will be quite different from results obtained before well casings were installed.

6) Vegetation potentials

Trees, shrubs, and grasses produce potentials which are commonly of order 10 mv. A technique used to reduce this noise, and noise due to varying soil moisture, involves making five measurements in a star about each observation point. Four readings are offset about 3 m north, south, east, and west of the central point. The five readings are then averaged.

7) Electrokinetic potentials from moving non-thermal water

Potentials generated by the flow of non-thermal surface and subsurface water, i.e. electrokinetic potentials, constitute a noise source in geothermal exploration, and may be a major cause of topographic noise (Corwin and Hoover, 1979) because of water flowing downhill beneath the surface.

tellurics

"---- the telluric method is mainly suitable for reconnaissance of horizontal resistivity variations. It is based on the assumption that telluric currents flowing in extensive sheets are affected by lateral variations in the resistivity structure, which can be caused, for example, by variations in geological structure or by hydrothermal systems. The method requires the simultaneous measurement of the telluric electric field at two stations. From the ratio of the amplitudes of the electric field at the two stations, inferences may be drawn about variations in the underlying resistivity structure. By keeping the base station fixed and moving a field station about, one can thus map resistivity variations in a qualitative way." (Palmason, 1975)

The method has been used in geothermal exploration by Beyer (1977), Isherwood and Mabey (1978), Jackson and O'Donnell (1980), and others. It appears to be a convenient method for regional surveys in order to detect areas worthy of more detailed exploration by resistivity methods (Palmason, 1975).

The method suffers from a number of problems which have already been described under MT/AMT as follows: random noise, geological noise due to overburden, lack of resolution, and effects of topography. Its worst problem is that it is a semi-quantitative method at best. However, Beyer (1977) advocated its use for northern Nevada because of its simplicity, low cost, and ease of interpretation.

heat flow

I will not dwell on the problems encountered with temperature gradient and heat flow measurements since that subject will be treated by D. D. Blackwell in this volume.

5.0 ILLUSTRATIVE RESULTS

Long Valley

Figure 3 presents the generalized geology of the Long Valley caldera, California from Bailey et al. (1976). I wish to draw attention to the resurgent dome depicted by early rhyolite flows,

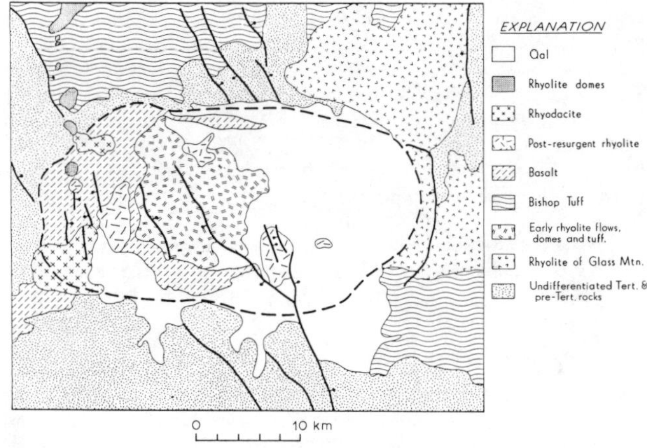

Figure 3.   Generalized geologic map of Long Valley caldera (from Bailey et al., 1976).

domes and tuff, and to the medial graben transecting the resurgent dome in a northwest direction. Heat flow has been measured in 29 drill holes from 50 to 300 m deep within the caldera. An additional 11 holes outside, but adjacent to the caldera, were drilled to 150 to 300 m and heat flow was measured in them (Lachenbruch et al., 1976a,b). Temperature at 10 m depth for the drill holes within the caldera is contoured in Figure 4. An outline of the area of

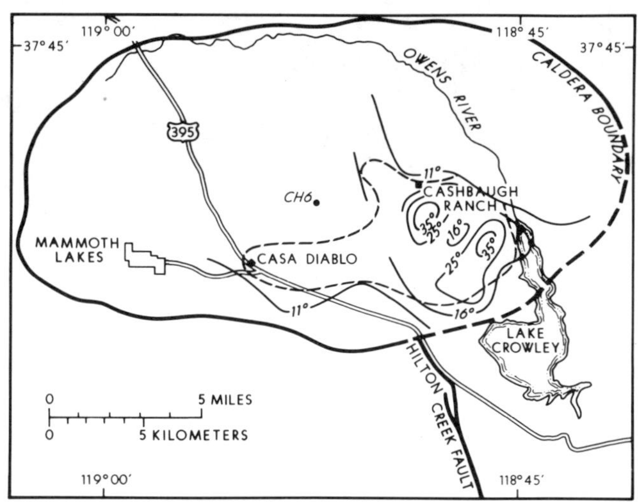

Figure 4.   Temperatures at 10 m depth, Long Valley caldera (from Lachenbruch et al., 1976a). The area of hot spring activity is superimposed as a dashed line.

hot springs is included in the figure. There is an obvious correlation between shallow temperature measurements and hot spring activity. Note that drill hole CH6 is the only drill hole within the resurgent dome; it was isothermal at 111°C from about 150 m to 300 m. Ground deformation, enhanced seismicity, and fumarolic activity in the last 5 years, and particularly in the last 2 years, testifies to a modern increase in tectonism and presumed magmatism. The enhanced seismicity appears to lie south of the resurgent dome. Figure 5 shows earthquakes occurring in the region

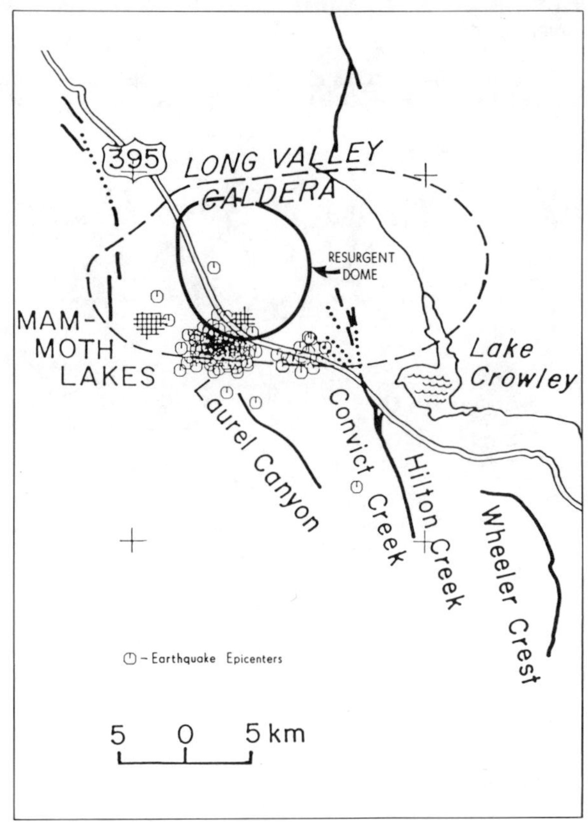

Figure 5.   Mammoth Lakes earthquakes July, 1981, (from Miller et al., 1982).

in July, 1981 (from Miller et al., 1982). A tongue of magma is believed to have risen to within 2 to 3 km of the surface beneath the epicenter of the earthquakes shown in Figure 5. Many more earthquakes than are shown in this figure have occurred in the last several years. Ryall and Vetter (1983) indicate that such earthquakes occur as far as 7 km north from the south rim of the caldera and as much as 15 km south of it.

The resurgent dome has bulged upward in recent years (Miller et al., 1982). Steeples and Iyer (1976a) reported a zone of significant teleseismic P-wave delay, beneath much of the resurgent dome. Hill (1976) found evidence for the roof of a magma chamber in late P-wave arrivals which he identified as reflections from a low-velocity horizon at a depth of 7 to 8 km. Ryall and Ryall (1981) found P- and S-wave attenuation which they attributed to a magma chamber beneath the Long Valley caldera. Ryall and Vetter (1983) reported on the absence of S waves at stations north of the caldera, for events occurring south of it; a magma chamber within the caldera is postulated to account for these observations; the postulated magma chamber roughly coincides with the area of hot spring activity shown in Figure 4.

Figure 6 (from Bailey et al., 1976) portrays a schematic east-west cross section through Long

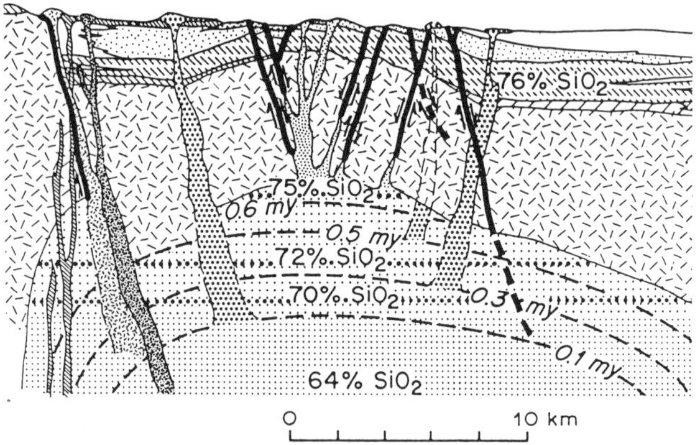

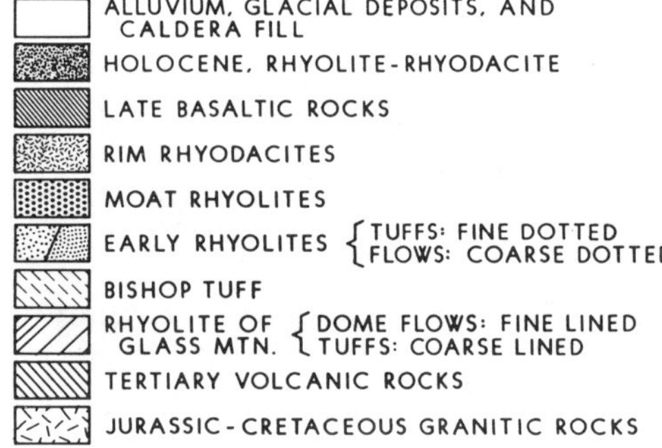

Figure 6. Schematic east-west cross section through Long Valley caldera and its underlying magma chamber (from Bailey et al., 1976).

Valley caldera. As noted by Muffler and Williams (1976) and by Sorey et al. (1978) the prevalent USGS view has been that the geothermal reservoir lies in the Glass Mountain rhyolites and the Bishop Tuff, the two earliest volcanics to infill the caldera.

The caldera fill above the Bishop Tuff consists of a variety of rhyolitic flows and tuffs, rhyodacitic flows, basaltic flows, and in the eastern half, lake, marsh, and periglacial sediments. These rocks are considered to be impermeable except along faults and hence do not form part of the reservoir (Muffler and Williams, 1976). The gravity data of Kane et al. (1976) and the seismic section of Hill (1976) have been used by USGS personnel to define the volume of the proposed reservoir to be 450 km^3.

Resistivity and AMT surveys mapped two near-surface low-resistivity zones associated with hydrothermally altered rocks and/or hot saline water (Hoover et al., 1976; Stanley et al., 1976).

Nine geothermal test wells were drilled near Casa Diablo and one well was drilled about 5 km farther east by Magma Power Company. One well was drilled by Republic Geothermal, Inc. about 4 km south of Cashbaugh Ranch. Union Geothermal drilled two wells, one near a clay pit in the north end of Little Antelope Valley, and one near Casa Diablo. All of the above wells have failed to discover a high temperature geothermal reservoir. None of the wells has been sited near the center of the resurgent dome, especially where transected by the medial graben. The self-potential data of Figure 7 draw attention to the whole of the area of the resurgent dome. W.R. Sill (pers. comm.) advises that cold water inflow from the west-northwest and outflow in the area of the thermal springs could account for the dipolar S.P. pattern. The cold water inflow could readily mask the center of a deep convective hydrothermal system possibly occurring beneath the resurgent dome.

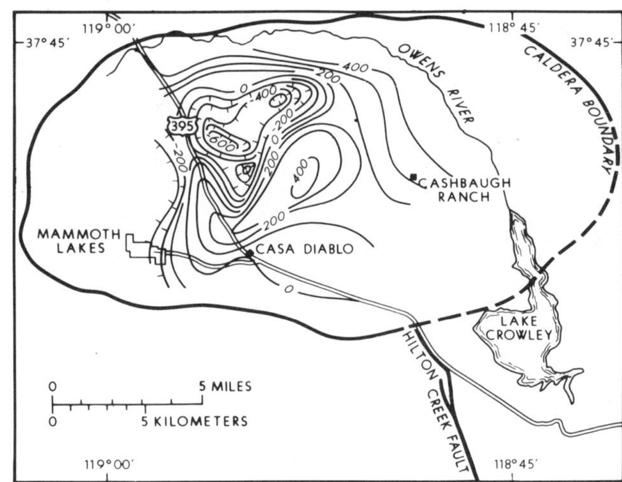

Figure 7. Self-potential map of Long Valley. (after Anderson and Johnson, 1976).

The magma chamber, as determined by the teleseismic P-wave delays, lies directly beneath the resurgent dome. The cross section of Bailey et al. (1976), Figure 6, schematically illustrates a potential fault-controlled reservoir within and below the caldera infill and beneath the resurgent dome. The problem with this notion is that for calderas elsewhere (e.g. Baca) the resurgent dome may be a poor exploration target because of lack of permeability (D.L. Nielson, 1983, pers. com.).

The resistivity and AMT data draw attention to areas south and east of the resurgent dome, while the aeromagnetic data draw attention to a small region of magnetite destruction around Casa Diablo. Surface manifestations of hydrothermal activity are abundant in these areas. I have downgraded shallow heat flow, magnetics, resistivity, and AMT at this geothermal prospect because drilling to date has not found a commercial reservoir where these methods have produced anomalies. I have up-graded self-potential on the speculation that a commercial reservoir may be found in the central part of the resurgent dome. Gravity and seismic refraction were very useful in delineating the Glass Mountain Rhyolites and Bishop Tuff in the floor of the caldera and were instrumental in quantifying the USGS model of the reservoir. Teleseismic P-wave delays, S-wave attenuation, and S-wave delays seem quite important and reliable indicators of a magma chamber in this setting. Earth noise so far has failed to help us delineate a reservoir. The earthquake epicenter map of Figure 5 may be delineating an east-west fracture zone up which a tongue of lava is believed to have recently intruded. What bearing this would have on a commercial high-temperature reservoir is unknown at present.

Evaluating geophysical methods at Long Valley is a questionable pursuit at present because we do not know where a commercial high-temperature reservoir is located if, indeed, it exists. However, the analysis has been made on the assumption that either the Bishop Tuff or the deeper parts of the medial graben will prove to be commercial. Note that the recent bulging of the resurgent dome would indicate that the medial graben may now be in extension.

Coso

Figure 8 portrays the generalized geology of the Coso geothermal area (Hulen, 1978). "The oldest rocks exposed at Coso are intermediate to mafic metamorphic rocks of uncertain age intruded by dikes and pods of quartz latite porphyry and felsite, and by a small stock of Late Cretaceous (?) granite. These rocks are locally overlain by Late Cenozoic volcanic rocks, which include the domes, flows, and associated pyroclastic deposits of the Coso rhyolite dome field."

"Principal structures in the geothermal area are older high-angle faults of uncertain displacement trending northwest, west-northwest, and east-northeast, and younger high-angle faults with a normal component of displacement trending north-northwest, north-northeast, and (subordinately) northeast. Active surface thermal phenomena and hydrothermal alteration are concentrated along the younger northerly-trending faults, especially where these faults intersect

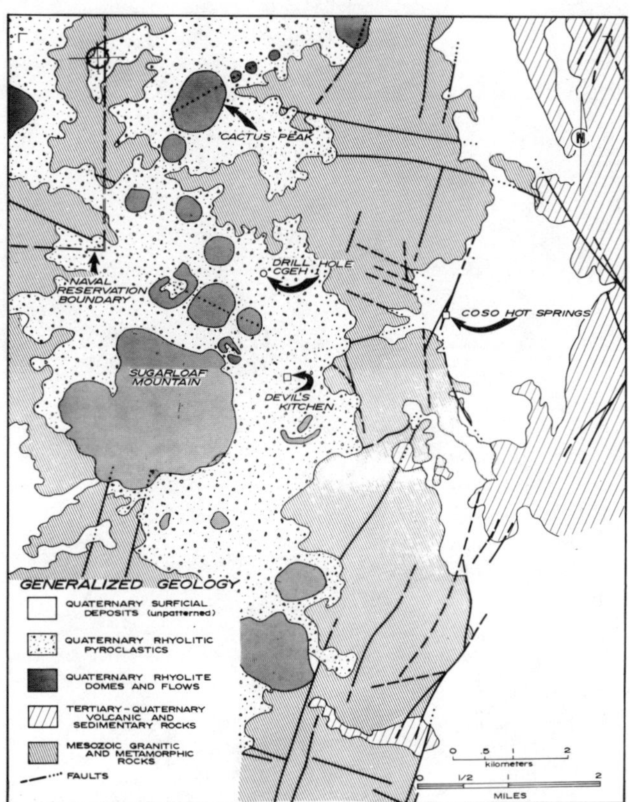

Figure 8. Generalized geologic map of Coso Hot Springs KGRA. (after Hulen, 1978).

older structures. Deep thermal fluid flow at Coso will be controlled entirely by structural permeability developed in otherwise tight and impermeable host rocks."

"Surface alteration at Coso is of three main types: (1) clay-opal-alunite alteration, (2) weak argillic alteration, and (3) stockwork calcite veins and veinlets, which are locally associated with calcareous sinter. Clay-opal-alunite and weak argillic alteration are typically developed around active thermal emissions. These are almost entirely restricted in distribution to an east-northeast-trending belt roughly one mile in width and four miles in length. Calcareous alteration is much more widely distributed, but is confined to a broad zone of anomalous geophysical response interpreted as evidence for a concealed geothermal reservoir" (Hulen 1978).

A low level aeromagnetic map of the area is given in Figure 9 (after Fox, 1978a). The anomalous low which extends southeast from Devil's Kitchen and Coso Hot Springs is attributed, in part, to magnetite destruction by hydrothermal fluids (Fox 1978a).

A resistivity low mapped in the vicinity of Devil's Kitchen and Coso Hot Springs (Figure 10) is attributed, in part, to hydrothermal alteration and hot brines in fractures (Fox, 1978b). The

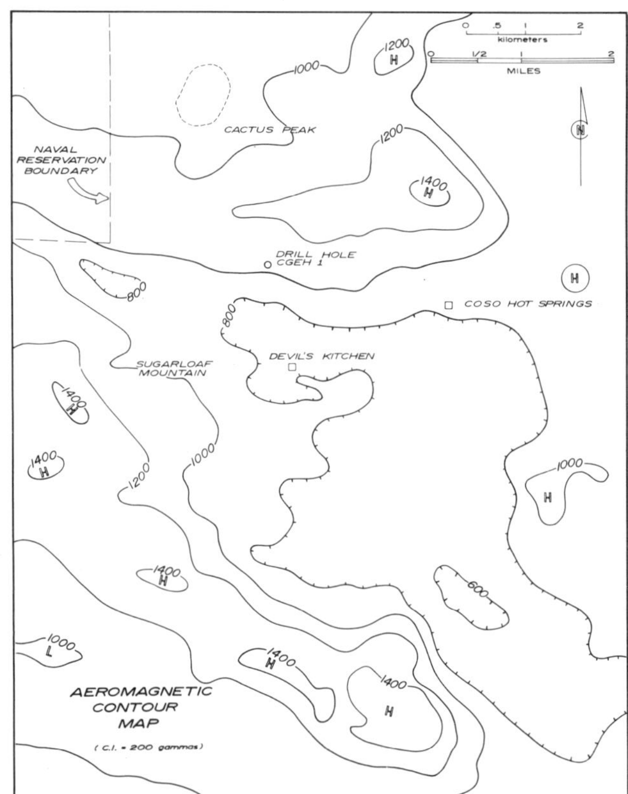

Figure 9.   Aeromagnetic residual total field intensity contour map of Coso. (after Fox, 1978a).

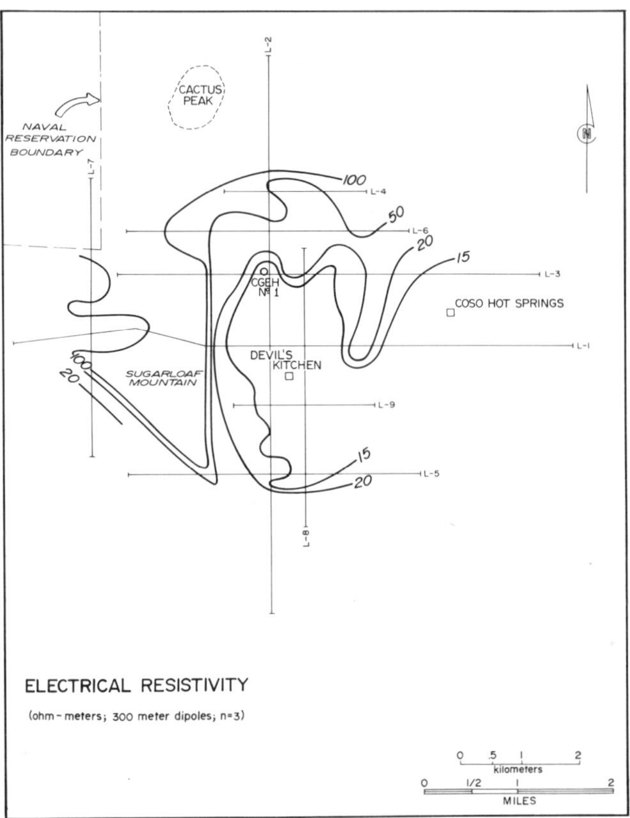

Figure 10.   Third separation apparent resistivity map of Coso obtained with 300 m dipole-dipole survey. (after Fox, 1978b).

dipole-dipole array with 300 m dipoles was used. Much the same zone of low resistivity was mapped by AMT earlier. Figure 11 displays the 7.5 Hz AMT apparent resistivity map produced by Jackson and O'Donnell (1980).   A telluric J-value map, also produced by these authors, was not nearly as effective in depicting the area of interest around Devil's Kitchen and Coso Hot Springs.

No one of the above geophysical techniques by itself created a confined target for drilling. However, when the resistivity and aeromagnetic data were combined with shallow heat flow data and a map of the hydrothermal alteration, then a target in the general vicinity of Devil's Kitchen was clearly delineated as shown in Figure 12 (from Hulen, 1978). Prior to Hulen producing Figure 12, drill hole CGEH-1 was drilled as a dry hole. Subsequently, all six wells have been producers. The importance of overlapping several data sets prior to spotting the first well on a geothermal prospect is vividly illustrated by this example.

I have not found that earth noise, teleseisms, gravity, or tellurics have contributed much to understanding the reservoir.   The earth noise data of Figure 13 (after Combs, 1980) could be aliased and noise power peaks near Coso Hot Springs may, in part, be due to Rayleigh waves in deep alluvium. The teleseismic evidence of a 0.2

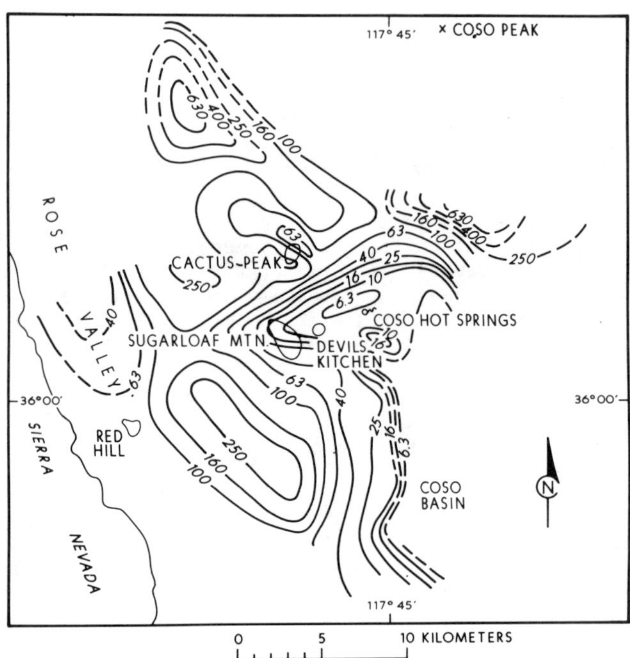

Figure 11.   The 7.5 Hz AMT apparent resistivity map of Coso. (after Jackson and O'Donnell, 1980).

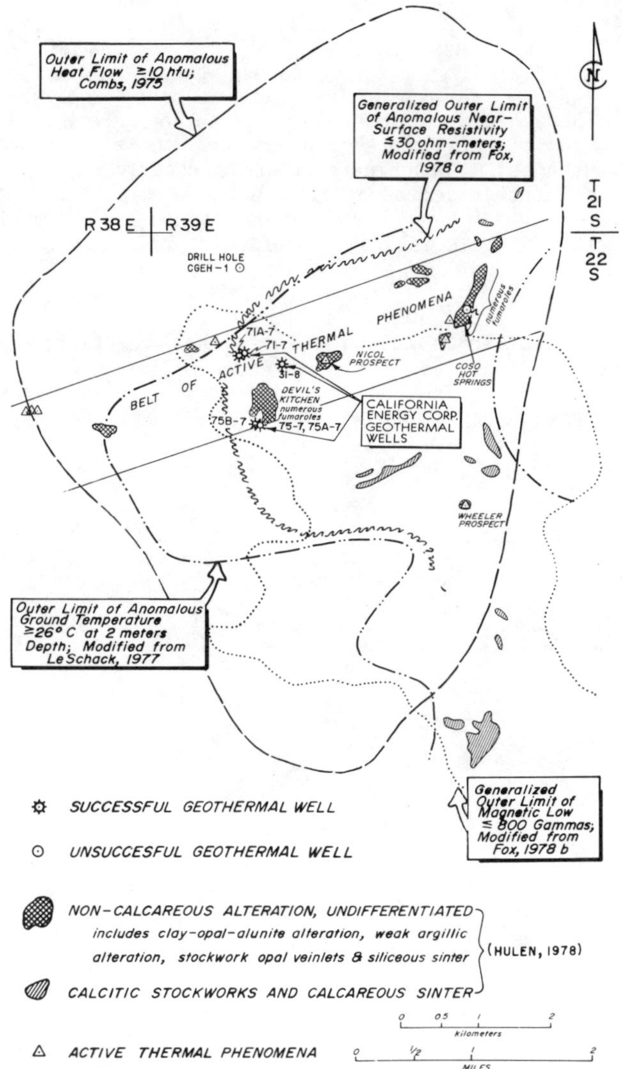

Figure 12.   Generalized alteration and geophysical map. (after Hulen, 1978).

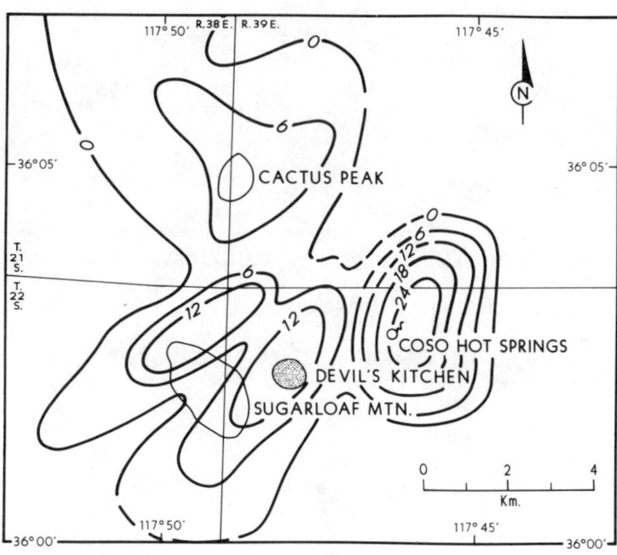

Figure 13.   Earth noise power expressed in decibels, Coso. (after Combs, 1980).

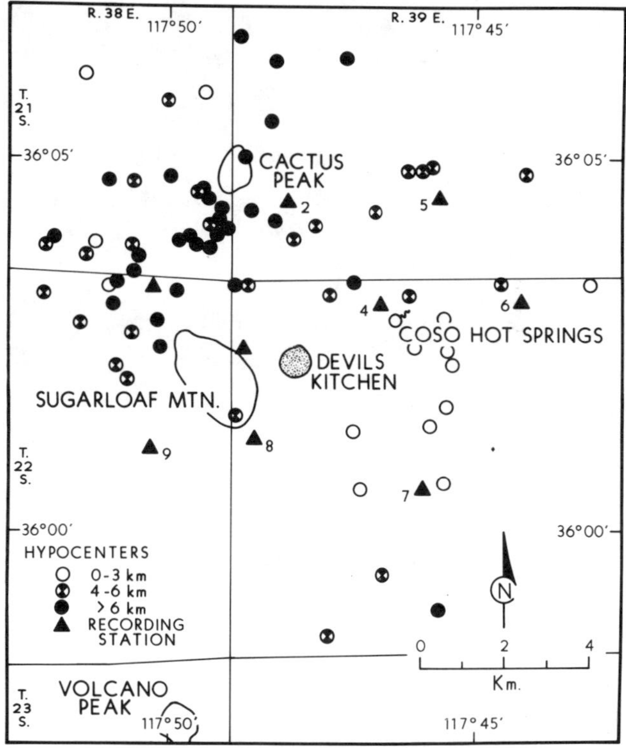

Figure 14.   Epicenters of 78 microearthquakes associated with the Coso geothermal area. (from Combs and Rotstein, 1975).

sec P-wave delay is not convincing. The gravity data are relatively featureless in the region between Devil's Kitchen and Coso Hot Springs. The telluric data are too much affected by Coso Basin to be of good quality in the region of interest. Microearthquakes were highly scattered throughout the area as Figure 14 illustrates (after Combs and Rotstein, 1975). No major faults were deduced from the data by Combs and Rotstein. However, the important contribution of microearthquakes at this prospect lies in calculation of a Poisson's Ratio of 0.16, an indicator of a vapor-dominated system. Based on current knowledge, the shallower parts of the reservoir appear to be vapor dominated (J.A. Whelan, 1983, pers. com.). The refraction method was useful at Coso in suggesting "the existence of a localized body of low velocity material at depth, possibly a magma chamber." (Combs and Jarzabek, 1977)

Roosevelt Hot Springs

Figure 15 portrays the generalized geology at the Roosevelt Hot Springs KGRA (after Sibbett and Nielson, 1980). From Ross et al. (1982) we extract the quotations which follow.

"The geothermal system is a high-temperature water-dominated resource, and is structurally controlled with permeability localized by faults and fractures cutting plutonic and metamorphic rocks."

"The oldest unit exposed in the area of the geothermal system is a banded gneiss which was formed from regionally metamorphosed quartzo-feldspathic sediments." "The rock is compositionally heterogeneous and contains thick sequences of quartzo-feldspathic rocks. The unit also contains metaquartzite and sillimanite schist layers which have been differentiated in the more detailed geologic study (Nielson et al., 1978)."

"The Mineral Mountains intrusive complex is the largest intrusive body exposed in Utah. Potassium-Argon dating and regional relationships suggest that the intrusive sequence is middle to

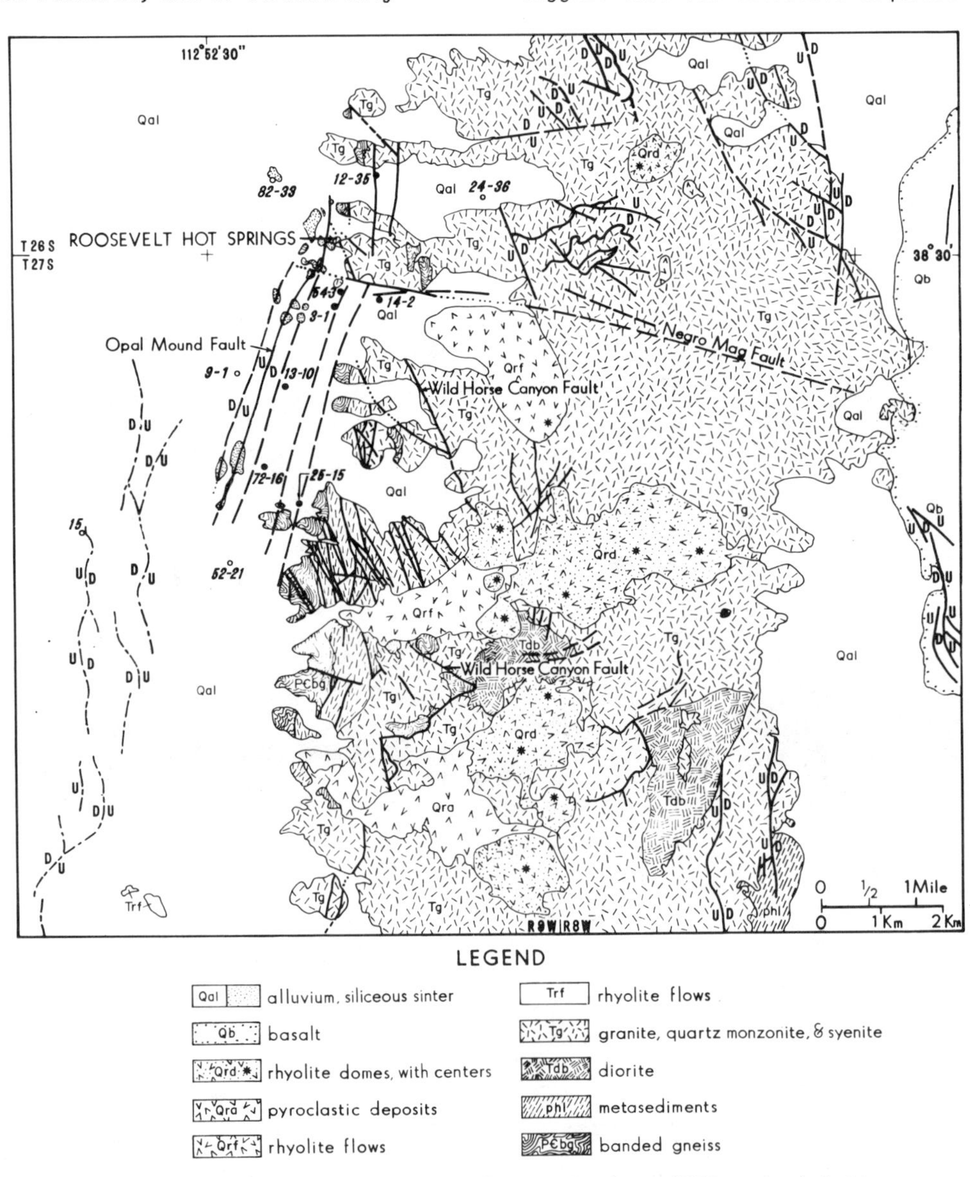

LEGEND

Qal — alluvium, siliceous sinter		Trf — rhyolite flows
Qb — basalt		Tg — granite, quartz monzonite, & syenite
Qrd — rhyolite domes, with centers		Tdb — diorite
Qrd — pyroclastic deposits		phl — metasediments
Qrf — rhyolite flows		PCbg — banded gneiss

Figure 15. Geologic map of Roosevelt Hot Springs KGRA and vicinity. (after Sibbett and Nielson, 1980).

late Tertiary in age. In the vicinity of the geothermal system, the lithologies range from diorite and granodiorite through granite and syenite in composition.

"Rhyolite flows, pyroclastics, and domes were extruded along the spine of the Mineral Mountains 800,000 to 500,000 years ago (Lipman et al., 1977). The flows and domes are glassy, phenocryst-poor rhyolites. The pyroclastic rocks are represented by air-fall tuff and nonwelded ash-flow tuffs."

"Hot spring deposits in the vicinity of the geothermal system have been mapped as siliceous sinter, silica-cemented alluvium, hematite-cemented alluvium, and manganese-cemented alluvium. The principal areas of hot-spring deposition are along the Opal Mound fault and at the old Roosevelt Hot Springs. In both of these areas, the deposits consist of both opaline and chalcedonic sinter."

The hot spring deposits are depicted in Figure 16. A soil survey within the area of

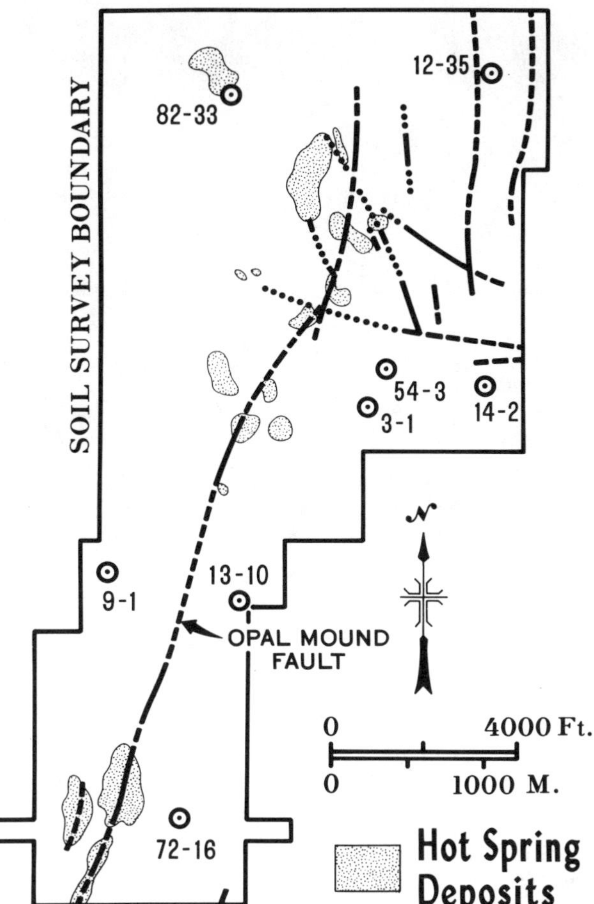

Figure 16. Hot spring deposits, Roosevelt Hot Springs. (after Bamford et al., 1980).

Figure 16 revealed anomalous mercury and arsenic. Hydrothermal alteration in the geothermal system and the adjacent Mineral Mountains is localized along faults and fractures.

"The hydrothermal alteration produced assemblages of quartz + chlorite + epidote + hematite. Hematite is commonly found as specularite veinlets and, where genetic relationships can be observed, hematite mineralization follows sulfide mineralization."

"The hydrothermal alteration assemblages associated with the present geothermal system are crudely zoned with depth. The uppermost assemblage, occurring around the hot-spring deposits and fumaroles, is characterized by quartz, alunite, kaolinite, montmorillonite, hematite, and muscovite. Parry et al. (1980) have studied the near-surface alteration and suggest that these minerals have formed above the water table by downward-percolating acid sulfate waters. Upward-convecting geothermal brines have produced, with increasing depth, alteration assemblages characterized by montmorillonite + mixed layer clays + sericite + quartz + hematite and chlorite + sericite + calcite + pyrite + quartz + anhydrite (Ballantyne, 1978). Thermochemical calculations and petrologic observations suggest that the brines are in equilibrium with the alteration assemblages produced by the upward-migrating fluids (Capuano and Cole, 1982)."

Figure 17 shows the heat flow contours, expressed in mWm^{-2}, over the main prospect area at Roosevelt Hot Springs KGRA. The southern lobe, up to Negro Mag fault, is a legitimate expression of the geothermal reservoir as it is currently known. The northern lobe, northwest of the intersection of the Negro Mag fault and the Opal Mound fault, may in part be an expression of leakage of hot water northwestward out of the reservoir. Wells 82-33 and 24-36 are dry, while well 12-35 is productive.

The resistivity contours of Figure 18 correspond with the heat flow high very nicely. The resistivity low is primarily due to surface conductivity of clay minerals produced in hydrothermal alteration (Ward, S.H. and Sill, 1976). Leakage of brine northwestward out of the reservoir would appear to be saturating valley alluvium leading to low resistivities beyond the heat flow high. The CSAMT apparent resistivities at 32 Hz, shown in Figure 19, also directly correlate with the zone of high heat flow. Quantitative two-dimensional modeling of the resistivity and CSAMT data has yielded high-resolution targets for drilling; such resolution is not afforded by modeling the heat flow data.

The self-potential map of Figure 20 reveals a number of closures, some of which are positive and some negative. Quantitative interpretation of the self-potential data has not yet been made and its contribution to understanding the reservoir is not known.

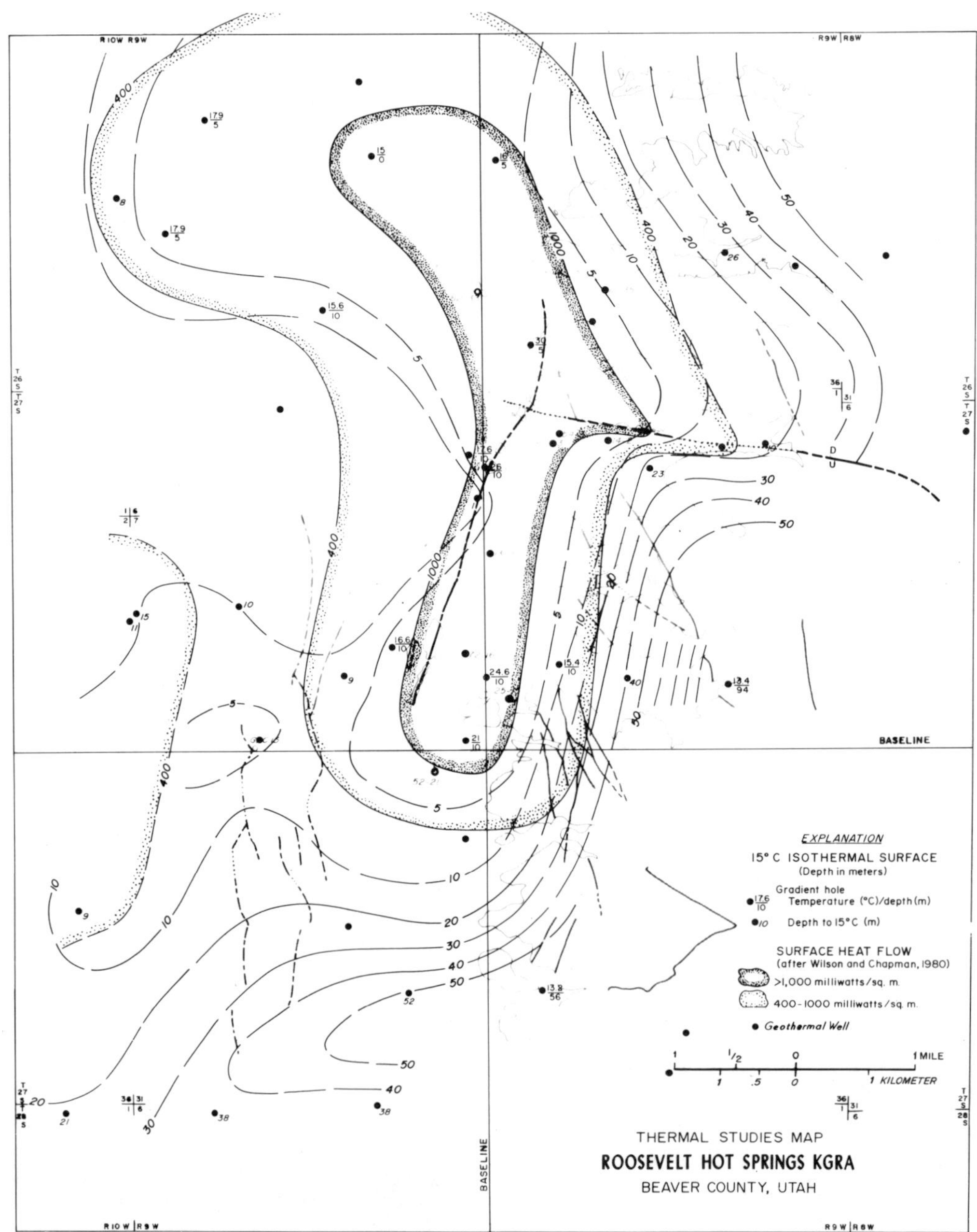

Figure 17. Thermal studies map, Roosevelt Hot Springs KGRA. (from Ross et al., 1982).

Very few microearthquakes have been found beneath wells currently known to be productive. Activity in the region is episodic. During one episode of increased activity, the numerous hypocenters of Figure 21 were located; most occur at a depth of about 5 km. They seem to depict an east-west fault along B-B'. Whether the hypocenters lie on a southward dipping Negro Mag fault, or whether they are associated with another east-west fault is unknown. Research on these microearthquakes and their structural significance is continuing. The Wadati diagram of Figure 22 was developed from the origin times and S- and P-arrival times. A Poisson's Ratio of 0.25 results. Hypocenters occurring beneath the reservoir as currently known, while few in number, yielded a Poisson's Ratio of 0.24. Thus the Poisson's Ratio does not reflect the fact that the reservoir is liquid dominated. It is too early to say whether microearthquakes will contribute to knowledge of the reservoir; to this date they have not.

Douze and Laster (1979) detected no earth

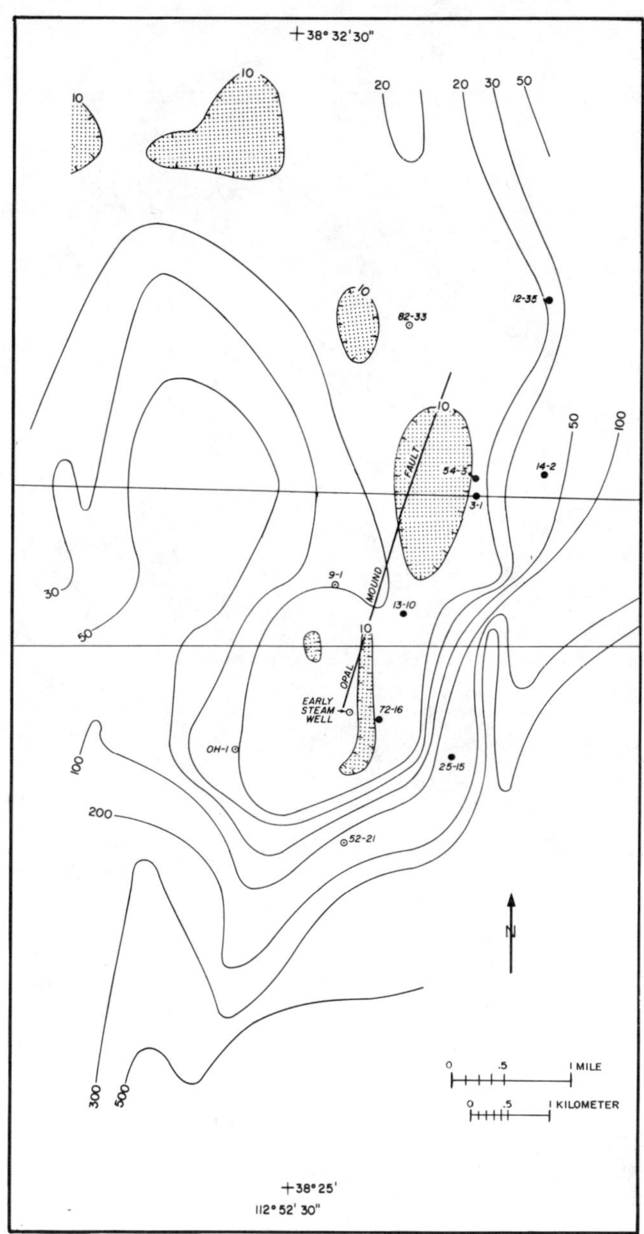

Figure 18. First separation resistivity map from 300 m dipole-dipole survey, Roosevelt Hot Springs KGRA. (after Ward and Sill, 1976).

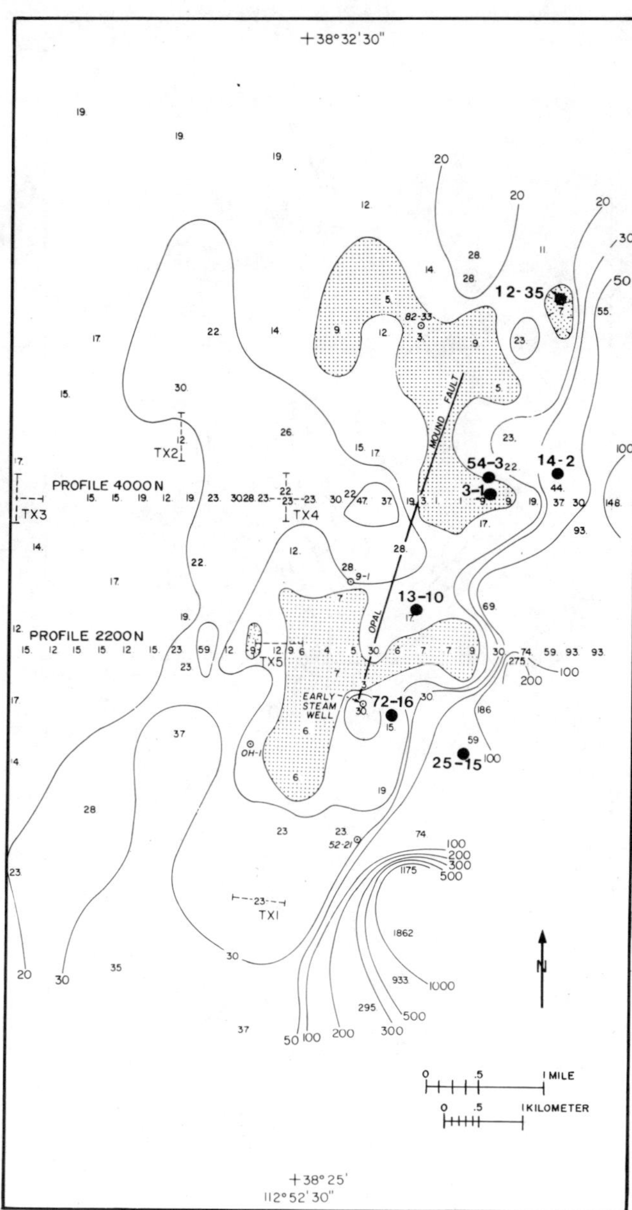

Figure 19. The CSAMT 32 Hz apparent resistivity map of Roosevelt Hot Springs KGRA. (from Sandberg and Hohmann, 1982).

noise that could be related to the geothermal reservoir. Robinson and Iyer (1981) make a modest claim that there is teleseismic evidence for a partial melt at about 15 km depth. Wechsler and Smith (1979) cast doubt on this interpretation and I am inclined to agree with them. Reflection and refraction surveys have yielded information on range front faulting and on depth of valley alluvium, but they have not given any useful information on the reservoir (Ross et al., 1982).

Gravity and magnetic methods have aided in

understanding the structure in the vicinity of Roosevelt Hot Springs, but their contribution to understanding the reservoir is nil.

The magnetotelluric method detected the shallow resistivity structure but at great expense. It also detected a partial melt in the upper mantle. Research is continuing to delineate the lateral extent of the partial melt and to determine whether or not it has any significance to the occurrence of the geothermal reservoir. No deep geothermal system and no magma body were

141

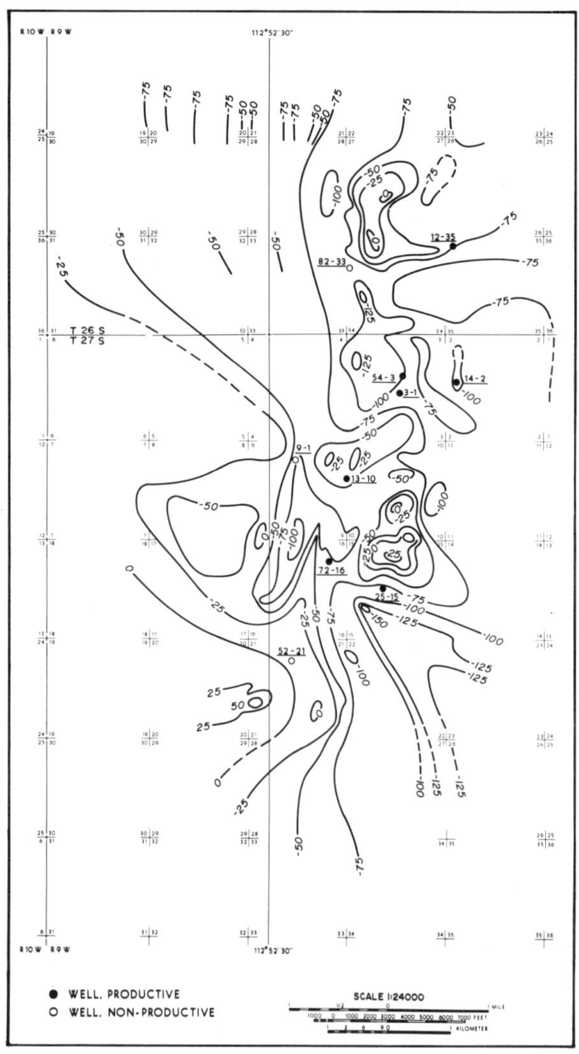

Figure 20. The self-potential map of Roosevelt
Hot Springs KGRA (courtesy W. R. Sill).

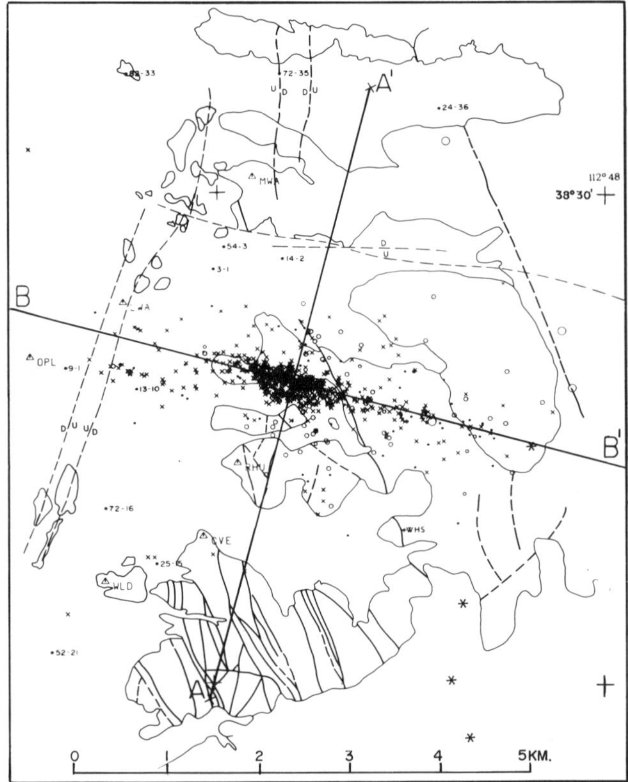

Figure 21. Microearthquakes occurring during
swarm, July 1981. (after Zandt et al.,
1982).

detected by the MT survey. Detectability of magma
chambers by MT is a questionable enterprise as
Newman et al. (1983) have demonstrated.

Tuscarora

Pilkington et al. (1980) and Sibbett (1982)
have described the geology of the Tuscarora
geothermal prospect in Elko County, Nevada.
Figure 23 portrays the simplified representation
of the geology of the prospect according to
Sibbett, who states, "The Tuscarora geothermal
prospect is located at the north end of
Independence Valley in northern Nevada. Thermal
springs issue from Oligocene tuffaceous sediments
near the center of an area of high thermal
gradient. The springs are associated with a large
siliceous sinter mound and are currently
depositing silica and calcium carbonate. Measured

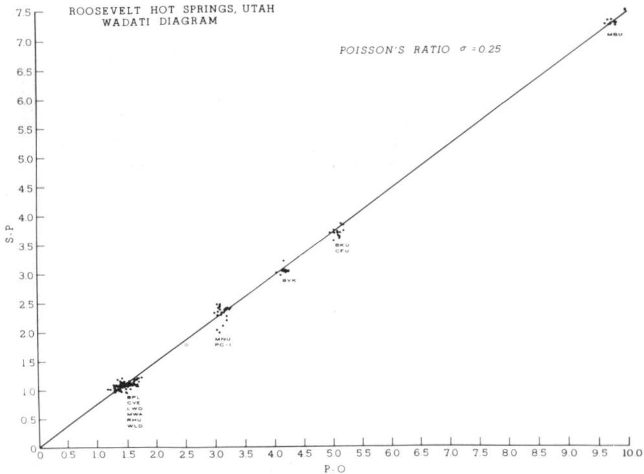

Figure 22. Wadati diagram derived from earthquakes occurring at Roosevelt Hot
Springs KGRA during July, 1981.

142

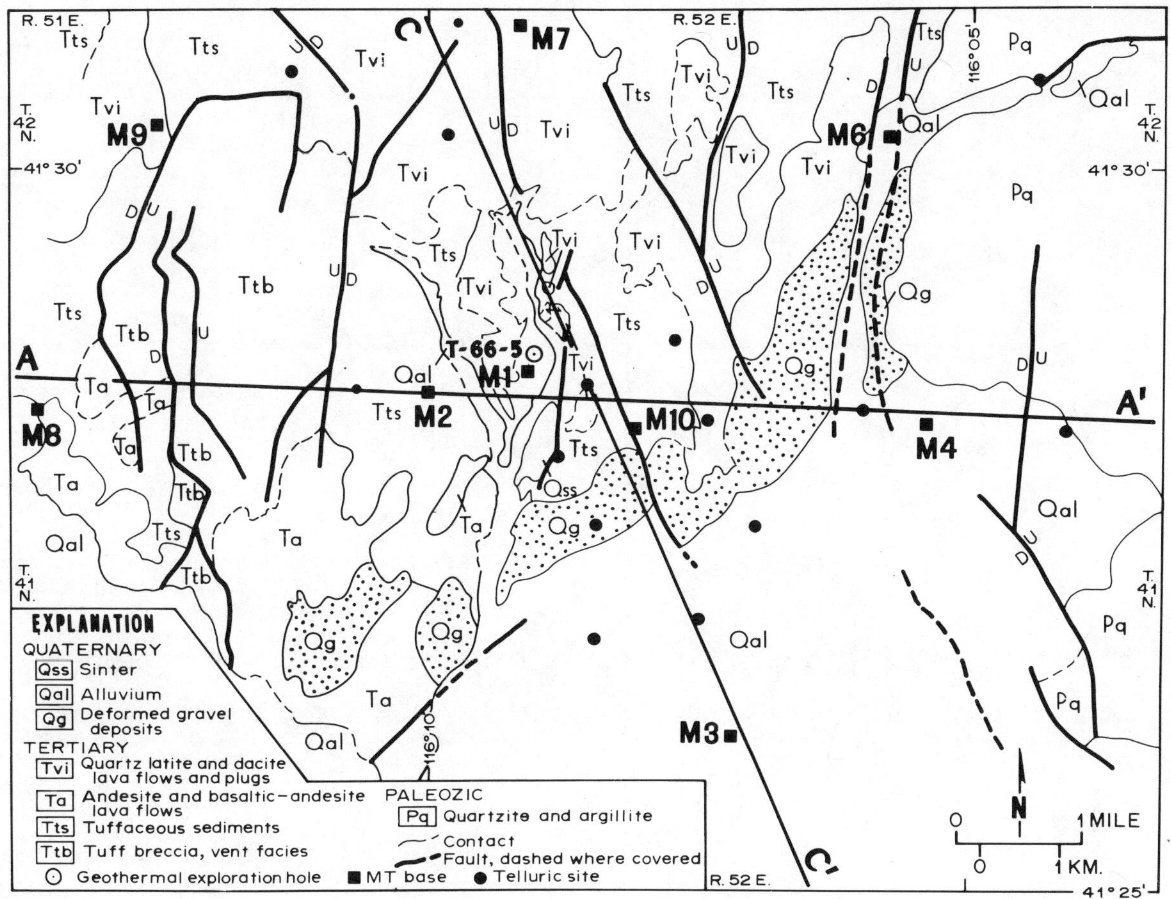

Figure 23. Generalized geology, Tuscarora. (from Sibbett, 1982).

fluid temperatures range up to 95°C, and chemical geothermometers indicate a reservoir temperature of 216°C. The Independence Valley contains 35- to 39-m.y.-old tuffs and tuffaceous sediments which overlie Paleozoic clastic and volcanic rocks and are overlain by Miocene lava and pyroclastic flows. The rocks have been deformed by normal faults trending north-south and northwest and by folds trending north-south which have been active in the Pleistocene."

The heat flow anomaly at the Tuscarora prospect is centered on Hot Sulphur Springs as Figure 24 illustrates (from Pilkington et al., 1980). The north-south faulting is evident in the gravity map of Figure 25 from Pilkington et al., 1980). Meidav and Tonani (1975) observed "that both microearthquake activity and thermal spring occurrences are more commonly associated with that side of the basin which has the steeper gravity gradient". If this is an observation upon which we can rely, let us test it with the available data. From Figure 25 it is evident that the east side of Independence Valley has the steepest gradient of gravity. The range lines in Figures 23 and 25 permit ready correlation of the gravity

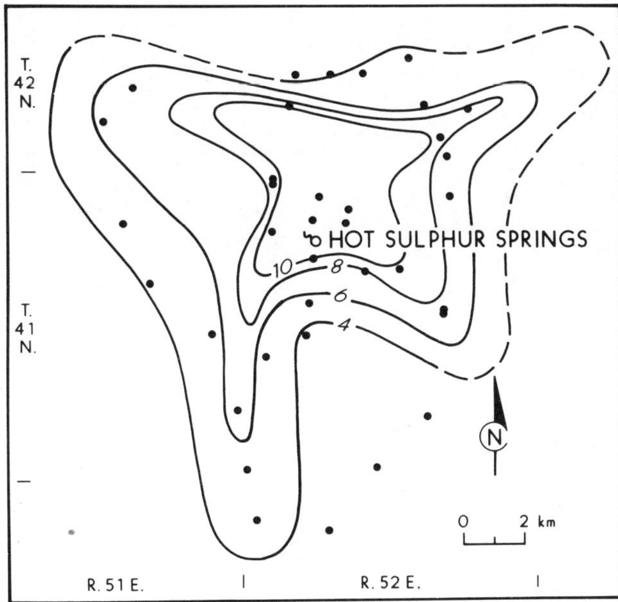

Figure 24. Contours of heat flow at Tuscarora (in HFU). (from Pilkington et al., 1980).

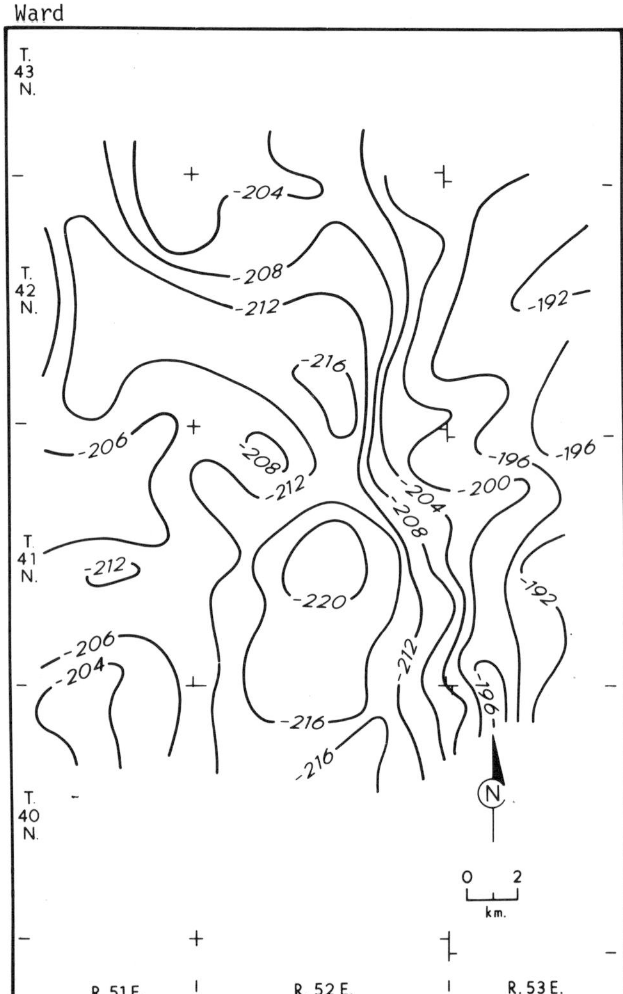

Figure 25. Contours of Bouguer gravity at Tuscarora. (from Pilkington et al., 1980).

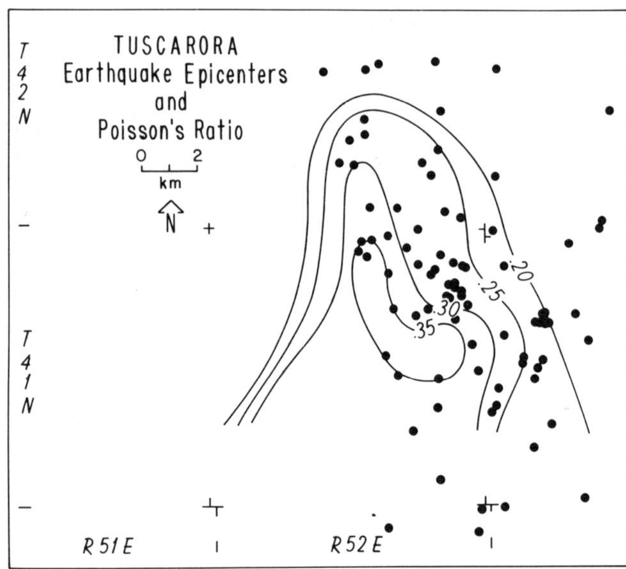

Figure 26. Distribution of microearthquakes and
contours of Poisson's ratio. (from
Pilkington et al., 1980).

data with the geology. The microearthquakes of
Figure 26, as documented by Pilkington et al.
(1980), are scattered but clearly follow a range
front fault suggested by the steep gravity
contours of Figure 25. Contrariwise, Hot Sulphur
Springs occur well west of this presumed fault
line.

As concluded by Sibbett (1982),

"The surface expression of the Hot Sulphur
Springs thermal system is controlled by a fault
zone trending N20°E. Exposed argillic alteration
produced by the thermal system is limited to the
spring area. Quartz-sericite alteration which
predates the present thermal system is present
along the fault zone."

"The subsurface character of the geothermal
system is not known, but the geophysical and
geological data are consistent with an
interpretation that the reservoir is 3 to 5 km
southeast of the hot springs. In this model,
meteoric water circulates down along the range-
front fault system and is heated at depth. The
thermal waters rise along major fractures, perhaps
the intersection of the N10°E and N30°W fault
zones, into either a solution reservoir in the
lower-plate carbonates or a fracture reservoir in
the overlying Valmy Group quartzite. The fracture
reservoir and feeder channelways may have been
formed by brecciation along the thrust fault and
by formation of the deep graben. The reservoir
cap consists of the incompetent and less permeable
Tertiary tuffs and tuffaceous sedimentary rocks,
the base being 1,200 m or more below the
surface. Some of the thermal fluids migrate up
major fractures within the Paleozoic shale, chert
and greenstone unit which overlies the Valmy Group
quartzite. The fluids probably move updip to the
northwest along gravel aquifers either at the base
of or within the tuffaceous sedimentary rocks,
ultimately reaching the surface along the faults
at the hot springs. Cold water aquifers in the
thick quartzite gravel overlying the tuffaceous
sedimentary rocks apparently mask the thermal
anomaly directly above the reservoir."

Drilling on this prospect has been confined
to the general vicinity of the heat flow high
surrounding Hot Sulphur Springs. If Sibbett's
model is correct, and resistivity data
(Mackelprang, 1982) would tend to support it, then
drilling might recommence along the zone of the
eastern range front depicted by the gravity and
microearthquake data. In this region, the
contoured values of Poisson's Ratio shown in
Figure 26 are probably not indicative of the
characteristics of the reservoir but rather of the
fluid saturated alluvium of Independence Valley
(Berkman and Lange, 1980).

Beowawe

Figure 27 depicts the generalized geologic
map at Beowawe (from Sibbett, 1983). The Beowawe
Geysers have formed a 850 m long sinter terrace at
the base of the Malpais scarp.

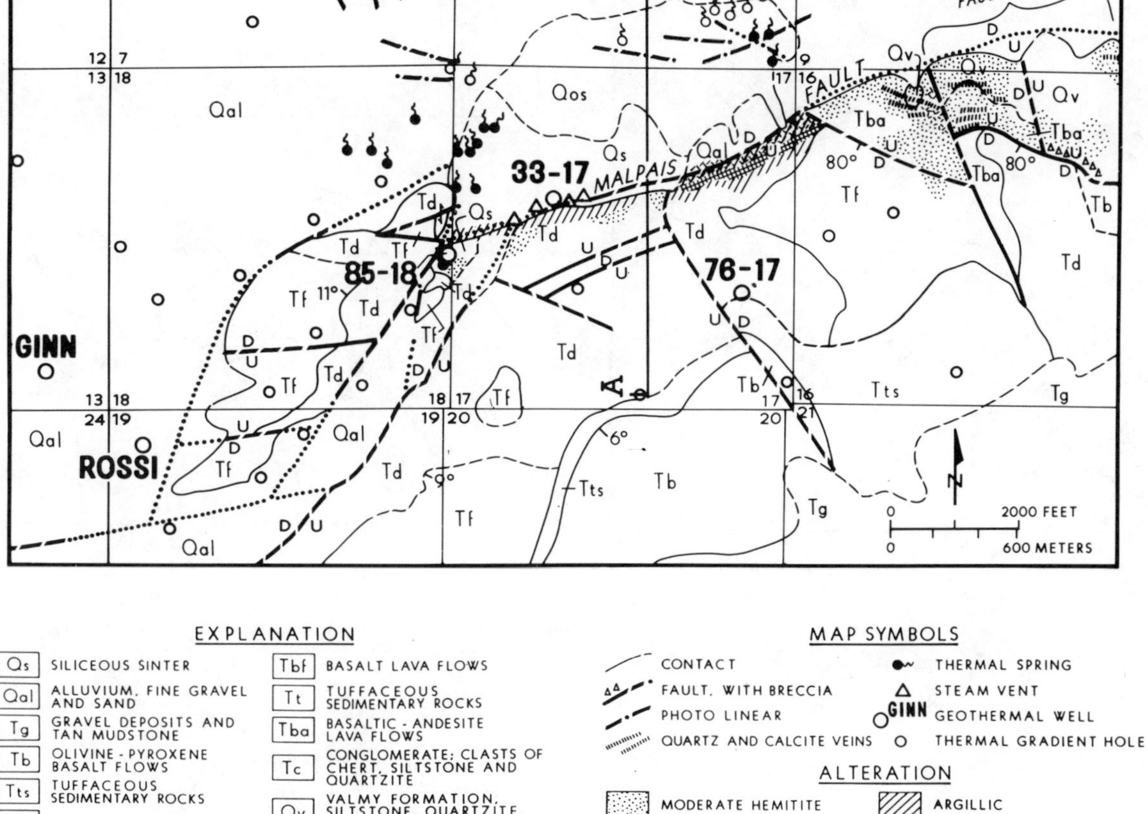

EXPLANATION

Qs	SILICEOUS SINTER
Qal	ALLUVIUM, FINE GRAVEL AND SAND
Tg	GRAVEL DEPOSITS AND TAN MUDSTONE
Tb	OLIVINE - PYROXENE BASALT FLOWS
Tts	TUFFACEOUS SEDIMENTARY ROCKS
Tf	FELSITE LAVA FLOWS
Td	PORPHYRITIC DACITE LAVA FLOWS
Tbf	BASALT LAVA FLOWS
Tt	TUFFACEOUS SEDIMENTARY ROCKS
Tba	BASALTIC - ANDESITE LAVA FLOWS
Tc	CONGLOMERATE; CLASTS OF CHERT, SILTSTONE AND QUARTZITE
Ov	VALMY FORMATION, SILTSTONE, QUARTZITE, CHERT AND ARGILLITE

MAP SYMBOLS

CONTACT
FAULT, WITH BRECCIA
PHOTO LINEAR
QUARTZ AND CALCITE VEINS
THERMAL SPRING
STEAM VENT
GINN — GEOTHERMAL WELL
THERMAL GRADIENT HOLE

ALTERATION

MODERATE HEMITITE
STRONG HEMATITE STAIN
ARGILLIC
SILICIFICATION AND MINOR SINTER

Figure 27. Generalized geology, Beowawe. (from Sibbett, 1983).

"The Beowawe geothermal system in northern Nevada is a structurally controlled, water-dominated resource with a measured temperature of 212°C (414°F). Surface expression of the system consists of a large, active opaline sinter terrace that is present along a Tertiary to Quaternary normal fault escarpment. The thermal system appears to be controlled by the subsurface intersection of the east-northeast trending, north dipping Malpais fault with a pre-existing northwest trending fault which dips south and has 884 m of vertical displacement.

Surface alteration associated with the geothermal system is vertically zoned along the Malpais escarpment with, from base to top: hematite stained, argillized rock along the fault trace; silicification and quartz veining; and argillic, acid leach zone at the top. Subsurface alteration generally increases with depth in the volcanic rocks and is most intense in basaltic-andesite lava flows which are capped by tuffaceous sedimentary rock." (Sibbett, 1983)

Figure 28 presents the contoured heat flow values, which provide a focus for attention to the flexure in the Malpais Fault near well 85-18. Smith (1983) discusses the thermal hydrology and heat flow at Beowawe. Swift (1979) outlined the area of low resistivity, as in Figure 29, and this outline passes through the center of the heat flow high. A dipolar self-potential anomaly, shown in Figure 30, best delineates the convective system at Beowawe (Swift, 1979).

According to Swift (1979), gravity and magnetic data delineate the Malpais Fault and the important north-northwest structures which may control the reservoir at depth. Seismic reflection data are too noisy, because of the interbedded volcanic section, to provide definitive structural information. In contrast, Swift (1979) reports that the Geysers area at Beowawe appears as an earth noise source. Magnetotelluric data seem to have yielded only regional information at Beowawe, based on Swift's remarks.

6.0 COMPARATIVE CASE HISTORIES

An evaluation has been made, in Table 3, of the contribution made by each of 14 geophysical

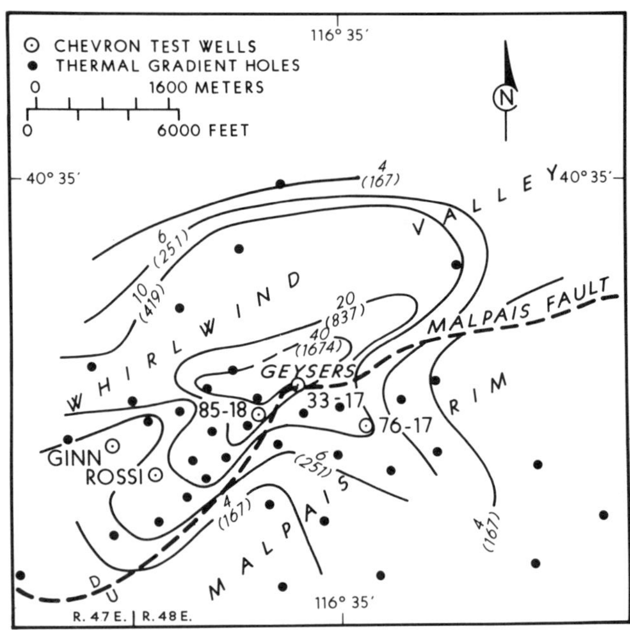

Figure 28.   Contours of heat flow at Beowawe (in HFU). (courtesy B. S. Sibbett).

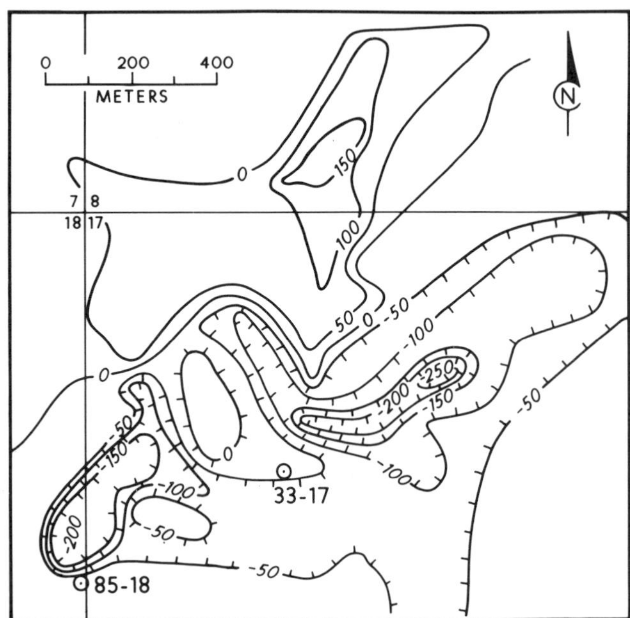

Figure 30.   Contours of self-potential at Beowawe. (from DeMoulley and Corwin, 1980).

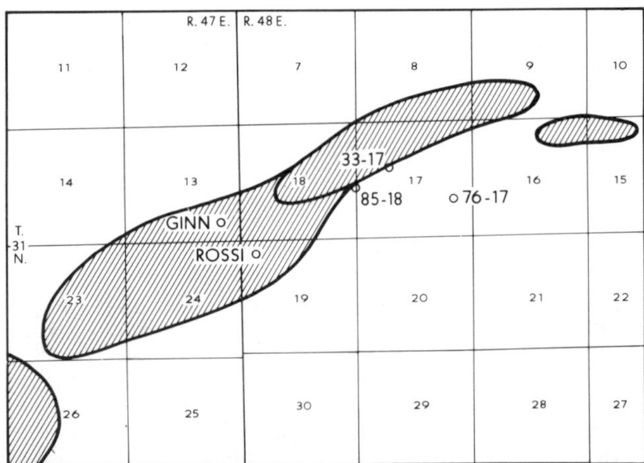

Figure 29.   Outline of zone of low resistivity at Beowawe. (after Swift, 1979).

methods to understanding the known or postulated reservoir at each of 13 geothermal prospects. Each method has been rated from 1 (good) through 4 (poor). The rating is subjective, but a serious attempt has been made to apply the rating uniformly throughout the matrix of Table 3. The surprising results of this analysis are that: a) geophysical methods are uniformly inconsistent in performance, b) no geophysical method ranks one, five methods rank about two, eight rank between two and one half and three, while none ranks four, c) no combination of any four methods has been successful at more than one site where "successful" means a ranking of one or two, d) the most useful of the methods, judging by their mean rankings, are heat flow, microearthquakes, gravity, resistivity, and self-potential, e) the least effective methods are earth noise, magnetics, and magnetotellurics. The CSAMT method has been included with CSEM for purposes of the comparative study. There are 23 entries of good (1), 29 of fair (2), 48 of questionable (3), and 10 entries of poor (4) in Table 3. Thus 53% of the time the results of geophysical surveys in geothermal exploration are either questionable or poor.

Overall, this study provides a somewhat discouraging picture. I would tend to use heat flow, microearthquakes, gravity, resistivity (or CSAMT), and self-potential methods at all prospects. Once these data were interpreted and correlated, I would then decide whether or not additional geophysical surveying was justified. All geophysical surveys should be designed with one or more conceptual geological models in mind and the density and extent of the geophysical coverage should be compatible with the range of conceptual geologic models.

Clearly the reader will want to know why each method has performed poorly or questionably under some circumstances. To attempt to satisfy the reader's curiosity, I have prepared Table 4 in which I have indicated my assessment of why a method rated a 3 or a 4 in Table 3. Six categories of answers seem to be sufficient. These are:

TABLE 3 CONTRIBUTION TO UNDERSTANDING RESERVOIR

METHOD	COSO*	LONG VALLEY+	ROOSEVELT HOT SPRINGS*	BALTAZOR	BEOWAWE*	COVE FORT+	DESERT PEAK*	GRASS VALLEY	McCOY	SAN EMIDIO	SODA LAKE+	TUSCARORA	RAFT RIVER	MEAN
	IGNEOUS RELATED													
EARTH NOISE	3	3	4		1			4	3	3				3.0
MICROEARTHQUAKES	1	3	3	3		2		1	1			1		1.9
TELESEISMS	2	1	3						3			4		2.6
REFRACTION	2	2	3					3					2	2.4
REFLECTION			3		3	4		1		2	2			2.5
GRAVITY	4	2	4	1	2	3	3	1	3	1	2	1	2	2.2
MAGNETICS	1	3	4	3	3	2		4	3		4	3	3	3.0
RESISTIVITY	2	3	1	1	2	1	4	2	3	2	3	2	3	2.2
CSEM & CSAMT		3	1					3	3		2			2.4
SCALAR AMT	2	3		2									3	2.5
MT / AMT			3		3		3	3	3		3	3	3	3.0
SELF POTENTIAL		1	2	3	1			2	2			3	3	2.1
TELLURICS	3				3			2	2					2.5
HEAT FLOW	1	3	1	2	1	2	1	2	3	3	1	3	2	1.9

* EXPECTED TO BE COMMERCIAL IN NEAR TERM
+ EXPECTED TO BE COMMERCIAL IN LONG TERM

CODE: 1 GOOD
2 FAIR
3 QUESTIONABLE
4 POOR

TABLE 4 REASONS FOR POOR PERFORMANCE OF GEOPHYSICAL METHODS

METHOD	COSO*	LONG VALLEY+	ROOSEVELT HOT SPRINGS*	BALTAZOR	BEOWAWE*	COVE FORT+	DESERT PEAK*	GRASS VALLEY	McCOY	SAN EMIDIO	SODA LAKE	TUSCARORA	RAFT RIVER	SUMMARY
	IGNEOUS RELATED													
EARTH NOISE	T_1	T_1	N						N	R		T_1		3T_1,2N,1R
MICROEARTHQUAKES		R	R	R										3R
TELESEISMS	I		I							R			R	2R,2I
REFRACTION			N					N						2N
REFLECTION			N		T_2	T_1								1T_1,1T_2,1N
GRAVITY	N		N		N	N				R				4N,1R
MAGNETICS		R	N	R	R				N	R	R	D	R	6R,2N,1D
RESISTIVITY		R					T_1			D		R	R	3R,1T_1,1D
CSEM & CSAMT		R							I	I				2I,1R
SCALAR AMT		R											R	2R
MT / AMT			N		N		I	I	I		N	I	R	4I,3N,1R
SELF POTENTIAL			T_1									R	R	2R,1T_1
TELLURICS	R		D											1R,1D
HEAT FLOW		R							R	R		R		4R

T_1 - TECHNOLOGY DEFICIENT, HAS BEEN IMPROVED
T_2 - TECHNOLOGY DEFICIENT, HAS NOT BEEN IMPROVED
I - INTERPRETATION PROCEDURE QUESTIONABLE
D - DATA SET INCOMPLETE
N - NO RECOGNIZABLE SIGNATURE OVER RESERVOIR
R - RELATIONSHIP TO RESERVOIR UNCERTAIN

T_1 - technology deficient, but has since been improved,

T_2 - technology deficient, and has not yet been improved,

I - interpretation procedure questionable,

D - data set incomplete,

N - no recognizable signature over reservoir, and

R - relationship to reservoir uncertain.

Table 4 contains 6 T_1, 1 T_2, 12 N, 28 R, 8 I, and 3 D entries. The T_1, I and D entries total 17 (29%), indicating that improvement in equipment or interpretation has or can be made. Unfortunately, we can do nothing about the poor ratings associated with the 41 entries (71%) reporting T_2, N, and R.

The following comments follow on each method:

earth noise

The technology has improved so that the method should perform better when advantage is taken of f-k processing.

microearthquakes

Some geothermal reservoirs do not yield sufficient numbers of microearthquakes to permit the method to be successful.

teleseisms

The method does not always produce useful results.

refraction seismology

The method is not always applicable to reservoir delineation. It should be used late in the exploration sequence.

reflection seismology

Data processing advances in reflection seismology may ultimately permit us to work in volcanic environments and efforts should be directed toward that end. However, the method will not always be applicable to reservoir delineation and should be used late in the exploration sequence.

gravity and magnetics

The main use of the gravity and magnetic methods will continue to be as aids in geological mapping of structure. Gravity seems to be quite useful in mapping range front faults with which some geothermal reservoirs are associated. The magnetic method occasionally will prove useful in mapping zones of magnetite destruction.

resistivity

The technology of data acquisition, processing, and interpretation has been much improved for Schlumberger and dipole-dipole resistivity surveys. Algorithms for 2D modeling of resistivity data are now used routinely. Algorithms for 3D modeling are available. The bipole-dipole method has not proven itself to be satisfactory for reservoir delineation in most cases.

CSEM/CSAMT

These methods have not been sufficiently tested in geothermal environments to be certain of their future. Interpretation procedures for CSAMT are capable of encompassing 2-D and 3-D earth models, although there has been a reluctance on the part of industry to use them. Reliable computer algorithms to permit 2D modeling of CSEM data are only now emerging. Algorithms for 3D modeling are still on the drawing boards.

scalar AMT

Scalar AMT doesn't always produce results that are as good as we prefer and its use should be restricted to reconnaissance surveys.

MT/AMT

There is a tendency to use 1D inversion, only, in interpretation of MT/AMT data. Algorithms for 2D and 3D interpretation are available but industry seems reluctant to use them. This is the weakest link in application of the method. Data acquisition and processing have improved greatly in recent years, to the extent that there is no excuse for data of poor quality.

There is also a tendency by industry to use MT/AMT when it is not warranted. It should be reserved for special situations, usually late in the exploration sequence.

self-potential

The self-potential method shows great promise, if the new quantitative interpretation schemes work as expected. The method doesn't always produce meaningful signatures over geothermal systems.

tellurics

Since it lacks a truly quantitative scheme for interpretation, the method should be used only in reconnaissance.

heat flow

As Blackwell (1983) has indicated, heat flow as measured at surface is not always a reliable indicator of a high quality geothermal resource. Reference should be made to Blackwell's paper for cognizance of the advantages and limitations of heat flow in geothermal applications.

7.0 CONCLUSIONS

The performance of 14 different geophysical methods used at 13 high temperature geothermal sites in the Northern Basin and Range Province has been somewhat disappointing. Heat flow, microearthquakes, gravity, resistivity and self-potential methods appear to be the most consistently useful, although none of them performs up to expectations all of the time. Recent improvements in interpretation procedures are expected to benefit the resistivity and self-potential methods. The least effective methods at the present time are earth noise, reflection seismology, magnetics, magnetotellurics, and tellurics. Recent improvements in survey design will make the earth noise method much more effective for those reservoirs which emit earth noise. Accomplished improvements in interpretation of magnetotelluric data, if adopted by industry, will improve that method's performance. However, the magnetotelluric method ought to be reserved for those special prospects to which it is applicable. Improvements in processing reflection seismic data in volcanic covered areas may make that method more generally applicable in the late stages of an exploration sequence. The teleseismic method, developed to a reasonable degree of sophistication, may occasionally contribute, in a broad sense, to understanding the reservoir. The refraction seismic method probably ought to be considered a subset of combined reflection-refraction surveys for optimum resolution of subsurface velocity structures. However, refraction seismic surveys should be considered also as an adjunct to microearthquake surveys since they can provide 3D velocity distributions necessary for interpretation of microearthquake data. Controlled source electromagnetic and controlled source audiomagnetotelluric methods require much more exposure in geothermal exploration before they can be evaluated properly. However, the controlled source audiofrequency magnetotelluric method looks very promising, and it may replace resistivity surveying if its cost-effectiveness can be established as is expected. Scalar AMT and tellurics seem destined for use as inexpensive reconnaissance techniques which will find occasional application.

There is evidence of need for improvement in survey design, data acquisition, data processing, and data interpretation in geothermal exploration in the Northern Basin and Range Province. Yet the predominant problems with application of geophysical methods in this environment are related to fundamental limitations of the methods as have been described in the text.

Of the five case histories presented for comparison in this article, we find the following results.

Long Valley

While no high temperature reservoir has been defined, there is a chance that ultimately the teleseismic, microearthquake and self-potential methods may prove to be valuable in delineating a postulated fracture-controlled reservoir beneath the medial graben in the resurgent dome. The existence of such a reservoir probably requires regional faulting through the resurgent dome following the graben, and subsequent extension produced by recent bulging of the dome. Refraction seismology and gravity have been exceptionally useful in quantifying the potential reservoir, advocated by the USGS, to be the Bishop Tuff within the Long Valley caldera. The heat flow anomaly at this site may be misleading.

Coso

The combination of resistivity, scalar AMT, shallow heat flow, and airborne magnetic data, when added to the knowledge of hydrothermal alteration provided by careful geological mapping, have clearly defined the drilling targets. Teleseismic P-wave delays and refraction seismology are methods which have provided information of secondary importance.

Roosevelt Hot Springs

The heat flow pattern at Roosevelt Hot Springs has been fundamental in outlining the reservoir. However, careful application of the resistivity and CSAMT techniques has provided the detail necessary to site wells. Self-potential data at this site have not yet been interpreted quantitatively but ultimately it may prove to be useful in understanding and delineating the reservoir.

Tuscarora

Microearthquakes and gravity draw attention to the eastern margin of Independence Valley, while heat flow and surface manifestations draw attention to a region around Hot Sulphur Springs about 4 km to the west. Drilling has been concentrated in the vicinity of Hot Sulphur Springs whereas our speculation dictates that the reservoir might be along the range front fault bounding the eastern margin of Independence Valley. Dipole-dipole resistivity interpretations would tend to confirm this notion. The heat flow data at this site may be misleading.

Beowawe

The self-potential method has been the best confirmation of the heat flow anomaly which seems to be centered over the shallow reservoir. Dipole-dipole resistivity has provided supporting evidence for the location of the drilling target. Gravity and magnetics have helped to define the Malpais Fault zone while application of MT/AMT was necessary to attempt to delineate the deep reservoir.

From the above comments it should be evident that there is no common denominator of geophysical methods which might lead us to a better record of discovery and delineation of geothermal reservoirs. Exploration for igneous-related geothermal systems is no different from those geothermal systems of no obvious igneous relationship. Exploration for geothermal resources of any type, those with surface manifestations or those without, is difficult.

8.0 ACKNOWLEDGEMENTS

This review article has only been made possible by my reliance upon the library facilities of the Earth Science Laboratory of the University of Utah Research Institute. I am grateful to C. M. Swift, Jr. and Dennis L. Nielson, for excellent reviews which have markedly improved the manuscript. P. M. Wright kindly provided a final editing of the manuscript. Connie Pixton, Doris Cullen, Pat Daubner, Sandra Bromley, and Paul Onstott prepared the illustrations. Joan Pingree dutifully and superbly turned out draft after draft on her word processor. I am indebted to all of the above mentioned people.

This report has been prepared under DOE Contract No. DE-AC07-80ID12079 with the U.S. Department of Energy.

REFERENCES

Ackerman, H.D., 1979, Seismic refraction study of the Raft River geothermal area, Idaho: Geophysics, v. 44, p. 216-225.

Aiken, C.L.V., and Ander, M.E., 1981, A regional strategy for geothermal exploration with emphasis on gravity and magnetotellurics: J. Volc. and Geotherm. Res., v. 9, p. 1-27.

Anderson, L.A., and Johnson, G.R., 1976, Application of the self-potential method to geothermal exploration in Long Valley, California: J. Geophys. Res., v. 81, p. 1527-1532.

Applegate, J.K., Goebel, V.S., Kallenberger, P., and Rossow, J., 1981, The use of seismic reflection techniques in geothermal areas throughout the U.S.: extended abstract, 5th Annual International Meeting and Exposition, Society of Exploration Geophysicists, Los Angeles, Oct. 11-15.

Arnorsson, S., Bjornsson, A., Gislason, G., and Gudmundsson, G., 1975, Systematic exploration of the Krisuvik high-temperature area, Reykjanes Peninsula, Iceland: Proc. Second U.N. Symposium on the Development and Use of Geothermal Resources, San Francisco, v. 2, p. 853-864.

Bailey, R.A., Dalrymple, G.B., and Lanphere, M.A., 1976, Volcanism structure and geochronology of Long Valley caldera, California: J. Geophys. Res., v. 81, p. 725-744.

Ballantyne, J.M., 1978, Hydrothermal alteration at the Roosevelt Hot springs thermal area, Utah: modal mineralogy and geochemistry of sericite, chlorite, and feldspar from altered rocks, Thermal Power Company well Utah State 14-2: Univ. of Utah, Dept. of Geol. and Geophys. Rept., 42 p.

Bamford, R.W., Christensen, O.D., and Capuano, R.M., 1980, Multielement geochemistry of solid materials in geothermal systems and its application, Part I: The hot-water systems at Roosevelt Hot Springs KGRA, Utah: Univ. of Utah Res. Inst., Earth Sci. Lab., Rept. ESL-30.

Bannister, P.R., 1969, Source distance dependence of the surface-impedance conductivity measurement technique: Geophysics, v. 34, p. 785-788.

Banwell, C.J., 1970, Geophysical techniques in geothermal exploration: U.N. Symposium on the Development and Utilization of Geothermal Resources, Pisa, Geothermics Spec. Issue 2, v. 2, pt.1, p. 32-56.

Banwell, C.J., and Macdonald, W.J.P., 1965, Resistivity surveying in New Zealand thermal areas: Eighth Commonwealth Mining and Metallurgical Congress, Australia and New Zealand, New Zealand Section, p. 1-7.

Baudu, R., Bernhard, J., Georgel, J.M., Griveau, P., Rugo, R., 1980, Application of d.c. dipolar methods in the Upper Rhinegraben: Advances in European and Geothermal Research, Dordrecht, Holland, D. Reidel Co., p. 823-832.

Benoit, W. R., and Butler, R. W., 1983, A review of high temperature geothermal developments in the northern Basin and Range Province: this volume.

Berkman, F., and Lange, A.L., 1980, Tuscarora geophysics - preliminary report: Amax Exploration, Inc. internal report, open filed by Univ. of Utah Res. Inst., Earth Sci. Lab., 7 p.

Berktold, A., 1982, Electromagnetic studies in geothermal regions: Proc. 6th Workshop on Electromagnetic Induction in the Earth and Moon, Dept. Physics, Univ. of Victoria, Canada.

Berktold, A., and Kemmerle, K., 1982, Distribution of electrical conductivity in the Urach geothermal area, a magnetotelluric and geomagnetic depth sounding investigation, The Urach Geothermal Project: Stuttgart, E. Schweizerdort'sche Verlagsbuchhandling, p. 289-300.

Beyer, J.H., 1977, Telluric and D.C. resistivity techniques applied to the geophysical investigation of Basin and Range geothermal systems: Univ. of California, Lawrence Berkeley Lab., Rept. LBL-6325, 461 p., 3 vol.

Beyer, J.H., Morrison, H.F., and Dey, A., 1975, Electrical exploration of geothermal systems in the Basin and Range valleys of Nevada: Proc. Second U.N. Symposium on the Development of Geothermal Resources, San

Francisco, p. 889-894.

Beyer, J.H., Dey, A., Liaw, A., Major, E., McEvilly, T.V., Morrison, H.F., and Wollenberg, H., 1976, Preliminary open file report, geological and geophysical studies in Grass Valley, Nevada: University of California, Lawrence Berkeley Lab., Rept. LBL-5262, 144 p.

Bibby, H.M., and Risk, G.F., 1973, Interpretation of dipole-dipole resistivity surveys using a hemispheroidal model: Geophysics, v. 38, p. 719-736.

Blackwell, D. D., 1983, Heat flow in northern Basin and Range Province: this volume.

Capuano, R.M., and Cole, D., 1982, Fluid-mineral equilibria in high temperature geothermal systems: the Roosevelt Hot Springs geothermal system, Utah: Geochim. et Cosmochim. Acta, v. 46, p. 1353-1364.

Cheng, W.T., 1970, Geophysical exploration in the Tatum volcanic region, Taiwan: Proc. U.N. Symposium on the Development and Utilization of Geothermal Resources, Pisa, Geothermics Spec. Issue 2, v. 2., pt. 1, p. 262-274.

Chu, J.J., Ward, S.H., Sill, W.R., Stodt, J.A., 1983, Induced polarization at Roosevelt Hot Springs Geothermal Area, Utah: unpublished manuscript.

Combs, J., 1980, Heat flow in the Coso geothermal area, Inyo County, California: J. Geophys. Res., v. 85, p. 2411-2424.

Combs, J., and Rotstein, Y., 1975, Microearthquake studies at the Coso geothermal area, China Lake, California: Proc. Second U.N. Symposium on the Development and Use of Geothermal Resources, San Francisco, p. 917-928.

Combs, J., and Jarzabek, D., 1977, Geothermal: State of the Art: Trans., Geoth. Res. Council, v. 1, p. 41-44.

Corwin, R.F., 1975, Self-potential exploration for geothermal reservoirs: Proc. Second U.N. Symposium on the Development and Use of Geothermal Resources, San Francisco, p. 937-946.

Corwin, R.F., and Hoover, D.B., 1979, The self-potential method in geothermal exploration: Geophysics, v. 44, p. 226-245.

DeMoully, G.T., and Corwin, R.F., 1980, Self-potential survey results from the Beowawe KGRA, Nevada: Trans., Geoth. Res. Council, v. 4, p. 33-36.

Dey, A., and Morrison, H.F., 1977, An analysis of the bipole-dipole method of resistivity surveying: Geothermics, v. 6, p. 47-81.

Douze, E.J., and Laster, S.J., 1979, Seismic array noise studies at Roosevelt Hot Springs, Utah, geothermal area: Geophysics, v. 44, p. 1570-1583.

Dupis, A., Marie, Ph., and Petian, G., 1980, Magnetotelluric prospecting of the Mont Dore area: Advances in European Geothermal Research, D. Reidel, Dordrecht, Holland, p. 935-943.

Edmiston, R.C., 1982, A review and analysis of geothermal exploratory drilling results in the northern Basin and Range geologic province of the U.S.A. from 1974 through 1981: Trans, Geoth. Res. Council, v. 6, p. 11-14.

Edquist, R.K., 1981, Geophysical investigation of the Baltazor Hot Springs known geothermal resource area and the Painted Hills thermal area, Humboldt County, Nevada: Univ. of Utah Res. Inst., Earth Sci. Lab., Rept. DOE/ID/12079-29, 89 p.

Fox, R.C., 1978a, Low-altitude aeromagnetic survey of a portion of the Coso Hot Springs KGRA, Inyo County, California: Univ. of Utah Res. Inst., Earth Sci. Lab., Rept. IDO/77.5.7., 19 p.

Fox, R.C., 1978b, Dipole-dipole resistivity survey of a portion of the Coso Hot Springs KGRA, Inyo County, California: Univ. of Utah Res. Inst., Earth Sci. Lab., Rept. IDO/77.5.6., 21 p.

Fox, R.C., Hohmann, G.W., Killpack, T.J., and Rijo, L., 1980, Topographic effects in resistivity and induced polarization surveys: Geophysics, v. 43, p. 144-172.

Frangos, W., and Ward, S.H., 1980, Bipole-dipole survey at Roosevelt Hot Springs thermal area, Beaver County, Utah: Univ. of Utah Res. Inst., Earth Sci. Lab., Rept. DOE/ID/12079-15, 41 p.

Gamble, T.D., Goubau, W.M., and Clarke, J., 1979a, Magnetotellurics with a remote reference: Geophysics, v. 44, p. 53-68.

Gamble, T.D., Goubau, W.M., and Clarke, J., 1979b, Error analysis for remote reference magnetotellurics: Geophysics, v. 44, p. 959-968.

Gamble, T.D., Goubau, W.M., Goldstein, N.E., and Clarke, J., 1980, Referenced magnetotellurics at Cerro Prieto: Geothermics, v. 9, p. 49-63.

Garcia, D.S., 1975, Geoelectric study of the Cerro Prieto geothermal area, Baja, California: Proc. Second U.N. Symposium on the Development and Use of Geothermal Resources, San Francisco, v. 2, p. 1009-1012.

Ward

Gertson, R.C., and Smith, R.B., 1979, Interpretation of a seismic refraction profile across the Roosevelt Hot Springs, Utah and vicinity: Univ. of Utah, Dept. Geol. and Geophys., Rept. IDO/78-1701.a.3, 116 p.

Goldstein, N.E., and Paulsson, B., 1979, Interpretation of gravity surveys in Grass and Buena Vista Valleys, Nevada: Geothermics, v. 7, p. 29-50.

Goldstein, N.E., Mozley, E., Gamble, T.D., and Morrison, H.F., 1978a, Magnetotelluric investigations at Mt. Hood, Oregon: Trans. G.R.C., v. 2, p. 219-221.

Goldstein, N.E., Norris, R.A., and Wilt, M.J., 1978b, Assessment of surface geophysical methods in geothermal exploration and recommendations for future research: Univ. of California, Lawrence Berkeley Laboratory, Rept. LBL-6815, 166 p.

Goldstein, N.E., Mozley, E., and Wilt, M., 1982, Interpretation of shallow electrical features from electromagnetic and magnetotelluric surveys at Mount Hood, Oregon: J. Geophys. Res., v. 87, p. 2815-2828.

Gupta, M.L., Singh, S.B., and Rao, B.V., 1975, Studies of direct current resistivity in the Puga geothermal field, Himalayas, India: Proc. Second U.N. Symposium on the Development and Use of Geothermal Resources, San Francisco, v. 2, p. 1029-1036.

Gupta, H.K., Ward, R.W., and Lin, T-L., 1982, Seismic wave velocity investigation at The Geysers - Clear Lake geothermal field, California: Geophysics, v. 47, p. 819-824.

Hamilton, R.M., and Muffler, L.J.P., 1972, Microearthquakes at The Geysers geothermal area, California: J. Geophys. Res., v. 77, p. 2081-2086.

Hatherton, T., Macdonald, W.J.P., and Thomson, G.E.K., 1966, Geophysical methods in geothermal prospecting in New Zealand: Bull. Volcanology, p. 485-497.

Hermance, J.F., and Pedersen, J., 1977, Assessing the geothermal resource base of the southwestern U.S.; status report of a regional geoelectromagnetic traverse: Geophysics, v. 42, p. 155-156.

Hermance, J.F., Thayer, R.E., Bjornsson, A., 1975, The telluric-magnetotelluric method in the regional assessment of geothermal potential: Proc. Second U.N. Symposium on the Development and Use of Geothermal Resources, San Francisco, v. 2, p. 1037-1048.

Hermance, J.F., and Peltier, W.R., 1970, Magnetotelluric fields of a line current: J. Geophys. Res., v. 75, p. 3351-3356.

Hill, D. G., Layman, E. B., Swift, C. M., Jr., and Yungul, S. H., 1979, Soda Lake, Nevada, thermal anomaly: Trans., Geoth. Res. Council, v. 3, p. 305-308.

Hill, D.P., 1976, Structures of Long Valley Caldera, California, from a seismic refraction experiment: J. Geophys. Res., v. 81, 5, p. 745-753.

Hochstein, M.P., and Hunt, T.M., 1970, Seismic, gravity, and magnetic studies, Broadlands geothermal field, New Zealand: Proc. U.N. Symposium on the Development and Utilization of Geothermal Resources, Pisa, Geothermics Spec. Issue 2. v. 2, pt. 1, p. 333-346.

Hohmann, G.W., and Jiracek, G.R., 1979, Bipole-dipole interpretation with three-dimensional models: Univ. of Utah Res. Inst., Earth Sci. Lab., Rept. DOE/ET/28392-29, 20 p.

Hoover, D.B., 1981, Self-potential investigations at Mt. Hood, Oregon, paper presented at the Self-Potential Workshop, Golden, Colorado, March 3-4, 1981.

Hoover, D.B., Frischknecht, F.C., and Tippens, C., 1976, Audiomagnetotelluric soundings as a reconnaissance exploration technique in Long Valley, Calif: J. Geophys. Res., v. 81, p. 801-809.

Hoover, D.B., and Long, C.L., 1976, Audiomagnetotelluric methods in reconnaissance geothermal exploration: Proc., Second U.N. Symposium on the Development and Use of Geothermal Resources, San Francisco, p. 1059-1064.

Hoover, D.B., Long, C.L., and Senterfit, R.M., 1978, Some results from audiomagnetotelluric investigations in geothermal areas: Geophysics, 43, p. 1501-1514.

Hulen, J.B., 1978, Geology and alteration of the Coso geothermal area, Inyo County, California: Univ. of Utah Res. Inst., Earth Sci. Lab., Rept. IDO/78-1701-b.4.1, 28 p.

Hutton, V.R.S., Dawes, G.J.K., Devlin, T., and Roberts, R., 1982, Magnetotelluric and magnetovariational studies in the Travale geothermal field: Report for the Commission of the European Communities, Directorate General for Science, Research, and Development.

Isherwood, W..F, 1976, Complete Bouguer gravity map of The Geysers Area, California: U.S. Geological Survey Open File Report, 76-357.

Isherwood, W.F., and Mabey, D.R., 1978, Evaluation of Baltazor known geothermal resources area, Nevada: Geothermics, v. 7, p. 221-229.

Iyer, H.M., and Hitchcock, T., 1975, Seismic noise as a geothermal exploration tool: techniques and results: Proc. Second U.N. Symposium on the Development and Use of Geothermal Resources, San Francisco, v. 2, p. 1075-1083.

Iyer, H.M. and Stewart, R.M., 1977, Teleseismic technique to locate magma in the crust and upper mantle, H.J.B. Dick, ed., Magma genesis; Oregon Dept. of Geol. and Min. Ind., Bull. 96, p. 281-299.

Iyer, H.M., Oppenheimer, D.H., and Hitchcock, T., 1979, Abnormal P-wave delays in The Geysers - Clear Lake geothermal area, California: Science, v. 204, p. 495.

Jackson, D.B., and Keller, G.V., 1972, An electromagnetic sounding survey of the summit of Kilauea Volcano, Hawaii: J. Geophys. Res., v. 77, p. 4957.

Jackson, D.B., and O'Donnell, J.E., 1980, Reconnaissance electrical surveys in the Coso Range, California: J. Geophys. Res., v. 85, p. 2502-2516.

Jacobson, J.J., Pritchard, J.I., 1975, Electromagnetic soundings in geothermal exploration: Proc. Second U.N. Symposium on the Development and Use of Geothermal Resources, San Francisco, p. 45-.

Jiracek, G.R., and Smith, C., 1976, Deep resistivity investigations at two known geothermal resource areas (KGRAs) in New Mexico: Radium Springs and Lightning Dock: New Mexico Geol. Soc. Spec. Pub. No. 6, p. 71-76.

Jiracek, G.R., Smith, C., Dorn, G.A., 1975, Deep geothermal exploration in New Mexico using electrical resistivity: Proc. Second. U.N. Symposium on the Development and Use of Geothermal Resources, San Francisco, p. 1095-1102.

Kane, M.F., Mabey, D.R., and Brace, R., 1976, A gravity and magnetic investigation of the Long Valley Caldera, Mono County, California: J. Geophys. Res., v. 81, p. 754-762.

Kauahikaua, J., 1981, Interpretation of time-domain electromagnetic soundings in the East Rift geothermal area of Kilauea volcano, Hawaii: USGS open-file report 81-979.

Keller, B.V., Furgerson, R., Lee, C.Y., Harthill, N., and Jacobson, J.J., 1975, The dipole mapping method: Geophysics, v. 40, p. 451-472.

Keller, G.V., 1970, Induction methods in prospecting for hot water: Proc. U.N. Symposium on the Development and Utilization of Geothermal Resources, Pisa, Geothermics Spec. Issue 2, vol. 2, pt. 1, p. 318-332.

Keller, G.V., and Frischknecht, F.C., 1966, Electrical methods in geophysical prospecting: New York, Pergammon Press, 517 p.

Keller, G.V., and Rapolla, A., 1974, Electrical prospecting methods in volcanic areas: Civetta, K., et al., eds., Physical volcanology: Amsterdam, Elsevier Sci., p. 133.

Keller, G.V., Taylor, K., and Santo, J.M., 1982, Megasource EM method for detecting deeply buried conductive zones in geothermal exploration: Geophysics, v. 47, p. 420 (abstract).

Klein, D.P., and Kauahikaua, J.P., 1975, Geoelectric-geothermal exploration, Hawaii Island, preliminary results: report, Hawaii Institute of Geophysics.

Lachenbruch, A. H., Sorey, M. L., Lewis, R. E., and Sass, J. H., 1976a, The near-surface hydrothermal regime of Long Valley caldera: J. Geophys. Res., v. 81, p. 763-768.

Lachenbruch, A. H., Sass, J. H., Munroe, R. J., and Moses, T. H., Jr., 1976b, Geothermal setting and simple heat conduction models for the Long Valley caldera: J. Geophys. Res., v. 81, p. 769-784.

Lange, A.L., and Westphal, W.H., 1969, Microearthquakes near The Geysers, Sonoma County, California: J. Geophys. Res., v. 74, p. 4377-4382.

Lange, A.L., 1980, The McCoy Nevada geothermal project: Paper delivered at the Fiftieth Annual Meeting of the Society of Exploration Geophysicists, Houston, Texas, 17 November.

Liaw, A.L., and McEvilly, T.V., 1979, Microseisms in geothermal exploration-studies in Grass Valley, Nevada: Geophysics, v. 44, p. 1097-1115.

Liaw, A., and Suyenaga, W., 1982, Detection of geothermal microtremors using seismic arrays: paper presented at 52 Annual International Meeting and Exposition, Society of Exploration Geophysicists, Dallas, Oct. 17-21.

Lipman, P.W., Rowley, P.D., Mehnert, H.H., Evans, S.H., Jr., Nash, W.P., and Brown, F.H., 1977, Pleistocene rhyolite of the Mineral Range, Utah: geothermal and archeological significance: U.S.G.S. J. Res., v. 6, p. 133-147.

Long, C.L., and Kaufman, H.E., 1980, Reconnaissance geophysics of a known geothermal resource area, Weiser, Idaho, and Vale, Oregon: Geophysics, v. 45, p. 312-322.

Lumb, F.T., and Macdonald, W.J.P., 1970, Near-surface resistivity surveys of geothermal areas using the electromagnetic method: U.N. Symposium on Development and Utilization of

Geothermal Resources, Pisa, Geothermics Spec. Issue 2, p. 311-317.

Macdonald, W.J.P. and Muffler, L.J.P., 1972, Recent geophysical exploration of the Kawerau geothermal field, North Island, New Zealand: New Zealand J. Geol. and Geophys., v. 18, p. 303.

Mackelprang, C.E., 1982, Interpretation of the dipole-dipole electrical resistivity survey, Tuscarora geothermal area, Elko County, Nevada: Univ. of Utah Res. Inst., Earth Science Lab., Rept. DOE/ID/12079-59

Madden, T.R., and Nelson, P.H., 1964, A defense of Cagniard's magnetotelluric method: Massachusetts Inst. of Technology, Geophys. Lab., Rept. NR-391-401.

Majer, E.L., 1978, Seismological investigations in geothermal regions: Univ. of California, Lawrence Berkeley Lab., Rept. LBL-7054, 225 p.

Majer, E.L. and McEvilly, T.V., 1979, Seismological investigations at The Geysers geothermal field: Geophysics, v. 44, p. 246-249.

Mansure, A.J., and Brown, G.L., 1982, A forecast of geothermal drilling activity: Geothermal Energy, v. 10, p. 8-18.

Martinez, M., Fabrial, H., and Romo, J.M., 1982, Magnetotelluric studies in the geothermal area of Culiacan, Mexico: Sixth Workshop on Electromagnetic Induction in the Earth and Moon, IAGA., Victoria, British Columbia, Dept. of Physics, Univ. of Victoria (abstract).

McNitt, J.R., 1975, Summary of United Nations geothermal exploration experience, 1965 to 1975: Proc. Second U.N. Symposium on the Development and Use of Geothermal Resources, San Francisco, p. 1137-1134.

Meidav, T., and Furgerson, R., 1972, Resistivity studies of the Imperial Valley geothermal area, California: Geothermics, v. 1, p. 47-62.

Meidav, T., and Tonani, F., 1975, A critique of geothermal exploration methods: Proc. Second U.N. Symposium and the Development and Use of Geothermal Resources, San Francisco, p. 1143-1154.

Miller, C.D., Mullineaux, D.R., Crandell, D.R., and Bailey, R.A., 1982, Potential hazards from future volcanic eruptions in the Long Valley-Mono Lake area, east-central California and southwest Nevada - A preliminary assessment: U.S. Geol. Survey, Circular 877, 10 p.

Morrison, H.F., Goldstein, N.E., Hoversten, M.,

Oppliger, G., and Riveros, C., 1978, Description, field test, and data analysis of a controlled-source EM system (EM-60): University of California, Lawrence Berkeley Laboratory, Rept. LBL-7088, 150 p.

Morrison, H.F., Lee, K.H., Oppliger, G., and Dey, A., 1979, Magnetotelluric studies in Grass Valley, Nevada: Univ. of California, Lawrence Berkeley Lab., Rept. LBL-8646, 50 p.

Muffler, L.J.P., and Williams, D.L., 1976, Geothermal investigations of the U.S. Geological Survey in Long Valley, California, 1972-73: J. Geophys. Res., v. 81, p. 721-724.

Musmann, G., Gramkow, B., Lohr, V., and Kertz, W., 1980, Magnetotelluric survey of the Lake Laach (Eifel) volcanic area: Advances in European Geothermal Research: D. Reidel Co., Dordrecht, Holland, p. 904-910.

Newman, G.H., Wannamaker, P.E., and Hohmann, G.W., 1983, A two- and three-dimensional magnetotelluric model study with emphasis on the detection of magma chambers in the Basin and Range: Univ. of Utah, Dept. of Geol. and Geophys. Rept.

Ngoc, P.V., 1980, Magnetotelluric survey of the Mount Meager region of the Squamish Valley (British Columbia): Rept. of the Geomagnetic Service of Canada, Earth Physics Section, Dept. of Energy, Mines, and Resources, Ottawa, 26 p.

Nicholl, J.J., and Lange, A.L., 1981, Passive seismic results near the Tuscarora prospect, Nevada; Trans., Geoth. Res. Council, v. 5, p. 197-200.

Nielson, D.L., 1978, Radon emanometry as a geothermal exploration technique; theory and an example from Roosevelt Hot Springs KGRA Utah: Univ. Utah Res. Inst., Earth Sci. Lab Rept. No. 14, 31 p.

Nielson, D.L., Sibbett, B.S., and McKinney, D.B., Moore, J. N., and Samberg, S., 1978, Geology of Roosevelt Hot Springs KGRA, Beaver County, Utah: Univ. of Utah Res. Inst., Earth Sci. Lab., Rept. 12, 121 p.

Nur, A., and Simmons, G., 1969, The effect of saturation on velocity in low porosity rocks: Earth Plan. Sci. Letters, v. 7, p. 183-193.

Palmason, G., 1975, Geophysical methods in geothermal exploration: Proc. Second U.N. Symposium on the Development and Use of Geothermal Resources, San Francisco, p. 1175-1184.

Parry, W.T., Ballantyne, J.M., Bryant, H.L., and Dedolph, R.E., 1980, Geochemistry of hydrothermal alteration at the Roosevelt Hot Springs thermal area, Utah: Geochim. et

Cosmochim. Acta, v. 44, p. 95-102.

Patella, D., Quarto, R., and Tramacere, A., 1980, Dipole-dipole study of the Travale geothermal field: Advances in European Geothermal Research, Dordrecht, Holland, D. Reidel Co., p. 833-842.

Patella, D., Rossi, A., Tramacere, A., 1979, First results of the application of the dipole electrical sounding method in the geothermal area of Travale-Radicondoli (Tuscany): Geothermics, v. 8, p. 111-134.

Peltier, W.R., and Hermance, J.F., 1971, Magnetotelluric fields of a Gaussian electrojet: Canadian Jour. of Earth Sci., v. 8, p. 338-346.

Pilkington, H.D., Lange, A.L., and Berkman, F.E., 1980, Geothermal exploration at the Tuscarora prospect in Elko County, Nevada: Trans., Geoth. Res. Council, v. 4, p. 233-236.

Plouff, D., and Isherwood, W. F., 1980, Aeromagnetic and gravity surveys in the Coso Range, California: J. Geophys. Res. v. 85, p. 2491-2501.

Razo, A., Arellano, F., and Fouseca, H., 1980, CFE resistivity studies at Cerro Prieto: Geothermics, v. 9, p. 7-14.

Reasenberg, P., Ellsworth, W., and Walter, A., 1980, Teleseismic evidence for a low-velocity body under the Coso geothermal area: J. Geophys. Res., v. 85, p. 2471-2483.

Risk, G.F., 1975a, Monitoring the boundary of the Broadlands geothermal field, New Zealand: Proc. Second U.N. Symposium on the Development and Use of Geothermal Resources, San Francisco, p. 1185-1190.

Risk, G.F., 1975b, Detection of buried zones of fissured rock in geothermal fields using resistivity anisotropy measurements: Proc. Second U.N. Symposium on the Development and Use of Geothermal Resources, San Francisco, p. 1191-1198.

Risk, G.F., Macdonald, W.J.P., and Dawson, G.B., 1970, D.C. resistivity surveys of the Broadlands geothermal region, New Zealand: Proc. U.N. Symposium on the Development and Utilization of Geothermal Resources, Pisa, Geothermics, Spec. Issue 2, v. 2, p. 287-294.

Robinson, R., and Iyer, H.M., 1981, Delineation of a low-velocity body under the Roosevelt Hot Springs geothermal area, Utah, using teleseismic P-wave data: Geophysics, 46, p. 1456-1466.

Ross, H.P., Nielson, D.L., and Moore, J.N., 1982, Roosevelt Hot Springs geothermal system, Utah - case study: Bull, Am. Assoc. Pet. Geol., v. 66, p. 879-902.

Ryall, A. S., and Vetter, V. R., 1983, Seismological investigation of volcanic and tectonic processes in the western Great Basin, Nevada and eastern California: this volume.

Ryall, A., and Ryall, F., 1981, Attenuation of P and S waves in a magma chamber in the Long Valley Caldera, California: Geophys. Res. Letters, v. 8, p. 557-560.

Sandberg, S.K., and Hohmann, G.W., 1982, Controlled-source audiomagnetotellurics in geothermal exploration: Geophys, v. 47, p. 100-116.

Senturion Sciences, Inc., 1977, N. W. Nevada, Microearthquake survey report for Earth Power Corporation: DOE Industry Coupled Case Study, Univ. of Utah Res. Inst., Earth Sci. Lab., 61 p.

Sibbett, B.S., 1982, Geology of the Tuscarora geothermal prospect, Elko County, Nevada: Bull. Geol. Soc. America, v. 93, p. 1264-1272.

Sibbett, B.S., 1983, Structural control and alteration at Beowawe KGRA, Nevada: submitted to Trans. Geoth. Res. Council.

Sibbett, B.S., and Nielson, D.L., 1980, Geology of the central Mineral Mountains, Beaver Co., Utah: Univ. of Utah Res. Inst., Earth Sci. Lab., Rept. 33, 42 p.

Sill, W.R., 1982a, Self-potential effects due to hydrothermal convection, velocity cross-coupling: Univ. of Utah, Dept. of Geol. and Geophys., Rept. DOE/ID/12079-61, 15 p.

Sill, W.R., 1982b, Diffusion coupled (electrochemical) self-potential effects in geothermal areas: Univ. of Utah, Dept. of Geol. and Geophys., Rept. DOE/ID/12079-73, 31 p.

Sill, W.R., 1982c, A model for the cross-coupling parameters of rocks: Univ. of Utah, Dept. of Geol. and Geophys., Rept. DOE/ID/12079-69, 29 p.

Sill, W.R., 1983a, Self-potential modeling from primary flows: Geophysics, v. 48, p. 76-86.

Sill, W.R., 1983b, Electrical methods used during injection testing: Univ. of Utah Res. Inst., Earth Sci. Lab., Rept. in preparation.

Smith, C., 1983, Thermal hydrology and heat flow of Beowawe geothermal area, Nevada: Geophysics, v. 48, p. 618-626.

Smith, C., 1980, Delineation of electrical resistivity anomaly, Malpais area, Beowawe KGRA, Eureka and Lander Counties, Nevada: Univ. of Utah Res. Inst., Earth Sci. Lab., Rept. DOE/ID/12079-7, 25 p.

Sorey, M.L., Lewis, R.E., and Olmsted, F.H., 1978, The hydrothermal system of Long Valley caldera, California: U.S. Geol. Survey Prof. Paper 1044A, 60 p.

Souto, J.M., 1978, Oahu geothermal exploration: Trans. Geoth. Res. Council, v. 2, p. 605-607.

Stanley, W.D., 1982, Magnetotelluric soundings on the Idaho National Engineering Laboratory facility, Idaho: J. Geophys. Res., v. 87, p. 2683-2691.

Stanley, W.D., Jackson, D.B., and Zohdy, A.A.R., 1976, Deep electrical investigations in the Long Valley geothermal area, California: J. Geophys. Res., v. 81, p. 810-820.

Stanley, W.D., Boehl, J.E., Bostick, F.X., Jr., and Smith, H.W., 1977, Geothermal significance of magnetotelluric sounding in the eastern Snake River Plain - Yellowstone region: J. Geophys. Res., v. 82, p. 2501-2514.

Steeples, D.W., and Iyer, H.M., 1976a, Low-velocity zone under Long Valley as determined from teleseismic events: J. Geophys. Res., v. 81, p. 849-860.

Steeples, D.W., and Iyer, H.M., 1976b, Teleseismic P-wave delays in geothermal exploration: Proc. Second U.N. Symposium on the Development and Use of Geothermal Resources, San Francisco, v. 2, p. 1199-1206.

Stodt, J.A., 1983, Magnetotelluric data acquisition, reduction, and noise analysis: Proc. Workshop on "Electrical Methods in Oil and Gas Exploration", Univ. of Utah Res. Inst., Earth Sci. Lab., Salt Lake City, Utah, (January).

Studt, F.E., 1964, Geophysical prospecting in New Zealand's hydrothermal fields: Proc., United Nations Conference on New Sources of Energy, v. 2, pt. 1, p. 380.

Swift, C.M., Jr., 1979, Geophysical data, Beowawe geothermal area, Nevada: Trans., Geoth. Res. Council, v. 3, p. 701-703.

Tripp, A.C., Ward, S.H., Sill, W.R., Swift, C.M., and Petrick, W.R., 1978, Electromagnetic and Schlumberger resistivity sounding in the Roosevelt Hot Springs KGRA: Geophysics, v. 43, p. 1450-1469.

Wannamaker, P.E., 1983a, Interpretation of magnetotelluric data: Proc. Workshop on "Electrical Methods in Oil and Gas Exploration", Univ. of Utah Res. Inst., Earth Sci. Lab., Salt Lake City, Utah, (January).

Wannamaker, P.E., 1983b, Resistivity structure of the Great Basin and its tectonic implications: this volume.

Wannamaker, P.E., Ward, S.H., Hohmann, G.W., and Sill, W.R., 1980, Magnetotelluric models of the Roosevelt Hot Springs thermal area, Utah: Univ. of Utah Res. Inst., Earth Science Lab., Rept. DOE/ET/27002-8, 213 p.

Wannamaker, P.E., Ward, S.H., Hohmann, G.W., and Sill, W.R., 1983, Deep resistivity structure in S.W. Utah and its geothermal significance: Univ. of Utah Res. Inst., Earth Sci. Lab. Report DOE/ID/12079-89, 95 p.

Ward, P.L., Palmason, G., and Drake, C., 1969, Microearthquake survey and the Mid-Atlantic Ridge in Iceland: J. Geophys. Res., v. 74, p. 665-684.

Ward, P.L., and Bjornsson S., 1971, Microearthquake swarms and the geothermal areas of Iceland: J. Geophys. Res., v. 76, p. 3953-3982.

Ward, P.L., 1972, Microearthquakes: prospecting tool and possible hazard in the development of geothermal resources: Geothermics, v. 1, p. 3-12.

Ward, R.W., Butler, D., Iyer, H.M., Laster, S., Lattanner, A., Majer, E., and Mass, J., 1979, Seismic Methods: D.L. Nielson, ed., Program Review Geothermal Exploration and Assessment Technology Program, Rept. DOE/ET/27002-6, Univ. of Utah Res. Inst., Earth Sci. Lab., 128 p.

Ward, S.H., 1981, Gamma-ray spectrometry in geologic mapping and uranium exploration: Economic Geology, 75th Anniversary Volume, p. 840-849.

Ward, S.H., 1983, Controlled source electromagnetic methods in geothermal exploration: U. N. Univ. Geothermal Training Programme, Iceland, Rept. 1983-4, 46 p.

Ward, S.H., Parry, W.T., Nash, W.P., Sill, W.R., Cook, K.L., Smith, R.B., Chapman, D.S., Brown, F.H., Whelan, J.A., and Bowman, J.R., 1978, A summary of the geology, geochemistry, and geophysics of the Roosevelt Hot Springs thermal area, Utah: Geophysics v. 43, p. 1515-1542.

Ward, S.H., Ross, H.P., and Nielson, D.L., 1981, Exploration strategy for high-temperature hydrothermal systems in Basin and Range Province: Bull, Am. Assoc. Pet. Geol., 65, p. 86-102.

Ward, S.H., and Sill, W.R., 1976, Dipole-dipole resistivity delineation of the near-surface zone at the Roosevelt Hot Springs area: Univ. of Utah, Dept. of Geol. and Geophys., Tech Rept. 76-1, 7 p.

Ward, S.H., and Sill, W.R., 1982, Resistivity, induced polarization, and self-potential methods in geothermal exploration: Univ. of

Utah Res. Inst., Earth Sci. Lab., Rept. DOE/ID/12079-90, 100 p.

Ward, S.H., and Wannamaker, P.E., 1983, The MT/AMT electromagnetic method in geothermal exploration: U.N. Univ. Geothermal Training Programme, Iceland, Rept. 1983-5, 107 p.

Wechsler, D.J., and Smith, R.B., 1979, An evaluation of hypocenter location techniques with applications to Southern Utah: Regional earthquake distributions and seismicity of geothermal areas: Univ. of Utah, Dept. of Geol. and Geophys., Rept. IDO/DOE/ET/28392-32, 131 p.

Whiteford, P.G., 1975, Assessment of the audiomagnetotelluric method for geothermal resistivity surveying, Proc., Second U.N. Symposium on the Development and Use of Geothermal Resources, San Francisco, p. 1255-1261.

Williams, P.D., Mabey, D.R., Zohdy, A.R., Ackerman, H., Hoover, D.B., Pierce, K.L., Oriel, S.S., 1975, Geology and geophysics of the southern Raft River valley geothermal area, Idaho, USA: Proc. Second U.N. Symposium on Development and Use of Geothermal Resources, San Francisco, v. 2, p. 1273-1282.

Wilt, M.J., Goldstein, N.E., and Razo, A., 1980a, LBL resistivity studies at Cerro Prieto: Geothermics, v. 9, p. 15-26.

Wilt, M., Beyer, J.H., and Goldstein, N.E., 1980b, A comparison of dipole-dipole resistivity and electromagnetic induction sounding over the Panther Canyon thermal anomaly, Grass Valley, Nevada: Trans., G.R.C., 4, p. 101-104.

Wilt, M., Goldstein, N.E., Stark, M., and Haught, R., 1980c, An electromagnetic (EM-60) survey in the Panther Canyon area, Grass Valley, Nevada: Univ. of California, Lawrence Berkeley Lab., Rept. LBL-10993, 97 p.

Wilt, M. J., Haught, J. R., and Goldstein, N. E., 1980d, An electromagnetic (EM-60) survey of the McCoy geothermal prospect, Nevada: Univ. of California, Lawrence Berkeley Lab., Rept. LBL-12012, 115 p.

Wilt, M., Goldstein, N.E., Stark, M., Haught, J.R., and Morrison, H.F., 1981a, Experience with the EM-60 electromagnetic system for geothermal exploration in Nevada: University of California, Lawrence Berkeley Lab, Rept. 12618.

Wilt, M. J., Stark, M., Goldstein, N. E., and Haught, J. R., 1981b, Electromagnetic induction sounding at geothermal prospects in Nevada: Univ. of California, Lawrence Berkeley Lab, Rept. LBL-12100.

Wollenburg, H.A., 1975, Radioactivity of geothermal systems: Proc. Second U.N.

Symposium on the Development and Use of Geothermal Resources, San Francisco, p. 1282-1292.

Zandt, G., McPherson, L., Schaff, S., Olsen, S., 1982, Seismic baseline and induction studies, Roosevelt Hot Springs, Utah, and Raft River, Idaho: Univ. of Utah Res. Inst., Earth Sci. Lab., unnumbered report, 59 p.

Zoback, M. L., 1983, Style of Basin and Range faulting as inferred from seismic reflection data in the Great Basin, Nevada and Utah: this volume.

Zohdy, A.A.R., Anderson, L.A., and Muffler, L.J.P., 1973, Resistivity, self-potential, and induced polarization surveys of a vapor-dominated geothermal system: Geophysics, v. 38, p. 1130-1144.

BASIN AND RANGE GEOTHERMAL HYDROLOGY:
AN EMPIRICAL APPROACH

FRANK YEAMANS

Reno, Nevada

ABSTRACT

Basin and Range hydrothermal systems are dynamic flow systems rather than reservoir systems. The systems share underlying fundamental similarities such that they can be classified as a generic group. The systems are structurally controlled, occurring where the extensional tectonics of the province have created permeable flow paths. Possible recharge zones can be identified by stable isotope relationships and, if available, hydrologic and geologic information from deep wells. Where there are no cold meteoric waters to sample and few deep wells the location of recharge zones is speculative. Depths of circulation can be roughly estimated by use of predictive chemical geothermometers and measured temperature gradients but actual flow paths are unknown. Saturated thicknesses, porosity, lateral extent and boundaries all are either unknown or can be estimated only in the most favorable of circumstances. The discharge zone is the best understood part of the system. The discharge rate, including convective heat discharge, can be calculated with varying degrees of accuracy, depending upon the physical setting (surface, subsurface, or both) of the individual discharge zone.

Production zones are fracture or fault controlled and may be separated by significant blocks of impermeable rock. The effective porosity of production zones may be very low, perhaps less than 1-2% of the total thermal anomaly defined by drilling. The existence of deep reservoirs is problematical. Flow test data have been interpreted as indicating extremely large volumes of fluid in place yet deep drilling (>1000 meters) has been disappointing.

Empirical methods that do not require knowledge of the controlling hydrologic parameters can be used to evaluate the natural energy discharge and to predict production potential. Step-wise increases in the production rate allows development of empirical response curves that, with longer periods of record, will predict system response to increased production rates in the future.

INTRODUCTION

Previous work on Basin and Range hydrothermal systems has demonstrated a basic similarity in the geologic setting of these systems (Hose and Taylor, 1974; Olmsted et al,. 1975; Garside and Schilling, 1979). Studies of individual hydrothermal systems have viewed them as being within the overall framework of Basin and Range hydrothermal systems; Steamboat Springs (White et al., 1964), Roosevelt Hot Springs (Nielson et al., 1978), Long Valley (Sorey et al., 1978), Leach Hot Springs (Welch et al., 1981), Stillwater hydrothermal area (Morgan, 1982). Thus, just as it can be held that Basin and Range hydrothermal systems share sufficient common geologic characteristics to be classified as a generic group, it is thought the hydrologic regimen of the systems share an underlying fundamental similarity.

Continued research coupled with exploration efforts have increased the knowledge of these systems, particularly the intermediate and deeper levels. Basin and Range hydrothermal systems are liquid dominated dynamic flow systems transporting thermal energy from deeper levels to the shallow subsurface and to the surface. They are not, in the main, porous-medium reservoir or storage systems and thus are distinct in this respect from geothermal reservoirs in other provinces such as the Imperial Valley. Successful development of Basin and Range hydrothermal systems must be based on a clear understanding of the fundamental hydrologic regimen.

The first part of this paper reviews the common hydrologic components of Basin and Range hydrothermal systems. The systems discussed generally are the high temperature (>150°C) ones where exploration drilling and testing have led to an increased understanding of the deeper parts of the systems. The fundamental similarities of these systems should be applicable to low (<50°C) and moderate (50°-150°C) temperature systems in the Basin and Range province. The Steamboat Springs system frequently is used as an example as much research and exploration work has been done there.

The second part of this paper examines the geologic and hydrologic setting of some Basin and Range production wells. Many of the production wells have similar production zone characteristics and thus an understanding gained from one well is applicable to others. The geologic and hydrologic factors that control production from one Basin and Range hydrothermal system also operate, although perhaps in different proportions, to control production from other Basin and Range hydrothermal systems.

Lastly, a conceptual hydrologic model is presented. An empirical lumped parameter method is proposed as an appropriate approach for future geothermal exploration and development in the Basin and Range. An advantage of this approach is that it does not require detailed knowledge of such hydrologic characteristics as hydrologic boundaries, porosity, transmissivity, and storativity.

HYDROLOGIC CHARACTERISTICS

In a systems approach a Basin and Range hydrothermal system can be divided into several components that can be studied individually. Within a geologic setting the system is composed of:

1) recharge zone	— where the recharge water enters the subsurface;
2) flow paths	— the subsurface pathways the fluid follows;
3) depth of circulation	— the depth of circulation required to obtain the maximum temperature, assumed to be the maximum depth of circulation;
4) discharge zone	— channels and flow paths of upflow from the deepest depth of circulation, includes surface discharge as hot springs and subsurface discharge into thermal aquifers.

Geologic Setting

When discussing the geologic setting of a Basin and Range system it generally is the discharge zone that is placed within a geologic setting as it often is the only component that can be readily identified and studied. Discharge zones are associated with three structural settings.

The close spatial association of many Basin and Range hot springs with major range-bounding faults indicates the underlying portions of the discharge zones of the hydrothermal systems also are related to the faults. Production wells at Dixie Valley (Parchman and Knox, 1981) and Beowawe (Epperson, 1982) are thought to tap upflow zones along the faults. The Kemp thermal anomaly indicates thermal waters may circulate along the range-bounding faults even where there are no surface thermal manifestations (Flynn et al., 1982). Such flow probably is a common occurrence creating numerous small thermal anomalies along many of the range-bounding faults. At Long Valley many hot springs are located within the resurgent dome suggesting the volcanic feature is responsible for the location of the springs. However, within the resurgent dome many of the hot springs are found along the Hilton Creek fault, an active Sierra Nevada frontal fault bisecting the caldera's southern margin (Bailey et al., 1976). The association of the hot springs with the fault suggests the springs occur where there is perme-

ability along the fault (Bailey et al., 1976).

A second geologic setting for the discharge zone is within small, highly faulted, relatively low relief, horst blocks. Found here are Desert Peak (Benoit et al., 1982), Steamboat Springs (White et al., 1964), and Coso Hot Springs (Duffield and Bacon, 1981). These are among the hottest known systems in the western part of the province. These horst blocks are not the main, high-relief ranges of the province.

A third setting is within the basins at considerable distances from the ranges and with few surface indications of structural control. Examples include the Hyder Hot Springs that emerge in the middle of Dixie Valley and the Soda Lake hydrothermal system that is at least 18 km. from any major mountain block. There is evidence, however, that structural features are controlling the location of these thermal systems. The homogeneous water chemistry of the Hyder Hot Springs suggests a fault-controlled flow system separated from the cold water aquifers in the alluvium (Denton et al., 1980). The Soda Lake hydrothermal system is associated with a buried structural feature defined by the phreatic explosion craters at Soda Lake and the volcanic Upsal Hogback to the northeast (Olmsted et al., 1975; Hill et al., 1979).

Thus Basin and Range hydrothermal systems are structurally controlled. They are not confined to the main range-bounding faults, however, but may occur within other structural settings where the extensional tectonics of the province have created permeable flow paths. There is no geologic evidence of stratigraphic or lithologic control of laterally extensive hydrothermal systems at drilled depths.

Recharge

The oxygen-deuterium stable isotope relationship has been used to identify possible recharge zones for hydrothermal systems. Meteoric waters, particularily cold springs, with the same deuterium value as the oxygen-shifted thermal water can be considered to be located within a possible recharge area for the hydrothermal system (Craig et al., 1956). Basin and Range recharge studies include Coso Hot Springs (Fournier and Thompson, 1980), Steamboat Springs (Nehring, 1980), Roosevelt Hot Springs (Rohrs and Bowman, 1980), and Long Valley (Sorey et al., 1978).

Nehring (1980) concluded the major recharge zone for the Steamboat Springs hydrothermal system is in the Carson Range some 16 km. to the west of the hot springs. Similar deuterium values for cold springs and thermal waters suggests the recharge zone is at an average elevation of 2100 meters and limited to an area between Galena Creek and Evans Creek (Fig. 1). Nehring (1980) concluded recharge along major faults within the Carson Range moves deep enough to be heated by regional conduction or by a possible magma chamber beneath the Steamboat Hills (Fig. 2).

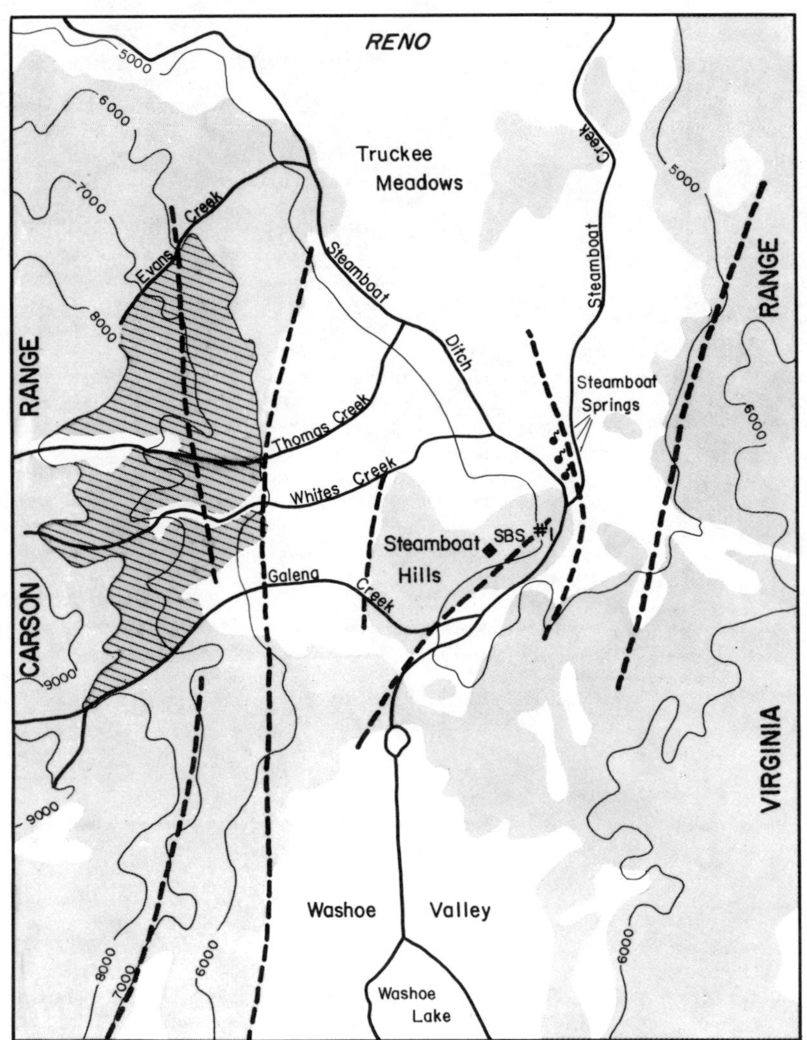
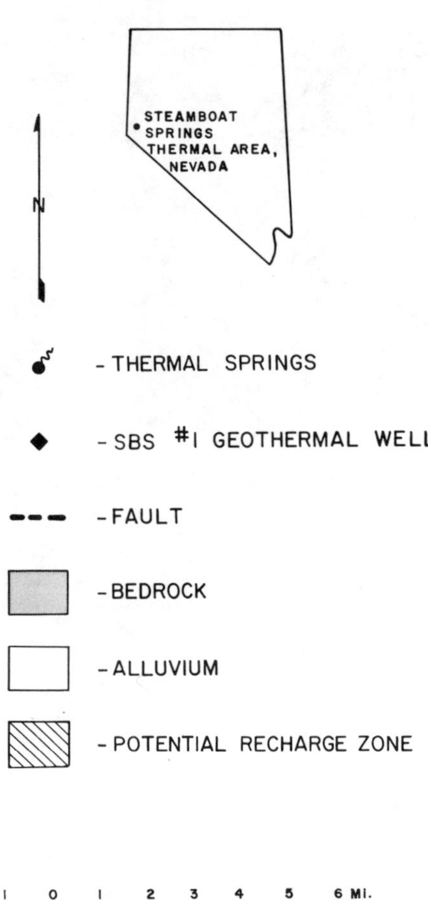

FIGURE I. Potential recharge zone to the Steamboat Springs hydrothermal system (modified from Nehring, 1980).

There also is hydrologic data to support this isotope recharge model. Skau (in preparation), using data from numerous drainage basins on the east front of the Sierra Nevada, defined an inverse relationship between streamflow and total linear length of mapped faults within a drainage basin. Skau's data indicates the Galena, Whites, and Thomas Creek drainages have extensive faulting within the mountain block and less than expected streamflow where the streams debouche onto the alluvial fan.

Identification of one potential recharge area does not preclude the possibility of other recharge areas or more than one flow path for the recharge water. At Long Valley the large areal variability in the isotopic composition of the thermal waters indicates the involvement of meteoric waters from at least two different sources (Sorey et al., 1978).

Some Basin and Range hydrothermal systems have very subtle recharge systems. Welch and others (1981) suggested discharge at the Leach Hot Springs either is from modern-day precipitation at an elevation higher than that in the immediate vicinity of the hot springs or it was recharged during an earlier, colder, climatic period. This conclusion is based on the fact deuterium values of the thermal waters are more negative than that of the surrounding cold waters. Elevation zones where precipitation can be expected to have the same deuterium values as that of the thermal waters are so far away (160 km.) that under a ground water flow rate of 10 meters/yr. the infiltration event could have occurred up to 16,000 years ago and thus the recharge water still would be considered "paleowater" (Welch et al., 1981). Young and Lewis (1980) suggested present-day discharge from the Bruneau-Grand View hydrothermal system in Idaho was recharged during late Pleistocene glacial advances when the climate averaged 3-5C° colder than at present.

The Desert Peak hydrothermal system is located in the arid Hot Springs Mountains where the small amount of precipitation is consumed by evapotranspiration and there is little or no ground water recharge except from rare, high-intensity storms that occur years or decades apart. No springs exist in the mountains nor are there a significant

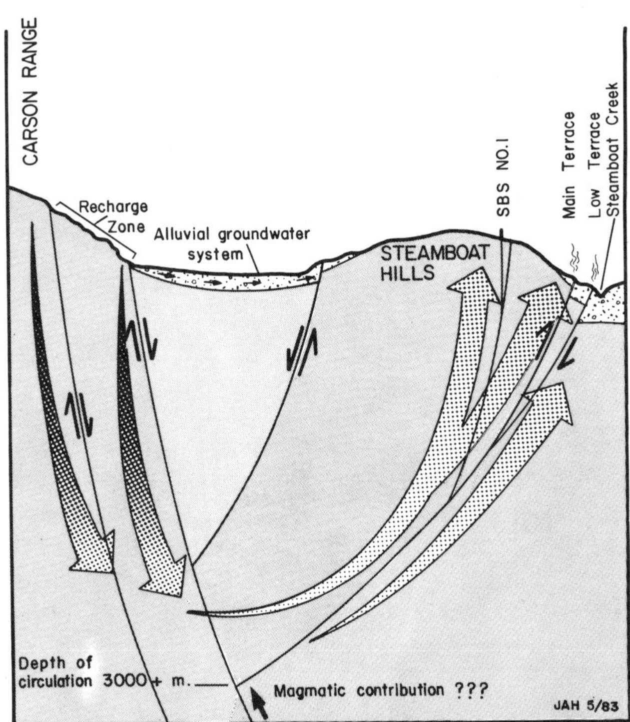

FIGURE 2. Diagrammatic cross-section showing postulated recharge to the Steamboat Springs hydrothermal system.

number of cold water wells that can be sampled for isotope values. In this instance another approach must be used to identify the recharge zone. A recharge mechanism is discussed later in the section on hydraulic gradients.

Delineation of recharge zones by isotope studies alone is circumstanial in those areas where there are meteoric waters to be sampled. Only in rare instances are there a sufficient number of deep wells from which hydrologic and geologic data can be collected to substantiate conclusions based on isotope data. In areas where there is no correlation with nearby meteoric waters the recharge mechanism of the hydrothermal system remains speculative. If recharge occurred at an earlier time under different climatic conditions is it still occurring and at what rate? If recharge is occurring at great distances from the discharge zone where is the recharge zone and what is the nature of the aquifer system that allows such a long flow path in the structurally discontinuous Basin and Range province?

Depth of Circulation and Flow Paths

The minimum depth of circulation required for conductively heated hydrothermal systems under steady-state conditions can be estimated from the following equation. Note that with flow paths of finite length greater depths of circulation would be required to attain the observed or estimated temperature. (Welsh, et al., 1981).

$$D = (T_s - T_r)/g$$

where
D = depth of circulation;
T_s = maximum temperature in system, predicted from geothermometers;
T_r = surface temperature of recharge water, generally taken as the average annual temperature;
g = temperature gradient.

A conductive temperature gradient of 76.4C°/km was measured in the fine-grained sediments of the Carson Sink (Olmsted et al., 1975). Using this gradient and assuming a 10°C temperature for the recharging water, the depth of circulation for the 204°C Desert Peak hydrothermal system would be at least 2.5 km. This estimate assumes the temperature gradient is constant with depth. If more thermally conductive rock is present above the 2.5 km. depth there would be a lower temperature gradient and thus greater depths of circulation would be required to obtain the same maximum temperature. For Leach Hot Springs Welch and others (1981) estimated a depth of circulation on the order of 3 km. to obtain temperatures of 180°C. Sammel and Craig (1981) calculated a circulation depth of at least 4 km. to obtain a temperature of 170°C for the Warner Valley hydrothermal system. Morgan (1982) calculated that for a maximum temperature of 160°C depth of circulation would be at least 2.5 km. for the Stillwater hydrothermal system.

The deepest production well in Dixie Valley is 2.8 km.(Benoit and Butler, 1983), within the range of the minimum depth of circulation required for conductive heating to what is thought to be a maximum temperature of over 200°C for the system. An earthquake swarm in Warner Valley produced seismic focal points along the western boundary fault at depths between 2 and 13 km. (Schaff, 1976 in Sammel and Craig, 1981). This indicates fault planes exist to depths of 13 km. in Warner Valley and such faults could be conduits for circulation within the hydrothermal system. Thus it appears permeable conduits exist at the depths required for conductive heating.

Fluids at or near the maximum predicted temperatures have been found at depths much shallower than the depth required for conductive heating; 204°C at 915 meters at Desert Peak (Benoit et al., 1982); 227°C at 760 meters at Steamboat Springs (Desormier, 1983). These high temperatures in the shallow subsurface are maintained by relatively high upflow rates that overcome conductive heat loss to the wallrock (Sorey, 1975).

It has been hypothesized that a magma chamber is present beneath the Steamboat Hills and serves as the heat source for the Steamboat Springs hydrothermal system (White, 1968). Depths of circulation could be much less than that required by regional conductive heating if the magma chamber were at a relatively shallow depth. However, if the alkali chlorides found in the thermal waters at Steamboat Springs are being transferred from the magma chamber, a hydrostatic pressure of at least 305 kg./cm.^{2} (3000 meters) is required for the existence of a dense high-pressure steam to transport the chlorides (White, 1968). Therefore, circulation may be to depths of a few kilometers even in the presence of a shallow, localized magmatic heat source.

Thus the extensional tectonics of the Basin and Range have created permeable flow paths at the necessary depths for conductive heating to the predicted and measured temperatures. Maximum depths of circulation remain unknown. It is interesting to note the maximum temperatures of the high temperature systems associated with conductive heat flow (Beowawe, Desert Peak, Dixie Valley, Humboldt House, and Soda Lake) generally are grouped together between 200°C and 225°C. This suggests they all may have a common maximum depth of circulation on the order of 2.5 to 3.0 km. There is no evidence that if a shallow localized heat source is present the depth of circulation is less than that for a regional conductive heat source. In any case fluids at or near the maximum predicted temperatures are found at relatively shallow depths due to rapid upflow along permeable conduits.

Hydraulic Gradients

Fluid flow in hydrothermal systems results from mechanical energy (potential differences) and thermal energy (density differences). Potential differences occur where recharge zones are higher than discharge zones. For hydrothermal systems in areas of high topographic relief recharge could occur in the mountain ranges and descend deeper than recharge to the shallow cold water aquifers. Flow direction, however, generally would be in the same direction (see Fig. 2).

Input of thermal energy creates a density difference when cold recharge water is heated. Depending on temperature differences, density differences can be a significant part of the overall hydraulic gradient. White (1968) calculated density differences could account for as much as 275 meters of the hydraulic head at Steamboat Springs. Recharge at an elevation of 2100 meters in the Carson Range would have an elevation advantage of 675 meters over the hot springs discharge zone.

At Desert Peak there is no evidence of a potential gradient from a recharge zone located at a higher elevation. It is possible fluid flow is solely the result of density differences. The Carson Sink to the east of and the Fernley Sink to the southwest of Desert Peak are sites of major shallow ground water discharge by evapotranspiration. Yet the sinks also may serve as recharge zones to the hydrothermal system. In a cross-section from the Carson Sink on the east through the well field at Desert Peak to the Fernley Sink on the southwest, the computed static water levels for the well field, referenced back to a column of water at 15.6°C, are lower than those of the surrounding sinks (Fig. 3). Although there is the possibility of a potential gradient from the Fernley Sink to

the Carson Sink of 43 meters it also is reasonable to visualize recharge to the hydrothermal system coming from one or both of the sinks with the energy gradient required for flow being density differences created by the thermal energy of the heat source. Such a flow system requires a diversion of the discharge from the hydrothermal system (see Fig. 3). Part of the hydrothermal discharge flows down gradient to discharge in the sinks and part moves downward under the density gradient to become recharge to the hydrothermal system.

It is difficult to separate the potential and density differences of the total gradient. Both potential and density differences have both horizontal and vertical components (Olmsted et al., 1975). Determination of the vertical component requires at least two observation wells at different depths at a single location in the hydrothermal system. Determination of the density differences requires data on the entire depth and temperature distribution of the discharge zone. This generally is unknown but reasonable estimates can be made using the predicted maximum temperature and depth of circulation of the system.

Determination of the static pressure of an individual production zone also is difficult. Deep exploration-production wells generally are completed with several hundred meters of uncased open hole. If separate production zones with different static pressures are intersected the borehole pressure will be intermediate between the highest and lowest pressured zones intersected.

Discharge

The discharge zone is the best understood part of a Basin and Range hydrothermal system. Hot springs are the easiest place to collect data on heat and mass transfer and to predict maximum subsurface temperatures by using chemical geothermometers. Hot springs represent discharge from the deeper parts of the system and as such are an intergal component in exploration.

The total convective heat flow of a system provides a first approximation of the minimum rate at which energy can be withdrawn (Benseman, 1959, and Bodvarsson, 1964 in White, 1968). Thus understanding the discharge zone may be the key to evaluating the production potential of the hydrothermal system. Olmsted and others (1975) proposed two discharge systems, a leaky system where thermal discharge is into shallow ground water aquifers and a non-leaky system where the upflow conduits are sealed off from the shallow aquifers and all fluid discharge is via hot springs. All systems studied apparently have some subsurface discharge. For systems that have both surface and subsurface fluid discharge the ratio between the two can vary widely. White (1968) calculated hot spring discharge at Steamboat Springs constitutes only 6% of the total thermal fluid discharge of 71.2 liters/sec., the rest being discharged into the shallow ground water system and eventually into Steamboat Creek.

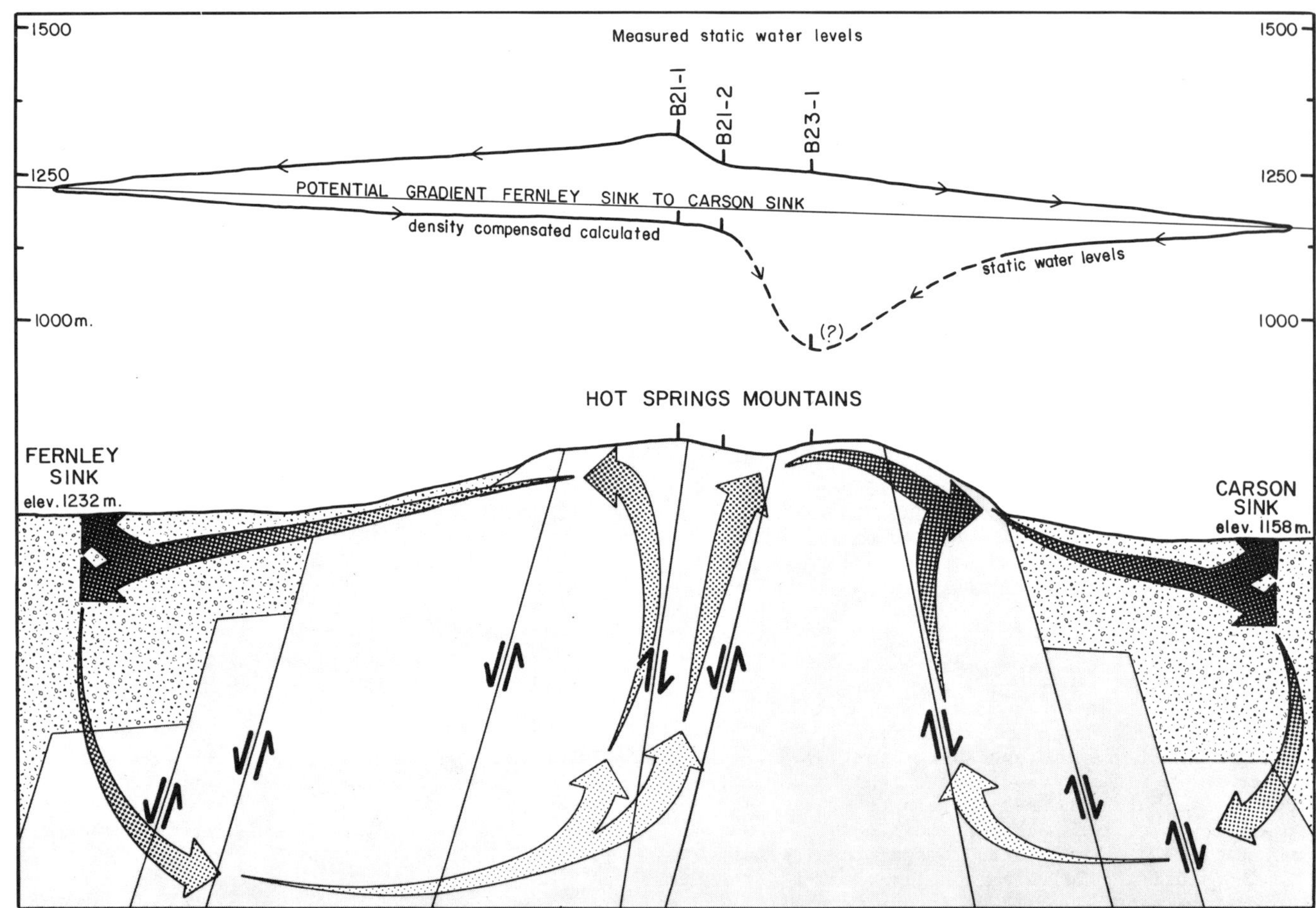

FIGURE 3.   Diagrammatic cross-section through the Desert Peak hydrothermal system showing postulated flow direction and measured and calculated static water levels.

At Long Valley 80% of the total thermal fluid discharge of 248 liters/ sec. occurs as hot spring discharge within Hot Creek Gorge (Sorey et al.,1976). Some systems are totally leaky with all discharge into the subsurface. Roosevelt Hot Springs and Humboldt House are examples where spring flow has ceased and all discharge now is into the shallow subsurface.

Where there is no surface discharge and little or no other surface evidence of a hydrothermal system such as fossil hot spring deposits or altered and bleached rocks, the system is referred to as being "blind" (Blackwell and Czang, 1973). These systems can be identified by the thermal "halo" created by leakage into shallow thermal aquifers (Benoit et al., 1982). The shallow thermal aquifers have anomalously high temperature gradients above them (often >300C°/km.) and generally have a temperature reversal below the aquifer.

Calculation of total heat and mass discharge can be based on predicted maximum temperature by geothermometers and the measurement of conservative chemical constituents of the thermal fluid as found in nearby surface waters into which the thermal waters drain (White, 1968; Sorey et al., 1978). If there are no surface waters that collect the thermal discharge then calculations can be made of the amount of heat convected into the shallow thermal aquifers

(Olmsted et al., 1975; Benoit et al., 1982).

A brief summary of the knowledge concerning the hydrologic setting of Basin and Range hydrothermal systems is as follows.

1) recharge zone – Where cold meteoric waters are available for sampling, possible recharge zones and elevations can be identified. Where there are no cold meteoric waters to sample, location of the recharge zone is speculative.

2) depth of circulation – Can be roughly estimated by use of predictive chemical geothermometers and measured temperature gradients; actual flow paths are unknown.

3) aquifer geometry – Thickness, porosity, lateral extent, boundaries, all are either unknown or can be only roughly estimated in the most favorable of circumstances.

4) discharge zone – Discharge rate, including convective heat discharge, can be calculated with varying degrees of accuracy depending on the physical setting (surface, subsurface, or both) of the individual discharge zone.

164

Production Zones

Deep drilling in the Basin and Range province has intersected the higher temperature parts of hydrothermal systems in fractures in a "hardrock" environment. Commonly an isothermal zone in a static equilibrated temperature profile is interpreted as the production thickness. In a porous, liquid dominated sedimentary hydrothermal reservoir an isothermal zone may accurately reflect thermal convection within the production zone. This interpretation, however, often is not correct for Basin and Range wells. One interpretation is that by Urban and Diment (1982) who suggested the over 1480-meter thick isothermal section measured in well B23-1 at Desert Peak might result from hot water from a shallow, higher pressure thermal zone flowing down hole to exit in deeper, lower-pressure permeable zones. This flow would mask the undisturbed thermal regimen of the lower portion of the hole and would suggest an exaggerated thickness to the pay zone.

At both Steamboat Springs and Beowawe measured wellhead temperatures and pressures did not agree with predicted values from computer simulations of test conditions (Baza, 1981; Epperson, 1982). At Beowawe later geologic data indicated the well was producing from two intervals with the deeper interval having a lower production temperature (Epperson, 1982). At Steamboat Springs the ambiguity between predicted and measured values was resolved when a temperature survey was run while injecting cold water. The injection temperature survey identified a second production zone at the predicted temperature located below the maximum temperature isothermal zone (see Fig. 4).

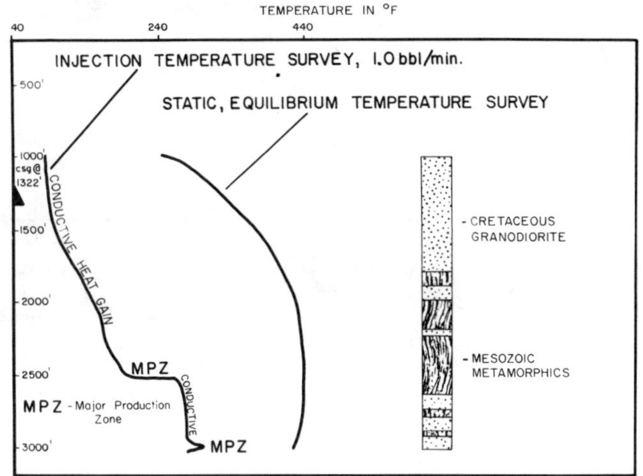

FIGURE 4. Static and injection temperature survey, S.B.S. no. 1, Steamboat Springs, Nevada.

The average depth of all successful high-temperature geothermal wildcat wells in the northern Basin and Range province from 1974 through 1981 was a relatively shallow 1030 meters (Edmiston, 1982). This depth is far less than the depths of circulation estimated by use of chemical geothermometers and background temperature gradients. The relatively shallow depths of production support the earlier statement that the existence of high

temperature fluids at relatively shallow depths results from rapid upflow within permeable channels. Figure 5 is a generalized geologic map of the Humboldt

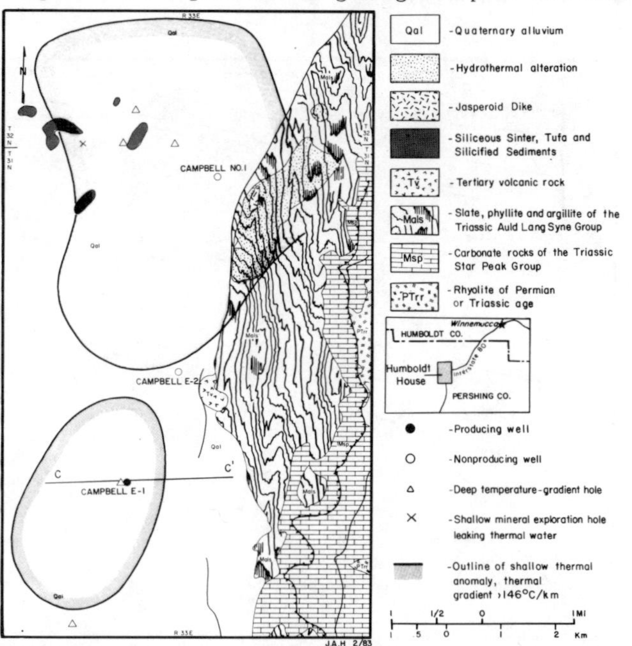

FIGURE 5. Generalized geologic map, Humboldt House Area (after Silberling and Wallace, 1967).

House prospect. It is of the main range-bounding fault type. Data from two deep wells, Campbell No. 1 and Campbell E-2, were used to construct the idealized geologic cross-section for the Campbell E-1, the only producing well in the prospect (Fig. 6).

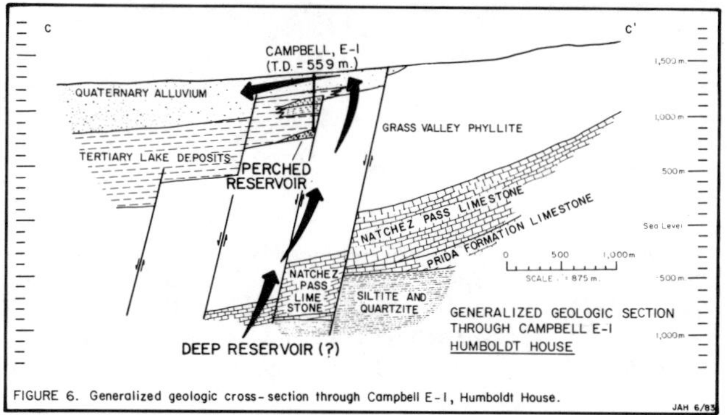

FIGURE 6. Generalized geologic cross-section through Campbell E-1, Humboldt House.

The well produces from a Tertiary limestone "boulder" deposit (Desormier, 1979) that can be considered a perched reservoir of limited volume. It is thought to be recharged from a deeper zone via a main range-bounding fault that is an upflow conduit. The recharge rate to the perched reservoir and thus the sustained or long-term production rate of the system is controlled by the least transmissive section of the upflow conduit (Welch et al. 1981).

At Desert Peak a temperature cross-section through the well field indicates the permeable upflow channels are nearly vertical and are separated by significant blocks of impermeable rock (Fig. 7). The elevated temperature contours in the vicinity of wells B21-2 and B23-1 result from upflow along

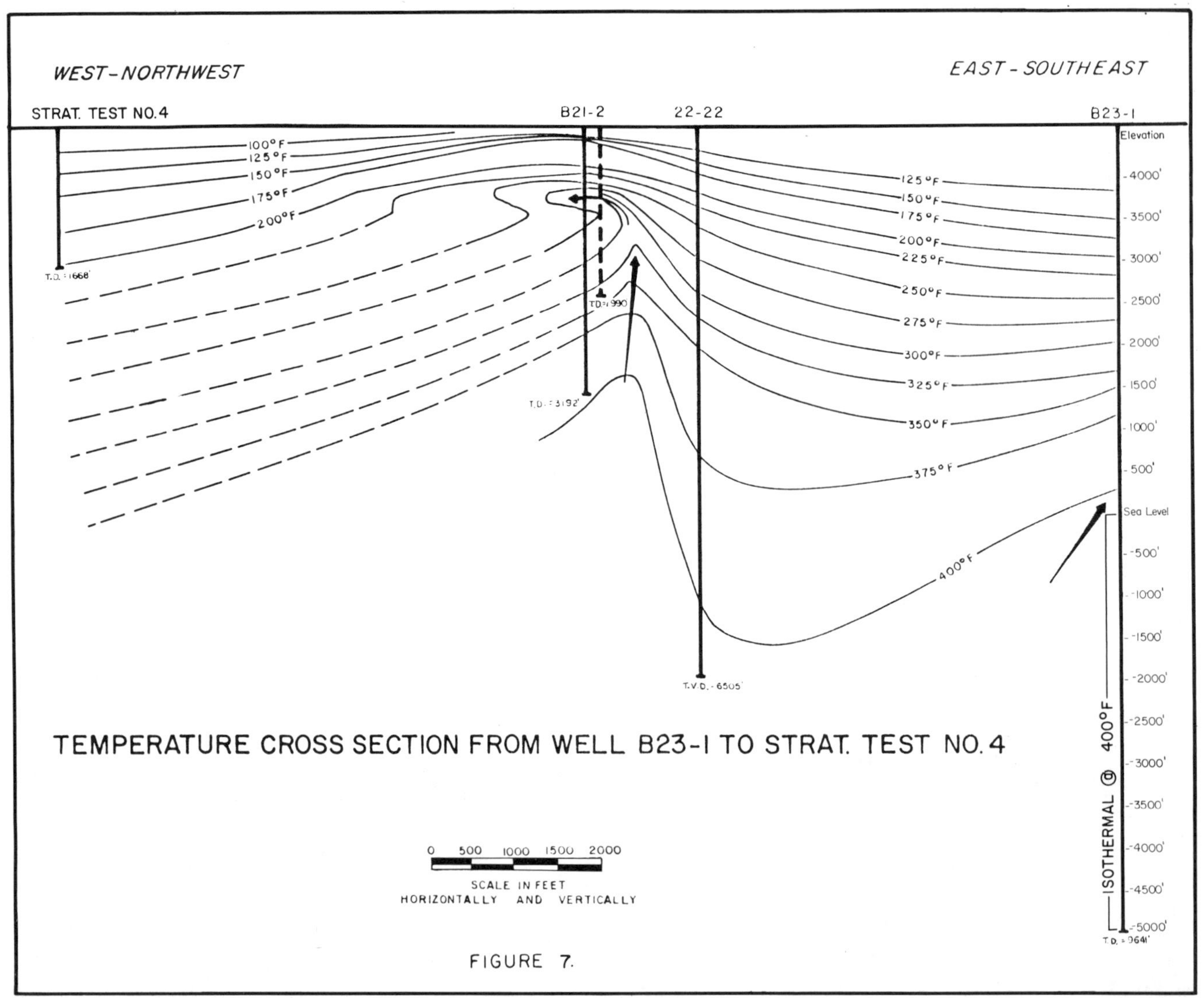

TEMPERATURE CROSS SECTION FROM WELL B23-1 TO STRAT. TEST NO. 4

FIGURE 7.

the discharge zones that are the production zones. Well 22-22, not tested at this time, is at some distance from a discharge zone and is within a zone of low permeability as indicated by the relatively depressed temperature contours.

Thus it seems reasonable to assume Basin and Range production zones, if not directly in discrete and singular faults and/or fractures (Desert Peak), are the result of upflow of thermal waters along such zones (Humboldt House). Fracture density is not thought great enough to consider the production zone to be a fractured rock matrix with any significant (>10 meters) saturated thickness and the effective porosity of the production zone may be very low, perhaps less than 1-2% of the total thermal anomaly defined by drilling.

Flow Testing and Analysis

Benson (1982) described a flow test of a 120°C Basin and Range geothermal well, the results of which should not go unnoticed. The well, WEN-1, is located on the eastern side of the Honey Lake Valley, California, near Wendel and Amedee hot springs. From a depth of 1545 meters to the TD of 1780 meters the well is completed open hole in granitic basement rock. From pre-test temperature and spinner surveys it was determined 80% of the flow comes from one major fracture zone. The production zone is under confined conditions and the well was tested by single phase artesian flow at four flow rates.

During testing the productivity index, discharge/ drawdown, decreased with increased flow rates, going from 0.45 liters/sec. per kN/m^2 at 13.9 liters/sec. to 0.20 liters/sec. per kN/m^2 at 42.9 liters/sec. In a 100% efficient well with laminar (Darcian) flow in the formation the productivity index is a constant, independent of flow rate. In reality well bore damage or enhancement from drilling results in a skin value. For this test an increase in the flow rate resulted in a proportionate increase in the calculated apparent skin value. Under the conventional definition of skin effects the skin value of a well should not be a function of the flow rate. Using a formulation (Ramey, 1965) that included non-Darcian flow in the production zone, Benson calculated a true skin value near zero for the well. It was concluded non-Darcian flow in the fracture contributed significantly to the draw-down, resulting in the varying productivity index. Benson then developed an empirical relationship between the flow rate and drawdown for the well tested. It is possible this non-Darcian flow is a common occurrence in high temperature, high dis-charge-rate Basin and Range geothermal wells producing from narrow fracture zones. The calcu-lated transmissivity of the production zone will vary inversely with the flow rate.

Responses in observation wells can be simul-taneously quick and at distances greater than 2500 meters. At the Beowawe field responses within one hour of start of flow were observed at distances up to 2000 meters from the production well (Epperson, 1982). Responses also indicate the anisotropic nature of the hydrothermal system. At Steamboat Springs one observation well 2573 meters from the SBS # 1 production well responded to testing within 48 hours after kick-off and had a 0.8 meter drawdown after 15 days of flow. A second well at a distance of 2598 meters did not show a response to testing. At Desert Peak the observation wells for the testing of well 86-21 were three deep wells that intersected the hydrothermal system at roughly the same depth. Responses in these wells indicate the production zone not only is anisotropic but also is hetero-geneous (Goyal, personal communication). Thus at a given point in the system hydraulic parameters of the system (permeability, transmissivity) are directionally dependent (anisotropic) and the values of the hydraulic parameters vary from point to point within the system (heterogeneous).

Limited testing of Basin and Range geothermal systems commonly results in the calculation of large volumes of fluid being affected by the testing. At Beowawe it was calculated a fluid volume of 1.59×10^{13} liters was affected by a 27-day flow test at a rate of 463,050 kg./hr. (Epperson, 1982). Other flow test data at Beowawe "indicate fluid volumes in excess of 10^{12} barrels (1.59×10^{15} liters); in other words, a volume larger than the Prudhoe Bay Oil Field" (Epperson, 1982). It is thought the major part of the system at Beowawe is in lower Paleozoic carbon-ates at depths approaching 6100 to 9100 meters but none of the wells drilled to date have reached the deep carbonates (Epperson, 1982). Such a proposed reservoir model must be examined to see if it is realistic. Using the calculated fluid volume of 1.59×10^{15} liters, a reservoir with 10% porosity would have a total rock and fluid volume of 15,900 km^3. One must ask if this is a realistic model, especially in light of the disappointing results from deep drilling in the Basin and Range (Edminston, 1982). Thus, although flow test data have been inter-preted to suggest deep, large-volume hydrothermal reservoirs, their existence still is problematical.

CONCEPTUAL MODELS

Reservoir Model

Traditional models for ground water and petro-leum resources have treated these systems as static volumes. Development of a ground water or petroleum reservoir is based on obtaining knowledge of the physical and hydraulic parameters that control the

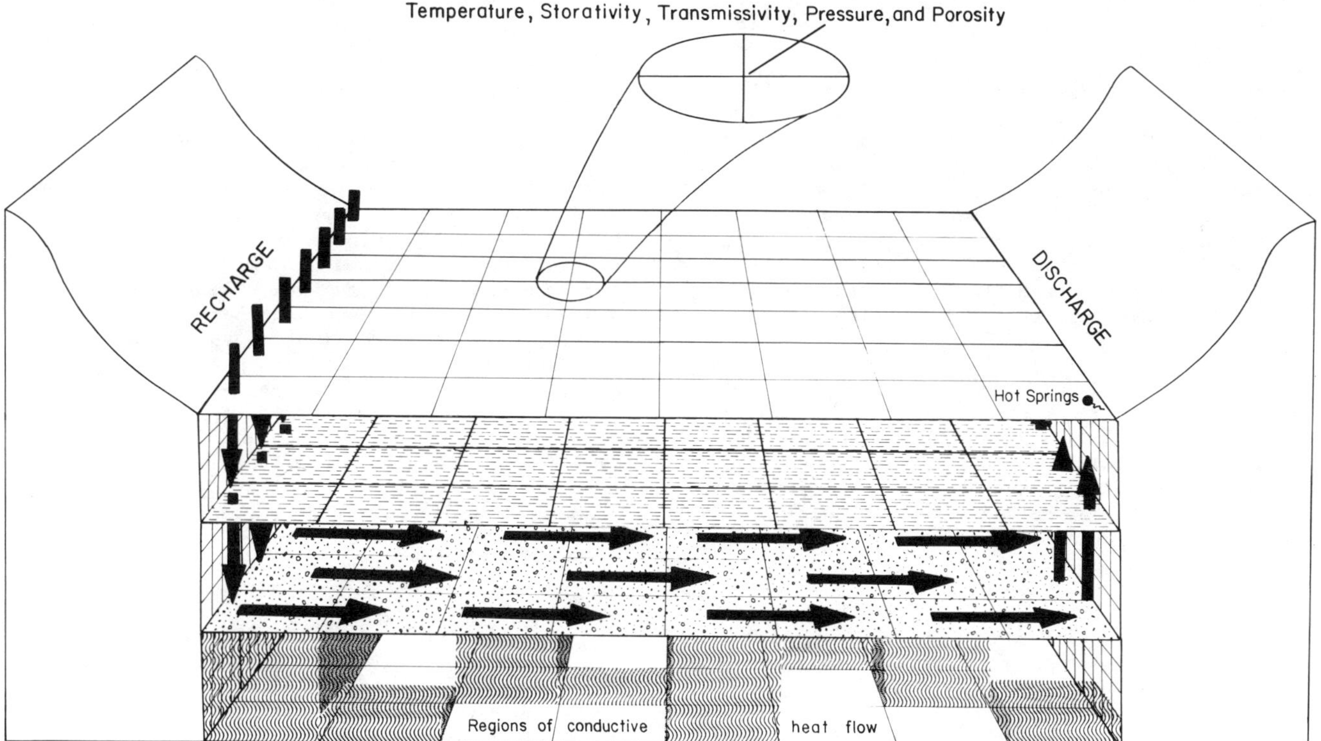

FIGURE 8. Idealized cross-section for a laterally continuous hydrothermal reservoir with overlay of grid required for a distributed parameter model (modified from Welch et. al., 1981).

production rate from the reservoir; transmissivity, storativity, viscosity, porosity and saturated thicknesses. Assuming a geologic setting (formation, lithology, and/or structure) that contains within it the hydraulic continuum of the reservoir, drilling and testing are performed to determine the physical and hydraulic parameters of the reservoir and to develop an analytical solution to predict future reservoir production potential. Figure 8 is an idealized hydrothermal system of the reservoir type that might be found in the Imperial Valley or at Cerro Prieto. A distributed parameter grid is overlain showing the reservoir parameters that would be defined for each node in the grid. However, drilling results to date have shown Basin and Range hydrothermal systems to be much more complex than the simplified model presented here.

In one respect development of a petroleum reservoir is different from that of a ground water system even though the underlying hydraulic principles are the same. Production from an oil or gas reservoir is an extraction process to remove as completely as possible the oil or gas that exists within a closed system (no "recharge" of oil or gas). Ground water development, however, involves taking water from storage at a faster rate than natural recharge. If pumping stops, over a long enough time recharge will replace the water removed.

Development of a Basin and Range hydrothermal system is closer to that for a ground water resource than that for the development of a petroleum reservoir. Using the Beowawe system as an example, it is uneconomical and impractical to drill wells to depths of 6100 to 9100 meters in order to "mine" the projected 10^{15} liters of fluid and contained energy in the hypothesized deep reservoir. Fluid produced from existing shallow production wells in the Malpais Fault discharge zone comes from the elastic compression of the reservoir as the pressure declines and from the expansion of the fluid (Domenico, 1972). As long as production is from the discharge zone the deeper portions of the systems will never be "mined" or dewatered. Productivity of the system will be governed by its storativity and transmissivity. In this situation "reservoir volumes" calculated from flow test data are an inappropriate measure of the production potential of the system.

Two Basin and Range hydrothermal systems have been modeled using numerical methods for assumed reservoir systems. The Long Valley system was modeled by Sorey and others (1978). The model was a three-dimensional distributed parameter numerical model that attempted to represent what were considered to be the actual physical parameters of the system. A later numerical model by Welch and others (1981) for the Leach Hot Springs was a generalized model that examined the constraints on the age, depth and lateral extent of the system without the formulation of a detailed physical model of the system.

Dynamic Flow Model

Hot spring and shallow subsurface discharge imply a dynamic flow system with recharge. If discharge zones and the associated upflow conduits are discrete zones separated by large volumes of impermeable rock with low or nonexistent porosity, then produced fluids cannot be coming from storage in the shallow part of the discharge zone. What occurs is that production, with its subsequent pressure decline, increases the flow rate up the upflow conduit in order to reach a new equilibrium with the new discharge rate, the natural discharge plus the production discharge. As explained in the prior section on the reservoir model, produced fluid comes from compression of the system and expansion of the fluid as system pressures decline. The system is not dewatered.

Use of a dynamic flow model to describe a Basin and Range geothermal system is analogous to using a general systems synthesis to describe surface water systems. In surface drainages there are numerous physical components that control streamflow from a precipitation event, e.g., soil type and moisture content, underlying bedrock, slope orientation, vegetation, humidity, wind speed, duration and intensity of the precipitation. To establish an analytical relationship between each of the components and also for the overall system is difficult, if not impossible. The general systems synthesis method develops empirical relations between the physical components that control the input-output relationship (Crawford and Linsley, 1964). The basis for the method is a lumped-parameter ordinary differential equation.

$$I(t) - O(t) = \frac{dS}{dT}$$

where
I = input
O = output
S = change in storage
t = time (Domenico, 1972).

A lumped-parameter system is one where the inputs and outputs can be measured or estimated but the intermediate processes that interrelate them are unknown or unobservable. The unknown intermediate processes are important because of their combined effect on the input-output relationship but detailed knowledge of the process is not necessary as the intermediate processes are "blackboxed" in the analysis (Domenico, 1972).

There are several advantages to using a lumped-parameter method to describe a Basin and Range hydrothermal system. One is that the unknown intermediate processes that include location and elevation of the recharge zone, rate of conductive heat flow beneath the system, depths of circulation, transmissivity and storativity, rock/fluid ratios and energy gradients are all included in the "blackbox". The input would be the cold water recharge which could be calculated by knowing the energy (heat and mass) output of the system, assuming the system is in a state of equilibrium.

A second advantage is that a space coordinate system is not required when a lumped-parameter approach is used in problem formulation and solution in geothermal hydrology. Thus knowledge of flow paths, aquifer geometries and boundary locations is not necessary.

A third feature of lumped-parameter analysis is that such methods are used extensively in optimization studies where the emphasis is on extreme value problems such as maximizing economic returns with production rates. Such studies would be applicable in determining the maximum potential of a system where there are trade-offs with increased production, such as decreasing enthalpy with reinjection.

APPLICATION OF THE DYNAMIC FLOW MODEL

Exploration

Exploration efforts, while focusing on locating a productive target, also should be directed towards collecting data that aids in determining the total energy flux of the system. The natural energy flux can be considered the minimum rate at which energy can be extracted (Bodvarsson, 1964). This would include predicting the maximum temperature of the system by use of chemical geothermometers. Hydrologic studies, coupled with drill hole data, would be used to calculate energy flux by convection, conduction, advection, and radiation. A complete energy budget would place that of the hydrothermal system within the background energy flux of the region (Olmsted et al., 1975; Welch et al. 1981). Ordinarily this would be beyond the scope of an exploration program as would be a detailed survey of the conductive and radiative heat flow of the system. The convective energy is what would be exploited during production and is the factor of most concern to an exploration program. Table 1 presents the net heat and mass

TABLE I. Heat and mass discharge rates for selected Basin and Range hydrothermal systems.				
	Max. Predicted Temp. in Reservoir, °C	Temp. Recharge Water °C	Net Enthalpy of Water, cal/g	Thermal Water Discharge, kg/s	Net Heat Discharge, cal/s $\times 10^6$
STILLWATER (1.)	159	11	148	95	14
STILLWATER (2.)	159	11	148	55	8.2
GERLACH (1.)	171	11	160	34	5.4
SULPHUR HOT SPRINGS (1.)	186	8	180	8.9	1.6
LEACH HOT SPRINGS (1.)	155	9	147	12	1.7
LEACH HOT SPRINGS (3.)	180	10	170	9.0	1.5
BRADY'S (1.)	200	11	193	42	8.1
BUFFALO VALLEY HOT SPRINGS(1.)	125	9	116	8.0	.93
STEAMBOAT SPRINGS (4.)	228	10	218	71	13
LONG VALLEY CALDERA (5.)	210	10	200	300	69
	282	10	272	190	69

(1.) Olmsted et.al.,1975 (2.) Morgan, 1982 (3.) Welch et.al.,1981 (4.) modified from White,1968 (5.) Sorey et.al.,1978

discharge for several hydrothermal systems in the Basin and Range province. The reader is strongly urged to read Welch and others (1981) and Morgan (1982) for examples of the detailed evaluation of the energy budget for two Basin and Range hydrothermal systems.

Testing

Flow testing must be done with an understanding of the potential complex nature of the production zone. Allman and others (1979) noted that for testing at the Raft River hydrothermal project "complex hydrologic conditions must be presumed, until proven otherwise, to result in drawdowns that do not vary directly with the flow rate." It is possible there are multiple production zones with different static pressures, thus the borehole static pressure is an intermediate one. One may have to assume the production zone is both anisotropic and heterogeneous. Non-Darcian flow may contribute significantly to drawdown.

An empirical approach such as that used by Benson (1982) or Allman and others (1979) can be used in analyzing flow test data. This approach will not yield values for the hydraulic parameters of transmissivity and storativity. Such data, however, are not necessary for the empirical methods used as tools to predict overall system performance. Multirate tests should be done to determine the specific capacity at varying flow rates. Data from these tests can be analyzed to determine the amount of drawdown attributed to non-Darcian flow. Multiple well tests can be run to determine in a preliminary manner interference patterns during the initial stage of production.

An intuitive knowledge of the transmissivity of the production zone can be gained by simple observation of the well response during both production and recovery segments of the test. Note the observations made during the testing of WEN-1. Benson (1982) noted at "each flow rate the downhole pressure quickly stabilized and showed little or no change for the duration of the flow period". It also was noted when "the well is shut-in, the pressure immediately increases by nearly 95% of the total pressure drop" (Benson, 1982). Such responses are indicative of a highly transmissive production zone.

Development and Production

Prediction of production potential is filled with uncertainties making development of a Basin and Range hydrothermal system a high risk operation. The risk is partially the result of little being known of the deeper parts of the systems. If the greater amount of fluid is at great depths (1 to 5 km. below the production zone) and upflow to the production zone is along a fault plane, the production rate and pressure decline will be governed by the least transmissive section along the fault plane flow path as was shown in the geologic cross-section for the Humboldt House hydrothermal system (Fig 6). If the porosity and volume of the system are unknown it is impossible to accurately estimate the stored thermal energy that can be recovered by a sweep process of either natural flow or injection.

On the other hand, however, if the recharge zone is at a relatively great distance from the discharge zone and there is an extensive and deep network of "feeder" conduits, then the highly transmissive fractures in the production zones may be capable of sustained high yields.

Thus whether Basin and Range hydrothermal systems are viewed as reservoir systems or as flow systems, the actual yield or production rate will be governed by the transmissivity of the production zone.

It is impractical and uneconomical to drill-out and define all of the physical and hydrologic parameters of a Basin and Range hydrothermal system. What is required are empirical methods that operate without knowledge of the various intermediate parameters of the system. Such methods are those developed from the empirical production-decline curves developed by Arps (1945). Production-decline curves were developed to predict production rate declines and total oil recoverable for a given oil field. The general method is to use historical production rate-time data to fit a curve to predict future production rate declines.

Two of the basic assumptions are:

1) the extrapolation procedure is strictly empirical and a mathematical expression of the curve based on physical considerations can be set up only for a few simple cases and,

2) whatever factors governed the trend of a curve in the past will continue to govern its trend in the future in a uniform manner (Gentry and McCray, 1978).

The method "black-boxes" unknowns such as the physical characteristics of the system, fluid characteristics and the drive mechanism. Figure 9 illustrates the empirical "black-box" approach to a Basin and Range hydrothermal system.

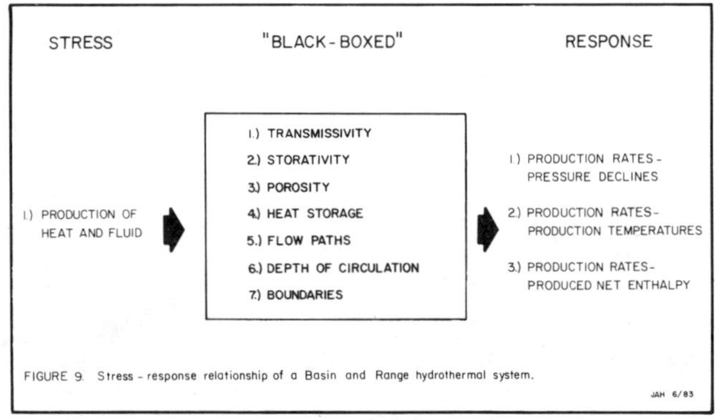

FIGURE 9. Stress - response relationship of a Basin and Range hydrothermal system.

Rivera (1977) applied decline-curve analysis to historical production data from the Cerro Prieto field. Figure 10 shows a production rate decline of the harmonic curve type. Future production rates can be estimated by either extrapolating the general trend of curve (1) or by an exponential extrapolation based on data from the last year, curve (2) (Rivera, 1977). The latter results in a more conservative prediction. Rivera (1977) also noted that when coupled with flowing pressure and temperature surveys the decline curves provided an auxiliary tool in diagnosing some well problems.

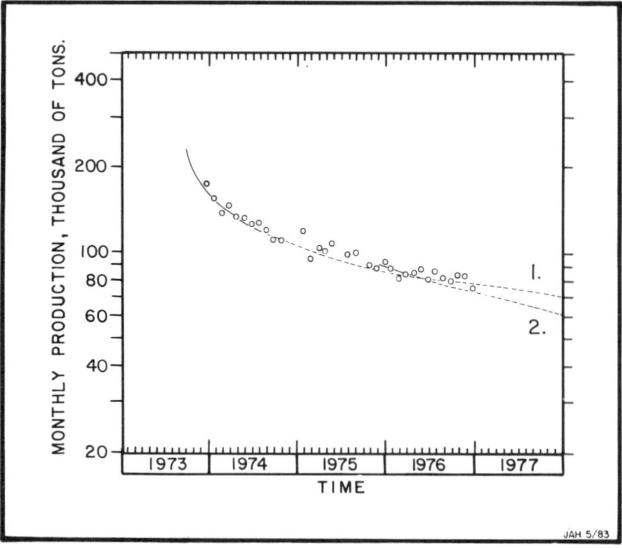

FIGURE 10. Production rate versus time for well B, Cerro Prieto (from Rivera, 1977).

Zais and Bodvarsson (1980) used a nonlinear least squares program to fit an exponential equation to production rate data from Wairakei (Fig. 11). They concluded many wells and fields can be considered to be declining exponentially.

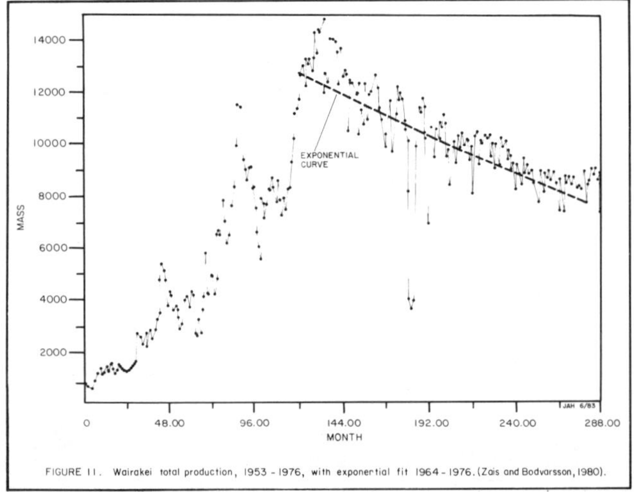

FIGURE 11. Wairakei total production, 1953 - 1976, with exponential fit 1964 - 1976. (Zais and Bodvarsson, 1980).

Zais and Bodvarsson (1980) also applied Coats' (1964) influence function method to the Wairakei data. The method is a black-box one that, given historical production and pressure data, can calculate an influence function and extrapolate it without proposing a specific reservoir model. Thus, either production or pressure can be predicted, given the other. For the Wairakei data, a one year extrapolation resulted in a calculated pressure drop of 539 psi since the beginning of production whereas the observed pressure drop was 543 psi. Thus it is not surprising Castanier and Sanyal (1980) suggested such empirical techniques coupled with modern methods of linear programming are promising tools for future use.

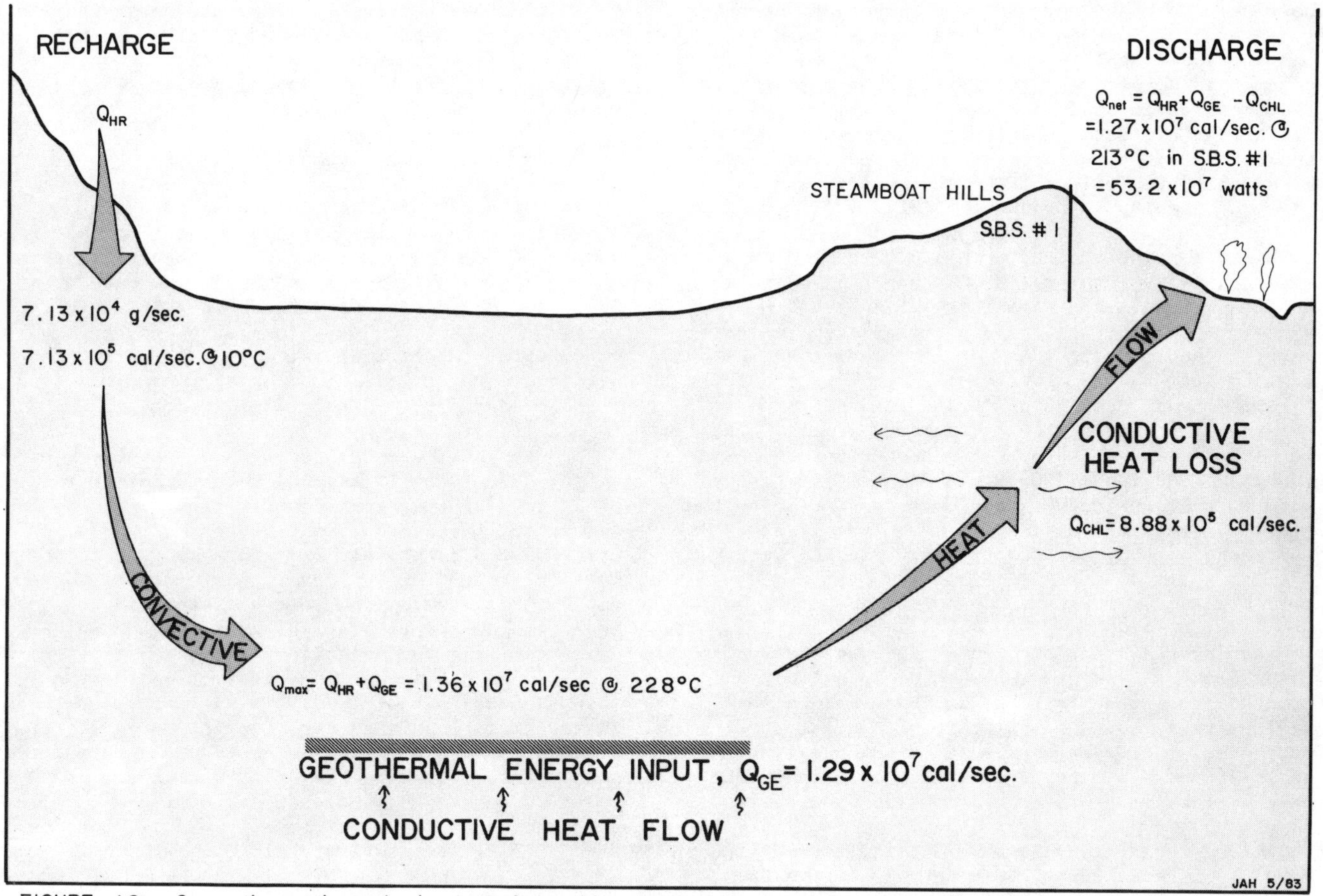

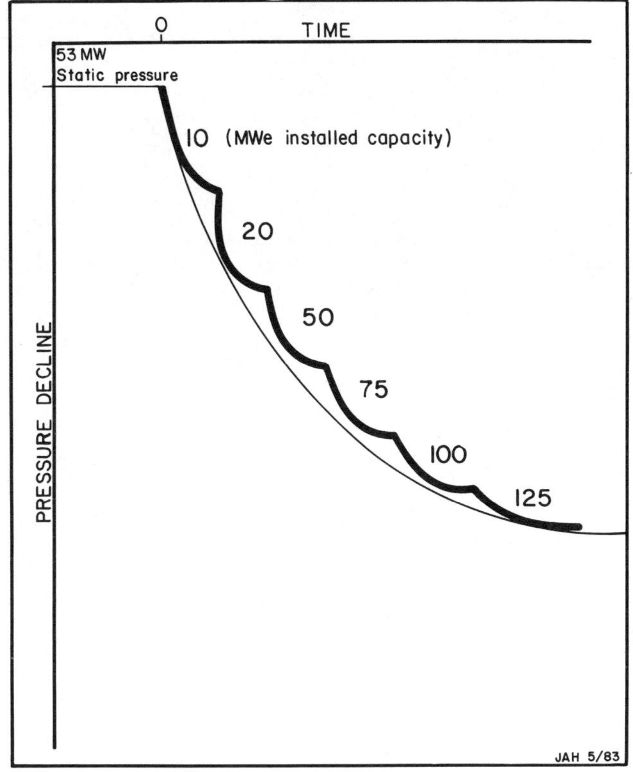

FIGURE 12. Convective and conductive heat flow within the Steamboat Springs hydrothermal system.

Steamboat Springs can be used as an example of how the empirical method could be applied to Basin and Range hydrothermal systems. White (1968) calculated total thermal fluid discharge at 71 liters/sec. From testing by Phillips Petroleum Company chemical geothermometers indicate a maximum temperature of 228°C. Using a recharge temperature of 10°C, the geothermal energy input is 1.29×10^7 cal/sec. (see Fig. 12). A calculated energy discharge is 53.2×10^7 watts or 1.27×10^7 cal./sec. at 213°C in Steamboat Springs #1. Development of the resource can be viewed as a stress (pumping) applied to the system which responds (pressure decline). Ideally the system would be stressed or developed to the maximum level possible without adverse effects in its response. Conditions that might set upper limits on production rates include:

1) reduction in enthalpy due to cold water influx;
2) pressure declines such that flashing occurs within the formation, resulting in scaling outside the borehole;
3) pressure declines such that deeper wells would have to be drilled to maintain the same production rate.

Development of a hydrothermal system for electricity may occur in stages. The initial plant might

FIGURE 13. Possible production-pressure decline curve for step-wise increases in generating capacity.

have a 10 Mw capacity. After a number of years of successful production additional wells would be drilled and additional generating capacity added. In essence this development is similar to a multiple rate flow test on an individual well. With several steps for historical data empirical relationships between production rates and pressure declines should give a reliable prediction of the system response to the next step-up in production (Fig. 13).

Limited prior production history results in a wider range of interpretations than that based on a longer period of historical data (Genty and McCray, 1978). This is true for both the distributed parameter method and the lumped parameter method. With a historical record of production the distributed parameter model can be calibrated and verified by reasonably adjusting the hydraulic parameters until the model response matches the historical response (Wang and Anderson, 1982). With empirical methods such as the Arps curve a new curve is fitted to the additional historical data to more closely establish the equation that defines the curve in the future.

Injection

The uncertainties of predicting system response to production also are to be found in injection in fracture dominated hydrothermal systems. Horne (1982), in a review of injection histories of five geothermal fields in Japan, noted the major impact of injection was a rapid interference with production wells. Among his conclusions were as follows:

1) Where interwell flow occurs, thermal interference can be very detrimental to the performance of the production wells.

2) While only one field benefitted by pressure maintenance, three had reduced performance by thermal interference.

3) An estimate of recoverable energy based on percentage of in-place heat will not be correct if short circuiting takes place between injection and production wells.

Later, in a summary of worldwide experiences in water injection into fractured geothermal systems, Horne (1982) made the following conclusions, among others.

1) Nonproducing fields tend to show slower rates of tracer return than producing fields, creating a difficulty in the prediction of production reinjection breakthrough from pre-production tests.

2) There appears to be a correlation between tracer return rates and subsequent thermal breakthrough in the field.

It is thought these findings are very significant in respect to Basin and Range hydrothermal systems. It is highly recommended anyone planning an injection program in the Basin and Range read the last two references.

SUMMARY

In spite of a decade of intense research and exploration much remains unknown concerning specific components of Basin and Range hydrothermal systems, either individually or as a group. To date no single Basin and Range system has been identified in its entirety. Recharge zones, flow paths and depths of circulation can be identified only by circumstantial evidence or by calculations using reasonable assumptions concerning the individual system.

Enough knowledge has been gained, however, to define these systems in broad conceptual terms. Controlled by the extensional tectonics of the Basin and Range, the hydrothermal systems are dynamic flow systems with relatively limited reservoir capacity. Thus in evaluating the development potential the amount of energy stored in the systems is not as important as the rate at which energy can be extracted from the relatively shallow discharge zone. Because Basin and Range hydrothermal systems are flow systems and not reservoir systems, petroleum engineering techniques for reservoir analysis are inappropriate, both from a conceptual and an engineering standpoint. Standard flow test analysis probably results in an exaggerated estimate of fluid in place and ignores the more important aspect of the "deliverability" of the systems.

Using the concept of a dynamic flow system, Basin and Range hydrothermal systems can be evaluated by empirical methods not requiring a specific detailed model of the system. Large-scale development of and production from a Basin and Range hydrothermal system will depend on the unknown transmissive characteristics of the deeper parts of the system. Realistic evaluations can be made using empirical curves that relate system response to production stress. As with many predictive methods, a longer period of record will result in a more accurate prediction of future system response.

ACKNOWLEDGMENTS

The author wishes to acknowledge the support given him by Phillips Petroleum Company during the preparation of this paper. Dick Benoit of the Reno office was liberal with his suggestions and comments and the author thanks him for his continuous interest. The author also expresses his appreciation to Frank Olmsted and Ed Sammel of the USGS for their technical review of the paper. Views and opinions expressed in this paper are solely those of the author.

BIBLIOGRAPHY

Allman, D. W., Goldman, D., and Niemi, W. L., 1979,
Evaluation of testing and reservoir parameters
in geothermal wells at Raft River and Boise,
Idaho: Geothermal Resources Council, Transactions, v. 3, pp. 7-10.

Arps, J. J., 1945, Analysis of decline curves: Am.
Inst. of Mining and Metallurgical Engineers,
Pet. Div., Transactions, v. 160, pp. 228-247.

Bailey, R. A., Dalrymple, G. B., and Lanphere,
M. A., 1976, Volcanism, structure and geochronology of Long Valley caldera, Mono County,
California: Jour. Geophysical Research, v. 81,
no. 5, pp. 725-744.

Baza, J. R., 1980, Steamboat Springs flow test
evaluation: report prepared for Phillips
Petroleum Company, 15 p.

Benoit, W. R., and Butler, R. W., 1983, A review
of high-temperature geothermal developments in
the northern Basin and Range province: Geothermal Resources Council, Basin and Range
Symposium, this publication.

____________, Hiner, J. E., and Forest, R. T.,
1982, Discovery and geology of the Desert Peak
geothermal field: a case history: Nev. Bureau
of Mines and Geology, Bull. 97, 82 p.

Benseman, R. F., 1959, Estimating the total heat
output of natural thermal regions: Jour. Geophys. Research, v. 64, no. 8, pp. 1057-1062.

Benson, S. M., 1982, Well test data analysis from
a naturally fractured liquid-dominated hydrothermal system: Geothermal Resources Council,
Transactions, v. 6, pp. 237-240.

Blackwell, D. D., and Czang-Go Baaz, 1973, Heat
flow in a "blind" geothermal area near Marys-
ville, Montana: Geophysics, v. 38, no. 5,
pp. 941-956.

Bodvarsson, Gummar, 1964, Utilization of geothermal energy for heating purposes and combined
schemes involving power generation, heating,
and/or by-products: United Nations Conf. New
Sources Energy, Rome 1961, Proc., v. 3
(Geothermal Energy 2), pp. 429-448.

Castanier, L., and Sanyal, S. K., 1980, Geothermal
reservoir modeling - a review of approaches;
Geothermal Resources Council, Transactions,
v. 4, pp. 313-315.

Coats, L. R., McCord, J., and Drews, W., 1964,
Determination of aquifer influence functions
from field data: Jour. of Pet. Tech., Dec.

Craig, H., Boato, G., White, D. E., 1956,
Isotopic geochemistry of thermal waters,
chapter 5 in Nuclear processes in geologic
settings: National Research Council Comm.
Nuclear Sci., Nuclear Sci. Ser. Rept. 19,
pp. 29-38.

Crawford, N. H., and Linsley, R. K., 1964, A
conceptual model of the hydrologic cycle:
Inter. Assoc. Sci. Hydrology, Publ. 63, pp.
573-587.

Denton, J. M., Bell, E. J., and Jodry, R. L.,
1980, Geothermal reservoir assessment case
study - northern Dixie Valley, Nevada: U.
S. Dept. of Energy DOE/ET/27006-1, 117 p.

Desormier, W., 1979, Report on the Humboldt
House geothermal project, Pershing County,
Nevada: report prepared for Phillips
Petroleum Company, 15 p.

____________, 1983, Steamboat Springs Geothermal
Project: Geothermal Resources Council Field
Trip Guidebook for the Role of Heat in the
Development of Energy and Mineral Resources
in the Northern Basin and Range Province, 7 p.

Domenico, P. A., 1972, Concepts and models in
groundwater hydrology: McGraw-Hill, 405 p.

Duffield, W. A., and Bacon, C. R., 1981, Geologic map of the Coso volcanic field and
adjacent areas, Inyo County, California:
U. S. Geol. Survey Misc. Invest. Map I-1200.

Edmiston, R. C., 1982, A review and analysis of
geothermal exploration drilling results in the
northern Basin and Range geologic province of
the USA from 1974 through 1981: Geothermal
Resources Council Transactions, v. 6, pp.
11-14.

Epperson, I. J., 1982, Beowawe, Nevada, well
testing; history and results: Geothermal
Resources Council, Transactions, v. 6, pp.
257-280.

Flynn, T., Trexler, D. T., and Koenig, B. A.,
1982, The Kemp thermal anomaly; a newly discovered geothermal resource in Pumpernickel
Valley, Nevada: Geothermal Resources Council,
Transactions, v. 6, pp. 121-124.

Fournier, R. O., and Thompson, J. M., 1980, The
recharge area for the Coso, California geothermal system deduced from deuterium and
oxygen-18 in thermal and non-thermal waters
in the region: U. S. Geol. Survey Open-File
Rpt. 80-454, 24 p.

Garside, L. J., and Schilling, J. H., 1979, Thermal waters of Nevada: Nev. Bur. of Mines and Geology, Bull. 91, 163 p.

Gentry, R. W., and McCray, A. W., 1978, The affect of reservoir and fluid properties on production decline curves: Jour. of Pet. Tech., Sept. pp. 1327-1341.

Goyal, Kishav, 1982, personal communication, Phillips Petroleum Company.

Hill, D. G., Layman, E. B., Swift, C. M., and Yungul, S. H., 1979, Soda Lake, Nevada, thermal anomaly: Geothermal Resources Council Annual Meeting, 1979, Transactions, v. 3, pp. 305-308.

Horne, Roland N., 1982, Geothermal reinjection experience in Japan: Jour. of Pet. Tech., March, 1982, pp. 495-503.

_____________, 1982, Effects of water injection into fractured geothermal reservoirs a summary of experience worldwide: Geothermal Resources Council Special Report no. 12, pp.47-63.

Hose, R. K., and Taylor, B. E., 1974, Geothermal systems of northern Nevada: U. S. Geol. Survey Open-File Rpt. 74-271, 27 p.

Morgan, D. S., 1982, Hydrogeology of the Stillwater geothermal area, Churchill County, Nevada: U. S. Geol. Survey Open-File Rpt. 82-345, 95 p.

Nehring, N. L., 1980, Geochemistry of Steamboat Springs, Nevada: U. S. Geol. Survey Open-File Rpt. 80-887, 61 p.

Nielson, D. L., Sibbett, B. S., McKinney, D. B., Hulen, J. B., Moore, J. N., and Samberg, S. M., 1978, Geology of Roosevelt Hot Springs KGRA, Beaver County, Utah: Univ. of Utah Research Institute ESL-12, 125 p.

Olmsted, F. H., Glancy, P. A., Harrill, J. R., Rush, F. E., and Van Denburgh, A. S., 1975, Preliminary hydrogeologic appraisal of selected hydrothermal systems in northern and central Nevada: U. S. Geol. Survey Open-File Rpt. 75-56, 274 p.

Parchman, W. L., and Knox, J. W., 1981, Exploration for geothermal resources in Dixie Valley, Nevada; a case history: Geothermal Resources Council Bull., v. 10, no. 5, pp. 3-6.

Ramey, H. J., Jr., 1965, Non-Darcy flow and wellbore storage effects in pressure build-up and drawdown of gas wells: Jour. of Pet. Tech., v. 17, no. 2, pp. 223-233.

Rivera, R. J., 1977, Decline curve analysis a useful reservoir engineering tool for predicting the performance of geothermal wells: Geothermal Resources Council, Transactions, v. 1, pp. 257-259.

Rohrs, D. T., and Bowman, J. R., 1989, A light stable isotope study of the Roosevelt Hot Springs thermal area, southwestern Utah: Univ. of Utah Topical Rpt. 78-1701, DOE/DGE contract DE-ACO7-78ET28392, 89 p.

Sammel, E. A., and Craig, R. W., 1981, The geothermal hydrology of Warner Valley, Oregon: A reconnaissance study: U. S. Geol. Survey Prof. Paper 1044-I, 47 p.

Schaff, S. C., 1976, The 1968 Adel, Oregon earthquake swarm: M. A. Thesis, University of Nevada, Reno, 63 p.

Silberling, N. J., and Wallace, R. E., 1967, Geologic map of the Imlay quadrangle, Pershing County, Nevada: U. S. Geol. Survey, map GQ-666.

Skau, C. M., and Brown, John, in preparation, The quality of natural waters in the east side Sierra Nevada: Desert Research Institute, Univ. of Nevada, Reno, 206 p.

Sorey, M. L., 1975, Numerical modeling of liquid geothermal systems: U. S. Geol. Survey Open-File Rpt. 75-613, 66 p.

_____, M. L., and Lewis, R. E., 1976, Convective heat flow from hot springs in the Long Valley caldera, Mono County, California: Jour. of Geophy. Research, v. 81, no. 5, pp. 785-791.

_____, M. L., Lewis, R. E., and Olmsted, F. H., 1978, The hydrothermal system of Long Valley caldera, California: U. S. Geol. Survey Prof. Paper 1044-A, 60 p.

Urban, T. C., and Diment, W. H., 1982, An interpretation of precision temperature logs in a deep geothermal well near Desert Peak, Churchill County, Nevada: Geothermal Resources Council, Transactions, v. 6, pp. 317-320.

Wang, H. F., and Anderson, M. P., 1982, Introduction to groundwater modeling: W. H. Freeman and Company, 237 p.

Welch, A. H., Sorey, M. L., and Olmsted, F. H., 1981, The hydrothermal system in southern Grass Valley, Pershing County, Nevada: U. S. Geol. Survey Open-File Rpt. 81-915, 193 p.

White, D. E., 1968, Hydrology, activity, and heat flow of the Steamboat Springs thermal system, Washoe County, Nevada: U. S. Geol. Survey Prof. Paper 458-C, 109 p.

White, D. E., Thompson, G. A., and Sandberg,
C. H., 1964, Rocks, structure, and geologic
history of Steamboat Springs thermal area,
Washoe County, Nevada: U. S. Geol. Survey
Prof. Paper 458-B, 63 p.

Young, H. W., and Lewis, R. E., 1980, Hydrology
and geochemistry of thermal ground water in
southwestern Idaho and north-central Nevada:
U. S. Geol. Survey Open-File Rpt. 80-2043,
40 p.

Zais, E. J., and Bodvarsson, G., 1980, Analysis of
production decline in geothermal reservoirs:
Lawrence Berkeley Laboratory, Univ. of California, LBL-11215, 75 p.

THERMOGENICS AND HYDROCARBON RESOURCES

HYDROCARBONS IN THE NORTHERN BASIN AND RANGE,
NEVADA AND UTAH

Louis C. Bortz

Amoco Production Company
Denver, Colorado

ABSTRACT

Occurrences of surface and subsurface hydrocarbons in the northern Basin and Range suggest that oil and gas have been generated in several areas in this province. Documented surface occurrences include: 1) oil in ammonites found in Triassic shales in the Augusta Mountains northeast of Dixie Valley, 2) the Bruffey oil and gas seeps and asphaltite dikes in Pine Valley, 3) Diana's Punch Bowl (probable gas seep) in Monitor Valley, 4) in the ranges surrounding Railroad and White River valleys, droplets of oil are found in goniatites (Mississippian Chainman shale) and part of the Sheep Pass formation is oil stained at one locality, 5) oil shale occurs in the Tertiary Elko formation near Elko and the Ordovician Vinini formation in the Roberts Mountains, 6) numerous outcrops have a petroliferous odor and a few are oil stained.

Subsurface oil and gas shows are more widespread, but most have been found in the same general area as the surface shows. However, there are some important exceptions.

To date all commercial and noncommercial oil and gas fields in the northern Basin and Range are located near the sites of the surface hydrocarbons. This relationship emphasizes the importance of source rock studies to exploration in this province. Prospective areas that lack surface hydrocarbons might be delineated by source rock studies.

A total of eleven oil and gas fields have been discovered in this province of which only three or four can be classed as commercial fields. All of these fields are located in Neogene basins--no fields have been found in an exposed mountain range. The significant fields have some additional common characteristics: 1) the traps are associated with a Tertiary unconformity, 2) the reservoirs have a relatively thick oil column, 3) fractures usually enhance the reservoir quality. Fields in Railroad Valley and the Great Salt Lake are used to illustrate these and other characteristics.

INTRODUCTION

Numerous occurrences of hydrocarbons in the Northern Basin and Range province are direct evidence that oil and gas have been generated in many places within the province. Because of the great diversity of the geology and geologic history associated with these oil and gas "shows", it can be concluded that oil and gas have been generated from a variety of source beds within the northern Basin and Range province.

The scope of this paper is to briefly discuss the oil and gas fields, describe the important surface and subsurface hydrocarbon occurrences, suggest possible hydrocarbon source rocks for some of the areas, and summarize the characteristics common to the significant fields.

NORTHERN BASIN AND RANGE PROVINCE

This province as defined in this paper includes the area from Reno, Nevada, east to Salt Lake City. From north to south it extends from the northern Utah and Nevada state boundaries to a few miles south of the south end of Railroad Valley. (The province extends south to at least the Las Vegas shear zone, but this area was excluded because of "figure" format.) Figure 1 is a regional map of this area generalized from the Nevada (Stewart and Carlson, 1978) and Utah (Hintze, 1980) state geologic maps. This map shows the major Neogene-Quaternary basins. The intervening areas are the ranges, and in some places, late Tertiary volcanic rocks. The Tertiary volcanic rocks in northern Nevada are basalts that range in age from 6-17 m.y. (Stewart and Carlson, 1978). The approximate east edge of the province is shown by the hatchured line that extends north and south from Salt Lake City.

OIL AND GAS FIELDS

There are eleven oil and gas fields in this province; however, only four of these fields are currently producing. These fields are shown on Figure 1 and pertinent data for each field (or producing well) are shown on Table I. The two best fields are both in Railroad Valley. Eagle Springs, discovered in 1954 by Shell, has produced 3,570,206 BO

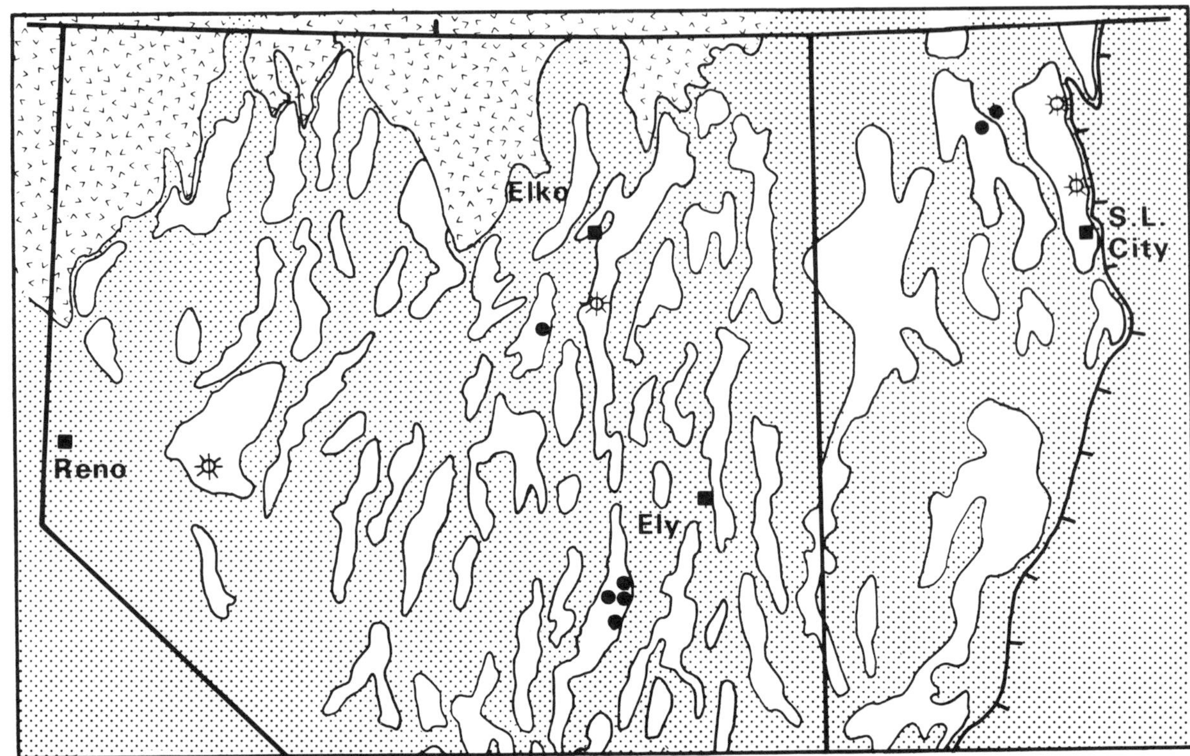

Figure 1.   Northern Basin and Range province. ☐ – Major Neogene and Quaternary basins; ▣– Mostly pre-Neogene rocks; ☷ – Upper Tertiary volcanic rocks (6-17 m.y.); ⟋ – East edge of Basin and Range province; ● – Oil field; ☼ – Gas field.

TABLE I

Oil and Gas Fields

Field	Location	Disc. Date & Co.	Producing Fms.	Prod. Depth	Cumulative BO thru 1982	Remarks
NEVADA						
Fallon Area	T17-18N, R28-30E	1920's - ?	Quaternary	160'±	Unknown	97-98% CH_4 Tr $C_2 H_6$+
Eagle Springs	T9N-R57E	1954 - Shell	Oligo. Volcanics Eocene Sheep Pass Paleozoics	5780'-7256'	3,570,206	26-29° API 65-80°F Pour point
Trap Spring	T9N-R56E	1976 - NW Expl.	Olgio. Volcanics	3330'-4865'	4,602,874	21-25° API 0-5° F Pour point
Currant	T10N-R57E	1979 - NW Expl.	Eocene Sheep Pass	6856'-7080'	635 - SI	95°F Pour point 15° API
Bacon Flat	T7N-R57E	1981 - NW Expl.	Paleozoics	5316'-5354'	60,127	28° API 10° F Pour point
Jiggs	T29N-R55E	1980 - Wexpro	Tert. Elko	9096'-9420'	None - SI	IP 93 MCFD 558' Oil on DST
Blackburn	T27N-R52E	1982 - Amoco-Getty-North Central	"Tite Hole"	?	12,434	27° API
UTAH - BASIN AND RANGE						
Farmington	T3N, R1W	1892 - ?	Quaternary	500'±	150,000 MCF (Abn)	BTU 833
Rozel Point	T8N, R8W	Early 1900's - ?	Pliocene Basalt	125'-300'	3000+ BO Since 1956	9° API
Brigham City	T9N, R3W	1920's - ?	Quaternary	400'-700'	Unknown	
West Rozel	T8N, R8W	1978 - Amoco	Pliocene Basalt	2100'-2400'	28,000 BO (Abn)	4-6° API

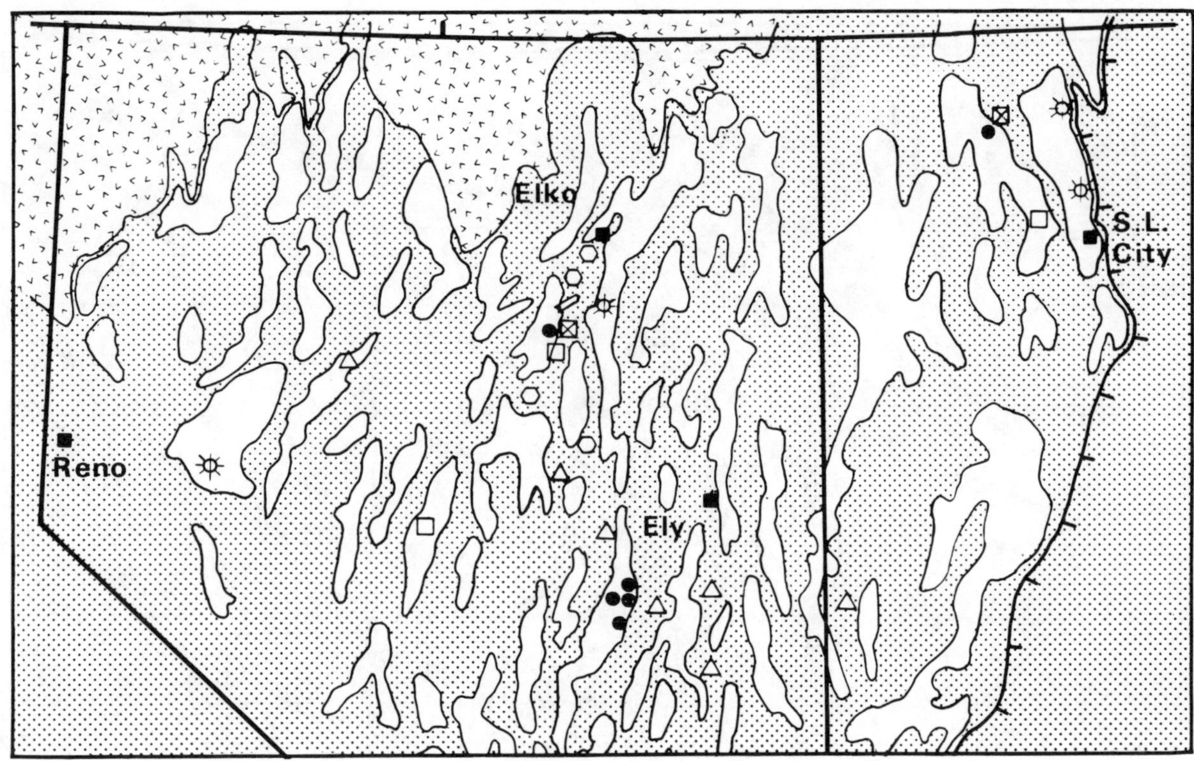

Figure 2.   Documented surface oil, gas and other hydrocarbon occurences. ⊠ – Oil seep;
            □ – Gas seep; △ – Oil stain or droplets; ⧄ – Asphaltite dike; ⬡ – Oil shale
            locality.

TABLE II

Surface Hydrocarbons

	Area	Location	Reference
NEVADA			
Bruffey oil seep	Pine Valley – Sulphur Spring Range	T27N, R52E	Foster, et al., (1979)
Bruffey gas seep	Pine Valley – Sulphur Spring Range	T27N, R52E	Foster, et a., (1979)
Asphaltite Dikes	Pinon Range	T29N, R52E	Smith & Ketner (1975)
Gas Seep – Diana's Punch Bowl	Monitor Valley	T14N, R47E	Garside & Schilling (1979)
Bitumen and liquid oil in voids-Dev. Woodruff Fm.	Southern Fish Creek Range	T15-16N, R52E	Desborough, et al., (1979)
Oil in ammonites	Augusta Mtns. – Dixie Valley	T25-27N, R39E	Nichols & Silbering (1977)
Oil in goniatites	Railroad – White River Valleys	T13N, R56E T6N, R63E	Youngquist (1959)
Oil stain in Sheep Pass Fm.	Egan & Grant Ranges	T10N, R62-63E T9N, R58E	Winfrey (1959)
Oil Shale – Vinini Fm.	Roberts Mtns.	T23N, R51E	Merrian & Anderson (1942)
Oil Shale – Newark Canyon Fm.	S. Diamond Range	T20N, R54E	Foster, et al., (1979)
Oil Shale – Elko Fm.	Near Elko	T23N, R55E	Winchester (1923)
Oil Shale – Dev. Woodruff Fm.	Pinon Range	T32N, R52E	Desborough, et al., (1981)
UTAH			
Rozel Point oil seeps	Great Salt Lake	T8N, R8W	Eardley (1963)
Oil in fossils; Dead oil stain	Northern Needle Range	T25-26S, R19W	Gould (1959)
Gas Seeps	Great Salt Lake – So. Arm	Unknown	---

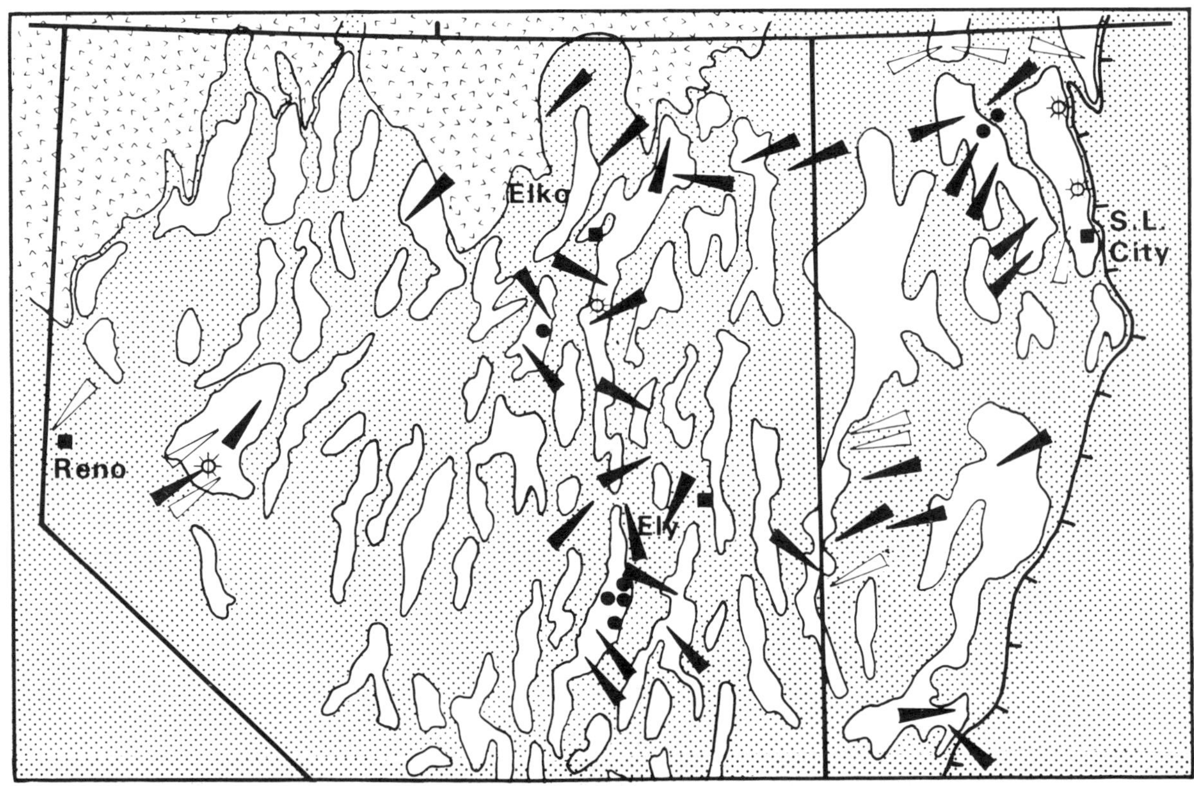

Figure 3. Subsurface shows of hydrocarbons. / - Oil and gas shows; / - gas shows only.

through October, 1982. Trap Spring was found in 1976 by Northwest Exploration and their partners and has produced 4,602,804 BO through October, 1982.

All of the fields produce from Tertiary or Quaternary sedimentary or volcanic rocks except for the Bacon Flat field which produces only from Paleozoic carbonates. Eagle Springs has one well that produced a small amount of oil from fractured Paleozoic carbonates (Bortz and Murray, 1979) and some of the wells in the eastern part of the field may also produce from Paleozoic rocks. The producing formation(s) in the Blackburn oil discovery has not been released.

All of these fields are located within a Neogene-Quaternary basin. Several wells have been drilled in the ranges or areas of pre-Miocene outcrop but only minor oil and gas shows have been encountered (Lintz, 1957; Shilling and Garside, 1968). Neogene and younger lacustrine and playa sediments apparently form effective reservoir seals in many of these late Tertiary basins. It is certainly possible for traps not associated with the Neogene basins to have economic reserves of oil and gas, but the chance of trap preservation is reduced because of tectonic activity prior to mid-Miocene time.

SURFACE HYDROCARBON OCCURRENCES

Documented surface oil, gas and other hydrocarbon occurrences are shown on Figure 2 and listed with references on Table II. Many outcrops within this province have sedimentary rocks that have a fetid or petroleum smell when freshly broken. These occurrences have not been included in Table II. No doubt there are other surface shows in this area that should be included in this list.

Surface shows of oil and/or gas are found in the vicinity of most of the producing areas. For example, the Bruffey oil and gas seeps are about four miles east of the Blackburn oil discovery in Pine Valley; the West Rozel oil field in the Great Salt Lake is 4 miles southwest of the Rozel Point oil seep; near the Railroad Valley oil fields, where surface shows are less obvious, droplets of oil are present in goniatites in a few local areas and oil stain is found in parts of the Eocene Sheep Pass formation in a few localities.

Oil shale is found in four different formations (Ordovician Vinini, Devonian Woodruff, Cretaceous Newark Canyon and Tertiary Elko) in an area from Elko to Eureka, Nevada. These organic-rich sediments and associated sediments may be possible oil and gas source rocks for this area.

Additional discussion of the surface hydrocarbons listed on Table II may be found in the section "Specific Areas for Discussion."

SUBSURFACE SHOWS OF OIL AND GAS

Subsurface shows of oil and gas in this province are shown on Figure 3 and are more widespread than surface shows. Data for this figure is from Lintz (1957), Schilling and Garside (1968), Garside, et al., (1977), commercial well reports and some Amoco data. It is very difficult to obtain complete and accurate data for all wells drilled in this province; some of these shows may not be legitimate in wells that were drilled on "promotional highs." Another problem is that many of these wells were drilled as "tite holes," and much of this valuable information has been lost or buried in company files. No list of subsurface shows is included in this paper.

Most of the subsurface shows will be discussed as part of the "Specific Areas for Discussion." Three areas which have had significant subsurface shows outside of these detailed areas will be discussed briefly in this section:

1) In the Fallon area there are several minor gas shows that are associated with the shallow gas field. The gas is probably biogenic, but several wells have reported minor "asphalt" or oil shows associated with the gas shows. An analysis by the U.S.G.S. (Casper Wyo. No. 47-G-8) of gas from a water well (Sec. 1, T19N, R30E) has a surprising amount of heavy hydrocarbons:

	Percent	
Carbon Dioxide	- 1.65	
Oxygen	- 0.0	
Methane	- 40.21	BTU - 689
Ethane+	- 15.88	
Nitrogen	- 42.26	

Possibly, there could be oil source rock in some of the western Nevada Neogene-Quaternary basins. Alternatively these oil shows may be the result of local oil generation near igneous activity centers.

2) North and northeast of Elko in the "Humboldt Basin" there is a concentration of documented oil shows in the Lower Tertiary Elko formation. Possible source rocks for this area include the Elko formation, Mississippian Chainman shale and Ordovician Vinini formation.

3) In west-central Utah, from the Confusion Range south to the Needles Range, several shows of oil and gas have been found in lower Paleozoic formations. Parts of the lower Paleozoic

section, the Devonian Pilot shale and the Chainman shale could be the source for these shows.

FREE OIL AND GAS SHOWS

There are surprisingly few free oil and gas shows in this province that can documented. The ones that I feel are legitimate are shown on Figure 4 and listed with pertinent information in Table III. There may be additional wells that have had shows of free oil or gas.

Somewhat surprising is the fact that in Railroad Valley, outside the limits of the oil fields, there is not one well that has recovered free oil on DST, production tests or on the pits while drilling.

Most of the free oil shows have been found close to the known fields and will be described in the "Specific Areas for Discussion" portion of this paper. Two exceptions to this are two wells NE of Elko and a well in the Snake Valley SE of Ely.

NE of ELko the Gulf No. 1 Wilkins Ranch (Sec. 21, T38N, R61E) had a show of oil on the pits while drilling at 7700 feet in the Elko formation. Several zones in this well had live oil staining but no free oil was recovered on DST. In the same township, Shell Oil at the 1 Mary's River Federal (Sec. 30, T38N, R61E) recovered 275 feet of GCM and 25 to 35 MCFD on a DST of the Tertiary from 3515 to 3545 feet. The sample chamber of this DST had 2100 cc mud/oil emulsion and about 50 cc of 39° API oil. A well that was recently abandoned, Sun No. 3-13 S.P. (Sec. 13, T39N, R65E) is rumored to have recovered free oil on at least 2 DSTs. The formation is unknown but it is probably the Elko formation.

In Snake Valley, Commodore No. 1 Outlaw-Federal (Sec. 1, T10N, R70E) recovered three feet of heavy viscous oil during a swab test of perforations from 12,081 to 12,090 feet in the Ordovician Pogonip.

There have been very few free gas recoveries outside of the gas fields. The oil fields in Railroad Valley produce virtually no gas. Gas shows near the Fallon, Brigham City and Farmington fields were not included in the map or list because very little reliable data is available. For additional information on the Fallon area see Lintz (1957) and Schilling and Garside (1968).

The only free gas shows on the map and list (Figure 4 and Table III) were reported from the Shell Oil 1 Mary's River Federal (previously mentioned) and the Amoco No. 1 Jiggs (Sec. 19, T29N, R56E) that had GTS, TSTM on a DST of 4685 to 4807 feet in Tertiary sediments. This dry hole is three miles east-southeast of the Wexpro Jiggs gas well.

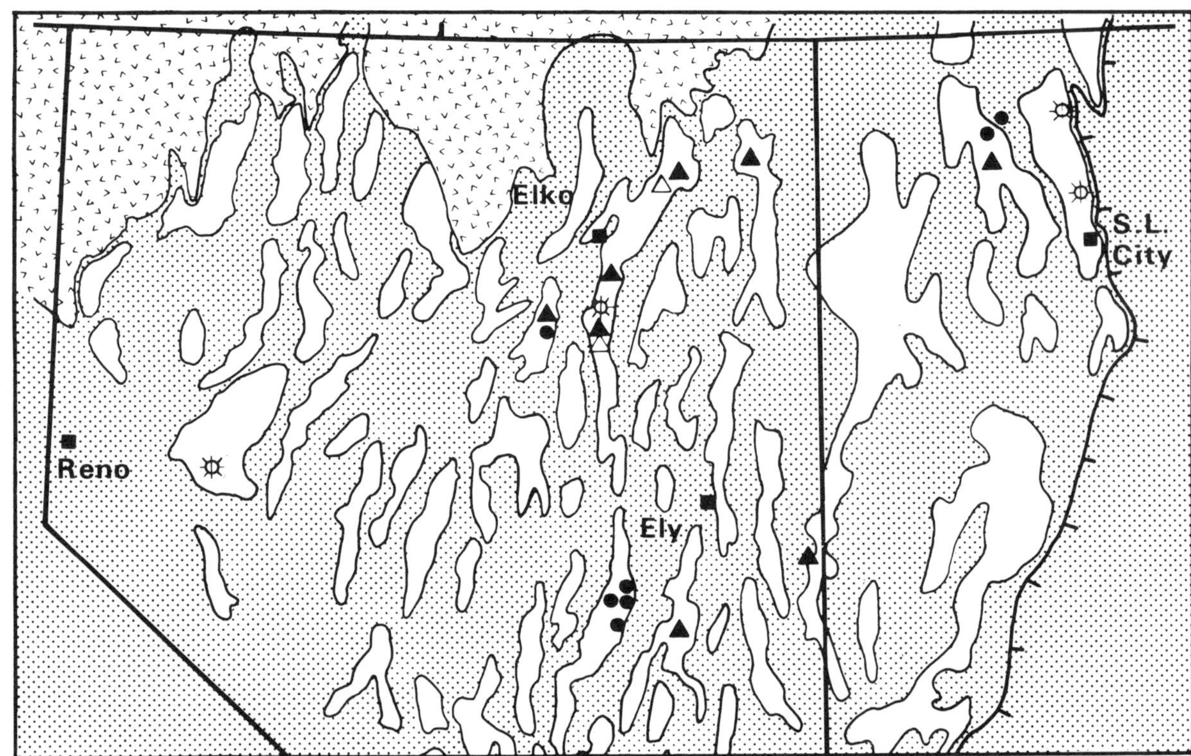

Figure 4. Documented free oil and gas shows. ▲ - Oil; △ - Gas

TABLE III

Free Oil and Gas Shows

Well	Location	Area	Formation	Show	Depth	Remarks
NEVADA						
NW Expl. 6 WRV	7N-61E	White River Valley	Tov	1 cu. in - DST	2190'-2225'	Pour point 115°F; BHT 95° F
Commodore 1 Outlaw-Fed.	10N-70E	Snake Valley	Pogonip Gp.	3 feet in tubing	12,081'-12,090'	High pour point
Beyerbach 1 Fee	27N-52E	Pine Valley	Qal	F 5 BO in 1 day	112'-242'	Near Bruffey seep
Amoco 1 Jiggs	29N-56E	Huntington Valley	Elko Formation	Oil on pits DST-Rec 587' OCM&W	6550'-6652'	GTS TSTM @ 4685'-4807'
Amoco 1 Huntington	31N-56E	Huntington Valley	Elko Formation	Swbd 8 BO + wtr in 8 days	5760'-5830'	Sli S/G
Gulf 1 Wilkins Ranch	38N-61E	Wells	Elko Formation	Show oil on pits	7705'-7772'	DST Rec 90' mud with oil spots
Shell 1 Mary's River	38N-61E	Wells	Tert.	50 cc oil, 39° API	3515'-3545'	GTS 26 to 39 MCFD
Sun 3-13 SP	39N-65E	Pequop	Tert. (?)	Rumored--2 DSTs rec. free oil	?	
UTAH						
Amoco 1 State "P"	8N-8W	Great Salt Lake "E. Gunnison"	Pliocene Sediments	DST 2 - Rec. 19' oil DST 3 - Rec. 1 gal. oil	4880'-5004'	High pour point oil

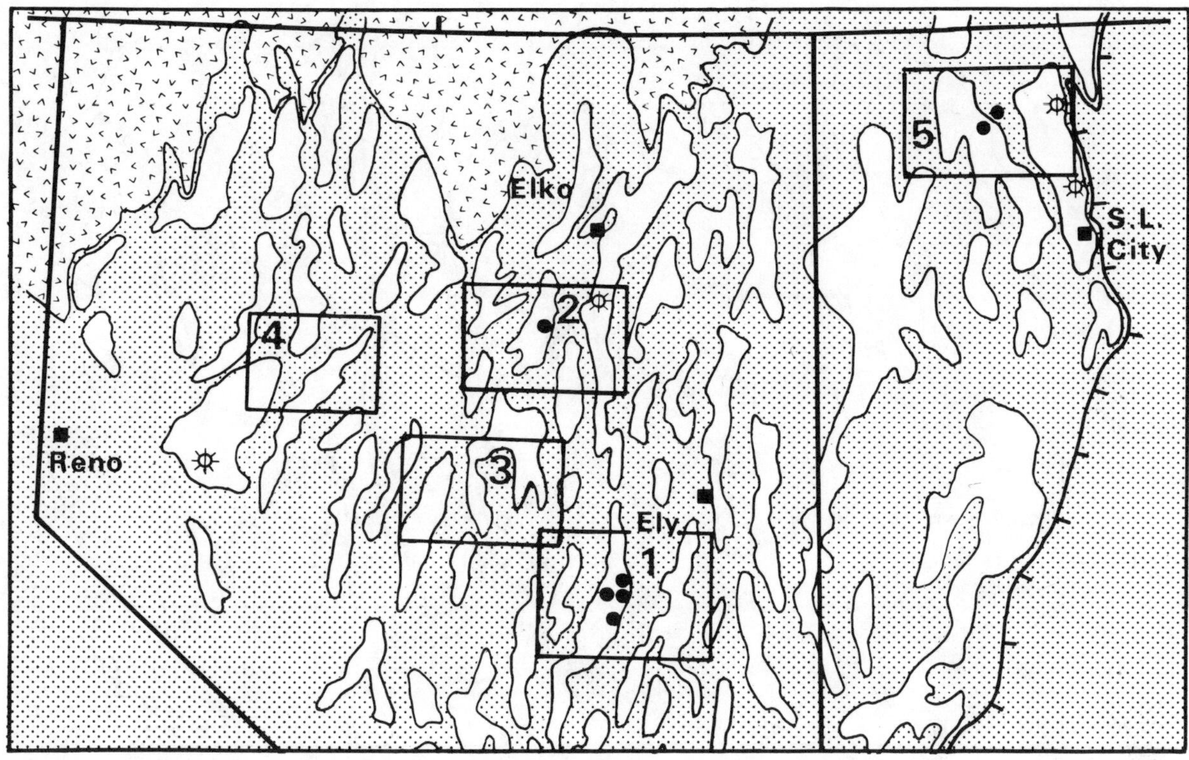

Figure 5. Map showing specific areas for discussion. 1 – Sheep Pass Basin; 2 – Pine Valley; 3 – Monitor Valley (Diana's Punch Bowl); 4 – Dixie Valley; 5 – North Great Salt Lake.

SPECIFIC AREAS FOR DISCUSSION

Figure 5 shows five areas that will be described in some detail to show the relationship between hydrocarbon occurrences and the geology. The five areas are:

1) Sheep Pass Basin (Railroad and White River Valleys)
2) Pine Valley
3) Monitor Valley (Diana's Punch Bowl)
4) Dixie Valley
5) North Great Salt Lake

SHEEP PASS BASIN (RAILROAD AND WHITE RIVER VALLEYS)

Railroad Valley, located within the Sheep Pass Basin (Winfrey, 1960), has been the center of oil exploration in the Basin and Range province since Shell Oil discovered the Eagle Springs field in 1954. Eagle Springs and the Trap Spring fields have produced virtually all of Nevada's cumulative total of 8,246,278 BO through October, 1982. The petroleum geology of Railroad Valley is discussed by Duey in this volume.

Figure 6 is a generalized geologic map of the Sheep Pass basin modified after Stewart and Carlson, 1978. The valley areas shown by the QT outcrops form the Neogene Basins which were initiated during the Miocene. Beneath the Miocene sediments the older Tertiary and Paleozoic outcrop pattern is approximately the same as can be seen in the adjacent ranges (Bortz and Murray, 1979). (This principle may apply to all five of the detailed areas.)

Tertiary intrusives are present in the Grant and White Pine Ranges east of the producing fields. The deepest well in the Eagle Springs field TD'd in an intrusive below Paleozoic rocks (Bortz and Murray, 1979). Obviously, the presence of intrusive rocks in this and other areas within this province does not destroy its petroleum potential.

Surface occurrences of oil in this area include oil in fossils and oil stain in the Sheep Pass formation. Free oil occurs in late Mississippian goniatites in black limestone lenses in the upper part of the Chainman formation (Youngquist, 1949). Two documented localities shown on Figure 6 are: 1) northern Railroad Valley in Section 17, T13N, R56E, and 2) southern Egan Range, Section 7, T6N, R63E. Only a small percentage of these fossils have free oil when cracked open.

The Sheep Pass formation is oil stained in an 800-foot interval in the type section in the Egan Range. Figure 7 is the type section as described by Winfrey (1960) that has oil

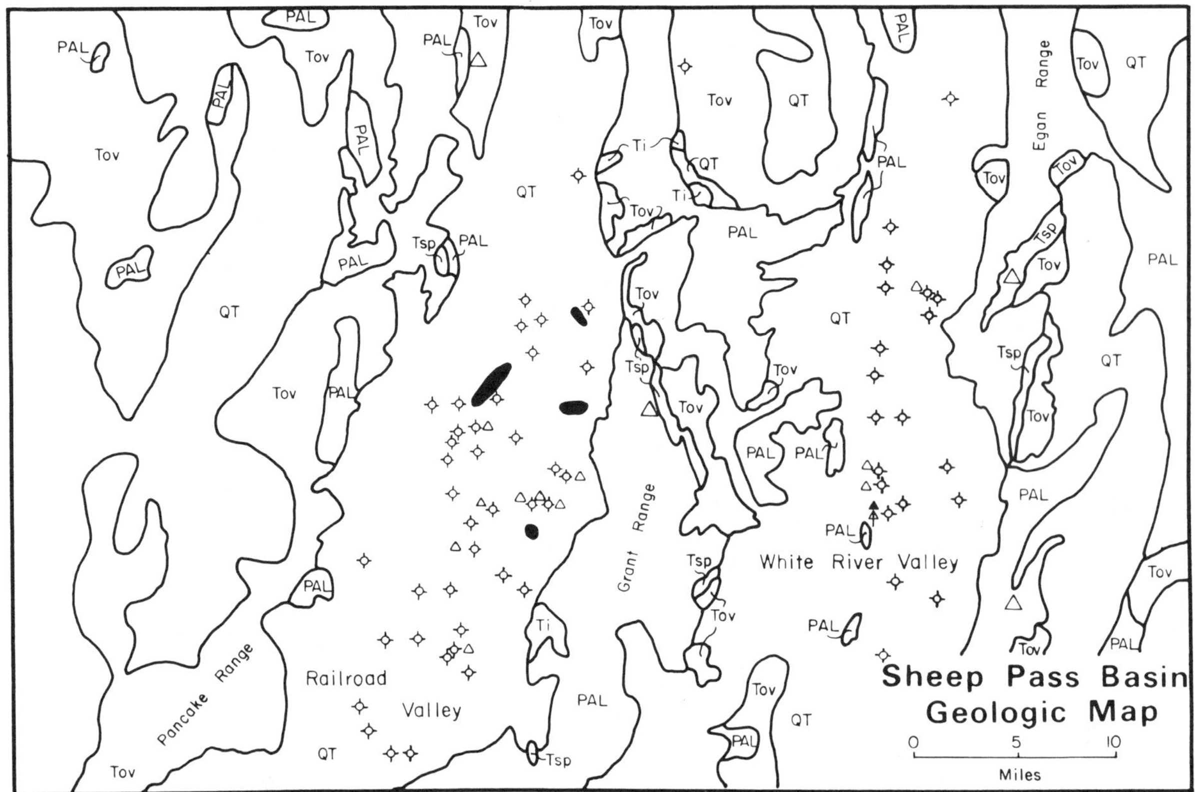

Figure 6.   Sheep Pass basin.   QT - Miocene to Recent volcanic and sedimentary rocks;   Tov -
Oligocene volcanic rocks;   Tsp - Cretaceous - Eocene Sheep Pass Formation;   PAL -
Paleozoics undifferentiated;   Ti - Tertiary intrusive; ⬬ - Oil field; ▲ - Free
oil show; △ - Minor oil show; ⬥ - Minor gas show; △ - Surface oil show.

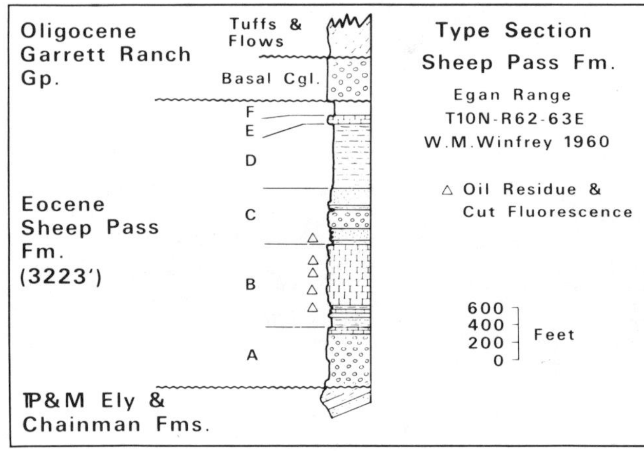

Figure 7.   Type section, Sheep Pass formation.

stain, residue and cut fluorescence in the "B"
and "C" members. The "B" member is probably
equivalent to the Sheep Pass reservoir in the
Eagle Springs field. Minor oil staining in
Sheep Pass limestones has been observed in the
Grant Range due east of the Eagle Springs
field.

Possible source rocks for the Sheep Pass basin
area are the Mississippian Chainman shale,
Devonian Pilot shale and the Eocene Sheep Pass
formation (Poole, et al., this volume).

Subsurface shows of oil and gas are shown on
Figure 6 for all wells completed through 1982.
Some of the dry holes near the producing
fields have been omitted from the map; however, many of these wells did not have any
significant shows of oil or gas. Only a small
percentage of the wells drilled in these two
valleys had oil and gas shows. Most of the
shows were found in the Sheep Pass formation
and the Oligocene volcanic rocks. A few shows
have been reported from Paleozoic formations
and the Miocene to Recent sediments. Typical
shows are scattered oil stain and minor
amounts of gas detected by mudlogging units.

No free oil recoveries have been reported outside of the field areas in Railroad Valley.
In White River Valley, one cubic inch of heavy
oil was recovered on a DST (2190-2225 feet) of
Oligocene volcanic rocks in the Northwest
Exploration No. 6 White River Valley (Sec. 10,
T7N, R61E). The small amount of oil recovery
can be attributed to the pour point of the oil
(115°F) being higher that the temperature of
the DST interval (95°F).

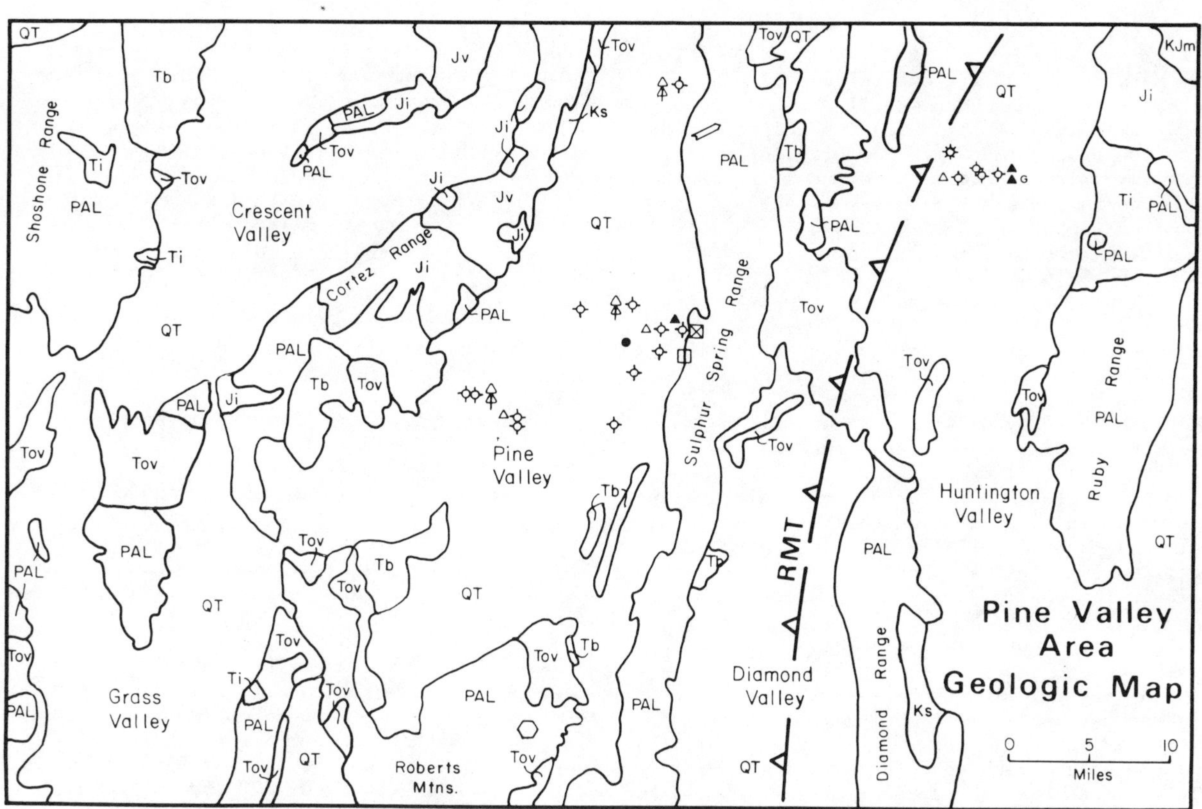

Figure 8. Pine Valley area. QT – Miocene to Recent volcanic and sedimentary rocks; Tb – Basalt; Tov – Oligocene volcanic rocks; Ks – Lacustrine and fluvial rocks; Jv – Volcanic sandstone, tuffs, rholite and rhyodacite; PAL – Paleozoics undifferentiated; T, Ji – Intusives; KJm – Metamorphic rocks; ⊻ – Approximate leading edge of Robert's Mtns. thrust; ● – Oil discovery (Amoco No. 3 Blackburn); ☼ – Gas discovery (Wexpro No. 10-1 Jiggs); ⊠ – Oil seep; □ – Gas seep; ⬌ – Asphatite dike; ⬡ – Oil shale locality; ▲ – Free oil show; △ – Minor oil show; ▲G – Free gas show; ⚲ – Minor gas show.

Figure 9.   Bitumen-impregnated conglomerate at Bruffey oil seep.

Figure 10.   Bruffey Gas seep.

PINE VALLEY AREA

The generalized geologic map (Figure 8) of the
Pine Valley area has been modified after Stewart
and Carlson (1978). This area was active during
the Antler orogeny in which the Ordovician Vinini
shales and siliceous rocks were thrust eastward
over Devonian and older miogeosynclinal
carbonates (Roberts, et al., 1958). The eastern
edge of the Roberts Mountains thrust is somewhere
between the Sulphur Springs Range and the Diamond
Range. Thus, in Pine Valley and the western part
of Huntington Valley, the Vinini formation may be
present beneath the Neogene and younger sediments
and is potentially an additional source rock.

Pine Valley has Nevada's best known oil and gas
seeps, the Bruffey Ranch seeps, that have been
described in detail by Foster, et al., (1979).
The seeps are located in Quaternary sediments and
Ordovician Vinini shales and cherts in
Sections 11 and 14, T27N, R52E. The seeps are
associated with the main boundary fault zone that
separates the east flank of Pine Valley from the
Sulphur Springs Range. Foster, et al., state
that "the seep consists of black tar- or
bitumen-impregnated conglomerate composed of
limestone and dolomite pebbles in the clay and
sandstone matrix." Figure 9 is a photograph of
the bituminous seep. About 100 feet west of this
seep is another oil seep in an outcrop of Vinini
shale that has been observed by Charles H. Thorman
(personal communication, 1983). This seep is also
within the boundary fault zone. The original seep
was discovered over 50 years ago when Mr. R. V.
Bruffey enlarged a hot spring for an irrigation
development and oil began to ooze out upon the
surface in black tarry masses. Foster, et al.,
submitted a sample of the bituminous material to a
major company (unnamed) for analysis. The subsequent report stated that the material is strongly
degraded oil, possibly of parafinic base.

A mile south of the oil seeps is a thermal
spring from which a large number of gas
bubbles rise from mud on bottom and break to
the surface (Figure 10). Foster, et al.,
collected a gas sample which was analyzed by
Hager Laboratories, Inc., of Denver, Colorado.
The following is a summary of the quantitative
analysis:

ppm by volume

Methane	8300
Ethane	1.5
Propane	0.3
n-Butane	0.2
n-Pentane	0.8
n-Hexane	0.2
z-Methyl-Heptane	0.1

This analysis containing fractions of C_1
through C_7 hydrocarbon compounds suggests that
origin of gas was thermogenic rather than biogenic.

Two other surface hydrocarbon occurrences
shown on Figure 8 are: the oil shales of the
Vinini formation exposed along Vinini Creek in
the Roberts Mountains and reported by Merrian
and Anderson (1942) to give assay yields in
excess of 25 gallons per ton, and the asphaltite dike in the Pinon Range (NE/4 Sec. 1,
T29N, R52E) where solid bituminous material
occurs as a narrow vein-like body and as fracture fillings in Chainman shale (Smith and
Ketner, 1975). Exposures of this dike are
very poor.

Possible source rocks capable of generating
oil and gas in the Pine Valley area are:

> Ordovician Vinini shales and cherts
> Devonian Woodruff formation
> Mississippian Chainman shale
> Cretaceous-Paleocene Newark Canyon forma-
> tion

Foster, et al., (1979) discuss the source rock
potential of the Vinini and Newark Canyon formations.

On January 3, 1983, Amoco Production Company
announced the completion of an oil discovery
in Pine Valley. The Amoco-Getty-North Central
Blackburn Unit No. 3 (SW SE 8-T27N-R52E) was
completed pumping 346 BO and 767 BWPD. The
oil has an API gravity of 27° and is being
trucked to a refinery in Tonopah, Nevada.
Through October, 1982, the Blackburn discovery
has produced 12,434 BO. The producing forma-
tion(s) and perforated intervals (two zones)
have not been released by Amoco.

The Blackburn discovery is 3 1/2 miles west of
the Bruffey oil and gas seeps. Two offset
wells, 1/2 mile west and 1/2 mile south, have
been completed as dry holes.

Several wells have been drilled near the
Bruffey seeps in Pine Valley, the boundary
fault zone and in the Sulphur Spring Range
(Foster, et al., 1979). The best show of oil
was found in the Beyerbach and Black well
(NE NW SE 11-T27N-R52E) which was drilled in
the fault zone to a depth of 1205 feet in
Quaternary sediments. Oil shows were encountered at 112 and 242 feet and reportedly the
well flowed 5 BO for one day in 1960.

Shows of oil and gas have been found in other
parts of Pine Valley. The Getty No. 1 NOST
(NW NE 32-T28N-R52E) drilled in 1977 to a TD
of 10,505 feet in Tertiary or Cretaceous volcaniclastics encountered only minor shows of
oil (oil stain) and gas. Aminoil No. 1-23 SP
(Sec. 23, T30N, R52E) drilled in 1979 to a TD
of 7145 feet in the Paleozoics also found only
minor increases of methane on the mud log and
at 5430 feet asphaltic material had a yellow
cut fluorescence. Four wells have been
drilled about 8 miles southwest of the Blackburn discovery, and two of these wells

reported oil and gas shows, but no description of the type of shows is available.

In Huntington Valley, Wexpro completed the No. 10-1 Jiggs (Sec. 10, T29N, R55E) flowing 93 MCFPD from perforations between 9096 and 9420 feet in the Teritary Elko formation. A DST of 9391 to 9440 feet recovered 558 feet of heavy oil and some natural gas, TSTM. The Wexpro well is probably not commercial; however, this well proves that hydrocarbons have been trapped in this area. Four other wells have been drilled nearby and shows of oil and gas were reported from two of them. The best show was in the Amoco No. 1 Jiggs (NE NE 19-T29N-R56E) which had a show of oil on the pits while drilling between 6550 and 6652 feet in the Tertiary Elko formation. A DST of this interval recovered 587 feet of oil cut mud and water. GTS TSTM was recovered on a DST at 4685-4807 feet.

<u>MONITOR VALLEY AREA</u> (DIANA'S PUNCH BOWL)

As can be seen on the generalized geologic map (Figure 11) modified after Stewart and Carlson (1978), this area is similar to Pine Valley in that the leading edge of the Roberts Mountain thrust is in eastern portion of the area. East of the thrust only pre-Antler eastern assemblage rocks are present, but to the west, both western and eastern assemblage rocks are present. Thus, the Paleozoic source rocks for the two areas are similar.

Diana's Punch Bowl is the only dramatic occurrence of surface hydrocarbons in this area. The best reference to Diana's Punch Bowl, located in Section 22, T14N, R47E, is Garside and Schilling (1979). They describe the bowl "...as a cup-shaped depression approximately 50 feet in diameter at the top of a domelike hill of travertine approximately 600 feet in diameter. Warm water in the pool of the bowl is 30 feet below the top of the rim while the top of the hill is about 75 feet above the level of Monitor Valley." Figure 12 is a photograph of Diana's Punch Bowl from the road. The hot spring was originally called the Devil's Punch Bowl in the early 1900's because occasional flames were seen and more gas than at present was emitted. This was described by J. J. Butler in Spurr (1905).

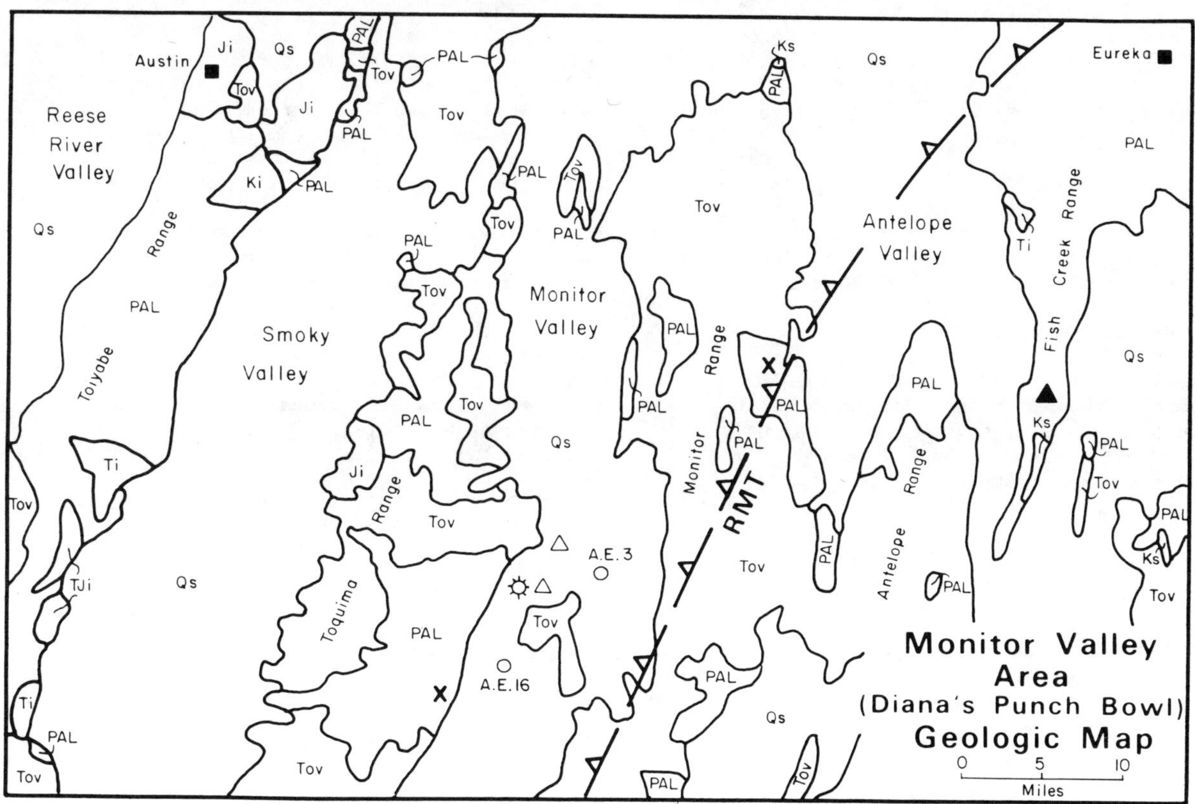

Figure 11. Monitor Valley area. Qs - Quaternary alluvial and playa deposits; Tov - Oligocene volcanic rocks; Ks - Cretaceous to Eocene lacustrine and fluvial rocks; PAL - Paleozoics undifferentiated; Ti, Tji, Ki, Ji - Intrusives; ⊻ - Approximate edge of Robert's Mtns. thrust; ☼ - Diana's Punch Bowl; △ - Gas seep; ▲ - surface oil show;○A.E. - A.E.C. well; ✕ - Petroliferous odor in rocks.

Figure 12. Diana's Punch Bowl.

No analysis of the gas from Diana's Punch Bowl is available to the writer. The gas could be biogenic and insignificant to oil and gas exploration. However with the presence of C_1 through C_7 in the Bruffey gas seep, there is a possibility that the gas in Diana's Punch Bowl is similar and that there may be significant source rocks for oil and gas in this area.

At Potts' Ranch four miles north of Diana's Punch Bowl there are several hot springs (Sec. 2, T14N, R47E) including one that emits some gas (Garside and Schilling, 1979). Mariner, et al., (1975) report an analysis of this gas in volume percent: $O_2 + A_r = 4$, $N_2 = 93$, $CH_4 < 1$, $CO_2 = 3$. This gas may be biogenic since methane is the only hydrocarbon present; however, the composition of this gas is probably different than the gas at Diana's Punch Bowl because this mixture with less than 1 percent methane probably would not ignite.

In an unnamed Devonian limestone about 7 miles southwest of Diana's Punch Bowl (X on Figure 15), McKee (1976) has described 100 feet or more of thin-bedded dark petroliferous-smelling limestone and black shale. In the Monitor Range (X on Figure 11), Bortz (1959) reported strong petroliferous odor in brown-black calcareous mudstone at the base of the Ordovician Hanson Creek formation. Many of the Paleozoic formations in the Basin and Range province contain limestones and shales that have a fetid or petroliferous odor. These occurrences have not been mentioned in other areas because they are so common. If there were no units in the Monitor Valley area with petroleum odor, its potential for significant source rocks would be reduced.

In the southern Fish Creek Range (T15-16N, R52E) Desborough, et al., (1979) have described oil shows in the Devonian Woodruff formation. They state, "In fresh rock solid bitumen and liquid oil fill voids and microfractures." The Woodruff formation is a possible source rock in the eastern part of the Monitor Valley area.

No wells have been drilled for oil and gas in the Monitor Valley area (Figure 15). Two wells were drilled in Monitor Valley by the A.E.C. as test holes for nuclear devices (Hoover, et al., 1969). One well (UCe 3) 5 miles east of Diana's Punch Bowl encountered only welded tuffs from surface to a TD of 2000 feet. The other well (UCe 16) 5 miles south of Diana's Punch Bowl penetrated 1100 feet of alluvium, then drilled volcanic tuffs and flows to a TD of 4353 feet. No oil or gas shows were reported in either well.

DIXIE VALLEY AREA

This area is the only detailed area in western Nevada and the generalized geologic map (Figure 13) shows that the geology is quite different from the other detailed Nevada areas. (This map is also modified after Steward and Carlson, 1978.) There are no Paleozoic rocks in the Dixie Valley area; the oldest rocks are Triassic sediments and intrusives. The valley areas are also Neogene to Recent depositional basins.

Since 1978 at least 12 geothermal wells have been drilled in Dixie Valley ranging in depth from 3010 to 12,500 feet. Some of these wells are geothermal producing wells and others are temperature gradient and observation wells. These wells are in the general area of the three wells shown on Figure 13 and are in T24N, R36 and 37E. See Edmiston (1982) for a review of geothermal activity in the northern Basin and Range from 1974 through 1981.

The petroleum industry is attracted to this area because of possible Triassic source rocks in the ranges flanking the northern part of Dixie Valley. Nichols and Siberling (1977) describe the Fossil Hill member of the Middle Triassic Favret formation as a 200-meter sequence of dark-gray calcareous shale with interbeds of fossiliferous lime mudstones in the Augusta Mountains (T25N, R39E). When broken, chambers of ammonoids from concretions commonly yield liquid hydrocarbons (Figure 14). Figure 15 shows an outcrop of dark-gray calcareous shale within the Fossil Hill member in the Augusta Mountains. This is part of the "Triassic "C" outcrop shown on Figure 13.

In the Tobin Range at the north end of Dixie Valley between the Augusta Mountains and the Stillwater Range, Burke (1973) found similar hydrocarbon-bearing cephalopods within the higher part of the Favret-Prida section (T26 and 27N, R39E). This section consists of 300 feet of poorly exposed dark mudstone, fine-grained limestone and fine-grained calcarenite. All of these rocks have a strong petroliferous odor when broken and liquid hydrocarbon was found in the hollow chambers of some cephalopod shells.

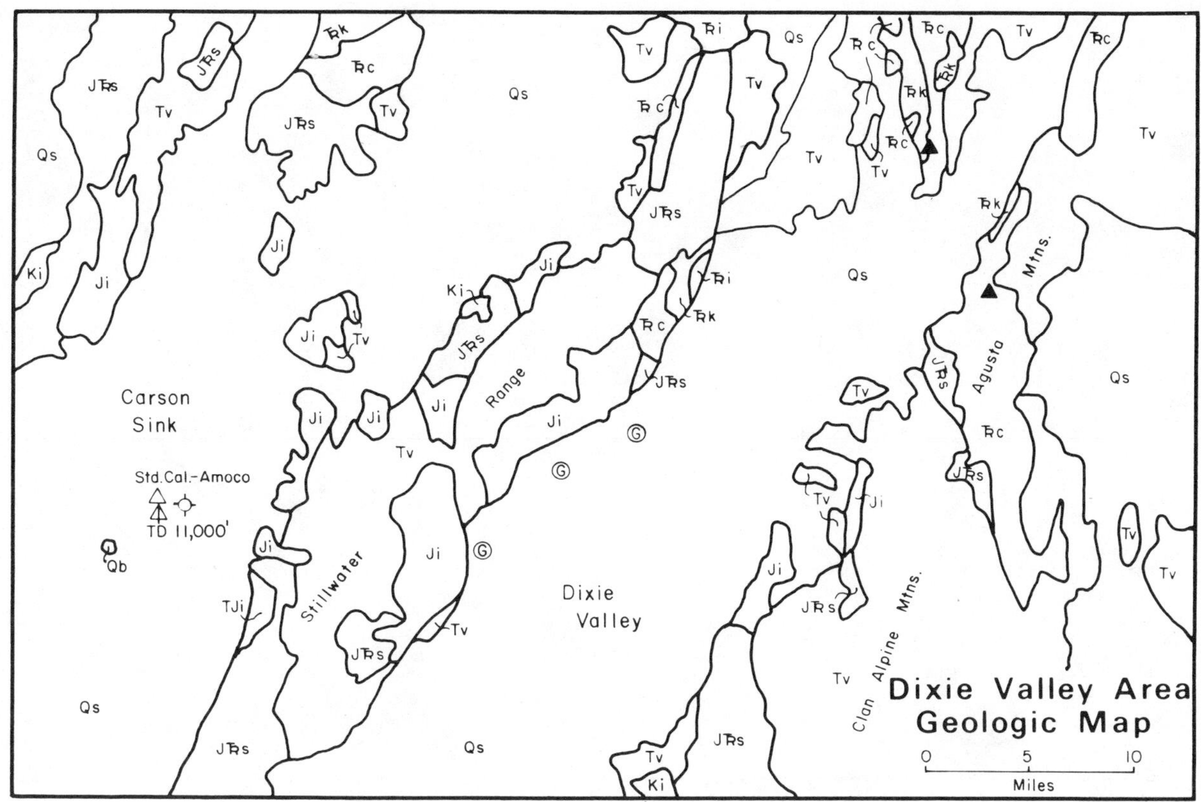

Figure 13.   Dixie Valley area.   Qs – Quaternary alluvial and playa deposits;   Qb – Quaternary basalts;   Tv – Tertiary volcanics;   JRs – Upper Triassic and Lower Jurassic sediments and volcanic rocks;   Rc – Lower, Middle, and Upper Triassic (Tobin, Dixie Valley, Favret and Augusta Mtns. fms.);   Rk – Lower Triassic Koipato volcanic and clastic rocks;   Tji, Ki, Ji, Ri – Intrusives;   Ⓖ – Geothermal well or test; △ – Minor oil show; ⬥ – Minor gas show; ▲ – Surface oil show.

Figure 14.   Concretion of calcareous mudstone from Triassic Favret formation in the August Mountains – chambers of this ammonite contained liquid hydrocarbons.

Figure 15.   Outcrop of the Fossil Hill member of the Favret formation in the Augusta Mountains.

Oil-bearing ammonoids have only been reported from the Favret and Prida formations in the ranges north and east of the geothermal wells in Dixie Valley. The Favret and Prida formations are included in the Triassic "C" outcrop shown on Figure 13. Thus, if the Favret and Prida are significant source rocks for oil and gas, they may be present beneath the Neogene sediments in Dixie Valley northeast of the geothermal wells. If traps are present, this part of Dixie Valley is an attractive prospective area for oil and gas.

The only well on Figure 13 that was drilled as an oil and gas exploratory well is the Standard-Amoco No. 1 S.P. Land Co. (Sec. 33, T24N, R33E) located in the northeastern part of the Carson Sink. Hastings (1979) states that this well penetrated 11,000 feet of Tertiary sediments and volcanic rocks. Oil and gas shows were reported by Hastings:

"Free oil was present in vugs at the top of a core of calcite cemented basalt breccias taken in the interval 8,168-8,198 feet. A laboratory analysis described the oil as moderately mature and paraffinic. The only

other hydrocarbon shows of significance consisted of weak cut fluorescence from sidewall samples taken in an interval above an igneous sill at 5,130 feet and strong methane shows in the drilling mud in the interval between 1000 and 4200 feet."

Hastings also summarizes the Tertiary source rock potential for this well. From the surface to 6,900 feet, the lacustrine-playa sediments are high in organic content (up to 5 1/2%) but immature. He concludes that rocks in the lower part of the section are too low in organic content to be good source rocks.

NORTH GREAT SALT LAKE

The Great Salt Lake is one of the remnant Pleistocene lakes in the Basin and Range province. Aside from the lake, this area is similar to the other areas that have been previously discussed. The Great Salt Lake covers several major Neogene depositional basins; one of these basins is essentially the area that includes the wells within the lake shown on Figure 16. Smaller subsidiary basins flank most of the major basin areas; an example of a smaller basin is the "Rozel graben" north and

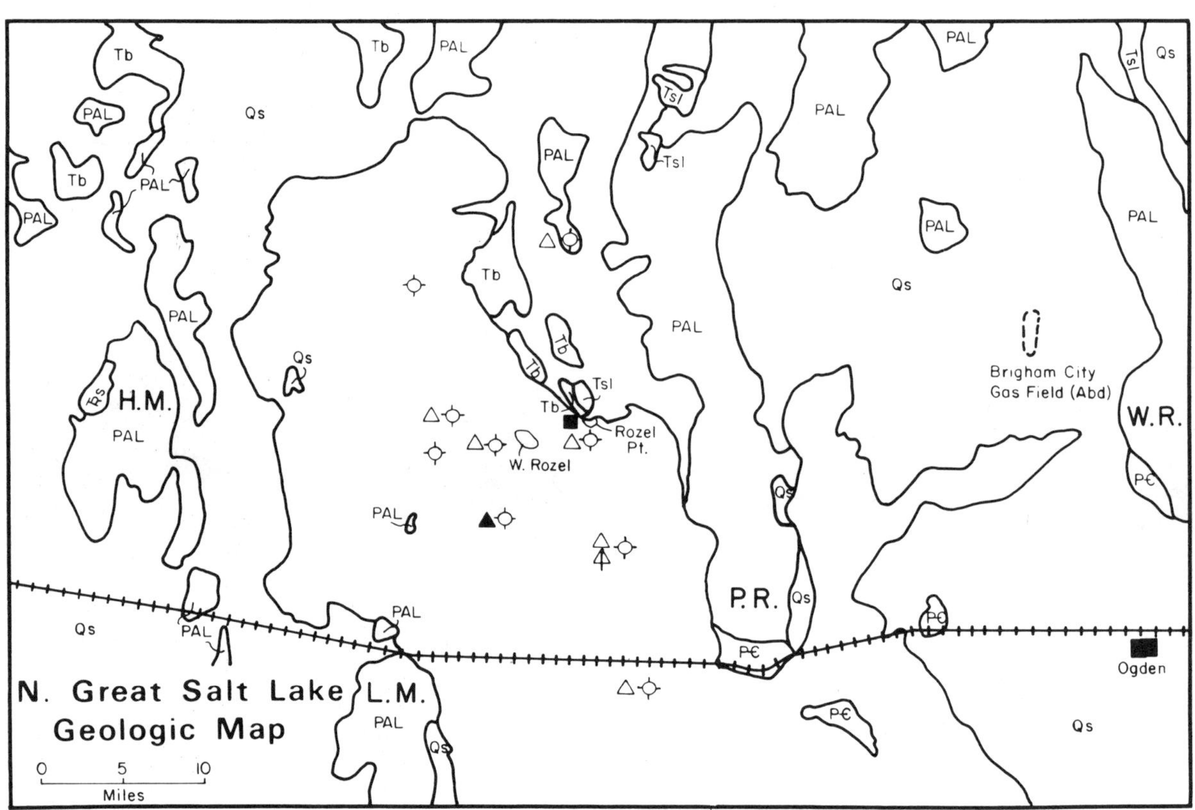

Figure 16. North Great Salt Lake. Qs – Quaternary lacustrine and fluvial sediments; Tsl – Miocene – Pliocene Salt Lake Group; Tb – Tertiary basalts; ℞s – Triassic sediments; PAL – Paleozoics undifferentiated; P€ – Precambrian; ▲ – Free oil show; △ – Minor oil show; ⚲ – Minor gas show.

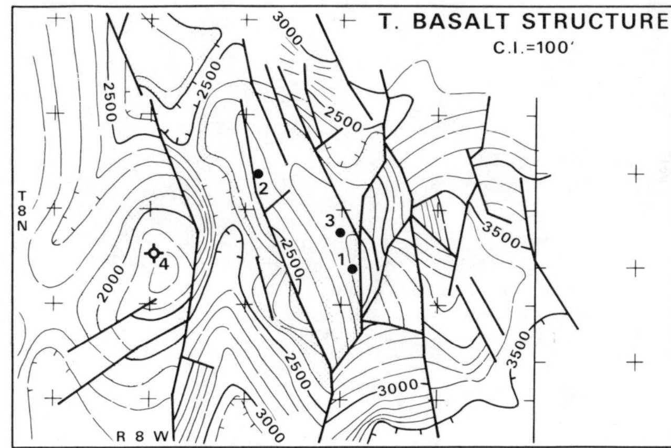

Figure 17. Top of basalt structure map in the
West Rozel Field area - datum is the
surface of the Great Salt Lake.

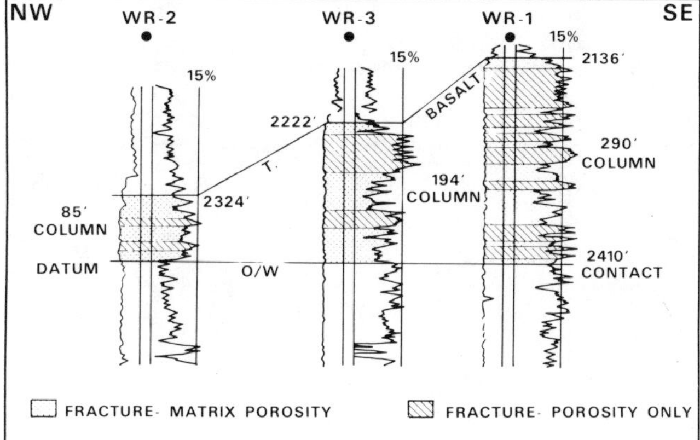

Figure 18. Structure section at West Rozel showing basalt reservoir. Closures are
shown by stippled line. See figure
17 for well locations.

east of Rozel Point. Figure 16 is a generalized geologic map of the area modified after
Hintze (1980).

The Rozel Point oil seeps flow heavy, viscous
oil at the south end of Rozel Point
(Sections 8, 9 and 16, T8N, R7W) along a
probable fault zone which has uplifted the
Rozel Hills relative to the lake to the southwest (Eardley, 1963). Some of the oil seeps
are associated with springs that emit oil and
water and build small circular mounds of travertine; these can be observed only during
periods when the lake level is low.

Since the early 1900's there have been many
attempts to produce oil from shallow wells
drilled near the seeps and from surface pools
of oil at the seeps. The area is currently
leased and intermittently produced. Wells
encounter the oil in a Pliocene basalt reservoir from 150 to 300 feet. A high percentage
of water is produced with the oil. Twenty or
thirty shallow wells have been drilled in and

around the seeps and also four or five holes
1500 to 2800 feet deep have been drilled.
(These wells are not shown on Figure 16).

An offshore Great Salt Lake exploration program by Amoco commenced in June, 1978, and
drilled a total of 15 wells before the program
was completed in January, 1981. The oil discovery at West Rozel (Sec. 23, T5N, R8W),
4 1/2 miles southwest of Rozel Point, was completed in Pliocene basalts from perforations
at 2270 to 2410 feet flowing intermittantly 1
to 5 BOPH using a nitrogen lift. API gravity
is 4-6 degrees with 13% sulphur. The second
well drilled at West Rozel was equipped with a
submersible hydraulic pump and produced at
rates up to 90 BOPH for short periods from
open hole at 2340 to 2384 in the same Pliocene
basalt reservoir. A third well was drilled
and produced from the basalt (2305 to
2348 feet) continuously for 64 days. During
the last five days of the test, the well
averaged 311 BO and 675 BW. Total field production was 28,000 BO. Economics for
developing the field were not favorable at
that time and the three wells were plugged and
abandoned.

The West Rozel structure is a faulted, closed
anticline with approximately 300 feet of
vertical closure covering 2300 acres
(Figure 17). Figure 18 is a structure section
that shows the basalt reservoirs of the three
producing wells.

Of the twelve exploratory wells drilled in
this program all but four encountered some
shows of oil and gas. Commonly oil shows were
stain and globules of heavy oil in the Neogene
sediments and volcanic rocks. Gas shows were
usually small increases in the mud in Tertiary
rocks; however, one well near the south end of
the lake (Sec. 4, T1S, R4W) had a minor gas
blowout at 435 feet.

In the Amoco No. 1 East Gunnison (Sec. 10,
T7N, R8W) high pour point oil was recovered on
2 DSTS (19' oil and 1 gallon oil) of Pliocene
limestones and tuffaceous shales from 4880 to
5004 feet.

Source rock studies by Amoco on Paleozoic and
Mesozoic surface and subsurface samples found
no significant oil source rocks in these
units. Their potential as a gas source was
poor. Since no pre-Neogene Tertiary rocks
were penetrated in the Great Salt Lake, the
Neogene fluvial and marginal lacustrine rocks
may be the source for the oil at Rozel Point
and West Rozel. Neogene to Recent sediments
and volcanic rocks are about 15,000 feet thick
in the basin west of the south end of the
Promontory Range. If high heat flow existed
in the northern Great Salt Lake during and
after the extrusion of the Pliocene basalts at
West Rozel and Rozel Point, then some of the
organic-rich Neogene sediments could be the
source of this heavy, viscous oil.

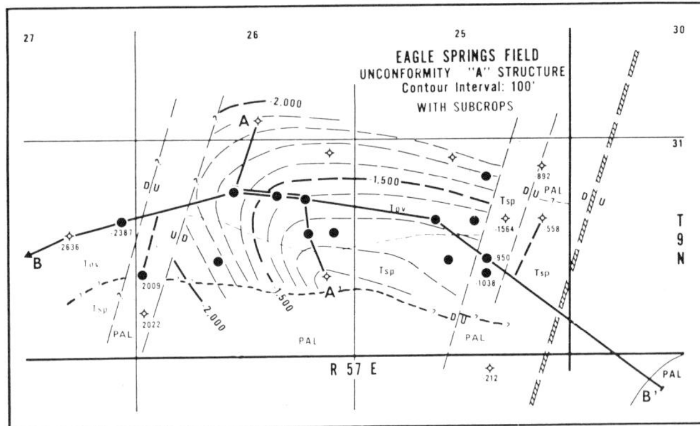

Figure 19. Eagle Springs field structure map. Discovery well in shown by circle around well symbol. Symbols the same as Figures 20 and 21.

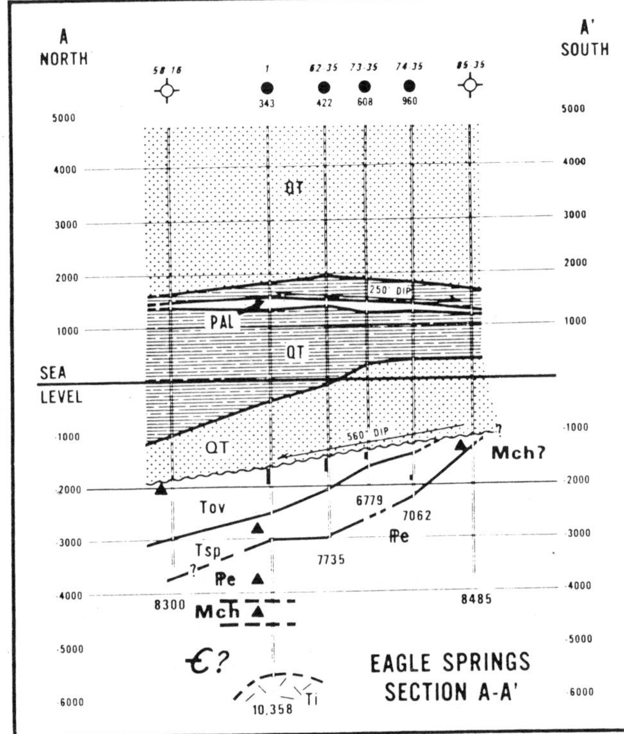

Figure 20. North-south 1:1 structural-stratigraphic section. Solid bars - perforated intervals; solid triangles - oil shows; $\in$? - Cambrian; Mch - Chainman Shale; Pe - Ely Group; Tsp - Eocene Sheep Pass Formation; Tov - Oligocene volcanic rock; Ti - Tertiary Intrusion; QT - Miocene to Recent sediments. Within the QT, stippled area shows coarse clastic facies and dashed lined area shows fine clastic facies. Well number, initial production, and total depth of each well are show. See Figure 19 for location of section.

COMMON CHARACTERISTICS OF BASIN AND RANGE OIL FIELDS

Three fields--Eagle Springs, Trap Spring and West Rozel--will be used to discuss the following common characteristics of Basin and Range oil fields:

- Most traps are associated with a Tertiary unconformity
- Oil columns are relatively thick
- Fractures usually enhance reservoir quality
- Most fields have volcanic rocks as reservoirs
- Faults form part of the trap

Eagle Springs Oil Field. The unconformity "A" structure is shown on Figure 19, and two true-scale structure sections are shown on Figures 20 and 21. (For detailed information on the Eagle Springs field see Duey, this volume, and Bortz and Murray, 1979.) Unconformity "A" is the base of the "valley fill" sediments probably mid-Miocene or younger in age at Eagle Springs. At the base of the valley fill claystones and siltstones provide the seal for the three oil reservoirs-- Oligocene volcanics, Eocene Sheep Pass formation and Paleozoics. The gross oil column is over 1500 feet and can best be seen on Section B-B' (Figure 21). The top of the pay in the easternmost producing well is at a datum of -959 feet and the base of the producing interval in the westernmost well is -2498 feet for total column along Section B-B' of 1539 feet. Bortz and Murray (1979) have evaluated the reservoir characteristics of the

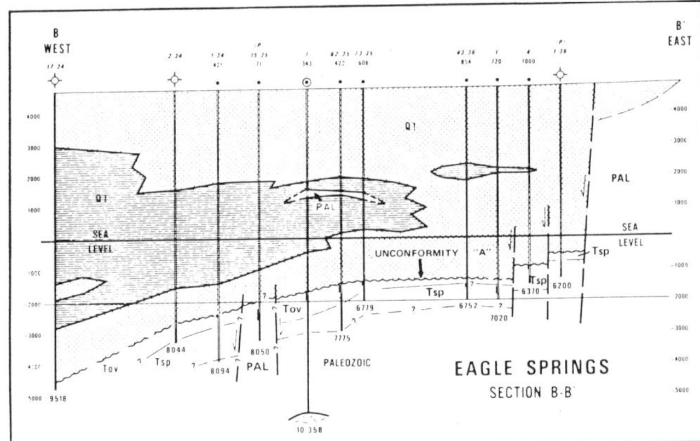

Figure 21. West-east 1:1 structural-stratigraphic section. Symbols used are same for Figure 20, plus PAL for Paleozoics undifferentiated. Section illustrates proximity of field to Grant Range on east. Fault separating Grant Range from field is diagrammatic. See Figure 19 for location of section.

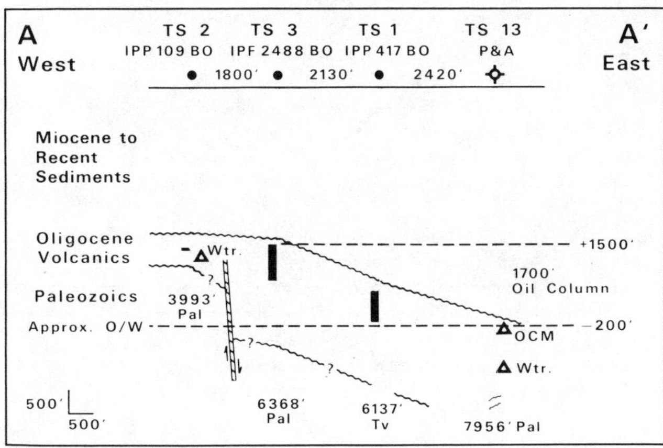

Figure 23. Trap Spring field true-scale cross section A-A'. See Figure 22 for line of section.

three producing formations and conclude that porosity and permeability in the Oligocene volcanic rocks is attributed primarily to fractures. The best producing zones in the Sheep Pass limestones have matrix porosity and permeability; however, some zones are fractured. The Pennsylvanian Ely limestone in the Shell No. 15-35 produced oil from a fractured and brecciated zone. Figure 21 shows that the east side of the trap is formed by down-to-the-basin normal faults.

Trap Spring Oil Field. Figure 22 is a current plat of the field showing the producing oil wells, dry holes, and the line of Section A-A' (Figure 23). See Duey (this volume and 1979) for a more complete discussion of this field. Unconformity "A" is at the top of the Oligocene volcanics in the Trap Spring field area. At the base of the "valley fill", thin beds of fine clastic sediments provide a top seal for the volcanic reservoir. An additional seal is formed by the "ash zone" at the top of the volcanic rocks. The oil column is about 1700 feet thick along Section A-A' which is probably the maximum column. There is very little effective matrix porosity in the volcanic rocks at Trap Spring which is an ignimbrite sequence. Duey (1979) states that the fractures resulted from ignimbrite cooling and subsequent faulting. The fault between TS 2 and TS 3 on Section A-A' is evidence that faulting forms part of the trap. The water recovery below the perforations in TS 2 suggests that this well is in a separate fault

block west of the main producing area. West of Section A-A', Duey (1979) has mapped a major fault which probably forms part of the west side of the Trap Spring trap.

West Rozel Oil Field. The Tertiary unconformity associated with the trap at West Rozel is the top of the basalt on Figure 18. The unconsolidated Upper Pliocene claystones and siltstones immediately above the unconformity form the top seal for trap. The maximum oil column shown on this section is 290 feet in the West Rozel No. 1. The maximum column is estimated to be a little over 300 feet. As can be seen on Figure 18, much of the basalt has high porosity (greater than 15% density log porosity); however, most of this is not effective porosity since the basalt contains a high percentage of vugs and vesicles. Numerous fractures are present in cores of the basalt reservoir and in surface exposures of the basalt in the Rozel Hills. Also, high formation permeabilities determined from DST and production pressure data suggest that this reservoir has an extensive fracture system. Shown on Figure 18 is my interpretation of those zones in the basalt that have only fracture porosity and those zones with both fracture and matrix porosity. The only faults in Figure 17 which form part of the trap are those faults east of WR No. 3 that are down to the east. These faults form a side seal in that they place impermeable clays and silts against the basalt reservoir.

CONCLUSIONS

Oil and gas production in the northern Basin and Range province is closely related to surface and subsurface hydrocarbons shows. Some of these shows are quite obvious as in the Pine Valley area with the Bruffey oil and gas seeps and in the Great Salt Lake with the Rozel Point oil seep. The Railroad Valley-

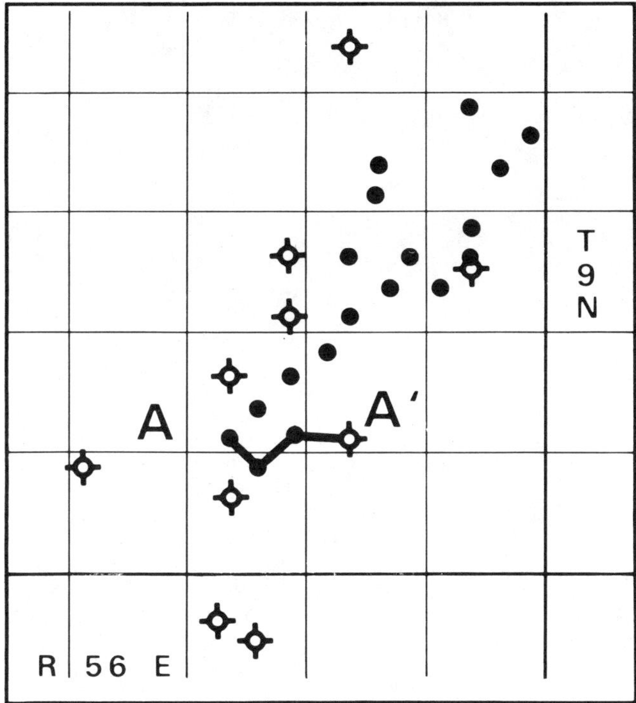

Figure 22. Trap Spring field plat

Sheep Pass basin may be more typical with less dramatic surface and relatively few subsurface shows. Monitor Valley (Diana's Punch Bowl) and Dixie Valley are two examples of areas that are prospective for oil and gas reserves because of surface shows of oil and gas.

Surface and subsurface shows are only a reflection of the source rocks present in a given area. Source rock studies must be an integral part of any exploration program in this province. Certain areas with favorable source rocks that lack surface and subsurface oil and gas shows may be prime prospective areas.

The most important common characteristic of the significant producing fields is the thick oil column. This province has giant oil field potential if good reservoir rocks can be found within a thick oil column.

ACKNOWLEDGEMENTS

My sincere thanks to Amoco Production Company for permission to publish the paper, to Harry Veal, Eugene Howard, Charles Thorman, Dave Drowley, Susan Laule, China Leonard, and John Graham for their help in preparing and editing the paper, to Keith Farmer and Marilyn Cowhick for drafting, Bob Lynn and Pat Currey for photography, and to Natalie Hook and Deborah Skelton for typing.

REFERENCES

Bortz, L. C., 1959, Geology of the Copenhagen Canyon area, Monitor Range, Eureka County, Nevada: Unpublished University of Nevada MS thesis.

Bortz, L. C. and Murray, D. K., 1979, Eagle Springs Oil Field, Nye County, Nevada: RMAG and UGA Basin and Range Symposium, p. 441-453.

Burke, D. B., 1979, Reinterpretation of the Tobin thrust: pre-Tertiary geology of the southern Tobin Range, Pershing County, Nevada: unpublished Stanford University Ph.D. thesis.

Desborough, G. A., and others, 1979, Metals in kerogenous marine strata at Gibellini and Bisoni properties in southern Fish Creek Range, Eureka County, Nevada: USGS open-file report 79-530.

Desborough, G. A., and others, 1981, Metalliferous oil shale in central Montana and northeastern Nevada: USGS open-file report L978.

Duey, H. D., 1979, Trap Spring oil field, Nye County, Nevada: RMAG and UGA Basin and Range Symposium.

Eardley, A. J., 1963, Oil seeps at Rozel Point: Utah Geol. and Min. Survey Special Studies 5, 32 p.

Edmiston, R. C., 1982, A review and analysis of geothermal exploratory drilling results in the northern Basin and Range geologic province of the USA from 1974 through 1981: Geothermal Resources Council, Transactions Vol. 6, p. 12-14.

Foster, N. H., et al., 1979, The Bruffey oil and gas seeps, Pine Valley, Eureka County, Nevada: RMAG and UGA Basin and Range Symposium, p. 531-540.

Garside, L. J., et al., 1977, Oil and gas developments in Nevada, 1968-1976: Nevada Bureau of Mines and Geology Report 29.

Garside, L. J. and Schilling, J. H., 1979, Thermal waters of Nevada: Nevada Bureau of Mines and Geology Bull. 91.

Gould, W. J., 1959, Geology of the northern Needle range, Millard County, Utah: BYU Research Studies Geology Series, Vol. 6, No. 5.

Hastings, D. D., 1979, Results of exploratory drilling northern Fallon basin, western Nevada: RMAG and UGA Basin and Range Symposium.

Hintze, L. F., 1980, Geologic map of Utah, Utah Geological and Mineral Survey, 1:500,000.

Hoover, D. L., et al., 1969, Lithology of the UCe-2, UCe-3, and UCe-16 drill holes, Stone Cabin and Monitor valleys, Nye County, Nevada: Central Nevada-27, Special Projects Branch, USGS 474-9.

Lintz, Joseph, Jr., 1957, Nevada oil and gas drilling data, 1906-1953: Nevada Bur. Mines Bull. 52.

McKee, E. H., 1976, Geology of the northern part of the Toquima Range, Lander, Eureka and Nye Counties, Nevada: USGS Prof. Paper 931.

Mariner, R. H., et al., 1975, Minor and trace elements, gas, and isotope compositions of the principal hot springs of Nevada and Oregon: USGS open-file report.

Merrian, C. W. and Anderson, C. A., 1942, Reconnaissance survey of the Roberts Mountains, Nevada: GSA Bull., v. 53, no. 12, pt. 1, p. 1675-1726.

Nichols, K. M. and Silberling, N. J., 1977, Stratigraphy and depositional history of the Star Peak group (Triassic) northwestern Nevada: GSA Special Paper 178.

Roberts, R. J., et al., 1958, Paleozoic rocks of north-central Nevada: AAPG Bulletin, vol. 42, no. 12, p. 2831-2857.

Schilling, J. H. and Garside, L. J., 1968, Oil
    and gas developments in Nevada, 1953-1967:
    Nevada Bur. Mines Rept. 18.

Smith, J. F. and Ketner, K. B., 1975, Strati-
    graphy of Paleozoic Rocks in the
    Carlin-Pinon Range area, Nevada:
    USGS PP 867-A.

Spurr, J. E., 1905, Geology of the Tonopah
    mining district, Nevada: USGS Prof.
    Paper 42.

Stewart, J. H. and Carlson, J. E., 1978,
    Geologic map of Nevada, USGS 1:500,000.

Winchester, D. E., 1923, Oil shale of the
    Rocky Mountain region: USGS Bull 729,
    204 p.

Winfrey, W. M., Jr., 1960, Stratigraphy,
    correlation, and oil potential of the Sheep
    Pass formation, east-central Nevada: IAPG
    and ENGS Guidebook to the geology of
    east-central Nevada.

Youngquist, Walter, 1949, The cephalopod fauna
    of the White Pine shale of Nevada: Jour.
    Paleontology, v. 23, p. 276-305.

LCB:das
072583
RPT534

OIL GENERATION AND ENTRAPMENT IN RAILROAD VALLEY, NYE COUNTY, NEVADA

Herbert D. Duey

Northwest Exploration Co.
Denver, Colorado

ABSTRACT

Railorad Valley is a graben block in the Basin and Range Structural province. Topographically it is basically flat with recent playa deposits on the surface. There are two structural deeps in the valley. Four oil fields are associated with the northern deep. All oil fields are related to faulting.

Oil has been generated from Tertiary Sheep Pass and Mississippian shales. This generation is probably due to recent local heating of the valley by intrusive rocks. Temperature gradients are as low as 0.9°F to has high as 7.3°F per hundred feet.

8,000,000 barrels of oil with no significant quantity of gas has been produced from the fields. The seals on the fields are imperfect and any gas generated, and much oil, has probably leaked into the overlying valley fill. Trap Spring and Eagle Springs fields are hydrostatically pressured while the smaller fields of Bacon Flat and Currant are slightly over-pressured.

The concept of immature source rocks occurring near a valley with high heat flow may improve exploration success.

INTRODUCTION

Oil was discovered in the first well drilled for oil in Railroad Valley. This well, the Shell #1 Eagle Springs Unit, Sec. 35-T 9N-R57E, was completed from the Oligocene Garrett Ranch ignimbrites for 343 BOPD. Since that time a total of 100 wildcat and development wells have been drilled. There are now four oil fields, Eagle Springs (1954), Trap Spring (1976), Currant (1978), and Bacon Flat (1981) in the valley.

This presence of commercial quantities of oil in the Basin and Range is unique to Railroad Valley. This paper is designed to point out the geological setting of the fields, the oil sources, and maturation of the oil, with hope that this knowledge will help further exploration in the Basin and Range Province.

Railroad Valley is an intermountain valley in the Basin and Range Geomorphic Province (figure 1). Geographically it is in east-central Nevada, some 65 miles southwest of the town of Ely. The only commercial establishment is the village of Currant where limited services are available.

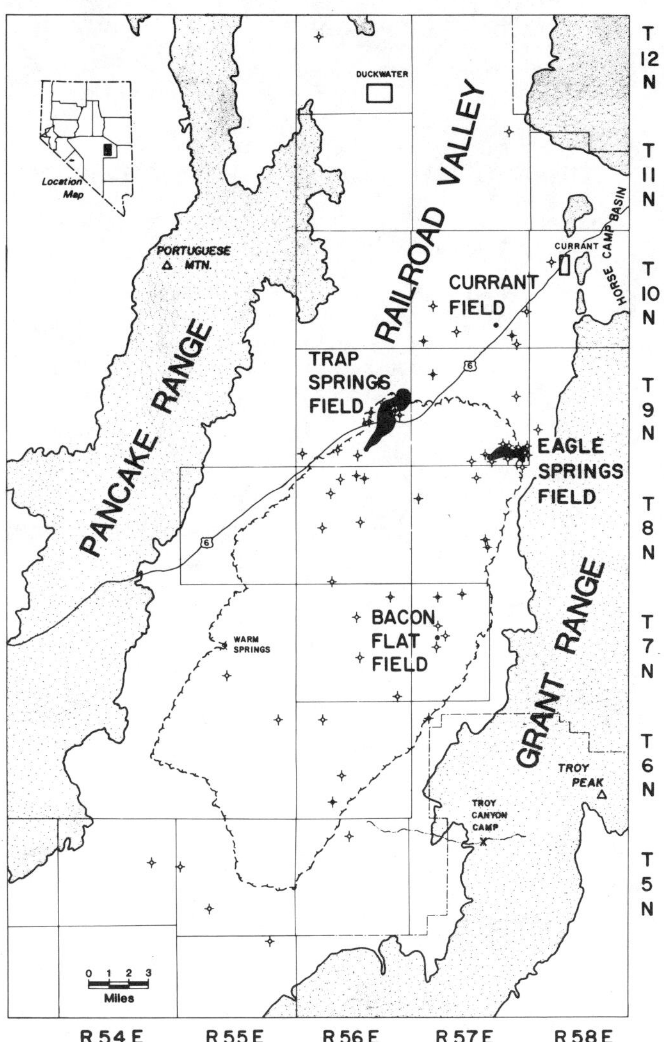

INDEX MAP OF RAILROAD VALLEY &
+5000' CONTOUR, APPROXIMATE PRESENT
LIMIT OF PLAYALAKE - FIGURE 1

DUEY

Railroad Valley, for the purposes of this paper
has a southern limit at the drainage divide at
approximately the middle of Township 4N. The
valley trends NNE-SSW, is 55 miles long and
approximately 14 miles wide. The surface of the
valley has an elevation of 4600-5000' with the
mountain on the east as high as 11,298' (Troy
Peak), and on the west as high as 9249'
(Portuguese Mtn.).

The surface of the valley is recent alluvium
with one large dry playa lake. The alluvium
includes the fans and bajadas from the nearby
mountains. The playa lake includes the surface
clays stabilized sand dunes and beach sand and
gravels.

STRUCTURE

Railroad Valley is a graben in the true sense,
as it is a down dropped block adjacent to
upthrown blocks on either side, but it is
asymmetric and very complex. (See figures 2 &
3) The most movement occurred on the east side
of the valley where over 12,000 feet of total
displacement has taken place. The west side of
the valley is broken by smaller faults with even
some secondary horsts which allow for a
broadening of the valley. The structure of the
valley can only be described in relation to the
time markers identifiable for mapping. These
time markers are unconformities at the base of
the valley fill, base of the ignimbrites and the
base of the Sheep Pass Fm. Often a younger
unconformity erodes an older one which masks
older structural movement.

The valley has two structural lows, one to the
north and to the south. While the northern deep
is structurally lower, on top of the volcanics,
the solution one may be structurally lower on
top of the Paleozoic rocks since the volcanics
thicken to the south.

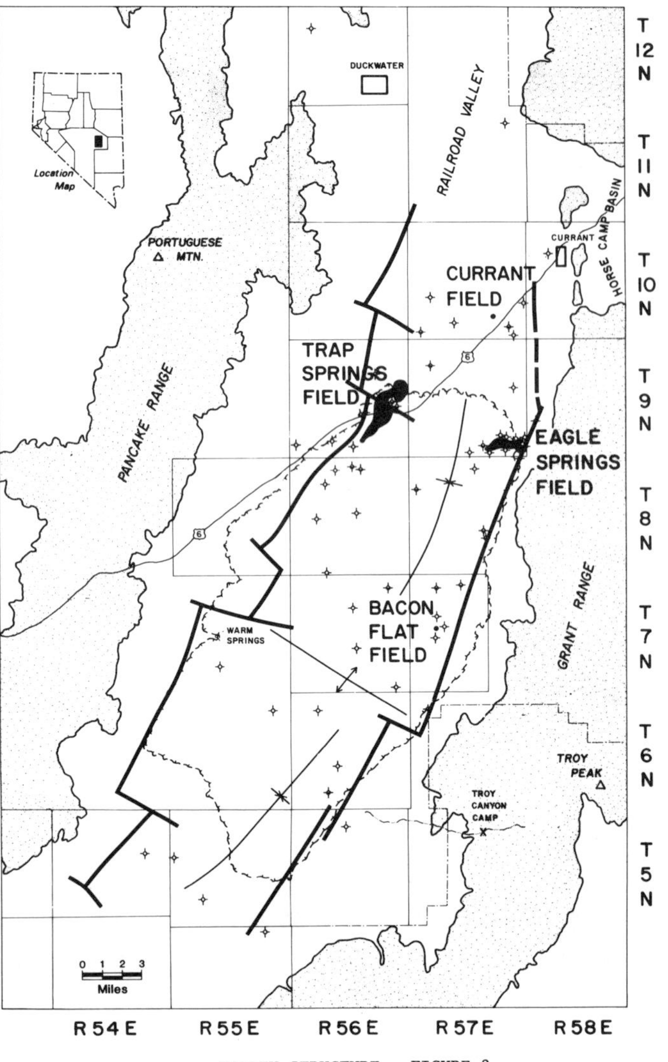

VALLEY STRUCTURE - FIGURE 2

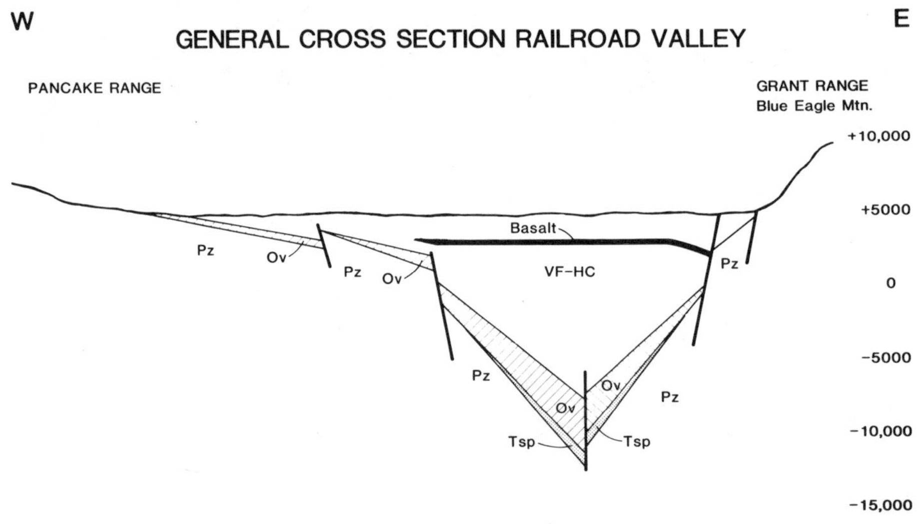

FIGURE 3

STRATIGRAPHY

The generalized stratigraphy of the valley is as follows, from youngest to oldest (fig. 4):

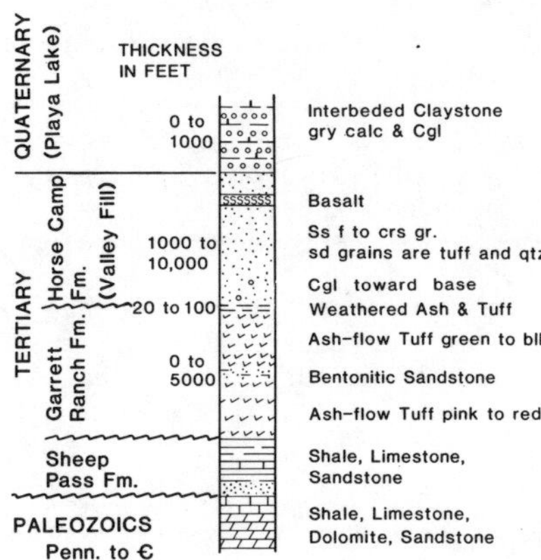

FIGURE 4

<u>Playa lake deposits,</u> probably Pleistocene to present in age. Usually clays and marlstones with mixed sand dunes and beach sands and gravels.

<u>Valley file (Horse Camp Fm.?),</u> Pliocene sands and gravels composed of all the material eroded from the nearby uplifted mountains, i.e. on east side of valley many Paleozoic fragments are included. Occasional basaltic lava flows are present in the subsurface of the valley in this unit. The relationship of Horse Camp Formation is unclear but often is synonomous with valley fill. Lithologically the Horse Camp Fm. may be more indurated and may represent only the oldest portion of valley fill.

The Horse Camp formation is identified in the Horse Camp Basin on the NE flank of Railroad Valley and may have no relation to Railroad Valley. Its relation may be only in the type of sedimentation.

The valley fill lies unconformably on the underlying rocks.

<u>Garrett Ranch Formation</u> Oligocene, used to include all ignimbrites, inter volcanic sediments and rhyolitic flows (French 1979). It includes the Windus Butte, Pritchard Station and Stone Cabin ignimbrites.

The Garrett Ranch lies unconformably on the underlying rocks.

<u>Sheep Pass Formation,</u> Cretaceous ? through Eocene shale, limestone and minor sandstone. This formation unconformably overlies the rocks below.

<u>Paleozoic sediments Pennsylvanian to Cambrian:</u>

Pennsylvanian	Ely limestone
Mississippian	Chainman shales and sandstones
	Joanna limestone
Devonian	Pilot shale
	Guilmette limestone and dolomite
	Simonson dolomite
	Sevy dolomite

Ordovivian, Silurian, and Cambrian dolomite, dolomites with lesser amounts of sandtone, shale and limestone.

HYDROLOGY-

There are two water systems in the valley. The shallowest is in the near surface lacustrine sediments immediately below the surface. In many places, during most seasons, holes as shallow as 1' deep will fill with water. In the rainy season the playa may be covered with water for short periods.

The deep water system is artesian, trapped beneath the lacustrine sediments. This system is fed through the alluvial fans and bajadas. All of the water is basically fresh with high concentrations of sodium and calcium carbonates and bicarbonates. The water of the artesian system fills the valley sands in the fill, and extends into the Paleozoic sediments and ignimbrites below the lacustrine sediments.

There is no surface external drainage from the valley. Price (1979) indicates that subsurface waters flow into the valley from the west (from Hot Creek and Little Smokey Valleys), and drains out of the valley to the southwest.

This subsurface movement of water from the valleys indicates a pervasive permeability system with moving ground water. A fresh supply of ground water enhances growth of bacteria which destroy hydrocarbons by supplying nutrients and oxygen.

Moving ground water enhances hydrocarbon migration, but in a dynamic system hydrocarbon traps must also be more effective than in a static system to keep oil and gas in place.

HEAT FLOW

Heat flow represented by temperature gradients is extremely variable (fig 5). Highest temperature gradient measured is at the Shell #1 Coyote, Sec. 28-T7N-R55E where a temperature of

166°F was measured at 1770 feet. This is a temperature gradient of 7.3°F per 100'. The well is near Warm Spring which flows water at 140°F.

The lowest temperature gradient is at the southern end of Trap Spring field, where at 4500' the temperature is 95°F, or a temperature gradient of 0.94°F per 100'. The highest temperature measured in the valley is 309°F at 5350 feet. The measurement was at the Bacon Flat (Sec 17-T7N-R57E) well while pumping with a high capacity downhole hydraulic pump. With this pump, bottom hole pressure was reduced enough to allow water to turn to steam which then restricted the pump capacity. Most of the valley has a temperature gradient in the range of 2°F per 100 feet.

The immediate sources of heat appear to be hot ground waters. Intrusives have been encountered in the Paleozoic rocks at Trap Spring, Eagle Spring and at the Pan Am #1 USA McDonald, Sec. 2-T11N-R57E. These intrusives did not appreciably affect the temperature gradient where encountered.

Hot groundwaters are probably confined to certain rock layers prior to mixing with the cold valley fill waters. This is evidenced by the deep temperatures being colder than those measured in shallow horizons in the Buckhorn 16-1 Federal Adobe, Sec. 1-T6N-R54E.

OIL MATURATION

Temperature, pressure, time, and possibly catlysts are the factors necessary to produce crude oil from organic material.

Organic material from which oil can be produced is present in the Mississippian Chainman Shale and Eocene Sheep Pass shales. In the vicinity of Trap Spring field the Chainman shales are immature, based both on the maturation of the organic matter and conodonts. The Sheep Pass shale, which outcrops near Eagle Springs field is also thermally immature. Both the Chainman and Sheep Pass are probably mature at depth.

An indication of the maturity of the oils is indicated by the pristane-phytane ratios seen in the table below:

PRISTANE-PHYTANE RATIO

Bacon Flat	1.45
Trap Spring	1.24
Eagle Spring	0.83
Currant	0.52

Pristane and phytane are isoprenoids found in living organisms, and as organic material is maturated into petroleum; Pristane is destroyed faster than Phytane. This is then a measure of maturity.

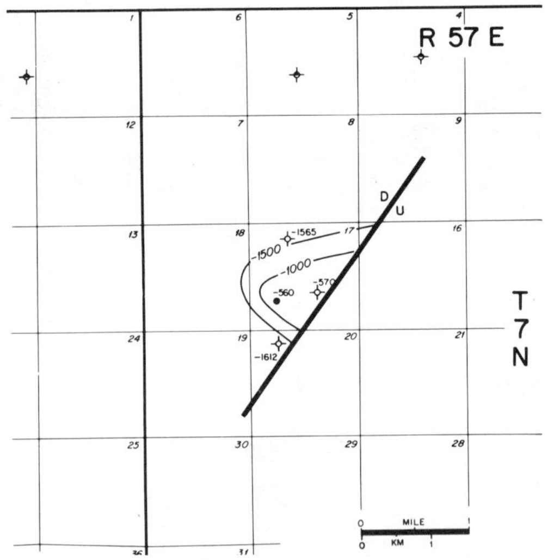

TEMPERATURE GRADIENT
ISOGRADIENT 0.5°F/100'
FIGURE 5

It is seen from these ratios that Currant field is the least mature and Bacon Flat field is the most mature.

The oil found at Currant Field has an even carbon number molecule preference. This factor is not found in the other fields. This even carbon preference is unusual but elsewhere it is associated with oils that have a carbonate source rock.

BACON FLAT FIELD

Geology

>   Producing formation: Devonian Guilmette Formation ?
>   Trap type: structural
>   Thickness and lithology of reservoir rock: 134', Dolomite
>   Oldest formation penetrated: Devonian Simonson

Reservoir Data

>   Productive area: 160 Ac ?
>   Number of producing wells: 1
>   Number of abandoned wells: 0
>   Number of dry holes: 3
>   Porosity: 8 - 12% + fractures
>   Seal: weathered ignimbrite above Paleozoic subcrop
>   Oil/water contact: unknown
>   Vertical oil column: 134' +
>   Associated water analysis - Na 1186, Ca 29, Mg 18, Fe 5.8, Cl 435, HCO_3 2183, SO_4 226 mg/1, R_w
>   Type of drive: water ?
>   Total production to 12-30-82: 64,740 BO

Discovery Well   DUEY

>   Northwest Exploration #1 Bacon Flat (#8 Railroad Valley)
>   SW SW Sec. 17-T7N-R57E
>   Date of completion: July 5, 1981
>   Total depth: 5450'
>   Initial production: 200 BOPD + 1050 BWPD
>   Perforations: 5316 - 5333'

EAGLE SPRINGS FIELD

Geology

>   Producing formations: Oligocene Garrett Ranch Volcanics, Eocene Sheep Pass Formation, Pennsylvanian Ely ? limestone
>   Trap type: structural - stratigraphic
>   Thickness and lithology of reservoir rock: thickness max 275', ignimbrites with vugs and fractures, fractured limestone, sandstone
>   Oldest formation penetrated: Cambrian

Reservoir Data

>   Productive area: 640 Ac
>   Approved spacing: none
>   Number of producing wells: 11
>   Number of abandoned wells: 3
>   Number of dry holes: 10
>   Porosity: ignimbrites 13.5%, Sheep Pass Formation 16%
>   Seal: indurated valley fill (Horse Camp)
>   Oil/water contact: -2000'
>   Vertical oil column: 1500'
>   Gas analysis: unknown, but low GOR

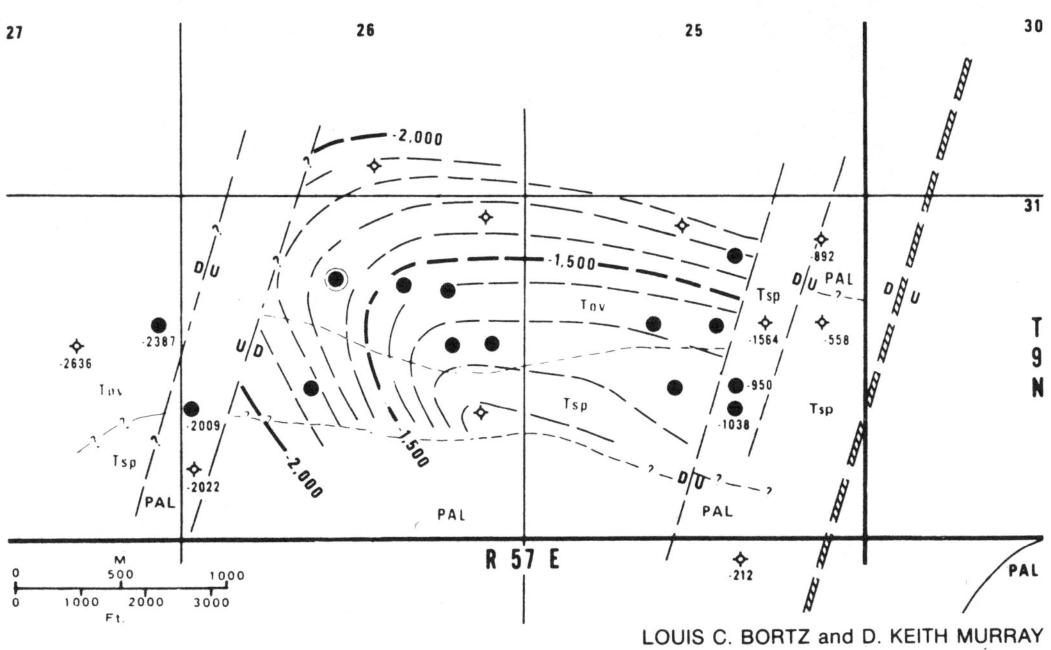

FIGURE 7 — EAGLE SPRINGS FIELD UNCONFORMITY "A" STRUCTURE CONTOUR INTERVAL: 100' (30M) WITH SUBCROPS

DUEY

Associated water analysis: 24,298 mg/1 C1;
  R_w = 0,32 ohms @ 70°F
Type of drive: water and gravity
  drainage
Total production to 12-30-82: 3,580,383

Discovery Well

Shell #1 Eagle Springs Unit
E/2 NW/4 Sec. 35-T9N-R57E
Date of completion: July 5, 1954
Total Depth: 10,358'
Initial production: 343 BOPD
Perforations: none - open hole 6450 - 6730'

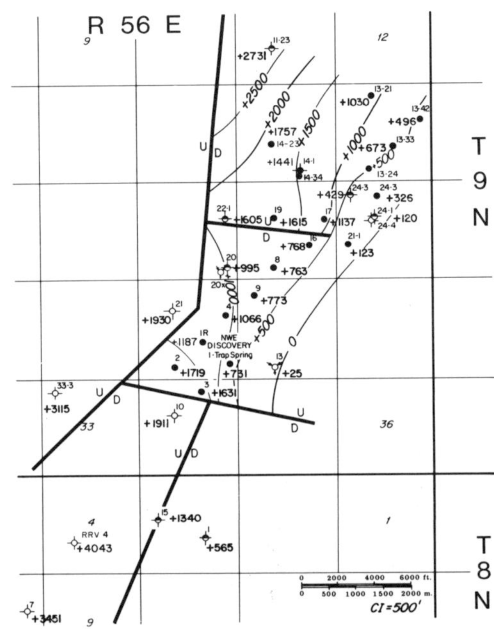

TRAP SPRING FIELD, NYE COUNTY, NEVADA
TOP UNCONFORMITY
"A" STRUCTURE
FIGURE 8

TRAP SPRING FIELD

Geology

Producing formation: Oligocene Garrett
  Ranch Volcanics
Trap type: structural - stratigraphic
Thickness and lithology of reservoir rock,
  847' thick, fractured ignimbrite (welded
  ash flow tuff)
Oldest formation penetrated: Devonian
  Guilmette

Reservoir Data

Productive area: 2080 acres
Approved spacing: 160 acres (state),
  variable in Munson Ranch Unit
Number of producing wells: 16
Abandoned wells: 1
Number of dry holes: 8
Porosity: unknown, fractures

Seal: weathered ignimbrite below
  unconformity
Oil/water contact: -200'
Vertical oil column: 1700'
Gas: none
Associated water analysis: Ca 23, Mg 6,
  Na 2100, Fe 0, C1 3020, HCO_3 488, SO_4
  26 mg/1; R_w: 1.1 @ 68°F.
Type of drive: water
Total production to 12-30-82: 4,692,842 BO

Discovery Well

Northwest Exploration #1 Trap Spring
200'/E & 800'/S Sec. 27-T9N-R56E
Date of completion: November 30, 1976
Total depth: 6,137'
Initial potential: 417 BOPD
Perforations: none, open hole 4220 - 4853"

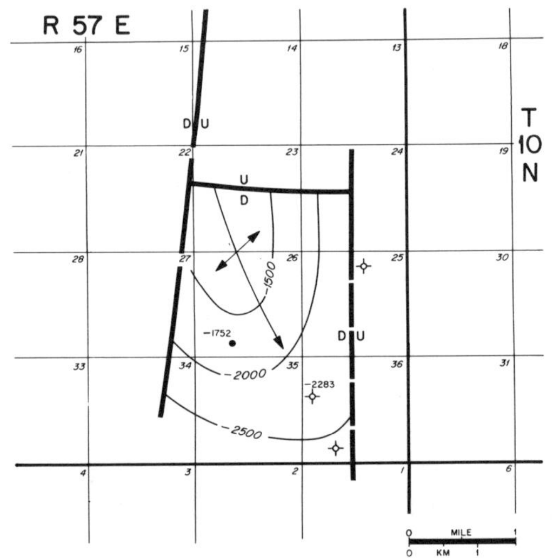

CURRANT FIELD, NYE COUNTY, NEVADA
STRUCTURE
TOP OF SHEEP PASS
C1:500' - FIGURE 9

CURRANT FIELD

Geology

Producing formation: Eocene Sheep Pass
Trap type: structural ?
Thickness and lithology of reservoir
  rocks: 380' thick, shaly limestone
Oldest formation penetrated:
  Devonian Sevy Dolomite

Reservoir Data

Productive area: 40 Ac ?
Number of producing wells: 1
Number of abandoned wells: 0
Number of dry holes: 0
Porosity: 6%
Seal: shales is Sheep Pass Formation
Oil/water contact: unknown
Vertical oil column: unknown
Gas analysis: unknown, low GOR

Associated water: none recovered
Type of drive: unknown
Total production to 1-30-83: 635 BO

Discovery Well

Northwest Exploration #1 Currant
SE SW Sec. 26-T10N-R57E
Date of completion: October 21, 1978
Total depth: 7800'
Initial production: 10 BOPD
Perforations: 6856 - 6994", 7038 - 7080'

The comparison of the oil viscosities with that of water indicates that water is much more easily moved than the oil. Also the oil at Eagle Springs is the easiest to move in relation to water. The oil at Currant is significantly different than that in the other fields. The oil is probably from Sheep Pass shales. Eagle Springs oil may be a mixture of oil generated from Sheep Pass shales and Mississippian shales.

SUMMARY

In comparing the oil fields, it is seen that in three, Trap Spring, Eagle Spring and Bacon Flat, oil occurs immediately beneath the unconformity below the valley fill. Two fields, Currant and Eagle Springs, have some oil in traps not associated with the unconformity. Other parameters such as reservoir rocks, water salinity, depth or datum to oil water contact have little or no similarity.

CONCLUSIONS

It is the authors opinion that the oil in Railroad Valley can be found migrated into any type reservoir (fracture, intergranular, vugular, etc.) and that only the seal and trap are critical. The rock type, and reservoir type are not important.

The high local heat flow and great depth of recent burial of the valley has allowed oil to be generated from Eocene and Mississippian source rocks.

Traps for oil and gas formed prior to basin and range deformation in Paleozoic rocks will be difficult to find due to the large amount of fracture, fold and fault deformation. However, seals for Paleozoic traps in highly deformed rocks could be plastic shales such as the Mississippian Chairman or more recent clays in Tertiary rocks.

ACKNOWLEDGMENTS

No paper can do without infusion of ideas from others. Don French, while at Northwest Exploration was an excellent mentor, but others, such as Lou Bortz and Harvey Pokorny also helped. Kay Theimer and Bill Boden with Northwest were strong arms in typing and drafting.

BIBLIOGRAPHY

Bortz, Louis C. and Murray, D. Keith, 1979, Eagle Springs Oil Field, Nye County, Nevada, Rocky Mountain Association of Geologists - Utah Geological Association, 1979 Basin and Range Symposium pages 441-454.

Duey, Herbert D., 1979, Trap Spring Oil Field, Nye County, Nevada, RMAG-UGA, 1979 Basin and Range Symposium pages 469-476.

French, Don E. and Freeman, Kevin J., Tertiary Volcanic Stratigraphy and Reservoir Characteristics of the Trap Springs Field, Nye County, Nevada, RMAG-UGA, 1979 Basin and Range Symposium pages 487-502.

Price, Don, 1979, Summary Appraisal of the Water Resources in the Great Basin, RMAG-UGA, 1979 Basin and Range Symposium pages 353-360.

OIL CHARACTERISTICS

	Bacon Flat	Currant	Trap Spring	Eagle Spring
Gravity	28°	15°	21-25°	26-29°
Pour Point	10°F	80-95°F	0-40°F	65-80°
Color	Black	Black-Brown	Black	Black
Sulphur	0.4%	3.9%	0.8%	1.7%
Saybolt Viscosity @ 100°F	173	28,500	90	66
Saybolt Viscosity of: Oil at reservoir temperature (T_r)	45	520	90	5
Saybolt Viscosity of Water @ T_r	0.34	0.35	0.75	0.34

MAJOR EPISODES OF PETROLEUM GENERATION IN PART OF THE NORTHERN GREAT BASIN

Forrest G. Poole, George E. Claypool, and Thomas D. Fouch

U.S. Geological Survey, Denver, Colorado 80225

ABSTRACT

Petroleum source beds in the northern Great Basin include marine Paleozoic rocks and nonmarine upper Mesozoic and lower Cenozoic rocks. The organic matter, which is mostly sapropel, ranges from thermally submature through postmature in the Paleozoic rocks, but is mostly submature to mature in the Cretaceous through Tertiary rocks. A discontinuity in thermal maturity occurs between Paleozoic and Cretaceous-Tertiary rocks in uplifted terrains, whereas a more continuous kerogen-maturation profile exists across the Paleozoic and Cretaceous-Paleogene rock boundary in east-central Nevada where these strata are deeply buried beneath Neogene rocks. These data indicate at least two episodes of petroleum generation in the Great Basin region.

INTRODUCTION

Oil and gas shows have been reported in many exploratory wells in the northern Great Basin. Known petroleum accumulations occur in sedimentary and (or) volcanic-rock reservoirs in Railroad Valley in east-central Nevada, in Pine Valley and Huntington Valley in north-central Nevada, and at Rozel Point in northwestern Utah. Most of the petroleum found prior to 1982 is in fractured-rock reservoirs in Paleogene ash-flow tuffs, and in sedimentary beds composing downfaulted blocks that are buried beneath Neogene and Quaternary deposits of intermontane basins. The downthrown blocks that contain petroleum are contiguous with rocks rich in organic matter.

This report briefly describes probable, possible, and potential source beds and their depositional settings, their organic maturation stages, and their rock-temperature history. These data indicate multiple episodes of petroleum generation in part of the northern Basin and Range Province (fig. 1). The conclusions in this report are based on our published and unpublished data.

SOURCE BEDS AND DEPOSITIONAL SETTINGS

Figure 2 shows the principal formations in the Roberts Mountains allochthon in the western part of the area that contain potential or

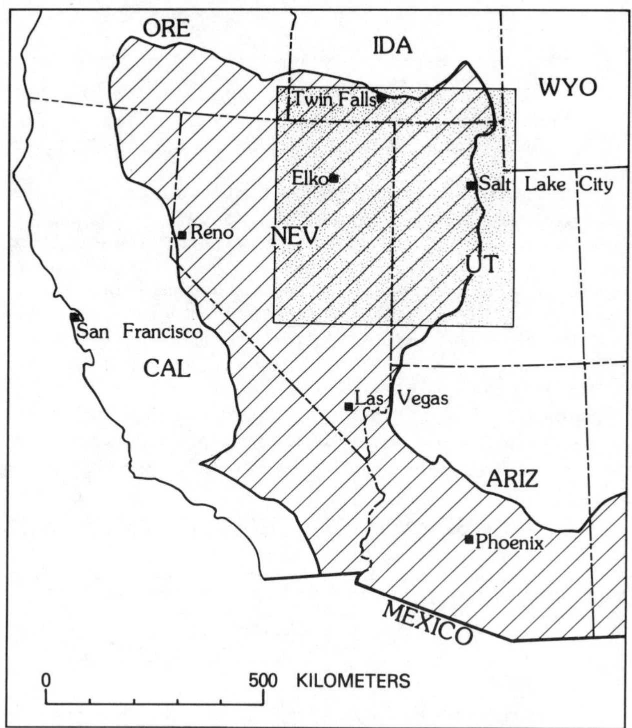

Figure 1. Index map showing extent of the Basin and Range Province (slanted-line pattern) in the Western United States and the area (stippled pattern) that contains evidence of multiple episodes of petroleum generation.

possible petroleum source beds. Solid circles after the formation names indicate submature sapropelic organic-rich marine beds. Oil-shale units occur within the Vinini and Woodruff Formations in central and northern Nevada. The eugeosynclinal Ordovician Vinini and Devonian Woodruff Formations are believed by us to be continental-rise deposits. The depositional environment and structural setting of the eugeosynclinal-like Lower Mississippian Webb Formation has not been established. Most of the Webb, which was deposited during the early stages of the Antler orogeny, is lithologically similar to pre-Mississippian continental-rise deposits exposed in the Antler orogenic belt.

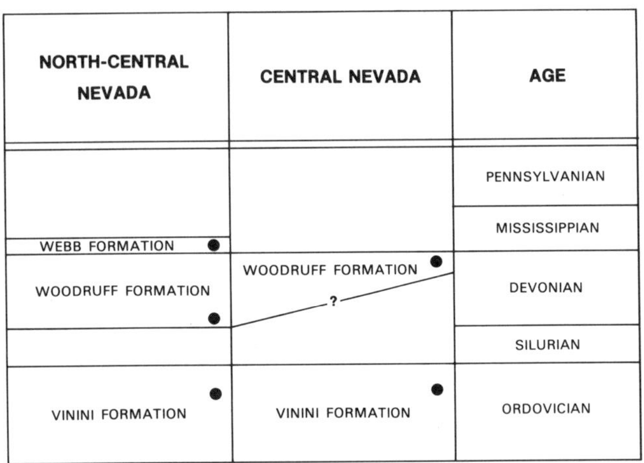

Figure 2. Chart showing principal formations in the Roberts Mountains allochthon that contain petroleum source rocks (solid circles).

Figure 3 shows principal Mississippian formations in the region that contain possible or probable (that is, correlated with oil) petroleum source beds. Solid circles indicate submature to supermature sapropelic to humic organically rich marine beds. In the left column of figure 3, the Chainman Shale and Dale Canyon Formation represent flysch-trough deposits in the western part of the Mississippian Antler foreland basin (Poole and Sandberg, 1977). In the right-center and right column of figure 3 the lower phosphatic members of the Chainman Shale and of the Deseret Limestone represent starved-basin deposits related to the westward prograding Lower Mississippian carbonate bank (Poole and Sandberg, 1977). These starved-basin deposits are along the west side of the carbonate bank in the eastern part of the Antler foreland basin. The concretionary shale member of the Chainman in the right-center column on figure 3 represents starved-basin deposits related to the westward prograding carbonate bank deposits of the Upper Mississippian Great Blue Limestone. By very Late Mississippian time, prodelta deposits in the upper Chainman Shale covered much of the central and eastern part of the Antler foreland basin between the delta systems on the west and the carbonate bank on the east (Poole, 1981). In eastern Nevada (fig. 3, left-center column), the prodelta Chainman Shale contains several organically rich units below the delta-front Scotty Wash Quartzite. Many dense limestone concretions and beds in the Chainman Shale below the Scotty Wash contain oil in open fractures, vugs, and fossil cavities.

Figure 4 shows the principal Cretaceous and Tertiary formations in the area that contain potential or possible petroleum source beds. Solid circles indicate submature sapropelic organic-rich lacustrine beds (Fouch, Hanley, and Forester, 1979). A 3-cm-thick layer of oil shale in the Tertiary Elko Formation near the old Catlin retort site south of Elko, Nevada, yielded 75 gallons of oil per ton of rock.

CENTRAL NEVADA	EASTERN NEVADA	WESTERN UTAH		CENTRAL UTAH		AGE
ELY LIMESTONE (PT)	ELY LIMESTONE (PT)	ELY LIMESTONE (PT)		MANNING CANYON SHALE (PT)		
	SHALE	JENSEN MBR¹ ●				
	SCOTTY WASH QTZITE					
DIAMOND PEAK FORMATION		WILLOW GAP MBR¹	CHAINMAN SHALE	GREAT BLUE LIMESTONE		CHESTERIAN
	SHALE ●	MIDDLE MBR				
		CONCRETIONARY SH. MBR ●				
		NEEDLE SILTSTONE MBR.		HUMBUG FORMATION		
●	NEEDLE SILTSTONE MEMBER			UNCLE JOE MBR	DESERET LIMESTONE	MERAMECIAN
CHAINMAN AND DALE CANYON FORMATIONS				TETRO MBR		
		PHOSPHATIC MBR ●		PHOSPHATIC MBR ●		OSAGEAN
LIMESTONE AND SHALE	JOANA LIMESTONE	JOANA LIMESTONE		GARDISON LIMESTONE		
SHALE	PILOT SHALE (PT)	PILOT SHALE (PT)		FITCHVILLE FORMATION (PT)		KINDERHOOKIAN

Figure 3. Chart showing principal units in the Mississippian Antler foreland-basin sequence that contain petroleum source rocks (solid circles).

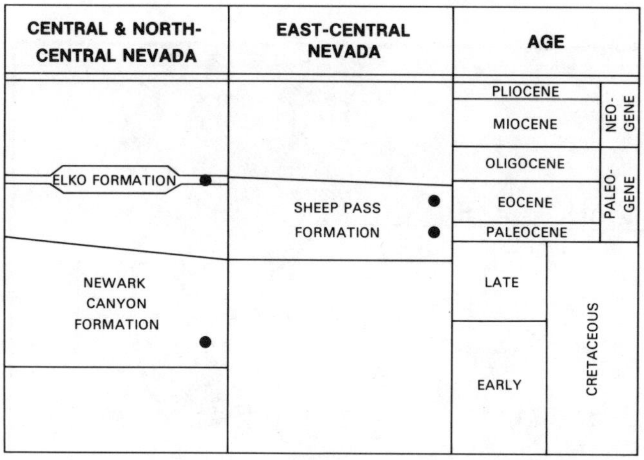

CENTRAL & NORTH-CENTRAL NEVADA	EAST-CENTRAL NEVADA	AGE	
		PLIOCENE	NEO-GENE
		MIOCENE	NEO-GENE
		OLIGOCENE	PALEO-GENE
ELKO FORMATION ●	SHEEP PASS FORMATION ●	EOCENE	PALEO-GENE
	●	PALEOCENE	PALEO-GENE
NEWARK CANYON FORMATION ●		LATE	CRETACEOUS
		EARLY	CRETACEOUS

Figure 4. Chart showing principal formations in Cretaceous and Tertiary lake-basin deposits that contain petroleum source rocks (solid circles).

ORGANIC MATURATION STAGES AND ROCK-TEMPERATURE HISTORY

Figure 5 is a generalized map showing the maximum temperatures that some outcropping and subsurface Paleozoic rocks have been subjected to since deposition. Temperature data are not available for Paleozoic rocks in many areas. All available organic maturation data on Ordovician to Permian rocks have been integrated on the map (fig. 5). Although maturation stages in upper Paleozoic and lower Mesozoic formations generally record a lower temperature history than lower Paleozoic formations, there are certain localities where apparent reversals occur. Some of these reversals may reflect different heat-flow regimes for Paleozoic and lower Mesozoic formations during and prior to tectonic transport; however, discussion of these aberrant localities is beyond the scope of this report. Stipple pattern depicts relatively cold rocks (<100°C), cross-hachuring relatively hot rocks (>300°C), and diagonal hachuring intermediate-temperature rocks (100-300°C). Estimation of Paleozoic rock temperatures is based on a combination of organic geochemical data and visual estimates of organic maturity using alteration colors of palynomorphs (R. M. Kosanke, written commun., 1981), conodonts (Harris and others, 1980), and kerogen in limestones and shales. Conodont color alteration, which is a relatively new method for assessing organic metamorphism (Epstein, Epstein, and Harris, 1977), was used in conjunction with organic geochemical data to estimate maximum rock temperature. Field and laboratory experiments by Anita Harris and her associates of the U.S. Geological Survey have shown that color alteration in conodonts from limestone beds correlates with fixed carbon, vitrinite reflectance, and palynomorph translucency. Composition of extractable organic matter in rocks with regard to temperature history is, in general, compatible with elemental composition of solid organic matter and thermal alteration color of recognizable spores and conodonts in sediments. The method of maximum paleotemperature assignment is discussed by Epstein, Epstein, and Harris (1977) and assumes that other thermal maturation indices can be used in the same manner as conodont color alteration.

The map (fig. 5) is a type of heat-flow record of the Phanerozoic. Of special interest are the rocks in the stippled areas, where much of the organic matter is thermochemically submature to early mature. Adequate rock temperature increase in the stippled areas could generate petroleum. The cross-hachured areas contain organic matter that is supermature and indicate earlier petroleum generation and possible migration. The diagonal-hachured areas or intermediate temperature areas contain organic matter that is mature to supermature. Within this area of generally moderate rock temperatures, some local cold and hot spots have been recognized.

Two major east-directed thrust systems and a strike-slip fault (Wells fault) are shown on figure 5. The western thrust is the Early Mississippian Roberts Mountains thrust with as much as 100 km displacement and the eastern thrust represents the Cretaceous Sevier belt with 50-100 km displacement. The Roberts Mountains allochthon or Antler orogen contains marine source rocks (fig. 2) near its eastern margin (fig. 5). The area between the two thrust systems contains marine source beds in the Mississippian Antler foreland basin (fig. 3) and nonmarine source beds in Cretaceous and Tertiary lake basins (fig. 4). The solid circles represent drill holes that are known to contain oil. The cluster of drill holes in the lower left on figure 5 are in Railroad Valley. Plus signs (+'s) indicate Ordovician and Devonian outcrops in the Antler orogen that contain oil and x's indicate Mississippian foreland basin outcrops that contain oil.

The rectangular area in the lower left part of figure 5 outlines the larger scale map of figure 6. This east-central Nevada area is important because it contains evidence for at least two episodes of petroleum generation in the northern Great Basin (Poole, Fouch, and Claypool, 1979). Mountains are stippled and valleys are unpatterned. Maturity or stage of thermochemical alteration of organic matter in samples taken from the Chainman Shale at several localities is shown by solid circles, diamonds, and triangles. Circles represent upper-range submature organic matter, diamonds mature, and triangles supermature or overcooked organic matter. Determination of organic maturity at each locality is based on analysis of several samples.

Marine Chainman Shale obtained from drill cores and cuttings of Railroad Valley wells southwest of the village of Currant, and from outcrops in several mountain ranges contains a mixture of marine-sapropel and detrital terrestrial-plant organic matter. The organic matter ranges from submature to supermature in

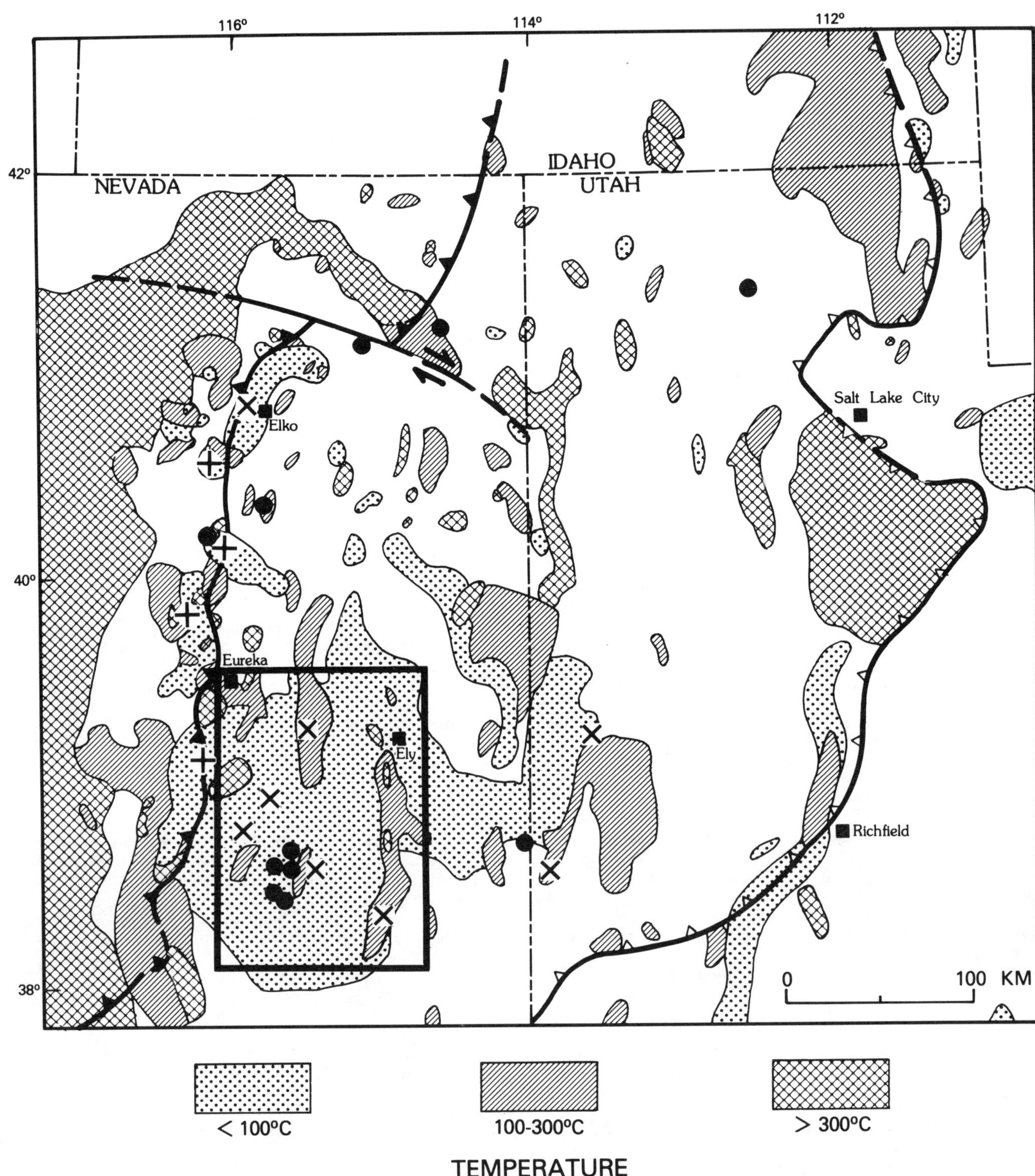

Figure 5. Generalized map of maximum temperatures of some outcropping and subsurface Paleozoic strata in part of the northern Great Basin. Solid circles indicate wells containing oil, +'s indicate Ordovician and Devonian outcrops containing oil, and x's indicate Mississippian outcrops containing oil. See text for explanation of faults.

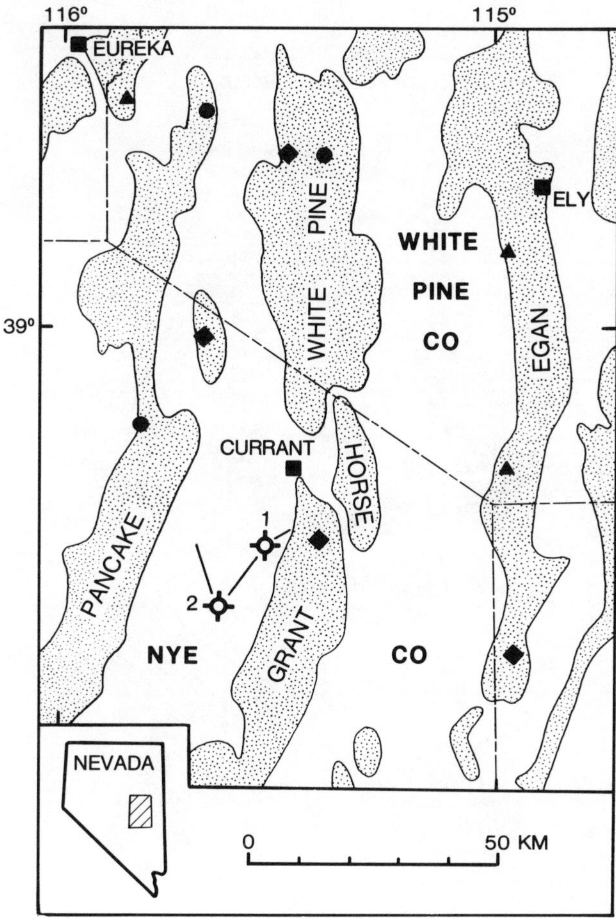

Figure 6. Map of east-central Nevada showing
maturity stage of organic matter in the
Mississippian Chainman Shale at several
localities. Solid circles indicate submature
stage, solid diamonds indicate mature stage,
and solid triangles indicate supermature
stage. Symbols 1 and 2 indicate wells shown
in cross section of figure 7. Mountains are
represented by stippled pattern.

that temperatures sufficient for oil generation
were reached during burial and prior to uplift.

Figure 7 is a diagrammatic subsurface section
across Railroad Valley from the Trap Spring oil
field on the west to the Eagle Springs oil field
on the east (line of section shown on fig. 6).
Shell Unit 2 east of Trap Spring is shown as well
number 2. Shell Unit 1 in the Eagle Springs field
is shown as well number 1. Chainman cores from
Shell Units 1 and 2 were studied in detail as was
shale of the Sheep Pass Formation from Shell Unit
1. Major oil-producing zones are shown by the
solid wedges. Vitrinite reflectance values (N. H.
Bostick, written commun., 1978) are given for the
Chainman cores in both Shell holes and for the
shale of the Sheep Pass in well number 1.
Vitrinite reflectance values indicate mature
organic matter in both the Chainman and Sheep Pass
Formations in Shell Unit 1 and show a uniform
kerogen-maturation profile in the subsurface.
Both the Chainman and Sheep Pass formations
contain mature to marginally mature organic matter
and rank as excellent source rocks (Fouch,
1979). In Railroad Valley, and elsewhere where
these formations are buried deeper, they should
have reached optimum temperatures for petroleum
generation.

Several samples of hydrocarbon extracts from
Chainman and Sheep Pass drill cores and cuttings,
and crude oil from Tertiary reservoir rocks in the
Eagle Springs and Trap Spring fields have been
analyzed to evaluate hydrocarbon compositions and
to test for genetic relationships among samples.
Crude oil compositions, including ratios of sulfur
to nitrogen and pristane to phytane and isotopes
of sulfur and carbon, indicate that they were
derived from different organic source materials
(Claypool, Fouch, and Poole, 1979). Hydrocarbon
extracts from the Chainman Shale are identical in
many respects to Trap Spring oil. Extracts of
available Sheep Pass Formation samples are unlike
either Eagle Springs oil or Trap Spring oil
(Claypool, Fouch, and Poole, 1979).

Data from Sheep Pass and Chainman formation
cores and cuttings from drill holes in Railroad
Valley indicate that immature and mature organic
matter in many Sheep Pass and Chainman rocks in
this region is now being thermochemically altered
and is generating oil.

An earlier cycle of petroleum generation is
indicated by geochemical and conodont color
alteration data from outcrop samples of Chainman
Shale and associated rocks in the Great Basin. In
east-central Nevada, organic matter in outcrop
samples of marine Chainman Shale range from
submature to postmature as shown on figure 6,
whereas all organic matter in outcrop samples of
lacustrine Tertiary rocks were found to be
submature. In some stratigraphic sections,
submature organic matter in Tertiary rocks
directly overlie Mississippian rocks containing
mature to supermature organic matter. This

thermochemical evolution as indicated by organic
geochemical data from kerogen and from rock
hydrocarbon extracts and oil, by vitrinite
reflectance, and by alteration colors of
palynomorphs and conodonts. Hydrocarbon contents
range from a few parts per million to 2,000 ppm
and organic-carbon contents range from almost nil
to 7 percent by weight. The presence of
significant amounts and different types of both
immature and mature organic matter in many
Chainman samples in east-central Nevada indicates
a history of moderate temperatures. Liquid oil of
varying viscosities found in fractures, voids, and
invertebrate-fossil cavities of dense limestone
concretions and beds in the Chainman Shale at
several localities in east-central Nevada indicate

211

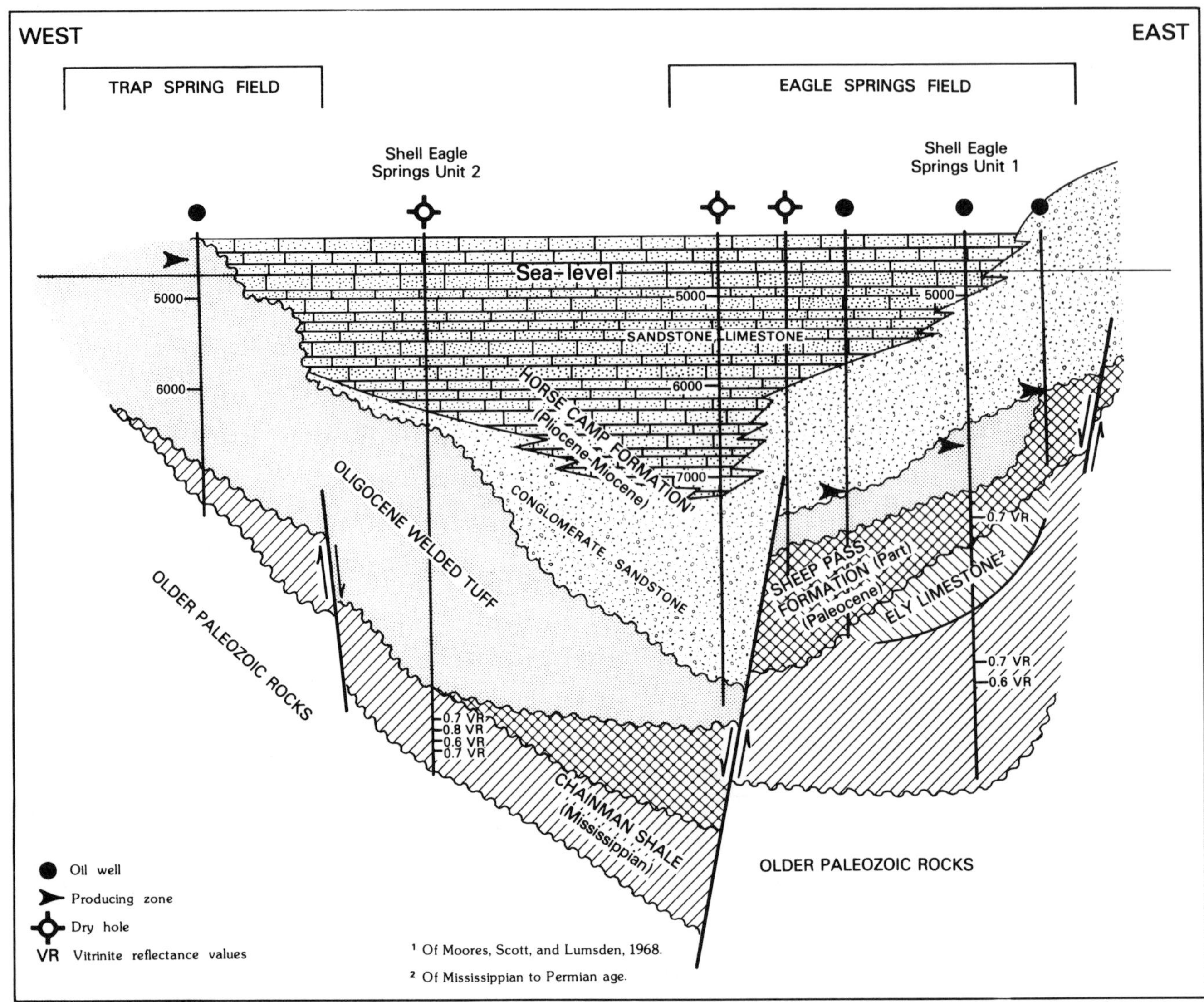

Figure 7. Generalized cross section of Railroad Valley showing rock units and vitrinite reflectance values obtained on cores from two wells.

discontinuity indicates an earlier heating event of the Chainman. The first episode of petroleum generation began probably in early Mesozoic time when the Chainman was buried beneath upper Paleozoic and lower Mesozoic rocks. Depth of burial and hence degree of thermochemical maturation varied with late Paleozoic and Mesozoic folding, faulting, and erosion. Organic matter in the Chainman is supermature in many localities where higher paleotemperatures occurred owing to subcrustal hot spots or deeper burial. In the Diamond Mountains in the northwestern corner of the map area, submature nonmarine Cretaceous rocks overlie supermature marine Mississippian flysch.

Analysis of surface and subsurface data indicates that although Chainman rocks were thermochemically degraded and generated some hydrocarbons in Mesozoic time, as evidenced by oil trapped in fractures, voids, and fossil cavities in dense limestone concretions and beds; petroleum generation was incomplete where rocks were buried at moderate depths. Uplift of buried Chainman Shale prior to the late Mesozoic arrested the first episode of petroleum generation in thermochemically submature and mature rocks.

Many organically submature and mature Chainman rocks in this region are now subjected to a second episode of thermochemical degradation and renewed oil generation in Neogene basins where valley fill has been adequate and temperature has increased. Rocks of the Chainman Shale probably are the major source of petroleum in the Trap Spring and Bacon Flat (about 12 km southeast of Trap Spring field) fields in Railroad Valley where oil has accumulated in fractured welded ash-flow tuffs of Oligocene age. Rocks of the Sheep Pass Formation may be the major source of petroleum in the Eagle Springs and Currant (about 10 km north of Eagle Springs field) fields. Oil is inferred to have accumulated in other valleys in eastern Nevada and western Utah where similar geologic conditions exist.

Oil samples analyzed to date suggest the possibility of several distinct genetically related Paleozoic and Tertiary crude oil/source rock families in the northern Great Basin. Further work to document these genetic relationships and to develop the implications for exploration is still in progress.

REFERENCES

Claypool, G. E., Fouch, T. D., and Poole, F. G., 1979, Chemical correlation of oils and source rocks in Railroad Valley, Nevada: Geological Society of America Abstracts with Programs for 1979, v. 11, no. 7, p. 403.

Epstein, A. G., Epstein, J. B., and Harris, L. D., 1977, Conodont color alteration--an index to organic metamorphism: U.S. Geological Survey Professional Paper 995, 27 p.

Fouch, T. D., 1979, Character and paleogeographic distribution of Upper Cretaceous(?) and Paleogene nonmarine sedimentary rocks in east-central Nevada, _in_ Armentrout, J. M., Cole, M. R., and TerBest, Harry, eds., Cenozoic Paleogeography of the Western United States, Pacific Coast Paleogeography Symposium 3: Pacific Section, Society of Economic Paleontologists and Mineralogists, p. 97-111.

Fouch, T. D., Hanley, J. H., and Forester, R. M., 1979, Preliminary correlation of Cretaceous and Paleogene lacustrine and related nonmarine sedimentary and volcanic rocks in parts of the eastern Great Basin of Nevada and Utah, _in_ Newman, G. W., and Goode, H. D., eds., 1979 Basin and Range Symposium: Rocky Mountain Association of Geologists and Utah Geological Association, p. 305-312.

Harris, A. G., Wardlaw, B. R., Rust, C. C., and Merrill, G. K., 1980, Maps for assessing thermal maturity (conodont color alteration index maps) in Ordovician through Triassic rocks in Nevada and Utah and adjacent parts of Idaho and California: U.S. Geological Survey Miscellaneous Investigations Series Map I-1249.

Moores, E. M., Scott, R. B., and Lumsden, W. W., 1968, Tertiary tectonics of the White Pine-Grant Range region, east-central Nevada, and some regional implications: Geological Society of America Bulletin, v. 79, no. 12, p. 1703-1726.

Poole, F. G., 1981, Molasse deposits of the Antler foreland basin in Nevada and Utah: Geological Society of America Abstracts with Programs for 1981, v. 13, no. 7, p. 530, 531.

Poole, F. G., Fouch, T. D., and Claypool, G. E., 1979, Evidence for two major cycles of petroleum generation in Mississippian Chainman Shale of east-central Nevada, _in_ AAPG Rocky Mountain Section Meeting Abstracts, 1979, Casper, Wyoming: American Association of Petroleum Geologists Bulletin, v. 63, no. 5, p. 838.

Poole, F. G., and Sandberg, C. A., 1977, Mississippian paleogeography and tectonics of the Western United States, _in_ Stewart, J. H., Stevens, C. H., and Fritsche, A. E., eds., Paleozoic Paleogeography of the Western United States: Society of Economic Paleontologists and Mineralogists, Pacific Section, Pacific Coast Paleogeography Symposium 1, p. 67-85.

Sadlick, Walter, 1965, Biostratigraphy of the Chainman Formation (Carboniferous), eastern Nevada and western Utah: Utah University, unpublished Ph. D. thesis, 228 p.

ORGANIC MATTER MATURATION AND PETROLEUM GENESIS:-GEOTHERMAL versus HYDROTHERMAL

Bernd R. T. Simoneit

Oregon State University

I. ABSTRACT

Sedimentary organic matter is derived from the catabolic and diagenetic residues of primary biological carbon fixation and in some cases also from recycled organic detritus of older geologic formations. This organic matter undergoes maturation following diagenesis and is easily converted to gas and petroleum under the catagenetic and metagenetic temperature regimes of increasing burial.

Organic matter in immature, recent sediments is comprised of minor amounts (based on total organic carbon content) of biogenic gas (CH_4 and CO_2, sometimes H_2S), significant lipid residues of terrigenous and/or marine origins and a major macromolecular fraction consisting of fulvic and humic acids and particulate detritus (eg. biopolymer fragments, cell membranes and miscellaneous carbonaceous matter) the pseudokerogen. The lipids and macromolecular material undergo diagenetic (including microbiological) alteration according to the environmental conditions during transport and in the depositional sinks, ie. oxidative degradation in high-energy, oxygenated environments and reductive in anaerobic environments.

Maturation of organic matter commences after cessation of diagenesis with increasing burial and concomitant rise in the geothermal gradient. This produces some low temperature cracking products from the kerogen, such as gas (CH_4-C_8+) and bitumen (C_8-C_{40+}), which are superimposed on the endogenous biogenic gas and lipid residues. Maturation of pseudokerogen occurs via molecular rearrangements and addition of geomonomers by copolymerization. Further thermal stress during catagenesis generates additional bitumen and gas, which far exceeds the original concentrations of endogenous lipids and gas, thus erasing their compositional signatures and resulting in the characteristic distributions of petroleum compounds. The spent kerogen remains as amorphous carbon. The late stages of catagenesis and the subsequent high-temperature phase called metagenesis (very deep burial) generate primarily methane from both bitumen and kerogen, and H_2S can also be formed, especially in carbonate sequences.

Hydrothermal processes can also act on sedimentary organic matter and result in "instantaneous" diagenesis and catagenesis of recent biogenic detritus and thus produce analogous petroleum products. Gas (CH_4-C_{8+}, CO_2 and H_2S) and bitumen (C_8-C_{40+}) are cracked from the pseudokerogen and superimposed on the endogenous gas and lipids. Additionally, products characteristic of elevated thermal processes (eg. olefins, PAH, stabilized molecular markers, etc.) are also found in the bitumen. The spent kerogen remains as amorphous "activated" carbon.

Pressure, temperature and time effects on the chemistry of the organic matter are all interrelated. Temperature and time primarily effect petroleum genesis and are coupled with mineralogical interactions and the elemental compositions of kerogens. Migration processes are understood least. In sedimentary basins migration appears to occur by diffusion and in solution ($CH_4/CO_2/ H_2O$ solvent). In hydrothermal areas migration proceeds by the previous processes and also by thermally driven diffusion, advection and mass transport as oil and/or emulsions, although the overall result may be the same for both regimes.

II. INTRODUCTION

Organic matter of sedimentary sinks, usually marine and either of Recent or geologically old origin, is derived from the syngenetic residues of posthumus biogenic debris (Simoneit 1978a; 1982a). This detritus is composed of both autochthonous detritus and allochthonous residues derived from continental sources (Simoneit, 1982a). Aquatic sediments receive organic detritus primarily by river washin and eolian transport, with ice-rafting and sediment recycling as minor contributing processes (Simoneit, 1975; 1978a).

The preservation of organic matter in sediments depends on the initial diagenetic processes involving microbial degradation and chemical conversion, coupled with the environmental conditions of acidity and redox potential (eg., Didyk et al., 1978; Demaison and Moore, 1980). Then, during subsequent sediment maturation and lithification, the organic matter is modified by the effects of temperature, pressure and petrology (eg., Hunt, 1979; Simoneit, 1978a; Tissot and Welte, 1978; Tissot et al., 1971).

The analytical techniques of organic geochemistry are ideally suited to examine the character of such organic matter in terms of its structural and compositional makeup (Simoneit, 1978a). The sources, the diagenetic and catagenetic histories and the migration mechanisms of this organic matter can then be evaluated from such data. It should be noted that in the following discussion organic matter is comprised of gas, lipids (bitumen), humic substances (with fulvic substances, or asphaltenes) and kerogen (with "pseudokerogen") as outlined in Table 1, which summarizes the equivalent organic fractions in contemporary (Recent) versus ancient samples.

Various basic organic geochemical procedures with only minor modifications have been in routine use for recent times to fractionate organic matter prior to instrumental analyses. They are described in detail elsewhere (eg. Reed, 1977; Boon et al., 1978; Deroo et al., 1978; Simoneit et al., 1979; 1980; 1981; Simoneit, 1978a; 1981; 1982a,b; Stuermer and Simoneit, 1978; Stuermer et al., 1978; Philp et al., 1978; van de Meent et al., 1980).

TABLE 1. Organic Matter of Sediments and Petroleum
____

SEDIMENTS:

Contemporary and Recent (generally biogenic components):

Gas -	Lipids -	Humic and Fulvic substances -	Carbonaceous detritus (humin, pseudokerogen)-
(CH_4, CO_2, H_2S)	$(C_8-C_{40}+)$	(macromolecular	(macromolecular,
minor amount	minor amount	M.W. $\sim 10^3$ to $>10^6$)	>humates)
of total C_{org}	(max. $\sim$ 10%)	variable amount	major amount

Geologically Ancient (generally geogenic components):

Gas -	Bitumen -	Asphaltenes (sometimes humates) -	Kerogen -
(CH_4-C_8, CO_2, H_2S)	$(C_8-C_{40}+)$	(macromolecular,	(macromolecular,
minor amount	minor amount	M.W. $\sim 10^4$ to $>10^6$)	>alphaltenes)
of total C_{org}	(max. $\sim$ 10%)	variable amount	major amount

Table 1. (continued)

PETROLEUM:

Composition (liquid products from mature sedimentary organic matter):

Gas (and gasoline range) -	Bitumen -	Asphaltenes -	Kerogen -
$(CH_4-C_8+, CO_2, H_2S, N_2)$	$(C_8-C_{40}+)$	(macromolecular,	(absent)
significant amount of	major amount	M.W. $\sim 10^4$ to $> 10^6$)	spent kerogen
total C_{org}		major and variable	remained in
		amount	source rocks

III. NATURE OF ORGANIC MATTER IN IMMATURE SEDIMENTARY SINKS

Organic matter that accumulates in contemporary sediments represents the residues from primary biological carbon fixation and its degradation (remineralization).

Gas:

Interstitial gas in recent sedimentary environments consists primarily of methane, carbon dioxide and sometimes hydrogen sulfide (Claypool and Kaplan, 1974). The biogenic hydrocarbon gases usually have $CH_4/(C_2H_6+C_3H_8)$ ratios greater than 1000, while those of a thermogenic origin have ratios less than 50 (Bernard et al., 1976). For example, the $C_1/(C_2+C_3)$ ratios for shallow sediment gases from Guaymas Basin, Gulf of California range from 41-150 and thus indicate a mixed origin of biogenic (CH_4) and thermogenic (C_2-C_8) hydrocarbons (Simoneit et al., 1979). The depth range where biogenic gas can be found is variable but shallow ($\sim$100 m) and depends on microbial production and environmental conditions in the sediments.

Lipids:

The lipids extractable from recent sediments, especially the hydrocarbons, have been examined most extensively. The compound classes which are commonly found as lipid components and which are reasonably stable over geologic times are: hydrocarbons (normal, iso-, anteiso-, alkene, aromatic and isoprenoid), fatty acids (also normal, iso-, anteiso-, unsaturated and isoprenoid), fatty alcohols, ketones, wax esters, steroids, terpenoids (sesqui-, di-, sester-, tri-, and tetra-), and tetrapyrrole pigments (Simoneit, 1978a; 1982a; Cranwell, 1982; Mackenzie et al., 1982a). The following additional compound classes are analyzed in recent sediments only, since their geologic half-lives are short: amino acids and peptides, purines and pyrimidines, and carbohydrates (Simoneit, 1978a). The homolog distributions and molecular markers of the first category of compound classes will be briefly illustrated with some examples and further details can be found elsewhere (Simoneit, 1978a and Cranwell, 1982 and references cited therein).

Some typical examples of homologous compound distributions in lipids of generally immature sediments are shown in Figure 1. DSDP sample 36-330-10-1, 95-102 cm, which is of Jurassic age from the South Atlantic, exhibits n-alkanes maximizing at n-C_{17} and n-C_{19} (Fig. 1A) and even carbon numbered n-fatty acids with a maximum at n-C_{16} (Fig. 1E). Both these distributions are indicative of primary bacterial lipid residues of a marine origin and there are only traces of allochthonous

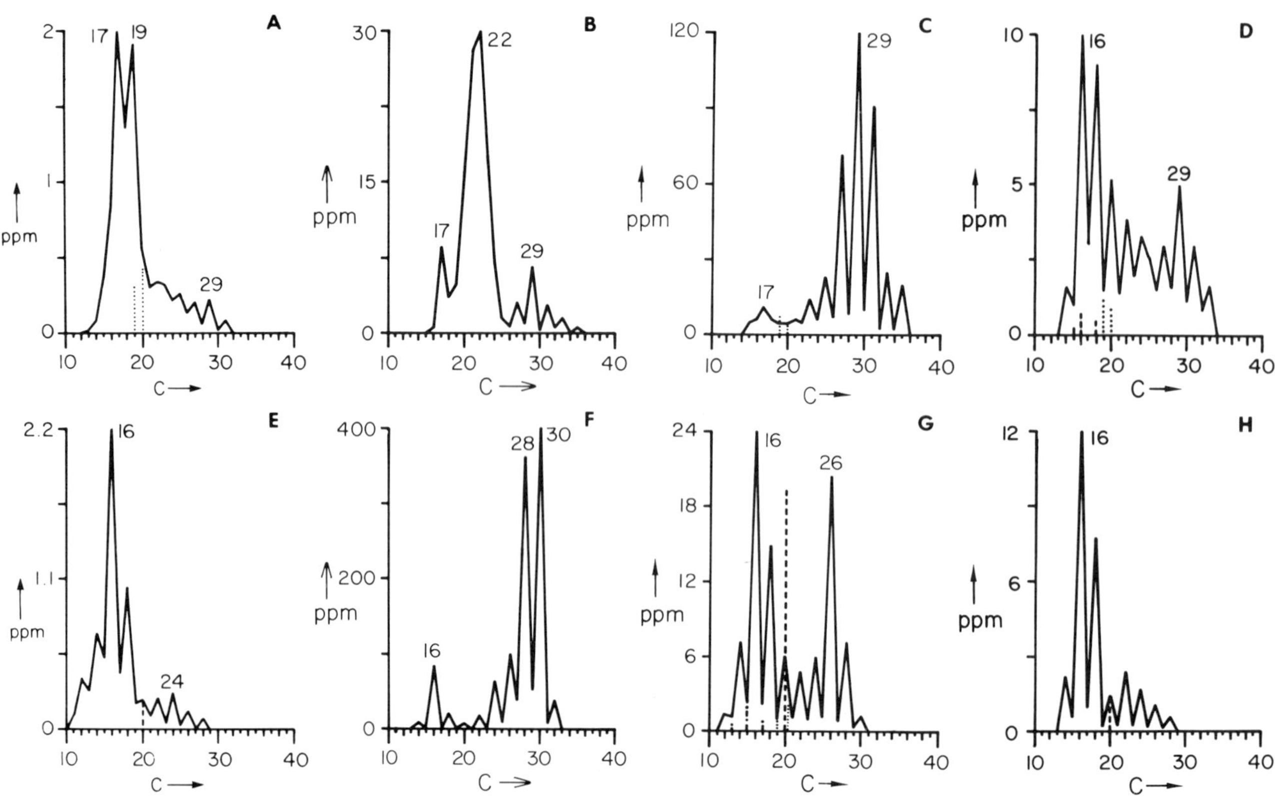

Fig. 1: Distribution diagrams (concentration versus carbon number) of n-alkanes (A-D) and n-fatty acids (E-H)(...isoprenoids,--- diterpenoids):

(A,E)   Sample 36-330-10-1, 95-102 cm, S. Atlantic (Simoneit, 1980)
(B,F)   Sample ML71-2-23, 8.0 m, Mangrove Lake, Bermuda (Hatcher et al., 1982)
(C,G)   Sample AII49-1462K, 5.0 m, Black Sea (Simoneit 1977b)
(D,H)   Sample 18-175-2-2, 45-47 cm, N.E. Pacific (Simoneit, 1977a, 1978).

homologs > C_{21} (Simoneit, 1980).   The n-alkanes are superimposed on a minor unresolved complex mixture (hump) which maximized in the GC retention region of C_{17}.   Such humps have been found for bacterial residues in surface sediments at natural gas emanations in Chile (Simoneit and Didyk, 1978).

Sample ML71-2-23, 8.0m is from Mangrove Lake, Bermuda, of recent origin in a highly productive sapropelic environment (Hatcher et al., 1977; 1982).   The n-alkanes (Fig. 1B) show maxima at n-C_{17} from primary algal synthesis, at n-C_{22} from microbial degradation of algal detritus and a minor one at n-C_{29}, with a strong odd carbon number predominance > C_{26}, derived from higher plant wax (eg. Simoneit, 1975; 1977a,b; 1978a).   The maximum at n-C_{22} with no carbon number predominance is superimposed on an unresolved hump which also maximizes in that retention region.   The n-fatty acids (Fig. 1F) exhibit a bimodal distribution where the predominant maximum at C_{28} and C_{30} is of a plant wax origin and the minor homologs < C_{20} of a marine derivation (Simoneit, 1977a,b; 1978a; 1982a).

The Black Sea is a sink for terrigenous lipids from plant waxes as is indicated by the example (Fig. 1C, G).   The n-alkanes exhibit a maximum at n-C_{29} with a strong odd-to-even carbon number predominance, typical of higher plant wax (Simoneit, 1974; 1977b; 1978b).   This is corroborated by the n-fatty acids with their maxima at n-C_{16} and n-C_{26} and strong even carbon number predomin- ance, where the homologs > C_{20} are of an allochthonous terrigenous origin (Simoneit, 1977b; 1978b).

DSDP sample 18-175-2-2, 45-47 cm is an example of a hemipelagic reducing microenvironment in the Northwest Pacific, where the $\underline{n}$-alkanes (Fig. 1D) exhibit a strong even-to-odd carbon number predominance for the homologs $< C_{24}$ and the distribution follows that of the n-fatty acids $<C_{24}$ (Fig. 1H). These even $\underline{n}$-alkanes $<C_{24}$ may be derived from reduction of fatty acids or of olefins from fatty alcohols (Simoneit 1977a; Tissot and Welte, 1978). The $\underline{n}$-alkanes $>C_{24}$ with the odd carbon number predominance and maximum at $\underline{n}\text{-}C_{29}$ are again derived from plant waxes.

Strong euxinic conditions of sedimentation are reflected in the saturated hydrocarbons by an excess of phytane (Ph) > pristane (Pr) as for example for DSDP sample 30-330-10-1 (Fig. 1A). A Pr/Ph ratio of less than unity, coupled with the presence of significant amounts of pigments and sulfur have been utilized as indicators for anoxic sedimentary conditions (Didyk $\underline{et}$ $\underline{al}$., 1978).

The distribution patterns and the carbon preference index (CPI*) of homologous compounds in lipids can be used to make partial assessments of the genetic origins of those compounds. This approach can be further strengthened by coupling it with the analysis and interpretation of the molecular marker compounds in the lipids.

Molecular markers are indicator compounds which can be utilized in correlations of genetic sources (Simoneit, 1978a; 1981; 1982a). Such molecules have definitive chemical structures that can be related either directly or indirectly via a set of diagenetic changes to their source. Such sources can be biogenic, geologic or synthetic (Simoneit, 1978a). The terpenoids and steroids are the compound groups which have had the greatest utility as specific biomarkers in geological applications.

Sesquiterpenoids have been identified in the marine environment as primarily cadalene (I, all chemical structures are shown in Appendix I), with various tetrahydro analogs. They are of both a marine (algal) and terrigenous origin (Simoneit and Kaplan, 1980).

The diterpenoids that have been characterized fit into two classes, those derived from terrigenous sources (Simoneit, 1977a) and extended tricyclic terpanes of probably a marine origin. The terrigenous diterpenoids consist of a large number of compounds and the most abundant analogs are dehydroabietic acid (II), dehydroabietin (III), dehydroabietane (IV), simonellite (V) and retene (VI). The extended tricyclic terpanes (VII) were first identified in shale and petroleum (eg. Reed, 1977), where they are ubiquitous, and then in recent sediments.

The triterpenoids found in most marine sediments are of an autochthonous origin and are usually comprised of the hopane and to a lesser extent the moretane series. The homologs consist of trisnorhopane (VIII, R=H), norhopane (VIII, R=C_2H_5), hopane (VIII, R=C_3H_7) and extended hopanes ranging from C_{31} to C_{35} (IX) with minor amounts of the corresponding moretanes (X). The 17β(H) stereochemistry predominates in Recent, immature sediments (Dastillung and Albrecht, 1976) and various triterpenes (eg. diploptene, XI) are also present. Recent sediments from the Gulf of California (Fig. 2A) illustrate this point in that only 17β(H)hopanes and hopenes are present (Simoneit $\underline{et}$ $\underline{al}$., 1979). The extended 17β(H)-hopanes occur as only the 22S diastereomer in Recent sediments as the direct markers of biosynthesis.

*CPI - Carbon Preference Index: for hydrocarbons it is expressed as a summation of the odd carbon number homologs over a range divided by a summation of the even carbon number homologs over the same range (Simoneit, 1978a; Cooper and Bray, 1963); for fatty acids and alcohols it is the same ratio only inverted to have even-to-odd homologs (Simoneit, 1978a; Kvenvolden, 1966).

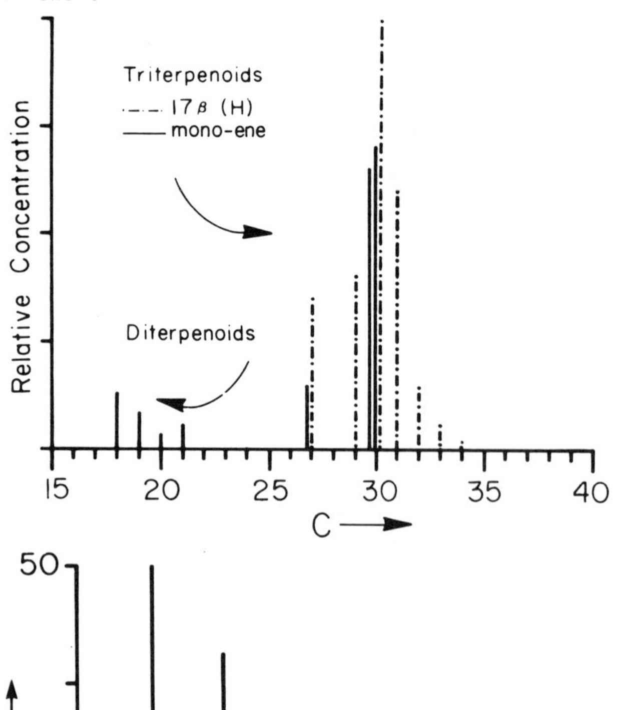

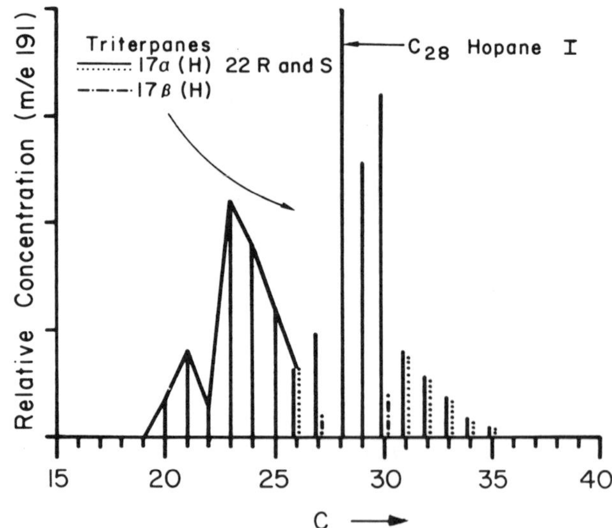

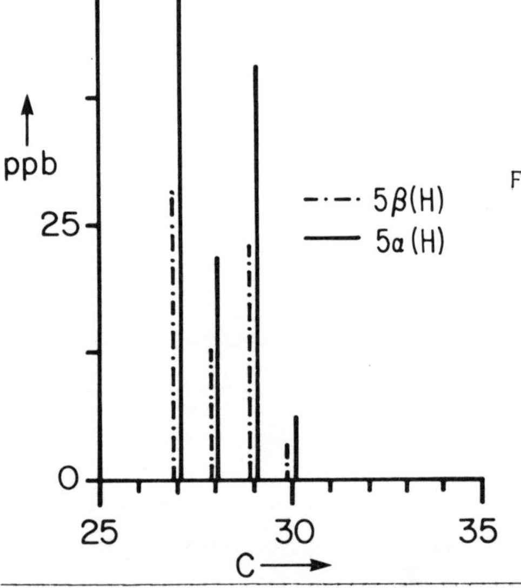

Fig. 2: Relative distribution histograms for diterpenoids, triterpenoids (based on the m/z191 mass chromatograms or gas
gas chromatographic response) and steranes (based on m/z
217) of some examples. The R and S diastereomers are also
indicated and C_{28} Hopane I is $17\alpha(H),18\alpha(H),21\beta(H)$-28,30-
bisnorhopane.

(A) Terpanes, sample 30G-I (102-105 cm), Guaymas Basin,
Gulf of California (Simoneit et al., 1979).
(B) Terpanes, sample 193 (25-31 cm), Santa Barbara Coastal
Area, Southern California Bight (Simoneit and Kaplan,
1979).
(C) Steranes, sample 36-330-4-2, 120-126 cm, South
Atlantic (Simoneit, 1979).

Steroidal compounds are widespread in sediments and they undergo complex diagenetic reactions
yielding various series of hydrocarbons, alcohols and ketones (XII) (Mackenzie et al., 1982a,b; Lee
et al., 1977, 1979; Huang and Meinschein, 1979). The saturated hydrocarbons are comprised of the
steranes [XIII, eg., cholestane, 5α(H) vs. coprostane, 5β(H)], diasteranes (XIV) and traces of
other isomers. An example of a sterane distribution is shown in Fig. 2C for a Cretaceous sample
from the South Atlantic. The 5α stereomers are predominant over the 5β (Simoneit, 1980). Steroid
residues in Recent sediments are unsaturated hydrocarbons, found mainly as ster-2-enes, ster-4enes,
diasterenes and monoaromatic diasteranes, and functionalized analogs found as stan-3-ols and stan-
3-ones (Simoneit, 1978a; Cranwell, 1982; Mackenzie et al., 1982a,b).

Other minor components that can be utilized as molecular markers are tetraterpenoids (Watts
and Maxwell, 1977), iso- and anteiso- alkanes or fatty acids, hydroxy fatty acids (Boon et al.,
1977), wax esters (Boon and deLeeuw, 1979), tetrapyrrole pigments and aromatic hydrocarbons
(Windsor and Hites, 1979). Tetrapyrrole pigments (chlorins) have been extensively studied and they
are excellent molecular markers (Baker and Smith, 1974).

Pseudokerogen, detritus, humic and fulvic substances:

Both humic and fulvic substances (solubles in aqueous base) are mixtures of complex macromole-
cules, and the latter are of lower molecular weights and thus soluble in dilute HCl. Little is

known about the detailed structures of the compounds concerned and the analytical data consist of their various bulk properties (eg., Schnitzer and Khan, 1972; Stuermer, 1975; Stuermer et al., 1978; Huc, 1973, 1978; Huc and Durand, 1977; Harvey et al., 1983). It has been shown that humic and fulvic substances decrease in concentration with depth of burial or geologic age and are probably incorporated into the kerogenous material (eg., Nissenbaum and Kaplan, 1972; Huc and Durand, 1977; Stuermer and Simoneit, 1978).

Kerogens are also complex mixtures of high molecular weight moieties of various, essentially unknown structures. The end members of kerogen sources are coals for terrigenous and alginites for marine origins, but most kerogens of sediments are admixtures of all input sources (eg., Tissot et al., 1974; van de Meent et al., 1980). Kerogen is an invaluable endogenous paleoenvironmental marker for the origin of the bulk of the sedimentary organic matter. Recent sediments (eg., algal mats) yield a "pseudokerogen" which is a lipoid macromolecular material, constitutionally less complex than ancient kerogen, but related to it (Philp and Calvin, 1976, 1977). In addition, recent sediments can receive terrigenous influx of plant detritus (eg. lignin, pollen, etc.), which then constitute the more oxygenated pseudokerogen components (Hedges and Parker, 1976; Hedges and Mann, 1979). Recycled detritus as for example charcoal can also be incorporated into Recent sediments (Griffin and Goldberg, 1975; 1983).

Diagenesis:

The microbiological, chemical and physical transformations of organic matter in sedimentary systems are termed diagenesis. These processes tend to approach equilibrium under mild conditions and shallow depths of burial, thus affecting primarily the lipids, humic/fulvic substances and the pseudokerogen. This degradation and alteration results in the polycondensation of some of the lower molecular weight organic residues, which upon further reaction with macromolecular detritus are insolubilized to generate the geopolymers (kerogen). Labile lipid components are also incorporated in part into the kerogen fraction and fulvic and humic substances condense further by elimination of H_2O, NH_3, etc. yielding moieties with even higher molecular weight ranges.

IV. NATURE OF ORGANIC MATTER IN MATURING BASINS

The processes of maturation, ie. catagenesis and metagenesis, will be examined in this section and how these affect sedimentary organic matter after diagenesis. Basically, the nature of the organic matter determines the resulting products in terms of gas versus oil provinces.

Gas:

Natural gas forms by a variety of processes and thus occurs with many different composition ranges. Methane (CH_4) is always a major constituent and higher homologs (wet gas), with carbon dioxide, hydrogen sulfide, nitrogen, hydrogen and noble gases may also be found in varying amounts. The origin of these constituents can be from organic matter and from atmospheric, volcanic and geothermal sources.

Organic matter derived from marine or lacustrine paleoenvironments (type I or II kerogen, Tissot et al., 1974) will ultimately yield wet gas associated with petroleum in the late catagenetic stage. More terrestrial and oxidized organic matter (type III kerogen) will generate dry gas (CH_4, CO_2 and N_2) during the cata- and metagenetic stages. The stable carbon isotope compositions of CH_4 in natural gases range from $\delta^{13}C$ = -60°/oo (immature, low temperature genesis) to $> -30^o/oo$ (late metagenetic stage) (Galimov, 1975; Frank _et al._, 1974; Galimov and Simoneit, 1982a,b; Simoneit and Galimov, 1983). Thermogenic natural gases reflect their varied hydrocarbon composition in a range of $C_1/C_2 + C_3$ ratios usually < 50 (Bernard _et al._, 1974). Intermediate values of $C_1/C_2 + C_3$ ratios are usually interpreted as admixtures of thermogenic and biogenic gas (eg. Simoneit _et al._, 1979).

<u>Bitumen:</u>

The principal stage of oil formation proceeds under the catagenetic regime, as the geothermal stress increases with greater burial. Hydrocarbons are cracked from the kerogen and are superimposed on the original bitumen of the source rock. Most of these new hydrocarbons have a medium ($<C_{20}$) to low molecular weight range and thus skew the _n_-alkane distribution to low carbon numbers (eg. Fig. 3). The CPI approaches unity and a major envelope of unresolvable branched and cyclic hydrocarbons (N,S,O compounds, naphthenes, or hump) is also generated (Vassoevich _et al._, 1974). This petroleum genesis also yields significant amounts of gas. The oil genesis window has been estimated to commence at temperatures in the range of 50 to 120^oC (Tissot _et al._, 1975).

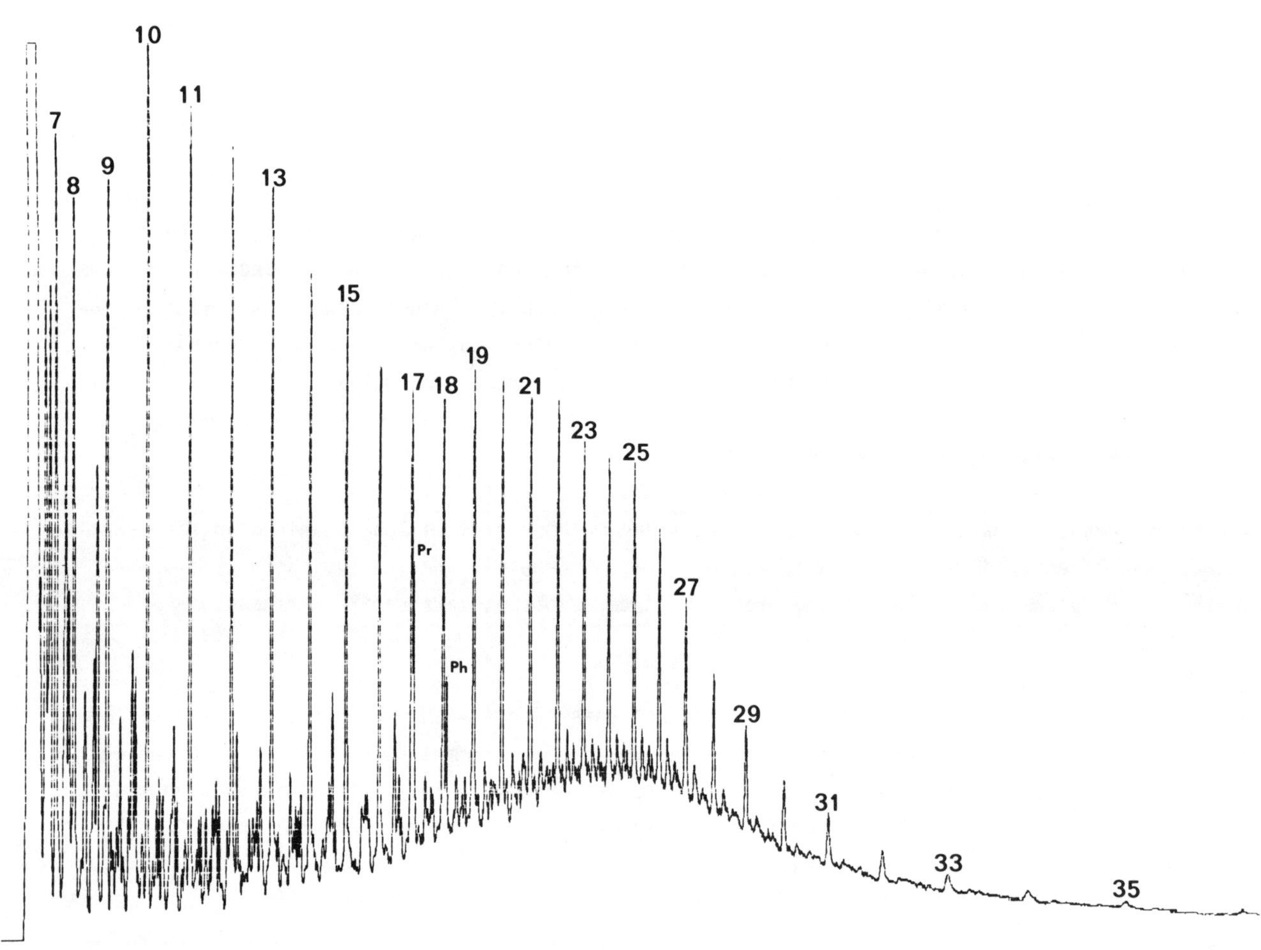

Fig. 3: Capillary gas chromatographic trace of an example of a petroleum (Bradford crude oil, Pennsylvania). The carbon numbers of the _n_-alkanes are indicated and Pr = pristane and Ph = phytane.

Molecular marker geochemistry has been extensively applied to assess maturity and for oil-source rock and oil-oil correlations (eg. Welte _et al._, 1975; Seifert, 1978; Seifert and Moldowan, 1978; 1980; Seifert _et al._, 1979). In ancient sediments or crude oils, the hopanes (VIII, C_{27}, C_{29} and C_{30}), occur as the $17\alpha(H)$ stereomers and the extended $17\alpha(H)$-hopanes (IX, C_{31}-C_{35}) are found as 22R and 22S diastereomeric pairs, where full maturity is indicated by an S/R ratio of about one (Dastillung and Albrecht, 1976; Simoneit and Kaplan, 1980). For example, a seep oil from the Southern California Bight off Santa Barbara shows the typical triterpane distribution of petro-leum consisting of mainly the $17\alpha(H)$-hopane series (eg. Fig. 2b). Also, $17\alpha(H),18\alpha(H),21\beta(H)$-28,30-bisnorhopane, a $C_{28}H_{48}$ triterpane is a major analog in this sample (Seifert _et al._, 1978; Simoneit and Kaplan, 1980). Extended tricyclic terpanes (VII) are present in this sample and they are usually coupled with the occurrence of the $17\alpha(H)$-hopane series of triterpanes (Simoneit and Kaplan, 1980). These extended tricyclic terpanes range from $C_{19}H_{34}$ to $C_{26}H_{48}$ and sometimes to $C_{29}H_{54}$, with very similar distributions and maximum at $C_{23}H_{42}$ (Fig. 2b). Steroidal hydrocarbons are widespread in mature source rocks and in petroleum. The steranes (XIII), usually with the $5\alpha(H) > 5\beta(H)$ stereochemistry, and diasteranes (XIV) are commonly present (Seifert and Moldowan, 1979; Ensminger _et al._, 1978). An example of a sterane distribution in a Cretaceous shale is shown in Fig. 2c (Simoneit, 1980). Monoaromatic steranes, common as minor constituents, have also been utilized as maturation temperature indicators (eg. Mackenzie _et al._, 1982b).

Other minor components that can be utilized as molecular markers are tetraterpenoids, _iso-_ and _anteiso-_ alkanes, tetrapyrrole pigments and alkylated aromatic hydrocarbons. The tetrapyrrole pig-ments (porphyrins) have been extensively studied and they are excellent molecular markers as well as geothermal sensors (Baker and Smith, 1974; Mackenzie _et al._, 1980).

Kerogen:

Kerogen is an invaluable endogenous paleoenvironmental marker for the origin of the bulk of the organic matter and for the thermal effects during maturation. This can be enhanced by charac-terizing its bulk structure, and thus its genetic origin. The elemental composition and atomic H/C and O/C ratios of kerogen can be correlated using van Krevelen (1961) diagrams to assess the magni-tude of the allochthonous influx and the extents of diagenesis and catagenesis (eg., Deroo _et al._, 1978). Using stable carbon isotope ratios in a correlation versus H/C (eg., Figure 4) of kerogens of Recent to Cretaceous age reveals a clustering which reflects the genetic origin, i.e., terrigen-ous, marine or a mixture of both.

Kerogen is sensitive to thermal stress from either the normal gradient due to depth of burial (eg., Tissot _et al._, 1971; Tissot and Welte, 1978) or from transients such as intrusions or hydro-thermal activity (Simoneit _et al._, 1978; 1981; Simoneit and Philp, 1982). This catagenesis results in the generation of bitumen (petroleum) from the kerogen, and this kerogen thus becomes more aro-matic or remains as spent amorphous carbon (Simoneit, 1982b). This is illustrated in Figure 5, where an igneous sill intrusion into Cretaceous shale resulted in the loss of volatiles and a de-crease in the H/C in the sill proximity (Simoneit _et al._, 1978; 1981). Additional parameters and techniques that can be utilized in kerogen evaluation are vitrinite reflectance, Rock-Eval pyroly-sis, pyrolysis GC and pyrolysis GC-MS and maceral analysis by microscopy. Thus, the bulk parame-ters of kerogen can be good indicators of the thermal history of the source rock and its organic matter.

V. NATURE OF ORGANIC MATTER IN HYDROTHERMAL SYSTEMS

The effects of hydrothermal activity on sedimentary organic matter will be illustrated here with the extensively studied case of the Guaymas Basin in the Gulf of California. This is an example

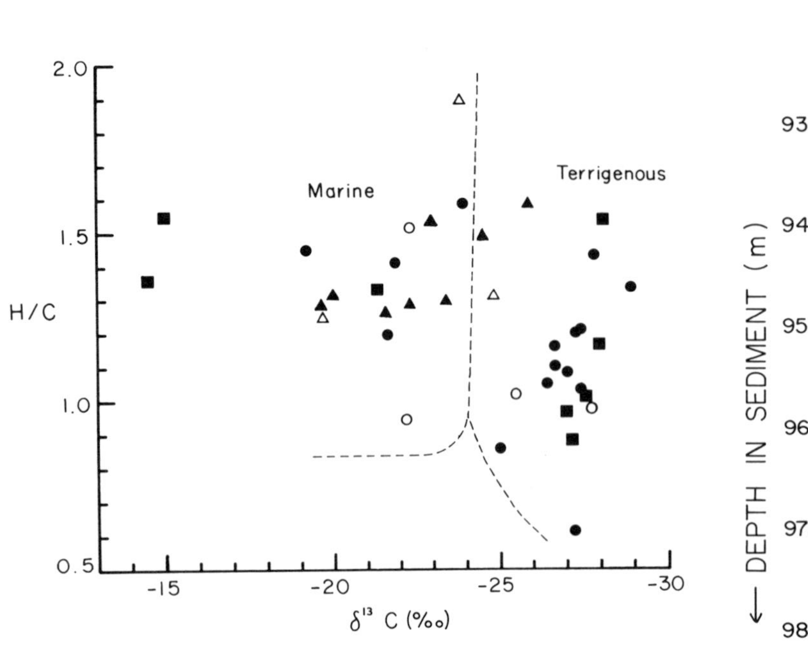

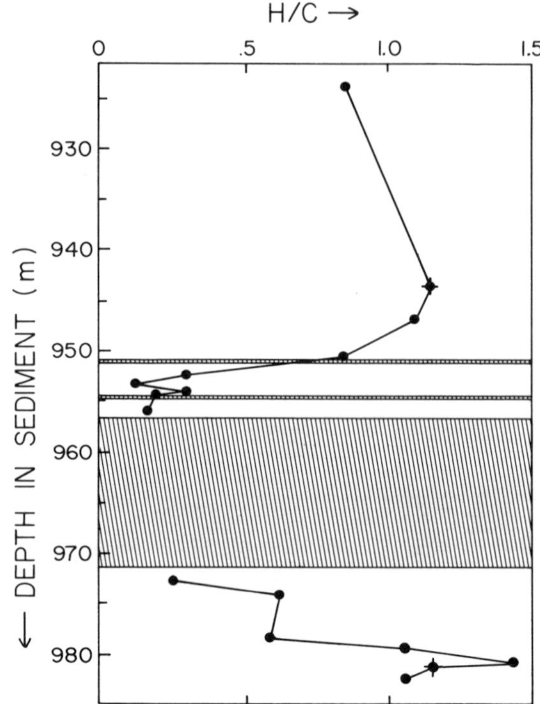

Fig. 4. Correlation diagram of H/C versus δ^{13}C for some kerogens and humates (▲ Recent kerogen; ● Cretaceous kerogen; ■ Recent kerogen from work of Stuermer et al., 1978; △ Recent and ○ Cretaceous humates.

Fig. 5: Plot of the atomic H/C of the kerogens at DSDP site 41-368 versus depth in the sediment (Simoneit et al., 1978; 1981). The intrusions are indicated by the hatched areas.

of a ridge-crest hydrothermal system with a sediment cover (maximum water temperature measured = 315°C at ∿200 atmospheres) and it differs from hydrothermal systems depositing more of the noble metals (Barnes, 1979). However, it is ideal in terms of petroleum genesis.

Geologic Setting:

Guaymas Basin (Fig. 6) is an actively-spreading oceanic basin, which is part of the system of spreading axes and transform faults that extend from the East Pacific Rise to the San Andreas fault (Curray et al., 1979, 1982). The processes of ocean plate accretion result in high conductive heat flow (locally exceeding 1.2 Wm^{-2}) and dike and sill intrusions into the unconsolidated sediments (Williams et al., 1979; Einsele et al., 1980; Curray et al., 1982). Sediments accumulate at a rate of more than 1 m/1000 yrs. and have covered the rift floors to a depth of up to 400 m (Curray et al., 1979; 1982).

Organic matter of these Recent hemipelagic sediments is comprised of various operational fractions as defined by the analytical procedures. The fractions that have been analyzed here are interstitial gas, lipids, fulvic and humic substances and detrital carbon (pseudokerogen). All of these carbonaceous fractions are very sensitive to thermal stress, as for example from an intrusion or a deep-seated heat source and are thus easily pyrolyzed, as is the case here, especially in the southern rift (Fig. 6).

Experimental:

The samples described in this summary are derived from gravity coring (30G, Simoneit et al., 1979), Deep Sea Drilling Project, Leg 64 coring (Curray et al., 1979, 1982; Simoneit et al., 1982), dredging operations (7D, Simoneit and Lonsdale, 1982), piston coring (LaPaz 9P, 13P and 15P) and sampling with the submersible Alvin (Fig. 6). The salient analytical results for all samples are collected in Table 2.

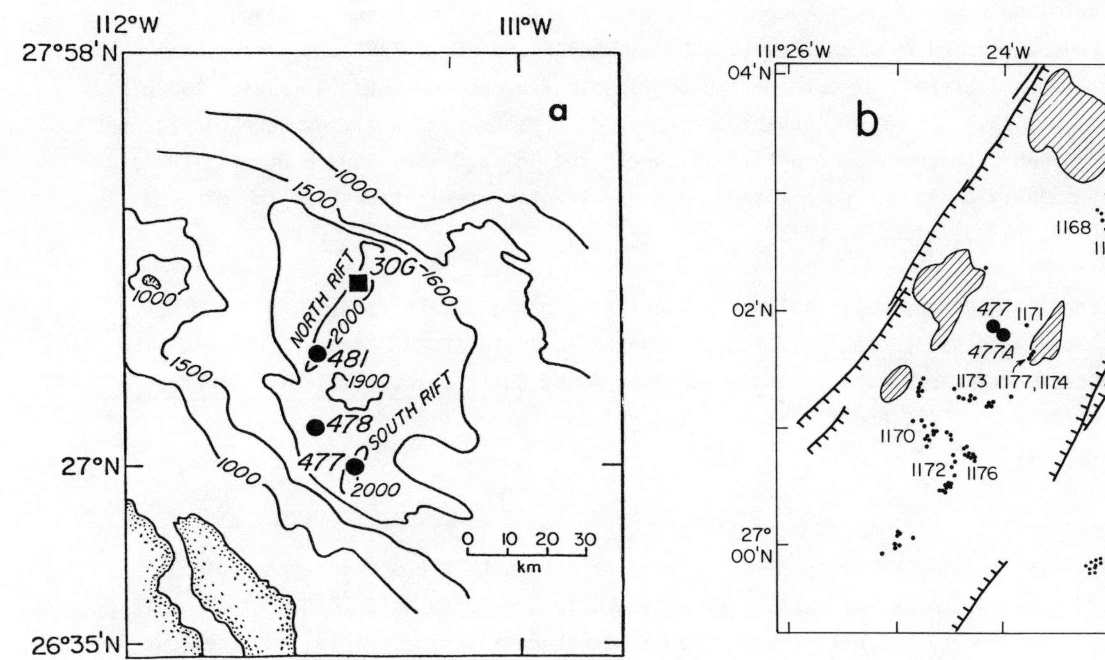

Fig. 6: Location maps of the Guaymas Basin area in the Central Gulf of California, showing the sampling locations. (a) Overall basin with DSDP sites, and (b) portion of Southern Trough showing locations of DSDP Site 477, dredge 7D and the mound patches visited by DSRV Alvin.

TABLE 2. Summary of analytical results for sedimentary bitumen and hydrothermal petroleum from Guaymas Basin, Gulf of California.

Samples	Total Organic Carbon (%)	Total Extract Yield (μg/g dry weight of sample)[1]	Total Hydrocarbon Yield (μg/g dry sample)[2]	CPI[3]	Reference
30G (∿3 m)	1.7-2.5	n.d.	1-24 (60-1000)	1.4-6.0	Simoneit et al. (1979)
64-477-5-1	2.6	2160 (83,000)	78 (3000)	1.9	Simoneit and Philp (1982)
64-477-17-3	0.8	1230 (153,750)	430 (53,700)	1.03	Simoneit and Philp (1982)
64-477A-9-1	0.4	2380 (600,000)	14 (3500)	1.02	Simoneit and Philp (1982)
7D-2B	n.a.	1075	27	1.03	Simoneit and Lonsdale (1982)
7D-4A,B	n.a.	70,400	4850	1.2	Simoneit and Lonsdale (1982)
9P (12.9-14.4 m)	n.d.	2110	41	1.05	Simoneit (1983)
13P (10.5-12.4 m)	n.d.	3040	17	1.7	Simoneit (1983)
15P (11.2-11.9 m)	n.d.	5000	16	1.2	Simoneit (1983)
1172-2 (DSRV Alvin)	n.a.	76,300	n.d.	1.0	Simoneit (1983)
1172-4 (DSRV Alvin)	n.a.	552,000	n.d.	n.d.	Simoneit (1983)
1177-3 (DSRV Alvin)	n.a.	55,000	n.d.	0.88	Simoneit (1983)

[1] Values include entrapped bitumen liberated after mineral removal with HF; values in parentheses are μg/g C_{org}.

[2] Values in parentheses are hydrocarbon yield in μg/g C_{org}.

[3] Carbon Preference Index (Simoneit, 1978a).

n.d. - not determined

n.a. - not applicable

Analyses for gasoline range (C_4-C_9) hydrocarbons were carried out on canned or bagged samples by the methods described (Simoneit et al., 1979; Whelan and Hunt, 1982). Interstitial gases in vacutainers were analyzed for composition and stable isotope contents (Simoneit, 1982b; Galimov and Simoneit, 1982a,b). The extractable bitumen and protokerogen analyses were carried out by the well defined organic geochemical practice (Simoneit and Philp, 1982; Jenden et al., 1982) and the petroleum was analyzed by the same methods after extractive separation from the minerals (Simoneit and Lonsdale, 1982; Simoneit, 1983).

The various organic fractions were analyzed by capillary gas chromatography (GC) and computerized gas chromatography/mass spectrometry (GC/MS)(Simoneit and Philp, 1982; Simoneit, 1983). The pseudokerogen from sediments was separated and then analyzed by Curie point pyrolysis, electron spin resonance spectrometry (ESR), and for stable isotope and elemental compositions (Simoneit and Philp, 1982; Jenden et al., 1982).

Hydrothermal Petroleum:
The first indication that thermogenic products were diffusing to the seabed from depth in Guaymas Basin was found in a gravity core taken at Site 30G in the north rift (Fig. 6a)(Simoneit et al., 1979). The bulk of the organic matter was of an autochthonous marine origin, but the lower sections of the core contained significant concentrations of gasoline range hydrocarbons. Similar migration of hydrocarbons was observed in other shallow (9P,13P and 15P, cf. Table 2) cores from this rift (Simoneit, 1983).

Leg 64 of the Deep Sea Drilling Project encountered intrusives and hydrothermal alteration at depth at Sites 477, 478 and 481 (Fig. 6a) (Curray et al., 1982). Thermogenic gas and H_2S and CO_2 were identified for all sites based on composition and stable carbon isotope data (Simoneit, 1982b; Galimov and Simoneit, 1982a,b). At shallow depths, the data indicated a typically biogenic pattern (Sites 481 and 478, and also 30G, cf. Simoneit et al., 1979), however, with increasing depth, the $\delta^{13}C$ values became heavier indicating the removal (by diffusion and/or distillation) of the lighter $^{12}CH_4$ due to the thermal stress from intrusive and conductive heat sources. The CH_4 at Site 477 was heaviest, reflecting the highest temperature effects, and the data for Site 481 between the sills indicated various thermal stresses.

The total hydrocarbon fractions of the lipids were evaluated to compare these effects and two examples are shown in Fig. 7, one of unaltered biogenic lipids and one of thermogenic petroliferous bitumen. The unaltered sample exhibits n-alkanes ranging from C_{14}-C_{35}, with a strong odd carbon number predominance, especially >C_{23} (terrestrial plant wax, eg., Simoneit, 1978a), and subordinate amounts of C_{20} and C_{25} natural cyclic olefins and triterpenes. The thermally altered sample exhibits n-alkanes with essentially no carbon number predominance and a range from C_{15}-C_{31}. Primary olefins and elemental sulfur are also dominant components.

The lipids in the sediments from shallow depths of Sites 477, 478 and 481 are primarily of an autochthonous marine origin, with a minor influx of terrestrial plant wax (Simoneit and Philp, 1982). The paleoenvironmental conditions of sedimentation were partially euxinic, probably as a result of the high deposition rates. Lipids are thermally altered close to and below the sills (Simoneit and Philp, 1982), which is indicated by the loss of the carbon number predominance of the n-alkanes, the appearance of a broad hump of unresolvable complex material, the thermodynamic equilibration of certain stereoisomers of the hopane molecular markers, and the presence of large

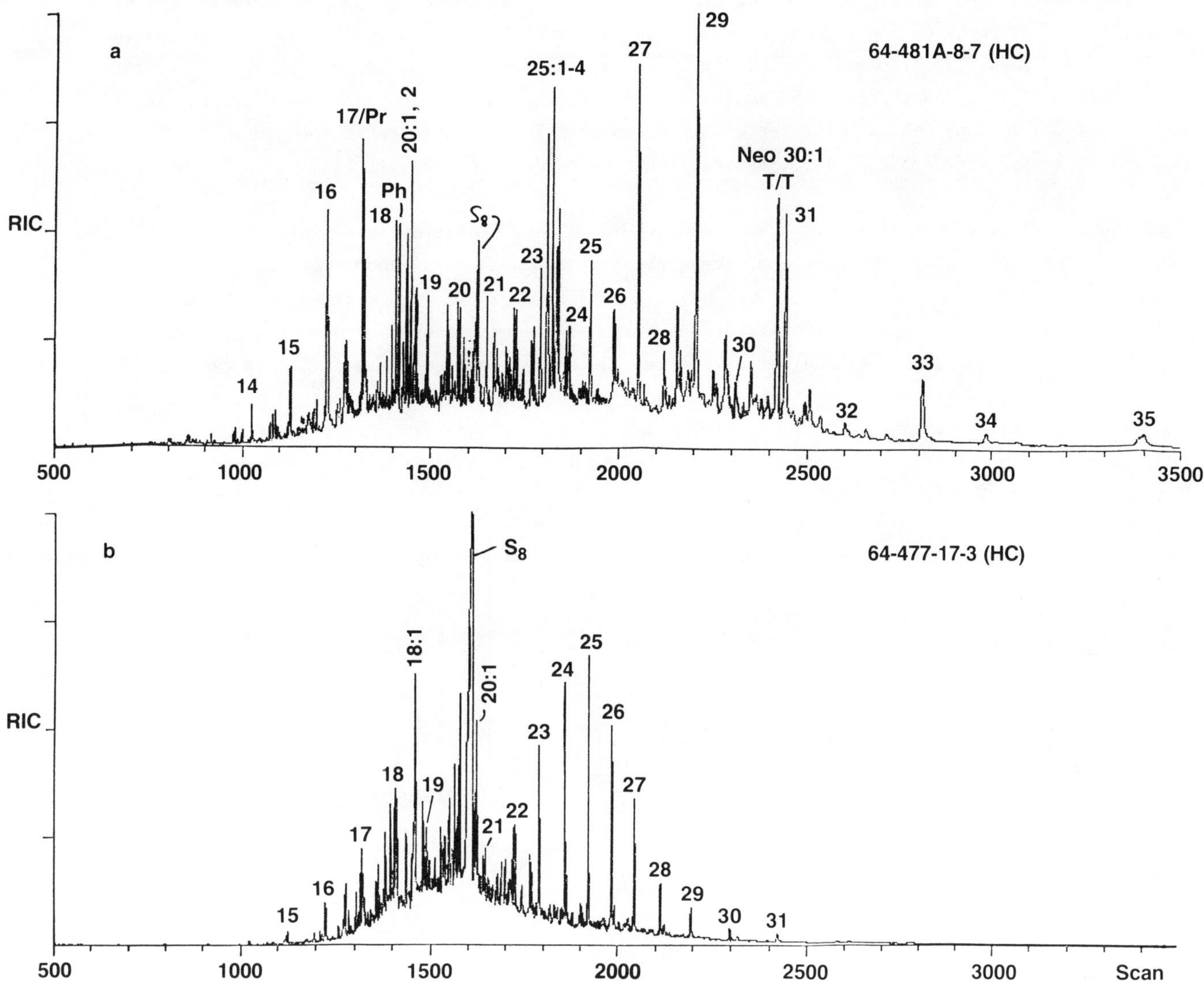

Fig. 7: GC-MS total ion current traces for total hydrocarbon fractions from (a) an unaltered, immature sample and (b) a thermally altered sample (a: DSDP 64-481A-8-7; b: 64-477-17-3; numbers indicate n-alkanes).

amounts of primary olefins and elemental sulfur. This thermal alteration of organic matter is most severe at Site 477 and intermediate at Sites 478 and 481 (Simoneit et al., 1982).

Numerous hydrothermal mounds rising 20-30 m above the basin floor (water depth about 2000 m) have been mapped in the Southern Trough and some have active hydrothermal plumes (Lonsdale, 1980). A dredge haul (7D, Fig. 6b) across a 50 m-wide patch of sinter deposits (forming a mound) recovered claystones, massive sulfides, barite, talc and other hydrothermal minerals, together with tube-worm specimens. Many of the fragments in the dredge were stained with a petroleum-like oil and had a strong odor similar to diesel fuel.

Gasoline range hydrocarbons (C_5-C_{10}) were analyzed in two dredge samples (Simoneit and Lonsdale, 1982). The relative distributions of all hydrocarbons in this range were very similar for both samples, and the large concentrations with the associated structural diversity confirmed their origin by thermal generation from the sedimentary organic matter (Whelan and Hunt, 1982).

Simoneit

The total bitumen (lipid) extract of sample 7D-2B is light amber in color and black for sample
7D-4A,B; both extracts of the samples exhibit blue fluorescence in solution typical for petroleum
condensates. These two bitumens are clearly different (Simoneit and Lonsdale, 1982). The aliphatic fraction (F1) is lowest for both samples and most of the bitumen is comprised of asphaltic (F3)
and then aromatic/naphthenic (F2) material. The aromatic to aliphatic ratio for sample 7D-2B is
9.3 and it is 4.7 for sample 7D-4A,B, typical for many crude oils (Hunt, 1979).

The GC traces of the aliphatic hydrocarbons (Fig. 8) are dramatically different. Sample 7D-2B
exhibits a pattern typical of petroleum, where the dominant $\underline{n}$-alkanes range from C_{12}-C_{33} with
no carbon number predominance (CPI_{12-33}= 1.03) and a maximum at $\underline{n}$-C_{21}. Pristane and phytane
are about equal (Pr/Ph = 1.06). The mixture of branched and cyclic hydrocarbons (i.e., hump)
ranges from C_{11} to C_{31}, also typical of petroleum. On the other hand, sample 7D-4A,B exhibits
a much narrower GC profile, skewed to lower carbon numbers. The dominant resolved peaks are various mono- and diolefins ranging from C_{12} to C_{19} and the hump ranges from C_9 to about C_{21}.
This is very much unlike typical petroleum, as olefins are not found in mature crudes (Hunt, 1979;
Tissot and Welte, 1978).

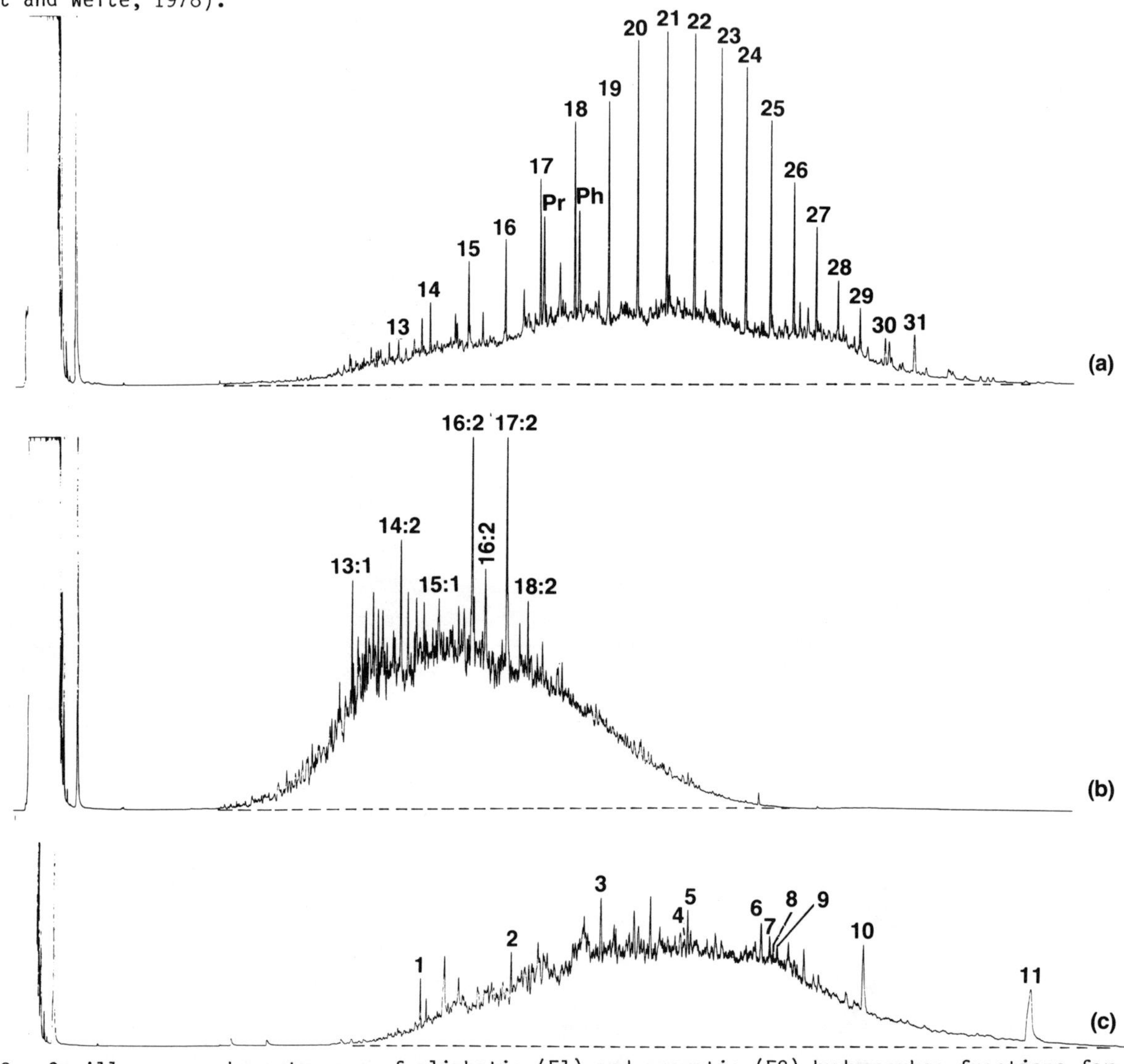

Fig. 8: Capillary gas chromatograms of aliphatic (F1) and aromatic (F2) hydrocarbon fractions for dredge
samples (a and c) 7D-2B and (b) 7D-4A,B (Simoneit and Lonsdale, 1982): (a) sample 7D-2B, F1
(the carbon chain length of the $\underline{n}$-alkanes is indicated by the arabic numerals, Pr = pristane,
Ph = phytane); (b) sample 7D-4A,$\overline{B}$, F1 (olefins are indicated by chain length: double bond
equivalent); (c) sample 7D-2B, F2 (1 = tetramethylbenzene, 2 = phenanthrene, 3 = pyrene,
4 = benz(a)anthracene, 5 = chrysene, 6 = benzofluoranthenes, 7 = benzo(e)pyrene, 8 = benzo(a)-
pyrene, 9 = perylene, 10 = benzoperylene, 11 = coronene).

The n-alkanes of sample 7D-4A,B were determined by GC/MS analysis to be present as minor components, ranging from C_{10} to C_{23}, with a minor odd carbon number predominance (CPI_{10-23} = 1.20) and a maximum at C_{15}. Pristane is more abundant than phytane (Pr/Ph = 1.6). The GC/MS data for both samples confirmed the presence of mono- and diolefins ranging from C_{12} to C_{19}. These compounds are not terminal olefins as were identified near the sills of DSDP Site 481A (Simoneit and Philp, 1982). They are slightly more stable "in-chain" olefins with methyl branching, similar to those in Bradford crude oil (cf. Fig. 3) (Hoering, 1977).

The major diagnostic molecular markers in the hydrocarbon fraction (7D-2B,F1) consist of triterpenoids, extended tricyclic terpanes and steranes with their rearranged analogs (diasteranes) (Simoneit and Lonsdale, 1982). The extended tricyclic terpanes range from C_{20} to C_{29} in a similar distribution pattern as observed for other mature petroleum samples (Simoneit and Kaplan, 1980). They are, however, not present in the unaltered surface sediments. The triterpenoids (Fig. 9) are essentially "mature"; they are for the most part in their thermodynamically more stable form, completely different from those in unaltered lipids of surface sediments (Simoneit and Philp, 1982; Simoneit et al., 1979). The triterpenoids are comprised primarily of the $17\alpha(H),21\beta(H)$hopane series ranging from C_{27} (no C_{28}) to C_{35} and the homologs from C_{31}-C_{35} are present as 22-S and R diastereomeric pairs in a ratio of about unity (Fig. 9). This series constitutes the stable mature form of these compounds (Dastillung and Albrecht, 1976; Ensminger et al., 1974) and appears to have been generated by the hydrothermal activity. The compounds are not found in the unaltered sediments, where the biogenic markers with the $17\beta(H),21\beta(H)$ stereochemistry and various triterpenes predominate. Minor amounts of $17\beta(H),21\alpha(H)$moretanes(C_{29}, C_{30} and C_{31}), $17\beta(H)$, $21\beta(H)$-hopanes (C_{27} and C_{30}), iso-hop-13(18)-ene and $17\beta(H)$-moret-22(29)-ene are also present in sample 7D-2B. These are precursor relics and intermediates from the thermal conversion process.

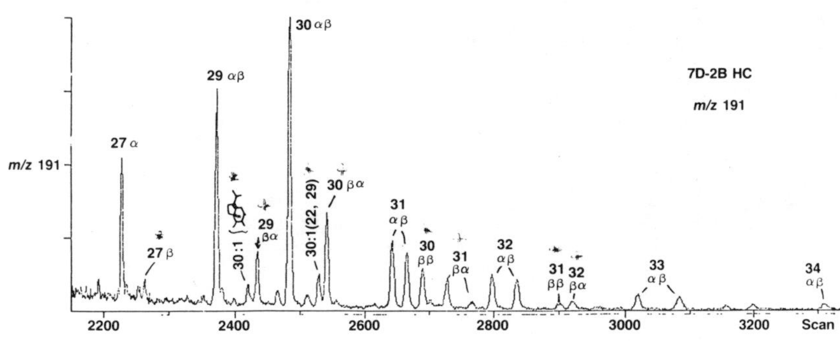

Fig. 9: Mass chromatogram (GC-MS, m/z 191) showing the triterpenoid distribution of dredge sample 7D-2B (Simoneit and Lonsdale, 1982). The predominant $17\alpha(H)$-hopanes are indicated by the arabic numerals followed by α, moretanes by $\beta\alpha$, $17\beta(H)$-hopanes by β and olefins by 30:1 (*indicates not present in mature crude oils, + not common in mature crude oils).

The steroidal markers consist of primarily $5\alpha(H),14\alpha(H),17\alpha(H)$-steranes (20 R), with lesser amounts of $5\beta(H),14\alpha(H),17\alpha(H)$steranes (20 R) and diasteranes [mainly the $13\beta(H),17\alpha(H)$-20 R or S series]. The steranes ranged from C_{26} to C_{30}, with cholestane as the major homolog. The distribution pattern indicates a marine autochthonous origin and fits best with similar data for DSDP Site 477 (Simoneit and Philp, 1982). The large $5\alpha(H)$-sterane concentration is a result of the elevated thermal stress which probably converted other steroidal compounds to these hydrocarbons.

An example of a GC trace of the aromatic/naphthenic fraction (F2) of sample 7D-2B is shown in Fig. 8c. The GC/MS data indicate that the major resolved peaks are polynuclear aromatic hydrocarbons (PAH), another group of compounds uncommon in petroleums but ubiquitous in higher temperature pyrolysis residues (Geissman _et al._, 1967; Hunt, 1979; Blumer, 1975; LaFlamme and Hites, 1978; Ishiwatari and Fukushima, 1979). The dominant analogs for both samples are the pericondensed aromatic series as for example pyrene, benzopyrenes, perylene, benzoperylene and coronene. A further indication for a pyrolytic origin is the presence of five-membered alicyclic rings (eg., acenaphthene, fluorene, fluoranthene, etc.), which are found in all pyrolysates from organic matter, since once formed they do not easily revert to pericondensed aromatic hydrocarbons (Blumer, 1975, 1976; Scott, 1982). It should also be noted that this fraction contains significant amounts of toxic PAH. The benzopyrenes are major components and the highly toxic benzo(a)pyrene is present in the two samples at levels of 25 and 16 ng/g of total bitumen, respectively. Perylene is present in these fractions and it is the predominant PAH in the unaltered sedimentary lipids of samples deposited in the Gulf from oxygen-minimum environments (Simoneit and Philp, 1982; Simoneit, 1982b). Thus, the chemical composition of this fraction indicates a source from pyrolysis with rapid quenching by hydrothermal removal and subsequent condensation at the seabed.

This process of pyrogenesis, hydrothermal removal and transport to the seabed appears to be operative at most of the mounds explored by subsequent diving with the deep submersible R.V. _Alvin_ during January, 1982 (Fig. 6b). Most of the areas with lower temperature regimes that were sampled at the mounds consisted of hydrothermal minerals cemented by solid petroleum. This material liquified upon warming on deck, thus liberating the characteristic volatile and odorous components. The data indicate that hydrothermal petroleum has become incorporated into seven of the nine areas sampled by the D.S.R.V. _Alvin_ and one vent was discharging a water-oil emulsion. The boiling ranges and compositions of the various oil samples are quite variable (eg., Fig. 10), but reflect

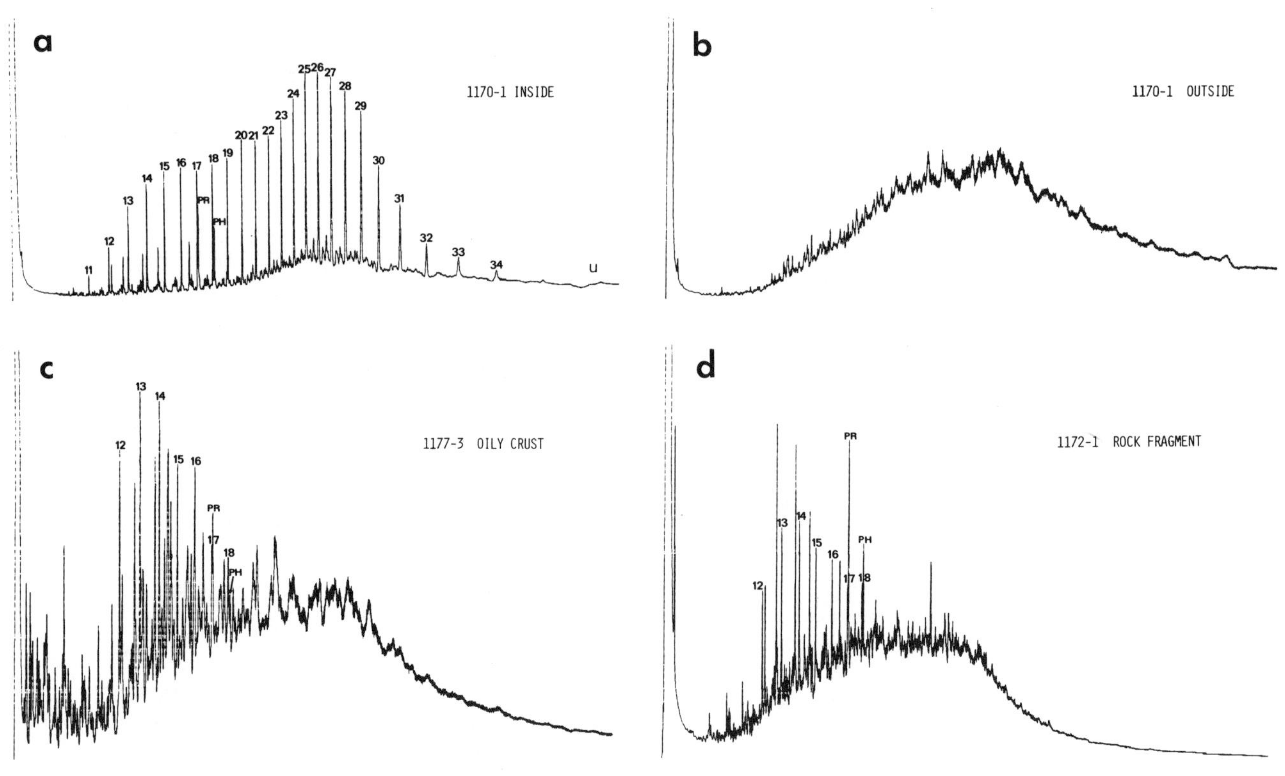

Fig. 10: Capillary gas chromatographic traces of total solvent extracts from samples recovered by D.S.R.V. _Alvin_ in Guaymas Basin (normal alkanes are indicated by the arabic numerals, Pr = pristane, Ph = phytane): (a) sample 1170-1 (interior); (b) sample 1170-1 (exterior, exposed); (c) sample 1177-3 (oily crust); (d) sample 1172-1 (rock fragment).

their rapid hydrothermal genesis and similarity with the samples described from the dredging operations. The oil on exterior surfaces of the mounds and in unconsolidated shallow sediments is weathered, leached and/or biodegraded, whereas interior samples are relatively unaltered.

<u>Pseudokerogen-amorphous carbon</u>:

Kerogen, the isoluble, detrital organic matter is an excellent <u>in situ</u> indicator of the effects of "instantaneous" thermal stress. For example, the organic nitrogen content at Site 477,
expressed as the carbon to nitrogen ratio in Fig. 11 can be used as an illustration. A C/N value
of 11-14 is typical for immature, marine organic matter (Simoneit, 1982b). At depths exceeding 150
m, the C/N ratio increases to infinity, indicating all organic nitrogen has been removed by pyrolysis (under laboratory simulation this requires temperatures in excess of 500°C). An increase in
the C/N is also observed above and below the upper sill, confirming that it is an intrusion and not
a flow, i.e., the organic matter was thermally stressed on both sides of the sill.

Pyrolysis-GC and pyrolysis-GC/MS has been utilized to characterize kerogens (van de Meent <u>et
al</u>., 1980) and the DSDP samples were analyzed by these same techniques to assess both their composition and degree of thermal catagenesis (Simoneit and Philp, 1982). Examples of pyrograms for
unaltered and thermally altered kerogen from Site 477 are given in Fig. 12. The shallow sample
exhibits a pyrogram that is virtually identical to those of surface samples from all the other
sites and is representative of typical unaltered marine organic matter (Simoneit and Philp, 1982).
The other (deepest) sample reflects a pattern of essentially complete expulsion of pyrolysate. The
<u>in situ</u> appearance of the kerogen in the zones of high thermal stress (at depth of Sites 477 and
478) resembles activated amorphous carbon (Simoneit, 1982b) and is representative of the spent
kerogen.

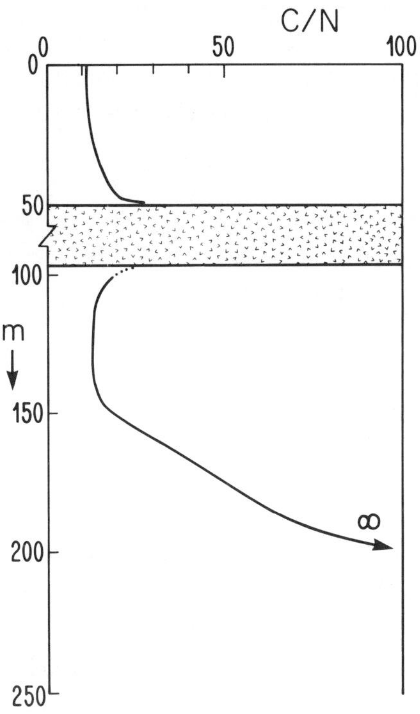

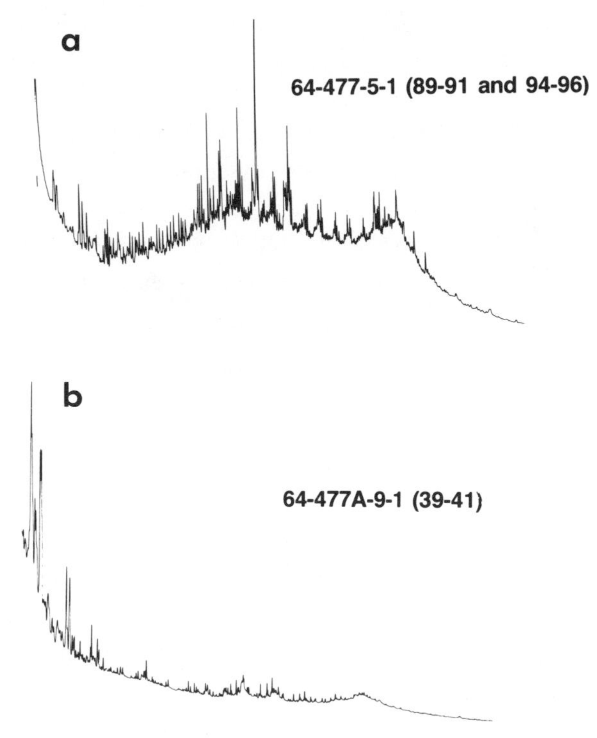

Fig. 11: Atomic ratios of organic carbon to
nitrogen versus depth at DSDP Site
477 (includes data for 477A)(Simoneit,
1982a).

Fig. 12: Examples of Curie point pyrolysis - GC traces
for kerogen concentrates from Site 477 (sub-
bottom depths are given in parentheses)
(Simoneit and Philp, 1982):
a) 64-477-5-1 (unaltered, 30 m)
b) 64-477A-9-1 (strong hydrothermal alteration, 240 m).

VI. EFFECTS OF PRESSURE, TEMPERATURE AND TIME ON CHEMISTRY OF ORGANIC MATTER

Pressure, temperature and time all affect the nature of sedimentary organic matter. The end results of these processes yield various grades of petroleums and cause its migration and interactions with the inorganic surroundings.

Petroleum genesis:

The principal zone of petroleum formation in sedimentary sequences under normal geothermal gradients commences at about 1 km to as low as 3 km (eg. Tissot and Welte, 1978; Hunt 1979). This corresponds to a temperature range of 50°-120°C and is dependent on the geologic age of the sediments (duration of heating). The effect of pressure on this process is significant, but less important and needs more supportive data (Tissot and Welte, 1978). The cracking of organic matter to natural gas is believed to take place at elevated temperatures of 150°-250°C (eg. Hunt, 1979; Vassoevich et al., 1974; Kartsev et al., 1971). These proposed temperature regimes for the oil and gas "windows" need some adjustment in consideration of more recent data.

The "instantaneous" petroleum genesis in Guaymas Basin occurs at temperatures approaching a maximum of 315°C. In this case, the lack of extensive organic matter destruction can be interpreted by the rapid removal of the thermogenic products from the hot zone. The formation of this hydrothermal petroleum appears to commence in low temperature areas, first generating products from weaker bonds (eg. ether, sulfide, carbonyl, tertiary carbon linkages), and later as the temperature regime rises, products from more refractory and even "resynthesized" (eg. PAH) organic matter.

The organic matter associated with deeper hydrothermal systems (eg. epithermal ores in volcanic terranes) is usually more asphaltic with a high PAH content. Such organic matter is widely distributed and for example has been studied from the California mercury deposits (idrialite, Blumer, 1975; Geissman et al., 1967) and other hydrothermal sulfide deposits (Germanov and Bannikova, 1972). The advent of deep well drilling (>7000 m) has yielded core materials (Cretaceous shales) which were at in situ temperatures of about 260°-300°C (eg. Price, 1982; Price et al., 1981). These samples had high concentrations of bitumen components and the kerogens still had a significant hydrocarbon generation potential. This indicates that in situ petroleum is stable at much higher temperatures as discussed above and over long geologic time periods. Metagenesis also appears to require much higher temperature conditions than normally believed.

Migration Processes:

Migration of petroleum in sedimentary sequences proceeds in solution and in the gas phase (supercritical?) from the source rocks to the traps (Hunt, 1979). The aqueous solubility of petroleums and various hydrocarbon fractions has been determined experimentally (Price, 1976). It was found that the petroleum solubility increased exponentially above 100°C to 180°C and these solubilities were high enough to account for the formation of petroleum reservoirs by the primary migration mechanism of molecular solution. Salinities of 150°/oo NaCl caused drastic exsolution of the petroleum and at 350°/oo essentially total "salt-out" was observed (Price, 1976). This finding supports the requirement for the exsolution of the petroleum from the migration solution in the salty waters of reservoir sands. In addition, it has been demonstrated that methane in the presence of water is an even better carrier for petroleum than each alone (Price et al., 1983). Both increases in pressure (to about 1800 atmospheres) and temperature (to 250°C) raised the solubility of petroleum. Cosolubility was found at rather mild conditions (eg. $100^{\circ}C$ at 1000 atm., $200^{\circ}C$ at 500 atm.). The addition of other gases (eg. CO_2, ethane) to this mixture also has a positive effect. Thus, primary migration appears to proceed by gaseous and aqueous solution.

In the case of the Guaymas Basin hydrothermal system the petroleum products have migrated by advection, diffusion, distillation and hydrothermal circulation away from the heat sources, also in gaseous but mainly in aqueous solution upward to the seabed. There the petroleum condenses according to the ambient temperatures in the conduits and vugs of the hydrothermal mineral mounds. PAH and sulfur condense in the hot vents; waxes crystallize in intermediate temperature regions ($\sim20-80^{\circ}C$); and the volatile petroleum partially collects in cold areas ($0^{\circ}C$) and emanates into the ambient sea water (Simoneit, 1983).

<u>Inorganic Interactions:</u>

Processes involving organic matter contribute to the formation of a variety of ore deposits (Saxby, 1976). The deposition of metal carbonates and phosphates is a process with a definite biogenic origin, whereas the deposition of sulfides represents a mineralization at the oxidative expense of organic matter. Metal-organic complexing has been invoked to concentrate metals and such entities can be derived from biogenic precursors (eg. porphyrins) or be generated <u>de novo</u>.

During mineral diagenesis and metamorphism under non-oxidizing conditions the organic matter composition changes progressively to more aromatic and asphaltic residues by the expulsion of volatile components (eg. CO_2, CH_4, H_2O, etc.). The inferred residuum is graphite and in the Guaymas Basin hydrothermal system the spent kerogen remaining in the altered sediments (at about $300^{\circ}C$) consisted of amorphous, activated carbon (Simoneit, 1982b). This higher temperature aromatic and asphaltic organic matter is often associated with heavy metal enrichments as for example uranium (eg. Schidlowski, 1981) or Carlin-type gold and silver ores (eg. Radtke and Scheiner, 1970). Heavy aromatic hydrocarbons (PAH) are a product of high temperature alteration and thus may be good indicators for such processes in the periphery of sulfide ore bodies (Germanov and Bannikova, 1972).

VII. SUMMARY AND CONCLUSIONS

The sequences of organic matter transformations in the lithosphere have been described briefly. Organic matter occurs as gaseous (volatile), liquid and solid components (Table 1) and is usually analyzed in these categories (Table 3). Recent immature sediments are the receptables of posthumus biogenic detritus, which upon deposition undergoes diagenetic and additional microbial alteration.

Increasing burial in sedimentary basins results in the onset of organic matter maturation, which generates some volatile products from the kerogen (easily cracked moieties) that become superimposed on the endogenous lipid residues. This is the beginning of petroleum formation. As the depth of burial (ie. temperature) keeps increasing catagenesis commences and here major petroleum generation takes place. At still greater depths of burial the metagenetic stage is envisaged, where extensive cracking, disproportionation and reforming of the organic matter, both petroleum and kerogen residues, occur to yield primarily gases and amorphous carbon (with heavy tars).

Simoneit

In the case of hydrothermal systems the previous processes are compressed into an "instantaneous" geological time frame. Hydrothermal systems operative below a sediment blanket (eg. Guaymas Basin) generate petroleum from that sedimentary organic matter, which migrates upward and leaves behind a spent carbonaceous residue. This same process of organic matter pyrolysis and movement appears to have occurred in hydrothermal regions where ore deposits formed and the organic matter content of the country rocks is low. Organic matter associated with such minerals reflects the temperature of their formation.

Migration of petroleum is aided by the presence of gases (CH_4, CO_2, etc.) and water, with elevated temperature and pressure regimes. High salinities result in essentially complete salting-out of dissolved petroleum. Heavy aromatics and asphalts can migrate during mineral metamorphism under non-oxidizing conditions and they can thus be associated with heavy metal enrichments.

In conclusion, gases, lipids (bitumen) and kerogens are ideal carbonaceous fractions that complement each other in providing information about the sources and thermal history of sedimentary organic matter (Table 3). Kerogen is a sensitive _in situ_ indicator for thermal stress, and bitumen (petroleum-asphalt) represents the product mixture of that stress - products that may have remained _in situ_ or migrated.

TABLE 3. Overview for the Analysis and Data Interpretation of Sedimentary Organic Matter.

I. INTERSTITIAL GAS:
- Analog identification and quantitation by gas chromatography (FID, TCD), GC-MS
- Range   a) C_1-C_7 hydrocarbons and other gases (CO_2, H_2S, etc).

    b) Gasoline range hydrocarbons (C_4-C_{15})
- Other methods (eg. stable isotope compositions, $\delta D, \delta^{13}C$...)
- Results
    - Evaluation of biogenic versus thermogenic gas
    - Commercial exploration and drilling safety

II. BITUMEN (LIPID) ANALYSIS:
- Chemical separations (into functional group or compound class fractions, sometimes with derivatization)
    - Liquid chromatography (column)
    - Thin layer chromatography
    - High pressure LC
- Instrumental analysis
    - Gas chromatography - homologies, quantitation, hump
    - GC-MS - homologies, hump, molecular markers
    - High resolution MS - absolute compositions
    - Stable isotope MS - element isotope fractionation
- Ancillary techniques
    - NMR - for individually isolated compounds
    - IR and UV - functional and structural determination
    - ORD - asymmetry
- Results
    - Compositional survey (concentration of homologs and analogs in lipids)
    - Source and degradation correlations

Table 3. (continued)
- Molecular markers - product/precursor relationships
- stereochemistry - thermal stress
- identity - sources and transport

III. KEROGEN ANALYSIS (Asphaltenes, fulvic and humic acids, and humin):

- Bulk properties
  - Elemental composition (C,H,O,N,S)
  - Stable isotopes ($\delta^{13}C$, $\delta^{15}N$, δD, $\delta^{34}S$)
  - ESR, NMR
- Results
  - Parameter correlations
  - Indicative of sources (autochthonous vs. allochthonous)
- Specific analyses
  - Vitrinite reflectance
  - Petroleum generating potential
  - Pyrolysis GC and py-GC-MS
- Results
  - Evaluations of maturity
  - Thermal history
  - Resource potential
  - Oil versus gas
- Other methods
  - Microscopy (visual classification and identification)
  - Chemical degradation (some indications of molecular origin)

VIII. ACKNOWLEDGEMENTS

I thank the Deep Sea Drilling Project and the National Science Foundation for access to DSDP samples, the NSF for my participation on the D.S.R.V. _Alvin_, Pluto 6 cruise; Dr. P. Lonsdale for the dredge and piston core samples, and Dr. E.M. Galimov, Dr. R.P. Philp, Mr. P. Jenden, Ms. M.A. Mazurek, Mr. E. Ruth and Mr. O.E. Kawka for data and assistance. Funding from the National Science Foundation, Division of Ocean Sciences (Grant OCE81-18897) is gratefully acknowledged.

IX. REFERENCES

Baker, E. W. and G.D. Smith (1974). Pleistocene changes in chlorophyll pigments. In: _Advances in Organic Geochemistry 1973_, B. Tissot and F. Bienner, eds., Editions Technip, Paris, pp. 649-660.

Barker, C. (1974). Pyrolysis techniques for source-rock evaluation. _Amer. Assoc. Petrol Geol. Bull. 58_, 2349-2361.

Barnes, H.L. (1979). _Geochemistry of Hydrothermal Ore Deposits_, 2nd ed., John Wiley and Sons, New York, 798 pp.

Bernard, B.B., J.M. Brooks and W.M. Sackett (1976). Natural gas seepage in the Gulf of Mexico. _Earth and Planet. Sci. Lett. 31_, 48-54.

Blumer, M. (1975). Curtisite, idrialite and pendletonite, polycyclic aromatic hydrocarbon minerals: Their composition and origin. _Chem. Geol. 16_, 245-256.

Blumer, M. (1976). Polycyclic aromatic compounds in nature. _Scient. Amer., 234(3)_, 34-45.

Boon, J.J., R. deLange, P.J.W. Schuyl, J.W. deLeeuw and P.A. Schenck (1977). Organic geochemistry of Walvis Bay diatomaceous ooze-II. Occurrence and significance of the hydroxy fatty acids. In: _Advances in Organic Geochemistry 1975_, R. Campos and J. Goni, eds., ENADIMSA, Madrid, pp. 255-272.

Boon, J.J. and J.W. deLeeuw (1979). The analysis of wax esters, very long mid-chain ketones and sterol ethers isolated from Walvis Bay diatomaceous ooze. Mar. Chem. 7, 117-132.

Claypool, G.E. and I.R. Kaplan (1974). The origin and distribution of methane in marine sediments. In: Natural Gases in Marine Sediments, I.R. Kaplan, ed., Plenum Publishing Corp., New York pp. 99-139.

Colombo, U., F. Gazzarrini, R. Gonfiantini, E. Tongiorgi and L. Caflisch (1969). Carbon isotopic study of hydrocarbons in Italian natural gases. In: Advances in Organic Geochemistry 1968, P.A. Schenck and I. Havenaar, eds., Pergamon Press, Oxford, pp. 499-516.

Cooper, J.E. and E.E. Bray (1963). A postulated role of fatty acids in petroleum formation. Geochim. Cosmochim. Acta 29, 1113-1127.

Cranwell, P.A. (1982). Lipids of aquatic sediments and sedimenting particulates. Prog. Lipid Res. 21, 271-308.

Curray, J.R., D.G. Moore, J.E. Aguayo, M.P. Aubry, G. Einsele, D.J. Fornari, J. Gieskes, J.C. Guerrero, M. Kastner, K. Kelts, M. Lyle, Y. Matoba, A. Molina-Cruz, J. Niemitz, J. Rueda, A.D. Saunders, J. Schrader, B.R.T. Simoneit and V. Vacquier (1979). Leg 64 seeks evidence on development of basins. Geotimes 24(7), 18-20.

Curray, J.R., D.G. Moore, J.E. Aguayo, M.P. Aubry, G. Einsele, D.J. Fornari, J. Gieskes, J.C. Guerrero, M. Kastner, K. Kelts, M. Lyle, Y. Matoba, A. Molina-Cruz, J. Niemitz, J. Rueda, A.D. Saunders, H. Schrader, B.R.T. Simoneit and V. Vacquier (1982). Initial Reports of the Deep Sea Drilling Project, Vol. 64, Parts I and II, U.S. Govt. Printing Office, Washington, D.C., 1314 pp.

Dastillung, M. and P. Albrecht (1976). Molecular test for oil pollution in surface sediments. Mar. Poll. Bull. 7, 13-15.

Demaison, G.J. and G.T. Moore (1980). Anoxic environments and oil source bed genesis. Org. Geochem. 2, 9-31.

Deroo, G., J.P. Herbin, J.R. Roucaché, B. Tissot, P. Albrecht and M. Dastillung (1978). Organic geochemistry of some Cretaceous claystones from site 391, Leg 44, Western North Atlantic. In: Initial Reports of the Deep Sea Drilling Project, Vol. 44, W.E. Benson, R.E. Sheridan et al., U.S. Govt. Printing Office, Washington, D.C., pp. 593-598.

Didyk, B.M., B.R.T. Simoneit, S.C. Brassell and G. Eglinton (1978). Organic geochemical indicators of paleoenvironmental conditions of sedimentation. Nature 272, 216-222.

Einsele, G., J. Gieskes, J. Curray, D. Moore, E. Aguayo, M.P. Aubry, D.J. Fornari, J.C. Guerrero, M. Kastner, K. Kelts, M. Lyle, Y. Matoba, A. Molina-Cruz, J. Niemitz, J. Rueda, A. Saunders, H. Schrader, B.R.T. Simoneit and V. Vacquier (1980). Intrusion of basaltic sills into highly porous sediments and resulting hydrothermal activity. Nature 283, 441-445.

Ensminger, A., G. Joly and P. Albrecht (1978). Rearranged steranes in sediments and crude oils. Tetrahedron Letters No. 18, 1575-1578.

Ensminger, A., A. Van Dorsselaer, C. Spyckerelle, P. Albrecht and G. Ourisson (1974). Pentacyclic triterpenes of the hopane type as ubiquitous geochemical markers: origin and significance. In: Advances in Organic Geochemistry 1973, B. Tissot and F. Bienner, eds., Editions Technip, Paris, pp. 245-260.

Frank, D.J., J.R. Gormley and W.M. Sackett (1974). Reevaluation of carbon-isotope compositions of natural methanes. Amer. Assoc. Petrol. Geol. Bull. 58, 2319-2325.

Galimov, E.M. (1975). Carbon isotopes in oil and gas geology, Nedra, Moscow, 1973, English translation by NASA, F-682, Washington, D.C.

Galimov, E.M. and B.R.T. Simoneit (1982a). Geochemistry of interstitial gases in sedimentary deposits of the Gulf of California, Leg 64. In: Initial Reports of the Deep Sea Drilling Project, Vol. 64, J.R. Curray, D.G. Moore et al., U.S. Govt. Printing Office, Washington, D.C., pp. 781-788.

Galimov, E.M. and B.R.T. Simoneit (1982b). Variations in the carbon isotope compositions of CH_4 and CO_2 in the sedimentary sections of Guaymas Basin (Gulf of California), Geokhimya, Acad. Nauk SSSR, 7, 1027-1034.

Geissman, T.A., K.Y. Sim and J. Murdoch (1967). Organic minerals. Picene and chrysene as constituents of the mineral curtisite (idrialite). Experientia 23, 793-794.

Germanov, A.I. and L.A. Bannikova (1972). Alteration of organic matter of sedimentary rocks during hydrothermal sulfide concentration. Dokl. Akad. Nak SSSR 203, 1180-1182.

Griffin, J.J. and E.D. Goldberg (1975). The fluxes of elemental carbon to coastal marine sediments. Limnol. Oceanogr. 20, 456-463.

Griffin, J.J. and E.D. Goldberg (1983). Impact of fossil fuel combustion on sediments of Lake Michigan: A reprise. Env. Sci. Techn. 17, 244-245.

Harvey, G.R., D.A. Boran, L.A. Chesal and J.M. Tokar (1983). The structure of marine fulvic and humic acids. Mar. Chem. 12, 119-132.

Hatcher, P.G., B.R.T. Simoneit and S.M. Gerchakov (1977). The Organic Geochemistry of a Recent Sapropelic Environment: Mangrove Lake, Bermuda. In: Advances in Organic Geochemistry 1975, R. Campos and J. Goni, eds., ENADIMSA, Madrid, pp. 469-484.

Hatcher, P.G., B.R.T. Simoneit, F.T. Mackenzie, A.C. Neumann, D.C. Thorstenson and S.M. Gerchakov (1982). Organic geochemistry and pore water chemistry of sediments from Mangrove Lake, Bermuda. Organic Geochemistry 4, 93-112.

Hedges, J.E. and D.C. Mann (1979). The lignin geochemistry of marine sediments from the southern Washington coast. Geochim. Cosmochim. Acta 43, 1809-1818.

Hedges, J.E. and P.L. Parker (1976). Land-derived organic matter in surface sediments from the Gulf of Mexico. Geochim. Cosmochim. Acta 40, 1019-1029.

Hoering, T.C. (1977). Olefinic hydrocarbons from Bradford, Pennsylvania, crude oil. Chem. Geol. 20, 1-8.

Huang, W.Y. and W.G. Meinschein (1979). Sterols as ecological indicators. Geochim. Cosmochim. Acta 43, 739-745.

Huc, A.Y. (1973). Contribution à l'étude de l'humus marin et de ses relations avèc les kerogènes, Thèse Docteur-ingenieur, Université de Nancy, France.

Huc, A.Y. (1979). Geochimie organique des schistes bitumineux du Toarcien du bassin de Paris, Thèse Docteur ès-Sciences, l'Université Louis Pasteur de Strasbourg, France, 137 pp.

Huc, A.Y. and B.M. Durand (1977). Occurrence and significance of humic acids in ancient sediments. Fuel 56, 73-80.

Hunt, J.M. (1979). Petroleum Geochemistry and Geology, W.H. Freeman and Company, San Francisco, 617 pp.

Ishiwatari, R. and K. Fukushima (1979). Generation of unsaturated and aromatic hydrocarbons by thermal alteration of young kerogen. Geochim. Cosmochim. Acta 43, 1343-1349.

Jenden, P.D., B.R.T. Simoneit and R.P. Philp (1982). Hydrothermal effects on protokerogen of unconsolidated sediments from Guaymas Basin, Gulf of California, elemental compositions, stable carbon isotope ratios and electron spin resonance spectra. In: Initial Reports of the Deep Sea Drilling Project, Vol. 64, J.R. Curray, D.G.Moore et al., U.S. Govt. Printing Office, Washington, D.C., pp. 905-912.

Kartsev, A.A., N.B. Vassoevich, A.A. Geodekian, S.G. Neruchev and V.A. Sokolov (1972). The principal stage in formation of petroleum. Proc. 8th World Petrol. Congr. 2, 3-11.

Kvenvolden, K.A. (1966). Molecular distributions of normal fatty acids and paraffins in some lower Cretaceous sediments. Nature 209, 573-577.

LaFlamme, R.E. and R.A. Hites (1978). The global distribution of polycyclic aromatic hydrocarbons in recent sediments. Geochim. Cosmochim. Acta 42, 289-304.

Lee, C., R.B. Gagosian and J.W. Farrington (1977). Sterol diagenesis in Recent sediments from Buzzard's Bay, Massachusetts. Geochim. Cosmochim. Acta 41, 985-992.

Lee, C., J.W. Farrington and R.B. Gagosian (1979). Sterol geochemistry of sediments from the western North Atlantic Ocean and adjacent coastal areas. Geochim. Cosmochim. Acta 43, 35-46.

Lonsdale, P. (1980). Hydrothermal plumes and baritic sulfide mounds at a Gulf of California spreading center (Abstract). EOS Trans. Am. Geophys. Union 61, pp. 995.

Mackenzie, A.S., S.C. Brassell, G. Eglinton and J.R. Maxwell (1982a). Chemical fossils: The geological fate of steroids. Science 217, 491-504.

Mackenzie, A.S., N.A. Lamb and J.R. Maxwell (1982b). Steroid hydrocarbons and the thermal history of sediments. Nature 295, 223-226.

Mackenzie, A.S., J.M.E. Quirke and J.R. Maxwell (1980). Molecular parameters of maturation in the Toarcian shales, Paris Basin, France - II. Evolution of metalloporphyrins. In: Advances in Organic Geochemistry 1979, A.G. Douglas and J.R. Maxwell, eds., Pergamon Press, Oxford, pp. 239-248.

Nissenbaum, A. and I.R. Kaplan (1972). Chemical and isotopic evidence for the in situ origin of marine humic substances. Limnol. Oceanogr. 17, 570-582.

Philp, R.P. and M. Calvin (1976). Kerogen structures in recently-deposited algal mats at Laguna Mormona, Baja California: A model system for the determination of kerogen structures in ancient sediments. In: Environmental Biogeochemistry, J.O. Nriagu, ed., Ann Arbor, Vol. 1, pp. 131-148.

Philp, R.P. and M. Calvin (1977). Kerogenous material in Recent algal mats at Laguna Mormona, Baja California. In: Advances in Organic Geochemistry 1975, R. Campos and J. Goni, eds., ENADIMSA, Madrid, pp. 735-752.

Philp, R.P.., M. Calvin, S. Brown and E. Yang (1978). Organic geochemical studies on kerogen precursors in recently-deposited algal mats and oozes. Chem. Geol. 22, 207-231.

Price, L.C. (1976). Aqueous solubility of petroleum as applied to its origin and primary migration. Amer. Assoc. Petrol. Geol. Bull. 60, 213-244.

Price, L.C. (1982). Organic geochemistry of core samples from an ultra-deep hot well (300°C, 7km). Chem. Geol. 37, 215-228.

Price, L.C., J.S. Clayton and L.L. Rumen (1981). Organic geochemistry of the 9.6 km Bertha Rogers No. 1 well, Oklahoma, Org. Geochem. 3, 59-77.

Price, L.C., L.M. Wenger, T. Ging and C.W. Blount (1983). Solubility of crude oil in methane as a function of pressure and temperature. Org. Geochem., in press.

Radtke, A.S. and B.J. Scheiner (1970). Studies in hydrothermal gold deposition (I). Carlin gold deposit, Nevada: The role of carbonaceous materials in gold deposition. Econ. Geol. 65, 87-102.

Reed, W.E. (1977). Molecular compositions of weathered petroleum and comparison with its possible source. Geochim. Cosmochim. Acta 41, 237-247.

Saxby, J.D. (1976). The significance of organic matter in ore genesis. In: Handbook of Strata-bound and Stratiform Ore Deposits: I, Principles and General Studies, Vol. 2, Geochemical Studies, K.H. Wolf, ed., Elsevier, Amsterdam, pp. 111-133.

Schidlowski, M. (1981). Uraniferous Constituents of the Witwatersrand Conglomerates: Ore-Microscopic Observations and Implications for the Witwatersrand Metallogeny, U.S. Geol. Survey Prof. Paper 1161-N, 29 pp.

Schnitzer, M., S.U. Khan (1972). Humic Substances in the Environment, Marcel Dekker, New York, 327 pp.

Scott, L.T. (1982). Thermal rearrangements of aromatic compounds. Acc. Chem. Res. 15, 52-58.

Seifert, W.K. (1978). Steranes and terpanes in kerogen pyrolysis for correlation of oils and source rocks. Geochim. Cosmochim. Acta 42, 473-484.

Seifert, W.K. and J.M. Moldowan (1978). Applications of steranes, terpanes and monoaromatics to the maturation, migration and source of crude oils. Geochim. Cosmochim. Acta 42, 77-95.

Seifert, W.K. and J.M. Moldowan (1979). The effect of biodegradation on steranes and terpanes in crude oils. Geochim. Cosmochim. Acta 43, 111-126.

Seifert, W.K. and J.M. Moldowan (1980). The effect of thermal stress on source-rock quality as measured by hopane stereochemistry. In: Advances in Organic Geochemistry 1979, A.G. Douglas and J.R. Maxwell, eds., Pergamon Press, Oxford, pp. 229-237

Seifert, W.K., J.M. Moldowan and R.W. Jones (1979). Application of biological marker chemistry to petroleum exploration. Proc. 10th World Petrol. Congr., Heyden and Son, London, pp. 425-440.

Seifert, W.K., J.M. Moldowan, G.W. Smith and E.V. Whitehead (1978). First proof of structure of a C_{28}-pentacyclic triterpane in petroleum. Nature 271, 436-437.

Simoneit, B.R.T. (1974). Organic analyses of Black Sea cores. In: The Black Sea-Geology, Chemistry and Biology, E.T. Degens and D.A. Ross, eds., Memoir 20, Amer. Assoc. Petrol. Geol., Tulsa, pp. 477-498.

Simoneit, B.R.T. (1975). Sources of organic matter in oceanic sediments, Ph.D. Thesis, University of Bristol, England, 300 pp.

Simoneit, B.R.T. (1977a). Diterpenoid compounds and other lipids in deep-sea sediments and their geochemical significance. Geochim. Cosmochim. Acta 41, 463-476.

Simoneit, B.R.T. (1977b). The Black Sea, A sink for terrigenous lipids. Deep-Sea Res. 24, 813-830.

Simoneit, B.R.T. (1978a). The organic chemistry of marine sediments. In: Chemical Oceanography, Vol. 7, J.P. Riley and R. Chester, eds., Academic Press, London, pp. 233-311.

Simoneit, B.R.T. (1978b). Organic geochemistry of terrigenous muds and various shales from the Black Sea, DSDP Leg 42B. In: Initial Reports of the Deep Sea Drilling Project, 42, Part 2, D. Ross, Y. Neprochnov et al., U.S. Govt. Printing Office, Washington, D.C., pp. 749-753.

Simoneit, B.R.T. (1980). Organic geochemistry of Mesozoic sediments from Deep Sea Drilling Project Site 330, Falkland Plateau. In: Initial Reports of the Deep Sea Drilling Project, Vol. 50, Y. Lancelot, E.L. Winterer, et al., U.S. Govt. Printing Office, Washington, D.C., pp. 637-642.

Simoneit, B.R.T. (1981). Utility of molecular markers and stable isotope compositions in the evaluation of sources and diagenesis of organic matter in the geosphere, The Impact of the Treibs' Prophyrin Concept on the Modern Organic Geochemistry, A. Prashnowsky, ed ., Bayerische Julius Maximilian Universität, Würzburg, pp. 133-158.

Simoneit, B.R.T. (1982a). The composition, sources and transport of organic matter to marine sediments-- the organic geochemical approach. In: Proc. Symp. Marine Chem. into the Eighties, J.A.J. Thompson and W.D. Jamieson, eds., Nat. Res. Council of Canada, pp. 82-112.

Simoneit, B.R.T. (1982b). Shipboard organic geochemistry and safety monitoring, Leg 64, Gulf of California. In: Initial Reports of the Deep Sea Drilling Project, Vol. 64, J.R. Curray, D.G. Moore et al., U.S. Govt. Printing Office, Washington, D.C., pp. 723-728.

Simoneit, B.R.T. (1983). Effects of hydrothermal activity on sedimentary organic matter: Guaymas Basin, Gulf of California - petroleum genesis and protokerogen degradation, NATO-ARI: Hydrothermal Processes at Sea-Floor Spreading Centers,K. Bostrom et al., eds., Plenum Press, in press.

Simoneit, B.R.T., S. Brenner, K.E. Peters and I.R. Kaplan (1978). Thermal alteration of Cretaceous black shale by basaltic intrusions in the Eastern Atlantic. Nature 273, 501-504.

Simoneit, B.R.T., S. Brenner, K.E. Peters and I.R. Kaplan (1981). Thermal alteration of Cretaceous black shale by basaltic intrusions in the eastern Atlantic. II: Effects on bitumen and kerogen. Geochim. Cosmochim. Acta 45, 1581-1602.

Simoneit, B.R.T. and B.M. Didyk (1978). Organic geochemistry of a Chilean paraffin dirt. Chem. Geol. 23, 21-40.

Simoneit, B.R.T. and E.M. Galimov (1983). Geochemistry of interstitial gases in Quaternary sediments of the Gulf of California. Chem. Geol., in press.

Simoneit, B.R.T., H.I. Halpern and B.M. Didyk (1980). Lipid productivity of a high Andean lake. Biogeochemistry of Ancient and Modern Environments, P.A. Trudinger, M.R. Walter and B.J. Ralph, eds., Australian Acad. Sci., Canberra, pp. 201-210.

Simoneit, B.R.T. and I.R. Kaplan (1980). Triterpenoids as molecular indicators of paleoseepage in Recent sediments of the Southern California Bight, Mar. Environ. Res. 3, 113-128.

Simoneit, B.R.T. and P.F. Lonsdale (1982). Hydrothermal petroleum in mineralized mounds at the seabed of Guaymas Basin. Nature 295, 198-202.

Simoneit, B.R.T., M.A. Mazurek, S. Brenner, P.T. Crisp and I.R. Kaplan (1979). Organic geochemistry of recent sediments from Guaymas Basin, Gulf of California. Deep-Sea Res. 26A, 879-891.

Simoneit, B.R.T. and R.P. Philp (1982). Organic geochemistry of lipids and kerogen and the effects of basalt intrusions on unconsolidated oceanic sediments: Sites 477, 478 and 481, Guaymas Basin, Gulf of California. In: Initial Reports of the Deep Sea Drilling Project, Vol. 64, J.R. Curray, D.G. Moore, et al., U.S. Govt. Printing Office, Washington, D.C., pp. 883-904.

Simoneit, B.R.T., C.P. Summerhayes and P.A. Meyers (1982). Sources, preservation and maturation of organic matter in Pliocene and Quaternary sediments of the Gulf of California: A synthesis of organic geochemical studies from Deep Sea Drilling Project Leg 64. In: Initial Reports of the Deep Sea Drilling Project, Vol. 64, J.R. Curray, D.G. Moore et al., eds., U.S. Govt. Printing Office, Washington, D.C., pp. 939-952.

Stuermer, D.H. (1975). The characterization of humic substances in seawater, Ph.D. Thesis, Massachusetts Institute of Technology and Woods Hole Oceanographic Institution, Woods Hole, 188 pp.

Stuermer, D.H., K.E. Peters and I.R. Kaplan (1978). Source indicators of humic substances and proto-kerogen: Stable isotope ratios, elemental compositions and electron spin resonance spectra. Geochim. Cosmochim. Acta 42, 989-997.

Stuermer, D.H. and B.R.T. Simoneit (1978). Varying sources for the lipids and humic substances at Site 391, Blake-Bahama Basin, DSDP Leg 44. In: Initial Reports of the Deep Sea Drilling Project, Vol. 44, W.E. Benson, R.E. Sheridan et al., U.S. Government Printing Office, Washington, D.C., pp. 587-591.

Tissot, B., Y. Califet-Debyser, G. Deroo and J.L. Oudin (1971). Origin and evolution of hydrocarbons in Early Toarcian shales. Amer. Assoc. Petrol. Geol. Bull. 55, 2177-2193.

Tissot, B., B. Durand, J. Espitalié and A. Combaz (1974). Influence of nature and diagenesis of organic matter information of petroleum. Amer. Assoc. Petrol. Geol. Bull. 58, 499-506.

Tissot, B. and J. Espitalié (1975). L'évolution thermique de la matière organique des sédiments: application d'une simulation mathématique. Inst. Fr. Pétr. Rev. 30, 743-777.

Tissot, B.P. and D.H. Welte (1978). Petroleum Formation and Occurrence, A New Approach to Oil and Gas Exploration, Springer-Verlag, Berlin, 538 pp.

van de Meent, D., S.C. Brown, R.P. Philp and B.R.T. Simoneit (1980). Pyrolysis-high resolution gas chromatography and pyrolysis gas chromatography-mass spectrometry of kerogens and kerogen precursors. Geochim. Cosmochim. Acta 44, 999-1013.

van Krevelen, D.W. (1961). Coal, Typology, Chemistry, Physics and Constitution, Elsevier Publishing Company, Amsterdam, 514 pp.

Vassoevich, N.B., A.M. Akramkhodzhaev and A.A. Geodekyan (1974). Principal zone of oil formation. In: Advances in Organic Geochemistry 1973, B. Tissot and F. Bienner, eds., Editions Technip, Paris, pp. 309-314.

Watts, C.D. and J.R. Maxwell (1977). Carotenoid diagenesis in a marine sediment. Geochim. Cosmochim. Acta 41, 493-497.

Welte, D.H., H.W. Hagemann, A. Hollerbach, D.Leythaeuser and W. Stahl (1975). Correlation between petroleum and source rock. Proc. 9th World Petrol. Congr. 2, 179-191.

Whelan, J.K. and J.M. Hunt (1982). C_1-C_8 in Leg 64 sediments, Gulf of California. In: Initial Reports of the Deep Sea Drilling Project, Vol. 64, J.R. Curray, D.G. Moore et al., U.S. Govt. Printing Office, Washington, D.C., pp. 763-779.

Williams, D.L., K. Becker, L.A. Lawver and R.P. von Herzen (1979). Heat flow at the spreading centers of the Guaymas Basin, Gulf of California, J. Geophys. Res. 84, 6757-6769.

Windsor, J.G., Jr. and R.A. Hites (1979). Polycyclic aromatic hydrocarbons in Gulf of Maine sediments and Nova Scotia soils. Geochim. Cosmochim. Acta 43, 27-33.

Appendix I -
Chemical Structures Cited

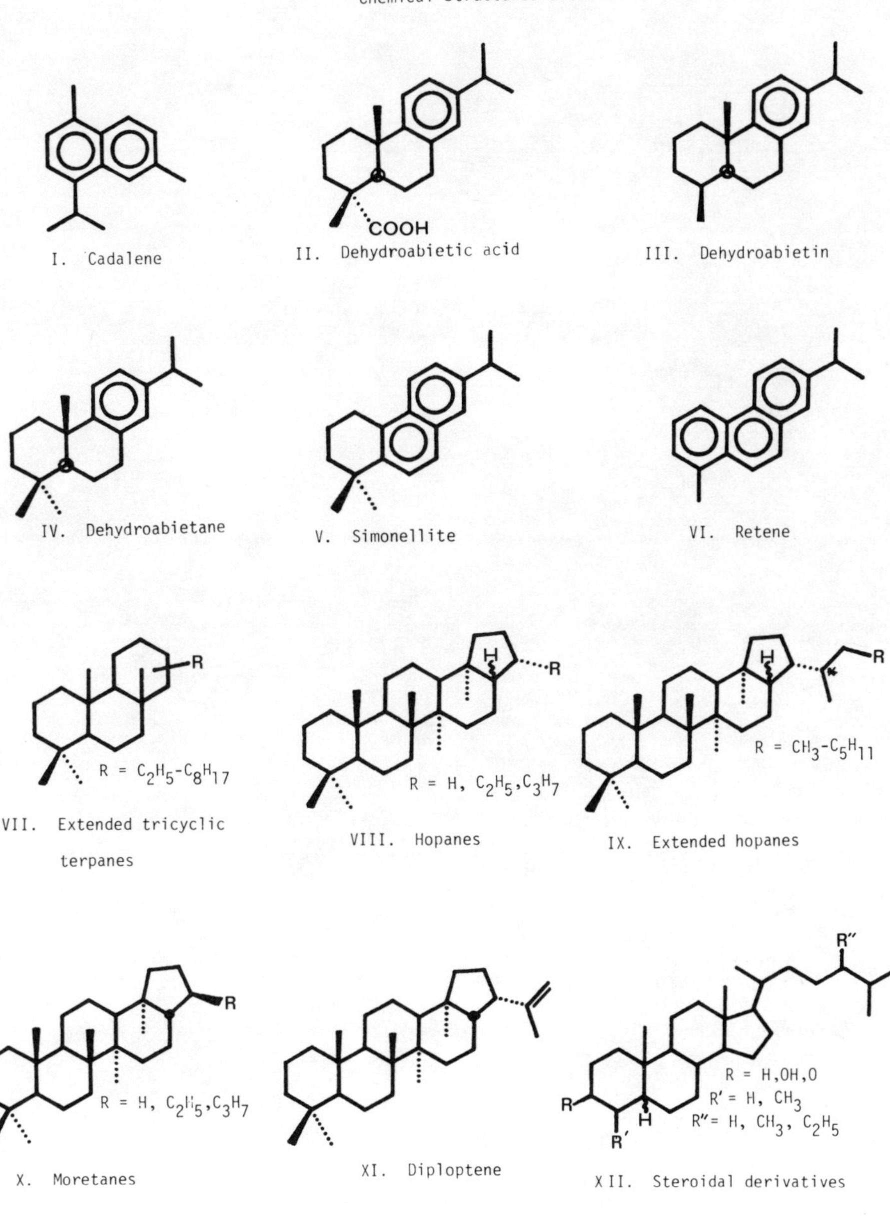

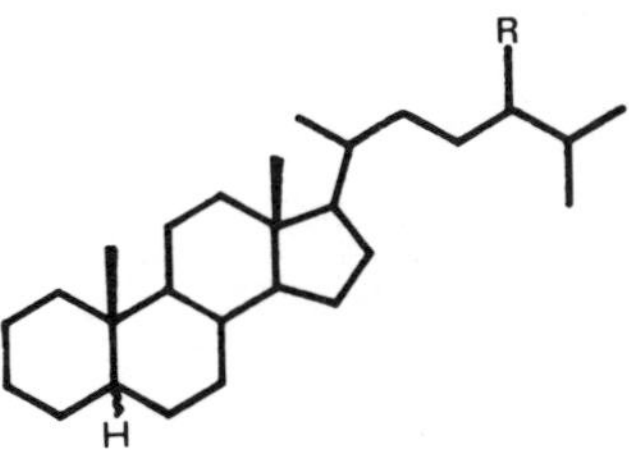

FOSSIL HYDROTHERMAL SYSTEMS

THE MAGMATIC-METEORIC TRANSITION

R. E. Beane

AMAX Exploration, Inc.

ABSTRACT

Circulation of fluids derived from both magmatic and meteoric sources in the shallow intrusive environment is documented by several empirical factors in porphyry copper deposits of western North America. Early in the cooling history of the shallow plutons, fluids circulating in rocks were endogenous. Following consolidation and pervasive fracturing of the central intrusion, the igneous rock was cooled by broad-scale convective circulation of meteoric waters. The transition from magmatic- to meteoric-derived fluid flow in the porphyritic intrusion is monitored by a shift in the hydrogen isotope ratios of silicate minerals, by a sharp decrease in fluid salinity as recorded by fluid inclusions, and by chloritization of earlier-formed biotite.

INTRODUCTION

Burnham and Ohmoto (1980) have defined a transitional stage during the cooling history of a magma that temporally bridges completely-magmatic and completely-hydrothermal events. During this transitional stage, aqueous fluids derived from the crystallizing magma interact with its crystalline and remaining-melt fractions. Another transition during pluton cooling involves the shift from a magmatic to a meteoric origin for fluids that circulate through and adjacent to the shallow intrusions. Several empirical parameters record changes in the behavior, origin, and characteristics of hydrothermal fluids related to the magmatic-meteoric transition. The information presented here is drawn chiefly from studies of porphyry copper deposits, a class of deposits for which abundant data is available to address the problem of magmatic and meteoric fluids in hydrothermal systems, and the transition between them.

In the western United States, porphyry copper systems are usually associated with granodioritic plutons. These subvolcanic intrusions were rapidly emplaced to depths as shallow as 1 to 2 km in an extensional tectonic regime, and some show evidence of volcanic venting (Figure 1). Several workers have

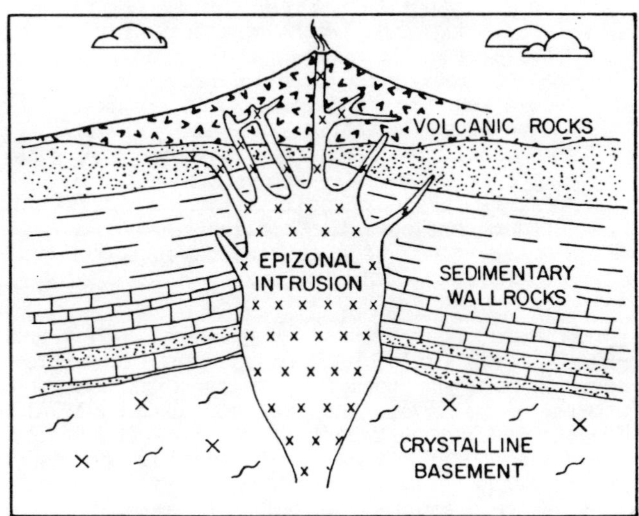

Figure 1. Vertical cross-section showing generalized geologic relations in the shallow intrusive environment of southwestern North America containing porphyry copper mineralization. The crystalline basement consists of Precambrian granite and metamorphic rocks. Sedimentary wallrocks comprise Paleozoic and/or Mesozoic clastic and carbonate rocks. Volcanic rocks are often cogenetic with the intrusion.

suggested two stages of fluid flow related to the development of porphyry copper deposits in this environment as shown schematically in Figure 2 (cf., Gustafson and Hunt, 1975; Gustafson, 1978). The first stage is characterized by circulation of magmatic fluids through a parent crystallizing magma or product igneous rock, concomitant with development of hypogene mineralization. The second stage, is termed "meteoric collapse", corresponds to the encroachment of meteoric fluids that produces late-stage argillic alteration. This convective circulation of meteoric waters through cooling plutons has been numerically predicted by several workers (eg., Norton and Knight, 1977; Cathles, 1977), but actual documentation of the presence of such fluids in fractured igneous rocks required application of stable isotope studies.

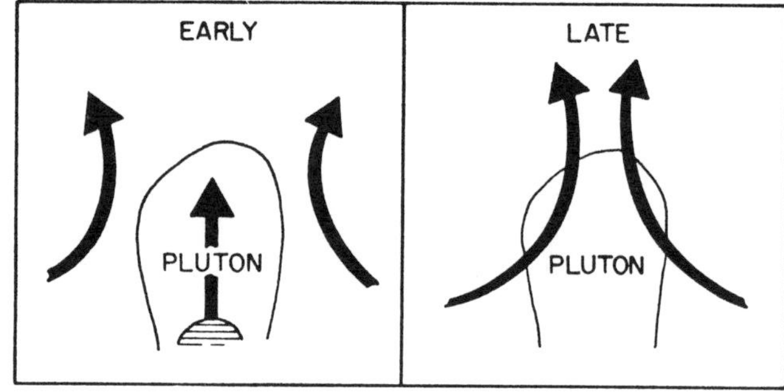

Figure 2. Vertical cross-section showing stages in the evolution of the
hydrothermal systems forming porphyry
copper deposits. Early fluids (left)
are endogenous, with magmatic fluids
traversing already-crystallized igneous rock and meteoric waters circulating in adjacent wall rocks.
Late-stage "meteoric collapse" (right)
consists of pervasive convection of
meteoric waters. Arrows show fluid
pathlines. The solid line in each
figure encloses the cooling pluton; a
small volume of melt remains at depth
during early circulation.

Figure 3a shows published D/H and
$^{18}O/^{16}O$ ratios of biotites and sericites from
several porphyry copper deposits (Sheppard and
others, 1969, 1971; Sheppard and Taylor, 1974).
The stable isotope ratios of biotites from early
alteration in numerous North American porphyry
copper deposits are essentially the same regard-
less of geographic location. For example, compare values for biotites from Santa Rita, New
Mexico, Ely, Nevada, Bingham, Utah, and Butte,

Montana, all of which fall within the range of
igneous biotites (Taylor, 1974). In contrast,
alteration sericites from Santa Rita, Ely, and
Butte show systematic variation in hydrogen
isotopes with geographic latitude. These variations are interpreted to reflect the formation of
alteration sericites from meteoric waters.

The influx and gradual dominance of meteoric
fluids flowing through igneous rocks during
progressive hydrothermal alteration is recorded
by systematic variations in the hydrogen and
oxygen isotope ratios of selected minerals.
Figure 3b shows the progressive variations in
these ratios in biotite and sericite at the
Butte, Montana deposit (Sheppard and Taylor,
1974). Among these data, igneous biotite
occupies the apex, biotite from early dark
micaceous alteration composes the variable
hydrogen leg, and later sericitic (and argillic)
alteration forms the variable-oxygen base of a
crude inverted "L"-shaped pattern. This pattern
resembles the characteristic variation in hydrogen and oxygen isotope ratios of biotite and
feldspar in batholitic igneous rock undergoing
increasing interaction with meteoric waters
(Taylor, 1977). The inverted "L" pattern
reflects the fact that igneous rocks are large
reservoirs of oxygen, but contain only small
amounts of hydrogen. The atomic oxygen to hydrogen ratios of igneous rocks are typically on the
order of 500 or 1000:1, whereas for water the
ratio is 8:1. As a result of this difference,
when meteoric waters flow through and exchange
with the igneous rock, it takes only a small
amount of water to effect the rock's hydrogen
isotope content. Large amounts of meteoric water
are required, on the other hand, before the oxygen isotopes of the rock are effected because the
rocks contain so much original igneous oxygen.

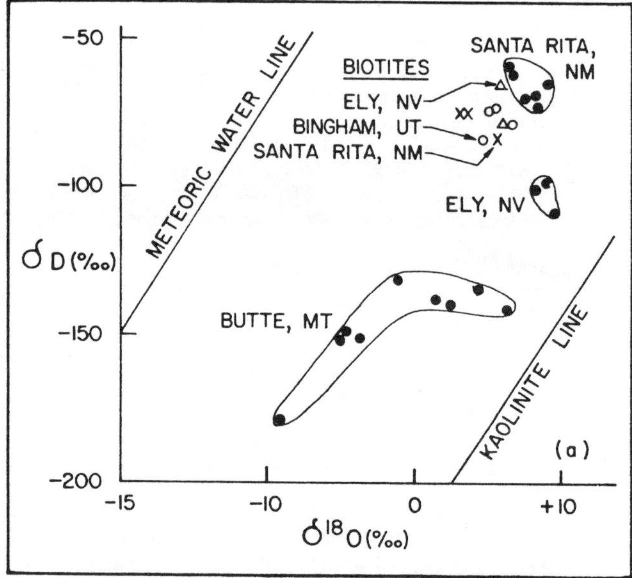

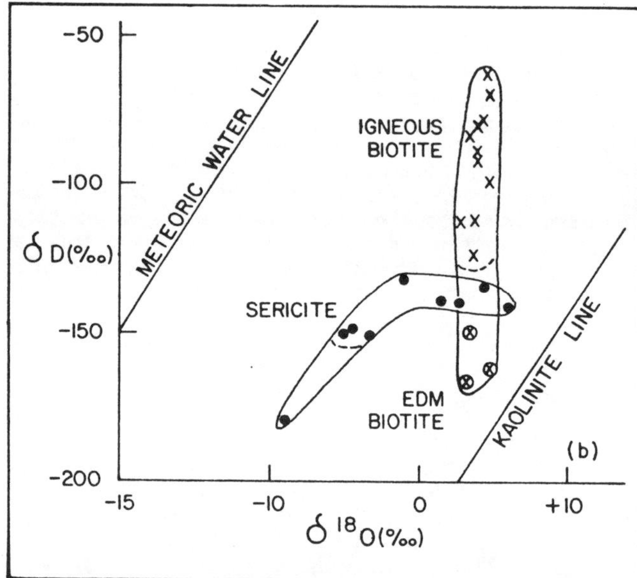

Figure 3. Hydrogen and oxygen isotope data for minerals from porphyry copper deposits (in parts
per thousand relative to standard mean ocean water). (a) Biotites from Bingham, Utah (circles);
Ely, Nevada (triangles); and Santa Rita, New Mexico (X's). Dots are sericites from Butte, Montana Main Stage alteration; Ely; and Santa Rita (Sheppard and others, 1971; Sheppard and Taylor,
1974). (b) Igneous biotite (X's), pre-Main Stage biotite from early dark micaceous (EDM) alteration (circles), and Main Stage sericite (dots) from Butte (Sheppard and Taylor, 1974).

MAGMATIC FLUIDS

Holland (1972) has pointed out that although several percent of dissolved H_2O are necessary to produce the amounts of hydrous phases (chiefly biotite and hornblende) seen in crystallized granodioritic rocks, chemical analyses typically reveal less than one weight percent H_2O in these igneous rocks (Nockolds, 1954). The conclusion is that the remaining magmatic H_2O originally present in the melts was exsolved during crystallization carrying with it the potential for forming hydrothermal ore deposits. Moreover, other workers have concluded that porphyry copper deposits were formed from magmatic fluids simply based on the spatial association of these mineralized systems with plutonic rocks.

The solubility of H_2O in granodioritic melts is directly related to pressure, but is essentially temperature-independent (Burnham, 1979). Dissolved H_2O can be released from a silicate melt in two ways. The first is the decrease from lithostatic to hydrostatic pressure on the magma as wall rocks are fractured during conductive heat transfer. The second is through early crystallization of anhydrous feldspar which causes H_2O to be concentrated in the remaining silicate liquid. When the H_2O content of the magma exceeds the saturation level for the prevailing pressure, an immiscible H_2O-rich fluid exsolves from the silicate melt. One result of this so-called "degassing" is a rapid crystallization of the silicate liquid surrounding the early feldspar phenocrysts, giving the igneous rocks a porphyritic texture. Also, because of the large volume increase resulting from H_2O exsolution at low confining pressures, there is a drastic release of mechanical energy which further fractures wall rocks and previously crystallized igneous rock (Burnham and Davis, 1971; Burnham and Ohmoto, 1980). The exsolution event will only cause quasi-plastic failure in igneous rocks at or slightly below the crystallization temperature, whereas magmas at temperatures well above the solidus are essentially unaffected. Diffusion of dissolved H_2O in magmas is slow, but exsolved bubbles of H_2O migrate rapidly through silicate liquid toward the top of the magma chamber, where development of a large low density bubble may lead to formation of collapse breccia pipes (Norton and Cathles, 1973). In crystalline portions of the igneous rock, exsolved fluids migrate through fracture networks generated by the strain-release associated with their exsolution.

The exsolved aqueous fluid is initially in chemical equilibrium with the crystallizing magma from which it was generated. If the temperature of the exsolved fluid does not decrease drastically below that of the parent magma, the fluid will precipitate a mineral assemblage comparable to the enclosing igneous rock, typically quartz, feldspar, biotite, and magnetite. This mineral assemblage will be deposited in fractures formed in response to thermal effects in previously-consolidated igneous rock and in immediately adjacent wall rocks.

The chemical composition and temperatures of hydrothermal fluids can be approximated using inclusions trapped in structural defects of minerals. Magmatic fluid inclusions can be identified by their mode of occurrence and by experimentally-estimated temperatures of their formation. When these fluid inclusions are observed with high magnification at room temperature, a number of included solid phases are seen. These daughter minerals indicate that the magmatic fluids are typically highly saline, containing up to approximately 75 weight percent dissolved sodium and potassium chloride; thus, they are essentially hydrous saline melts (Roedder, 1971). Identification of such daughter minerals using the electron microprobe indicates that such magmatic fluids transported significant amounts of sodium, potassium, calcium, iron, copper, chloride, and sulfur. Notably low in the magmatic fluid is magnesium, which appears to be fixed in mafic minerals of igneous or hydrothermal origin (Anthony and others, 1983).

Experimental studies have shown that chloride in silicate melts is strongly partitioned into a coexisting H_2O-rich phase (Burnham, 1979; Kilinc and Burnham, 1972). However, the available partition coefficients and the chloride contents of granitic rocks (cf., Holland, 1972) both indicate that exsolved H_2O-rich magmatic fluids should have no more than 5 to 10 weight percent dissolved NaCl. Consequently, fluid inclusions that contain seven to ten times this amount of dissolved NaCl suggest that the hypersaline magmatic fluids have undergone some modification following original exsolution from the crystallizing magma. The most straight-forward mechanism for concentrating NaCl in fluids is by selective partitioning between aqueous liquid and vapor. If an initially low salinity liquid undergoes boiling, the salt content of the remaining liquid may be significantly elevated. Conversely, a low salinity vapor may undergo condensation to yield a small fraction of high salinity liquid. Either of these concentration processes require coexistance of liquid and vapor fractions, with the mass and volume fractions of the vapor portion being far greater than that of the liquid portion. Although the phenomena of boiling and condensation in porphyry copper deposits, as evidenced by fluid inclusions, have been discussed by Cunningham (1978), Henley and McNabb (1978), Nash (1976) and a number of other workers, much additional work remains to be done in order to completely understand their role in these systems.

THE METEORIC INCURSION

The steep temperature gradient imposed upon wall rocks by magmatic emplacement causes a proportionate lateral gradient in fluid density which in turn drives relatively dilute meteoric waters into convective motion through permeable wall rocks. High-temperature, lower-density fluids nearest the contact are displaced upward by cooler, denser fluids further away. Initial permeabilities of wall rocks vary significantly among different lithologies. These initial values are then either augmented or reduced through a variety of thermal and mechanical processes during pluton cooling (Knapp and Norton, 1981). Magma emplacement deforms and fractures the enclosing host rocks by mechanical effects. Additional fracturing of wall rocks is then caused by thermal expansion of pore fluids owing to heating by the adjacent intrusion (Knapp and Knight, 1977). Also, conductive heat transfer across the magma-wall rock contact causes additional fracturing in the rigid wall rock. Early fracturing of this latter type in host rocks is maximized in the region above the cooling pluton.

The pluton itself is initially molten and deforms plastically. Following emplacement, heat loss and magma crystallization proceed most rapidly in the roof of the cooling magma chamber. Volume decrease associated with the rapid cooling and crystallization causes extensive fracturing of this roof zone (Knapp and Norton, 1981). When magma crystallization is complete, the resulting igneous rock becomes susceptible to fracturing caused by thermal effects and devolatilization of uncrystallized magma. Permeabilities of plutons caused by cooling vary in location, orientation, and intensity with time. Detailed studies of mineralized plutons by Titley and coworkers (Titley and others, 1978; Haynes and Titley, 1980) show that although early fracturing occurs throughout the igneous rock mass following consolidation, there is a methodical retreat of fracturing toward the pluton contact with advancing time.

Fracturing incorporates the pluton into the permeable system through which magmatic and meteoric fluids circulate. When the crystallized pluton attains the large fracture-permeability of its enclosing wall rocks, broad-scale convective circulation cools the igneous rock with meteoric waters. The classic pattern of fluid convection in rocks of homogeneous permeability is depicted by Figure 4a. The broken arrow on Figure 4a shows early circulation in permeable wall rocks while the intrusion is still molten; the solid arrow corresponds to later fluid circulation when the crystalline stock is also permeable. Convective circulation around a cooling pluton leads to the distribution of isotherms shown schematically by Figure 4b (Norton and Knight, 1977; Cathles, 1977). The extended isotherm over the intrusion represents a broad zone of low vertical temperature gradient. The dotted extension of the low temperature contour shows the effects produced by a plumbing system which is confined above by low permeability.

In nature, flowlines vary depending on inhomogeneities in rock permeability and

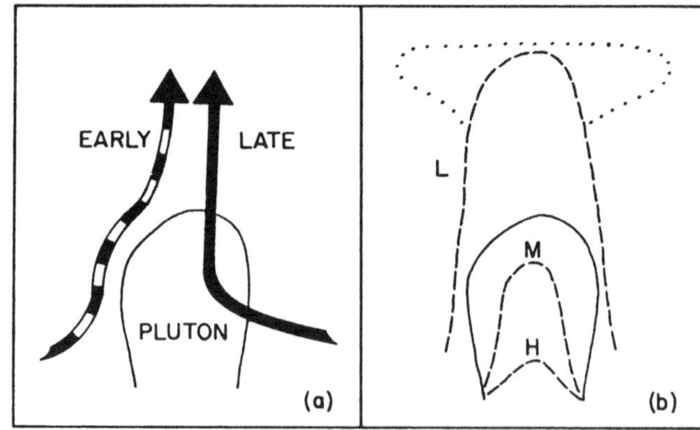

Figure 4. Positioning of fluid flow-lines (a) and isotherms (b) about a cooling pluton in uniformly permeable wall rocks. Flowlines, shown by arrows, are for early circulation (broken) when the central pluton is unfractured (impermeable), and later circulation (solid) when the stock becomes permeable through fracturing. Isotherms show relative positions for high (H), intermediate (M) and low (L) values. The dotted extension of the low isotherm depicts possible spreading as a result of sealing by an impermeable cap. The solid line in each figure encloses the cooling pluton.

differences in pluton geometry. Extensive fluid circulation during formation of porphyry copper deposits is evidenced by widespread isotopic exchange between minerals and hydrothermal fluids, and by wholesale chemical gains and losses in rocks (Norton, 1981). The changing position and abundance of flow permeability with time, coupled with the evolving temperature and composition of circulating fluids, is largely responsible for the zoning and paragenetic relations observed among alteration assemblages in porphyry copper deposits (Beane and Titley, 1982).

THE MAGMATIC-METEORIC TRANSITION

Fluid inclusions from porphyry copper deposits record the magmatic-meteoric transition through systematic temporal variations in their temperature and salinity. A detailed record of this transition at the Santa Rita, New Mexico deposit has been compiled by Reynolds (1980). The earliest fluids to circulate in the Santa Rita stock had high salinities, up to 75 weight percent NaCl, and temperatures in excess of 750°C (Type IIIA, Figure 5a). The next generation of hydrothermal fluids was significantly cooler, having temperatures in the range 250-525°C (Type IIIB, Figure 5a); however, their salinities remained high. The temperatures and salinities of the second-stage fluids both decreased with progressing time. The 250° to 525°C homogenization temperatures of

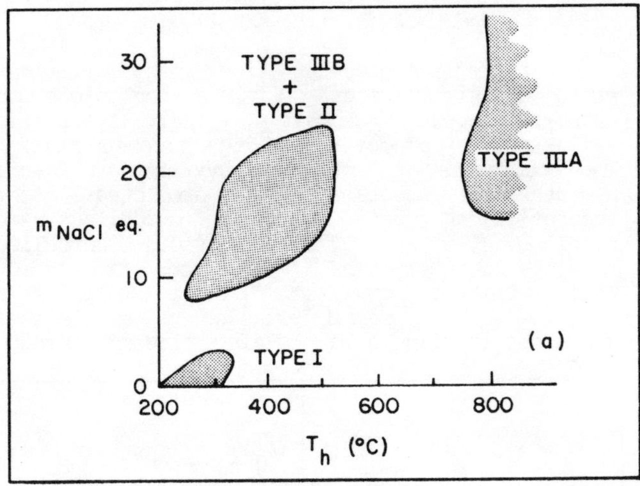

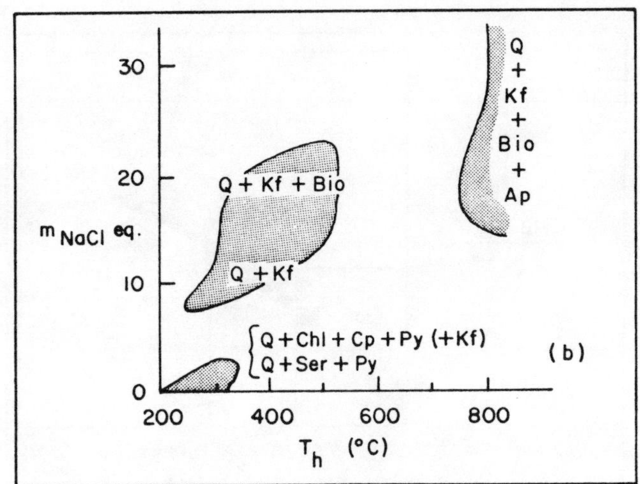

Figure 5. Thermal, salinity and alteration-mineral characteristics of different generations of hydrothermal fluids from the Santa Rita, NM porphyry copper deposit. (a) Measured homogenization temperatures (°C) and salinities (equivalent NaCl molality) of fluid inclusions. Type I = dilute, Type II = vapor-rich and Type III = hypersaline. (b) Mineral assemblages associated with each of the fluid types: Q = quartz, Kf = orthoclase, Bio = biotite, Ap = apatite, Chl = chlorite, Cp = chalcopyrite, Py = pyrite, Ser = sericite. From Reynolds and Beane (1983).

these fluid inclusions are characteristic of both liquid-rich and vapor-rich (Type II) varieties, indicating that the parent hydrothermal solutions were probably boiling. The third stage of hydrothermal fluids (Type I, Figure 5a) are characterized by significantly lower salinities, although their fluid inclusion homogenization temperatures are essentially the same as those of the immediately-preceeding Type IIIB inclusions.

Walker (1979) has documented a comparable change in fluid properties in the breccia pipes at Copper Creek, Arizona, and various workers have shown all three fluid types to have been present in other porphyry copper systems (Nash, 1976, Eastoe, 1978; Wilson and others, 1980). The majority of fluid inclusion studies of porphyry copper deposits, however, have not recognized the high-temperature, high-salinity type, and only report the two later intermediate-temperature varieties.

Mineral assemblages formed by the various fluid types at Santa Rita are shown in Figure 5b. The mineral assemblage in the earliest veins, namely quartz, K-feldspar, and biotite, is consistent with fluids exsolved from a crystallizing granodioritic pluton. Sheppard and others (1971) have shown that the hydrothermal biotite at Santa Rita is isotopically comparable to igneous biotite as discussed above. Also, the oxygen isotope ratio of quartz accompanying the biotite is comparable to that of quartz in igneous rocks (Reynolds and Beane, 1983). The temperatures of the earliest generation of fluids are clearly magmatic. Thus, isotopic, fluid inclusion and mineralogical evidence all indicate that these early Type IIIA fluids are of magmatic origin. Oxygen isotope ratios of quartz and the vein-filling mineral assemblage in the second high-salinity stage (Type IIIB, Figure 5a) are similar to the earlier Type IIIA fluids. These ratios suggest that the Type IIIB fluids were also originally exsolved from the pluton, but had cooled since generation at magmatic temperatures. The late Type I dilute fluids produced two alteration assemblages. The earliest, K-feldspar and chlorite with sulfides, formed the economic mineralization at the deposit; chlorite occurs as a deposited phase and as an alteration of earlier-formed biotite. This was followed by development of quartz-sericite-pyrite alteration. Stable isotope data indicate that this late sericite assemblage was formed from meteoric waters (Sheppard and others; 1969, 1971). The oxygen isotope ratio of quartz from the preceeding chlorite-sulfide stage is comparable to that of the later quartz-sericite-pyrite assemblage (Reynolds and Beane, 1983). The fluid inclusion homogenization temperatures and salinities of these two stages are also comparable, providing further evidence that the fluids producing the chlorite-sulfide veins were derived from the same source as the fluids producing late quartz-sericite alteration, namely of meteoric origin. The K-feldspar + chlorite assemblage carrying sulfide mineralization is typically referred to as retrograde modification of early K-feldspar + biotite alteration, and the mineralization is credited to this supposed predecessor. Textural features, however, indicate that chalcopyrite deposition accompanyied formation of hydrothermal chlorite.

Three mineralogically discrete alteration assemblages form at essentially the same temperature at Santa Rita as well as at some other porphyry copper deposits in southwestern North America (Beane and Titley, 1981). Precipitation of the earliest assemblage,

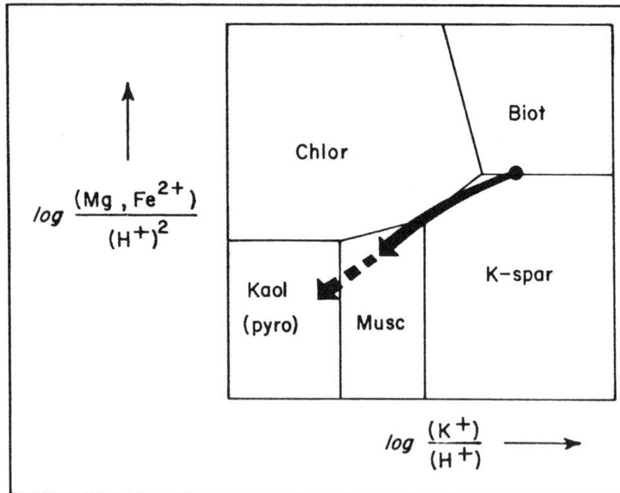

Figure 6. Schematic stability relations among silicate minerals in terms of ion-activity ratios in a co-existing aqueous phase. Phase relations are drawn based on observed mineral associations in porphyry copper deposits with formation temperatures near 350°C as determined using fluid inclusions. Alteration assemblages evolve as shown by the arrow. Mineral abbreviations: Biot(ite), Chlor(ite), Kaol(inite), Pyro(phyllite), Musc(ovite), and K-(feld)spar. From Beane and Titley (1981).

K-feldspar and biotite, is followed by K-feldspar plus chlorite deposition, and finally quartz and sericite are formed. The compositional evolution of a hydrothermal fluid required to precipitate these three assemblages at constant temperature is illustrated by the silicate-mineral stability diagram shown in Figure 6. This diagram depicts isothermal stability relations among K-feldspar, biotite, chlorite, muscovite (sericite), and kaolinite (or pyrophyllite) as a function of the ion activity ratios $(Mg^{2+})/(H^+)^2$ and $(K^+)/(H^+)$ in a coexisting aqueous fluid at quartz saturation. Although temperature is unspecified, the phase relations are consistent with vein assemblages observed in porphyry copper deposits which formed at temperatures in the range 300°-400°C (cf., Reynolds and Beane, 1983; Preece and Beane, 1982). The earliest K-feldspar + biotite assemblage was formed from high-salinity fluids, presumably of magmatic origin. This assemblage defines the starting point for successive reactions. When dilute fluids enter the system, as recorded by fluid inclusion salinities, the early minerals are progressively altered to K-feldspar plus chlorite and then to sericite as shown by the arrow in Figure 6. These later assemblages represent progressive reaction products as the meteoric fluids interact with, and eventually dominate, the original magmatic assemblage (Helgeson, 1970).

Detailed studies of time space relations among different kinds of fluid inclusions in porphyry copper deposits reveal systematic variations in properties of the trapped fluids. The occurrence of such variations in the fossil hydrothermal systems is predicted from theoretical concepts of fluid sourcelines and pathlines (Norton, 1978). A fluid sourceline connects the sequence of locations from which fluids flowing through some point of interest were derived. Several examples of sourcelines are shown in Figure 7a. Fluids flowing through

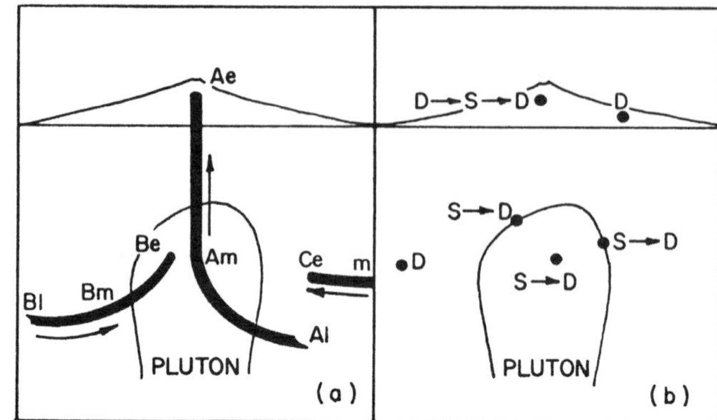

Figure 7. Vertical cross-section showing changes in sources and salinities of hydrothermal fluids with time in porphyry copper deposits. (a) Sourcelines A, B and C connect the sequence of sources of fluids which flow through these points at early (e), intermediate (m) and late (l) times. Arrows show fluid-flow pathlines. Modified from Norton (1978). (b) Order of occurrence of saline (S) and dilute (D) fluids with time, as indicated by the arrows, at different positions in the composite porphyry copper environment. From Beane and Titley (1981).

point B in a convecting system are drawn first from the area immediately surrounding that point (Be). As time progresses, the source location for fluids circulating through point B moves progressively further away from point Be outward into wallrocks (first Bm, then Bl), a directional trend nearly opposite to the fluid-flow pathlines shown by Figure 7a and the fluid flowlines discussed above. Thus the earliest fluids flowing through B are magmatic, while later-infiltrating fluids are meteoric. In contrast, early fluids that flow through rocks overlying the pluton are locally derived, hence meteoric in origin (Ae). As time progresses, high salinity magmatic fluids ascend to this point (Am). Still later, meteoric fluids that flowed from a wall rock source (Al) through the pluton, eventually circulate through this roof zone. Adjacent to the pluton, fluids circulate inward and upward, so magmatic fluids never reach point C.

Figure 7b shows time-space variations in fluid inclusion salinities that characterize several porphyry copper deposits (Beane and Titley, 1981). Assuming that magmatic fluids can be correlated with high-salinity fluid inclusions, and that meteoric fluids are indicated by the low-salinity variety, then the time-sequence of high- and low-salinity fluids at different positions in porphyry copper systems (Figure 7b) is consistent with the predicted evolution of fluid sourcelines in a convecting system (Figure 7a).

REVIEW

Thermal and compositional variations that characterize the transition from magmatic- to meteoric-dominated fluid flow during the cooling history of shallow plutons related to porphyry copper mineralization are traced schematically in Figure 8. Immediately following magma emplace-

TIME	EARLY MAGMATIC	LATE METEORIC	
MINERALS	Biotite	Chlorite	Sericite
TEMPERATURE	>750°		250°
SALINITY	>30m		2–3m
δD_{rk} (Biot.)			
$\delta^{18}O_{rk}$ (Feld.)			

Figure 8. Variations in hydrothermal parameters during evolution of porphyry copper deposits. The double vertical line denotes the magmatic-meteoric transition. Fluid temperatures (^{o}C) and salinities (equivalent NaCl molality) are determined from fluid inclusion studies. Decreases in hydrogen and oxygen isotope ratios are provisional based on concepts outlined by Taylor (1977) and measurements at Butte, Montana (Sheppard and Taylor, 1974).

ment, adjacent and overlying wall rocks are fractured while conductive heat transfer away from the still-molten magma causes convective flow of meteoric fluids through permeable wall rocks. This heat loss results in magma crystallization initially concentrated in the hood of the magma chamber. Once crystallized, the still-cooling igneous rock is itself fractured by thermal contraction.

Crystallization of the magma concentrates dissolved H_2O in the remaining melt until it reaches the saturation level, at which time an immiscible H_2O-rich phase, "magmatic water", is violently exsolved causing fracturing of already-crystallized igneous rock and wall rocks. The exsolved magmatic fluid concentrates chloride from the magma, and with it base metals which form stable complexes. Trapped inclusions of these fluids, recognized by near-magmatic homogenization temperatures, have high salinities and can be called hydrous saline melts. Although the manner in which these high salinities are obtained is uncertain, they do exist. The logical explanation is salt partitioning, either through boiling or condensation, but fluid inclusion evidence for these phenomona is sometimes lacking. The magmatic fluids circulating through newly crystallized igneous rock produce a "granitic" vein assemblage of quartz, K-feldspar, biotite, and magnetite. The vein-forming minerals exhibit an oxygen isotope signature indicative of their magmatic heritage.

When the pluton has completely crystallized, generation of magmatic fluids ceases. Fracturing from violent exsolution is finished, but the effects of continued thermal contraction persist. Lateral thermal and fluid-density gradients through the fractured pluton-wall rock system cause widespread convective fluid flow centered on the intrusion. The convective circulation introduces lower-salinity meteoric fluids into the cooling pluton. The earliest indicators of the meteoric incursion, as seen in Figure 8, are: 1) a significant decrease in fluid salinity as recorded by fluid inclusions, 2) a shift in the hydrogen isotope composition of potassic-altered rocks, and 3) chloritization of earlier-formed biotite. This incursion is not marked by drastic temperature decrease, because fluid inclusion data indicate that magmatic fluids have cooled to temperatures on the order of 300°–$400^{o}C$ before the meteoric influx. Neither do oxygen isotopes provide a contemporaneous indicator, because relatively large amounts of meteoric fluid are required to exchange with igneous rocks before their magmatic isotopic signature can be changed.

ACKNOWLEDGEMENTS

I am indebted to E. Versluis and D. Seiler for assistance in preparation of the manuscript, and to J. Johnson for helpful suggestions in review of the same.

REFERENCES

Anthony, E. Y., Reynolds, T. J. and Beane, R. E., 1983, Recognition of daughter minerals in fluid inclusions using scanning electron microscopy and energy dispersive analysis: ms.

Beane, R. E. and Titley, S. R., 1982, Porphyry copper deposits: Part II. Hydrothermal alteration and mineralization: Econ. Geol., 75th Anniv. Vol., p. 235-269.

Burnham, C. W., 1979, Magmas and hydrothermal fluids: in Barnes, H. L., ed., Geochemistry of hydrothermal ore deposits, 2nd ed.: New York, Wiley-Interscience, p. 71-136.

Beane

Burnham, C. W. and Davis, N. F., 1971, The role of H_2O in silicate melts: II. Thermodynamic and phase relations in the system $NaAlSi_3O_8-H_2O$ to 10 kilobars, $700°$ to $1100°C$: Am. Jour. Sci., v. 270, p. 54-79.

Burnham, C. W. and Ohmoto, H., 1980, Late-stage processes of felsic magmatism: Min. Geol., Spec. Iss. No. 8, p. 1-11.

Cathles, L. M., 1977, An analysis of the cooling of intrusives by ground-water convection which includes boiling: Econ. Geol., v. 72, p. 804-826.

Cunningham, C. G., 1978, Pressure gradients and boiling as mechanisms for localizing ore in porphyry systems: Jour. Research U. S. Geol. Survey, v. 6, p. 745-754.

Eastoe, C. J., 1978, A fluid inclusion study of the Panguna porphyry copper deposit, Bougainville, Papua New Guinea: Econ. Geol., v. 73, p. 721-748.

Gustafson, L. B., 1978, Some major factors of porphyry copper genesis: Econ. Geol., v. 73, p. 600-607.

Gustafson, L. B. and Hunt, J. P., 1975, The porphyry copper deposit at El Salvador, Chile: Econ. Geol., v. 70, p. 857-912.

Haynes, F. M. and Titley, S. R., 1980, The evolution of fracture-related permeability within the Ruby Star granodiorite, Sierrita porphyry copper deposit, Pima County, Arizona: v. 75, p. 673-683.

Helgeson, H. C., 1970, A chemical and thermodynamic model of ore deposition in hydrothermal systems: Mineral. Soc. America, Spec. Pap. 3, p. 155-186.

Henley, R. W. and McNabb, A., 1978, Magmatic vapor plumes and ground-water interaction in porphyry copper emplacement: Econ Geol., v. 73, p. 1-20.

Holland, H. D., 1972, Granites, solutions, and base metal deposits: Econ. Geol., v. 67 p. 281-301.

Kilinc, I. A. and Burnham, C. W., 1972, Partitioning of chloride between a silicate melt and coexisting aqueous phase from 2 to 8 kilobars: Econ. Geol., v. 67, p. 231-235.

Knapp, R. B. and Knight, J. E., 1977, Differential thermal expansion of pore fluids: Fracture propogation and microearthquake prediction in hot pluton environments: Jour. Geophys. Research, v. 82, p. 2515-2522.

Knapp, R. B. and Norton, D., 1981, Preliminary numerical analysis of processes related to magma crystallization and stress evolution in cooling pluton environments: Am. Jour. Sci., v. 281, p. 35-68.

Nash, J. T., 1976, Fluid inclusion petrology - Data from porphyry copper deposits and applications to exploration: U. S. Geol. Survey Prof. Pap. 907-D, 16 p.

Nockolds, S. R., 1954, Average chemical composition of some igneous rocks: Geol. Soc. America Bull., v. 65, p. 1007-1032.

Norton, D., 1978, Sourcelines, sourceregions, and pathlines for fluids in hydrothermal systems related to cooling plutons: Econ. Geol., v. 73, p. 21-28.

Norton, D., 1981, Transport processes related to copper-bearing porphyritic plutons - Fluid and heat transport in pluton environments typical of the southeastern Arizona porphyry copper province: in, Titley, S. R., ed., Advances in geology of the porphyry copper deposits of southwestern North America: Tucson, Univ. Arizona Press, p. 59-72.

Norton, D. and Cathles, L. M., 1973, Breccia pipes - products of exsolved vapor from magmas: Econ. Geol., v. 68, p. 540-546.

Norton, D. and Knight, J., 1977, Transport phenomona in hydrothermal systems: Cooling plutons: Am. Jour. Sci., v. 277, p. 937-981.

Preece, R. K., III and Beane, R. E., 1982, Contrasting evolutions of hydrothermal alteration in quartz monzonite and quartz diorite wall rocks at the Sierrita porphyry copper deposit, Arizona: Econ. Geol., v. 77, p. 1621-1641.

Reynolds, T. J., 1980, Variations in hydrothermal fluid characteristics through time at the Santa Rita porphyry copper deposit, New Mexico: Unpub. M.S. Thesis, Univ. Arizona, 52 p.

Reynolds, T. J. and Beane, R. E., 1983, Variations in hydrothermal fluid characteristics through time at the Santa Rita, New Mexico porphyry copper deposit: ms.

Roedder, E. R., 1971, Fluid inclusion studies of the porphyry-type ore deposits at Bingham, Utah, Butte, Montana, and Climax, Colorado: Econ. Geol., v. 66, p. 98-120.

Sheppard, S. M. F., Nielson, R. L. and Taylor, H. P., Jr., 1969, Oxygen and hydrogen isotope ratios of clay minerals from porphyry copper deposits: Econ. Geol., v. 64, p. 755-777.

Sheppard, S. M. F., Nielson, R. L. and Taylor, H. P., Jr., 1971, Hydrogen and oxygen isotope ratios in minerals from porphyry copper deposits: Econ. Geol., v. 66, p. 515-542.

Sheppard, S. M. F. and Taylor, H. P., Jr., 1974, Hydrogen and oxygen isotope evidence for the origins of water in the Boulder batholith and Butte ore deposits: Econ. Geol., v. 69, p. 926-946.

Taylor, H. P., Jr., 1974, The application of oxygen and hydrogen isotope studies to problems of hydrothermal alteration and ore deposition: Econ. Geol., v. 69, p. 843-883.

Taylor, H. P., Jr., 1977, Water/rock interaction and the origin of H_2O in granitic batholiths: Jour. Geol. Society, v. 133, p. 509-558.

Titley, S. R., Fleming, A. W. and Neale, T. I., 1978, Tectonic evolution of the porphyry copper system at Yandera, Papua New Guinea: Econ. Geol, v. 73, p. 810-828.

Walker, V. A., 1979, Relationships among several breccia pipes and a lead-silver vein in the Copper Creek mining district, Pinal County, Arizona: Unpub. M.S. Thesis, Univ. Arizona, 163 p.

Wilson, J. W. J., Kesler, S. E., Cloke, P. L., and Kelly, W. C., 1980, Fluid inclusion geochemistry of the Granisle and Bell porphyry copper deposits, British Columbia: Econ. Geol., v. 75, p. 45-61.

* * * * *

The Relationship of Alteration and Trace-Element Patterns in Epithermal Precious-Metal-Bearing, Fossil Geothermal Systems in the Great Basin

Byron R. Berger

U.S. Geological Survey, Box 25046
Denver Federal Center, MS 955
Denver, CO 80225

ABSTRACT

In the Great Basin, epithermal precious-metal deposits occur principally in carbonate and volcanic host-rock environments. Detailed study of these fossil geothermal systems indicates that similar ore mineral deposition processes and trace-element associations occur in both environments. The principal differences are in alteration processes and consequent mineralogies.

Carbonate-hosted deposits characteristically show early decarbonation and multiple episodes of silica deposition including the formation of a silica cap (jasperiod). Argillaceous and feldspathic material is altered to illite. During these alteration episodes, gold, silver, arsenic, and sulfur are the most significant minor elements deposited, along with lesser amounts of tungsten, molybdenum, mercury, thallium, antimony, and tellurium. Carbonaceous material is mobilized and enriched along fault zones and as rinds around intensely silicified zones. Superimposed on this early alteration and metallization are abundant arsenic, mercury, antimony, thallium, and fluorine in association with calcite and kaolinite. A late-stage hypogene argillization is common with the deposition of barite, kaolinite, montmorillonite, jarosite, and alunite.

High-level volcanic-hosted deposits display variable intensities of propylitization in the more mafic host rocks and weak propylitization in silicic host rocks. Phyllic alteration is generally restricted to vein selvages, and feldspar is primarily within the veins. A silicified or capping zone occurs in the upper parts of vein systems. Hypogene argillic alteration, when present, commonly occurs as an apron collapsing downward across the other alteration types adjacent to and beneath the silica cap. The precious and base metals occur in trace amounts in the silica cap, and are primarily restricted to quartz veins beneath the cap. Arsenic, antimony, mercury, and thallium are highest in the main fluid conduit areas as well as being strongly enriched in the hypogene argillic alteration. When base metals are present, cadmium is also high. Fluorine, bismuth, tellurium, selenium, and tungsten are highly variable from deposit to deposit.

EPITHERMAL GOLD/SILVER DEPOSITS: THE GEOTHERMAL CONNECTION

Harold F. Bonham, Jr.
David L. Giles

Nevada Bureau of Mines and Geology
Cimarron Exploration Inc.

ABSTRACT

Epithermal precious metal deposits in the western United States can be related to two main plate tectonic regimes: (1) active subduction zones of Mesozoic to Quaternary age with associated subduction-related magmatism and geothermal phenomena and (2) extensional tectonics of Tertiary-Quaternary age with associated rifting, magmatism and geothermal activity.

Three main deposit types and numerous subtypes of epithermal gold-silver deposits can be recognized in the western U.S.: (1) volcanic- or volcaniclastic-hosted vein, disseminated and stockwork deposits in or adjacent to volcanic centers; (2) sediment-hosted, stratigraphically-controlled replacement and stockwork deposits, typically in thin-bedded carbonaceous limestones or calcareous, carbonaceous shales (Carlin type), or in solution breccias, paleokarst horizons and channels (Sherman type), (3) hot spring-related stockwork and fissure systems, in which permeable rock horizons including volcanicalastics and hydrothermal explosion breccias are common host rocks for massive silicification and mineralization. Volcanic and subvolcanic rocks are common in the vicinity due to the association of volcanism with geothermal systems.

Mineralized fossil goethermal systems are directly analogous to a number of modern geothermal systems and the processes involved are essentially identical. Modern geothermal systems that contain significant amounts of base and precious metals (e.g. Taupo zone, New Zealand; Salton Sea, California; Steamboat Springs, Nevada; O-Yunuma and Sakurajima, Japan) are related to volcanic centers active within the Quaternary. Geothermal systems unrelated to volcanic centers, the deep circulation type related to high regional heat flow, typically contain very minor amounts of metals (e.g. Beowawe, Buffalo Valley and Darrough in central Nevada; Rio Grande rift in New Mexico).

EPITHERMAL GOLD/SILVER DEPOSITS: THE GEOTHERMAL CONNECTION

INTRODUCTION

Epithermal precious metal deposits in the western United States can be related to two main plate tectonic regimes: (1) subduction zones of Mesozoic to Quaternary age with associated subduction-related magmatism and geothermal phenomena, and (2) extentsional tectonics of Tertiary-Quaternary age with associated rifting, magmatism and geothermal activity. Deposits related to subduction-generated magmatism are associated with volcano-plutonic belts that range in magma chemistry from calcic through calc-alkaline, high potassium calc-alkalic, alkali-calcic and alkalic (Westra and Keith, 1981; Keith, this volume). The subduction-related deposits occur in subparallel, time-dependent belts and their metallogeny reflects the magma chemistry of associated igneous rocks (c.f. Keith, this volume). Deposits related to extensional tectonism are associated with rhyolites of the bimodal, basalt-rhyolite assemblage (McKee, 1971; Christiansen and Lipman, 1972). The rhyolites of this suite are high-silica, subalkaline or peralkaline types. No significant precious metal deposits are associated with volcanic centers in which only basalt or peralkaline rhyolite is present.

Three main types and numerous subtypes of epithermal gold/silver deposits can be recognized in the western U.S.: (1) volcanic and/or volcaniclastic-hosted vein, disseminated and stockwork deposits in or adjacent to volcanic centers; (2) sediment-hosted, stratigraphically-controlled replacement and stockwork deposits, typically in thin-bedded carbonaceous limestones or calcareous, carbonaceous shales (Carlin-type), but also in solution breccias, paleokarst horizons and channels (Sherman-type); (3) hot spring-related stockwork and fissure systems, in which permeable rock horizons including volcaniclastics and hydrothermal explosion breccias are common host rock for massive silicification and mineralization. Volcanic and subvolcanic rocks are usually associated with hot spring precious metal deposits, reflecting the common relationship between volcanism and metal-bearing geothermal systems (White, 1981).

EPITHERMAL Au-Ag DEPOSIT TYPES

The term epithermal (Lindgren, 1933) denotes Au-Ag deposits with the following generalized characteristics: deposition from relatively low salinity fluids (12% equivalent NaCl or less) at temperatures in the 150°–300°C range; limited vertical range of

Au-Ag deposition, typically 400 m or less and within 1000 m of the premineral surface; and, presence of As, Sb, Hg, Tl, Se, Te minerals associated with electrum, silver sulfides and sulfosalts with variable base metals. Au-Ag tellurides or Ag selenides are important ore minerals in some districts.

The source of the metals in epithermal Au-Ag deposits is an unresolved problem. Stable isotope data from many deposits indicate that the ore fluid was dominantly composed of highly exchanged meteoric water and sulfur derived from a sedimentary source. This has led many workers to propose that the metals were derived by hydrothermal leaching of source rocks, principally sedimentary rocks. Conversely, the close spatial association with intrusive and extrusive igneous rocks at nearly all deposits suggests to some that igneous rocks may have been the source of at least part of the metals present.

Continuing detailed geologic studies at a large number of deposits, involving stable isotopes, geologic mapping, geochemistry of the mineralized zones and related unmineralized sedimentary rocks and associated igneous rocks, and detailed fluid inclusion studies will be required in order to more closely define sources of metals.

Volcanic-Hosted Epithermal Au-Ag Deposits

Volcanic-hosted Au-Ag deposits include many of the famous bonanza deposits of the Circum-Pacific rim. In the western U.S. and Mexico they include such major districts as the Comstock Lode, Tonopah and Goldfield in Nevada, Cripple Creek in Colorado, and Guanajuato, Pachuca-Real del Monte and Tayoltita in Mexico. Deposits are typically in the 1-10 million ton range but often occur in districts with as many as thirty discrete orebodies. Grades vary widely up to a hundred ounces per ton silver and ten ounces per ton gold, but are generally less.

The observed ages of mineralization vary from late Mesozoic through the Cenozoic, coinciding with the ages of the volcanotectonic complexes with which they are genetically associated. Conceptually, older volcanic-hosted Au-Ag deposits will occur insofar as their supracrustal host rocks are preserved. Conversely, these deposits - as well the other epithermal types - are in the process of formation today at various locales around the Circum-Pacific region.

Three main subtypes of volcanic-hosted Au-Ag deposits can be distinguished by differences in alteration mineralogy, major and trace metal content, gold/silver ratios, and associated igneous rock types.

I. Enargite-gold subtype

One major type is characterized by advanced argillic alteration, gold associated with enargite-group minerals, high total sulfur content in both sulfides and sulfates, and by associated volcanic and subvolcanic igneous rocks ranging in composition from andesite to rhyodacite. Goldfield, Nevada, El Indio, Chile, and Summitville, Colorado are major examples of this deposit type. Silver-bearing tetrahedrite, galena and sphalerite underlie gold-enargite mineralization in some districts but, are virtually absent in others. Bismuth and tellurium are important trace metals. The gold mineralization commonly occurs in silicified hydrothermal breccias. Grades are often erratic and can be spectacularly high. Alteration, both hypogene and supergene, is widespread and typically extends far beyond economic mineralization.

Sillitoe (1983) has recently described enargite-bearing massive sulfide bodies with associated gold mineralization which occur in volcanic host rocks within zones of advanced argillic alteration. His enargite-bearing massive sulfide deposits, several of which he shows as genetically related to porphyry copper systems, are a variant of our enargite-gold deposit type.

II. Silver-base metal subtype

A second major subtype is associated with quartz-adularia-K-mica alteration, a generally high silver to gold ratio, a low total sulfur environment, geochemically anomalous amounts of Mo, W, Mn, F, and Se, variable amounts of Pb and Zn, and minor Cu. Genetically related igneous rocks are high and low-silica rhyolites. Deposits related to high-silica rhyolite systems typically contain more Mo, W, Sn, F and Mn than those associated with low-silica rhyolite systems. Many of the major bonanza, epithermal silver districts of western North America such as Creede, Colorado; Comstock and Tonopah in Nevada; Delamar, Idaho; and Tayoltita and Pachuca-Real del Monte in Mexico are examples of this deposit type.

Propylitic alteration is pervasive in these deposits, with potassic and phyllic alteration typically restricted to veins and immediately adjacent wall rocks. Quartz, adularia, fluorite, manganese carbonates and silicates, calcite and barite are the common gangue minerals. Argillic and phyllic alteration can form an alteration halo adjacent to and above the silver-gold ore bodies (Buchanan, 1981; Randall, 1979; Giles and Nelson, in press). The major precious metal ore minerals are electrum with silver sulfides, sulfosalts and/or selenides. The chief base metal sulfides are galena, sphalerite, and chalcopyrite with variable pyrite. Molybdenite is present in trace to minor amounts in many deposits, and huebnerite or wolframite occurs in several districts and may be present, but unidentified, in many more.

Economic silver-gold mineralization is underlain in some deposits by a zone of high base metals, chiefly Pb and Zn with some Ag. The silver-gold ore zone is typically exposed at the present erosion surface, but blind orebodies are present in some districts, e.g. Pachuca-Real del Monte and Tonopah. Structures above buried orebodies are typically narrow zones of quartz-calcite stringers which are essentially devoid of precious metals and sulfides. Anomalous amounts of Hg, As and Sb are usually present with sulfidic silicification above the precious metal ore shoots. If the water table at the time of alteration and mineralization did not intersect the paleosurface, there will be no hot spring sinters deposited, and hypogene solfataric alteration will take place. A productive precious metal system can be overlain by rock which contains no anomalous silver-gold mineralization, but may contain alunite, cinnabar, native sulfur and various sulfate minerals.

III. Gold telluride subtype

The third major deposit subtype is characterized by quartz-fluorite-adularia-carbonate alteration, a low silver to gold ratio with gold and silver occurring chiefly as tellurides, and a genetic relationship to alkalic igneous rocks. Fluorine and tellurium are diagnostic geochemical indicators for this type of deposit. Small amounts of base metal sulfides, stibnite and cinnabar may also occur with the precious metal ores. Cripple Creek, Colorado and Zortman-Landusky, Montana, are examples.

Gold telluride mineralization at Cripple Creek is hosted by quartz-fluorite-carbonate-adularia stockworks, veins, breccia shoots (Cresson Pipe) and sheeted zones which are surrounded by an envelope of propylitized volcanic rock. It seems probable that the Cripple Creek type of deposit is the volcanic and sub-volcanic expression of alkalic porphyry Cu-Au-Ag-Pt mineralization, such as that recently described by Werle and others (1982) in the Allard stock, Colorado.

Carbonate-hosted Epithermal Au-Ag Deposits

Deposits in this group are frequently referred to as "Carlin-type" deposits after the Carlin gold mine in Eureka County, Nevada. Host rocks for this group of deposits are typically thin-bedded, calcareous, carbonaceous, clastic sediments.

The carbonate-hosted deposits are hydrothermal, disseminated-replacement gold deposits. They are characterized by a high gold/ silver ratio and a geochemical association of Au, As, Sb, Hg, Ba and Tl. Anomalous amounts of W, Mo, Sn and F are usually present. Gold grades are typically in the 0.1 to 0.4 ounce/ton range with initial reserves ranging from a few hundred thousand tons to over one hundred million tons. Typical ore bodies range from five to fifteen million tons in size.

Host rocks are decarbonated in the ore horizon and variably silicified and argillized. Hydrocarbons are dissolved in hydrothermal fluids and reprecipitated adjacent to zones of acid alteration and in haloes around the ore zone. Jasperoid replacements occur in and adjacent to high-angle fault conduits. Jasperoid usually shows evidence of multiple periods of brecciation and may occur above, within or below the main ore horizon. Jasperoid is ore grade in some deposits, and almost always contain geochemically anomalous amounts of gold. In addition to the ubiquitous silicification, alteration minerals typically present include illite, montmorillonite, kaolinite, chlorite and sericite.

Spatially associated with nearly all Carlin-type deposits are stocks, dikes, sills and plugs ranging from granodiorite to rhyodacite in composition. Intrusive rocks are usually altered and mineralized and, in some deposits, contain economic gold mineralization. Primary controls of ore deposition in Carlin-type deposits are high-angle faults which transect a favorable host rock type, typically a thin-bedded, silty to sandy carbonaceous siltstone or carbonate. The gold in Carlin-type deposits is typically submicroscopic and has particle sizes in the micron to submicron range (Radtke, 1981). Gold occurs as free gold, coatings on pyrite, as gold-organic complexes dispersed on amorphous carbon grains, as discrete grains in realgar, and in solid solution in realgar and native arsenic (Radtke, op. cit.).

Sulfide minerals present in Carlin-type deposits include pyrite (2%) and, typically, highly variable amounts of cinnabar, realgar, orpiment, and stibnite. Small amounts of base metal sulfides including sphalerite, galena, chalcopyrite and molybdenite are usually present. Several rare thallium-bearing minerals have been identified in a few deposits. The amount of arsenic present in Carlin-type deposits is highly variable; some deposits such as Getchell, Nevada and Mercur, Utah contain abundant arsenic sulfides; others such as Northumberland and Alligator Ridge, Nevada contain only small amounts of arsenic.

Unoxidized primary ore at a number of Carlin-type deposits is overlain by a zone of hypogene oxidation produced by a late stage of boiling of the hydrothermal solutions. Supergene oxidation may be superimposed on the zone of hypogene oxidation, making separation of hypogene and supergene alteration stages difficult (Radtke, 1981; Wells and others, 1969).

The age of Carlin-type deposits is a subject of some controversy. Many workers prefer a late Tertiary age for all deposits of this type. Their argument is based on apparent control by Basin and Range faults of late Tertiary age, and an inferred relationship to late Tertiary intrusion and volcanism as a heat source for the hydrothermal fluids (Radtke and Dickinson, 1974; Radtke,

1981). Indeed, a remnant volcanic surface of the same age as the younger felsic dikes and sills associated with mineralization is frequently located within several kilometers of the deposits (Stevens and Hawkins, in press). On the other hand, K-Ar ages of hydrothermal sericite from several of the Carlin-type deposits give ages that range from at least Cretaceous to late Tertiary (e.g. Silberman and others, 1974; Norton and others, 1977; Berger 1980; Bonham and Silberman, unpub. data).

In some deposits, as at Getchell and Gold Acres, the geologic relationships further suggest a genetic connection between mineralization and older granodioritic or granitic plutons near the ore zone (Silberman and McKee, 1971; Wrucke and Armbrustmacher, 1975; Berger, 1980). Clearly, more work is needed to establish whether the ages of formation of Carlin-type deposits range from late Mesozoic through the Teritiary, or are confined to a distinct time span within the Tertiary.

Several published models for the formation of Carlin-type deposits have placed a strong emphasis on hot-springs model. This is somewhat misleading, because there does exist a class of disseminated gold-silver deposits (discussed later in this paper) which clearly formed in a hot springs environment, within 100 meters or less of the paleosurface. The Carlin-type deposits are certainly related to fossil geothermal systems, and some or all of these systems must have vented to the surface as hot springs. However, none of the Carlin-type deposits have as yet been conclusively shown to have formed in an actual hot spring environment.

Current estimates on the depth vary from as shallow as 200 meters to as deep as 2000 meters. Such estimates usually reflect the inferred connection - or lack thereof- of mineralization with a deeper plutonic body, shallow volcanic surface, and the like as discussed above. Proponents to the shallower environment cite as further evidence the apparent open and short lived nature of the mineral systems, probable boiling of the hydrothermal fluids, and abrupt changes in fluid Eh and pH as reflected in the mineral and alteration assemblages.

Disseminated, stratiform replacements in solution breccia (paleokarst) horizons developed on thick limestone units are another type of carbonate-hosted deposit. These gold-silver deposits have been informally termed "Sherman-type" after the Sherman mine in the Leadville district, Colorado. Paleokarst zones in upper lower Mississippian strata (e.g. Madison, Leadville) provide favorable ore locales throughout the Rocky Mountain region. Examples include the Warm Springs (Gilt Edge) and Kendall districts in Montana, the Aspen and Gilpin districts in Colorado, and possibly the Lake Valley

district in New Mexico. In all cases there is evidence of associated igneous activity.

The host in these deposits is dominantly paleokarst fill material, usually a heterogeneous mixture of limestone solution-breccia fragments and blocks in a carbonaceous, calcite-cemented siltstone and clay matrix. The old solution channels, combined with the porous and reactive infill, focus hydrothermal fluids and localize ore. Gold-/silver ratios vary widely within and between deposits, as does the accompanying base and trace metal assemblage. As examples, Gilman and Aspen are Pb-Zn-Ag districts with little gold; Sherman and Lake Valley are oxidized silver deposits; Gilt Edge and Kendall contain micron-size gold with pyrite, fluorite, minor base metals, and an As-Sb-Hg trace metal signature (Giles, 1983). The ore shoots are very irregular in plan, reflecting the primary ore control. The various central Montana deposits contained 1-2 million tons averaging around 0.4 ounces Au per ton.

Hot Spring-related Au-Ag Deposits

Hot spring lode gold deposits are silicified breccias and vein stockworks of quartz-sulfide mineralization found at or very near the pre-mineral surface. They represent an important and relatively new source of gold production. Examples currently include Kasuga, Iwato and Akeshi in Japan; Waihi and other deposits of the Hauraki goldfield in New Zealand; Cinola in British Columbia; Hasbrouck, Borealis and McLaughlin in the United States. Hydrofracted vein stockworks and mineralized explosion breccias are the main ore hosts.

Ascending fluids produce a typical mushroom-shaped cap of near surface silica replacement and flooding. The cap is often overlain by hot spring sinter. Ore grades are typically 0.05 to 0.5 oz per ton with local grades in confined volcanic throat areas and vein clusters up to 1 ounce per ton. Supergene processes may lead to local silver enrichment. The Au-Ag ratio is overall about 1:1, but is widely variable within and between deposits. Individual orebodies known to date range from 1.5 million tons (e.g. Kasuga) to more than 20 million tons (e.g. McLaughlin).

Ore is typically composed of micron-sized native gold and electrum. Associated minerals include fine-grained pyrite-marcasite and a large suite of silver sulfosalts. The gangue mineralogy is microcrystalline quartz and chalcedony with lesser calcite and adularia. Cinnabar and stibnite are zoned toward the surface; minor base metals are enriched at depth. Host rocks are typically hardened by a pre-ore episode of K-Na metasomatism and near-surface silicification. Sulfide ore is often enveloped by alunite, kaolinite and montmorillonite, and late acid supergene leaching often overprints the mineralization.

Hot spring environments contain the strongest enrichment in the epithermal geochemical suite, a response to shallow boiling and steep near-surface gradients in temperature. Mercury, antimony, thallium and arsenic are typically anomalous. The geochemical suite is identical to that identified in the bonanza and disseminated replacement environment, but concentrations are up to several orders of magnitude higher. Individual elements may be absent, e.g. mercury in some Japanese deposits, or additional elements may be present, e.g. tungsten at Cinola, British Columbia, depending on local source rock characteristics.

Hot spring lode gold deposits are associated with the sealed portions of fossil geothermal systems and may be considerably offset from currently active centers of geothermal activity. All deposits discovered to date are associated with explosive fracturing (hydrofracting) and hydrothermal brecciation. Explosive release of pressure and resultant high permeability contribute to shallow boiling of rapidly rising hydrothermal fluids and to repeated episodes of mineralization. Temperatures of precious metal deposition are under $200^{\circ}C$, and the fluids are meteoric and dilute.

SUMMARY AND CONCLUSIONS

Epithermal gold/silver deposits can be subdivided into a series of subtypes based upon mineralogy, alteration characteristics, host rock lithology, magma chemistry of associated igneous rocks, depth of formation and deposit form (vein, stockwork, breccia-hosted, etc.). They are essentially confined to arc and back-arc settings and are predominantly of Mesozoic or Cenozoic age. Depth of formation ranges from surface hot spring environments to at least 2000 meters. The fluids associated with ore deposition range from dilute alkali chloride waters to acid sulfate waters with temperatures ranging from $150^{\circ}C$ to $300^{\circ}C$.

This diversity in host rock lithology, ore fluid chemistry, tectonic setting and depth of formation, leads to an equal diversity in deposit form, ore mineralogy and wallrock alteration. Recognition of this diversity of epithermal gold/silver deposit types is a key factor in exploration. For example, gold/silver deposits in carbonate host rocks typically have inconspicuous wallrock alteration haloes, but engargite-gold deposits have very extensive wallrock alteration haloes.

The unifying factor in epithermal gold/silver deposits is their clear relationship to geothermal systems. The study of modern geothermal systems in the Circum-Pacific belt, in particular, has led to a better understanding of this diverse group of deposits. Future research in active geothermal systems will clearly lead to a better understanding of epithermal gold/silver deposits.

REFERENCES CITED

Berger, B.R., 1980, Geological and geochemical relationships at the Getchell mine and vicinity, Humboldt County, Nevada, in Society of Economic Geologists Epithermal Deposits Field Conference, Nevada, 1980, p. 111-135.

Buchanan, L.J., 1981, Precious metal deposits associated with volcanic environments in the southwest: Arizona Geological Society Digest, v. XIV, p. 237-262.

Christiansen, R.L. and Lipman, P.W., 1972, Cenozoic volcanism and plate tectonic evolution of the western United States-- Part II, Late Cenozoic: Philosophical Transactions of the Royal Society of London, v. 271, p. 249-284.

Giles, D.L., 1983, Gold mineralization in the laccolithic complexes of central Montana, in The genesis of Rocky Mountain ore deposits - changes with time and tectonics: Proceedings of the Denver Region Society Symposium, Denver, 1982; p. 157-162.

Giles, D.L. and Nelson, C.E., in press, Principal features of epithermal lode gold deposits of the Circum-Pacific rim, in Proceedings of the Circum-Pacific Energy and Minerals Resource Conference, Hawaii, 1982.

Lindgren, W., 1933, Mineral Deposits: McGraw-Hill Co., 930 p.

McKee, E.H., 1971, Tertiary igneous chronology of the Great Basin of western United States -- implications for tectonic models: Geological Society of America Bulletin, v. 82, p. 3497-3502.

Morton, J.L., Silberman, M.L., Bonham, H.F., Jr., Garside, L.J., and Noble, D.C., 1977, K-Ar ages of volcanic rocks, plutonic rocks and ore deposits in Nevada and eastern California - determinations under the USGS-NBMG Cooperative Program: Isochron West, No. 20, p. 19-29.

Radtke, A.S., 1981, Geology of the Carlin
Gold deposit, Nevada: U.S. Geological
Survey Open-file Report 81-97, 221p.

Radtke, A.S. and Dickinson, F.W., 1974, Genesis and vertical position of fine-
grained, disseminated, replacement-type
gold deposits in Nevada and Utah, USA,
in Problems of Ore Deposition, IAGOD
Symposium, Varna; v. 1, p. 71-78.

Randall, J.A., 1979, Structural setting and
emplacement of Veta Madre orebodies
using the Sirena and Rayas mines as
examples, Guanajuato, Mexico: Nevada
Bureau of Mines and Geology Report 33,
p. 203-212.

Silberman, M.L., Berger, B.R. and Koski,
R.A., 1974, K-Ar age relations of grano-
diorite emplacement and tungsten and
gold mineralization near the Getchell
Mine, Humboldt County, Nevada: Economic
Geology, v. 69, No. 5, p. 646-656.

Sillitoe, R.H., 1983, Enargite-bearing massive sulfide deposits high in porphyry
copper systems: Economic Geology, v.
78, No. 2, p. 345-352.

Stevens, D.L. and Hawkins, R.B., in press,
A comparison of the gold mineralization
at Jerritt Canyon, Nevada, with other
disseminated gold deposits of the Basin
Range province, in Proceedings of the
Circum-Pacific Energy and Minerals Resource Conference, Hawaii, 1982.

Wells, J.D., Stoisor, L.R. and Elliott, J.E.,
1969, Geology and geochemistry of the
Cortez gold deposit, Nevada: Economic
Geology, v. 64, No. 5, p. 525-537.

Werle, J.L., Ikramuddin, M. and Mutschler,
F.E., 1982, Allard stock, La Plata
Mountains, Colorado -- a porphyry
copper-precious metals deposit in
potassic alkaline rocks: Geological
Society of America, Abstracts with
Programs, v. 14, No. 7, p. 645.

Westra, Gerhard and Keith, S.B., 1981,
Classification and genesis of stockwork
molybdenum systems: Economic Geology,
v. 76, No. 4, p. 844-873.

White, D.E., 1981, Active geothermal systems
and hydrothermal ore deposits: Economic
Geology -- Seventy-fifth Anniversary
Volume, p. 392-423.

Wrucke, C.T. and Armbrustmacher, T.J., 1975,
Geochemical and geologic relations of
gold and other elements at the Gold
Acres open-pit mine, Lander County,
Nevada: U.S. Geological Survey
Professional Paper 860, 27 p.

ACTIVE HYDROTHERMAL SYSTEMS AS ANALOGUES OF FOSSIL SYSTEMS

Robert O. Fournier

U. S. Geological Survey, Menlo Park, Calif.

ABSTRACT

The physical and chemical characteristics of many diverse active hydrothermal systems have been determined from exploratory drilling and commercial production of geothermal resources. Fluid pressures and temperatures vary widely, depending on the distribution of permeable rocks and their specific permeability, the position of the water table, the source of recharge water, the salinity and gas content of the hydrothermal fluids, and the nature of the heat source. In convecting hot-water systems, the maximum temperatures attainable at given depths are given by boiling-point curves that are appropriate for hydrostatic conditions. In vapor-dominated systems, liquid water is present in pore spaces within the rock, but vapor (steam and gas) fills open fractures throughout much of the system. Temperatures and pressures vary little within vapor-dominated zones, and these systems are underpressured with respect to normal hydrostatic systems. Lithostatic fluid pressures have been encountered in deep sedimentary basins. There is reason to believe that exceptionally high fluid pressures might be encountered in other environments with temperatures higher than about 350°C. Increases in fluid pressure from hydrostatic to lithostatic can occur only where a permeability barrier prevents free movement of liquid from the high- to the low-pressure region. With increasing fluid pressures, boiling temperatures increase, or supercritical conditions might be attained. In either event, a relatively steep temperature gradient may develop across a thin impermeable barrier. Sudden rupturing of such a barrier, and the accompanying drop in confining pressure, could result in violent boiling, brecciation of the overlying rock, and simultaneous deposition of minerals.

Fluid compositions vary widely in presently active systems. Total dissolved cations adjust to the total available anions, whereas dissolved silica, cation ratios, and pH are fixed by temperature-dependent mineral-solution reactions. Most hot-spring waters at depth have pH values in the range 6-7. As these neutral to slightly acidic waters ascend they usually become more alkaline owing to loss of CO_2. The importance of contributions from magmatic emanations to dissolved-metal concentrations in presently active systems is difficult to assess because little, if any, magmatic water appears to be present in the systems that have been well studied to date. It is possible that water evolving from crystallizing magma may become trapped in brines that underlie presently active systems at deeper levels than drilling has yet reached. Magmatic water may yet be found in some of the newly discovered acid systems associated with active or very young andesite volcanism in the South Pacific.

Evidence for underground boiling, and for mixing of waters with different compositions and temperatures, is commonly found in active systems. These processes are likely to cause deposition of minerals, such as quartz, K-feldspar, calcite, sulfides, and gold. Where ascending hot water (>100°C) flows too rapidly to be cooled entirely by conduction, the water heats the surrounding rock to the boiling temperature of the solution at the prevailing hydrostatic pressure (on the boiling-point curve), and the deposition of minerals is likely to be chemically and physically uniform (without banding). Banded mineral deposits may result from seasonal mixing of different waters, or from intermittent changes in water table that disrupt steady-state temperatures and pressures, causing massive and widespread boiling and even hydrothermal brecciation.

Hydrologic characteristics of presently active systems show great variations. Some systems discharge large amounts of hot fluid to the surface while others discharge very little. Many systems show evidence of long periods of inactivity between periods of active convective discharge. Decreased convective flow results mainly from self-sealing, but also may be influenced by changes in water table. Seismic activity, hydraulic fracturing, or a sudden decrease in water table that results in hydrothermal explosions may reopen self-sealed rock and allow convective flow to resume.

Stages in the evolution of hydrothermal activity within an intermediate to silicic volcanic system might include: (1) early venting of gas from magma directly to the surface; (2) formation of a layer of ground water (from condensation of steam and rain water) above the gas which increases the fluid pressure near the the magma; (3) acid alteration near the base of the thickening condensate layer and formation of brine adjacent to the magma; (4) development of a relatively dilute, but chloride-rich, convecting system fed by meteoric water above the brine; (5) formation of a vapor-dominated system within rock previously heated by the convecting hot water system; and (6) final influx of cool water throughout the system as the heat source wanes.

INTRODUCTION

Investigations of fossil hydrothermal systems provide information about processes that occurred throughout the lives of those systems. Such investigations, however, seldom provide an unambiguous picture of the physical and chemical conditions throughout the system at a given time, the kind of information required to develop more than crude conceptual models. Studies of active hydrothermal systems do provide information about their physical and chemical nature at a given time, but not enough for complete understanding of the history of the system. Studies of fluid inclusions and hydrothermal alteration products, found in cores and cuttings retrieved from wells drilled in active systems, provide additional information about the previous hydrothermal history.

Fossil hydrothermal systems are of great interest because they commonly contain ores. Processes that may lead to the transport and deposition of metals in hydrothermal systems include heating and cooling, pressure changes, partitioning of volatiles between liquid and gas during boiling, mixing of different fluids, and reactions between fluids and wall rocks. Examples of all of the above processes have been found in presently active hydrothermal systems.

Relationships between hot-spring activity and ore deposition have been described and summarized in a series of papers by White (1955, 1967, 1968a, 1974, 1981). The deposition of mercury ore within the presently active hot-spring system at Sulphur Bank, California, is well documented (White and Robinson, 1962; White, 1981). Ore-grade precipitates of Au and high concentrations of Ag, Sb, Hg, and Tl have been found in sinters presently being deposited at Steamboat Springs, Nevada (White, 1955, 1981), and at Broadlands and Waiotapu, New Zealand (Weissberg, 1969; Ewers and Keays, 1977; Weissberg and others, 1979). Naboko (1974) described Hg-Sb-As mineralization and gold and polymetals being deposited from thermal waters at Uzon Caldera, Kamchatka. Ore-grade precipitates of base metals have been found in sediments beneath the Red Sea hot brine pools (Bischoff, 1969; Brewer and Spencer, 1969; Hendricks and others, 1969; Shanks and Bischoff, 1977), and precipitated from hot, deeply circulating ocean waters at oceanic spreading centers (Edmond and others, 1979; Francheteau

and others, 1979; Koski and others, 1982; Normark and others, 1982; Vidal and others, 1978). High concentrations of dissolved metals have been found in high-temperature (>300°C) brines in the Salton Sea geothermal system (White and others, 1963; White, 1968a), in low-temperature brines at Cheleken, U.S.S.R. (Lebedev, 1967, 1973, 1975), and in oil-field brines in Alberta, Canada (Billings and others, 1969) and central Mississippi (Carpenter and others, 1974). To date, however, ore-grade deposits of base metals have not been encountered in wells drilled into hot, presently active hydrothermal systems in silicic volcanic environments, although ore minerals have been reported in the vapor-dominated system at The Geysers, California (Sternfeld, 1981), in the low to moderately saline hot-water systems in New Zealand (Browne, 1969, 1971), in the highly saline system at the Salton Sea, California (McKibben, 1979), and elsewhere (summarized by Weisberg and others, 1979). This suggests that in our exploration of active hydrothermal systems we have not drilled in the right places, or deep enough, or, perhaps, at the right time to find ore-grade mineralization. Studies of fossil systems have shown that barren stages of hydrothermal activity commonly precede as well as follow ore-forming stages. In assessing the reasons why ore-grade mineralization has not been found, it should be remembered that active systems have been drilled to find hot water, not base and precious metals.

It is of practical importance to gain information about all processes that occur in active systems, and to determine whether some conditions always lead to hydrothermal alteration with ore mineralization and other conditions always lead to alteration without ore deposition; and, if so, how the alteration products differ. That sort of information might decrease the amount of exploration drilling within barren fossil hydrothermal systems, or lead to successful location of ore within a system that at first seemed unpromising. Unfortunately, the alteration products found in the cuttings and core from fruitless exploration drilling in fossil systems are seldom detailed in the literature. However, many studies of hydrothermal alteration products found in active systems have been summarized by Browne (1978). It is not the intent of this presentation to repeat that summary, but rather to focus upon the physical and chemical characteristics of the hydrothermal fluids in active systems and their possible significance in ore-forming processes.

FLUID PRESSURES AND TEMPERATURES IN ACTIVE SYSTEMS

The physical characteristics of many active hydrothermal systems in diverse geologic environments have been determined through exploratory drilling and commercial production of geothermal resources; excellent summaries are presented by Ellis and Mahon (1977), Ellis (1979), and Henley and Ellis (1983). Most geothermal wells drilled for production of electricity are 1 to 3 km deep. Fluid pressures and temperatures vary widely, depending on the distribution of permeable rocks and their specific permeability, the position of the water table, the source of recharge water, the salinity and gas content of the hydrothermal fluids, and the nature of the heat source. In silicic volcanic environments, reservoir temperatures commonly are in the 230° to 260°C range and temperatures up to about 320°C are not unusual. A temperature of 360°to 370°C was measured in a 2000-m well drilled near Ibusuki, Japan (H. Sakai, oral communication, 1982), and a temperature above 419°C (the melting temperature of zinc) has been reported in a geothermal well near Naples, Italy (A. Ten Dam, written communication, 1982).

Active hydrothermal systems are now generally subdivided into two main categories, hot-water and vapor-dominated (White and others, 1971). In convecting hot-water systems, liquid fills most of the pore spaces and open fractures within the rock (Fig. 1). Although scattered

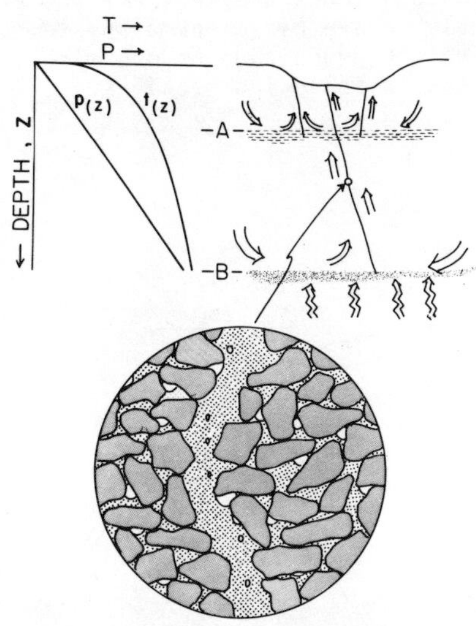

Figure 1. Schematic model of conditions in a hot-water-dominated geothermalsystem where boiling temperatures prevail through a steeply dipping structure filled with liquid water (from Fournier, 1981).

gas or steam bubbles may be present, liquid water is essentially the continuous phase in fractures leading from depth to the surface, and the maximum temperatures attainable at given depths are given by boiling-point curves that are appropriate for hydrostatic conditions (Figs. 1 and 4). In vapor-dominated systems, liquid water is generally present in pore spaces, but vapor (steam and gas) fills open fractures throughout much of the system (Fig. 2). Temperatures and pressures may change very little within vapor-dominated zones (generally about 240°C and 33.4 bars), and these

systems are underpressured with respect to normal hydrostatic systems through vertical

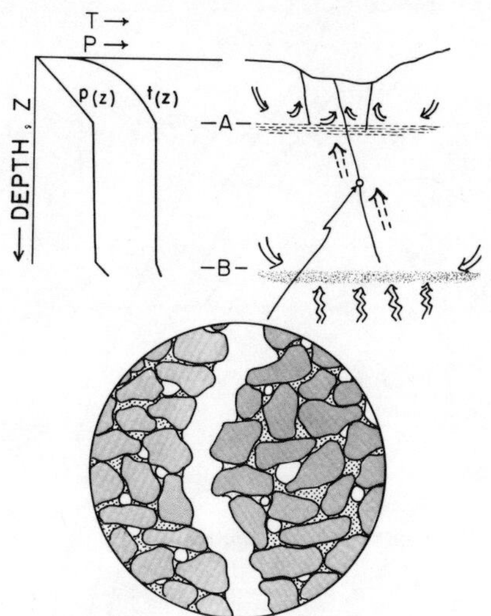

Figure 2. Schematic model of conditions in a vapor-dominated geothermalsystem (from Fournier, 1981).

distances of hundreds to a few thousands of meters. The existence of a hydrostatically underpressured system requires that a region of relatively low permeability surround the vapor-dominated zone that prevents the free movement of cold ground water into the system.

The buoyant force that drives convection in hot-water systems results from the greater density of relatively cold recharge water compared to hot discharging water; the "thermal-artesian" pressure of Studt (1958). Densities of pure water and aqueous NaCl solutions at the vapor pressures of the solutions are shown as functions of temperature in Fig. 3. White (1968b) constructed theoretical boiling-point curves for hydrostatic pressures controlled by the weight of an overlying column of water everywhere at its boiling temperature and showed that the maximum temperatures measured at given depths in wells drilled at Steamboat Springs, Nevada, closely agreed with that theoretical curve. Haas (1971) constructed similar boiling-point curves for water and aqueous NaCl solutions containing up to 25 weight percent salt (Fig. 4). Where partial pressures of dissolved gases are large, boiling-point curves may be significantly depressed (Sutton and McNabb, 1977; Mahon and others, 1980). A partial pressure of CO_2 of 10 bars should lower the boiling point curve of water by about 150 meters (Fig. 4).

Where maximum underground temperatures (the boiling point curve) are controlled by the weight per unit area of the overlying column of hot water, the hot outflowing part of the system must be relatively more open than the inflowing, recharge part of the system. This appears to be

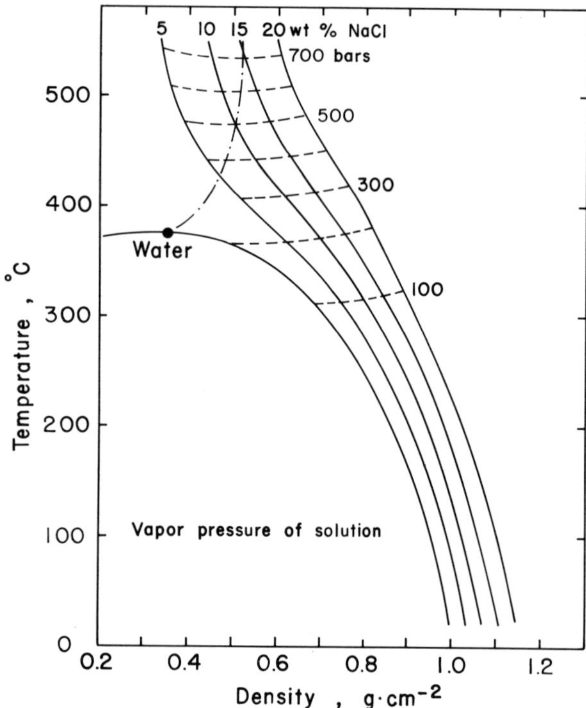

Figure 3. Density-temperature relations in the system NaCl-H$_2$O at the vaporpressures of the solutions. The dashed lines are isobars. The dot-dashed line is the critical curve. Densities are from an unpublished correlation of literature values by J. Tanger.

the situation in shallow parts of the hot spring system at Steamboat Springs, Nevada (White, 1968b). Where the outflow is restricted by impermeable strata, or by self-sealing resulting from mineral deposition, hydrostatic pressure may be controlled by the weight per unit area of the relatively cold recharge water. In many geyser basins at Yellowstone National Park, White and others (1975) measured fluid pressures in geothermal wells significantly above those that would be expected if the overlying hot water controlled pressure (Fig. 5). Theoretical depth-pressure curves are shown in Fig. 6 for boiling water and various boiling NaCl solutions, with pressure controlled by the weight of the overlying, freely discharging solution (Haas, 1971); for comparison, the depth-pressure curve for a cold column of water also is shown. The pressure exerted by the cold column of pure water is significantly greater at given depths than that exerted by a free-standing column of 20 weight percent aqueous NaCl everywhere at its boiling temperature; therefore, inflow of cold, dilute, meteoric water into the deep part of a hydrothermal system can cause upward convection of highly saline hot fluids, possibly with as much as 25 weight percent dissolved salts.

Boiling-point curves are calculated relative to the position of the water table that controls hydrostatic pressure. In many places that water table is far below the earth's surface. In other places, artesian systems are present in which the pressure-controlling water table is elevated in distant hills or mountains. In still other places, hot springs discharge onto the floors of lakes or the ocean, and boiling-point curves adjust to the overlying column of lake or ocean water. Thus, the 350°C hot-spring waters discharging on the ocean floor at 21°N in the East Pacific (Edmond and others, 1979) are below boiling temperature because of their great depth.

Fluid pressures approaching lithostatic have been encountered in deep sedimentary basins (Kharaka and others, 1978). There is reason to believe that mineral deposition and self-sealing in deep parts of hydrothermal systems in other environments, where temperatures higher than about 350°C are attained, might allow the development of fluid pressures much greater than hydrostatic (Fournier, in press).

The consequences of the above generalizations about temperatures and pressures in presently active hydrothermal systems, relative to the deposition of ore minerals, will be discussed in following sections.

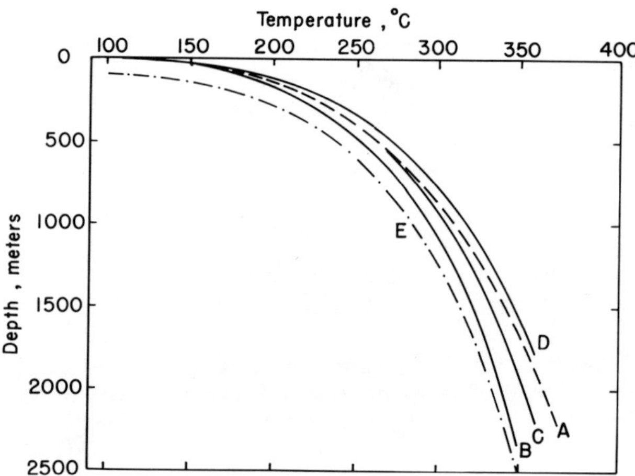

Figure 4. Depth-temperature relations for boiling solutions. Depth-pressure relations for curve A fixed by the weight per unit area of a free-standing column of cold water extending to the surface. Depth-pressure relations for curves B to E fixed by the weight of per unit area of free-standing columns of the given solutions everywhere at their boiling temperatures, and extending to the surface. Curves A and B for pure water, curve C for 10 weight percent aqueous NaCl, curve D for 20 weight percent aqueous NaCl, and curve E for water plus a partial pressure of CO$_2$ of 10 bars.

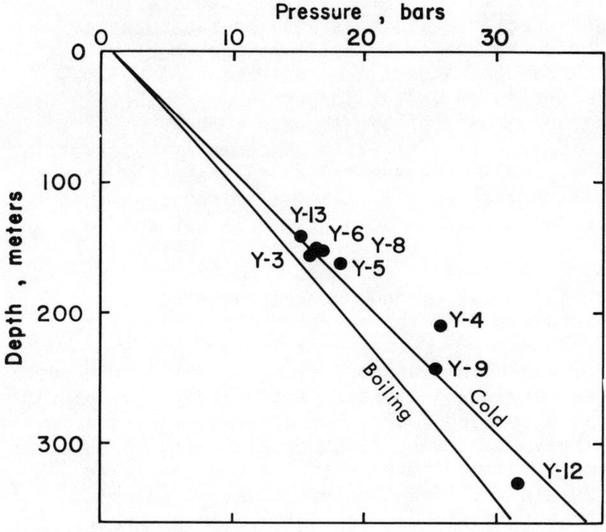

Figure 5.   Depth-pressure relations for boiling and cold columns of pure water. Numbered dots show depth-pressure relations at the bottoms of shut-in wells in Yellowstone National Park, measured by White and others (1975).

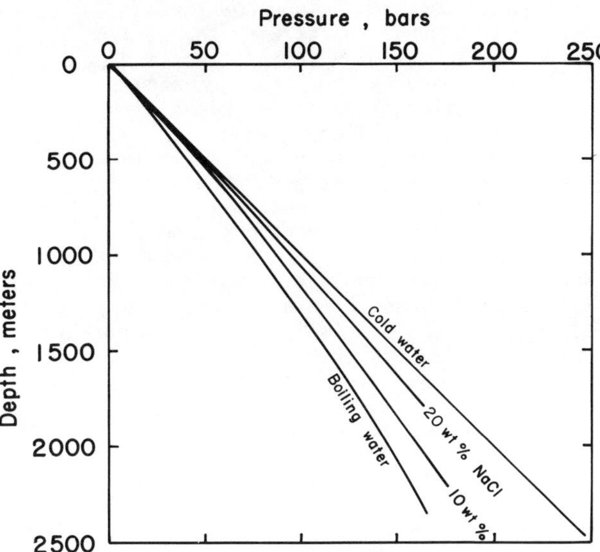

Figure 6.   Depth-pressure relations for boiling and cold columns of pure waterand boiling aqueous NaCl solutions.

COMPOSITIONS OF FLUIDS IN ACTIVE HYDROTHERMAL SYSTEMS

Fluids in explored parts of presently active hydrothermal systems are dominated by meteoric or ocean water that has changed composition during underground movement in response to changing temperature, pressure, and rock type, as well as mixing of different waters. Wooding (1963) and Henley and McNabb (1978) emphasized the importance of mixing of cool and hot fluids on the margins of upward moving convection plumes. Meteoric water is identified by isotopic evidence (Craig, 1963; White, 1968a; Truesdell and Hulston, 1980) and ocean water by isotopes, salinity, and ratios of dissolved constituents, such as Cl/Br (White, 1965). Connate and metamorphic waters have been identified by isotopic methods in thermal waters at Wilbur Springs and Sulphur Bank, California, respectively (White and others, 1973). Magmatic water has not yet been recognized in presently active systems, although some water must evolve from magmas as they crystallize. Present isotopic techniques cannot detect less than about 5 percent magmatic water in a hydrothermal fluid. Therefore, the presence of a small proportion of magmatic water and other "magmatic" constituents, such as chloride, sulfur, and metals, in presently active systems cannot be ruled out.   According to Ohmoto and Rye (1974), isotopic data show that up to 25 weight percent of magmatic water could have been involved in the formation of Kuroko deposits. Evolving magmatic water might be incorporated or trapped in highly saline brines (>30 weight percent dissolved salt) that form at high temperatures and at moderate depths around crystallizing magmas. Convective circulation of these brines to shallow levels, as part of a hydrothermal system recharged by meteoric water, might not be possible because of their high densities. In order for mixing of dilute and highly saline water to occur at a rate faster than by diffusion, the two fluids should have about the same density or the rising fluid be less dense than the overlying fluid. When cold, dilute water has about the same density as hot brine, mixing might occur by flow of the cold water into the hot brine across a nearly vertical fluid composition boundary. At 300 bars, dilute water at 150°C (point A in Fig. 7) has about the same density as a 20 weight percent NaCl solution at 305°C (point B). It is possible that very saline brines, possibly containing significant proportions of magmatic water, will be encountered when geothermal wells are drilled to greater depths and higher temperatures (McNabb, 1975; Truesdell and Fournier, 1976; Griffiths, 1978).

From a comparison of the C/S ratios of volcanic and geothermal gases, Giggenbach (1977) estimated that only about 5 percent of the sulfur entering geothermal systems from magmatic sources reaches the relatively shallow level of present exploitation; the rest is fixed as sulfides in the deeper, hotter zones. This implies that we have not drilled deep enough in presently active hydrothermal systems to find commercial-grade base-metal sulfide deposits.

Changing temperature has a major effect upon the ratios of cations in solution and the

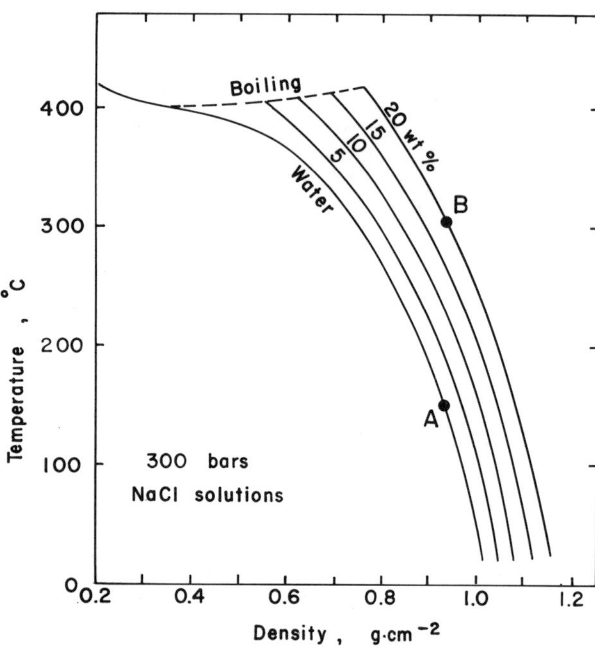

Figure 7. Density-temperature relations in the system NaCl-H$_2$O at 300 barspressure. Densities are from an unpublished correlation of literature values by J. Tanger.

concentration of dissolved silica. The effects of pressure are more variable. For non-carbonate minerals, changing pressure at constant temperature generally has little effect on solubilities when temperatures are below about 250°C, and great effect at higher temperatures. At 400°C the solubility of quartz in water is about 1000 mg/kg at 400 bars, 500 at 300 bars, and only 50 at 100 bars (Fournier and Potter, 1982). Changing pressure is an important factor wherever decompressional boiling occurs because of evaporative concentration and partitioning of volatile components between water and steam. Where boiling takes place deep underground, volatile components, such as CO$_2$ and H$_2$S, preferentially partition into the steam phase and move with it to the upper and marginal parts of hydrothermal systems, where steam condenses. There, some CO$_2$ and H$_2$S will dissolve in cooler ground water (Oki and Hirano, 1970; White and others, 1971; Kartokusumo and others, 1976; Mahon and others, 1980). These ground waters generally contain less chloride than do waters in the main part of the hydrothermal system because they are shallow and locally derived, or are mixtures of shallow and deeper waters. The redissolved CO$_2$ forms carbonic acid and attacks the wall rocks, resulting in calcium bicarbonate-rich solutions at low temperatures or sodium bicarbonate-rich solutions at temperatures above about 140°C. Calcium bicarbonate concentrations decrease at higher temperatures because of the decreasing

solubility of calcite as temperature is increased. H$_2$S is oxidized to H$_2$SO$_4$ in an oxygen-rich shallow environment. Therefore, the tops and margins of boiling systems become enriched in sulfate as well as bicarbonate. Where the buffer capacity of the rock is exceeded, the continued oxidation of H$_2$S may result in pH values less than 4 and the disappearance of bicarbonate. Thus, acid-sulfate alteration is commonly observed above boiling water tables in presently active hot-spring systems. The silicification that accompanies this acid attack is not due to influx of silica, but to strong leaching of alkalies that are flushed from the rock by condensed steam and meteoric water that percolate down to the water table.

Variations in rock type strongly influence the total salinity and particularly the chloride concentration that a hydrothermal solution is likely to attain. Geothermal waters in basaltic rocks in the interior of Iceland generally contain very low concentrations of chloride (Arnorsson and others, 1983). In contrast, geothermal waters in basalts near the seashore generally have chloride concentrations about the same as seawater. I found that geothermal waters in granites and highly metamorphosed rocks well inland from the ocean also generally contain very low concentrations of chloride. Bicarbonate is commonly the main anion in these low-chloride waters, although major amounts of sulfate may also be present. Bicarbonate concentrations are proportional to the partial pressure of CO$_2$. Dissolved CO$_2$ forms carbonic acid that reacts with the wall rock, liberating Na$^+$ at high temperatures and Ca^{+2} at low temperatures, as discussed above. Geothermal waters in silicic volcanic rocks (andesites to rhyolites) commonly contain a few hundred to a few thousand mg/kg chloride (Truesdell, 1976; Ellis and Mahon, 1977), even in interior regions of Iceland (Arnorsson and others, 1983). The highest chloride concentrations are found in geothermal waters that have come in contact with sedimentary rock; particularly those containing evaporites. At Cesano, Italy, a hot (>200°C) brine, rich in sodium sulfate with over 350,000 mg/kg total dissolved solids, was encountered in a 1435-m geothermal well (Calamai and others, 1976). Evidently, magma or very hot water reacted with gypsum that is known to be present in the underlying sedimentary section at Cesano, probably producing calcium silicates and a solution rich in sulfate. The conclusion seems inescapable that the compositions of waters in the relatively shallow parts of presently active hydrothermal systems are controlled mainly by leaching of the wall rocks by meteoric water. Experimental data also support this conclusion (Ellis and Mahon, 1964, 1967; Mahon, 1967; Kissen and Pakhomov, 1967; Ellis, 1968; Ewers, 1977).

High concentrations of dissolved metals can be obtained by reaction of chloride-rich brines with surrounding rocks (Barnes and Czamanske,

1967; Helgeson, 1967, 1968; Ellis, 1968; Carpenter and others, 1974; Bischoff and Dickson, 1975; Hajash, 1975; Seyfried and Bischoff, 1977; Hanor, 1979). However, in order for an ore deposit to form, those metals must precipitate within a relatively restricted volume of rock. Base-metal ore deposits require sulfide for their formation. Iron sulfides originally present in country rock can be transformed to Cu, Zn, and Pb sulfides through reaction with sulfide-deficient, metal-bearing brines; but the formation of large sulfide vein deposits is difficult to explain by this mechanism. Sulfides can be leached from rocks by saline solutions (Mottl and others, 1979), but it is questionable whether large enough quantities can be obtained this way to explain most observed ore deposits. A plausible mechanism for generating dissolved sulfide involves leaching of sulfate and later reduction of that sulfate by bacterial activity, ferrous iron (Mottl and others, 1979) or buried organic material. It has generally been assumed that reduction by bacteria activity could occur only at temperatures below 100°C. New data showing bacterial growth at temperatures of at least 250°C at high fluid pressure (Baross and Deming, 1983) suggest that reduction by bacteria activity might take place at high temperatures deep in hydrothermal systems. This has great implications for the interpretation of gas ratios and isotopic fractionation patterns of C, O, S, and H. Reduction of sulfate to sulfide should produce an alkaline solution,

$$SO_4^{-2} + 2Fe_3O_4 + H_2O = 3Fe_2O_3 + H_2S + 2OH^{-2}, \quad (1)$$

$$SO_4^{-2} + 2CH_2O = H_2S + 2CO_2 + + 2OH^{-2}, \quad (2)$$

while alteration associated with sulfide mineralization usually indicates acid conditions. Deposition of calcite and albitization (as a solution is heated) or K-feldspathization (as a solution is cooled) could be a consequence of reduction of sulfate to sulfide. However, a volcanic or magmatic source of sulfide would appear to be a very favorable circumstance for precipitation of base-metal ores.

Most waters in presently active hydrothermal systems are neutral or only slightly acid because pH is controlled by silicate hydrolysis reactions involving feldspars and micas or clays (Hemley and Jones, 1964; Meyer and Hemley, 1967; Ellis, 1970, 1979; Ellis and Mahon, 1977). Yet, as mentioned above, acid alteration commonly accompanies the formation of hydrothermal ores. Until very recently, drilling for production of geothermal energy in active hydrothermal systems had encountered extremely acid conditions at depth at only a few localities in zones of active volcanism, such as Matsao in Taiwan (Chen, 1970, 1975) and Onikobe (Yamada, 1976) and Matsukawa (Nakamura and others, 1970) in Japan. It now appears that there is deep, acid-chloride thermal water

and/or deep acid alteration (pyrophyllite plus quartz) in many active hydrothermal systems associated with active or relatively young andesitic volcanism. These include Biliran (Lawless and Gonzales, 1982), Nasuji-Sogonon (Seastres, 1982), Palimpinon (Leach and Bogie, 1982), and Baslay-Dauin (Harper and Arevalo, 1982) in the Philippines, and Suretimeat (Heming and others, 1982) in the New Hebrides. In these systems it is likely that much of the acidity comes from reactions with volcanic gases, including SO_2, H_2S, and HCl.

At Matsao and Onikobe, acid-chloride waters with pH values less than 2 were found at depths greater than 1000 meters at temperatures exceeding 275°C. The reservoir at Matsao is in quartzite, and at Onikobe it is in andesite altered to pyrophyllite and quartz. Waters collected at intermediate depths, and alteration products found in cuttings from wells, show that these deep, acidic waters are neutralized by mixing with shallow ground water and by reaction with overlying volcanic rocks as they rise toward the surface. Therefore, the acidity does not appear to be the result of downward movement of waters that had become acid by surface oxidation. Ellis (1977) attributed the deep acidity at Matsao to the reaction of water with deeply buried native sulfur deposits, producing sulfuric acid and hydrogen sulfide,

$$4S + 4H_2O = 3H_2S + H_2SO_4 . \quad (3)$$

However, as mentioned above, some or all of the acidity could result from interaction of water with gases evolved from a crystallizing magma at depth, or from hydrolysis reactions between salt and water that occur at high temperatures and low pressures.

Iwasaki and Ozawa (1960) and Saki and Matsubaya (1977) present evidence for the generation of acidity by the reaction,

$$4SO_2 + 4H_2O = 3H_2SO_4 + H_2S . \quad (4)$$

Relatively oxidized, sulfur-rich gases also may be evolved where gypsum or anhydrite are involved in hydrolysis reactions. For example, the 1982 eruption of El Chichon Volcano in Mexico contributed far more sulfuric acid to the atmosphere and stratosphere than is usual for comparably sized eruptions of other volcanoes, such as Mount St. Helens (B. Toon, oral communication, 1982). Gypsum beds occur in the sedimentary section beneath El Chichon, but are not present beneath Mount St. Helens. The isotopic composition of sulfur should be different in sulfate derived from gypsum compared to sulfate derived from volcanic SO_2.

The importance of HCl as a cause of acidity in hydrothermal systems should not be overlooked. Over 7000 mg/kg Cl as HCl was found in dry steam coming from a shallow well drilled at Hakone volcano in Japan (Kimio Noguchi, oral communication, 1970). Some or all of that HCl may have been generated by hydrolysis of NaCl at moderate to high temperatures and low pressures:

$$NaCl + H_2O = NaOH + HCl .\qquad(5)$$

Many investigators have found HCl in condensate after circulating dry steam over solid NaCl (Briner and Roth, 1948; Martynova and Samoilov, 1957; Galobardes and others, 1981). In experiments at 600°C I have found that significant amounts of HCl are generated by reaction 5 at pressures below about 350 bars, with more HCl produced at lower pressures. Addition of quartz to the system greatly increases the yield of HCl. This occurs because NaOH is removed from the solution by reaction with quartz, with precipitated sodium silicates as products. Solubilities of sodium silicates decrease with increasing temperature (Rowe and others, 1967). In natural systems, where aluminum is available in plagioclase and other minerals, albitization is likely to result from the hydrolysis of NaCl.

In some places acidity deep within a hydrothermal system does appear to result from downward movement of water that has become acid by oxidation of H_2S at and near the water table, as discussed previously. To the southeast of the Norris Geyser Basin in Yellowstone National Park, acid-sulfate waters are generated high on a hillside where the rocks have been extensively altered by fumarolic activity. Some of that acid water appears to percolate hundreds of meters underground where it mixes with high-temperature (~270°C) neutral water that is rich in chloride. The resulting "acid-chloride-sulfate" waters that issue as hot springs and geysers in Norris Geyser Basin have been extensively analyzed (Gooch and Whitfield, 1888; Allen and Day, 1935; Rowe and others, 1973), and exhibit widely ranging chloride and sulfate concentrations. However, the compositions of some individual "mixed-water" springs, such as Echinus, have remained remarkably constant since they were first analyzed in the late 1880's.

Truesdell (1976), Ellis and Mahon (1977), Ellis (1979) and Henley and Ellis (1983) summarize information about compositions of geothermal waters, and provide comprehensive reference lists.

MODELS OF HYDROTHERMAL SYSTEMS IN SHALLOW MAGMATIC ENVIRONMENTS

White (1973) published a schematic model of a convecting hydrothermal system within what appears to be a volcanic caldera (Fig. 8). That model shows an end-member situation in which the total convective flow is discharged at the surface; White (1973) described several other subtypes of hot-water systems, including ones with little or no surface discharge. In many hydrothermal systems it is likely that some or all of the fluid, after becoming cooled near the surface, recycles back downward in large convection cells, as shown in computer

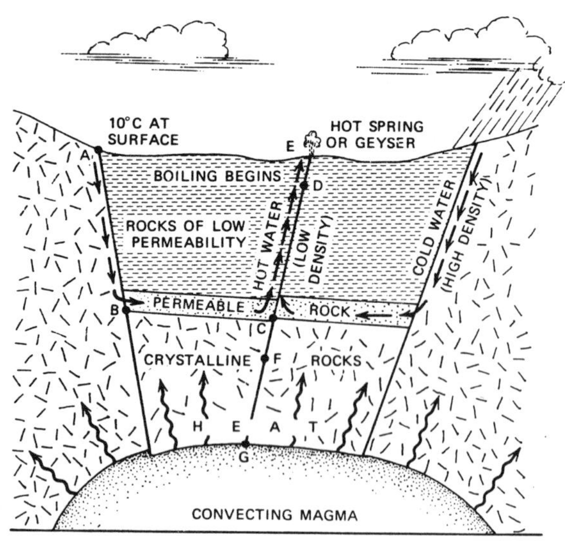

Figure 8. Schematic model of convective flow in a hydrothermal system showing effect of variations in permeability of the region on the flow, and with surface discharge of the total flow (from White, 1973).

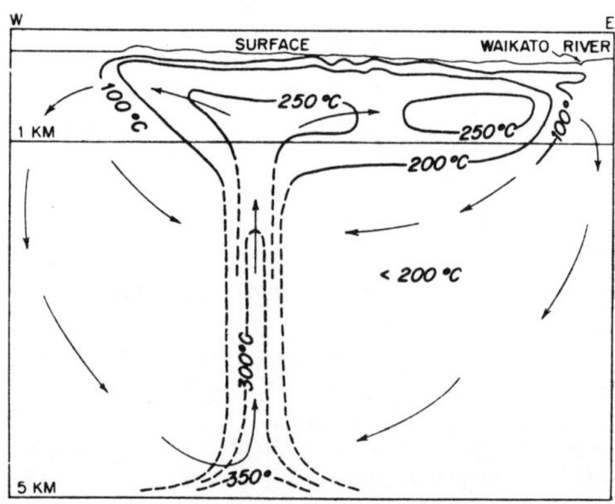

Figure 9. Cross section through the geothermal system at Wairakei, New Zealand (from Elder, 1965). The solid lines show isotherms derived by projecting measured temperatures onto the section, and dashed lines show estimated isotherms to a depth of 5 km. The approximate flow lines of meteoric water are shown by arrows.

simulations (Cathles, 1977; Norton and Cathles, 1979). Figure 9 is a cross section showing temperatures measured in wells at Wairakei, New Zealand. Note the mushroom shape of the thermal anomaly, indicating a lateral movement of hot water at shallow levels toward the topographically low Waikato River, and a postulated counter-flow of colder water beneath it. Ellis and Wilson (1955) calculated the natural discharge of chloride and heat from the Wairakei system into the Waikato River at 460 g Cl/sec and 82,000 kcal/sec. Using their data, the calculated natural discharge of thermal water and steam was 228 kg/sec. These figures do not include discharges from geothermal wells that were produced at the same time: 340 g Cl/sec, 62,000 kcal/sec, and 169 kg water/sec. For comparison, the calculated natural discharges at Yellowstone National Park are 1,319 g Cl/sec, 1,213,000 kcal/sec, and 3,200 kg water/sec (Fournier and others, 1976).

Lateral flow of thermal water appears to occur in many systems (Healy and Hochstein, 1973; Healy, 1976). The geothermal system at El Tatio, Chile (Fig. 10) is an excellent example

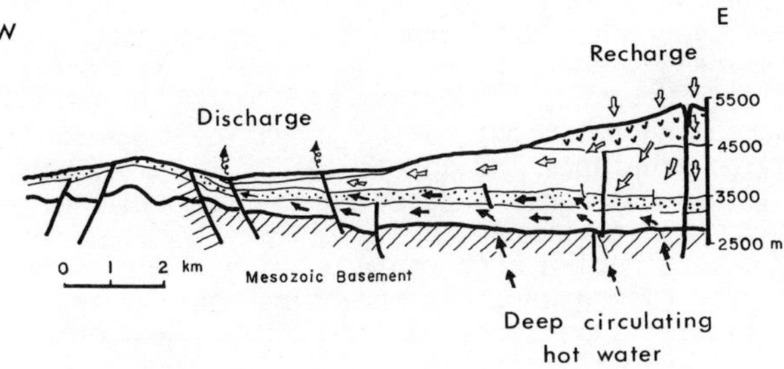

Figure 10. East-west cross section showing movement of geothermal fluids in the El Tatio sysyem (from Lahsen and Trujillo, 1976).

of one with major lateral flow (Lahsen and Trujillo, 1976; Ellis and Mahon, 1977). Based on isotopic data, recharge for the El Tatio geothermal field comes from the high Andes Mountains at least 10 km to the east. The heat source probably also lies beneath volcanoes to the east. Hot water produced from geothermal wells at El Tatio is drawn from a reservoir at 260-265°C. The water in the reservoir contains about 5,000 mg/kg chloride and measurable tritium. In one deep well, a brine saturated with salt and at a lower temperature (180°-200°C) was found underlying the high-temperature reservoir (Ellis and Mahon, 1977; W. A. J. Mahon, oral communication, 1982). From the above observations, it appears that little or none of the hot (5,600 mg/kg chloride) water flowing from east to west is recycled back into the system.

In many active hydrothermal systems that are associated with andesitic volcanoes, there appears to be underground flow of chloride-rich water laterally away from the volcanic edifice while gases rise more directly upward to the surface (Oki and Hirano, 1970; Heming and others, 1982; Harper and Arevalo, 1982; Muffler and others, 1982; Henley and Ellis, 1983).

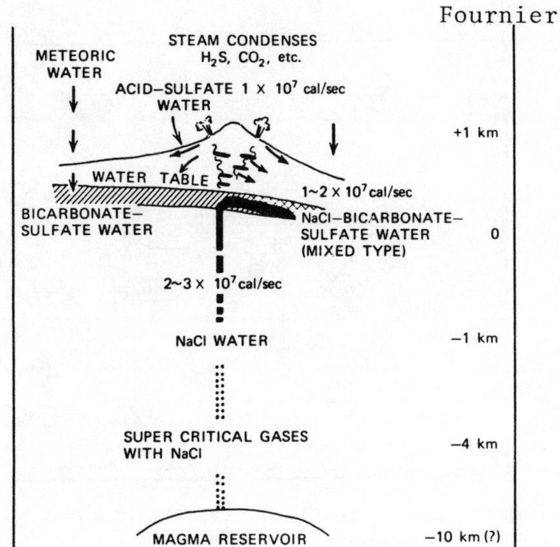

Figure 11. Schematic model for the geochemical development of fluids in the Hakone geothermal system (from Oki and Hirano, 1970).

Figures 11 and 12 show conceptual models of volcanic hydrothermal systems at Hakone, Japan (Oki and Hirano, 1970), and Lassen, California (Muffler and others, 1982), respectively. In Fig. 11 upward flow of hot and supercritical gas is shown above (and presumably from) a magma reservoir, and acidic fumaroles and springs emerge high on the slopes of the volcano. Acid-sulfate and bicarbonate-rich waters form in condensate zones and percolate downward and outward, mixing with deeper chloride-rich waters, as discussed previously. Henley and McNabb (1978) and Henley and Ellis (1983) showed how these models might be related to ore deposition. Beneath Lassen volcano, Muffler and others (1982) showed a relatively shallow vapor-dominated reservoir, underlain by 240°C water that is rich in chloride (Fig. 12). There, gases feeding the fumaroles and acid-sulfate springs on the summit and slopes of the volcano are thought to have been dissolved in upflowing chloride-rich (~2,300 mg/kg) water that boils at about 240°C. White and others (1971) suggested that mercury deposits may form above vapor-dominated systems and porphyry copper mineralization may occur in the zone of boiling brine below the vapor-dominated systems.

Many important variables govern the character and evolution of hydrothermal systems, and the time and place of ore deposition. These include the initial water, chloride, metal, and sulfur contents and oxidation state of the magma, the depth, size, shape and composition of the magma, the degree of "conditioning" of the overlying rocks by previous intrusions and extrusions that heated, fractured and altered those rocks, the position of the water table, the magnitude and distribution of porosity throughout the system, the pore pressures that are attained, and the compositions of the surrounding rocks. The evolution of water from

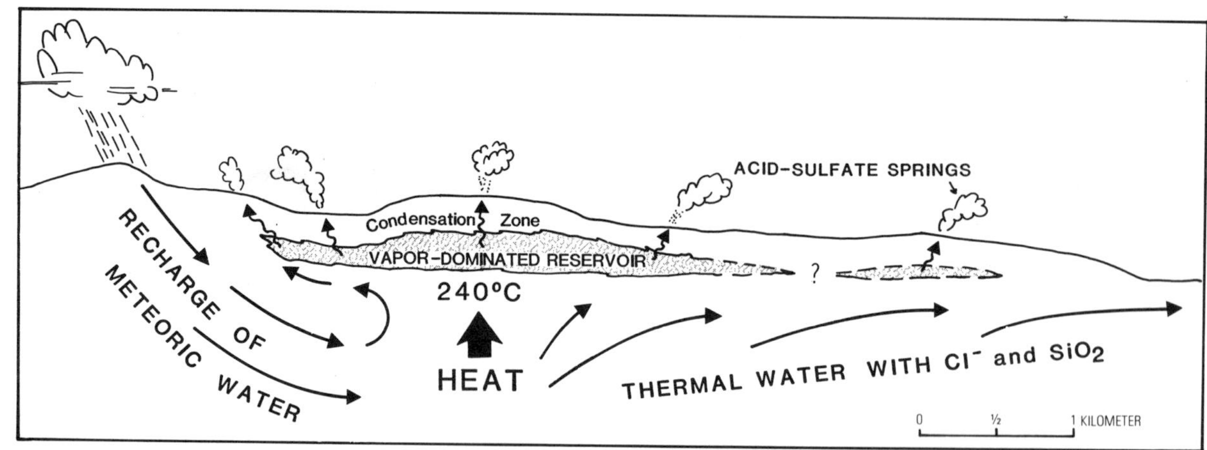

Figure 12. Schematic cross section of the Lassen geothermal system (from Muffler and others, 1982).

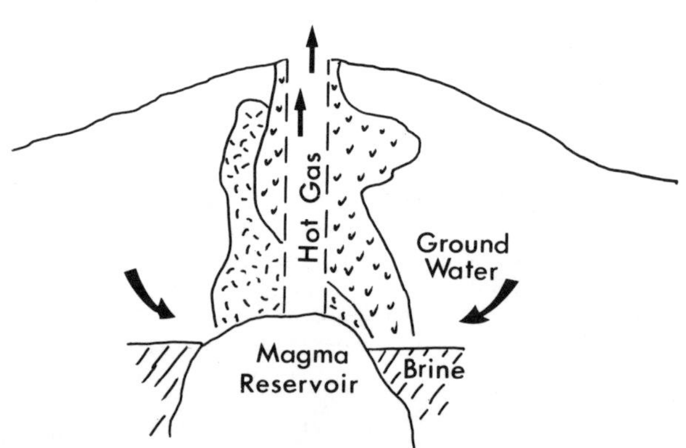

Figure 13. Schematic cross section of a volcanic system venting gas to the surface from a magma reservoir at moderate depth. See text for dicussion.

magmas has been discussed by many authors, including Morey (1922), Goranson (1931), Bowen (1933), Burnham (1967, 1979), Fournier (1968), and Whitney (1975), and the generation of chloride-rich gases and brines has been discussed by Sourirajan and Kennedy (1962), Ryabchikov and Hamilton (1971), Fournier (1972), Holland (1972), Kilinc and Burnham (1972), Carmichael and others (1973), Cunningham (1978), Henley and McNabb (1978), Burnham (1979), and Cloke and Kesler (1979).

Figure 13 shows a schematic cross section illustrating a situation in which very hot (500°-800°C) volcanic gases vent directly to the surface. Such gases usually are rich in H_2O, SO_2, H_2S, CO_2, and HCl, and may carry significant quantities of metals (Fenner, 1933; White and Waring, 1963; Menyailov and Nikitina, 1974; Tkachenko and Zotov, 1974; Gerlach and Nordlie, 1975; Graeber and others, 1982). Usually the parent magmas are visible, or thought to be very close to the surface within the volcanic edifice. Gases evolved from magmas a few to several kilometers deep are likely to cool appreciably by conduction, adiabatic expansion, and condensation of water as they rise. However, conductive cooling of upward streaming gas might not be significant where prior eruptions and intrusions of magma within the conduit zone have heated the surrounding rock to high temperatures. The heat stored in the surrounding rock could provide extra energy to the rising gas, counteracting cooling by adiabatic expansion. Volcanic necks and dikes which are still hot, but solidified and thoroughly cracked by shrinkage during solidification and cooling, would make excellent conduits to bring very hot gases to the surface from considerable depth.

From the bottom to the top of the conduit leading from the magma to the surface there may be steady or abrupt decreases in pressure. Decreasing pressure greatly decreases the ability of a water-rich gas to transport dissolved substances, such as silica and salts, as shown in Figs. 14-16. It is likely that precious metals, sulfides, and other minerals also would precipitate as a result of decreasing pressure and expansion of a gas, particularly where throttling occurs (Barton and others, 1961; Toulmin and Clark, 1967). Therefore, venting of magmatic gases (and of non-magmatic water and other volatiles heated to near magmatic temperatures) directly to the surface, as depicted in Fig. 13, is likely to be relatively short-lived, particularly where the magma is more than 1 or 2 km deep. Mineral deposition, either high in the system or very close to the magma could terminate this venting.

An estimate of the ability of hot, water-rich gases to transport dissolved salt can be obtained from the experimental results of Sourirajin and Kennedy (1962) in the system H_2O-NaCl. Figure 15 shows compositions of coexisting gas plus solid salt and gas plus brine at 700°C and at various depths, with

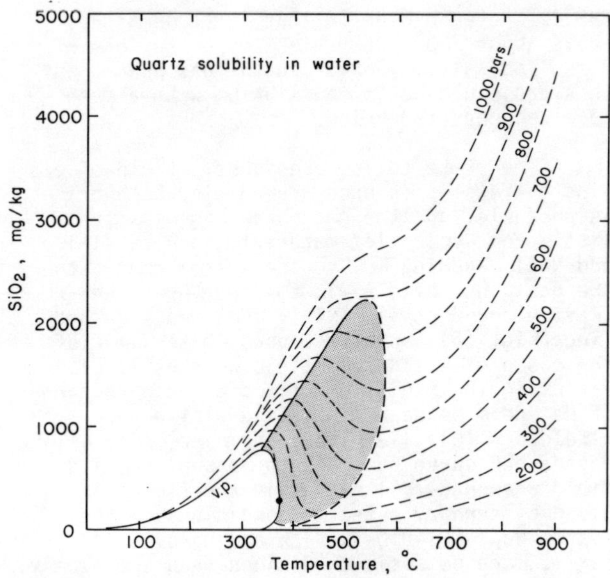

Figure 14. Solubilities of quartz in water up to 900°C at the indicated pressures, calculated using the equation of Fournier and Potter (1982). The shaded area emphasizes a region of retrograde solubility.

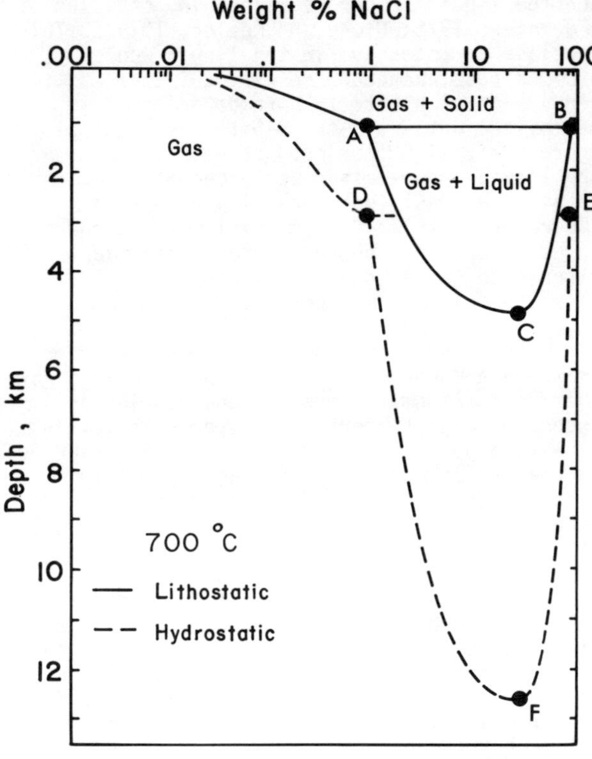

Figure 15. Compositions within the system NaCl–H₂O of coexisting gas plus solid salt and gas plus brine at 700°C and at various depths, with results for lithostatic fluid pressures contrasted with hydrostatic pressures (assuming hydrostatic pressure controlled by a cold column of dilute water).

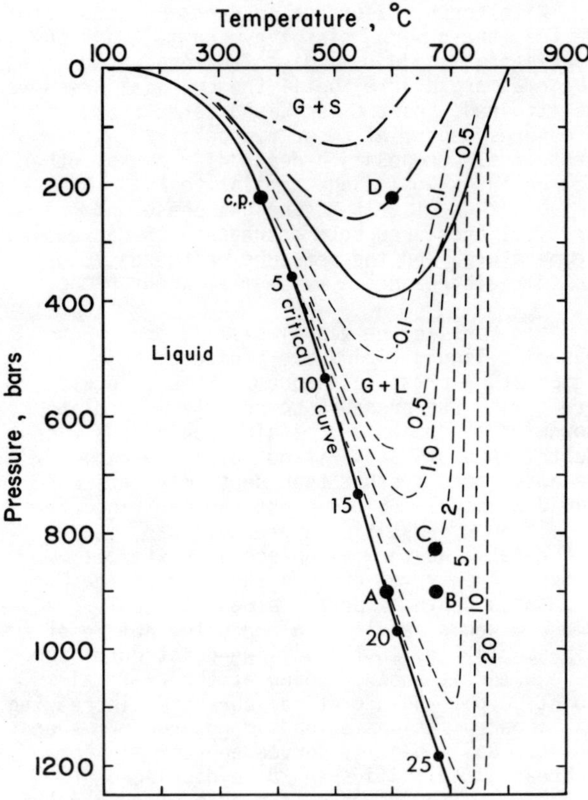

Figure 16. Some phase relations in the systems NaCl–H₂O and NaCl–KCl–H₂O projected onto a pressure-temperature diagram. The solid lines outline phase boundaries in the system NaCl–H₂O and short dashed lines show isopleths of NaCl solubility in steam. See text for discussion.

results for lithostatic fluid pressures contrasted with hydrostatic pressures (assuming hydrostatic pressure controlled by a cold column of dilute water). It is assumed in Fig. 15 that the partial pressure of water is equal to the total pressure. At 700°C a true NaCl brine may not exist at a pressure less than about 287 bars. Adding other salts to the system, such as KCl and CaCl₂, will allow brine to exist at significantly lower pressures, and will slightly increase the total amount of salt that can dissolve in the gas. Gas (steam) will dissolve increasing amounts of salt as pressure is increased (going to greater depths) from near atmospheric to about 287 bars. At a depth where 287 bars fluid pressure is reached, the solid salt will melt and that melt will dissolve a small amount of water, point B for lithostatic fluid pressures and point E for hydrostatic pressures. This melt may be called a brine, but this terminology can lead to confusion because at high temperatures and high pressures the gas phase too may dissolve enough salt to be called a brine. Liquid B would be in equilibrium with

273

gas at point A and liquid E with gas at point D. With further increase in pressure at greater depths, the gas can dissolve more salt and the coexisting liquid will dissolve more water (become more dilute) until the critical pressure is attained, point C for lithostatic fluid pressures and point F for hydrostatic pressures. Composition-depth diagrams at other temperatures would look similar to Fig. 15. Solubilities of salt in the gas phase and critical pressures both decrease with decreasing temperature, and the pressure of the gas plus solid field reaches a maximum at about 600°C.

In Fig. 16 the gas-plus-solid and gas-plus-liquid fields are projected onto a temperature-pressure diagram. Figure 16 was drawn with the pressure coordinate increasing downward to emphasize changing conditions with depth. A depth scale is not given in Fig. 16 because factors other than depth may influence fluid pressure. Fluid pressures may range from less than hydrostatic to greater than lithostatic at the same place at different times. A potrayal of more than one of the many plausable depth-temperature conditions in one diagram would result in a confusing jumble of lines. In Fig. 16 the boiling-point curve for pure water is shown, ending at the critical point, c.p., and a critical curve for increasing NaCl concentrations extends downward from that point. Boiling-point curves appropriate for increasing salinities would be displaced slightly to the right of the pure water curve and terminate at appropriate points on the critical curve. Termination points for 5, 10, 15, 20, and 25 weight percent NaCl are indicated. For a solution of given salinity, critical conditions can exist at only one temperature and pressure, such as point A in Fig. 16. To the left of the critical curve, and at pressures greater than the critical pressure, solutions are supercritical; but the fluids act more like liquids than gases in their ability to dissolve minerals, such as quartz. To the right of the critical curve is a field of liquid plus gas. The short dashed lines show isopleths of NaCl solubility in a gas phase that is in equilibrium with a liquid of specific composition (not shown), or solid salt (see Sourirajan and Kennedy (1962) for compositions of coexisting gas and liquid, and Fig. 15 for relations at 700°C). The critical point for an aqueous solution containing 2 weight percent NaCl is close to 395°C and 275 bars. A fluid containing 2 weight percent NaCl at 700°C and 900 bars (point B in Fig. 16) might be called a supercritical gas by many people. From a different point of view, however, that fluid is a gas that is unsaturated with salt. Decreasing the pressure to below 825 bars (point C) would cause the gas to become supersaturated, and it would separate into brine plus gas containing less dissolved NaCl, as discussed by Henley and McNabb (1978). Decreasing the temperature of a 2 weight percent NaCl solution from point B (675°C) to less than 595°C (point A) would cause the solution to become supercritical in the strict sense, until the temperature fell below

395°C. Note in Fig. 16 that at temperatures above about 700°C very high concentrations of dissolved salt may exist in the gas phase, but this gas could be in equilibrium with a much more concentrated brine.

At moderate to low pressures a field of gas plus solid salt is encountered, in which brine is unstable, and the gas phase can dissolve little "non-volatile" material, such as silica and NaCl. Adding KCl to the system contracts the gas-plus-solid field to lower pressures (Ravich and Borovia, 1949). The heavy dashed line (Fig. 16) shows the approximate limit of the gas-plus-solid salt field when NaCl/KCl ratios in the fluids are fixed by base exchange of Na and K between coexisting albite and K-feldspar (Orville, 1963; Fournier, 1976). The dot-dashed curve outlines the gas-plus-solid field when NaCl/KCl ratios in the fluids are given by ternary eutectic conditions in the system NaCl-KCl-H_2O (Ravich and Borovia, 1949), more potassic conditions than are likely to be found in most hydrothermal solutions. At 600°C and 220 bars two alkali feldspars can coexist with liquid plus gas of appropriate Na/K composition (point D in Fig. 16). Because the compositions of most natural waters are rich in NaCl compared to KCl, a decrease in pressure would cause NaCl to precipitate preferentially, and the liquid to become richer in potassium (Fournier, 1976; Cloke and Kesler, 1979). This increase in potassium in the liquid would promote base exchange, resulting in the conversion of albite to K-feldspar (or muscovite) until all the albite was used up. Thereafter, with continued decrease in pressure the liquid and feldspar would become more potassic as NaCl continued to precipitate (Fournier, 1976). This mechanism may account for much of the high-temperature potassic alteration found in porphyry-copper and porphyry-molybdenum deposits.

Where magmas are more than 1 or 2 km deep, or where a conduit has cooled appreciably before the current magmatic event, condensation of ascending high-temperature, water-rich gas is likely to occur within the vent. Acids carried in the gas at high temperatures are mainly associated complexes. Where condensation occurs, these acids will dissolve in the liquid, dissociate with decreasing temperature, and cause acid alteration of the surrounding rock. A neutral, chloride-rich solution, of moderate salinity and dominated by meteoric water, is likely to evolve above and at the sides of the acid-altered rock and hot gas region. This is essentially the model of Oki and Hirano (1970), depected in Fig. 11. In that figure a highly saline brine might be shown accumulating at the sides of the magma reservoir. Also, a small vapor-dominated system, similar to the one postulated at Lassen Volcano by Muffler and others (1982), could be incorporated into Fig. 11.

Figure 17 shows a schematic section through a hydrothermal system in which a self-sealed

envelope has developed in the country rock close
to a magmatic intrusion. In some systems the
self-sealed envelope might be totally within the
chilled part of the intrusive body. Fluid at
about lithostatic pressure fills the fractures
between the impermeable barrier and the magma.
This fluid is likely to be highly saline liquid
(>50 weight percent salt), but, depending on the
temperature, pressure, origin, and prior history
of the fluid, it could be a gas of moderate to
low salinity, or a brine capped by a gas. At a
depth between 4 and 5 km the confined fluid
pressure resulting from lithostatic load could
approach 1000 bars, and a gas at 650°C could
contain over 5 weight percent dissolved salt.

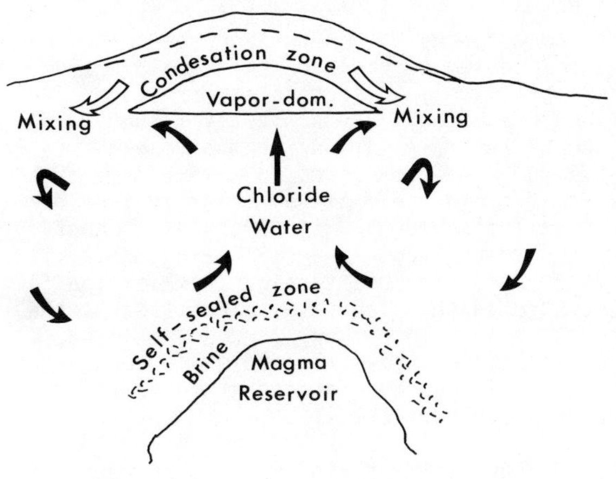

Figure 17. Schematic cross section of a
hydrothermal system with a self-sealed envelope
separating geopressured fluid from a
hydrostatically pressured convecting fluid. See
text for discussion.

A hydrostatically pressured hydrothermal system
of low to moderate salinity, dominated by
meteoric water, circulates at the sides and
above the self-sealed envelope. The maximum
temperature attained within this "meteoric"
system is likely to be about 350° to 450°C
(Fournier, 1977; in press). Both pressure and
temperature gradients across the self-sealed
zone of rock are likely to be large, and a small
amount of brine and gas may leak at slow rates
into the hydrostatic system. Deposition of
sulfides is likely to occur where hot brine and
cooler dilute water mix.

In the volcanic systems discussed above,
magma reservoirs were shown directly beneath
volcanoes, and hydrothermal activity of greatest
intensity was located above the tops of the
magma chambers. However, conduits from deep
magma chambers to volcanoes may be inclined.
They may also tap elongate bodies of magma,
rather than the spherical bodies that are
commonly shown. Movement of water at great
depth is likely to be controlled mostly by

faults and joints, and hydrothermal systems may
develop by deep circulation at the sides of a
magmatic bodies, rather than above the tops. In
the silicic caldera system at Yellowstone
National Park, centers of hydrothermal activity
are concentrated along the caldera ring
fracture, at intersections of faults resulting
from resurgent doming, and along radial faults
extending away from the caldera.

Evidence of hydrothermal conditions that
change with time has been found in several
active systems. On the basis of observed
self-sealing, chemical balance, and isotopic
evidence, Ellis (1979) concluded that brief
periods (of the order of 10^3 years) of major
flow of hot water from active systems commonly
alternate with long periods (10^4-10^5 years)
of conductive water heating and minor outflow.
Ellis (1979) also concluded that periods of
major flow may be triggered by tectonic activity
or by hydrothermal explosions. Fluid-inclusion
data show that there have been major changes in
temperature-depth profiles within the Kirishima
geothermal field in southern Kyushu, Japan
(Hayashi and others, 1981), and at Yellowstone
National Park (K. E. Bargar, oral communication,
1983). The changes in temperature at both
localities can be correlated with changing water
tables. At The Geysers, California, fluid
inclusion data and the high-temperature
character of the coexisting mineral assemblage
indicate that the present vapor-dominated system
(240°C) has evolved from an earlier and hotter
(350°C) hot-water system (Sternfeld, 1981;
McLaughlin et al., 1983).

BOILING AND BRECCIATION

Enthalpy-chloride relations found in
different parts of presently active hydrothermal
systems (Fournier, 1979) show that boiling
occurs deep in these systems as well as near the
surface. Boiling will change the composition of
a hydrothermal fluid by increasing the
concentrations of dissolved constituents that
remain in the residual liquid and by removing
dissolved gases that strongly partition into the
steam phase (Ellis, 1967). This partitioning of
volatiles, in turn, may significantly change the
pH of the system. All these factors are likely
to contribute to the deposition of ore
minerals.

Where ascending hot water (>100°C) flows too
rapidly to be cooled entirely by conduction, the
water heats the surrounding rock to temperatures
appropriate for boiling at the prevailing
hydrostatic pressure (the boiling-point curve).
With steady rates of fluid flow, the deposition
of minerals from solutions that are cooled
adiabatically is likely to be chemically and
physically uniform (without banding). Locally
intermittent geyser activity may temporarily
disrupt the steady-state temperature, pressure,
and chemical composition of the water, causing
deposition of compositionally banded minerals as
conditions oscillate from "normal" to "geyser".

Possibly more important but less widely recognized periodic boiling may result from seasonal changes in the water table. Where ascending hot water has heated the adjacent rock to a boiling-point curve appropriate for a high water table, a rapid drop in the water table will result in lower hydrostatic pressure at given depths throughout the system. Consequently, heat stored in the rock will cause vigorous and widespread boiling until the system boils dry or until rock and water temperatures decrease enough to correspond to a new boiling-point curve appropriate for the lower water table. Similarly, intermittent boiling might also occur within hot-spring systems that discharge into lakes or shallow oceanic environments, triggered by sudden changes in lake level (Muffler and others, 1971), or unusually low tides. In some convection systems, self-sealing at the outflow causes hydrostatic pressure to increase to the weight of the cold column of recharge water, as discussed previously. Intermittent seismic activity or hydraulic fracturing, induced by local excess gas pressure or heating of confined fluids, may break the sealed cap. This may allow hot water to flow freely with a drop in pressure throughout the hot-water column until mineral deposition reseals the outflow conduit, or the water supply is exhausted. This process may operate at time intervals from months to thousands of years.

Self-sealing within shallow, outflow channels of hydrothermal systems, followed by increased temperatures and pressures, and subsequently by sudden rupturing of the seal with explosive force, has been evoked as a mechanism to explain hydrothermal brecciation in many presently active systems (Lloyd, 1959, 1972, 1976; Skinner, 1966; Muffler and others, 1971; Grindley and Brown, 1976; Henley and McNabb, 1978; Keith and Muffler, 1978; Ellis, 1979; Henley and Thornley, 1979; Nairn and Wiradiradja, 1980). At Yellowstone National Park Muffler and others (1971) attributed the formation of hydrothermal eruption craters up to 0.6 km in diameter to decreased hydrostatic pressures resulting from sudden draining of glacial lakes.

Self-sealing may also occur deep in a hydrothermal system. Figure 14 shows that at constant pressure and increasing temperature, quartz has a solubility maximum (first reported by Kennedy, 1950) that extends from about 340°C at the vapor pressure of solution to 520°C close to 900 bars. The stippled area in Figure 14 shows a region of retrograde solubility in P-T space. Where water is heated at constant pressure less than about 900 bars, it will dissolve silica until either the solution starts to boil (at pressures below about 165 bars) or the solubility maximum is reached. With further heating that water will precipitate quartz. The precipitation of quartz in deep parts of a hydrothermal system may decrease the permeability to such an extent that convecting meteoric water can no longer attain temperatures much greater than

those shown by the quartz solubility maximum in Figure 14. Therefore, the time interval over which meteoric water may interact directly with a shallow intruded body of magma (or very hot solidified magma) may be limited to the early stage of development of the hydrothermal system, or may occur episodically thereafter with creation of new fractures by tectonic activity or thermal or hydraulic cracking (Phillips, 1973; Henley and McNabb, 1978; Ellis, 1979).

Deep hydrothermal explosion activity is a possible consequence of the deposition of an impermeable quartz seal (Henley and McNabb, 1978). Large and steep temperature and pore-pressure gradients are likely to evolve where an impermeable zone becomes established around a heat source. Although convective flow of meteoric water is cut off from the outside, the pore spaces within the zone between the quartz-sealed barrier and the remaining very hot rock are likely to contain gas or brine. This fluid may be entirely or partly meteoric or connate water, remaining from before the silica sealing became complete. However, some or all of that fluid could be volatiles evolved from a crystallizing magma. If volatiles do continue to be evolved from a crystallizing magma, it is easy to envision a situation in which the fluid pressure on the high-temperature side of the quartz seal becomes very large, (Phillips, 1973); sufficiently large to cause formion of a breccia pipe or even a conduit for a volcanic eruption (Morey, 1922).

Hydraulic fracturing will occur when the pore fluid pressure exceeds the confining pressure (the least principle stress) by an amount equal to the tensile strength of the rock. The confining pressure may range from less than normal hydrostatic to lithostatic, depending on whether open fissures are present, the nature of the fluid in those fissures, and permeability relations. Propogation of either a hydraulic or tectonic fracture through impermeable rock from a region of high fluid pressure into a region of lower fluid pressure may cause a significant decompression of the high-pressure fluid. If the thermal energy in the decompressing liquid and surrounding rock is large, massive flashing of water to steam may result. The expanding steam may explosively propel rock fragments into the air, where flashing occurs at relatively shallow levels, and into cavities and open fissures at deeper levels. Even without a magmatic contribution to the trapped fluids, pore pressures of those fluids could increase sufficiently to rupture the enclosing rock as a result of conductive heating.

Many of the conclusions in the above discussion are based on the solubility behavior of quartz in pure water. The effects of added salts can be modeled using NaCl solutions. Calculated solubilities of quartz in aqueous NaCl, using the method of Fournier (1983), show that adding dissolved salts should change the position of the quartz solubility maximum toward higher temperatures, and thus the extent of the field of retrograde quartz solubility shown in Fig. 14. However, the conclusion that quartz deposition may cause an impermeable barrier to form, preventing fluids at hydrostatic pressure from interacting directly with very hot rock or magma, is not changed by adding salt to the system (Fournier, in press).

Because the initial permeability several kilometers deep in a hydrothermal system is likely to be limited to a few widely spaced fractures or fractured zones of rock, an impermeable zone resulting from quartz deposition in those few fractures may go unrecognized as a significant feature. Also, in fossil hydrothermal systems where estimated temperatures at the time of vein formation are >340°C, it may be difficult to determine whether a given quartz vein was deposited as a result of increasing or decreasing temperature. If there is other hydrothermal alteration associated with the quartz deposition, that alteration may give an indication of the thermal history: for example, where a solution is heating, albite is likely to form in veins and after K-feldspar, and where a solution is cooling, K-feldspar or muscovite is likely to be deposited in veins and after plagioclase (Hemley et al., 1971).

Ores have been found in brecciated rocks and breccia pipes in too many deposits to enumerate here. Horikoshi (1969) and Henley and Thornley (1979) suggested that hydrothermal eruptions have initiated ore deposition in many Miocene Kuroko deposits. Hydrothermal explosive activity may affect ore deposition for various reasons, both physical and chemical. Brecciation greatly increases the permeability, providing easy access for later hydrothermal fluids that may deposit ores. The sudden and massive conversion of water to steam (boiling) may cause deposition of ore minerals for the reasons discussed above. Deposition of quartz and, to a lesser extent, K-feldspar is likely to occur wherever there is a sudden drop in pore pressure when initial temperatures exceed about 340°C. Also, the K-feldspar that forms as a result of a sudden drop in pressure is likely to be more potassium-rich than that which was in equilibrium with the fluid prior to the drop in pressure (Fournier, 1976). Therefore, hydrothermally brecciated rubbles that formed at very high temperatures are likely to be cemented by quartz, very potassium-rich feldspar, and a variety of other minerals, including sulfides. However, these minerals may also cement breccias that form at lower temperatures; quartz, adularia, and generally pyrite are the phases observed in the hydrothermal breccias that formed at Wairakei and Broadlands at about 200°-300°C (Grindley and Browne, 1976). Where exceptionally high degrees of silica supersaturation occur, particularly at lower temperatures, amorphous silica may precipitate and later alter to chalcedonic silica or quartz.

Sudden decreases in pore fluid pressure may also contribute to the formation of porphyry ores (Fournier, 1968, 1972; Henley and McNabb, 1978). Consider a fluid evolving from a small body of metal-rich magma that is crystallizing at a depth of about 1 to 4 km. If the pore pressure is near lithostatic, that fluid is likely to be a gas of moderate to high salinity that is unsaturated with salt. Metals initially in the melt will partition into the saline fluid. Slow, uniform crystallization of magma, without a drop in fluid pressure, will result in slow release of gas and most of the dissolved metals into the country rocks, where vein deposits might form if other circumstances are favorable. In contrast, a sudden drop in fluid pressure to hydrostatic or less will chill the magma by about 75°C, freeze phenocrysts in a fine-grained groundmass, and cause the gas to split into highly saline brine plus dilute gas, or gas plus solid salt. Some metals will enter the gas phase and be transported upward. However, most should precipitate as sulfides as the magma is chilled, or remain dissolved in any dense brine that forms. If the magma body is small enough to be chilled throughout its extent, and if most of its dissolved water escapes with the gas phase, there may be little further hydrothermal alteration until the rock cools enough to allow liquid at hydrostatic pressure into the system, or until there is another pulse of magmatic intrusion. Self-sealing by quartz deposition as water is heated to temperatures above 340°-400°C will tend to prevent movement of water into the system, as discussed above, but thermal cracking and tectonic movements may counteract this effect to some extent.

DISCUSSION AND CONCLUSIONS

The major- and trace- element compositions of thermal waters in explored parts of presently active hydrothermal systems appear to be controlled mainly by reaction of deeply circulating meteoric water with the surrounding rocks. There may be undetected magmatic contributions of water and other chemical constituents to these systems, particularly gases. However, the thermal energy that causes convective circulation of fluids with temperatures above about 200°C generally comes from a magmatic source within 10 km of the surface.

Mercury appears to be depositing in high enough concentration and sufficient quantity to form ore in altered rocks above some presently active systems, and mineable quantities of native sulfur have been formed by fumarole activity near the summits of many active volcanoes. Other types of epithermal ore deposits could form as a result of continued hydrothermal activity similar to that presently observed. Commercial grade concentrations of gold and silver, along with high concentrations of relatively volatile elements, such as S, Hg, Sb, As, Se, and Tl, have been found in precipitates deposited from some hot-spring waters.

Ore-grade deposits of base metals have not yet been encountered by wells drilled in active systems for production of geothermal resources, although iron sulfides, galena, sphalerite, chalcopyrite, and other base metal minerals have been found in cores and cuttings. In many of these systems the sulfides are precipitating from relatively dilute (less than a few thousand mg/kg total dissolved solids), neutral pH, chloride waters. With favorable hydrologic conditions that restrict the flow of most of the thermal water to a few relatively open channels, base-metal and precious-metal ore deposits could form at moderate to shallow depths in a system similar to those presently being investigated. Sulfides and precious metals are particularly likely to be deposited where ascending solutions boil, either continuously or episodically. Presently active systems with deep acidity, such as many that are now being investigated in regions of active and recently active andesitic volcanism in Japan and in the South Pacific, may provide new information about ore-forming processes.

It is possible that some presently active systems are underlain by highly saline brines that transfer heat by convection from very hot rock or magma to the overlying cooler, dilute system. Significant amounts of water and other constituents derived from magma (particularly sulfur and metals) may be trapped within that brine. Ore deposits may form where sulfur-rich gases, evolved from a crystallizing magma, flow into a surrounding brine that is rich in chloride and metals. Ores might also be deposited where the convecting brine boils or at interfaces between hot brine and cooler, dilute fluid. In some systems brines with successively higher salinities may underlie each other. At very high temperatures and relatively low pressures, gas plus solid salt could underlie a brine or a dilute hydrothermal system. Hydrologically this is an unstable situation that might be maintained by self-sealing that reduces the permeability of rock between the gas and the overlying liquid.

In some volcanic systems very hot sulfur- and chloride-rich gases vent directly from visible magma to the atmosphere. In other volcanic systems the hot gases that vent at the surface have near-magmatic temperatures and are assumed to come from magmas at undetermined depth. In some volcanic or subvolcanic systems the magma may be sufficiently deep that the escaping gases must bubble through (and react with) an overlying zone saturated with water before finally venting to the surface. The water in the saturated zone may be mostly condensed "magmatic" water or mostly meteoric water. The thickness of this water saturated zone may strongly influence the pore pressure where the gases are evolved from the magma. If the pore pressures are sufficiently low (<100-300 bars), brine is not likely to form (Fig. 16), and some metals are likely to be carried off with the gas as volatile complexes. Where these gases cool and condense, acid

alteration will occur and an ore deposit may form, possibly by mechanisms discussed by Lovering (1961) or Walker (1965). If pore pressures are in the range 300-600 bars, the gas evolved from a magma will probably split into gas plus liquid (brine) upon cooling to less than 650°-700°C, and metals initially carried in the gas are likely to partition strongly into the brine. Because that brine is relatively dense it is likely to drain downward and remain near the magma, while the gas phase rises, as in the model of Henley and McNabb (1978). At still higher pressures (generally at greater depths) the gas phase that evolves is likely to be a moderately saline fluid that is unsaturated with salt; the salinity is dependent on the initial concentrations of water and salt in the magma (Kilinc and Burnham, 1972; Holland, 1972). With cooling, that gas may become a supercritical fluid (Fig. 16), or with decreasing pressure it may separate into gas plus brine and eventually into gas plus solid salt.

The geologic record (and sometimes memories of local inhabitants) shows that some hot-spring systems have undergone great reductions in surface discharge by natural processes, and subsequent drilling has shown that large reservoirs of hot water remain at relatively shallow depths. Other hot-spring systems appear to have undergone repeated periods of decline and rejuvenation. Most of these changes appear to be the result of changing water tables and changing permeabilities as a result of mineral deposition, although waxing and waning heat sources will affect all hydrothermal systems. Self-sealing is a common phenomenon that occurs at and near the tops of hydrothermal systems. Increases in fluid pressure below the self-sealed zone may result, followed by hydrothermal explosions that brecciate the overlying rocks, opening the system to renewed flow. Ore deposition may accompany or follow the explosive activity. Self-sealing may also occur deep in hydrothermal systems, particularly near very hot bodies of rock. Narrow zones of self-sealed rock may separate geopressured brine from more dilute, hydrostatically pressured water. Seismic activity or hydraulic pressure may rupture this deep seal and decrease the fluid pressure within the brine, causing it to boil. The large pressure gradient and expanding gas phase may cause brine plus gas to squirt through the fractured region into the hydrostatically pressured system. Ore deposition may take place where the fluids mix and where boiling occurs. Deposition of quartz, sulfides, and other minerals is likely to re-establish the self-seal in a relatively short time.

The following sequence of changes might occur in a hydrothermal system that forms in a volcanic environment with a magma reservoir 1 to about 5 km deep. After volcanic eruptions cease, hot gases continue to vent to the surface. These gases may be derived in part from the remaining magma and in part from ground water (containing dissolved salts that may hydrolyze) that seeps into contact with the magma or with very hot rock (Fig. 13). In the shallowest magma systems, gas plus solid salt will be present in the hottest parts, and brine may form at slightly cooler margins where inflowing waters boil (Fig. 13). In deeper magma systems, brines may also form by dissociation of upflowing gas of moderate salinity into liquid of higher salinity plus gas of lower salinity. The liquid (brine) that forms will drain back downward, possibly interacting further with the magma or additional gas. If self-sealing keeps inflowing water away from magma and very hot rock and the outflow part of the system remains open, or if a relatively shallow layer of water does not soon become established in the path of the gas discharge, the magma may boil dry and cool at a relatively low vapor pressure. However, a shallow layer of ground water is likely to form above the gas region, derived in part from condensation of steam and in part from meteoric water. Acid alteration will occur at the base of the liquid water layer where condensation occurs. The acidity is likely to result from H_2S, SO_2, and HCl. Neutralization of HCl by reactions with rocks contributes to the chloride content of the overlying water. As the ground water layer becomes thicker, the fluid pressures in the underlying system increase; brines as well as gas may exist in contact with the more shallow magmas, and brines already formed in the deeper systems may become more dilute.

Eventually the hot-gas region will completely disappear, being displaced by relatively dilute but chloride-rich meteoric water that flows from the sides and above. The dilute, meteoric hydrothermal system may float directly on brine or be separated by a narrow self-sealed impermeable barrier (Fig. 17). Convective flow of this dilute hydrothermal system heats a large body of the overlying rock to temperatures in the $250°-350°C$ range. If the permeability at the deep margins of the convecting meteoric system then becomes restricted, and the heat source remains large, a vapor-dominated system might become established above the convecting dilute hot-water system (Figs. 12, 17). This vapor-dominated system may slowly expand downward as fluid pressures drop and heat previously stored in the rock is used to convert water to steam. Boiling of the hot-water part of the system during this process may result in formation of a second brine of intermediate salinity between the vapor-dominated zone and the underlying initial brine. When the available thermal energy is finally dissipated the vapor-dominated region will be displaced by relatively cool water.

The above sequence of events may be interrupted by renewed pulses of magmatic injection or by hydrothermal explosive activity at shallow or deep levels. In many places the stage during which there is relatively low fluid pressure immediately around the magma and venting of very hot gases directly to the atmosphere may be omitted. This may happen if the magma reservoir is relatively deep and the rocks in the vent are cool or impermeable. Fluid pressures near the magma also may remain relatively high if venting of gas or liquid occurs on the floor of the ocean or a deep lake.

It appears that different types of ore deposits may form at the same time in different parts of a hydrothermal system; for example, epithermal gold may occur in sinters and shallow veins, while base-metal sulfides may be deposited at deeper levels, possibly underlain by still deeper porphyry deposits. Ore deposition may also be episodic, occuring mostly during and shortly after events that open self-sealed systems to increased flow, or immediately after pulses of renewed igneous intrusion.

ACKNOWLEDGMENTS

My knowledge of the chemical and physical nature of active hydrothermal systems has been greatly expanded by discussions in the field and in the laboratory with many colleagues. In particular, I thank L. J. P. Muffler, M. Nathenson, A. H. Truesdell, and D. E. White within the USGS, and A. J. Ellis, R. W. Henley, and W. A. J. Mahon within the N. Z. DSIR for sharing their thoughts and observations. The manuscript benefited from technical reviews by W. C. Bagby and D. E. White.

REFERENCES

Allen, E. T., and Day, A. L., 1935, Hot springs of the Yellowstone National Park: Carnegie Inst. Washington Pub. 466, 525 p.

Arnorsson, S., Gunnlaugsson, E., and Svavarsson, H., 1983, The chemistry of geothermal waters in Iceland II. Mineral equilibria and independent variables controlling water compositions: Geochim. et Cosmochim. Acta, v. 47, p. 547-566.

Barnes, H. L., and Czamanske, G. K., 1967, Solubilities and transport of ore minerals, in Barnes, H. L., ed., Geochemistry of hydrothermal ore deposits: New York, Holt, Rinehart and Winston, p. 334-381.

Baross, J. A., and Deming, J. W., 1983, Growth of "black smoker" bacteria at temperatures of at least $250°C$: Nature, v. 303, p. 423-426.

Barton, P. B., Jr., and Toulmin, Priestley III, 1961, Some mechanisms for cooling hydrothermal fluids: U.S. Geol. Survey Prof. Paper 424-B, p. 348-352.

Billings, G. K., Kesler, S. E., and Jackson, S. A., 1969, Relation of zinc-rich formation waters, northern Alberta, to the Pine Point ore deposit: Econ. Geology, v. 64, p. 385-391.

Bischoff, J. L., 1969, Red Sea geothermal brine deposits: their mineralogy, chemistry, and genesis, in Degens, E. T., and Ross, D. A., eds., Hot brines and recent heavy metal deposits in the Red Sea: New York, Springer-Verlag, p. 368-401.

Fournier

Bischoff, J. L., and Dickson, F. W., 1975, Sea water–basalt interaction at 200°C and 500 bars: Implications for the origin of sea-floor heavy metal deposits and regulation of seawater chemistry: Earth and Planetary Sci. Letters, v. 25, p. 379–384.

Bowen, N. L., 1933, The broader story of magmatic differentiation, briefly told, in Ore deposits of the Western States (Lindgren Vol.); New York, Am. Inst. Mining Metall. Engineers, p. 106–128.

Brewer, P. G., and Spencer, D. W., 1969, A note on the chemical composition of the Red Sea brines, in Degens, E. T., and Ross, D. A., eds., Hot brines and recent heavy metal deposits in the Red Sea: New York, Springer-Verlag, p. 174–179.

Briner, E., and Roth, P., 1948, Recherches sur l'hydrolyse par la vapeur d'eau de chlorures alcalins seuls ou additionnés de divers adjuvants: Helvetica Chimica Acta, v. 31, p. 1352–1360.

Browne, P. R. L., 1969, Sulfide mineralization in a Broadlands geothermal drillhole, Taupo Volcanic Zone, New Zealand: Econ. Geology, v. 64, p. 156–159.

Browne, P. R. L., 1971, Mineralization in the Broadlands geothermal field, Taupo Volcanic Zone, New Zealand: Society Mining Geology Japan (Spec. Issue, 2) (Proc. IMA-IAGOD Meetings '70 Joint Symp. Vol.), p. 64–75.

Browne, P. R. L., 1978, Hydrothermal alteration in active geothermal fields: Annual Review of Earth and Planetary Sciences, v. 6, p. 229–250.

Burnham, C. W., 1967, Hydrothermal fluids at the magmatic stage, in Barnes, H. L., ed., Geochemistry of hydrothermal ore deposits: New York, Holt, Rinehart and Winston, p. 34–76.

Burnham, C. W., 1979, Magmas and hydrothermal fluids, in Barnes, H. L., ed., Geochemistry of hydrothermal ore deposits, 2nd ed.: New York, John Wiley and Sons, p. 17–136.

Calamai, A., Cataldi, R., Dall'Aglio, M., and Ferrara, G. C., 1976, Preliminary report on the Cesano hot brine deposit (Northern Latium, Italy), in Proceedings, 2nd U. N. Symp. on the Development and Use of Geothermal Resources, 1975, San Francisco: v. 1, p. 305–313.

Carmichael, I. S. E., Turner, F. J., and Verhoogen, J., 1973, Igneous petrology: New York, McGraw Hill, 739 p.

Carpenter, A. B., Trout, M. L., and Pickett, E. E., 1974, Preliminary report on the origin and chemical evolution of lead- and zinc-rich brines of central Mississippi: Econ. Geology, v. 69, p. 1191–1206.

Cathles, L. M., 1977, An analysis of the cooling of intrusives by ground-water convection which includes boiling: Econ. Geology, v. 72, p. 804–826.

Chen, C.-H., 1970, Geology and geothermal power potential of the Tatun volcanic region: Geothermics (Spec. Issue, 2), (Pt. 2), p. 1134–1143.

Chen, C.-H., 1975, Thermal waters in Taiwan, a preliminary study: Proc. Internat. Assoc. Hydrol. Sci., Grenoble, Publ. 199, p. 79–88.

Cloke, P. L., and Kesler, S. E., 1979, The halite trend in hydrothermal solutions: Econ. Geology, v. 74, p. 1823–1831.

Craig, H., 1963, The isotopic geochemistry of water and carbon in geothermal areas, in Tongiorgi, E., ed., Nuclear geology on geothermal areas, Spoleto, Italy, 1963: Consiglio Nazionale delle Ricerche Laboratorie di Geologia Nucleare, Pisa, p. 17–53.

Cunningham, C. G., 1978, Pressure gradients and boiling as mechanisms for localizing ore in porphyry systems: U.S. Geol. Survey Jour. Research, v. 6, p. 745–754.

Edmond, J. M., Craig, H., Gordon, L. I., and Holland, H. D., 1979, Chemistry of hydrothermal waters at 21°N on the East Pacific Rise: EOS (Am. Geophys. Union Trans.), v. 60, p. 864.

Elder, J. W., 1965, Physical processes in geothermal areas, in Terrestial Heat Flow: Geoph. Mon. 8, NAS-NRC noo. 1288, Am. Geophys. Union, p. 211–239.

Ellis, A. J., 1967, The chemistry of some explored geothermal systems, in Barnes, H. L., ed., Geochemistry of hydrothermal ore deposits: New York, Holt, Rinehart and Winston, p. 465–514.

Ellis, A. J., 1968, Natural hydrothermal systems and experimental hot-water/rock interactions: Reactions with NaCl solutions and trace metal extraction: Geochim. et Cosmochim. Acta, v. 32, p. 1356–1363.

Ellis, A. J., 1970, Quantitative interpretation of chemical characteristics of hydrothermal systems: Geothermics (Spec. Issue, 2), v. 2 (Pt. 1), p 516–528.

Ellis, A. J., 1977, Chemical and isotopic techniques in geothermal investigations: Geothermics, v. 5, p. 3–12.

Ellis, A. J., 1979, Explored geothermal systems, in Barnes, H. L., ed., Geochemistry of hydrothermal ore deposits, 2nd ed.: New York, John Wiley and Sons, p. 632–683.

Ellis, A. J., and Mahon, W. A. J., 1964, Natural hydrothermal systems and experimental hot-water/rock interactions: Geochim. et Cosmochim. Acta, v. 28, p. 1323–1357.

Ellis, A. J., and Mahon, W. A. J., 1967, Natural hydrothermal systems and experimental hot-water/rock interactions (Pt. 2): Geochim. et Cosmochim. Acta, v. 31, p. 519–539.

Ellis, A. J., and Mahon, W. A. J., 1977, Chemistry and geothermal systems: New York, Academic Press, 392 p.

Ellis, A. J., and Wilson, S. H., 1955, The heat from the Wairakei-Taupo thermal region calculated from the chloride output: New Zealand Jour. Sci. and Technology, Sect. B, v. 36, p. 622–631.

Ewers, G. R., 1977, Experimental hot-water/rock interactions and their significance to natural hydrothermal systems of New Zealand: Geochim. et Cosmochim. Acta, v. 41, p. 143–150.

Ewers, G. R., and Keays, R. R., 1977, Volatile and precious metal zoning in the Broadlands geothermal field, New Zealand: Econ. Geology, v. 72, p. 1337-1354.

Fenner, C. N., 1933, Pneumatolytic processes in the formation of minerals and ores, in Ore deposits of the western states: Rocky Mountain Fund Series, A.I.M.E., p. 58-106.

Fournier, R. O., 1968, Depths of intrusion and conditions of hydrothermal alteration in porphyry copper deposits (abs.): Geol. Soc. America Program with Abstracts, 1968 Annual Meeting, p. 101.

Fournier, R. O., 1972, The importance of depth of crystallization on the character of magmatic fluids (abs.): Internat. Geol. Cong., 24th, Montreal, 1972, Sect. 10, p. 214.

Fournier, R. O., 1976, Exchange of Na^+ and K^+ between water vapor and feldspar phases at high temperature and low vapor pressure: Geochim. et Cosmochim. Acta, v. 40, p. 1553-1561.

Fournier, R. O., 1977, Constraints on the circulation of meteoric water in hydrothermal systems imposed by the solubility of quartz: Geol. Soc. America Abstracts with Programs, v. 9, p. 979.

Fournier, R. O., 1979, Geochemical and hydrologic considerations and the use of enthalpy-chloride diagrams in the prediction of underground conditions in hot spring systems: Jour. of Volcanology and Geothermal Research, v. 5, p. 1-16.

Fournier, R. O., 1981, Application of water geochemistry to geothermal exploration and reservoir engineering, in Rybach, L., and Muffler, L. J. P., eds., Geothermal systems: principles and case histories: New York, John Wiley and Sons, p. 109-143.

Fournier, R. O., 1983, A method of calculating quartz solubilities in aqueous sodium chloride solutions: Geochim. et Cosmochim. Acta, v. 47, p. 579-586.

Fournier, R. O., in press, Silica minerals as indicators of conditions during gold deposition: U.S. Geol. Survey Circ.

Fournier, R. O., and Potter, R. W. II, 1982, An equation correlating the solubility of quartz in water from 25° to 900°C at pressures up to 10,000 bars: Geochim. et Cosmochim. Acta, v. 46 p. 1969-1973.

Fournier, R. O., White, D. E., and Truesdell, A. H., 1976, Convective heat flow in Yellowstone National Park, in Proceedings, 2nd U. N. Symp. on the Development and Use of Geothermal Resources, San Francisco, 1975: v. 1, p. 731-739.

Francheteau, Jean, Needham, D., Choukroyne, P., and others, 1979, Massive deep-sea sulfide ore deposits discovered on the East Pacific Rise: Nature, v. 277, p. 523-528.

Galobardes, D. R., Van Hare, D. R., and Rogers, L. B., 1981, Solubility of sodium chloride in dry steam: Jour. Chem. Eng. Data, v. 26, p. 363-366.

Gerlach, T. M., and Nordlie, B. E., 1975, The C-O-H-S gaseous system, Pt. I: Am. Jour. Sci., v. 275, p. 353-377.

Giggenbach, W. F., 1977, The isotopic composition of sulphur in sedimentary rocks bordering the Taupo Volcanic Zone, in Ellis, A. J., ed., Geochemistry 1977: New Zealand D.S.I.R. Bull., v. 218, p. 57-64.

Gooch, F. A., and Whitfield, J. E., 1888, Analyses of waters of the Yellowstone National Park, with an account of the methods of analysis employed: U. S. Geol. Survey Bull. v. 47, 84 p.

Goranson, R. W., 1931, The solubility of water in granite magmas: Am. Jour. Sci., v. 22, p. 481-502.

Graeber, E. J., Gerlach, T. M., and Hlava, P. F., 1982, Metal transport and deposition in high-temperature fumaroles at Mount St. Helens [abs.]: EOS Am. Geophysical Union Trans., v. 63, p. 1143.

Griffiths, R. W., 1978, Layered double diffusion convection in a porous medium: Australian Nat. Univ., Ph.D. Thesis, 55 p.

Grindley, G. W., and Browne, P. R. L., 1976, Structural and hydrological factors controlling the permeabilities of some hot-water geothermal fields, in Proceedings, 2nd U. N. Symp. on the Development and Use of Geothermal Resources, San Francisco, 1975: v. 1, p. 377-386.

Haas, J. L., 1971, Effect of salinity on the maximum thermal gradient of a hydrothermal system at hydrostatic pressure: Econ. Geology, v. 66, p. 940-946.

Hajash, Andrew, Jr., 1975, Hydrothermal processes along mid-ocean ridges: An experimental investigation: Texas A. and M. Univ., Unpub. Ph.D. thesis, 61 p.

Hanor, J. S., 1979, The sedimentary genesis of hydrothermal fluids, in Barnes, H. L., ed., Geochemistry of hydrothermal ore deposits, 2nd ed.: New York, John Wiley and Sons, p. 137-172.

Harper, R. T. and Arevalo, E. M., 1982, A geoscientific evaluation of the Basley-Danin prospect, Negros Oriental, Philippines, in Proceedings, Pacific Geothermal Conference, 1982, incorporating 4th New Zealand Geothermal Workshop, Pt. 1: p. 235-240.

Hayashi, M., Taguchi, S., and Yamasaki, T., 1981, Activity index and thermal history of geothermal systems: Trans. Geothermal Resources Council, v. 5, p. 177-180.

Healy, J., 1976, Geothermal fields in zones of recent volcanism, in Proceedings, 2nd U. N. Symp. on the Development and Use of Geothermal Resources, 1975, San Francisco: v. 1, p. 415-422.

Healy, J., and Hochstein, M. P., 1973, Horizontal flows in geothermal systems: J. Hydrol. (N.Z.), v. 12, p. 71-82.

Helgeson, H. C., 1967, Silicate metamorphism in sediments and the genesis of hydrothermal ore solutions: Econ. Geology, Monograph 3, p. 333-342.

Helgeson, H. C., 1968, Geologic and thermodynamic characteristics of the Salton Sea geothermal system: Am. Jour. Sci., v. 266, p. 129-166.

Fournier

Heming, R. F., Hochstein, M. P., and McKenzie,
W. F., 1982, Suretimeat geothermal system:
An example of a volcanic geothermal system,
in Proceedings, Pacific Geothermal
Conference, 1982, incorporating 4th New
Zealand Geothermal Workshop, Pt. 1: p.
247-250.

Hemley, J. J., Montoya, J. W., Nigrini, A., and
Vincent, H. A., 1971, Some alteration
reactions in the system
$CaO-Al_2O_3-SiO_2-H_2O$: Soc. Mining
Geol. Japan, Spec. Issue 2, p. 58-63.

Hemley, J. J., and Jones, W. R., 1964, Chemical
aspects of hydrothermal alteration with
emphasis on hydrogen metasomatism: Econ.
Geology, v. 59, p. 538-569.

Hendricks, R. L., Reisbick, F. B., Mahaffey, E.
J., Roberts, D. B., and Peterson, M. N. A.,
1969, Chemical composition of sediments and
interstitial brines from the Atlantis II,
Discovery, and Chain Deeps, in Degens, E.
T., and Ross, D. A., eds., Hot brines and
recent heavy metal deposits in the Red Sea:
New York, Springer-Verlag, p. 402-440.

Henley, R. W., and Ellis, A. J., 1983,
Geothermal systems ancient and modern: A
geochemical review: Earth-Science Reviews,
v. 19, 50 p.

Henley, R. W., and McNabb, A., 1978, Magmatic
vapor plumes and ground-water interaction in
porphyry copper emplacement: Econ. Geology,
v. 73, 19 p.

Henley, R. W., and Thornley, P., 1979, Some
geothermal aspects of polymetallic massive
sulfide formation: Econ. Geology, v. 74, p.
1600-1612.

Holland, H. D., 1972, Granites, solutions and
base metal deposits: Econ. Geology, v. 67,
p. 281-301.

Horikoshi, E., 1969, Volcanic activity related
to the formation of Kuroko-type deposits,
Japan: Mineralium Deposita, v. 4, p.
321-345.

Kartokusumo, W., Mahon, W. A. J., and Seal, K.
E., 1976, Geochemistry of the Kawah Kamojang
geothermal system, Indonesia, in
Proceedings, 2nd U. N. Symp. on the
Development and Use of Geothermal Resources,
1975, v. 1, p. 575-579.

Keith, T. E. C., and Muffler, L. J. P., 1978,
Minerals produced during cooling and
hydrothermal alteration of ash flow tuff
from Yellowstone drill hole Y-5: Jour. of
Volcanology and Geothermal Research, v. 3,
p. 373-402.

Kennedy, G. C., 1950, A portion of the system
silica-water: Econ. Geology v. 45, p.
629-653.

Kharaka, Y. K., Brown, P. M., and Carothers, W.
W., 1978, Chemistry of waters in the
geopressured zone from coastal
Louisiana--Implications for the geothermal
development: Trans. Geothermal Resources
Council, v. 2, p. 371-374.

Kilinc, I. A., and Burnham, C. W., 1972,
Partitioning of chloride between a silicate
melt and coexisting aqueous phase from 2 to
8 kilobars: Econ. Geology, v. 67, p.
231-235.

Kissen, I. G., and Pakhomov, S. I., 1967, Effect
of high temperatures on the chemical
composition of ground waters: Geochemistry
Int., v. 4, p. 295-308.

Koski, R. A., Normark, W. R., Morton, J. L., and
Delaney, J. R., 1982, Metal sulfide deposits
on the Juan de Fuca Ridge: Oceanus, v. 25,
p. 42-48.

Lahsen, A., and Trujillo, P., 1976, The
geothermal field of El Tatio, Chile, in
Proceedings, 2nd U. N. Symp. on the
Development and Use of Geothermal Resources,
1975: v. 1, p. 170-177.

Lawless, J. V., and Gonzalez, R. C., 1982,
Geothermal geology and review of
exploration, Biliran Island, in Proceedings,
Pacific Geothermal Conference, 1982,
incorporating 4th New Zealand Geothermal
Workshop, Pt. 1: p. 161-166.

Leach T. M., and Bogie, I, 1982, Overprinting of
hydrothermal regimes in the Palimpinon
Geothermal Field, Southern Negros,
Philippines, in Proceedings, Pacific
Geothermal Conference, 1982, incorporating
4th New Zealand Geothermal Workshop, Pt. 1:
p. 179-184.

Lebedev, L. M., 1967, Recent deposition of
native lead from Cheleken thermal brines:
Akad. Nauk SSSR Doklady, v. 174, p. 197-200.

Lebedev, L. M., 1973, Minerals of contemporary
hydrotherms of Cheleken: Geochemistry Int.,
v. 9, p. 485-504.

Lebedev, L. M., 1975, Ore-forming modern
hydrotherms (in Russian): Moscow, Nedra,
261 p.

Lloyd, E. F., 1959, The hot springs and
hydrothermal eruptions of Wairapu: New
Zealand Jour. Geology and Geophysics, v. 2,
p. 141-176.

Lloyd, E. F., 1972, Geology and Hot Springs of
Orakeikorako: New Zealand Geol. Survey
Bull. , v. 85, 163 p.

Lloyd, E. F., 1976, Waimungu hydrothermal field,
in Volcanic and geothermal geology of the
central North Island, New Zealand:
Internat. Geol. Cong., 25th, Sydney 1976,
Excursion Guides 55A and 56A, p. 23-26.

Lovering, T. S., 1961, Sulfide ores from sulfide
deficient solutions: Econ. Geology, v. 56,
p. 68-99.

Mahon, W. A. J., 1967, Natural hydrothermal
systems and the reaction of hot water with
sedimentary rocks: New Zealand Jour. Sci.,
v. 10, p. 206-221.

Mahon, W. A. J., Klyen, L. E., and Rhode, M.,
1980, Neutral sodium/bicarbonate/sulphate
hot waters in geothermal systems: Chinetsu
(Jour. Jpn. Geotherm. Energy Assoc.), v. 17,
p. 11-24.

Martynova, O. I., and Samoilov, Yu. F., 1959,
Dissolution of sodium chloride in an
atmosphere of water vapor of high parameters
(Trans.): Zhurnal Neorganisheskoi Khimii,
v. II, no. 12, p. 2829-2833.

McKibben, M. A., 1979, Ore minerals in the
Salton Sea geothermal system, Imperial
Valley, California, U.S.A.: Dept. of Earth

Sciences and Inst. of Geophysics and Planetary Physics, Univ. of Calif., Riverside, Report 79/7, 99 p.

McLaughlin, R. J., Moore, D. E., Sorg, D. H., and McKee, E. H., 1983, Multiple episodes of hydrothermal circulation, thermal metamorphism, amd magma injection beneath The Geysers steam field, California: Geol. Soc Amer. Abstracts With Programs 1983, p. 417.

McNabb, A., 1975, Geothermal physics: New Zealand D.S.I.R., Appl. Math. Div. Tech. Rep., 32, 35 p.

Menyailov, I. A., and Nikitina, L. P., 1974, Zinc and lead in gases and waters of the Ebeko Volcano and Pauzhetka deposit, in Naboko, S. I., ed., Hydrothermal mineral-forming solutions in areas of active volcanism: Translation, Published for the U.S. Dept. Interior and NSF, New Delhi, Amerind Publishing Co., 1982, p. 149-159.

Meyer, Charles, and Hemley, J. J., 1967, Wall rock alteration, in Barnes, H. L., ed., Geochemistry of hydrothermal ore deposits: New York, Holt, Rinehart and Winston, 670 p.

Morey, G. W., 1922, The development of pressure in magmas as a result of crystallization: Washington Acad. Sci. Jour., v. 12, p. 219-230.

Mottl, M. J., Holland, H. D., and Corr, R. F., 1979, Chemical exchange during hydrothermal alteration of basalt by seawater, II. Experimental results for Fe, Mn and S species: Geochim. et Cosmochim. Acta, v. 43, p. 869-884.

Muffler, L. J. P., Nehring, N. L., Truesdell, A. H., Janik, C. J., Clynne, M. A., and Thompson, J. M., 1982, The Lassen Geothermal System, in Proceedings, Pacific Geothermal Conference, 1982, incorporating 4th New Zealand Geothermal Workshop, Pt. 2: p. 349-356.

Muffler, L. J. P., White, D. E., and Truesdell, A. H., 1971, Hydrothermal explosion craters in Yellowstone National Park: Geol. Soc. America Bull., v. 82, p. 723-740.

Naboko, S. I., 1974, Metal content of the Uzon Caldera, in Naboko, S. I., ed., Hydrothermal mineral-forming solutions in areas of active volcanism: Translation, Published for the U.S. Dept. Interior and NSF, New Delhi, Amerind Publishing Co., 1982, p. 132-140.

Nairn, I. A., and Wiradiradja, S., 1980, Late Quaternary hydrothermal explosion breccias at Kawerau geothermal field, New Zealand: Bull. Volcanol., v. 43, 13 p.

Nakamura, H., Sumi, K., Katagiri, K., and Iwata, T., 1970, The geological environment of Matsukawa geothermal area, Japan: Geothermics (Spec. Issue, 2), v. 2 (Pt. 1), p. 221-231.

Normark, W. R., Lupton, J. E., Murray, J. W., Koski, R. A., Clague, D. A., Morton, J. L., Delaney, J. R., and Johnson, H. P., 1982, Polymetallic sulfide deposits and water-column tracers of active hydrothermal vents on the Southern Juan de Fuca Ridge: Marine Technology Society Journal, v. 16, p. 46-53.

Norton, Denis, and Cathles, L. M., 1979, Thermal aspects of ore deposition, in Barnes, H. L., ed., Geochemistry of hydrothermal ore deposits, 2nd ed.: New York, John Wiley and Sons, p. 611-631.

Ohmoto, H., and Rye, R. O., 1974, Hydrogen and oxygen isotopic compositions of fluid inclusions in the Kuroko deposits, Japan: Econ. Geology, v. 69, p. 947-953.

Oki, Y., and Hirano, T., 1970, The geothermal system at the Hakone Volcano: Geothermics (Spec. Issue, 2), v. 2, pt. 2, p. 1157-1166.

Orville, P. M., 1963, Alkali ion exchange between vapor and feldspar phases: Am. Jour. Sci., v. 261, p. 201-237.

Phillips, W. J., 1973, Mechanical effects of retrograde boiling and its probable importance in the formation of some porphyry ore deposits: Inst. Mining Metallurgy Trans., Sec. B, v. 82, p. B90-B98.

Ravich, M. I., and Borovia, F. E., 1949, Phase equilibria in ternary water-salt systems at elevated temperatures (in Russian): Akad. Nauk SSSR. Institut obshchei i neorganicheskoi. Izvestiia. Sektor fiziko khimicheskogo analiza, v. 19, p. 69-81.

Rowe, J. J., Fournier, R. O., and Morey, G. W., 1967, The system water-sodium oxide-silicon dioxide at 200, 250, and 300°: Inorg. Chem., v. 6, p. 1183-1188.

Rowe, J. J., Fournier, R. O., and Morey, G. W., 1973, Chemical analysis of thermal waters in Yellowstone National Park, Wyoming, 1960-1965: U.S. Geol. Survey Bull. 1303, p. 31.

Ryabchikov, I. D., and Hamilton, D. L., 1971, Possible separation of concentrated chloride solutions during crystallization of felsic magma: Akad. Nauk SSSR Doklady, v. 197, p. 219-220.

Seastres, J. S., Jr., 1982, Subsurface geology of the Nasuji-Sogongon sector, Southern Negros Geothermal Field, Philippines, in Proceedings, Pacific Geothermal Conference, 1982, incorporating 4th New Zealand Geothermal Workshop, Pt. 1: p. 173-178.

Seyfried, W., and Bischoff, J. L., 1977, Hydrothermal transport of heavy metals by sea-water: The role of seawater/basalt ratio: Earth and Planetary Sci. Letters, v. 334, p. 71-77.

Shanks, W. C. III, and Bischoff, J. L., 1977, Ore transport and deposition in the Red Sea geothermal system: a geochemical model: Geochim. et Cosmochim. Acta, v. 41, p. 1507-1519.

Skinner, D. N. B., 1966, Geology of Ngawha: New Zealand Geol. Survey, Rept. 16, p. 3-19.

Sourirajan, S., and Kennedy, G. C., 1962, The system H_2O-NaCl at elevated temperatures and pressures: Am. Jour. Sci., v. 260, p. 115-141.

Sternfeld, J. N., 1981, The hydrothermal petrology and stable isotope geochemistry of two wells in The Geysers geothermal field, Sonoma County, California: Dept. of Earth Sciences and Inst. of Geophysics and Planetary Physics, Univ. of Calif., Riverside, Masters Thesis, 201 p.

Stoiber, R. E., and Jepsen, A., 1973, Sulfur dioxide contributions to the atmosphere by volcanoes: Science, v. 182, p. 577-578.

Studt, F. E., 1958, The Wairakei hydrothermal field under exploitation: New Zealand Jour. Geology and Geophys., v. 1, p. 703-723.

Sutton, F. M., and McNabb, A., 1977, Boiling curves at Broadlands geothermal field, New Zealand: New Zealand Jour. Sci., v. 20, p. 333-337.

Tkachenko, R. I., and Zotov, A. V., 1974, Ultra-acidic therms of volcanic origin as mineralizing solutions, in Naboko, S. I., ed., Hydrothermal mineral-forming solutions in areas of active volcanism: Translation, Published for the U.S. Dept. Interior and NSF, New Delhi, Amerind Publishing Co., 1982, p. 126-131.

Toulmin, Priestly III, and Clark, S. P., Jr., 1967, Thermal aspects of ore formation, in Barnes, H. L., ed., Geochemistry of hydrothermal ore deposits: New York, Holt, Rinehart and Winston, p. 437-464.

Truesdell, A. H., 1976, Summary of section III: Geochemical techniques in exploration, in Proceedings, 2nd U. N. Symp. on the Development and Use of Geothermal Resources, 1975, v. 1, p. liii-lxxix.

Truesdell, A. H., and Fournier, R. O., 1976, Conditions in the deeper parts of Yellowstone National Park, Wyoming: U. S. Geological Survey Open-file Report 76-428, 29 p.

Truesdell, A. H., and Hulston, J. R., 1980, Isotopic evidence on environments of geothermal systems, in Fritz, P., and Fontes, J. C., eds., Handbook of environmental isotope geochemistry: New York, Elsevier, v. 1, p. 179-225.

Vidal, V. M. V., Vidal, F. V., Isaacs, J. D., and Young, D. R., 1978, Coastal submarine activity off northern Baja California (Mexico): Jour. Geophys. Research, v. 83, p. 1757-1774.

Walker, A. L., 1965, Some factors affecting gas phase ore transport: Econ. Geology, v. 60, p. 117-123.

Weissberg, B. G., 1969, Gold-silver ore-grade precipitates from New Zealand thermal waters: Econ. Geology, v. 64, p. 95-108.

Weissberg, B. C., Browne, P. R. L., and Seward, T. M., 1979, Ore metals in active geothermal systems, in Barnes, H. L., ed., Geochemistry of hydrothermal ore deposits, 2nd ed.: New York, John Wiley and Sons, p. 738-780.

White, D. E., 1955, Thermal springs and epithermal ore deposits: Econ. Geology, 50th Anniversary Volume, p. 99-154.

White, D. E., 1965, Saline waters of sedimentary rocks: Am. Assoc. Petroleum Geologists Mem. 4, p. 342-366.

White, D. E., 1967, Mercury and base-metal deposits with associated thermal and mineral waters, in Barnes, H. L., ed., Geochemistry of hydrothermal ore deposits: New York, Holt, Rinehart and Winston, p. 575-631.

White, D. E., 1968a, Environments of generation of some base-metal ore deposits: Econ. Geology, v. 63, p. 301-335.

White, D. E., 1968b, Hydrology, activity, and heat flow of the Steamboat Springs thermal system, Washoe County, Nevada: U.S. Geol. Survey Prof. Paper 458-C, 109 p.

White, D. E., 1973, Characteristics of geothermal resources, in Kruger, Paul, and Otte, Carol, eds., Geothermal Energy: Resources, Production, Stimulation: Stanford Univ. Press, Stanford, California, p. 69-94.

White, D. E., 1974, Diverse origins of hydrothermal ore fluids: Econ. Geology, v. 69, p. 954-973.

White, D. E., 1981, Active geothermal systems and hydrothermal ore deposits: Econ. Geol., 75th Anniversary Volume, p. 392-423.

White, D. E., Anderson, E. T., and Grubbs, D. K., 1963, Geothermal brine well/mile-deep drill hole may tap ore-bearing magmatic water and rocks undergoing metamorphism: Science, v. 139, p. 919-922.

White, D. E., Barnes, I., and O'Neil, J. R., 1973, Thermal and mineral waters of nonmeteoric origin, California Coast Range: Geol. Soc. America Bull. v. 84, p. 547-560.

White, D. E., Fournier, R. O., Muffler, L. J. P., and Truesdell, A. H., 1975, Physical results of research drilling in thermal areas of Yellowstone National Park, Wyoming: U.S. Geol. Survey Prof. Paper 892.

White, D. E., Muffler, L. J. P., and Truesdell, A. H., 1971, Vapor-dominated hydrothermal systems compared to hot water systems: Econ. Geol., v. 66, p. 75-97.

White, D. E., and Robinson, C. E., 1962, Sulphur Bank, California, a major hot-spring quicksilver deposit: Geol. Soc. America, Buddington Volume, p. 397-428.

White, D. E., and Waring, G. A., 1963, Volcanic emanations, in Data of geochemistry, 6th ed.: U.S. Geol. Survey Prof. Paper 440-K, p. K1-K29.

Whitney, J. A., 1975, Vapor generation in a quartz monzonite magma: A synthetic model with application to porphyry copper deposits: Econ. Geol., v. 70, p. 346-358.

Wooding, R. A., 1963, Convection in a saturated porous medium at large Rayleigh number or Peclet number: Jour. Fluid Mech., v. 15, p. 527-544.

Yamada, Eizo, 1976, Geological development of the Onikobe caldera and its hydrothermal system, in Proceedings, 2nd U. N. Symp. on the Development and Use of Geothermal Resources, 1975, San Francisco: v. 1, p. 655-672.

Distribution of Fossil Metallogenic Systems and Magma Geochemical Belts within the Great Basin and Vicinity from 145 Million Years Ago to Present

Stanley B. Keith
2748 East 9th Street
Tucson, AZ 85716

ABSTRACT

The southwestern United States (including the Great Basin) contains over 2000 Mesozoic-Cenozoic fossil hydrothermal systems with commercial metallic production. No significant correlation exists between metal content of ore deposit type and metal content of the enclosing tectonic terrane. For example, "Carlin-type" disseminated gold deposits with anomalous amounts of Hg, As, Sb, W, Tl occur in western Arizona in Precambrian Granite and lower Paleozoic quartzites and in the California coast range mercury belt in Franciscan graywacke-serpentinite hostrocks in addition to the more familiar Antler carbonate rocks of north-central Nevada. In contrast, a strong correlation exists between alkalinity (expressed by $K_{57.5}$ index) of time-related metaluminous igneous rock series and overall metal content of over 1000 worldwide sulfide systems (see Table below).

OVERALL METAL CONTENT OF MINERAL DEPOSITS ASSOCIATED WITH METALUMINOUS IGNEOUS ROCK SERIES[1]

SUBALKALINE [4] $K_{57.5} \leq 2.45$ [2]; PI ≥ 56 [3]; $Ca_{70} > 1.5$; F content of magmatic biotite $< 0.7\%$		ALKALINE $K_{57.5} \geq 2.45$; PI ≤ 56; $Ca_{70} < 1.5$; F content of magmatic biotite commonly $>0.7\%$	
CALCIC $K_{57.5} = 0.0-1.15$	CALC-ALKALI $K_{57.5} = 1.15-2.45$	ALKALI-CALCIC $K_{57.5} = 2.45-3.8$	ALKALIC $K_{57.5} = \geq 3.8$
Fe[5], Mn, V, Ti, Cu, Bn, Sn, Pb, Cd, Ag, Au, Platinoids, Co, Cr, Ni, S; Lack of U, Th, F, Li, Hg, Sb, Mo, Ba, and generally W, As (except for exhalative calcic gold systems) and Sn (except exhalative Cu-Zn massive sulfide facies systems).	Fe[5], Cu, Zn, Pb, Ag, Au, Sb, Hg, S, Ba, W, Mo, Mn, Bi, (As, Se, Tl); Lack of Ni, Cr, Sn, F, U, Th, Ti and generally Co and platinoids; W occurs strictly as scheelite; S is apparently maximized in this class; pyrite: pyrrhotite ratio is much higher than in calcic systems.	Fe[5], Mn, Pb, Zn, Ag, Cu, Au, Mo, Sn, Be, U, F, W, (Te); general lack of Hg (except peralkalic), Sb, Co, Ni, Ti, Cr, platinoids; lower S than in subalkalic systems; W occurs as huebnerite-wolframite series $\pm$ scheelite.	Cu[5], Pb, Zn, Au, Ag, F, U, Th, P, Mo, Fe, Mn, rare earths, Nb (Te); lacks same elements as alkali-calcic systems; W occurs mostly as huebnerite-wolframite series.

Notes: (1) Metaluminous igneous rocks series are where molecular $(Al_2O_3 / CaO + K_2O + Na_2O) < 1.0$ somewhere in the differentiation continuum. (2) $K_{57.5}$ index = K_2O content at 57.5% SiO_2 on a K_2O-SiO_2 variation diagram. (3) PI = Peacock alkali-lime index. (4) CA_{70} = CaO content at 70% SiO_2 on a CaO-SiO_2 variation diagram. (5) Nonunderlined elements have major production with respect to other classes; underlined elements have minor production with respect to other classes; double underlined elements have very minor production (by-product only) with respect to other classes; elements enclosed by parentheses are consistently occurring trace elements in anomalous but generally noncommercial amounts.

One interpretation of the above table is that the different metallic suites and their correlative alkali magma chemistry represent chemically distinct layered source regions in the upper 700 km of the earth's mantle with lower layers progressively more enriched in incompatible elements.

In Southwest North America, time-slice treatment of magmatic and metallogenic data for fossil hydrothermal systems reveals that in any time-slice (1) spatial patterns of magmatism form linear belts of equal alkalinity that generally parallel the inferred position of the paleotrench. (2) Alkalinity within the magmatic arc generally increases continentward away from the trench except for periods younger than 14-12 m.y. in Arizona-Southeast California and 8-6 m.y. in the Great Basin when truly basaltic magma chemistry and no regional alkalinity gradients are apparent. (3) Each alkalinity zone is systematically associated with a metallogenic belt that contains the metallic elements outlined in the above table. (4) Between 145 and 0 m.y. the magmatic arc migrates eastward (145-43 m.y. with a dramatic eastward acceleration at 85 m.y.) and then southwestward (43-0 m.y.). (5) Migration of the magmatic arc in space and time produces a cumulative overprinting of magma chemistry and metallogenic types. For example, the region around Tonopah, Nevada was the site of hypothermal tungsten (scheelite) skarn mineralization associated with calc-alkali granodiorite plutons 95-80 m.y. ago, lithophile-poor mesothermal Mo systems (Hall, Redlich) associated with calc-alkali granites about 85-70 m.y. ago, lithophile-poor tungsten (heubnerite-wolframite)-bearing veins associated with peraluminous muscovite-bearing granites 74 to 60 m.y. ago, epithermal Au-Cu mineralization (Goldfield) associated with 25-22 m.y. calc-alkali volcanism, epithermal silver, lead-zinc mineralization (Tonopah, Divide) associated with alkali-calcic volcanism 22-17 m.y. ago, and Au-F mineralization (Bullfrog) associated with alkalic volcanism 15-8 m.y. ago.

GEOCHRONOLOGY OF HYDROTHERMAL ALTERATION AND MINERALIZATION:
TERTIARY EPITHERMAL PRECIOUS METAL DEPOSITS IN THE GREAT BASIN

Miles L. Silberman

U.S. Geological Survey, Denver, Colorado 80225

ABSTRACT

Hydrothermal mineral deposits can be dated directly by the K-Ar method because a variety of potassium-bearing gangue and wall rock alteration minerals are formed by the hydrothermal fluids that deposit the ore metals. The minerals muscovite (sericite), adularia, alunite, natroalunite, biotite, amphibole, and chlorite seem to quantitatively retain argon produced by the radioactive decay of potassium. K-Ar ages of the hydrothermal alteration minerals have been used successfully to determine the ages of formation of precious metal vein and disseminated deposits, poly-metallic base metal vein deposits, porphyry copper, molybdenum and tin deposits, tungsten and base metal scarn deposits, mercury deposits, and thermal spring systems. The hydrothermal mineralization ages have provided useful information about the temporal relationships between ore deposition and spatially associated igneous rocks, and on the patterns of mineralization within tectonic evolution and magmatic and metamorphic timing on a regional scale.

Only a few geochronologic studies on ore deposits have been done in enough detail to yield conclusive evidence on the duration of an entire hydrothermal event or on the timing of individual ore deposition and alteration episodes within that event. The most conclusive data from volcanic-hosted precious metal vein and disseminated deposits, thermal spring systems, and porphyry copper deposits suggest that on average, the total time span of hydrothermal activity is about 1 m.y., although the range of activity is between 0.6 and 2.5 m.y. From textural and theoretical studies of volcanic hosted precious metal deposits a consensus is developing that ore deposition episodes within the overall framework of hydrothermal activity are short lived, transitory, and repetitive.

INTRODUCTION

The timing of hydrothermal alteration and mineralization is important to investigate for a number of academic and practical reasons in the context of comments to the following questions:

(1) How is the time of hydrothermal activity related to volcanism and plutonism? Geochronological studies of epithermal precious metal deposits and porphyry copper deposits have shown that alteration and mineralization can be related to specific stages of igneous activity of a very local nature (e.g. Silberman and others, 1972; Ashley and Silberman, 1976; Chivas and McDougal, 1978; Whalen and others, 1982). In epithermal precious metal deposits a consistent relationship between hydrothermal alteration and later stages of volcanic activity in a cycle can be demonstrated (Silberman and others, 1972; Ashley and Silberman, 1976).

(2) What is the duration of hydrothermal activity, including both alteration and mineralization? Estimates of the duration of hydrothermal activity in ore deposits or thermal spring systems vary from nearly 3 million years (Silberman and others, 1979) to considerably less than one million years (Whalen and others, 1982; Noble and Silberman, 1983, in press). It is still uncertain as to how long it takes for ore deposition to occur within overall hydrothermal phases of activity. Studies of the physical characteristics of hot spring type and vein type of Au, Ag deposits (Buchanan, 1981; Berger and Eimon, 1983, in press; Silberman, 1982b) and attempts at detailed geochronology in complex alteration-mineralization systems (Noble and Silberman, 1983, in press) suggest that the ore depositing episode or episodes are short lived. On the other hand, there are indications from geochronology that major precious metal deposits form in conjunction with hydrothermal systems where the overall period of activity is relatively lengthy, on the order of at least 1 and perhaps greater than 3 m.y. (Silberman and others, 1972; Silberman, unpubl. data; Ashley and Silberman, 1976), and are characterized by repetitive and episodic, short lived hydrothermal processes (Silberman, 1982a; Berger and Eiman, 1983, in press).

(3) What are the regional age relationships of ore deposits within the context of the tectonic framework, including structural evolution, igneous activity and metamorphism? Age determinations of hydrothermal alteration and mineralization have been successful in relating mineral deposits to

the tectonic framework; as examples, Silberman and others (1976), and Rowen and others (1983, in press), in the Great Basin, Mitchell and others (1981), Silberman and others (1980), and Wilson (1980), in southern Alaska, and Clark and others (1982), and Damon and others (in press) in Mexico, related the timing of mineralization to tectonic processes impacting the areas in question.

Perhaps the most significant hydrothermal deposit type that has not been studied conclusively as to its age of mineralization and timing of hydrothermal activity are the carbonate sedimentary rock-hosted (Carlin type) disseminated gold deposits. Silberman and others (1973) determined the age of mineralization at the Getchell deposit, Nevada, and reported ages from a similar deposit at Gold Acres, Nevada. Data exist for similar deposits at Northumberland and Pinson, Nevada (Silberman and others, unpub. data, 1973 to 1982), but in general attempts to interpret the age relations in these deposits unequivocally have been unsuccessful. Geochronological technology, particularly the widespread application of spectrum Ar-Ar dating, is adequate to determine the hydrothermal alteration and mineralization timing in these important deposits and an attempt to carry out the studies should be made.

ACKNOWLEDGEMENTS

Many of the data presented in this paper were produced or gathered during USGS sponsored research between 1967 and 1982. The ideas were crystallized during my work with Anaconda Minerals Co. in 1982 and 1983. I am appreciative of the assistance of many colleagues and co-authors of previous reports at the USGS, the Nevada Bureau of Mines, the California Division of Mines and Geology, the Mackay School of Mines, University of Nevada, and many mining companies whose properties I worked on, upon which this summary is based. I am particularly indebted to Paul Damon and Donald White for their pioneering work on geochronology of ore deposits and on the relationship of thermal spring systems and epithermal ore deposits, respectively.

METHODS AND MATERIALS--EPITHERMAL PRECIOUS METAL DEPOSITS

Several methods of isotopic age determination can be applied to dating hydrothermal alteration, but the most widely applied one has been conventional K-Ar analyses. Hydrothermal fluids that deposit ore metals also deposit K-bearing gangue and wall rock alteration minerals that can be dated by the method. The alteration mineralogy in any hydrothermal system depends on a variety of factors including original host rock mineralogy, hydrothermal fluid temperature, composition, pH, and eH (Meyer and Hemley, 1967; Rose and Burt, 1979).

The K-bearing minerals used to determine the age of hydrothermal alteration include hydrothermal K-feldspars (adularia) (Silberman and others, 1972; Koski and others, 1978), sericite (Silberman and others, 1973), (muscovite), alunite

and natroalunite (Ashley and Silberman, 1976; Mehnert and others, 1973), and jarosite (Erickson and others, 1978). All of these are common gangue and wall rock alteration minerals in epithermal vein and disseminated Au, Ag deposits. Adularia, sericite, and alunite are the most commonly dated alteration phases in epithermal deposits. Other minerals that have been used to date alteration directly are hydrothermal biotite and phlogopite (Moore and Lanphere, 1971), hydrothermal amphibole (Silberman and others, 1977), chlorite (Silberman and others, 1977), mariposite (Silberman and Dodge, unpublished data, 1979), and whole rock samples containing combinations of the K-bearing phases (Silberman and others, 1972; Morton and others, 1977; Silberman and others, 1979a; 1979b).

The second most widely applied method of geochronology of alteration has been fission track dating. Apatite, zircon, and sphene have been used successfully to study age relations in alteration systems (Ashley and Silberman, 1976; Lipman and others, 1976). The combination of K-Ar and fission track techniques has been particularly useful as the temperature of annealing of fission tracks in minerals is better known than that for diffusional loss of argon (Nasser and Faul, 1969). Combined fission track and K-Ar data allows a more complete thermal history of an alteration system to be constructed.

EXAMPLES OF ORE DEPOSIT GEOCHRONOLOGY

There are numerous examples of detailed geochronological studies of hydrothermal alteration and mineralization in the literature. I have chosen to summarize two here because of the large numbers of isotopic ages reported in them and because they were specifically designed to answer parts of the questions posed in (1) and (2) of the introduction. Silberman and others (1972) reported on the age of mineralization at Bodie, California in a paper that summarized results of 40 K-Ar analyses from rocks and veins within and nearby the district. Silberman and Ashley (1970) and Ashley and Silberman (1976) summarized the results of 55 K-Ar and fission track ages from the Goldfield district, Nevada.

Bodie mining district

Geology of the mining district

The Bodie mining district, which produced more than 34 million dollars worth of Au and Ag from fissure veins in volcanic rocks of intermediate composition, is near the eastern margin or the Bodie Hills, a volcanic massif of approximately 35 km^3 volume and thickness of up to 1300 feet (400 m). The area surrounding the Bodie Hills contitutes a major volcanic province (fig. 1) which has been subdivided on the basis of dominant overall lithology and age of activity (Kleinhampl and others, 1975; Chesterman, 1968; Gilbert and others, 1968).

Chesterman (1968) and Chesterman and Gray (1975) defined five volcanic formations of Tertiary age rocks in the Bodie Hills, based on

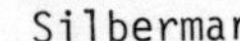

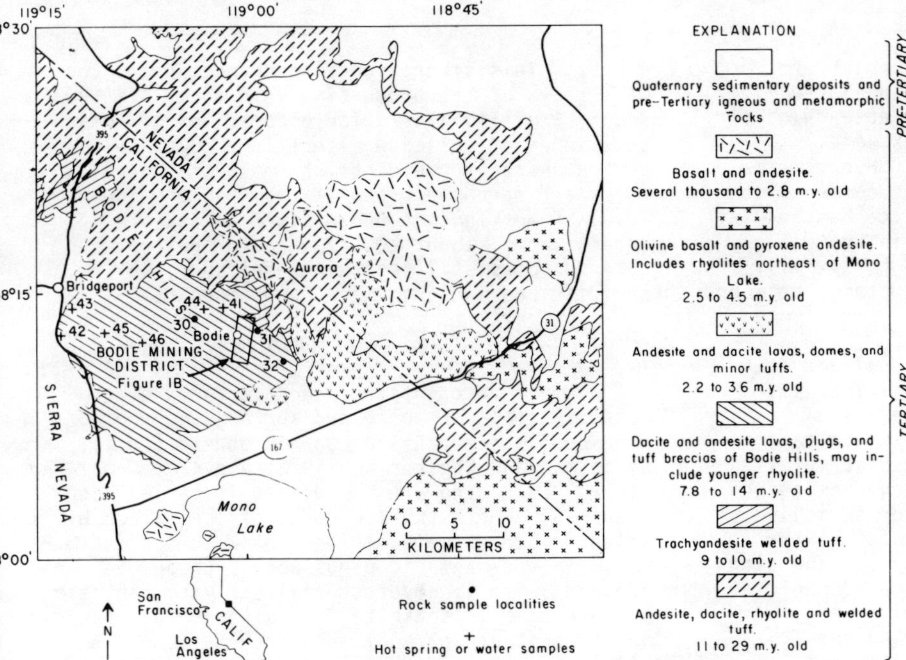

Figure 1. Regional geologic setting of the Bodie Hills and vicinity, California and Nevada, showing age and dominant composition of the Tertiary volcanic rocks (modified from Kleinhampl and others, 1975)

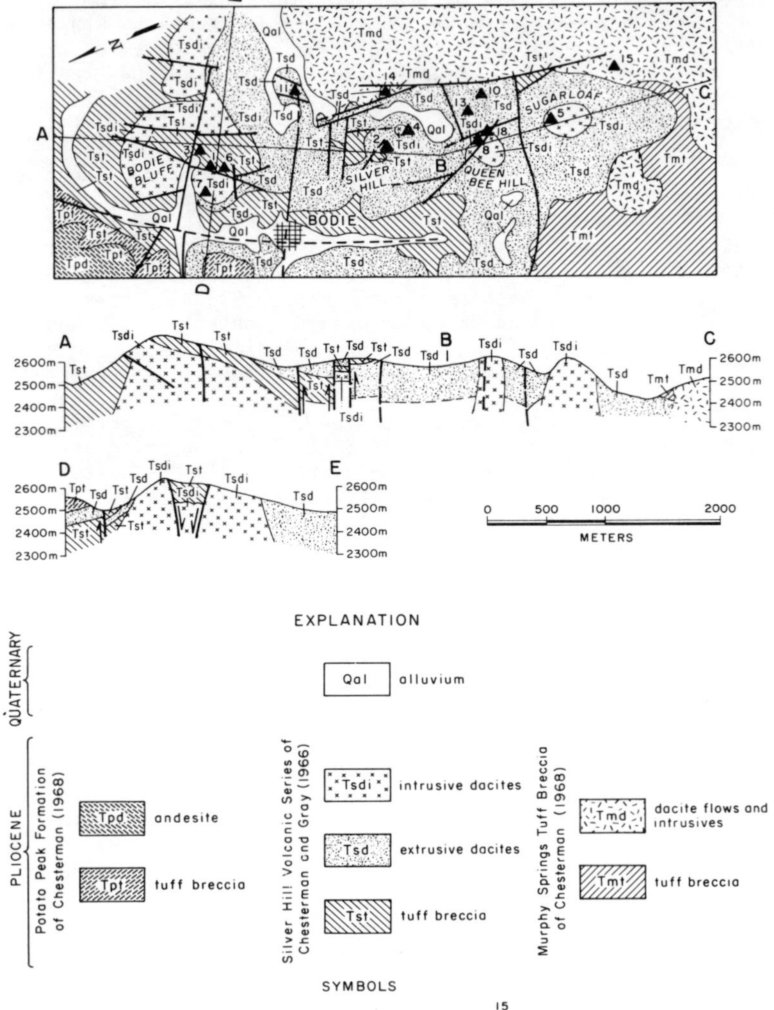

EXPLANATION

Figure 2. Generalized geologic map and cross sections of the Bodie mining district, Mono County, CA (modified from Chesterman and others, in press).

289

their proximity to eruptive centers which are characterized by complex structure and intrusive plugs. Many of the formations overlap each other in age as adjoining eruptive centers were frequently active at the same time. Hydrothermal alteration and mineral prospects are common, but the only major deposit located to date has been at Bodie. Although volcanic activity in the Hills spanned the time interval from 13.3 to 5.7 m.y. most of the eruptive material was emplaced between about 9.5 and 7.8 m.y.

The Bodie mining district is localized in dacite flows, tuff breccias and small intrusive plugs of the Silver Hill Volcanic series of Chesterman and Gray (1975), one of the volcanic formations of the Bodie Hills. The district is an eruptive center whose major structure consists of an irregular, faulted, north trending anticline (fig. 2) formed by intrusion and doming of the flows and tuff breccias by small plugs. The plugs occupy vents from which the extrusive volcanic rocks were erupted. Several sets of steeply dipping faults cut all units including the plugs. One prominent set strikes north to northeast, and another is normal to this (fig. 2). The major vein and fractures at Bodie also strike north to northeast parallel to one of the fault sets. Most of the production of the district comes from the vicinity of Bodie Bluff and Standard Hill in the north, in the vicinity of the graben of tuff breccia that was faulted down into the intrusive dacite during or shortly after emplacement (fig. 2).

The productive quartz veins cut both tuff breccia and intrusive rock. Several sets of quartz veins, which vary in thickness from about 1 to 90 ft are present, although most veins are not more than a few feet thick (Chesterman and others, 1983, in press). Adularia is a common constituent of the veins, sometimes forming crystals of up to 3 cm long coating open fractures. Ore minerals are principally native gold and silver, but argentite, pyrite, and sphalerite also occur. The latter increases with depth as the tenor of gold decreases. In the main productive zone near the graben, old records quoted by Chesterman and others (1983, in press) indicate that the ore averaged 1.7 opt (ounces per ton) Au and 3.1 opt Ag.

Wall rock alteration zoning at Bodie has not been studied in detail, although O'Neil and others (1973) reported on the isotopic and chemical effects of hydrothermal K-silicate alteration at Bodie Bluff. Chesterman and others (1983, in press) indicate that alteration zoning occurs both laterally and vertically. In the north, at Bodie Bluff, both the intrusive dacite and tuff breccia and flows are strongly silicified at the surface. Silberman (1982b) suggested that this silicified zone, a part of which has chalcedonic quartz vein stockworking represents a very shallow level in the system, probably just sub-surface. Below this near surface alteration, the rocks of the Bodie Bluff area are strongly K-silicate altered, largely recrystallized to the assemblage K-feldspar (adularia), K-mica (sericite) and

quartz. This alteration also characterizes the wall rocks of the productive vein zone. To the south, argillic alteration occurs, and at least some zones of sericite alteration. Outside of the central part of the district, most of the volcanic rocks are strongly propylitized. At the lowest level of workings at Bodie Bluff accessible in recent years, about 700 ft beneath the top, the altered rocks still contain K-feldspar, but sericite and kaolinite are more abundant.

Age relationships

Sample descriptions and data for the age determinations at Bodie and surrounding region were published by Silberman and others (1972), Silberman and Chesterman (1972) and Kleinhampl and others (1975) and are summarized in table 1 and figure 3. The extrusive rocks of the Silver Hill volcanic series were emplaced between 9.4 and 8.8 m.y. The co-magmatic plugs were intruded between 8.6 and 9.4 m.y. Hydrothermal activity, as dated by K-Ar ages of sericite from altered tuff breccia in the southern part of the district, and K-sillicate altered Bodie Bluff intrusive dacite, started at 8.6 m.y., which was just about at the end stage of volcanic activity in the district. Three K-Ar ages from adularia separated from mineralized quartz veins are in the range 8.0 to 7.1 m.y., with the youngest age from a vein occupying a structure that cuts others that hosted veins in the northern part of the district.

Volcanic rocks were emplaced within the volcanic center between 9.4 and 8.6 m.y., about a 1 m.y. duration for the volcanic activity. Hydrothermal alteration commenced at the end of or immediately after volcanism and continued for $1\frac{1}{2}$ m.y., during which several sets ofcross cutting veins (Chesterman and others, 1983, in press) were emplaced and a major precious metal deposit was formed. The alteration-mineralization at Bodie was nearly the last stage of igneous-hydrothermal activity associated with the development of the Bodie Hills volcanic field. It was not until nearly $2\frac{1}{2}$ m.y. later that minor volcanism again occurred, west of the district. To the east, major volcanic activity commenced again at 3.6 m.y. (table 1; fig. 1).

Goldfield mining district

Geology of the mining district

The Goldfield mining district (fig. 4) in the Goldfield Hills of western Nevada is underlain by a complex sequence of Oligocene and Miocene volcanic rocks, which cover a pre-Tertiary basement composed of Ordovician sedimentary rocks and Jurassic granitic rocks. The older Oligocene volcanic rocks (Tov of fig. 4) are silicic flows and tuffs of local origin from a caldera whose ring fracture zone is outlined by the faulting and alteration patterns. The pre-Tertiary rocks are exposed in the core of this caldera. Approximately 4 to 8 m.y. years after cessation of the silicic volcanic activity, a new pulse of dominantly intermediate flows, tuffs, breccias, and domes were emplaced in the early Miocene from

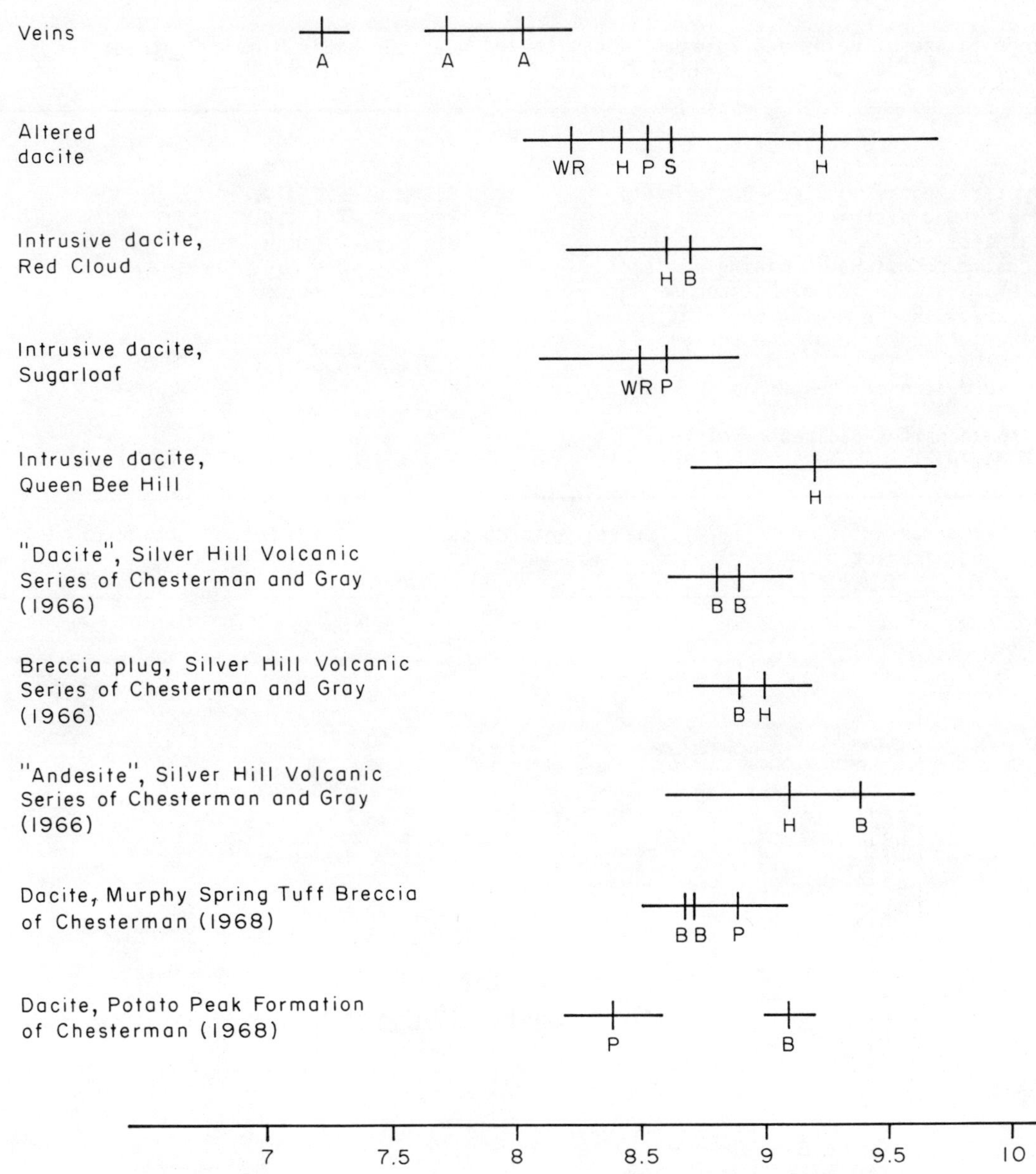

Figure 3. K-Ar ages of volcanic rocks, altered volcanic rocks and quartz-adularia veins at the Bodie mining district and vicinity, Mono County, California (modified from Silberman and others, 1972).

Table 1. Range in Age of Veins and Volcanic Rocks in and near the Bodie Mining District, Bodie Hills, Mono County, California

Younger postore sequence (east of Bodie mining district)	3.6 m.y.-250,000 m.y.
Younger rhyolitic rocks, western Bodie Hills	5.7 m.y.-5.3 m.y.
Veins, Bodie mining district	8.0 m.y.-7.1 m.y.
Hydrothermal activity	8.6 m.y.-7.1 m.y.
Dacite intrusive rocks, Bodie mining district	9.2 m.y.-8.6 m.y.
Silver Hill Volcanic Series of Chesterman and Gray (1976), Bodie mining district	9.4 m.y.-8.8 m.y.
Murphy Spring Tuff Breccia of Chesterman (1968)	8.9 m.y.-8.7 m. y.
[1]/Potato Peak Formation of Chesterman (1968)	9.1 m.y.-8.4 m.y.
Other basalts-andesites-dacites-rhyolites, western Bodie Hills	13.3 m.y.-7.8 m.y.

Modified from Silberman and others (1972). [1]/Volcanic rocks of the Bodie Hills outside of the Bodie mining district.

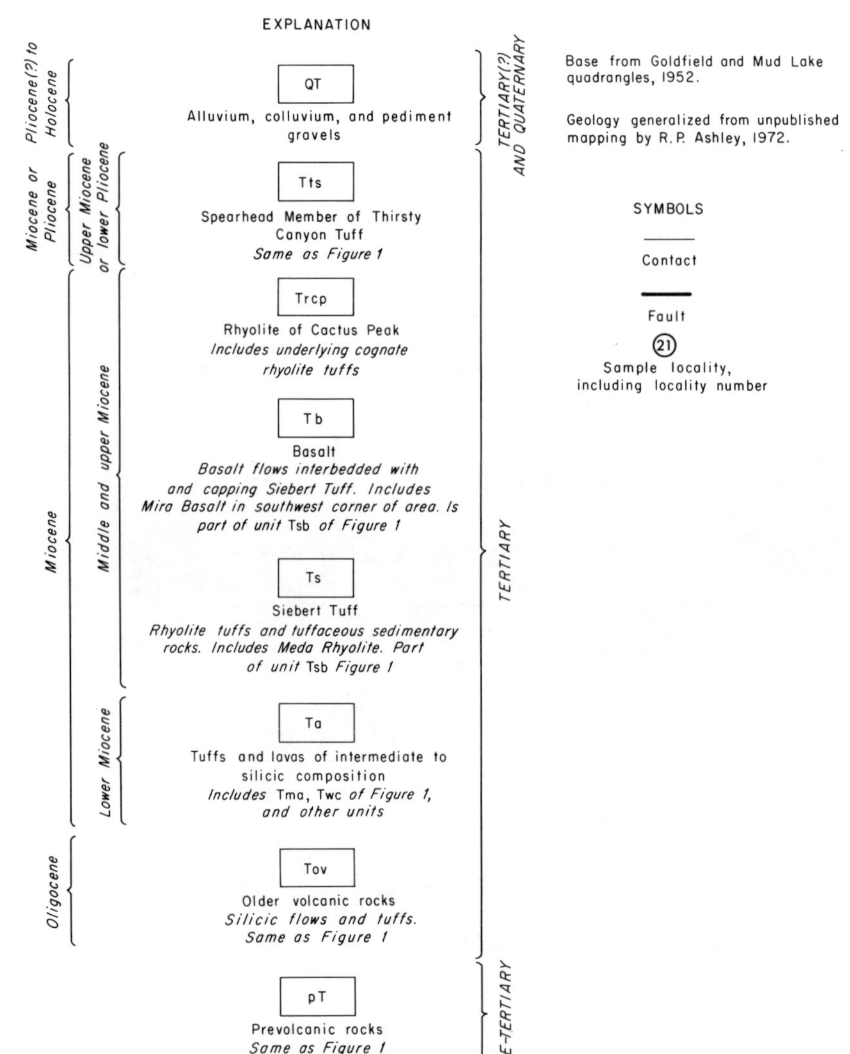

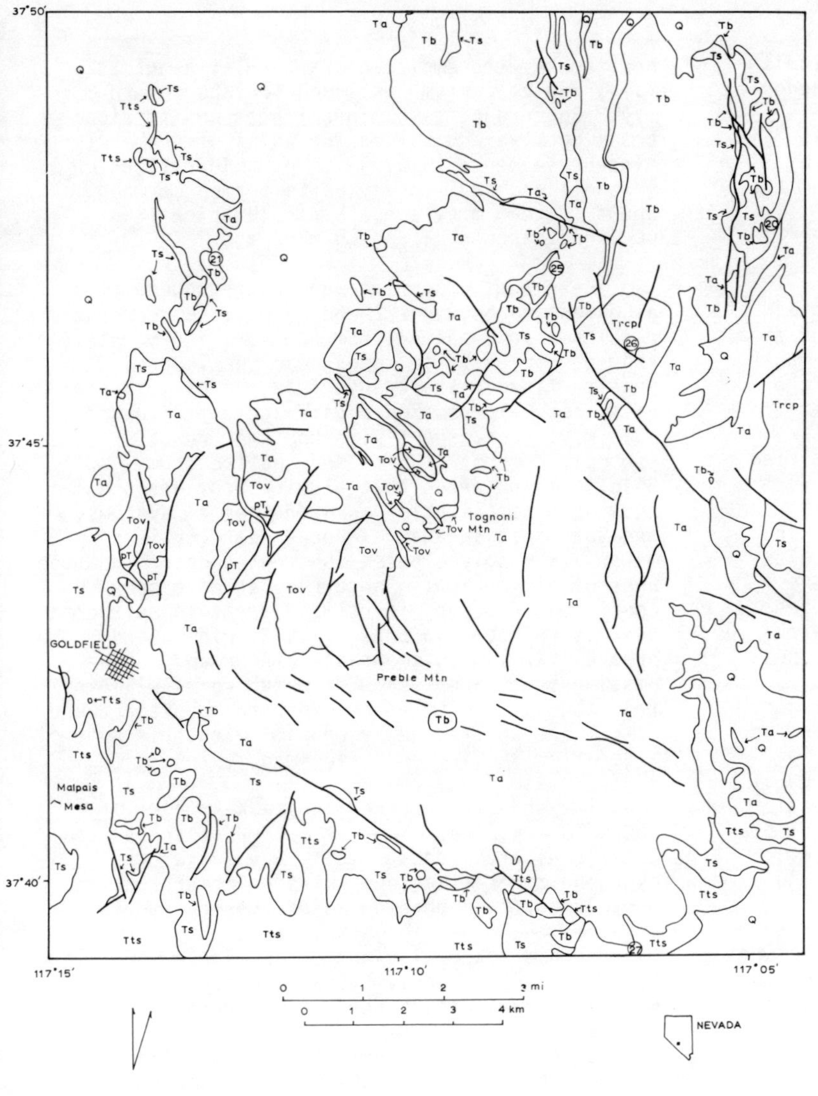

Figure 4. Simplified geologic map of the Goldfield mining district, Esmeralda and Nye Counties, Nevada (modified from Ashley and Silberman, 1976).

Figure 5. Schematic summary of volcanic history and timing of alteration and mineralization at Goldfield (modified from Ashley and Silberman, 1976).

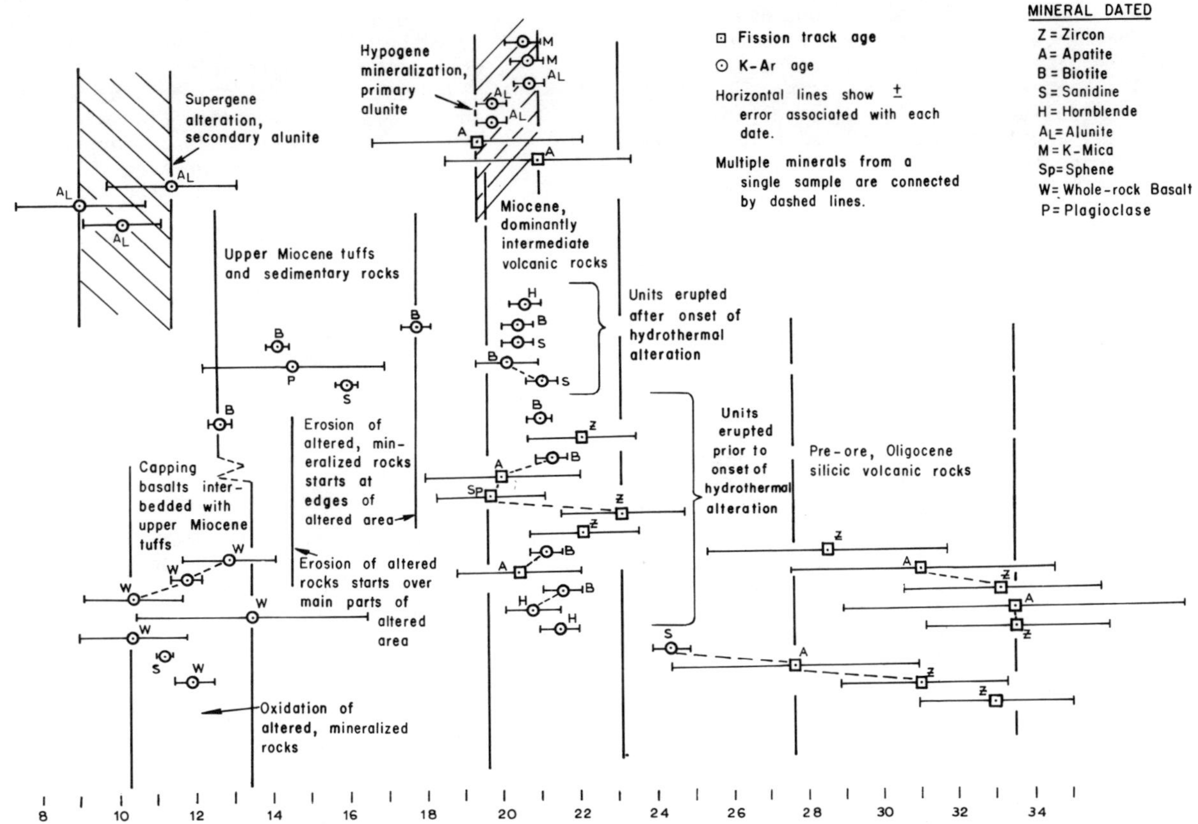

vents within and near the mining district (units Tma and Twc, fig. 4) (Ashley, 1974; Ashley and Silberman, 1976). Middle and upper Miocene sedimentary and volcanic rocks unconformably overlie the lower Miocene volcanics, and are largely from sources some distance away from the mining district (units Tsb and Tts, fig. 4), although some of the basalts in this younger sequence are erupted from local vents.

Hydrothermal alteration and mineralization at Goldfield are spatially and temporally associated with the dominantly intermediate Tma unit. Unit Twc was emplaced after onset of and during the hydrothermal alteration, but outside of the central part of the district (fig. 4). Mineralization at Goldfield occurs in silicified ledges consisting of prominent outcrops of fine grained quartz that occur throughout a 40 km^2 area of poorly exposed argillic and propylitic rocks. The ledges formed as replacement bodies along faults and fractures and may contain alunite, kaolinite, pyrophyllite and diaspore along with quartz. Limonite and pyrite (latter in unoxidized rocks) occur in all silicified zones. Productive mineralization in the ledges is restricted to a small area of 1.3 km^2, just north of Goldfield (fig. 4). Unoxidized ore formed open cavity fillings in brecciated silicified ledge material and consisted of varying proportions of native gold, and sulfide and sulfosalt minerals (Ashley, 1974). The overall average grade of ore at Goldfield was 0.54 opt (Albers and Stewart, 1972) but bonanza grades as high as 600 opt Au were reported (Ransome, 1909).

Alunite was an important hypogene constituent of the silicified ledges, including those hosting mineralization. It occurred as replacements of feldspar phenocrysts and pumice and as coarse hypogene veins cutting the silicified zones. Alunite deposition in the ledges persisted into early stages of mineralization (Ashley and Silberman, 1976). Supergene alunite occurs as fine grained veins cutting both oxidized silicified and argillized rocks. The zones peripheral to the quartz-alunite ledges were usually argillized, but at least in some areas, including the productive zone, quartz-sericite alteration mantles the ledges (Ashley and Silberman, 1976). The occurrence of the advanced argillic, alunite-containing and quartz-sericite alteration assemblages allowed the K-Ar ages of two alteration minerals to be determined.

Age relations

K-Ar ages of minerals from unaltered volcanic rocks, and from alunite and K-mica (sericite, $2M_1$ polymorph) were determined. In addition fission track ages from several units including the pre-Tertiary and older volcanic rocks (Tov) were obtained. The isotopic ages are summarized in figure 5 from Ashley and Silberman (1976). The older silicic units were propylitically altered and could not be dated by the K-Ar method. Zircon and apatites in these rocks were not affected by that alteration, and indicate emplacement ages of between 28 to 34 million years. The early Miocene host rocks were emplaced starting at about 22 m.y., and volcanism continued for about 1 to 2 m.y. Concordant K-Ar mineral ages and fission track ages were obtained for units of the early Miocene sequence (fig. 5). Post mineralization volcanic and sedimentary units formed between about 18 to 7 m.y. ago. Mineralization thus occurred between 22 and 18 m.y. ago.

K-Ar ages of replacement and hypogene vein alunites are between 21 and 20 m.y. Concordant results were obtained from the sericite samples from quartz-sericite zones near the ledges. Apatite fission track ages from pre-early Miocene units near altered zones gave ages of 21.0 and 19.6 m.y., concordant with the K-Ar results. The isotopic ages of alunite and apatite from Goldfield indicated that alunite provides accurate ages of mineralization in hydrothermal systems, and that apatite ages of rocks near (scale of meters) strongly altered zones also determined the ages of alteration. These results have been confirmed in other mineralized areas in epithermal systems (Morton and others, 1977; Lipman and others, 1976). In contrast, zircon and sphene fission track ages (some of which were collected from the same units as the apatite ages that were reset) are unaffected by proximity to this type of alteration (Ashley and Silberman, 1976).

The three supergene alunite K-Ar ages of about 10 m.y., represent oxidation of the sulfide bearing altered rocks some 10 m.y. after the hypogene mineralization. Fission track apatite ages demonstrate that no major thermal events affected the mineralized area after the early Miocene hydrothermal event. The middle and lower Miocene volcanic activity, apparently had no strong thermal effect here. Structural and stratigraphic data are in accord with erosion and oxidation starting to affect the mineralized area at about the time indicated by the supergene alunite ages. Thus at Goldfield, both hypogene hydrothermal alteration and supergene oxidation were dated by application of K-Ar analyses to alunites. The age relations of volcanic activity and alteration-mineralization at Goldfield are similar to those at Bodie. Several generalized conclusions can be drawn about the timing of igneous-hydrothermal activity from these studies.

(1) Hydrothermal alteration and mineraliza-tion starts late in and usually continues for some time after some significant local stage of volcanic activity. Pre- and syn-alteration volcanism lasts on the order of 1 m.y. or longer.

(2) The overall duration of hydrothermal activity indicated by the data at Bodie and Goldfield is 1 to 1.5 m.y. This is close to the average age of hydrothermal systems related to mineralization, as will be shown later.

(3) Volcanic activity can and does take place during and after hydrothermal mineralization, but generally outside of the zone of mineralization.

These generalizations have been confirmed by detailed study of other epithermal systems (Silberman and others, 1979; Noble and Silberman, 1983, in press).

DATA ON LIFETIMES OF HYDROTHERMAL SYSTEMS

A large amount of radiometric age data have become available on mineral deposits since the pioneering studies on porphyry copper deposits by Paul Damon and his associates (Damon and Mauger, 1966), and it is possible to compare the spans of hydrothermal activity at Bodie and Goldfield and their temporal relations to spatially associated igneous rocks with data from other hydrothermal mineral deposits and thermal spring systems. There are, unfortunately, few detailed studies on the chronology of hydrothermal systems, and even fewer that provide unequivocal estimates of the duration of hydrothermal activity.

Figure 6 summarizes data from a tabulation of geochronological studies of hydrothermal systems, including porphyry copper deposits, epithermal precious metal deposits and poly-metallic vein deposits (PM), and thermal hot spring systems (M. L. Silberman, unpubl. data, 1983).

Hydrothermal activity in porphyry copper deposits is generally found to have lasted about 1 m.y. or less, with the exception of the Bingham, Utah (Warnaars et al, 1978) and El Salvador, Chile (Gustauson and Hunt, 1975) systems, where I interpreted the data as indicated lifetimes on the order of 2 m.y. The few detailed studies of precious- and base-metal vein systems available suggest that hydrothermal activity commonly extends over periods of about 0.5 to 1.5 m.y. The longest-lived documented system is at Bodie, California (Silberman and others, 1972). The estimate of about 1.5 m.y. duration is relatively accurate, since the youngest age was obtained on vein material from a structure that cuts other mineralized structures and veins (Silberman and others, 1972). A similar lifetime is suggested for the Tui mine, New Zealand (Adams et al, 1974). Lifetimes of about 1, 0.7, and 0.5 are indicated for Goldfield and Divide, Nevada (Silberman and others, 1979b), and Summitville, Colorado (Mehnert and others, 1973), respectively. Data on other epithermal districts generally indicate a close association of igneous and hydrothermal activity, but are inadequate to infer the interval over which hydrothermal activity was taking place.

Figure 6 also depicts data on the lifetime of several active thermal spring systems. White (1955; 1974) has stressed the close similarity of many thermal springs and epithermal Au-Ag systems, and it now appears that several low-grade, large-tonnaged disseminated gold-silver deposits and prospects such as Round Mountain (Berger and Tingley, 1980), Borealis (Strachan and others, 1982), Sulfur (Wallace, 1980) and Hasbrouch Peak (Bonham and Garside, 1979) are fossil thermal spring systems. The age ranges for most of the thermal spring systems are the same as those for the porphyry and vein deposits, with the exception of Steamboat Springs, for which episodic activity appears to have been unusually long lived.

There do not seem to be significant differences in the lifetimes of these three types of systems, although the available data base is sparse. The average lifetime of hydrothermal activity based on this summary is 1.2 m.y., approximately the age of activity determined for Bodie and Goldfield.

PATTERNS OF GREAT BASIN MINERALIZATION

The relationship between the tectonic evolution of this region and the distribution of epithermal mineral deposits is well documented (Silberman and others, 1976).

Precious metal deposits are associated in space and time with several suites of volcanic rocks that were erupted in the Great Basin from mid to late Tertiary in response to interactions of the North American plate and various Pacific plates. The patterns of volcanic activity are thought to be related to variations in the dip of and the rate of convergence of the plates, and to transform faulting and migration of triple junctions (Atwater, 1970; Lipman and others, 1972; Christiansen and Lipman, 1972; Snyder and others, 1976). The distribution of volcanic rocks in the Great Basin in space and timed was summarized by Stewart and Carlson (1976). Figures 7 through 10 show simplified versions of the Stewart and Carlson maps along with locations of dated epithermal precious metal deposits (Silberman and others, 1976).

The entire pattern represented in figures 7 through 10 shows a generally outward, accurate progression of volcanic activity in time from the central Great Basin toward its margin. The figures also illustrate a similar pattern of outward migration for the hydrothermal mineral deposits. The overall distribution of dated hydrothermal Au-Ag deposits in the Great Basin is shown in figure 11. When the distribution and ages are considered in detail, several patterns and associations of interest are evident; the patterns differ in the northern and central Great Basin, and in the Walker Lane (western Great Basin).

<u>Northern and central Great Basin</u>

Figures 12 and 13 show dated ore deposits and their host rocks as well as the nature of volcanic activity that was taking place in the two regions of the Great Basin. The pattern for the central and northern Great Basin shows epithermal deposits associated with all stages of volcanic activity, with particularly vigorous hydrothermal activity immediately after the onset of crustal extension, in association with no strongly predominant host rock lithology. The area is characterized by a grouping of deposits in time, rather than with a host rock association. Of the 16 dated deposits in the northern and central area (figs. 11, 12) four formed during an early stage of intermediate volcanic activity and three during the period of

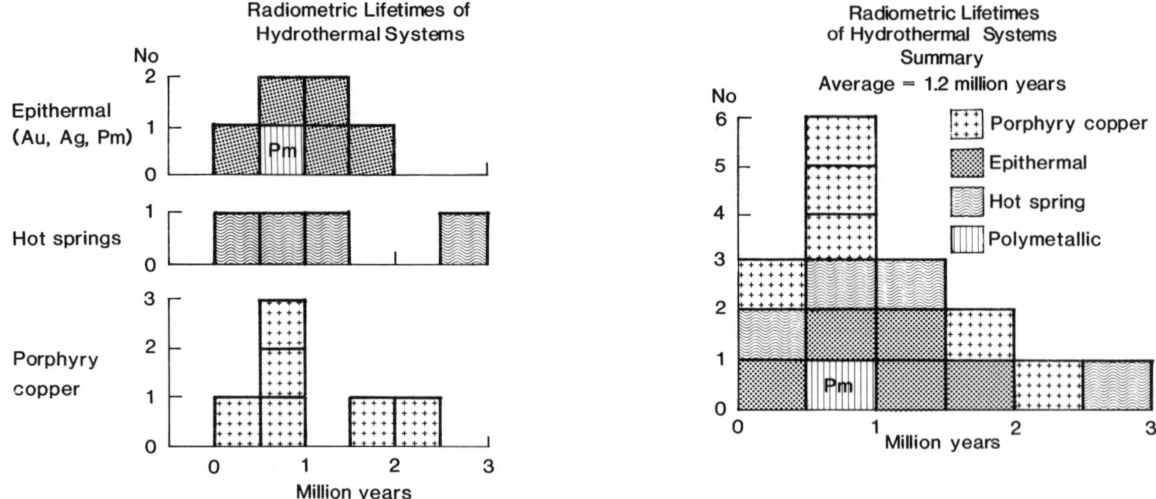

Figure 6. Histograms summarizing radiometric lifetimes of hydrothermal alteration-mineralization systems.

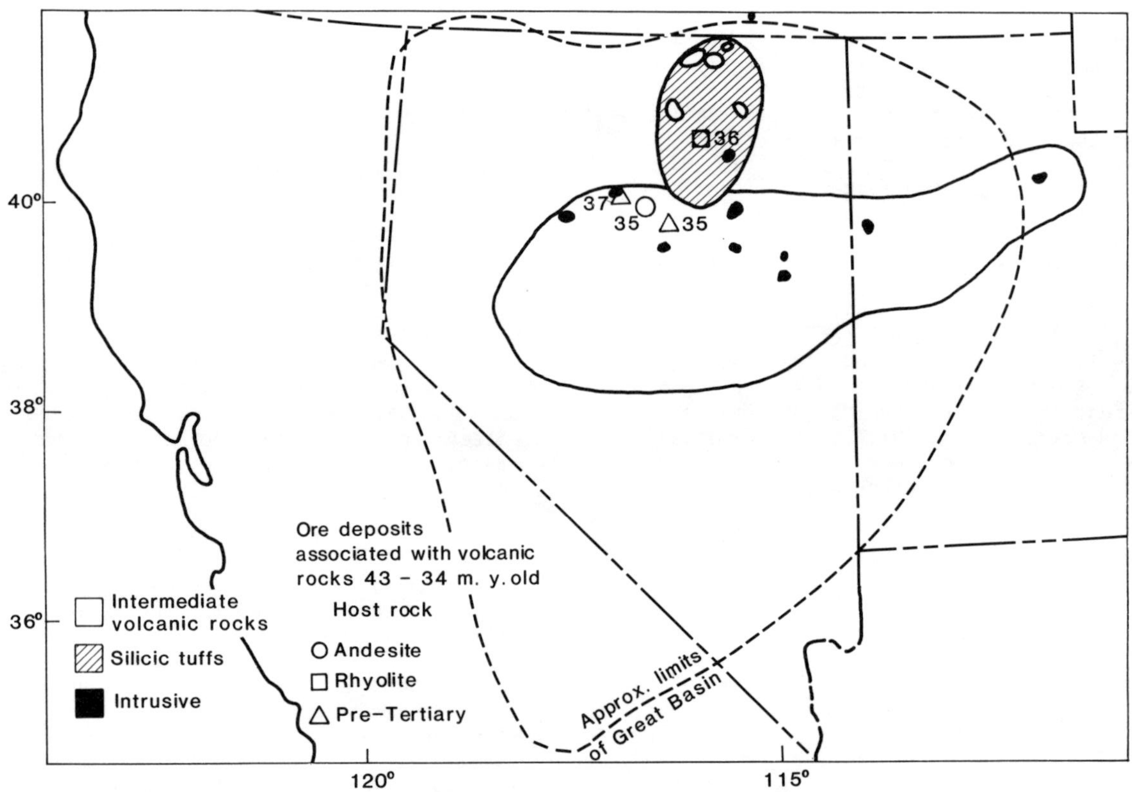

Figure 7. Distribution of volcanic rocks 43-34 m.y. old in the Great Basin and surrounding regions modified from Stewart and Carlson (1976). Location and age of ore deposits, and host rock lithologies are indicated.

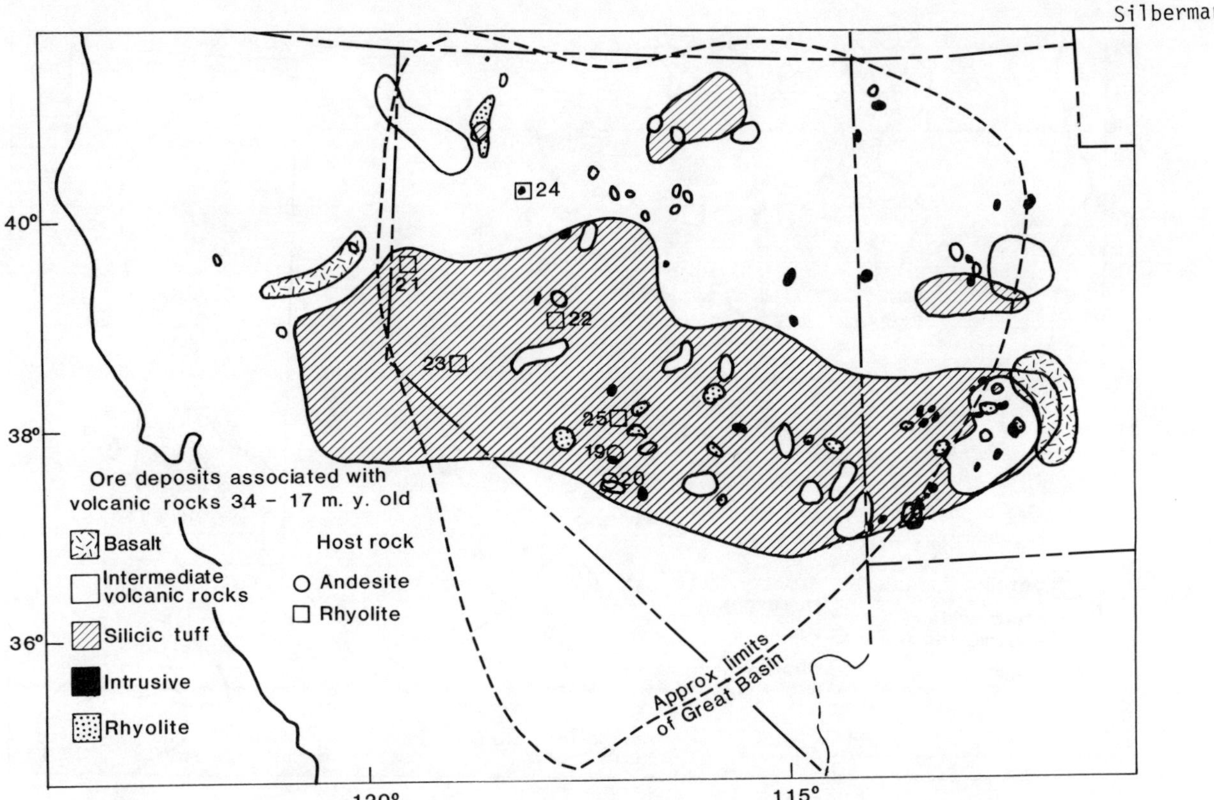

Figure 8. Distribution of volcanic rocks 34-17 m.y. old in the Great Basin and surrounding regions modified from Stewart and Carlson (1976). Location and age of ore deposits, and host rock lithologies are indicated.

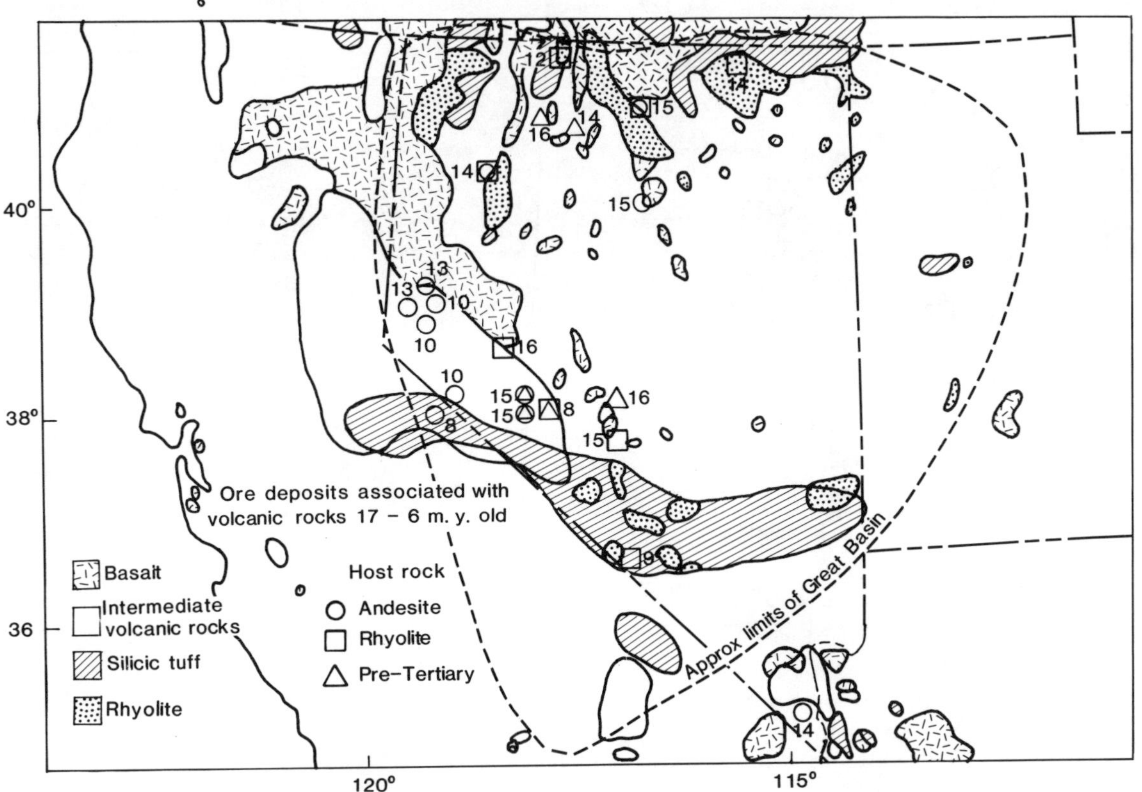

Figure 9. Distribution of volcanic rocks 17-6 m.y. old in the Great Basin and surrounding regions, modified from Stewart and Carlson (1976). Location and age of ore deposits, and host rock lithologies are indicated.

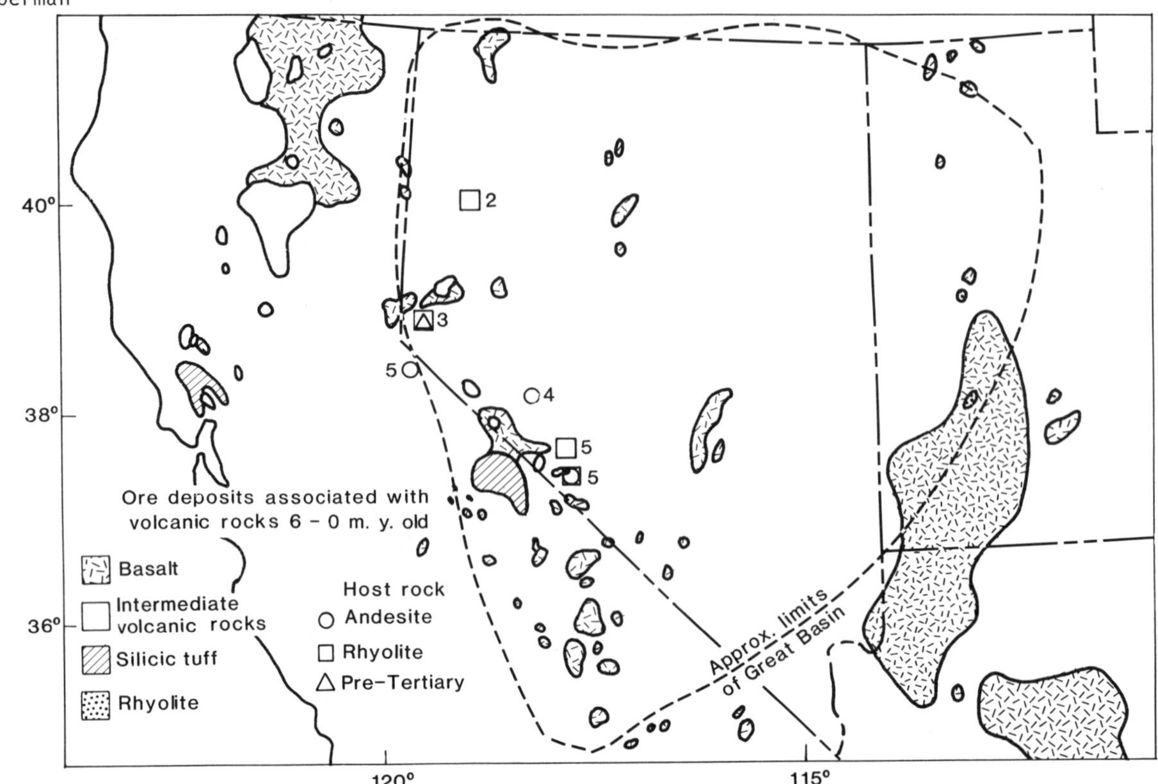

Figure 10. Distribution of volcanic rocks 6-0 m.y. old in the Great Basin and surrounding regions, modified from Stewart and Carlson (1976). Location and age of ore deposits, and host rock litholigies are indicated.

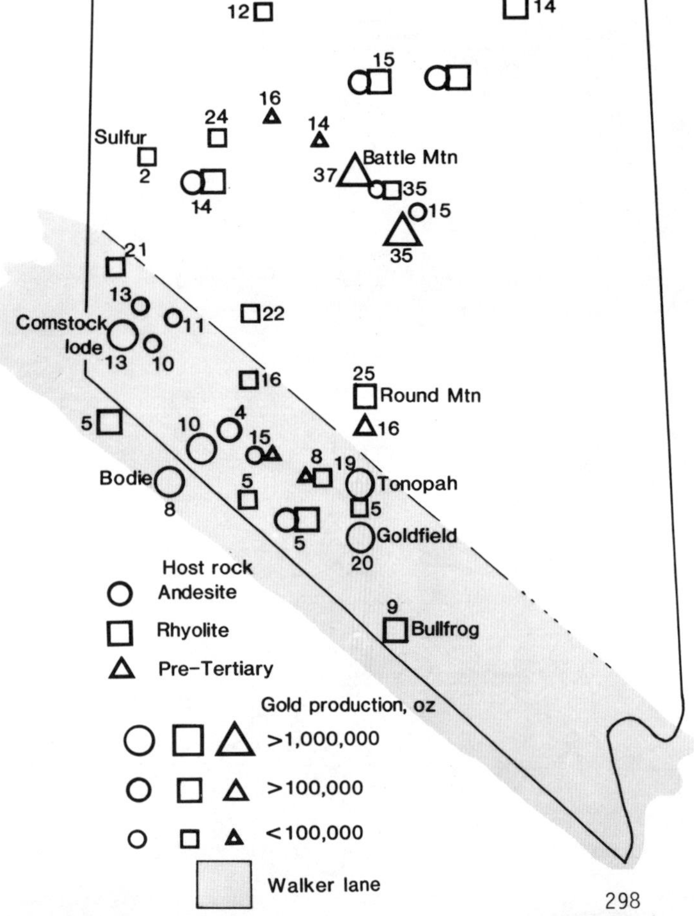

Figure 11. Distribution of ore deposits dated by the K-Ar method in the Great Basin, keyed to host rock lithology and approximate production of gold in troy ounces (modified from Silberman and others, 1976).

dominant ash flow-tuff eruption. All of these deposits are related to local intrusive or volcanic activity, and many are significant with production and or reserves of greater than 100,000 oz Au, and some with greater than 1,000,000 oz (e.g. Round Mountain and Cortez). A significant grouping of fissure vein deposits occurs between 12 and 16 m.y. ago, associated with the 17 and 6 m.y. mixed volcanic episode. Many of these are small deposits, less than 100,000 oz, and occur in faults or shear zones related to Basin and Range faulting, which followed the onset of crustal extension at 17 to 20 m.y. ago. The deposits are relatively independent of host rock type, and are not demonstrably related to local volcanic activity. Exceptions to this occur at Jarbidge and McDermitt, Nevada, where the deposits occur in voluminous volcanic sequences near the margins of the Great Basin. The two million year age on figure 11 and 12 is from the Sulfur prospect in northwest Nevada (M. L. Silberman and A. B. Wallace, unpub. data, 1980), and represents the youngest known Au containing system in the central Great Basin.

Walker Lane (western Great Basin)

The pattern of distribution of deposits along the Walker Lane is considerably different (fig. 13). There is no clear grouping in time, rather, mineral deposits occur throughout most of the time shown. Of 20 dated deposits, 9 are in andesitic host rocks, and 4 more are in the volcanic sequences that contain andesites as part of the host rock association. The major producers (greater than 1,000,000 oz Au) are all in andesites and include the Comstock Lode, Bodie, Aurora, Goldfield, and Tonopah. Production figures summarized in Silberman and others (1976) bear out the importance of andesites as a host rock for precious metal deposits in the Walker Lane. Most of the deposits are associated with local stages of volcanic or intrusive activity. Thus, the pattern for the Walker Lane shows no essential grouping in time, but a very strong association with a particular host rock lithology.

It is interesting to note that five deposits of 5 m.y. age or younger occur in the Walker Lane. The 2 m.y. old Sulfur deposit, although in the northern-central region, is close to the Walker Lane. Mineralization in the Great Basin has continued until nearly recent time, in spite of the fact that subduction affecting the area ended between 5 and 10 m.y. ago (Atwater and Molnar, 1973; Silberman and others, 1975).

TIMING OF STAGES OF ALTERATION-MINERALIZATION

None of the geochronological studies quoted above attempted to define the duration of a single stage of hydrothermal activity, or of a mineralizing event within an overall period of hydrothermal activity. There does appear to be a concensus developing among workers in epithermal systems that mineralizing events are short lived, transitory, and repetitive events within a longer context of hydrothermal activity (Silberman, 1982). Some evidence of this interpretation:

(1) Multiple stages of hydrothermal brecciation and stockwork veining, only some of which are associated with sulfide deposition and mineralization in bulk tonnage precious metal systems (Silberman, 1982; Berger and Eimon, in press).

(2) Banding of epithermal quartz veins in bonanza systems, with only a few of the bands containing sulfides. The banding and the brecciation referred to in (1) are believed to be related to boiling which occurs episodically in most systems (Buchanan, 1981).

(3) The occurrence of several alteration assemblages which succeed each other in a restricted area. In some systems, such as Bodie (Silberman and others, 1972; O'Neil and others, 1973) and Julcani (Noble and Silberman, 1983, in press; Peterson and others, 1977) mineralization occurs associated with one particular assemblage, and not others.

Epithermal mineral deposits are commonly found in close association with intrusive phases, where the heat of the crystallizing magma generates a convective hydrothermal system (White, 1974, 1981; Norton and Cathles, 1979). Cooling of the magma by conduction and convection, unless new sources of heat (magma) are provided will limit the period of time that the circulation of a hydrothermal system can occur. Factors such as permeability, salinity of the fluid, and recharge conditions (amount of fluid) will also affect this length of time. Norton and Cathles, (1979) suggest that permeability in hydrothermal systems is due to fracturing, and that cooling can be considered by a convective circulation model . Models of cooling plutons on the order of 2 km ± width suggests decay of a hydrothermal circulation cell well within about 100,000 years. This time period is short relative to the estimates discussed previously for overall hydrothermal lifetimes in epithermal systems. These observations suggest that magmatic bodies related to the generation of epithermal ore deposits are either much larger than the 1-3 km bodies frequently found at the present surface near such deposits (e.g. Bodie Bluff at Bodie; Mount Davidson at the Comstock Lode; rhyolite domes at Steamboat Springs, Nevada, etc.) or there must be successive, closely spaced pulses of magmatic influx that renew the sources of heat for the convective systems. In at least two areas, Steamboat Springs, Nevada (Silberman and others, 1979a), and Tonopah, Nevada (Silberman and others, 1979b), the presence of a regional batholith with several pulses of magma influx have been called upon to explain long lived, complex hydrothermal systems.

K-Ar data from the Julcani, polymetallic vein district of Peru, to be published shortly (Noble and Silberman, 1983, in press) support the interpretation that individual pulses of alteration and mineralization are short lived. K-Ar ages and geologic relationships at Julcani suggest that eight stages of interspersed volcanic and hydrothermal activity, including a pulse of

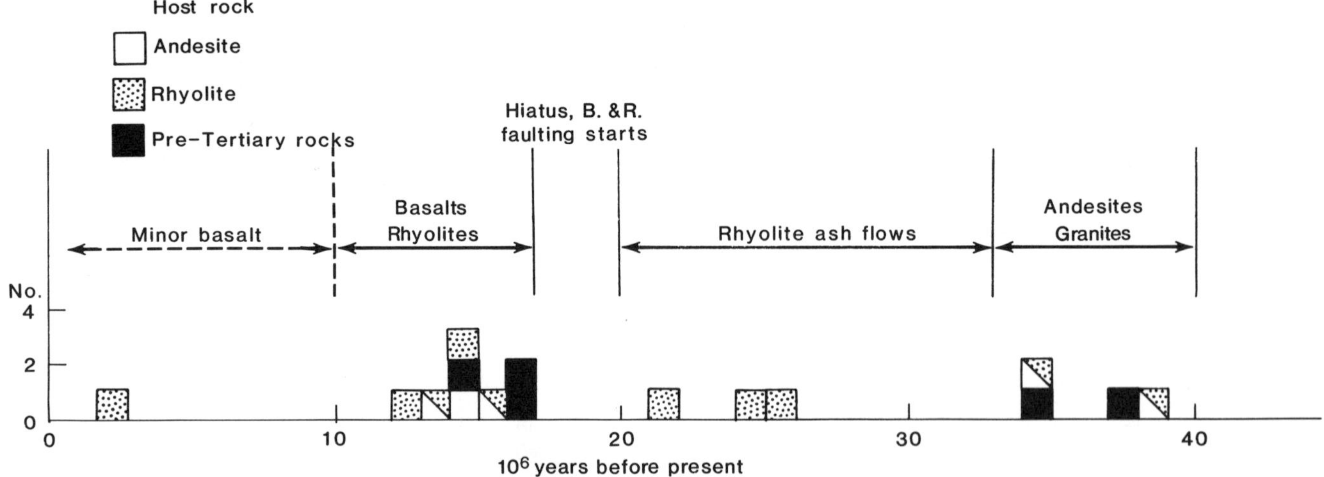

Figure 12. Histogram summarizing ages of ore deposits in the central and northern Great Basin with host rock lithology indicated.

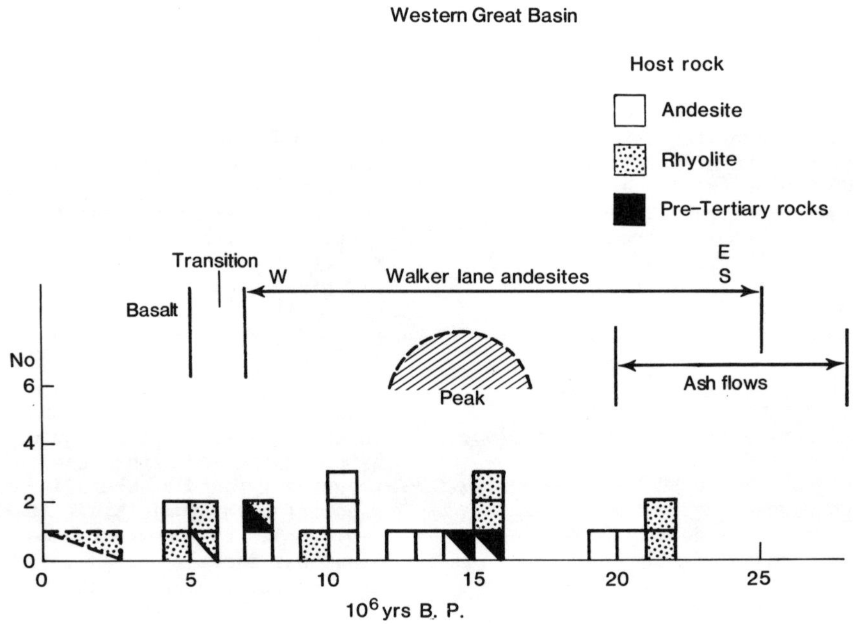

Figure 13. Histogram summarizing ages of ore deposits in the Walker Lane (western Great Basin) with host rock lithology indicated.

sulfide mineralization, occurred over a span of 700,000 years. Not all of the stages could be definitevely dated, but estimates of duration of the hydrothermal episodes range from 100,000 to 200,000 years. These time intervals are interestingly similar to the convective cooling periods for small plutons suggested by the Norton and Cathles (1979) models.

SUMMARY

This discussion of the geochronology of hydrothermal alteration and mineralization, particularly of the epithermal precious metal systems, has demonstrated the temporal relationships of hydrothermal processes to volcanic activity, and has suggested a range of timing of overall hydrothermal activity of 1/2 to 3 m.y. for many systems. It has also shown that regional patterns of hydrothermal mineralization occur, and that these patterns can be related to the timing of tectonic and volcanic evolution, although not always in a simple fashion. Finally, a start has been made in interpreting the timing of pulses of alteration and mineralization within a broader frame of overall hydrothermal activity. It is this topic that needs further investigation. Refinement of methods of isotopically determining the period of activity of alteration pulses and mineral deposition, and whether this fine scale tuning of the timing relationships has any correlation with the degree of mineralization (read economic significance) is yet to be accomplished.

REFERENCES CITED

Adams, C. J. A., Wodzicki, A., and Weissberg, B. G., 1974, K-Ar dating of hydrothermal alteration at the Tui mine, Te Aroha, New Zealand: New Zealand Jour. Sci., v. 17, p. 193-199.

Albers, J. P., and Stewart, J. H., 1972, Geology and mineral deposits of Esmeralda Co., Nevada: Nevada Bur. Mines Bull. 78, 80 p.

Ashley, R. P., 1974, Geoldfield mining district: Nevada Bur. Mines and Geol. Rept. 19, p. 49-66.

Ashley, R. P., and Silberman, M. L., 1976, Direct dating of mineralization at Goldfield, Nevada, by potassium-argon and fission-track methods: Econ. Geol., v. 71, p. 904-924.

Atwater, Tanya, 1970, Implications of plate tectonics for the Cenozoic tectonic evolution of western North America: Geol. Soc. America Bull., v. 81, no. 12, p. 3513-3536.

Atwater, Tanya, and Molnar, Peter, 1973, Relative motions of the Pacific and North American plates deducted from sea-floor spreading in the Atlantic, Indian, and South Pacific oceans, in Kovach, K. L., and Nur, Amos, eds., Proceedings of the conference on tectonic problems of the San Andreas fault system: Stanford Univ. Publ. Geol. Sciences, v. 13, p. 136-148.

Berger, B. R., and Eimon, P. I., 1983, Comparative methods of epithermal silver-gold deposits: AIME Trans. (in press).

Berger, B. R., and Tingley, J. V., 1980, Geochemistry of the Round Mountain gold deposit, Nye Co., Nevada: Society of Mining Engineers, AIME, Northern Nevada Section, Precious Metals Symposium, Abstracts, p. 18c.

Bonham, H. F., Jr., and Garside, L. R., 1979, Geologic map of the Tonopah, Lone Mountain, Klondike, and Northern Mud Lake quadrangles, Nevada: Nevada Bur. Mines and Geol. Bull. 92, 142 p.

Buchanan, L. J., 1981, Precious metal deposits associated with volcanic environments in the southwest, in Dickinson, W. R., and Payne, W. D., eds., Relations of tectonics to ore deposits in the Southern Cordillera: Arizona Geol. Soc. Digest, Vol. XIV, p. 237-262.

Chesterman, C. W., 1968, Volcanic geology of the Bodie Hills, Mono Co., California, in Coats, R. R., Hay, R. L., and Anderson, C. A., eds., Studies in volcanology, a memoir in honor of Howell Williams: Geol. Soc. America Mem. 116, p. 45-68.

Chesterman, C. W., and Gray, C. H., Jr., 1975, Geology of the Bodie 15-minute quadrangle: California Div. Mines and Geology Map Sheet 21, scale 1:48,000.

Chesterman, C. W., Chapman, R. H., and Gray, C. H., 1983, Geology and ore deposits of the Bodie mining district, Mono Co., California: California Div. Mines and Geology Bull. (in press).

Chivas, A. R., and McDougall, Ian, 1978, Geochronology of the Koloula porphyry copper prospect, Guadalcanal, Solomon Islands: Econ. Geol., v. 73, p. 678-689.

Christiansen, R. L., and Lipman, P. W., 1972, Cenozoic volcanism and plate-tectonic evolution of the western United States. II. Late Cenozoic: Royal Soc. London Philos. Trans. A, v. 271, p. 249-284.

Damon, Paul, and Mauger, R. L., 1966, Epeirogeny-orogeny viewed from the Basin and Range province: Am. Inst. Mining, Metall. and Petroleum Engineers Trans., v. 235, p. 99-112.

Damon, P. E., Shafiquallah, Muhammad, and Clark, K. F., 1983, Geochronology of the porphyry copper deposits and related mineralization of Mexico: Canadian Jour. Earth Sci. (in press).

Erickson, R. L., Silberman, M. L., and Marsh, S. P., 1978, Age and composition of igneous rocks, Edna Mountain quadrangle, Humboldt County, Nevada: U.S. Geol. Survey Jour. Research, v. 6, no. 6, p. 727-743.

Gilbert, C. M., Christensen, J. N., Al-Rawi, Yehya, and Lajoie, K. R., 1968, Structural and volcanic history of Mono Basin, California-Nevada, in Coats, R. R., Hay, R. L., and Anderson, C. A., eds., Studies in volcanology, a memoir in honor of Howell Williams: Geol. Soc. America Mem. 116, p. 275-329.

Gustafson, L. B., and Hunt, J. P., 1975, The porphyry copper deposit at El Salvador, Chile: Econ. Geol., v. 70, p. 857-912.

Kleinhampl, F. J., Silberman, M. L., Chesterman, C. L., Chapman, R. H., and Gray, C. H., Jr., 1975, Aeromagnetic and limited gravity studies and generalized geology of the Bodie Hills region, Nevada and California: U.S. Geol. Survey Bull. 1384, 38 p.

Koski, R. A., Chesterman, C. W., Silberman, M. L., and Fabbi, B. P., 1978, Hydrothermal adularia at Bodie, Mono Co., California: U.S. Geol. Survey Open-File Rept. 78-942, 24 p.

Lipman, P. W., Fisher, F. S., Mehnert, H. H., Naeser, C. W., Luedke, R. G., and Steven, T. A., 1976, Multiple ages of mid-Tertiary mineralization and alteration in the western San Juan Mountains, Colorado: Econ. Geol., v. 71, no. 3, p. 571-588.

Lipman, P. W., Prostka, H. J., and Christiansen, R. L., 1972, Cenozoic volcanism and plate-tectonic evolution of the western United States. I. Early and Middle Cenozoic: Royal Soc. London Philos. Trans. A, v. 271, p. 217-248.

Mehnert, H. H., Lipman, P. W., and Steven, T. A., 1973, Age of mineralization at Summitville, Colorado, as indicated by K-Ar dating of alunite: Econ. Geol., v. 68, p. 365-401.

Meyer, Charles, and Hemley, J. J., 1967, Wall rock alteration, in Barnes, H. L., ed., Geochemistry of hydrothermal ore deposits: New York, Holt, Rinehart, and Winston, p. 166-235.

Mitchell, P. A., Silberman, M. L., and O'Neil, J. R., 1981, Genesis of gold vein mineralization in an Upper Cretaceous turbidite sequence, Hope-Sunrise district, Southern Alaska: U.S. Geol. Survey Open-File Rept. 81-103, 20 p.

Moore, W. J., and Lanphere, M. A., 1971, The age of porphyry-type copper mineralization in the Bingham mining district, Utah--refined estimate: Econ. Geol., v. 66, p. 331-334.

Morton, J. L., Silberman, M. L., Bonham, H. F., Jr., Garside, L. J., and Noble, D. C., 1977, K-Ar ages of volcanic rocks, plutonic rocks and ore deposits in Nevada and eastern California--Determinations run under the USGS-NBMG cooperative program: Isochron/West, no. 20, p. 19-29.

Naeser, C. W., and Faul, Henry, 1969, Fission track annealing in apatite and sphene: Jour. Geophys. Research, v. 74, no. 2, p. 705-710.

Noble, D. C., and Silberman, M. L., 1983, Volcanic and hydrothermal evolution and K-Ar chronology of the Julcani district, Peru: Econ. Geol. (in press).

Norton, Dennis, and Cathles, L. M., 1979, Thermal aspects of ore deposition, in Barnes, H. L., ed., Geochemistry of hydrothermal ore deposits, Second Edition: New York, John Wiley and Sons, p. 611-683.

O'Neil, J. R., Silberman, M. L., Fabbi, B. P., and Chesterman, C. W., 1973, Stable isotope and chemical relations during mineralization in the Bodie mining district, Mono County, California: Econ. Geol., v. 68, p. 765-784.

Petersen, Ulrich, Noble, D. C., Arenas, M. J., and Goodell, P. C., 1977, Geology of the Julcani mining district, Peru: Econ. Geol., v. 72, p. 931-949.

Ransome, F. L., 1909, The geology and ore deposits of Goldfield, Nevada: U.S. Geol. Survey Prof. Paper 66, 258 p.

Rose, A. W., and Burt, D. M., 1979, Hydrothermal alteration, in Barnes, H. L., ed., Geochemistry of hydrothermal ore deposits, Second Edition: New York, John Wiley and Sons, p. 173-235.

Rowen, L. C., Kingston, M. J., and Heald-Wetlaufer, Pamela, 1983, The Humboldt and Walker Lane structural zones: post-Oligocene belts of mineralization in Nevada: Econ. Geol. (in press).

Silberman, M. L., 1982a, Geochronology of hydrothermal mineralization [abs.]: Geol. Soc. America Abs. with Programs, v. 14, no. 7, p. 617.

Silberman, M. L., 1982b, Hot-spring type, large tonnage, low-grade gold deposits, in Erickson, R. L., ed., Characteristics of mineral deposit occurences: U.S. Geol. Survey Open-File Rept. 82-795, p. 131-143.

Silberman, M. L., MacKevett, E. M., Jr., Connor, C. L., and Matthews, Alan, 1980, Metallogenic and tectonic significance of oxygen isotope data and whole-rock potassium-argon ages of the Nikolai Greenstone, McCarthy quadrangle, Alaska: U.S. Geol. Survey Open-File Rept. 80-2019, 31 p.

Silberman, M. L., and Ashley, R. P., 1970, Age of ore deposition at Goldfield, Nevada from K-Ar dating of alunite: Econ. Geol., v. 65, no. 3, p. 352-354.

Silberman, M. L., Chesterman, C. W., Kleinhampl, F. T., and Gray, C. H., Jr., 1972, K-Ar age of volcanism and mineralization, Bodie mining district and Bodie Hills volcanic field, Mono County, California: Isochron/West, no. 3, p. 13-22.

Silberman, M. L., Berger, B. R., and Koski, R. A., 1973, K-Ar age relations of granodiorite, tungsten, and gold mineralization near the Getchell mine, Humboldt County, Nevada: Econ. Geol., v. 69, p. 646-656.

Silberman, M. L., Noble, D. C., and Bonham, H. F., Jr., 1975, Ages and tectonic implications of the transition of calc-alkaline andesitic to basaltic volcanism, in the western Great Basin and the Sierra Nevada [abs.]: Geol. Soc. America Abs. with Programs, v. 7, p. 375.

Silberman, M. L., Stewart, J. H., and McKee, E. H., 1976, Igneous activity, tectonics, and hydrothermal precious-metal mineralization in the Great Basin during Cenozoic time: Soc. Mining Engineers Trans., v. 260, p. 253-263.

Silberman, M. L., White, D. E., Keith, T. E. C., and Docker, R. D., 1979a, Duration of hydrothermal activity at Steamboat Springs, Nevada, from ages of spatially associated volcanic rocks: U.S. Geol. Survey Prof. Paper 458-D, p. D1-D14.

Silberman, M. L., Bonham, H. F., Jr., Garside,
L. J., and Ashley, R. P., 1979b, Timing of
hydrothermal alteration--mineralization and
igneous activity in the Tonopah mining
district and vicinity Nye and Esmerald
Counties, Nevada: in Ridge, J. D., ed.,
papers in Mineral Deposits of Western North
America, IAGOD Fifth Quadrennial Symposium,
Proceedings, Vol. II: Nevada Bur. Mines and
Geol., Rept. 33, p. 119-126.
Smith, R. L., and Bailey, R. A., 1968, Resurgent
cauldrons, in Coats, R. R., Hay, R. L., and
Andersen, C. A., eds., Studies in
volcanology: Geol. Soc. America Mem. 116,
p. 613-662.
Snyder, W. S., Dickinson, W. R., and Silberman,
M. L., 1976, Tectonic implications of space
time patterns of Cenozoic magmatism in the
western United States: Earth and Planetary
Sci. Letters, v. 32, p. 91-106.
Stewart, J. H., and Carlson, J. E., 1976, Cenozoic
rocks of Nevada: Nevada Bur. Mines and Geol.
Map 52.
Strachan, D. G., Pettit, Paul, and Reid, R. F.,
1982, The geology of the Borealis gold
deposit, Mineral Co., Nevada [abs.]: Geol.
Soc. America Abs. with Programs, v. 14,
no. 7, p. 627.

Wallace, A. B., 1980, Geology of the Sulphur
district, southwestern Humboldt Co.,
Nevada: Soc. Econ. Geologists, 1980 Field
Conference, Road Log and articles (preprint),
p. 80-91.
Warnaars, F. W., Smith, W. H., Bray, R. W.,
Lanier, George, and Shafiquallah, Muhammed,
1978, Geochronology of igneous intrusions and
porphyry copper mineralization at Bingham,
Utah: Econ. Geol., v. 73, p. 1242-1249.
Whalen, J. B., Britten, R. M., and McDougall, Ian,
1982, Geochronology and geochemistry of the
Frieda River Prospect area, Papua, New
Guinea: Econ. Geol., v. 77, p. 592-616.
White, D. E., 1955, Thermal springs and epithermal
ore deposits: Econ. Geol., Fiftieth Ann.
Vol., Pt. I, p. 99-154.
White, D. E., 1974, Diverse origins of
hydrothermal ore fluids: Econ. Geol., v. 69,
p. 954-973.
White, D. E., 1981, Active geothermal systems and
hydrothermal ore deposits: Econ. Geol.,
Seventy-fifth Ann. Vol., p. 392-423.
Wilson, F. H., 1980, Late Mesozoic and Cenozoic
tectonics and the age of porphyry copper
prospects; Chignik and Sutwick Island
quadrangles, Alaska Peninsula: U.S. Geol.
Survey Open-File Rept. 80-543.

"This paper has not been edited for conformity
with Geological Survey standards and nomenclature."

REGIONAL GEOPHYSICS
OF THE NORTHERN GREAT BASIN

REGIONAL GRAVITY AND MAGNETIC ANOMALIES IN THE NORTHERN BASIN AND RANGE PROVINCE

Don R. Mabey, Howard W. Oliver and Thomas G. Hildenbrand

Utah Geological and Mineral Survey
U. S. Geological Survey

Abstract

The distribution of gravity stations in the northern Basin and Range Province is adequate to define regional anomalies and the more important features of most of the major local anomalies. These gravity data have been combined with digital topographic data to produce complete Bouguer, free-air and Airy-type isostatic maps. The Bouguer anomaly maps are dominated by the inverse relation with regional topography and by local gravity lows produced by thick accumulations of low-density Cenozoic rocks. Filtered versions of the Bouguer anomaly data enhance both regional and local anomalies. Free air and isostatic anomalies indicate that the region is in approximate isostatic equilibrium. Most of the individual basins and ranges are not locally compensated, but regional topographic features a few tens of kilometers in extent appear to be compensated. Isostatic highs are associated with the Lake Bonneville basin and a zone across northern Nevada.

Aeromagnetic data from numerous surveys have been merged to produce a map of the magnetic field at 3810 m above sea level. The map shows that the major magnetic anomalies over the northern Basin and Range Province are associated with Phanerozoic igneous rocks and only in a few areas is there evidence of an older magnetic basement. Several zones of magnetic anomalies are apparent, such as those reflecting west-trending belts of igneous rocks in western Utah and the Tertiary rift in north-central Nevada. The magnetic data delineate large crustal units with contrasting magnetic signature.

The regional gravity data in the northern Basin and Range Province for the most part reflect the distribution of late Cenozoic masses in the upper crust and the state of isostatic equilibrium. Gravity data can be combined with the seismic refraction data to provide control on the nature of isostatic compensation. The magnetic anomalies provide information on the formation of the crust and on temperature variations within the lower crust. Together the data indicate that the crust of the region is thinner, hotter, more mobile and generally less complexly magnetized than the crust of the craton to the east.

Introduction

For over thirty years gravity and aeromagnetic surveys have been an important part of many geologic programs in the northern Basin and Range Province. In the last few years, gravity and magnetic data from many of these programs have been merged into digital data sets that provide good regional coverage over most of the province. With these data sets computers and plotters can be used to prepare maps and other presentations of the data in a variety of forms. These data combined with digital topography and other sets of digital geological and geophysical data are powerful tools for studying the geology and resources of the province. Gravity and magnetic methods are useful in exploration programs by contributing to the definition of the present geology and also by helping to understand the geologic development of the region. Much work remains to be done with these important data sets. This report describes the major gravity and magnetic anomalies and discusses the possible geologic significance of some of the more prominent features.

The geophysical data come from many sources. The basic gravity set is from the U.S. Geological Survey and Department of Defense files and includes the data from surveys by hundreds of individuals and groups (Oliver and others, 1982; Snyder and others, 1982). The magnetic data set is built around a compilation of Nevada data by Sweeney and others (1978). The present set was compiled by Hildenbrand and others (1983). Most of the magnetic data are from surveys done by or for the U.S. Geological Survey. The original data used to develop both the gravity and magnetic data sets are of varying quality. Many of the earlier surveys were made with instruments that were not well calibrated and the observations were not always referenced to an absolute datum. Much work was involved in merging the individual surveys into the regional data sets. We (the authors) have been involved in all phases of the work and are convinced that the accuracy of the data sets is consistent with the presentations and interpretations in this report.

Various aspects of the regional geology of the northern Basin and Range Province are described in other reports in this volume and will not be repeated here. Both the gravity and magnetic fields over the northern Basin and Range Province reflect geologic events that occurred from very early geologic time to the present. Most of the larger gravity anomalies reflect mass anomalies related directly to the extensional tectonics that has dominated the province in late Cenozoic time. In the central and eastern part of the province the leading cause of the magnetic anomalies is Cenozoic intrusive and extrusive rocks, but in the west the pattern of magnetic anomalies is more complex and reflects rocks and structures with a much wider range of ages.

The gravity and magnetic anomalies associated with Cenozoic events are superimposed on anomalies related to earlier events, and the older anomalies are often difficult to identify. Not only have the older anomalies been submerged in the Cenozoic anomalies, but some anomalies are destroyed or severely altered by geologic processes. For example, isostatic processes work to reduce regional gravity anomalies, and heating of the lower crust above the Curie temperature of magnetite eliminates most of the magnetic anomalies from this zone. Identifiable gravity and magnetic anomalies might be found that relate to any of the major events in the formation of the crust of the northern Basin and Range Province, particularly those that involved igneous events and those that juxtaposed rocks of different density or magnetization. Also, gravity anomalies directly or indirectly reflecting the condition of the upper mantle should be discernible. However, such deep effects produce broad anomalies at the earth's surface that may be difficult to separate from those associated with isostasy.

Gravity Data

Gravity data from land areas are most commonly presented as simple or complete Bouguer anomaly maps or profiles. The complete Bouguer anomaly map (Fig. 1) was prepared by reducing the gravity data to a sea-level datum assuming a density of 2.67 g/cm^3 for the material above the datum. The resulting anomalies reflect all mass anomalies below sea level and any departures from 2.67 g/cm^3 density above sea level. The dominant regional Bouguer anomalies in the northern Basin and Range Province have a pronounced inverse correlation with regional topography. This correlation reflects the general isostatic equilibrium of the region. The mass anomalies reflected by the regional gravity anomalies include variations in crustal thickness and variations in the density of the crust and upper mantle. Although the correlation between gravity anomalies and regional topography was recognized by the first interpreters of data from the region, much work remains to be done before the full significance of this correlation is understood. The dominant local gravity anomalies are lows produced by thick accumulations of low-density Cenozoic rocks within the structural basins of the region and volcanotectonic depressions. In addition to providing a quick and relatively inexpensive technique for estimating the distribution and thickness of the Cenozoic rocks, gravity surveys provide important information on the Cenozoic structures. More subtle gravity anomalies are produced by Cenozoic igneous rocks and by mass anomalies within the older rocks.

The gross symmetry of Bouguer anomaly and regional topography over the northern Basin and Range Province was discussed by Eaton and others (1978). The low gravity and high topography of the Sierra Nevada on the west and the Wasatch Range and Colorado Plateau on the east are

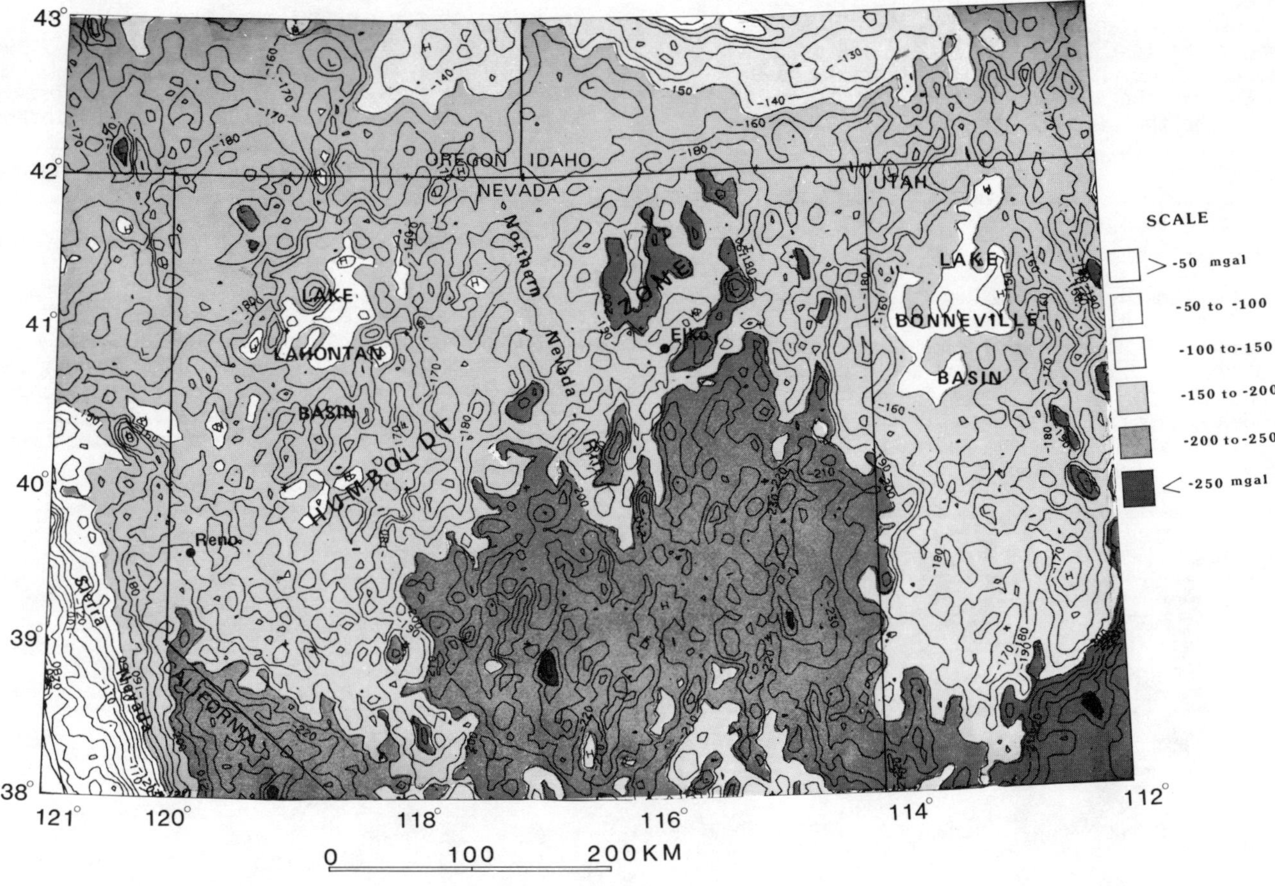

Figure 1. Complete Bouguer gravity anomaly of the northern Basin and Range Province - reduction density 2.67 g/cm^3

bordered by the high gravity and low elevations of the Lake Lahontan and Lake Bonneville basins. The symmetry is best developed in the south-central part of the northern Basin and Range Province where the regional topographic high and gravity low are divided by the north-trending axis of symmetry. To the north, this pattern is less well developed and subject to conflicting descriptions. Eaton and others (1978) proposed an eastward offset of the axis of symmetry in the north and a possible extension into central Idaho. However, the symmetry is severely disrupted by the northeast-trending Humboldt Zone and any extension to the north is poorly defined. Wave-length filtering of the Bouguer anomaly can

be used to emphasize anomalies of different extent (Hildenbrand and others, 1982). A long-wavelength anomaly map emphasizes the north- and north-northwest-trending anomalies involved in the symmetrical pattern, and the complementary short-wavelength anomaly map emphasizes the trends of the local anomalies produced by the Basin and Range structures and some of the mass anomalies in the pre-Tertiary rocks.

Isostatic anomalies are computed by removing a calculated gravity effect related to the regional topography. This can be done either by assuming a direct correlation between the regional topography and the measured Bouguer anomaly or by assuming a model for isostatic compensation and using the model to compute the gravity effect of isostatic compensation for the regional topography. An isostatic anomaly map shows the gravity effect of uncompensated masses and departures from the model assumed in computing the anomaly and helps to isolate small but broad intracrustal gravity anomalies that tend to be lost in the larger effects of isostasy. Recent studies in California have shown that the isostatic anomaly map reveals anomalies that are not apparent on the Bouguer anomaly map (Oliver and Robbins, 1982, fig. 2; Jachens and Griscom, 1982) and provides a basis for comparing anomalies that occur in widely differing topographic regimes (Oliver and others, 1982, table 1).

The isostatic map included here (Fig. 2) is based on the Airy model assuming a crustal thickness of 25 km at sea level and a density contrast of 0.4 g/cm^3 between the crust and upper mantle. Varying these assumptions produces significant changes in the isostatic anomalies in the northern Basin and Range Province, although local gravity differences are insensitive to the assumed parameters. Free-air anomalies can also be used to investigate isostatic phenomena and may be thought of as isostatic anomalies with a sea-level depth of compensation. In the Basin and Range Province, however, the free-air anomaly is strongly dependent on the elevation of the gravity station, and regional free-air anomalies must be used with care to be certain that the observations provide a representative sample. Adjusting the free-air anomaly for the difference between the station elevation and the regional elevation results in a kind of isostatic anomaly (Mabey, 1966a).

The isostatic anomaly map shows most of the regional and local anomalies apparent on the Bouguer anomaly map, but the amplitudes of the regional anomalies are greatly reduced. Important features to note on the isostatic map are the prominent high over the Lake Bonneville basin, the high along the axis of symmetry and its extension to the north, and the generally high zone across northern Nevada excluding the northwest corner.

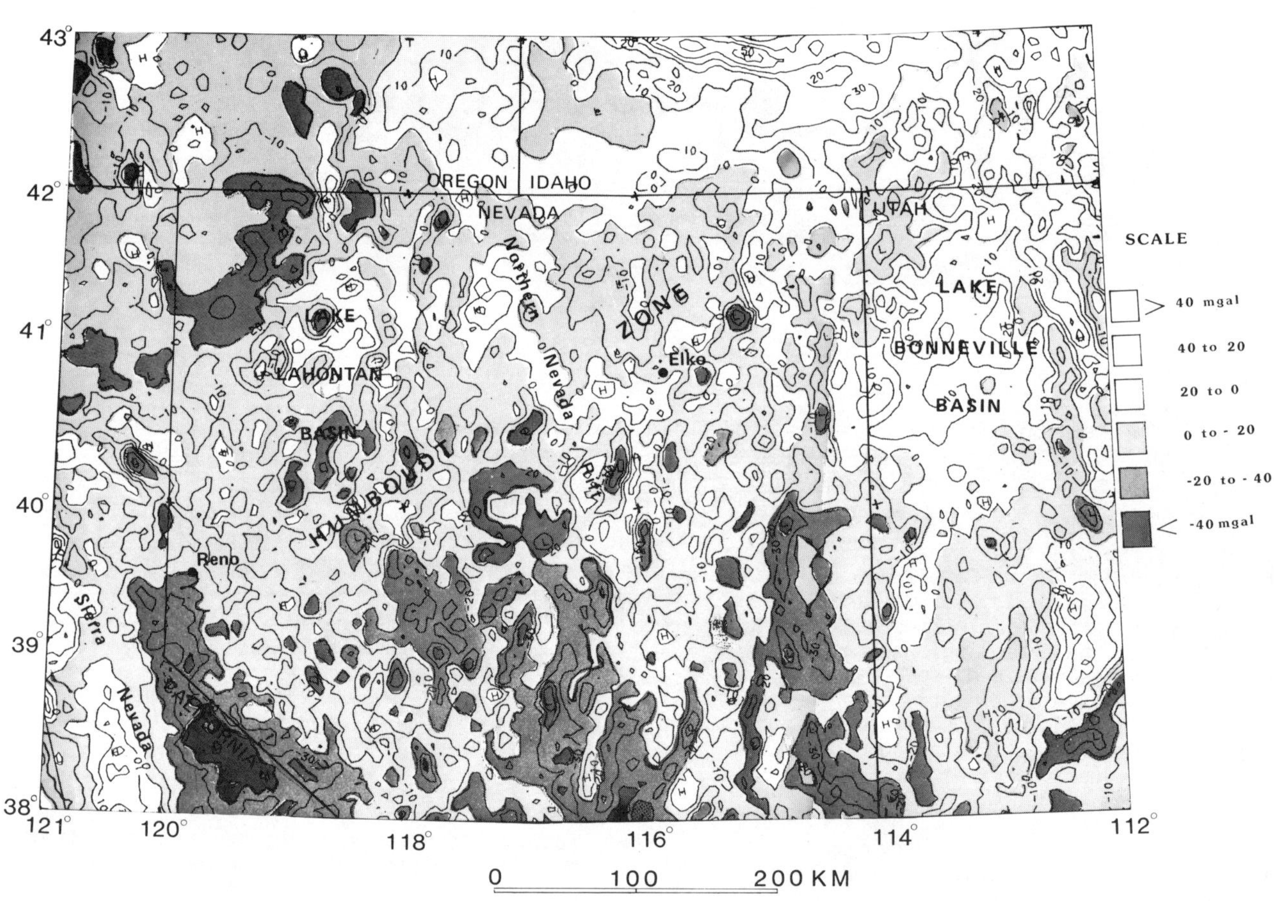

Figure 2. Isostatic anomaly map of the northern Basin and Range Province. Based on an airy model with reduction density of 2.67 g/cm^3, 25 km crust at sea level and a 0.4 g/cm^3 density contrast at the base of the crust.

Magnetic Data

The residual magnetic anomaly map (Fig. 3) portrays the total magnetic intensity at 3810 m above sea level with a reference field removed. Figure 3 at top of next page. This level is about 2000 m above the general land surface; thus, many of the local near-surface anomalies are effectively filtered out of the map. The residual anomaly map reflects large masses of contrasting magnetization lying above the Curie isothermal surface for magnetite. The Curie temperature of magnetite may range from 350°C to 580°C depending on the amount of titanium in the spinel structure (Nagata, 1961, fig. 3-10; Ade-Hall, 1964). The magnetic anomalies as they appear on the residual anomaly map and defined in the digital data do not reflect an internally consistent measure of the magnetic field. Differences in flight level and line spacing of the numerous surveys involved, the digitization and the projection of the data to a common plane produce systematic variations in the data that do not reflect geologic anomalies. Any operation on the data set or the map, such as filtering or quantitative analysis, must take this into account.

Over the northern Basin and Range Province magnetic anomalies reflecting the Precambrian crystalline basement are much less apparent than over the craton to the east of the province. This dearth of anomalies can be attributed to three factors: (1) The generally great depth of burial of the basement in the northern Basin and Range Province and the high level of projection (3810 m above sea level) combine to reduce the apparent magnetic effect of basement features. (2) Much of the basement underlying the province is younger and less complexly magnetized than the area to the east of the province. (3) High heat flow over most of the province has elevated the Curie isothermal surface and thinned the magnetized crust.

Most of the individual magnetic anomalies in the northern Basin and Range Province except in northwestern Utah are produced by Phanerozoic igneous rocks; at the level of the residual map, intrusive rocks are the dominant source. In northwest Utah several magnetic highs appear to reflect Precambrian basement rock. South of this zone of Precambrian anomalies the most prominent magnetic features are west-trending zones of magnetic anomalies (MB, fig. 3) produced by

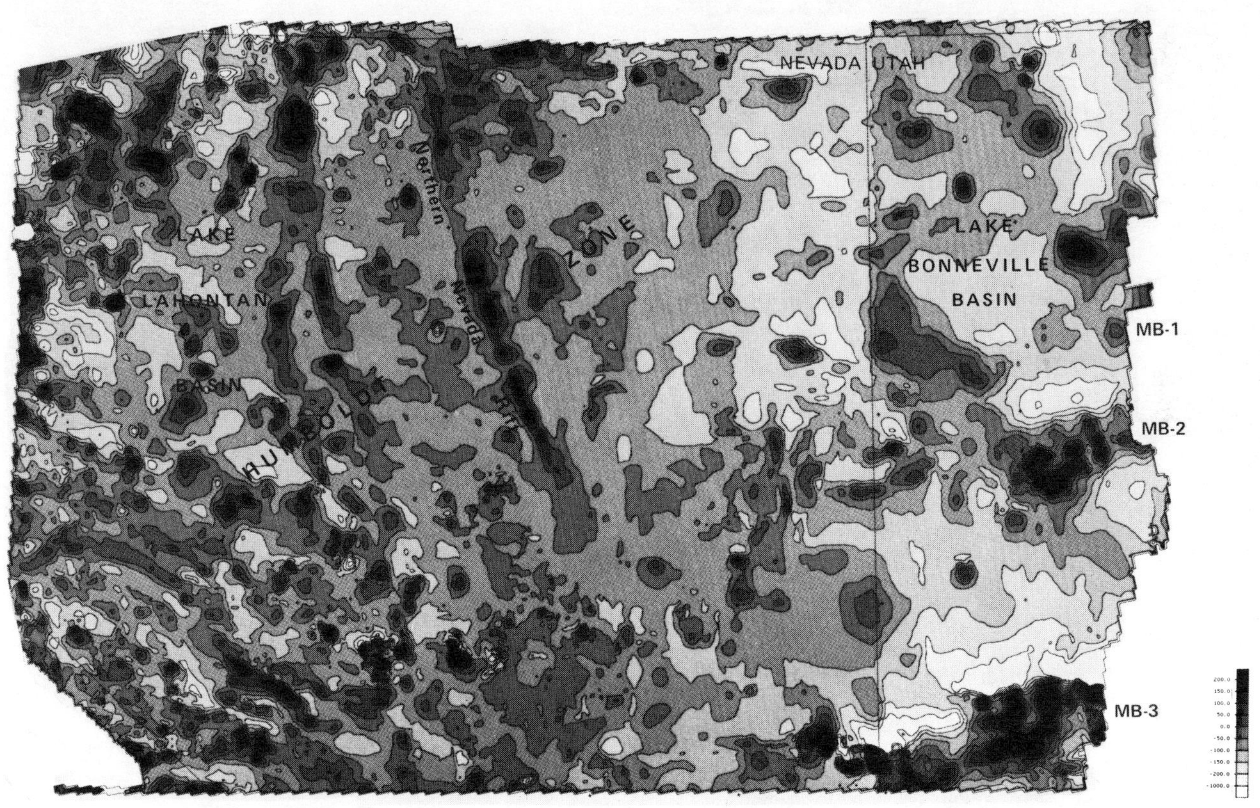

Figure 3. Residual aeromagnetic map of northern Nevada and northwest Utah. Based on numerous surveys merged to a common datum and projected to 3810 meters above sea level.

Tertiary intrusive rocks in major mineral belts. West of these anomalies is the "quiet zone" of Stewart and others (1977, fig. 5) with few large anomalies. One of the most prominent magnetic features is produced by a north-northwest- trending Tertiary rift in north-central Nevada. In the western part of the province the magnetic anomalies are produced by rocks with a wide range of ages, but the trend of many of the anomalies reflects, in large part, Cenozoic structures. (For discussions of the correlation of magnetic anomalies in this region to geology see Stewart and others, 1977, and Mabey and others, 1978).

Interpretation

The eastern margin of the northern Basin and Range Province is approximately coincident with the Cordilleran hingeline. The Cordilleran hingeline, Wasatch line, and hingeline are different terms referring to the zone that formed the eastern margin of the Cordilleran geosyncline and has persisted as a major geologic boundary (Stokes, 1976). The hingeline is expressed in a number of ways throughout Phanerozoic time, and although geologic evidence for a major structure at this location in Precambrian time is not compelling, the structure was well developed in Cambrian time and likely began to develop earlier. The hingeline and the province boundary are marked by a large eastward decrease in Bouguer anomaly values and a rise in regional elevation. Both primarily reflect an eastward thickening of the crust (Thompson and Zoback, 1979). East of the hingeline are large regional magnetic anomalies produced by the Precambrian basement rocks of the North American craton. This contrast suggests a fundamental difference in the basement across the hingeline. A notable exception is in northwest Utah where a band of moderately magnetized Archean rocks extends into the Basin and Range Province. Strontium-isotope ratios indicate that the western edge of the sialic Precambrian basement lies near the center of the province (see Kistler and others, this volume); however, the magnetic data indicate that the crust lying between the hingeline and the edge indicated by the isotopic data has had a different magnetic history from the area to the east.

The western edge of the sialic Precambrian crust indicated by the isotopic data is approximately coincident with the western edge of the quiet magnetic zone in eastern Nevada. That part of the continental crust accreted since Precambrian time is more complexly magnetized than the crust immediately to the east. No pronounced gravity anomaly is coincident with the inferred edge of the Precambrian crust. Apparently, any major difference in crustal thickness or density across the edge is in approximate isostatic equilibrium. Most of the changes in crustal thickness that are reflected in the Bouguer

anomaly probably relate to crustal thinning in Cenozoic time, but older contrasts in the composition and structure of the crust may control the amount and nature of thinning that has occurred in response to the regional tension.

Major igneous events during Paleozoic time were confined to the western part of the northern Basin and Range Province while sedimentary rocks were deposited over most of the province. Part of the complex magnetic pattern in the west may reflect these igneous events, but few magnetic anomalies that can be attributed to Paleozoic events can be identified elsewhere in the province.

Most of the igneous and sedimentary rocks of Paleozoic age in the northern Basin and Range Province are well indurated. The volcanic and carbonate units are generally more dense than the clastic units. Where Paleozoic rocks of contrasting density are juxtaposed by structures, such as the Roberts Mountains thrust, gravity anomalies exist, but they are low in amplitude and difficult to isolate from the much higher amplitude anomalies produced by younger rocks and structures.

As the North American plate began to move westward in Mesozoic time, a volcano-plutonic terrane developed in an arc environment in the western part of the province, and large batholiths related to the Sierra Nevada batholith developed. Numerous magnetic anomalies are produced by these igneous rocks. Some of the batholiths produce large gravity lows and the smaller igneous units produce both gravity highs and lows depending on the density contrast with adjacent rocks. Some of the Mesozoic intrusions produce large magnetic highs.

In the central and eastern part of the province are scattered igneous intrusions and extensive nonmarine sedimentary rocks of Mesozoic age, including evaporites. The Sevier orogeny near the end of the Mesozoic involved major thrusting with generally older, denser rocks moving relatively eastward over younger rocks. Gravity anomalies associated with the thrust faults and related tear faults are generally low amplitude but can be identified in regional surveys. To the west of the thrust belt a zone of metamorphic core complexes developed. The core complexes are usually topographic highs with associated regional Bouguer anomaly lows and modest magnetic highs. The Sevier orogeny probably thickened the crust along the Cordilleran hingeline and produced both a regional Bouguer anomaly low and isostatic anomaly high. Parts of these anomalies may have persisted to the present.

During early and middle Cenozoic time the eruption of silicic volcanic rocks was widespread over the northern Basin and Range Province. The volcanism started in the north and generally progressed southward over a period of about 20 million years (Stewart and Carlson, 1978). In western Utah the progression was in three clearly defined steps that produced three igneous belts (MB1, MB2, MB3, fig. 3). These belts, which are clearly expressed on the regional anomaly magnetic maps, contain abundant intrusive rock and related mineral deposits. Farther west the distribution of igneous rocks and related magnetic anomalies is more diffuse. However, the east-west grain is apparent across Nevada consisting primarily of "interruptions" of magnetic anomalies (Fuller, 1964 and Mabey and others, 1978). Some of these linear features coincide with structural lineaments described by Ekren and others (1976), who concluded that the lineaments reflect pre-Oligocene structural trends. Clearly, deep east-trending structures have controlled Tertiary igneous events. The best geophysical evidence for these east-trending structures is the magnetic anomalies; however, there is subtle expression of some of them in the gravity anomalies. Important mineralization is related to this extensive igneous activity over much of the province. The northern Basin and Range Province probably developed as a crustal unit during this time, but the rapid extension that has characterized the province in Neogene time does not appear to have been widespread then.

Regional extension of the northern Basin and Range Province became widespread after about 17 m.y. ago. At the beginning of this extension the crust was probably thicker and more competent than now. Numerous older structures including large uncompensated mass anomalies controlled response to the extension. One of the first features to form in response to the regional extension was the northern Nevada rift.[1] The rift is expressed by one of the most prominent magnetic anomalies in the province, which is produced by basalt flows, a near-surface swarm of basalt dikes, and a deeper large dikelike mass. A low-amplitude regional gravity high is coincident with the rift and extends south beyond the magnetic and surface expression of the rift. Eaton and others (1978) recognized that the southern part of the axis of symmetry of the Bouguer anomalies and topography coincided with part of the rift but concluded that to the north the axis was offset eastward from the rift. The positive mass anomaly associated with the rift is reflected by an isostatic high. We propose that the isostatic high defines the axis of symmetry and that it is coincident with the northern Nevada rift throughout the length of the rift.

[1]The magnetic evidence of a Tertiary rift in northern Nevada was first reported by Mabey (1966b) and modeled by Robinson (1970), but it was not named in those reports. Mabey and others (1978) applied the name Cortez rift to the feature, and Zoback and Thompson (1978) called it the northern Nevada rift. Mabey here proposes that the name northern Nevada rift be accepted.

Later in Miocene time thick sections of non-marine sedimentary rocks accumulated in local basins. In the northeast part of the province the distribution of the sedimentary rocks over some of the ranges suggests that the Miocene basins were wider than the present basin and range basins. If the brittle part of the crust was thicker in the Miocene than now, the Miocene normal faults likely penetrated the crust to a greater depth and at a higher angle than do the younger basin and range faults. As the crust thinned, mass anomalies that had formed earlier but had been held in vertical position by the strength of the crust now moved vertically to achieve a degree of local isostatic equilibrium. Most topographic features more than about 25 km wide are in approximate isostatic equilibrium. The Lake Bonneville basin may be an example of a positive mass anomaly that, in response to isostatic forces, subsided relative to the surrounding areas. The sensitivity of this basin to isostatic forces is demonstrated by the subsidence and rebound when the water load of Lake Bonneville was applied and then removed (Crittenden, 1963).

The local gravity and magnetic anomalies in the northern Basin and Range Province are two important sources of information on the basin-and-range structure. The gravity lows produced by the low-density basin fill indicate the approximate depth and configuration of the basins and the location of the major normal faults. The basin fill is often several kilometers thick and produces gravity lows of more than 70 mgal (Oliver and others, 1982). Density variations within the basin fill and problems in isolating the gravity effect of the fill from effects produced by the underlying older rocks often determine the limit to which a quantitative analysis can be carried.

Magnetic data usually indicate if the basin fill includes abundant volcanic rock and if normal faults displace a magnetic basement. Idealized gravity and magnetic profiles across basin-and-range structures in the eastern and western parts of the province are shown in Figure 4. In the east the Bouguer gravity anomalies primarily reflect two sources: the near-surface basin-and-range structure and a deeper source that is the compensating mass for the regional topography. Generally, the basin-and-range structures in the east are not apparent in the magnetic data. This pattern of gravity and magnetic anomalies is suggestive of a system of listric faults that do not displace major density or magnetization horizons. In the west both the gravity and the magnetic patterns are more complex. In addition to gravity anomalies produced by the shallow and deep structures, there is evidence of intermediate- depth sources. Some basin-and-range structures are also reflected in the magnetic data with highs over the ranges. The normal faults appear to juxtapose pre-Miocene units of differing

NORTHERN BASIN AND RANGE

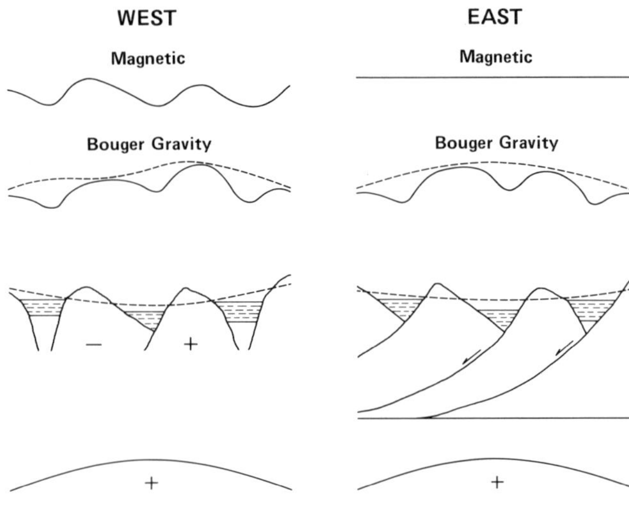

Figure 4 - Diagram of generalized gravity and magnetic anomalies over basin and range structures in northern Basin and Range Province. Positive mass anomalies indicated by + and negative by - .

density and magnetization. Density and magnetization differences are common in pre-Tertiary rocks of this part of the province. The pattern of the gravity and magnetic anomalies suggests but does not prove that the faults penetrate to greater depths than the faults in the east. In northwest Nevada the northeast structural grain of the Humboldt zone (Mabey and others, 1978) influenced the trend of the faults which are generally northeasterly rather than normal to the direction of extension and thus involve considerable strike-slip movement, a style of faulting that also argues against a listric character. In southwest Nevada complex interaction has occurred with deformation related to the San Andreas fault system.

Most of the known hydrothermal convection systems in the northern Basin and Range province are controlled, at least in part, by normal faults. Deep circulation within normal fault zones is commonly thought to be an essential element of the systems. Many of the systems are located where there is evidence for anomalous geometry of the fault zone. For example, the hot springs along the Wasatch Range are at salients where the faults may depart from the listric pattern that is inferred to be characteristic elsewhere.

Only a few igneous events have been associated with the last few million years of extension of the northern Basin and Range Province. Important exceptions are the southwest and southeast edges of the province. Here major eruptions of silicic volcanic rocks and basalt have occurred, and some of these igneous systems appear to still be active. Some of the volcano-tectonic depressions produce large gravity lows, and magnetic anomalies are produced by the extrusive rocks. These young igneous events are important in the development of the geothermal resources of the province.

Conclusions

Gravity and magnetic anomalies in the northern Basin and Range Province can be related to the development of the upper crust. Understanding the development of the geophysical anomalies contributes to understanding the geologic processes that have operated in the province. Quantitative interpretation of the anomalies often helps define the present geology. The digital data sets have opened important opportunities for using the geophysical data in a variety of investigations that will make major contributions to the knowledge of the geology of the province. These include the use of filtered maps to isolate anomalies with distinctive trends and wave-lengths and the preparation of isostatic maps with a range of compensation depths and densities.

REFERENCES

Ade-Hall, J.M., 1964, The magnetic properties of some submarine oceanic lavas: Royal Astrom. Soc. Geophys. Jour., v. 9, p. 85-92.

Crittenden, M.D., Jr., 1963, New data on the isostatic deformation of Lake Bonneville: U.S. Geol. survey Prof. paper 454-E. p. E1-E3.

Eaton, G.P., R.R. Wahl, H.J. Prostka, D.R. Mabey, and M.D. Kleinkopf (1978), Regional gravity and tectonic patterns: their relation to late Cenozoic epeirogeny and lateral spreading in the western Cordillera, in Cenozoic Tectonics and Regional Geophysics of the Western Cordillera: R.B. Smith and G.P. Eaton, eds., Geol. Soc. America Mem. 152. p. 51-92.

Ekren, E.B., R.C. Bucknam, W.J. Carr, G.L. Dixon, and W.D. Quinlivan, 1976, East-trending structural lineaments in central Nevada: U.S. Geol. Survey Prof. Paper 986, 16 p.

Fuller, M.D., 1964, Expression of E-W fractures in magnetic surveys in parts of the U.S.A.: Geophysics v. 29, p. 602-622.

Hildenbrand, T.G., R.P. Kucks, and R.E. Sweeney, 1983, Colored digital magnetic anomaly map of the Basin and Range Province: U.S. Geol. Survey Open-File Report 83-198.

Hildenbrand T.G., R.W. Simpson, R.H. Godson, and M.F. Kane, 1982, Digital colored residual and regional Bouguer gravity maps of the conterminous U.S. -- Wavelength cut-offs of 250 km and 1000 km: U.S. Geol. Survey Geophysical Investigations Map GP-953-A, scale 1:7,500,000.

Jachens, R.C., and Andrew Griscom, 1982, An isostatic residual gravity map of California: A residual map for interpretation of anomalies from intracrustal sources (Expanded Abs.): Society of Exploration Geophysicists Program, 1982 Annual Meeting, Dallas, TX., p. 299-301.

Mabey, D.R., 1966a, Relation between Bouguer gravity anomalies and regional topography in Nevada and the eastern Snake River Plain, Idaho: U.S. Geol. Survrey Prof. Paper 550-B, p. 108-110.

Mabey, D.R., 1966b, Regional gravity and magnetic anomalies in part of Eureka County, Nevada: Soc. Exploration Geophysicists Mining Case histories, v. 1, p. 77-83.

Mabey, D.R., I. Zietz, G.P. Eaton, M.D. Kleinkopf, 1978, Regional magnetic patterns in part of the Cordillera in the western United States, in Cenozoic Tectonics and Regional Geophysics in the Western Cordillera, R.B. Smith and G.P. Easton, eds., Geol. Soc. America Mem. 152, p. 93-106.

Nagata, Takesi, 1961, Rock magnetism: Maruzen Company, Ltd., Tokoyo, Japan, 350 p.

Oliver, H.W., and S.L. Robbins, 1982, Bouguer gravity map of California, Fresno sheet: California Division of Mines and Geology, scale 1;250,000, 23 p. interpretive text.

Oliver, H.W., R.W. Saltus, D.R. Mabey, and T.G. Hildenbrand, 1982, Comparison of Bouguer anomaly and isostatic residual maps of the southwestern Cordillera (expanded abs.): Society of Exploration Geophysicists Technical Program, 1982 Annual Meeting, Dallas, TX, p. 306-308.

Robinson, E.S., 1970, Relation between geological structure and aeromagnetic anomalies in central Nevada: Geol. Soc. America Bull., v. 81, p. 2045-2060.

Snyder, D.B., C.W. Roberts, R.W. Saltus, and R.F., Sikora, 1982, Magnetic tape containing approximately 64,000 gravity stations in the state of California: U.S. Geol. Survey Report, 9-track magnetic tape and 33 p. text, available from U.S. Department of Commerce, National Information Service, Springfield, VA, 22161, NTIS PB 82-168279 (text) and PB 82-168287 (tape).

Stewart, J.H., and J.E. Carlson, 1978, Generalized maps showing distribution, lithology, and age of Cenozoic igneous rocks in the western United States, in Cenozoic tectonics and Regional Geophysics in the western Cordillera: R.B. Smith and G.P. Eaton, eds., Geol. Soc. of America Mem. 152, p. 263-264.

Stewart, J.H., W.J. Moore, and I. Zeitz, 1977, East-west patterns of Cenozoic igneous rocks, aeromagnetic anomalies, and mineral deposits, Nevada and Utah: Geol. Soc. America Bull. v. 88, p. 67-77.

Stokes, W.L., 1976, What is the Wasatch Line? in Geology of the Cordilleran Hingeline: Rocky Mtn. Assoc. of Geol., p. 11-25.

Sweeney, R.S., R.H. Godson, J.H. Hassemer, D.D. Bansereau, B.K. Bhattachryya, 1978, Composite aeromagentic map of Nevada: U.S. Geol. Survey Open-File Report 78-695.

Zoback, M.L., and G.A. Thompson, 1978, Basin and Range rifting in northern Nevada - clues from a mid-Miocene rift and its subsequent offset: Geology, v. 6, n. 2, p. 111-116.

Thompson, G. A., and M. L. Zoback, 1979, Regional geophysics of the Colorado Plateau: Tectonophysics, v. 61, p. 149-181.

Analysis of Lineaments in the Great Basin: Relationship to Geothermal Resources

Lawrence C. Rowan, Terry W. Offield and Melvin H. Podwysocki
U.S. Geological Survey
National Center, MS 327
Reston, VA 22092

ABSTRACT

Linear features, mainly streams, ridge crests and escarpments, mapped on Lansat Multispectral Scanner (MSS) images of the Great Basin were analyzed along with diverse geological, geophysical, and geochemical data in order to delineate potentially fractured zones that might be favorable for subjacent magmatic intrusion. In Nevada, analysis of linear features $\geq$ 50 km length indicated the presence of nine lineaments trending dominantly northeast, northwest, and east-northeast. Analysis of fault density, aeromagnetic gravity, and heat-flow data indicate that seven of these lineaments are the morphological expressions of four structural zones that are 100 to 200 km wide.

In southern Nevada, three structural zones were delineated: the northwest-trending Walker Lane zone, east-northeast-trending Southern Nevada zone, and northeast-oriented Pahranaget zone. The Walker Lane structural zone contains the well-documented Walker Lane right-lateral strike-slip faults, and extends southeastward to include the Las Vegas shear zone. Northwest-trending mapped faults are concentrated within the Walker Lane zone but are relatively sparse elsewhere in Nevada. East-northeast-trending faults are abundant locally within the Southern Nevada structural zone, and field evidence indicates that left-lateral strike-slip, as well as dip-slip, displacement is characteristic. Aeromagnetic anomalies and lineaments in MSS images of Utah suggest extension of this structural zone eastward to the Wasatch front. The Pahranaget structural zone, which is marked by northeast-oriented left-lateral strike-slip faults, transects the Southern Nevada structural zone near the Utah-Nevada border and continues for approximately 175 km into Utah. A zone of high heat flow parallels part of this structural zone.

Northern Nevada is dominated by the northeast-oriented Humboldt structural zone. This zone is 150-200 km wide and characterized by numerous northeast- and north-northeast-trending faults— many of which are Pliocene or younger, disruptions in the trends of aeromagnetic and Bouguer gravity anomalies, the Battle Mountain heat-flow high, numerous hydrothermal convection systems with temperatures$>$150^{0}C, and broad arching during the last 10,000 years. Left-lateral strike-slip, as well as dip-slip, displacement has been documented along faults within this zone. The zone continues northeastward along the southeastern border of the Snake River Plains, across the Yellowstone caldera and into central Montana. It is interrupted in north-central Nevada by the north-northwest oriented Northern Nevada rift.

The northeast-trending linear features marking the Humboldt and Pahranaget structural zones are interpreted as left-lateral strike-slip fractures formed during the last 10 m.y. when the minimum principal stress direction was oriented west-northwest, as it is at present. These fractures provided the conduits for magmatic intrusions which are responsible for the anomalously high heat flow within parts of these zones. The left-lateral strike-slip faults that characterize the Pahranaget zone probably formed under these regional stress conditions.

The north-northwest-trending right-lateral strike-slip fractures anticipated during the west-northwest-oriented extension of the Great Basin have not been noted along any of the structural zones. We believe that northwest- rather than north-northwest-oriented right-lateral strike-slip faults formed in the Walker Lane structural zone owing to the presence of northwest-oriented faults that formed prior to 10 m.y. ago in response to west-southwest-directed extension. Although only a few high heat-flow anomalies are scattered along the northwestern part of the Walker Lane structural zone, late Cenozoic volcanic rocks and mineralized areas are abundant within the central and northeastern parts.

In the Southern Nevada structural zone, the generally east-orientation of the left-lateral strike-slip faults do not correspond to any of the primary shear directions resulting from the post- and pre-10 m.y. old-extension directions. However, the orientation and sense of displacement of these faults are consistent with those expected for second-order features in a west-northwest-directed extensional field.

The Walker Lane, Pahranaget, and Southern Nevada structural zones occur at the margins of the Great Basin and, therefore, probably reflect zones of compensation between this extensional province and adjacent areas having less extension during the past 10 m.y. The Humboldt structural zone separates the highly extended region to the south from the area to the north, where extension has been active during the past 10 m.y. but the magnitude has been less.

CRUSTAL STRUCTURE IN NORTHERN NEVADA FROM SEISMIC REFRACTION DATA

by Douglas A. Stauber

U.S. Geological Survey
345 Middlefield Rd.
Menlo Park, CA 94025

Abstract

Three seismic refraction profiles were
recorded in and around the Battle Mountain heat
flow high in north-central Nevada. Models of P
and S velocity structure derived from these
profiles together with previously published
profiles reveal a substantial variation in
crustal thickness in this region (from 23 km to
35 km). The crust is thinnest in a band between
the Battle Mountain-Winnemucca area and Reno,
and thickens to the northwest, east and south.
The expected gravity anomaly associated with the
variation in crustal thickness matches the shape
of the observed Bouger gravity field well, but
acceptable fit to the gravity data requires
either an unusually small density contrast (0.16
g/cm^3) between the crust and uppermost mantle
or a partially compensating mass somewhere in
the crust or upper mantle. A likely candidate
for the latter option involves variation in the
thickness of the lithosphere itself.

Introduction

Large amplitude long wavelength anomalies in
the heat flow and gravity fields in northern
Nevada as well as published seismic refraction
work indicate significant lateral variations
occur in the crust in this region. The Bouguer
gravity field in northern Nevada, Figure 1,
shows a large rectangular region of high values
in northwestern Nevada corresponding to the Lake
Lahonton topographic low and a gravity low over
the topographically high region in the center of
Nevada. Seismic P velocity models for two
reversed refraction lines in northern Nevada
have been published by Eaton (1963) and Hill and
Pakiser (1966). Locations are shown as
long-dashed lines in Figures 1 and 2. These
models have a variation in crustal thickness of
10 km (24 to 34 km) with the thin crust
occurring near Fallon, Nevada where the
refraction line is in the gravity high. Several
other unreversed refraction lines in northern
Nevada, using nuclear tests occurring in
southern Nevada as sources, have been
interpreted for crustal thickness variations by
Stauber and Boore (1978) and Priestly and others

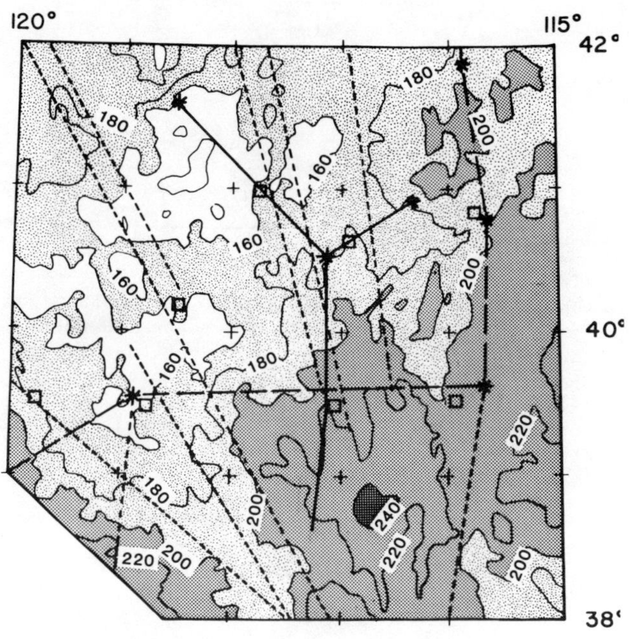

Figure 1. Locations of refraction lines in
northern Nevada superimposed on Bouguer gravity
map for the region from Eaton (1978). Gravity
values are in mgal. Contour interval is 20
mgal. Solid lines are described in this paper.
Long dashed lines are published reversed
refraction lines and short dashed lines are
published unreversed refraction lines. See
figure 2 for references to published lines.

(1982). Locations of these profiles are shown
with short dashed lines on Figures 1 and 2.
These interpretations have a thin (20 to 24 km)
crust in the region of the gravity high and a
thick crust (30 to 35 km) under the gravity low
in central Nevada. The interpretation of
unreversed refraction lines is not unique so
verification of these interpretations is
necessary.

319

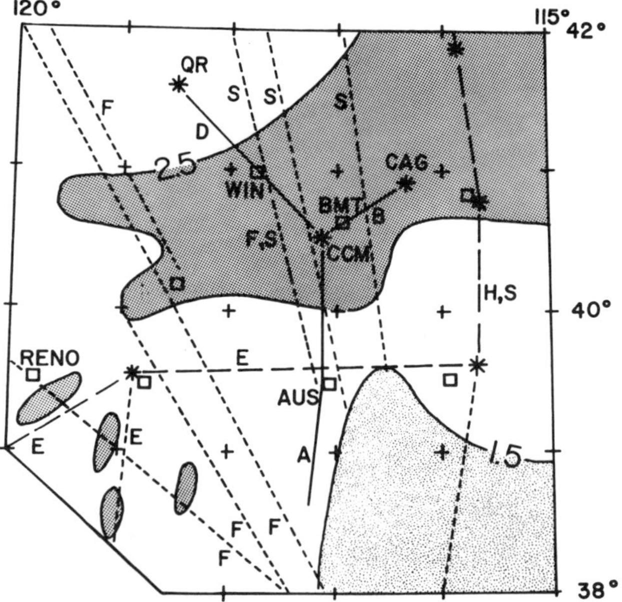

Figure 2.   Locations of refraction lines superimposed on heat flow map of Lachenbruch and Sass (1978). Dark pattern shows the HFH where the conductive heat flow is greater than 100 mW/m^2. Light pattern shows location of heat flow low in central Nevada where heat flow is less than 60 mW/m^2. Place name abbreviations are; WIN=Winnemucca; BMT= Battle Mountain; AUS= Austin. Shot points are shown with stars. Key to refraction lines ; A,B,D are described in this paper; E = Eaton (1963); F = Priestley and others (1982); H = Hill and Pakiser (1967); S = Stauber and Boore (1978).

In addition to the large gravity anomalies and seismic velocity structures of northern Nevada, there is a large heat flow anomaly.

The Battle Mountain Heat Flow High, HFH, described by Lachenbruch and Sass (1978) is a large region in which the conductive heat flux is greater than 100 mW/m^2, compared to the more typical, but still high, Basin and Range values near 80 mW/m^2. The location of the HFH is shown in Figure 2 along with the published refraction line locations. Stauber and Boore (1978) noted a correlation in the position of the southern boundary of the HFH and the crustal thickness change, thinner within the HFH than to the south, on the four eastern north-south profiles. Partial melting in the crust of the HFH as well as the tectonic extension models of Lachenbruch and Sass (1978) could produce large variations in the seismic velocity structure of the crust and mantle in the HFH. More detailed, reversed refraction lines are required to search for these effects and to constrain the interpretation of the unreversed profiles.

Experiment Description

To accomplish these goals three detailed reversed refraction lines in north central Nevada were recorded and interpreted. The locations of these lines are shown in Figures 1 and 2 as solid lines. These lines span the HFH and the gravity anomalies discussed above. Regular quarry blasts at the Copper Canyon Mine (CCM), 40.54° N, 117.13° W, and at the Carlin Gold Mine (CAG), 41.61° N, 116.32° W, served as the primary sources for these detailed refraction lines. In addition, a 2000 lb shot detonated by the U. S. Geological Survey in northwestern Nevada near the Quinn River (QR), 41.61° N, 118.52° W, was recorded. Line A, running south from CCM was partially reversed by recordings of Nevada Test Site, NTS, events. Additionally, line A crosses the east-west reversed profile of Eaton (1963), which provided control for dip along line A. Line A crosses both the HFH boundary and the gravity anomaly. Line B, between CCM and CAG, is entirely within the HFH. Both shotpoints were recorded on this line. Reversed coverage on line D was obtained by recording both the QR and CCM shotpoints.

Recording was performed with several types of seismographs. Most of the detailed refraction lines were recorded with 5 portable analog FM recorders built at Stanford University (Stauber, 1980). The QR shot was also recorded with U. S. Geological Survey 5-day analog FM recorders (Criley and Eaton, 1978). Sprengnether MEQ-800 smoked paper recorders were used to record NTS events and as reference stations on lines A and B. 1 Hz geophones were used as sensors for all the recording systems.

Recording site spacing of 2 to 10 km was obtained with 5 recorders by moving the recorders along the lines and recording many blasts. One smoked drum station remained fixed at a "reference" site to obtain shot time and amplitude information. For each reference site, a few shots at the appropriate quarry were recorded with a geophone at the edge of the shot hole pattern to provide an accurate shot time and travel time to the reference site. For subsequent shots, the arrival time measured at the reference station was used to obtain shot time and corrections for small shot location changes in the large quarries by the method described by Stauber (1980). Amplitudes of the P waves were measured from the reference records and used to correct for shot size varations in the record sections on Lines B and D (Stauber, 1980). Because of variation in waveform of different shots and instrumental variations, the amplitude normalization is considered accurate within a factor of 2. No amplitude normalization was attempted for line A because several reference sites were used at various

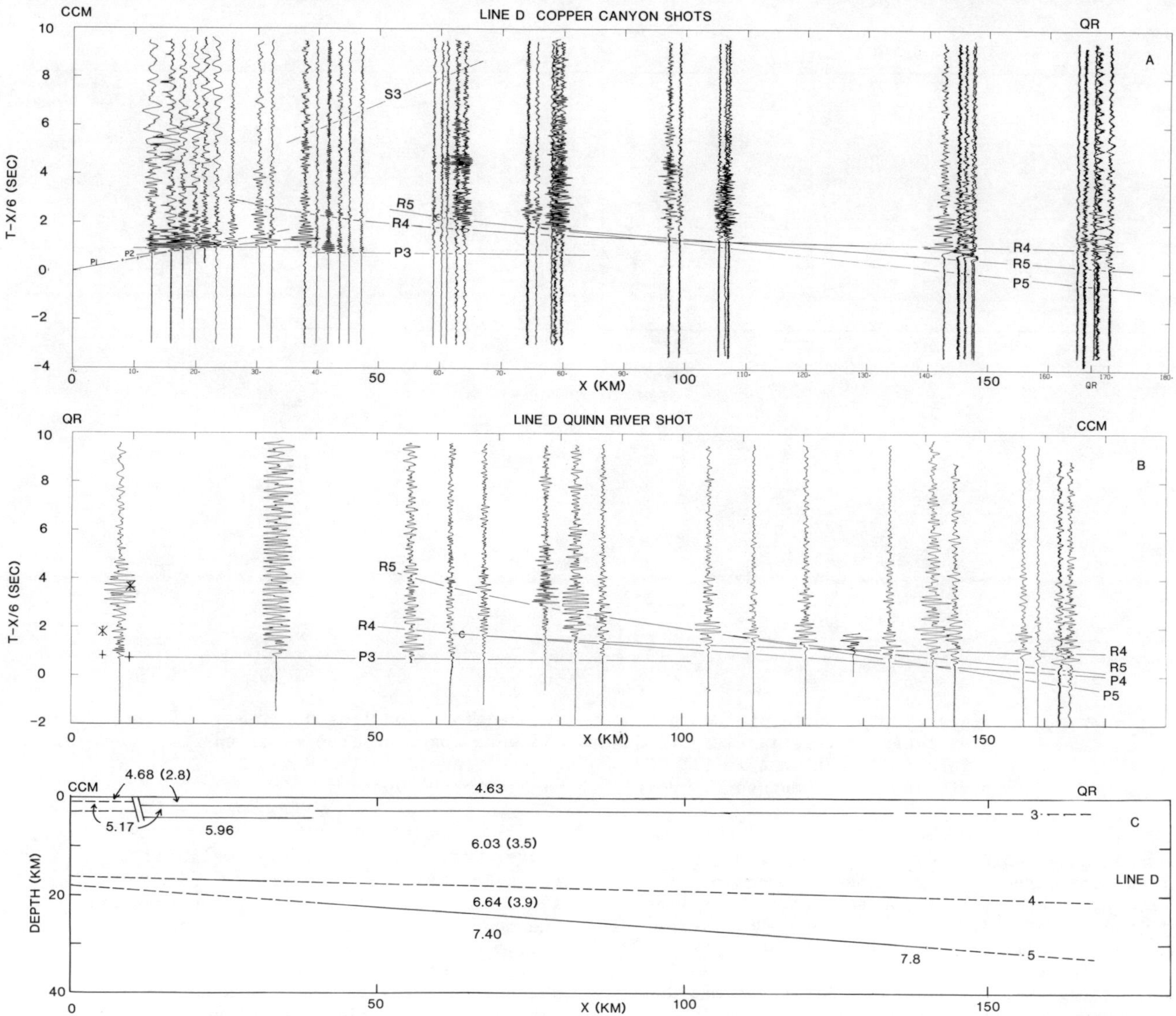

Figure 3.   A) Vertical component record section for CCM shots on line D. Travel-time curves are computed from model in figure 3C. B) Vertical component record section for QR shot on line D.
C) Velocity model for line D. P velocities and S velocities (in parenthesis) are in km/sec. The solid portions of the interfaces in the model indicate the parts of the interfaces for which observations of refracted and reflected waves exist.

times on this line. Vertical component record sections for lines D, B and A with reducing velocity of 6.0 km/sec are shown in Figures 3, 5 and 7.

First motion times and amplitudes were measured on the record section for all recordings where it was possible. For some of the weaker signals, a tracing of a clearer example at a nearby station was overlayed on the weak trace and used to estimate the first motion time. An uncertainty was assigned to each measurement as it was made. Later phases were also identified on all the record sections. In identifying the later phases I relied mainly on an abrupt (within one cycle) increase in signal amplitude as well as coherence across more than three or four records. I was further biased

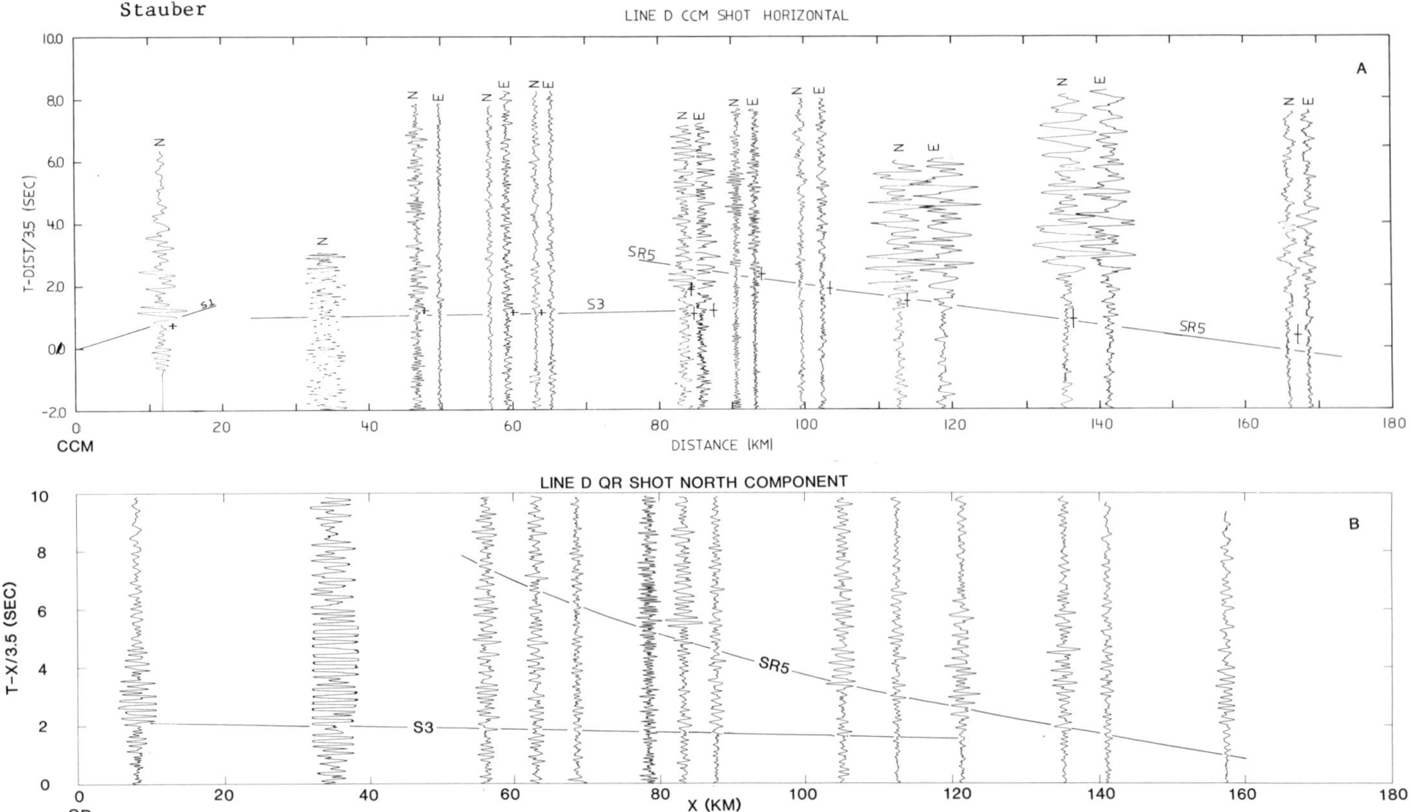

Figure 4.   Horizontal component record sections for line D.   The travel-time curves labeled S1, S3 and SR5 were computed from model in figure 3C.   Section A is for CCM shots.   The traces labeled N and E are north and east components.   Section B is for the QR shot.

toward phases which could be fit by smooth traveltime curves such as those produced by refractions and reflections from planar portions of interfaces. Finally, since many shots were used on each line, I required that later phases be seen on records from several shots before considering them significant. Arrival times and amplitudes were measured and assigned an uncertainty for these later phases. A scatter of arrival times of $\pm$ 0.1 sec about potential smooth traveltime curves was commonly found ·for these later phases.

Shear waves were observed on all the lines and identified as such by their phase velocities, less than 4.5 km/sec. Record sections with a reducing velocity of 3.5 km/sec were produced to enhance the visibility of S waves for several vertical and horizontal component profiles. They are shown in Figures 4, 6 and 7.

Interpretation of Seismic Refraction Data

The seismic velocity structure along the lines was modeled as uniform velocity layers separated by planar dipping interfaces. Slightly more complicated structure was allowed near the surface to account for the alluvial filled basins crossed by the profiles. The dipping layer models were fit to the observed

travel times by an iterative weighted, least-squares inversion program (Stauber, 1980). The reciprocals of the uncertainties in the measurements were used as weights so that poorly determined observations had only a weak influence on the model compared to well determined observatons. The S wave observations formed a less complete set of phases than the P wave observations. Consequently S wave velocities were obtained for the models by keeping the layer boundaries fixed at the positions determined by the P wave travel times and then solving for the S velocities within the layers with the same inversion program. Having discussed the data reduction and the interpretative method, we can proceed to discuss the results for the individual lines.

Line D is the longest, most fully reversed profile. Travel times from the dipping layer model, shown in Figure 3C, are plotted on the record sections in Figure 3 and 4. The first P arrivals near CCM, Figure 3A, are explained by three layers with velocities of 4.68, 5.17 and 5.96 km/sec as depth increases. 11 km northwest of CCM, Figuer 3A, the profile crosses from a range into a basin and the first arrivals are delayed by 0.10 to 0.14 sec. Assuming a valley fill with a velocity of 2.0 km/sec this corresponds to a fill thickness of 0.22 to 0.30

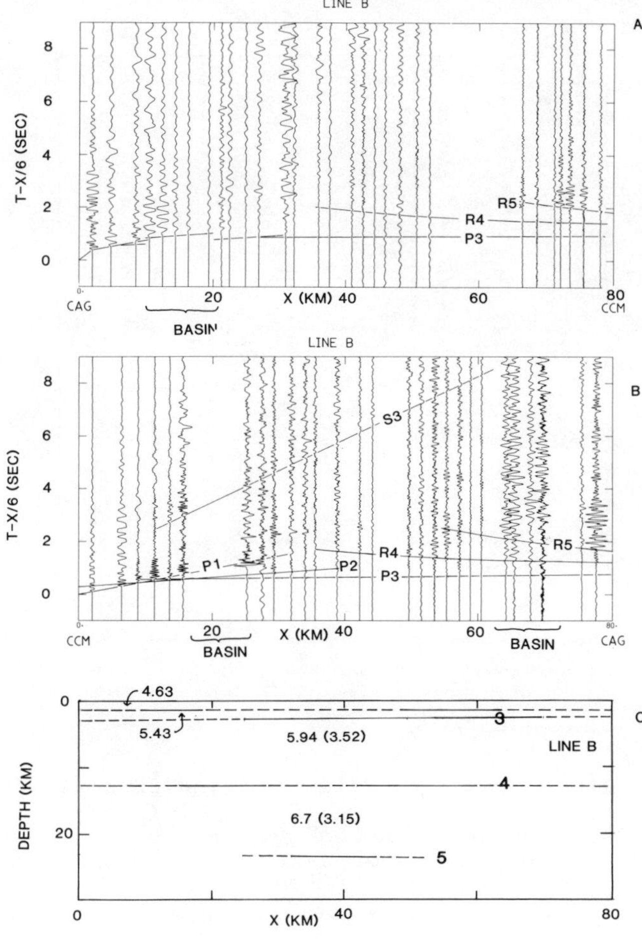

Figure 5.   Vertical component record sections
for line B.   Section A is for CAG shots and
section B is for CCM shots.   Travel-time curves
are computed from the model in figure 5C. P and
S (in parenthesis) velocities in figure 5C are
in km/sec.

km. At 40 km the first arrivals are offset
about 0.2 sec. This corresponds to a lateral
offset in the recording locations. The two
segments, less than 40 km and more than 40 km,
are radial to CCM but are about 15° apart in
azimuth. This offset represents a lateral
variation in the near surface structure. First
arrivals from 40 to 90 km and from 0 to 90 km
for the QR shot define apparent velocities of
6.0 km/sec. The phases marked R4 on the CCM and
QR record sections are interpreted as
reflections from the top of the 6.64 km/sec
layer, the fourth interface in the model. These
reflections control the position of the top of
this layer. The phases marked R5 are
interpreted as reflections from the bottom of
this layer, the fifth interface of the model.
The portion of these phases at large distance
controls the velocity (6.64 km/sec) in the layer

and the nearer shot portions control the depth
to the base. Small first arrivals near 160 km
for the QR shot in Figure 3B are interpreted as
refracted waves from below the 6.64 km/sec
layer. Using this interpretation, one obtains a
velocity of 7.4 km/sec below the fifth interface.

S waves were also observed and modeled on
line D. For the CCM shot, Figure 3A, a clear
impulsive low frequency (2-5 hz) phase labeled
S3 with an apparent velocity of 3.5 km/sec is
visible between 35 and 65 km. Record sections
for horizontal components recorded on Line D
from both shotpoints are plotted with a reducing
velocity of 3.5 km/sec in Figure 4.

For the QR shotpoint, Figure 4B, a tenuous
phase (labeled S3) with an apparent velocity of
3.5 km/sec and an intercept time near 2 seconds
is seen. A reflected phase, SR5, is clearly
visible, particularly at larger distances. For
the reverse direction, at CCM shotpoint in
Figure 4A, low frequency S3 and SR5 phases are
again visible. Assuming that these S phases
were generated at the source, and holding the
layer boundares fixed in the positions
determined by the P waves, the S velocities of
the layers were found by the least square
inverse program and are plotted, in parenthesis
on the model for line D, Figure 3C. The
travel-time curves for the observed S phases
produced by the model are plotted as solid lines
in the reversed sections of Figures 3 and 4.

Record sections and the model for line B are
shown in Figure 5. The amplitudes of the
seismograms have been scaled by distance
squared. The first arrivals define several
linear segments which are modeled by thin layers
with P velocities of 4.63, 5.43 over a thicker
layer with a P velocity of 5.94 km/sec. These
traveltime curves are offset when they cross two
basins along the profile, indicated in Figure
5. Between 10 and 20 km from the CAG quarry,
Figure 5A, the profile crosses Boulder Valley
and the P2 phase is delayed by 0.2 seconds.
Between 18 and 26 km from the CCM quarry, Figure
5B, the Reese River Valley is crossed near
Battle Mountain and the P3 phase from CCM is

delayed 0.4 seconds. Assuming a valley-fill
velocity of 2.0 km/sec, delays of 0.2 and 0.4
seconds correspond to fill thickness of 0.43 and
0.85 km. A reflection, labeled R4 on Figure 5
is visible at distances greater than 35 km and
has a similar character as the R4 reflection
observed on Line D. This reflection is modeled
by an increase in velocity to 6.7 km/sec below
the 5.94 km/sec layer. Since line B is not long
enough to see the wide angle R5 reflections
which would constrain the velocity of 6.7
km/sec, this value was assumed and held fixed.
6.7 km/sec is typical of the velocity found in
the lower crust of the other refraction lines in
the area. A strong reflection, labeled R5, is
visible beyond 60 km for both shotpoints. This
reflection constrains the base of the 6.7 km/sec
layer at a depth of about 23 km below the
surface.

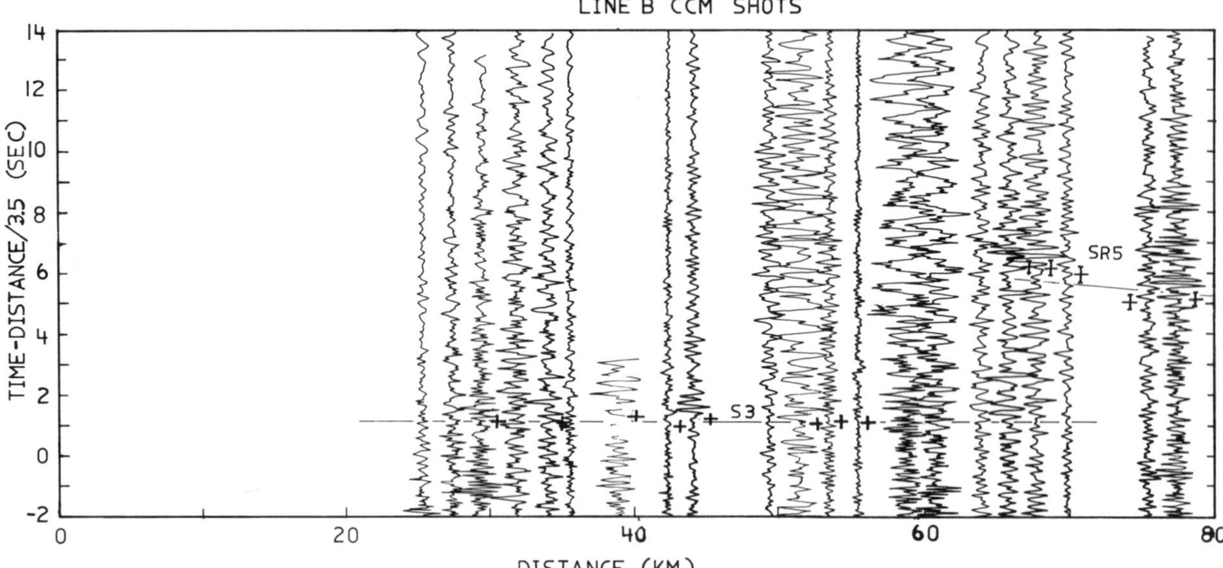

Figure 6.   Vertical component record section for CCM shots on line B
with a reducing velocity of 3.5 km/sec. S wave travel-time curves are
calculated from model in figure 5C.

Low frequency S waves with an apparent
velocity near 3.5 km/sec are visible for the CCM
shots, S3 on Figures 5B and 6. On Figure 6, a
high frequency reflection, SR5, is visible near
70 km at a reduced time near 5 seconds. No S
waves were observed on the comperable record
section from CAG shots. If these two phases are
interpreted as S waves generated at the source,
and the layer boundaries are fixed by the P wave
travel times, then the upper and lower crustal S
velocities are 3.52 and 3.15 km/sec. The
average shear velocity of 3.15 km/sec for the
lower crust, P velocity of 6.7 km/sec, is
unusually low.

Record sections and a model for line A are
shown in Figure 7. Amplitudes are scaled to
yield approximately constant maximum amplitude
on each trace. First arrivals between 10 and
110 km have an apparent velocity of 5.97
km/sec. Between 10 and 60 km the recording
sites were located in the sediment filled Reese
River Valley. Two stations, one near 21 km and
one at 39.5 km were located on bedrock outcrops
through the sediment fill. First arrivals at
these two stations and for stations beyond 60 km
are 0.15 to 0.20 seconds earlier than the
stations on the valley fill. Assuming a valley
fill velocity of 2.0 km/sec, about 0.32 to 0.42
km of valley fill occupies the Reese River
Valley south of CCM. Beyond 150 km, the first
arrivals have an apparent velocity of 7.8

km/sec. Starting at 60 km and a reduced time of
2.6 sec a large amplitude reflection with an
apparent velocity near 7.8 km/sec is clear and
was called the R5 reflection. An abrupt
increase in the low-frequency signal at reduced
times near 0.0 sec was picked between 180 km and

the end of the line and identified as R5.
Another reflection, R4, was picked between 90
and 160 km at reduced times near 1.4 sec and
with an apparent velocity near 6.0 km/sec. R4
arrivals between 60 and 90 km are also visible
in Figure 7A. They were based on an increase in
the amplitude of the signal except at the two
stations nearest 80 km where the signal level
abruptly decreased at the expected arrival
time. This is probably due to destructive
interference between the R4 signal and the codas
of earlier arriving phases.

Looking at figure 2, one can see that one of
the NTS profiles of Stauber and Boore (1978)
coincides with line A between CCM and Austin.
The two lines differ in azimuth by about 10
degrees. South of Austin, the two lines diverge
more strongly. An expanded view of the line
locations around and south of Austin is shown in
figure 8. The heavy solid line shows the
location of the recording sites for CCM shots
which were used to produce the record section in
figure 7A. The line between CCM and NTS is
shown with a dashed line in figure 8. Recording
sites for NTS events are shown with circles near
the CCM to NTS line. Six sites in the Toquima
Range along this line recorded CCM shots and
these are shown with triangles in figure 8.
Unfortunately, the shot time is not known for
the CCM recordings made in the Toquima Range so
only apparent velocity information can be
extracted from the seismograms. Reduced
travel-times from Stauber and Boore (1980) for
NTS events recorded on the four profiles
indicated in figure 2 are plotted in Figure 9.

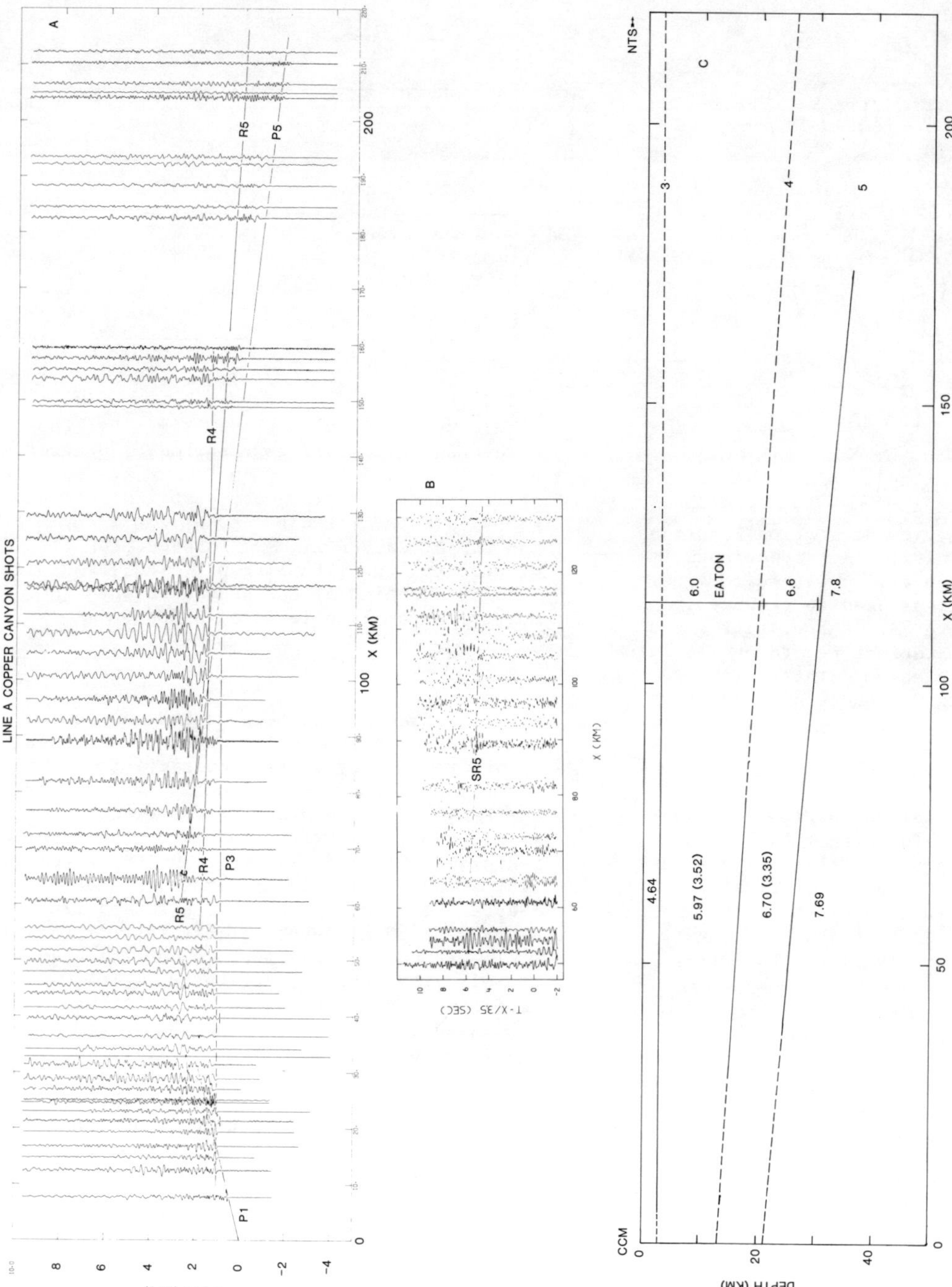

Figure 7.   Vertical component record sections and model for line A.
Travel–time curves are calculated from model in figure 7C. The reducing
velocity for section A is 6.0 km/sec and the reducing velocity for
section B is 3.5 km/sec. P and S (in parenthesis) velocities are in
km/sec. Velocity model from Eaton (1963) at intersection with line A is
shown at vertical bar near 126 km.

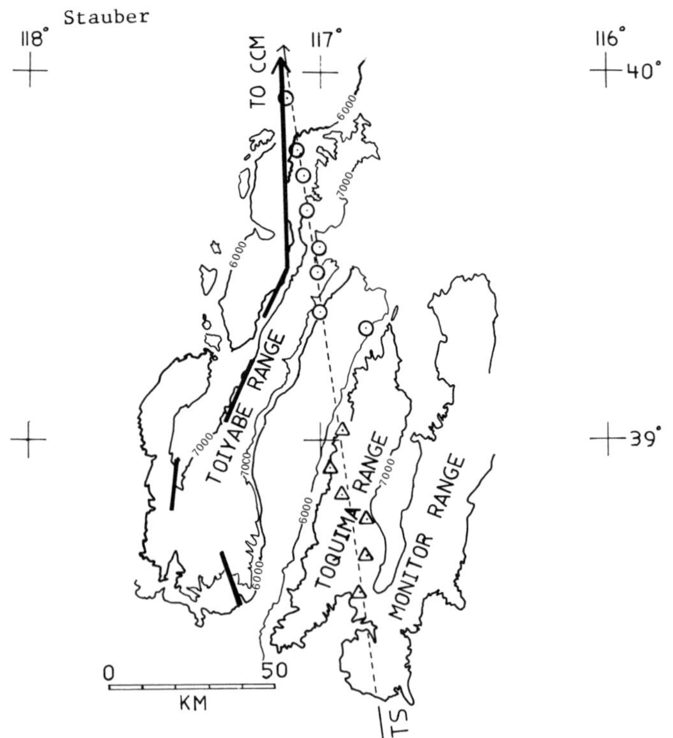

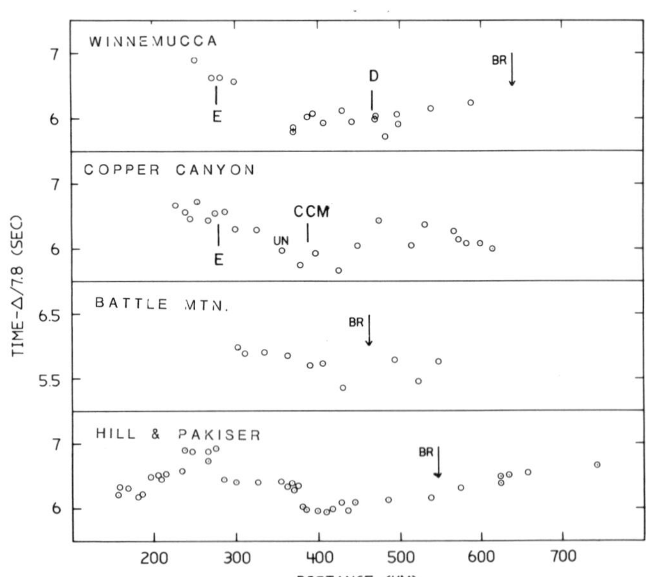

Figure 8.   Expanded view of station locations near the southern end of line A. Topographic contours outline the ranges. Heavy solid lines show locations of stations used to produce the record section in figure 7A. Dashed line shows the CCM to NTS line. Stations recording NTS shots are shown as circles and stations in the Toquima Range recording CCM shots are shown as triangles.

Figure 9.   Reduced travel-times for four unreversed NTS profiles reported by Stauber and Boore (1978), labeled S on Figure 2. Locations of intersections with east-west line of Eaton (1963) are indicated by "E" on the Winnemucca and Copper Canyon lines. Intersection of Winnemucca line with line D is indicated with "D" and location of CCM projected onto the Copper Canyon line is labeled "CCM".

The box labeled Copper Canyon in Figure 9 gives the data for the line coinciding with line A. Reduced travel-times, plus an arbitrary constant since the shot time is unknown, for first arrivals from the CCM shot recorded in the Toquima Range are plotted in the top of Figure 10. Six reduced travel-times from the Copper Canyon NTS profile for stations in the Toiyabe Range are reproduced in the bottom of Figure 10.

Treating line A from CCM shots and the Copper Canyon NTS line as a reversed line, ignoring for the moment the differences in location just discussed, and modeling them with dipping planar interfaces results in the model shown in Figure 7C. Like the structure on line D, the fifth interface dips away from CCM. At 126 km, line A crosses the east-west line of Eaton (1963). Eaton's structure is shown on Figure 7C at the intersection with line A, and the agreement is within 1 km in depths of interfaces and 0.1 km/sec in velocities. This is a good agreement, since the structure of Eaton (1963) was not used as a constraint on the inversion of the line A travel-times. The greatest discrepancy between the model and the observations is in the P5 phase from CCM (Fig. 7A). The apparent velocity that fits the first

arrival observations between 140 and 220 km is near 7.8 km/sec while the model predicts an apparent velocity of 7.26 km/sec. The discrepancy occurs for probably two reasons. The first is a departure of the fifth interface from the planar surface forced by the modeling technique. A fifth interface which is nearly flat at a depth of 24 km between 0 and 70 km, deepens to about 33 km between 70 and 120 km and stays near 33 km depth for the rest of the line would fit the P5 travel-times on line A more closely The second contributor is the difference in locations of the profiles discussed above. Line A and the Copper Canyon NTS line do not strictly form a reversed profile. Despite these problems, the aggreement between the models for line A and Eaton's 1963 profile at their intersection is a strong argument for the general validity of the model in Figure 7A.

The real P velocity below the fifth interface can be determined near Austin by using the data in Figure 10. The stations in the Toiyabe Range recording NTS shots and those in the Toquima Range recording CCM shots provide reversed coverage of a portion of the fifth interface. The apparent velocities resulting from least-square, linear fits give a real velocity below the fifth interface of 7.65 + 0.11 km/sec, close to the value of 7.69 km/sec obtained from the model in Figure 7C.

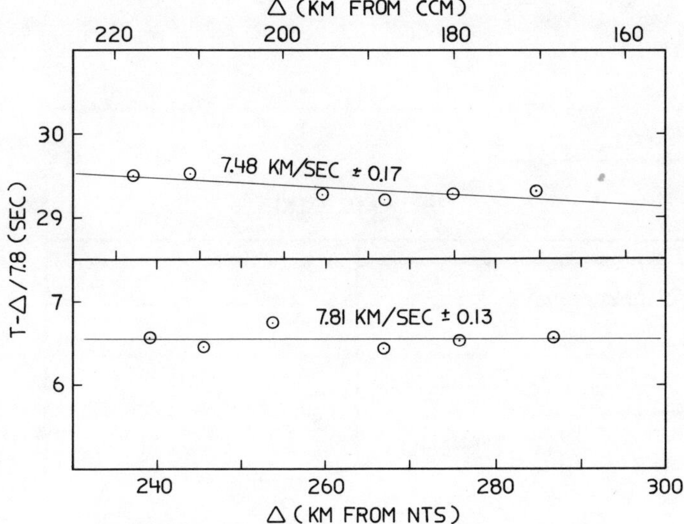

Figure 10.   Reduced travel times for stations located along the line between Copper Canyon and NTS. Top panel has times recorded in the Toquima Range for CCM shots. The locations of these stations are shown with triangles on figure 8. The bottom panel has reduced travel times for NTS shots recorded in the northern Toiyabe Range. The station locations are shown as circles in figure 8. Least square linear fits, their slopes and one standard deviation in the uncertainty of these slopes are shown.

Another estimate of the P5 velocity between Winnemucca and Austin can be obtained from the arrival times on the Winnemucca NTS profile, Figure 9. This profile crosses both Line D and the refraction line of Eaton (1963). Using the crustal velocities and thickness from these two cross lines and the travel times from NTS at the intersections, the upper mantle velocity between can be measured. The value that results is 7.73 km/sec. A change in crustal thickness on one of the cross lines of 2 km, within the uncertainty of the determinations, can change the derived velocity by 0.08 km/sec. Uncertainties in the travel times from NTS of 0.10 sec due to the scatter of points, not measurement accuracy, can change the computed velocity by 0.05 km/sec. Therefore, the error in the upper mantle velocity of 7.73 km/sec could be as large as 0.13 km/sec. This value of 7.73 $\pm$ 0.13 km/sec is close to the value 7.65 $\pm$ 0.11 km/sec obtained near Austin and 7.69 km/sec obtained from the line A model.

A record section was also plotted with a reducing velocity of 3.5 km/sec to facilitate the identification of shear waves (Fig. 7B). The vertical component recordings show a large increase in amplitude between 70 km, 5.5 sec, and 130 km, 4.5 sec. This is probably a near critical-angle shear wave reflection, possibly SR5, the shear wave equivalent of the R5 reflection. The picks are indicated by plus (+) symbols. The solid line shows part of the SR5

travel-time predicted by modeling discussed later. Because of the emergent nature of the SR5 reflection, the picked times are expected to be late and, by inspection of figure 7B, the true arrival times might reasonably be expected to be up to 1.5 sec earlier than the picks.

The travel times of the large-amplitude SR5 arrivals beyond 70 km in Figure 7B were inverted to obtain an average shear velocity in the lower crust along line A. The depth and dip of the interfaces from inversion of the P travel times were held fixed and the S3 shear velocity was held at 3.52 km/sec, the value from line B. The resulting velocity in the lower crust is 3.35 km/sec. This should be considered a minimum since the arrivals are emergent and one expects measured times to be biased toward large values. Subtracting 1.5 sec from all the SR5 travel times, corresponding to the earliest reasonable time, causes the velocity to increase by 0.30 km/sec to 3.65 km/sec. Because no horizontal component records are available, the emergent nature of the arrivals, and possible misidentification of the raypath, I am not confident of the lower crustal shear velocities obtained along lines A and B.

A crustal cross section with a vertical exaggeration of 2.5 constructed from the layered model interpretations of Lines A, B and D is given in Figure 11. The section bends at CCM with the line A structure extending south from CCM and the line D structure extending northwest from CCM. The structure from line B is projected to the cross section at the vertical bar below CCM. P and S velocities, the latter in parenthesis, are shown where determined. The structure proposed by Eaton (1963) is shown at the vertical bar below the location of Austin. The location of the heat flow high, HFH, is shown by the bar at the top. I interpret the region with P velocities greater than 7 km/sec to be mantle. The crust, then, varies in thickness from 34 km at the SE end to 23.5 km near CCM and 30.5 km just SE of QR. This structure confirms the previous interpretations by Stauber and Boore (1978) and Priestly and others (1982) that the early P_n arrivals in northern Nevada are due to a thinning of the crust in the Lake Lahonton depression and not because of an increase in the crustal or upper mantle P velocities.

Using the detailed refraction lines from this study and those of Eaton (1963) and Hill and Pakiser (1966) and the unreversed refraction lines of Stauber and Boore (1978) and Priestly and others (1982), a fairly detailed map of the elevation of the crust-mantle boundary can be constructed for northern Nevada, Figure 12A. To interpolate or extrapolate the crustal thickness, between and around the reversed

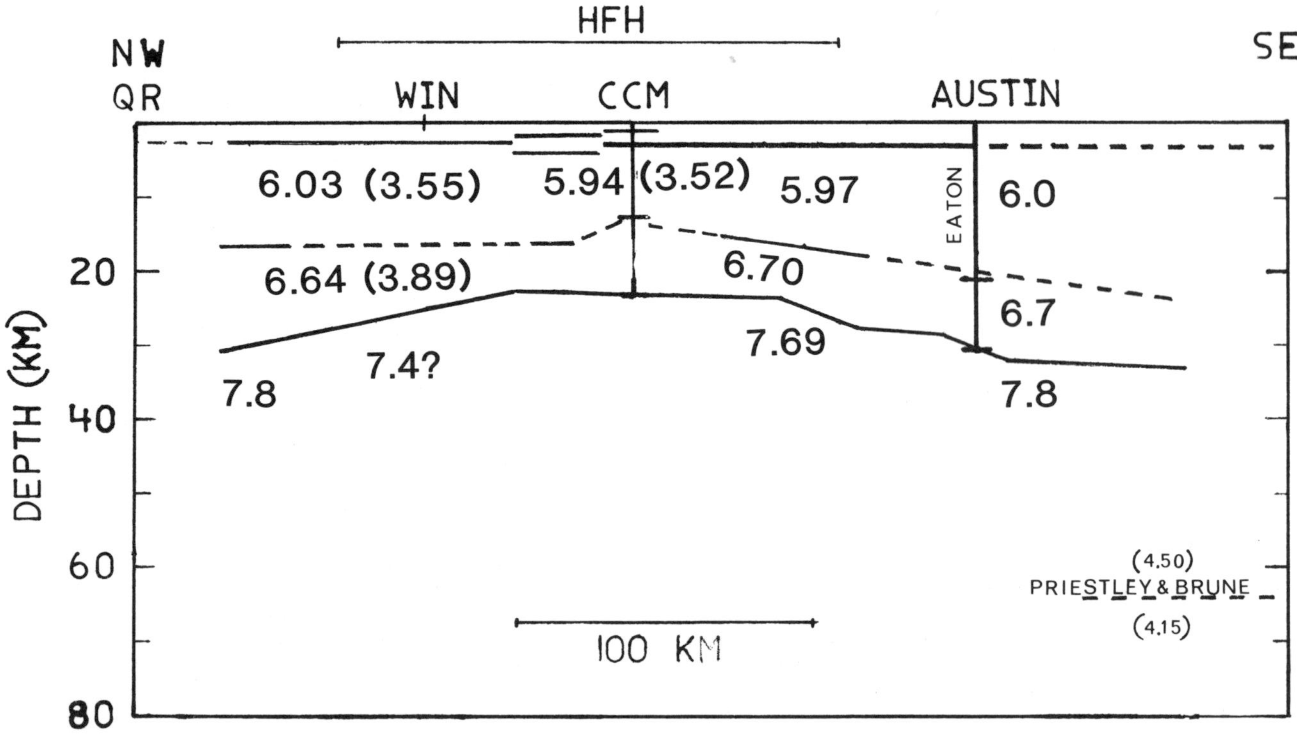

Figure 11.   Composite cross section along lines A and D. The profile
bends at CCM with the line A model shown to the right and the line D
structure shown on the left. The velocity model for line B and Eaton's
1963 line are shown along the vertical lines which mark the intersections
of these lines with the cross section. The position of the
lithosphere-asthenosphere boundary published by Priestley and Brune
(1978) is shown near the southeast end and the location of the Heat Flow
high of Lachenbruch and Sass (1976) is indicated with the horizontal bar
at the top. WIN marks the position of Winnemucca.

refracton lines using the NTS observations, an
upper mantle velocity of 7.8 km/sec was used.
Crustal thickness variations were calculated
from the deviations in arrival times, from the
7.8 km/sec refraction, by multiplying the
arrival time deviations by 10.0 for the four
profiles of Stauber and Boore (1978) and by 9.4
for the observations of Priestly and others
(1982). The difference in the two constants
results in a negligible difference in
elevation. The value of 10.0 approximates the
case where the relative proportion of upper and
lower crust remains constant as the total
crustal thickness changes and the value of 9.4
is appropriate when the lower crust maintains a
constant thickness. Where the data are sparse
the positions of some contours are estimated
with the aid of the Bouguer gravity field.

These contours are consistent with the seismic
observations, however. The most prominent
feature is the shallow crust-mantle boundary in
a region extending from Reno to the Battle
Mountain-Winnemucca area. The elevations reach
as high as -20 km. The elevations decrease to
between-30 and -35 km in central Nevada. The
pattern coincides with the regional elevation
map of northern Nevada, Figure 12C. The areas
of thin crust corresponds to a regional
topographic low and the thick crust to a
topographic high. The crust-mantle elevation
map also correlates with the regional Bouguer
gravity field, Figure 12B. The gravity field is
high where the crust-mantle boundary is shallow
and the gravity values are low in central Nevada
where the boundary is deep.

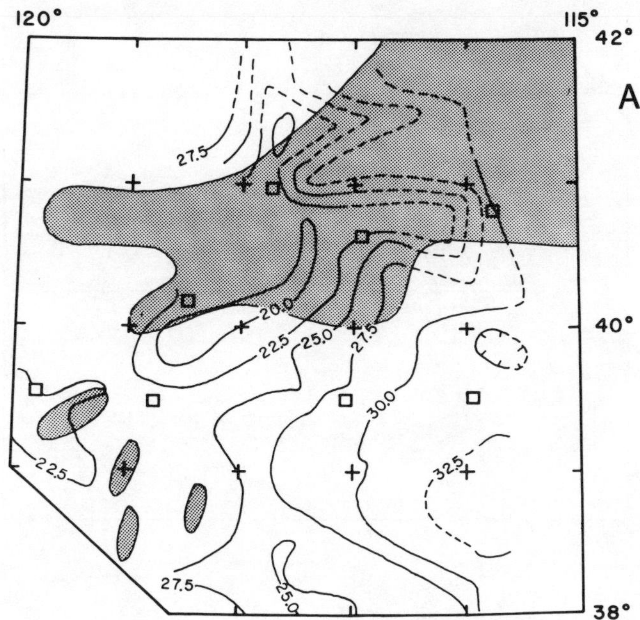

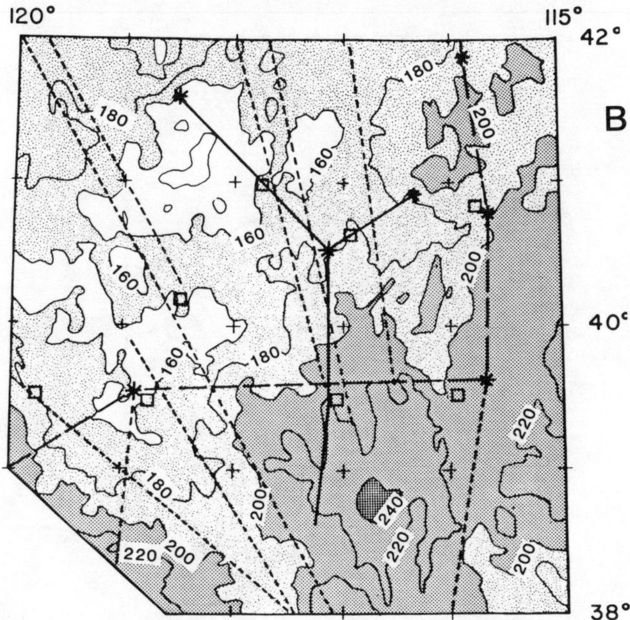

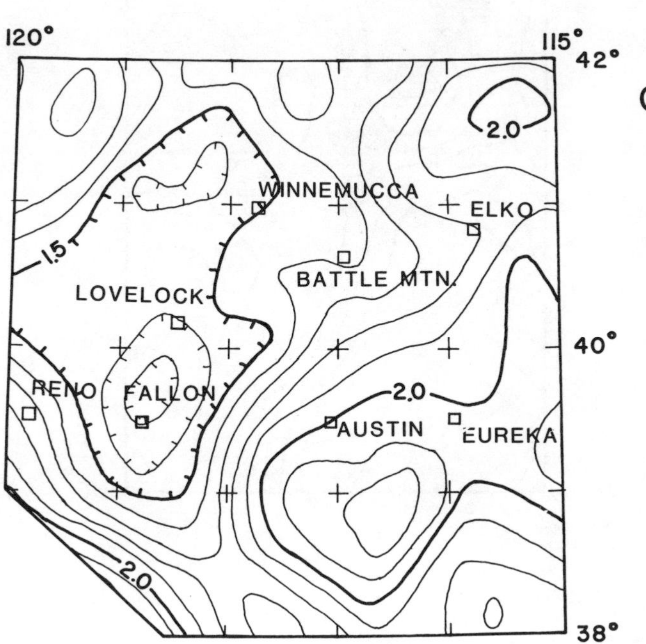

Figure 12.   A) Elevation below sea level, in km, of the crust-mantle boundary in northwest Nevada from all sources. Location of the Battle Mountain Heat Flow High from Lachenbruch and Sass (1978) is shown by stippled pattern. B) Bouguer gravity map of northern Nevada from Eaton (1978). Contour interval is 20 mgal. Shading density increases as anomaly decreases. C) Regional topography of northwestern Great Basin from Diment and Urban (1981). Contour interval is 0.2 km and values are in km.

Gravity Field Modeling

Since the regional crustal thickness and gravity maps exhibit the same large scale features it is reasonable to attempt to compare the observed gravity field with the gravity field calculated from the crust-mantle boundary map. This has been attempted by Stauber (1980) for north-central Nevada between 115° and 118° W longitude and 39° and 42° N latitude. A smoothed version of the regional gravity field is presented in Figure 13A. The gravity data used in producing Figure 13A come from Erwin

(1974), Erwin and Bittleston (1977), and Woollard and Joesting (1964). The lowpass smoothing filter had a corner wavelength of 35 km to smooth out the anomalies of individual basins and ranges (Stauber, 1980). The crust mantle elevation map was approximated by a set of polygonal boundaries shown on figure 13B. The polygonal boundaries were derived from an older version of figure 12A. The chief difference between the two versions is in the southwest corner of figure 12A. Use of the older polygonal boundaries results in a misfit between the observed and computed gravity in that corner which will be pointed out later. A sheet of excess mass north of 42° and east of 118° was added to approximate the effects of the Snake River Plain structure. The 3-D gravity program of Plouf (1975) was used to compute the gravity anomaly of the polygonal bodies. The gravity program also determined the density contrast for which the computed anomaly most closely matches, in a least-square sense, the smoothed obeserved field (all the polygonal boundaries have the same density contrast) . This density contrast was found to be 0.16 g/cm^3 and the gravity field computed with this contrast is plotted in Figure 13C. The residual field, the difference between the observed field in 13A and the computed field, 13C, is shown in Figure 13D.

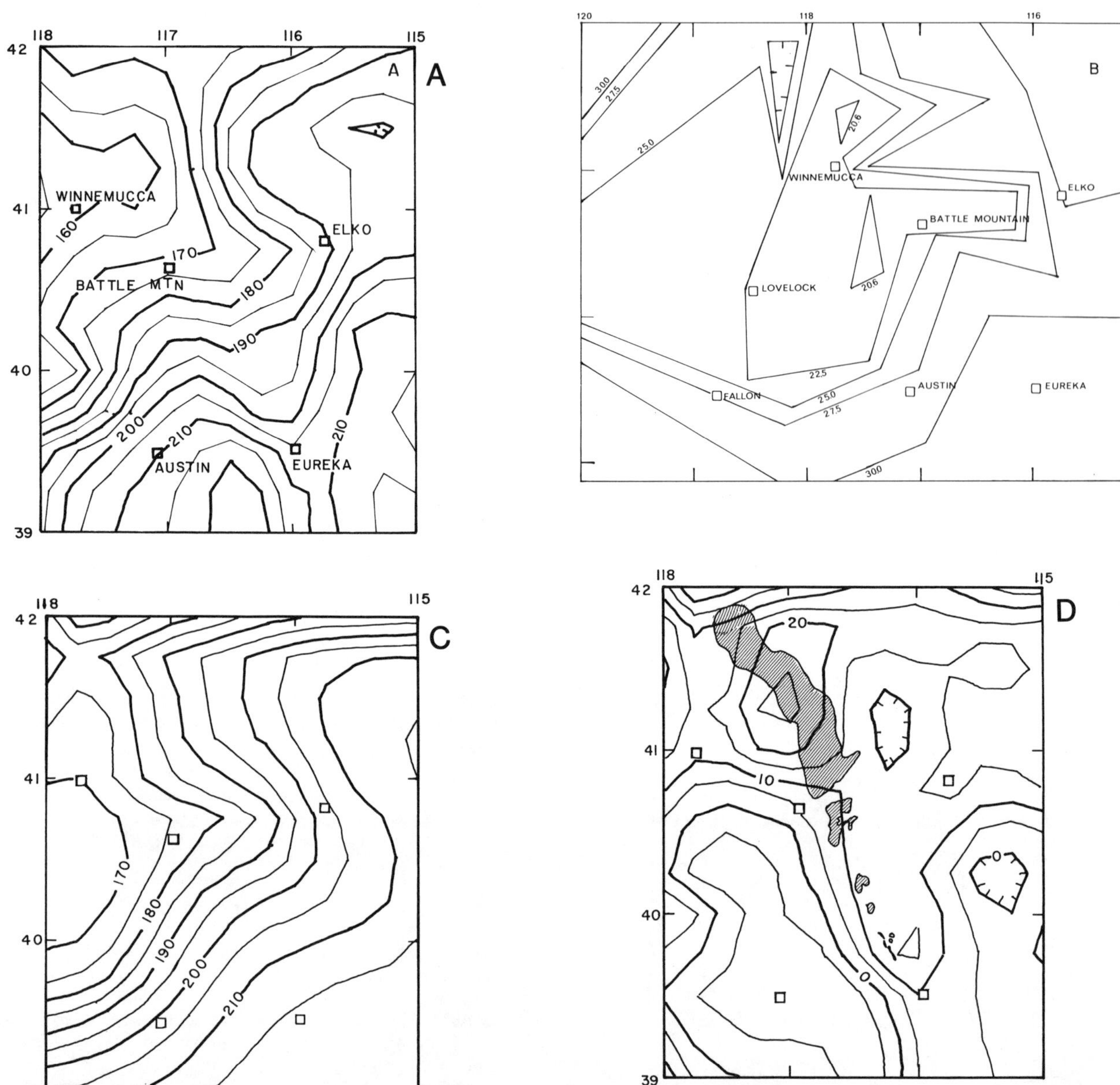

Figure 13   A)   Smoothed observed Bouguer gravity field.   Contours at
5 mgal intervals.
B) Polygonal boundaries used to approximate crust-mantle boundary
topography in figure 12A. Depths are in km.
C) Gravity anomaly calculated from 13B. Contour interval is 5 mgal.
D) Residual gravity field, figure 13A minus 13C. Contour interval is 5
mgal. Outcrops of volcanic rocks associated with the Miocene Northern
Nevada Rift, Zoback and Thompson (1978), shown with diagonal hatching.

The computed gravity field, figure 13C, for the polygonal model of Figure 13B which approximates the crustal model presented in this paper is similar to the smoothed observed gravity. The gravity high in the Winnemucca area relative to central Nevada near Eureka in the calculated gravity map is 50 mgal and is 60 mgal for the observed gravity maps. The most obvious residual visible in figure 13D is a NNW-SSE striking ridge of positive residuals coincident with the Miocene Northern Nevada rift zone (Zoback and Thompson, 1978), which is indicated on figure 13D. The gravity high along this rift zone may result from the presence of

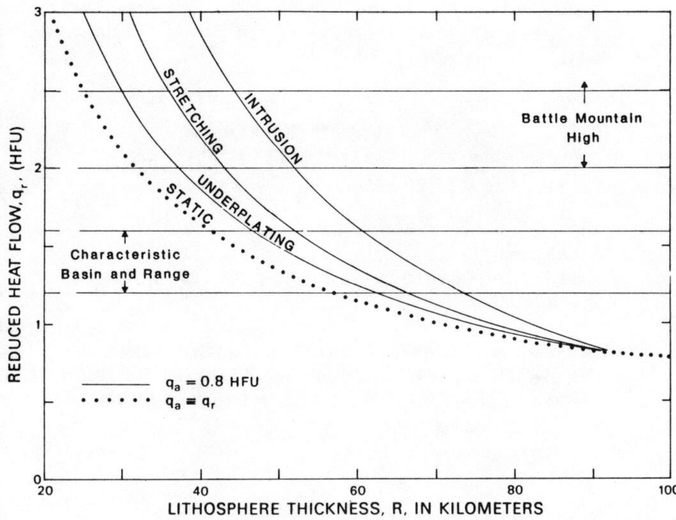

Figure 14.   Reduced heat flow vs lithospheric thickness for three models of lithospheric extension taken from Lachenbruch and Sass (1978). Reduced heat flow ranges appropriate for the Battle Mountain heat flow high and normal Basin and Range are indicated.

high density intrusive bodies in the crust which are associated with the Miocene rifting. A steep gradient along the north edge of the residual gravity map probably results from an inadequate representation of the Snake River Plain structure to the north. The north-south striking contours along the southwest edge of the residual gravity map results from the use of the older version of the Crust-Mantle boundary elevation map discussed earlier. Finally there is a gradient resulting in east-west striking contours at about the latitude of Battle Mountain. The width of this gradient zone is comparable to the cutoff wavelength of the smoothing filter used on the gravity data. Because of this one can only say that it's source is shallower than the mid-crust, about 17 km.

The density contrast of 0.16 g/cm^3 is rather small for the density contrast between the crust and mantle. This is especially troublesome since Figure 11 shows that the thickness of the lower crust is relatively constant and the main contrast should be between mantle rocks and upper crustal rocks. One solution to this problem is to postulate a larger density contrast between the crust and mantle and partially compensating low density anomaly below. One candidate is the lithosphere-asthenosphere boundary observed at a depth of 65 km in central Nevada (Priestly and Brune, 1978) and shown in Figure 11. A density contrast of 0.08 g/cm^3 at this boundary (Crough and Thompson, 1976), and a lithosphere thickness 20 to 30 km less in northwest Nevada than in

central Nevada would allow a larger, 0.3 to 0.4 g/cm^3, density contrast between the crust and mantle. Thermal-mechanical models for lithospheric extension in the Basin and Range Province proposed by Lachenbruch and Sass (1978) predict relationships between lithospheric thickness, extension rates and surface heat flux. Figure 14, reproduced from Lachenbruch and Sass (1978, Figure 11), shows the relationship between lithospheric thickness and reduced heat flow for three proposed lithospheric extension models. The ranges of reduced heat flow appropriate for normal Basin and Range and the Battle Mountain High are indicated. One can see that the lithospheric thickness decreases with increasing heat flow and that for all three extension modes, the lithosphere could be 15 to 25 km thinner in the heat flow high than in the normal Basin and Range. This is in the range which is consistent with gravity and seismic observations discussed above.

Summary

A combination of unreversed refraction lines utilizing NTS events as sources and reversed refraction lines using quarry blasts and borehole shots results in a roughly north-south crustal cross section across the Battle Mountain Heat Flow High, near Battle Mountian, Nevada, shown in figure 11, and a contour map of the crust-mantle boundary elevation in northern Nevada, figure 12A. The dominant features observed are the area of thin crust coinciding with the Lake Lahonton basin in northwest Nevada and the thick crust occurring in the central part of Nevada. On the profile in figure 11, the heat flow high corresponds to the area of thin crust. The correlation between the spatial extent of the heat flow high, described by Lachenbruch and Sass (1978) shown in figure 2, and the thin crust as shown in figure 12A is not perfect.

The long wavelength Bouguer gravity field correlates well with the topography of the crust-mantle boundary in northeastern Nevada with high gravity values occurring where the crust-mantle boundary is high. If the topography of the crust-mantle boundary is the only source of the Bouguer gravity field, then a low density contrast, 0.16 g/cm^3, between the average crustal and upper mantle densities is required. A larger value for this density contrast is permitted if there is a partly compensating density contrast at another boundary having a long wavelength shape similar to the crust mantle boundary. One possible boundary is the lithosphere-asthenosphere boundary which occurs at about 65 km depth in central Nevada (Priestley and Brune, 1978). The magnitude of the gravity high observed in the area of thin crust could be explained by a lithosphere-asthenosphere boundary which is 20 to 30 km higher in the Lake Lahonton basin than in central Nevada and a larger, 0.3 to 0.4 g/cm^3 instead of 0.16 g/cm^3 density difference between the average crustal and the

upper mantle densities. Further three
dimensional gravity modeling using the updated
crust-mantle elevation map of figure 12A, which
covers a larger region of northwestern Nevada,
the observed Bouguer gravity field over this
larger region, and including upper mantle
density variations such as topography on the
lithosphere-asthenosphere boundary is necessary
to make this discussion more quantitative.
Since thermal-mechanical models of lithospheric
extension such as those proposed by Lachenbruch
and Sass (1978) predict relationships between
lithospheric thickness, extension rates, and
heat flow, more extensive gravity modeling of
the upper mantle in northwest Nevada is likely
to yield useful constraints on the extension
mechanism of the lithosphere in this region.

Acknowledgements

This work was funded by N.S.F. grant number
EAR75-04291 A02.

References

Criley, E. and Eaton, J., 1978, Five-day
recorder seismic system,U. S. Geol. Surv.,
open file report, No. 78-266.

Crough. S. T., and G. A. Thompson, 1976, Thermal
model of continental lithosphere,
J. Geophys. Res., v. 81,p. 4857-4861, 1976.

Diment, W. H., and T. C. Urban, 1981, Average
elevation map of the conterminous United
States (Gilluly averaging method),
Geophysical Investigations Map GP-933,
1:2,500,000 , U.S. Geological Survey.

Eaton, J. P., 1963, Crustal structure from San
Francisco, California to Eureka, Nevada,
from seismic-refraction measurements, Jour.
Geophys. Res., v. 68, p. 5789-5806.

Erwin, J. W., 1974, Bouguer Gravity map of
Nevada Winnemucca sheet, Nevada Bureau of
Mines and Geology, Map 47.

Erwin, J. W. and Bittleston, E. W., 1977,
Bouguer gravity map of Nevada Millet sheet,
Nevada Bureau of Mines and Geology, map 53.

Hill, D. P. and Pakiser, L. C., 1966, Crustal
structure between the Nevada Test Site and
Boise, Idaho, from seismic-refraction
measurements, Amer. Geophys. Union Monograph
10, p. 391-319.

Lachenbruch, A. H., Sass, J. H., 1978, Models of
an extending lithosphere and heat flow in
the Basin and Range province, Geophys. Soc.
Am., Memoir 152, p. 209-250.

Plouff, D., 1975, Derivation of formulas and
FORTRAN programs to compute gravity
anomalies of prisms, U.S. Geological Survey
Report, USGS-GD-75-015.

Priestley, K., Brune, J., 1978, Surface
waves and the structure of the Great Basin
of Nevada and western Utah, J. Geophys.
Res., v. 83, p. 2265-2272.

Priestley, K. F., A. S. Ryall and G. S. Fezie,
1982, Crust and upper mantle structure in
the northwest Basin and Rance Province,
Bull. Seis. Soc. Am., v. 72, p. 911-923.

Stauber, D. A. and Boore, D. M., 1978, Crustal
thickness in northern Nevada from seismic
refraction profiles, Bull., Seis. Soc. Am.,
68, v. 1049-1058.

Stauber, D. A., 1980, Crustal structure in the
Battle Mountain Heat Flow High in northern
Nevada from seismic refraction profiles and
Rayleigh wave phase velocities, PhD thesis,
Stanford University.

Woollard, G. P. and Joesting, H. R., 1964,
Bouguer gravity anomaly map of the United
States, Am. Geophys. Union, Special
Committee for Geophysical and Geological
Study of the Continents, G. P. Woolord,
Chairman, and the U. S. Geological Survey,
H. R. Joesting, Coordinator.

Zoback, M. L. and Thompson, G. A., 1978, Basin
and Range rifting in northern Nevada: Clues
from a mid-miocene rift and its subsequent
offsets, Geology, v. 6, p. 111-116.

SEISMOLOGICAL INVESTIGATIONS OF VOLCANIC AND TECTONIC PROCESSES
IN THE WESTERN GREAT BASIN, NEVADA AND EASTERN CALIFORNIA

Ute R. Vetter, Alan S. Ryall and Christopher O. Sanders

Seismological Laboratory
University of Nevada
Reno, NV 89557-0018

ABSTRACT

The western Great Basin is structurally complex, and seismogenic processes vary within it. A significant increase in seismicity around the "White Mountains seismic gap", north of the rupture zone of the 1872 (M = 8+) Owens Valley earthquake, suggests that the potential for a major earthquake in the gap is greater than it has been for several decades. Seismicity within the "Stillwater gap" of northern Dixie Valley (Wallace, 1978) is very low and earthquakes that do occur there are small and shallow, suggesting that the potential for a large shock there in the near future is low.

Earthquakes around the Adobe Hills Tertiary (ca. 3 m.y.) volcanic center show a tendency for temporal clustering, which appears to be characteristic of geothermally active areas in the Great Basin. Teleseismic P-wave traveltime residuals indicate a low-velocity zone, possibly a zone of partial melting, at shallow depth in the crust.

Since October 1978 a major earthquake swarm has been in progress near Mammoth Lakes, California. Shallow earthquakes of this sequence are distributed in in an irregular-shaped zone extending 30 km in a WNW-ESE direction and about the same distance from north to south. At the north end this zone extends 7 km into Long Valley caldera. Within this shallow zone the crust is undergoing complex brecciation, possibly accompanied by the formation of clusters of dikes or fissures. Events deeper than about 10 km are located in a northerly-trending zone 25 km long and about 5 km wide; mechanisms for these events are consistent with oblique or normal faulting along the Sierran front. Some of the Mammoth Lakes earthquakes are characterized by lack of S-waves at regional stations north of the caldera, and P-waves for the same station-event combinations are deficient in frequencies greater than about 2-3 Hz -- effects that can be explained by one or more shallow magma chambers. In July 1980 spasmodic tremor began to occur in one small area inside the caldera. Taken together with the observation of uplift (Savage and Clark, 1982) and new fumarole activity within the caldera, these effects are interpreted as evidence for adjustments within the cauldron block, caused by stress changes related to the earthquake sequence.

More than 150 P-wave fault-plane solutions for earthquakes in western Nevada and the eastern Sierra Nevada show a systematic change in mechanism with depth. Earthquakes shallower than 6 km are characterized by strike-slip motion, while most events deeper than 9 km have a strong normal-slip component. These observations are consistent with a model of lithospheric extension that involves oblique or normal faulting at mid-crustal depth and formation of vertical fissures or dikes in the shallow crust.

INTRODUCTION

From the geologic literature, active faulting in the western Great Basin appears to be rather evenly distributed over much of the region. For example, Figure 1 is from a photogrammetric study by Slemmons (1967) and shows faults in late Quaternary alluvium, glacial deposits or lake sediments of the playa, Bonneville or Lahonton type. Approximately a thousand faults are shown on the figure, ranging in length from about 1 to more than 100 km, striking from NW to N to NE. In contrast to this picture of relatively even fault distribution in the Nevada region, historic seismicity has been concentrated in the western half of Nevada. Figure 2 shows this seismicity from 1969 to 1978 in western Nevada and eastern California, on a generalized map of late Cenozoic structural features (Wright, 1976).

In the first part of this paper we present a general description of seismicity in the western Great Basin, with emphasis on (1) the Dixie Valley area, where the last major earthquakes in this region occurred in 1954 and where a number of industrial groups have focused considerable effort on geothermal exploration; (2) the Excelsior Mountains area where teleseismic P-residuals indicate the possibility of a magma chamber; and (3) Long Valley caldera, where a sequence of earthquakes and associated magmatic activity has been in progress for almost five years. In the second part of the paper we discuss results of a study (Vetter and Ryall, 1983) that indicates a systematic change in focal mechanism with depth in this region and leads to estimates of maximum and minimum principal stresses to mid-crustal depths.

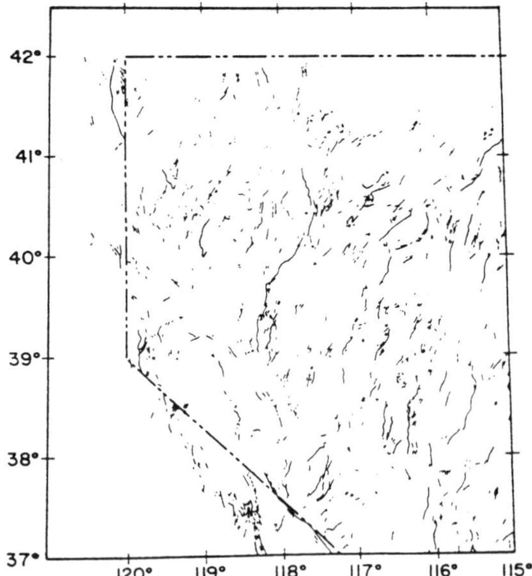

Figure 1. Map of faults that show photogrammetric evidence of Quaternary displacement (adapted from Slemmons, 1967).

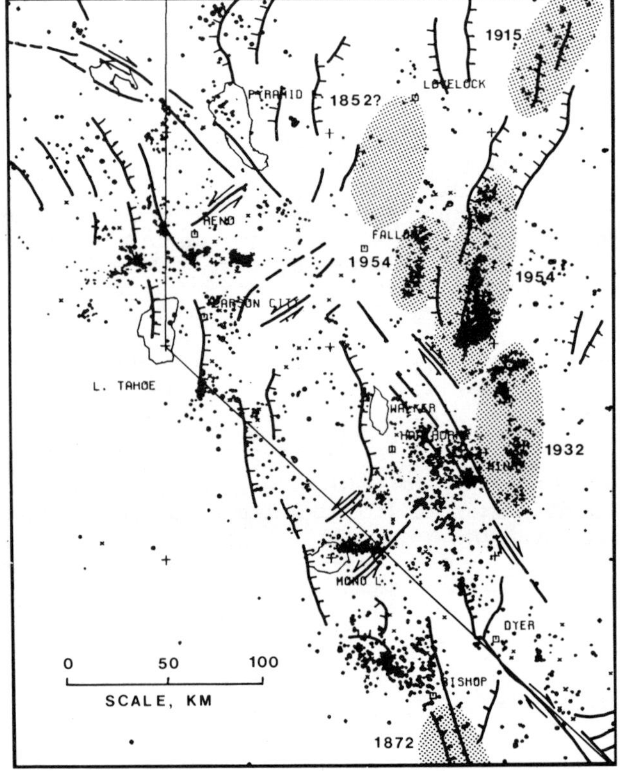

Figure 2. Generalized map of late Cenozoic structural features of the western Great Basin (Wright, 1976), together with epicenters for the period 1969-1978 (dots) and approximate rupture zones of major historic earthquakes (stippled areas with year of main shock). The "Stillwater seismic gap" is between the 1915 and 1954 rupture zones and the "White Mountains gap" is between the 1872 and 1932 zones -- respectively in the upper right and lower right parts of the figure.

General Description. From the rupture zone of the M = 8+ Owens Valley earthquake of 1872 (bottom of map in Figure 2), current seismicity spreads to the north through west-central Nevada and to the northwest along the eastern Sierra Nevada, forming two more-or-less separate zones. In the southern part of this region earthquake swarms occur frequently in the area north of Bishop, and a major swarm near Mammoth Lakes has been in progress since 4 October 1978. So far this sequence has produced four earthquakes with M = 6+ and uplift, spasmodic tremor and increased fumarole activity in Long Valley caldera have been associated with probable magma injection (Ryall and Ryall, 1981a, 1983; Savage and Clark, 1982; U. S. Geological Survey, 1982).

As shown by the stippled areas on Figure 2, major earthquakes during the historic period have occurred in a northerly-trending belt from the southern Owens Valley in southeastern California to Winnemucca in north-central Nevada. Wallace (1978) pointed out that three "gaps" within this belt are likely candidates for future large earthquakes. Two of these are the Stillwater gap between the 1954 and 1915 ruptures in the upper right part of Figure 2, and the White Mountains gap between the 1872 and 1932 zones in the lower right part of the figure. The third is the Southern Sierra Nevada gap south of the Owens Valley rupture zone, and will not be discussed in this paper.

In the region around the White Mountains gap since 4 October 1978 more than 50 earthquakes with ML 4.0-5.4 have occurred in the following zones: a NE-trending zone between Mono Lake, California and Luning, Nevada, at the north end and east side of the White Mountains; and in the northern Owens Valley (Figure 3). In addition, more than 100 shocks with ML 4.0-6.3 have occurred in the Mammoth Lakes area. In comparison only 18 events with ML 4.0 or greater were observed in this entire region (including the Mammoth Lakes area) during the previous nine-year period. Thus, the level of moderate seismicity since 1978 is about six times higher than the previous decade if the Mammoth Lakes sequence is excluded, and almost twenty times higher if it is included. The similarity between this pattern and that observed by Mogi (1969) before large earthquakes in Japan suggests that the potential for a major earthquake in the White Mountains gap is substantially higher now than it has been for at least the last two decades (Ryall and Ryall, 1983).

The belt of recent seismicity extends north through the 1932 Cedar Mountains (M = 7.3) rupture zone, the 1954 Dixie Valley-Fairview Peak rupture zone (M = 6.9, 7.1, respectively) and the 1915 Pleasant Valley zone (M = 7.6). A second zone, of frequent moderate earthquakes but no major historic shocks, extends NNW from the Mono basin along the eastern boundary of the Sierra Nevada. In the northern part of this zone near Reno (Figure 4), clusters of earthquakes are observed in the following areas: (1) on the California-Nevada border west of Reno, where an ML 6 event occurred in 1948; (2) north of Truckee, California, where an ML 5.7 shock occurred in 1966; (3) south of Reno in the Steamboat Hot Springs area; and (4) in the Virginia Range SE of Reno.

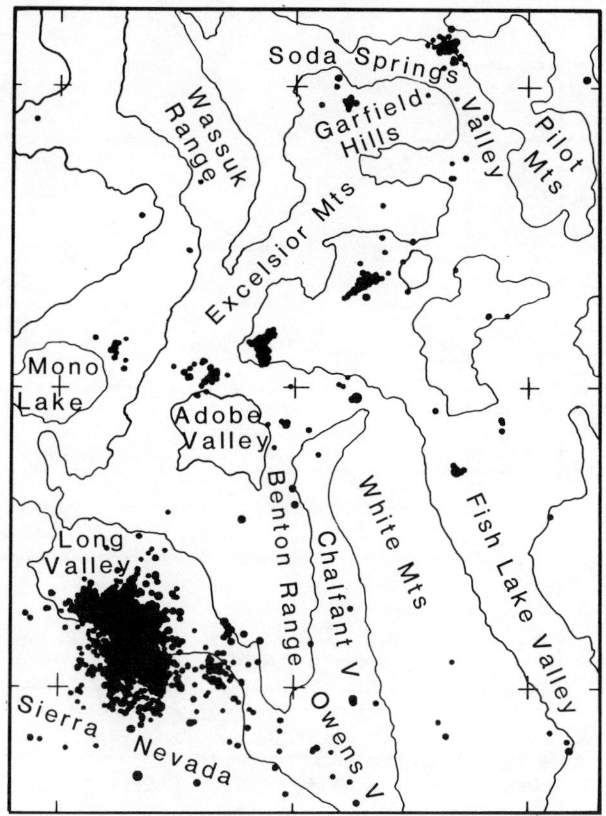

Figure 3. Seismicity north of the 1872 Owens Valley rupture zone, 1978-1982.

Dixie Valley Area. In recent years Dixie Valley, in north-central Nevada, has been the focus of geothermal exploration involving surface and drilling investigations by a number of oil companies. In 1980 and 1981 the University of Nevada Seismological Laboratory operated an eleven-station seismic network in and around northern Dixie Valley, the area of most intensive exploration. Figure 5 shows the seismic network used in the two-year study, together with 1,128 earthquakes analyzed for the period from 1970 to 1981. As the figure indicates, almost all of the seismicity was in a 90 km-long zone extending through southern Dixie Valley and along the east side of Fairview Peak (dense cluster of earthquakes south of station DIX) -- within the rupture zone of two major earthquakes that occurred within four minutes of each other on 16 December 1954 (Slemmons, 1957; Romney, 1957). Detailed analysis of epicenter maps for different periods of time indicates the existence of northerly-trending (NW to NE) fracture zones, with an average trend of about N5-10' E. Since the seismicity is almost entirely confined to the 1954 rupture zone, we conclude that it represents a decaying aftershock sequence. Indeed, a plot of earthquake occurrence in this zone as a function of time (Figure 6) shows a gradual decrease in activity over the 12-year period of observation.

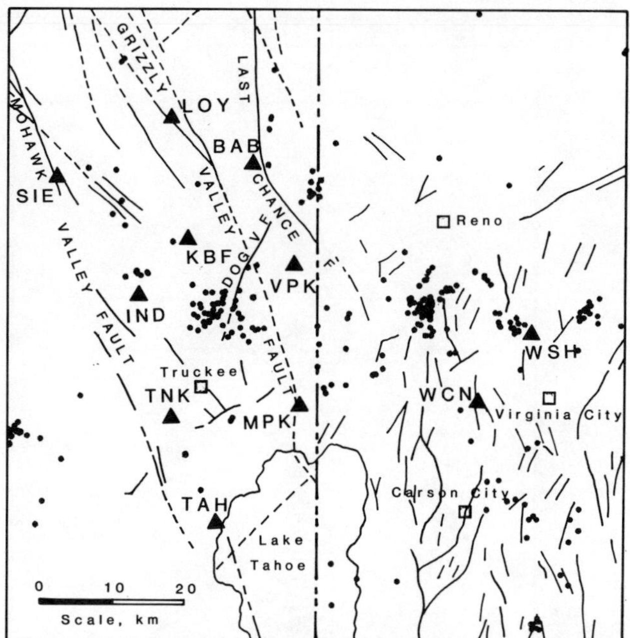

Figure 4. Map of the Reno-Truckee-Lake Tahoe area (39.0° - 39.8°N, 119.5°- 120.5°W), showing seismic stations (triangles), faults, and earthquakes with location quality "C" or better for the period 1975-1982.

A histogram of focal depths for earthquakes in the Fairview Peak-southern Dixie Valley zone is shown on Figure 7. The plot has a peak in the depth range 10-12 km. The few earthquakes that we recorded in northern Dixie Valley all had depth less than 7 km, and all of them were quite small. Meissner and Strehlau (1982) and Sibson (1982) studied earthquake depths on a global scale, based on laboratory tests with quartz-bearing rocks and different assumptions of water content. For a given rock type and water content they predicted the depth of occurrence of earthquakes in a specified region as a function of heat flow. In the western Great Basin the depth distribution of earthquakes agrees well with the theoretical distribution calculated for estimated conditions of water saturation in the crust.

Evidence related to the possibility of a large earthquake in northern Dixie Valley is ambiguous. On the one hand we noted that Wallace (1978) considers that area to represent a seismic gap, with the potential for an M = 7+ earthquake in the near future, and his conclusions are supported by our observation of very low seismicity north of the 1954 rupture zone. On the other, the few earthquakes we recorded in northern Dixie Valley were very shallow and could even be associated with listric faulting in a shallow crustal section overlying a zone in which deformation does not involve brittle fracturing. In a previous study, Richins (1974) noted that earthquakes in NW Nevada, a region characterized by high heat flow and geothermal activity, tend to be shallower and smaller than those in the major earthquake zones of central Nevada. He concluded that the maximum magnitude of

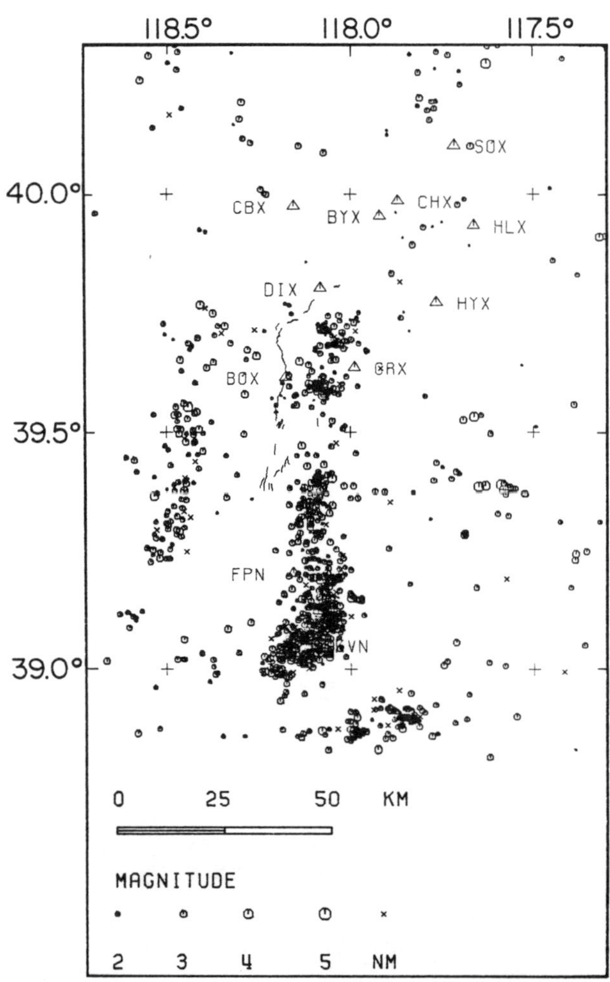

Figure 5.   Earthquakes in the Dixie Valley-Fairview
Peak area, 1970-1981.

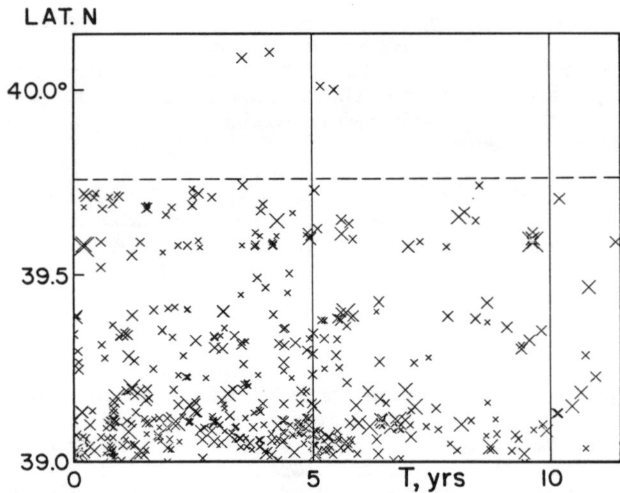

Figure 6.   Distribution of earthquakes in the Dixie
Valley-Fairview Peak zone as a function of time,
1970-1981. Only events with ML $\geq$ 2 are shown.
Dashed line -- northern limit of 1954 rupture zone.

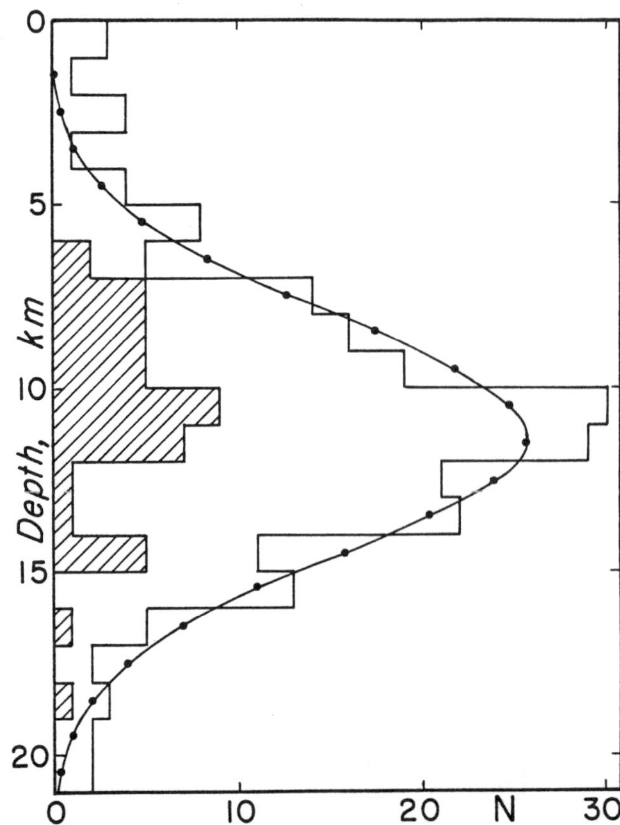

Figure 7.   Histogram showing depth distribution for
217 events with quality "C" or better (unshaded
area) and 42 events with quality "A" or "B" (shaded
area).   Smooth curve is normal distribution for
mean depth 11.3 km, standard deviation + 3.22 km.

earthquakes in geothermally active regions may be
only 5-3/4 to 6, as a result of the weakening of
crustal rocks in the vicinity of intrusive bodies,
or by the effects of stress corrosion and leaching
due to geothermal fluids.

To study the stress pattern in the Fairview
Peak-Dixie Valley region, P-wave fault-plane solutions were determined for eleven earthquakes in the
magnitude range 3.0-4.5. These solutions, shown on
Figure 8, were based primarily on Pg arrivals, but
clear Pn arrivals were used. We found strike-slip,
oblique-slip and normal-slip mechanisms, with the
axis of minimum compressive stress in all cases
oriented roughly NW-SE in agreement with the known
direction of lithospheric extension in the western
Great Basin. Of particular interest was the observation that the deeper events had a strong component of normal slip, while the shallower events
were all strike-slip. For the shallow events, the
two planes of the fault-plane solution were essentially vertical, but for the deeper shocks the
planes were inclined with dips of 40-60°. Our
interpretation of this change, following a paper by
McGarr (1980), was that the change in mechanism
could be due to increasing overburden pressure with
depth. This topic is discussed further in the section on focal mechanisms.

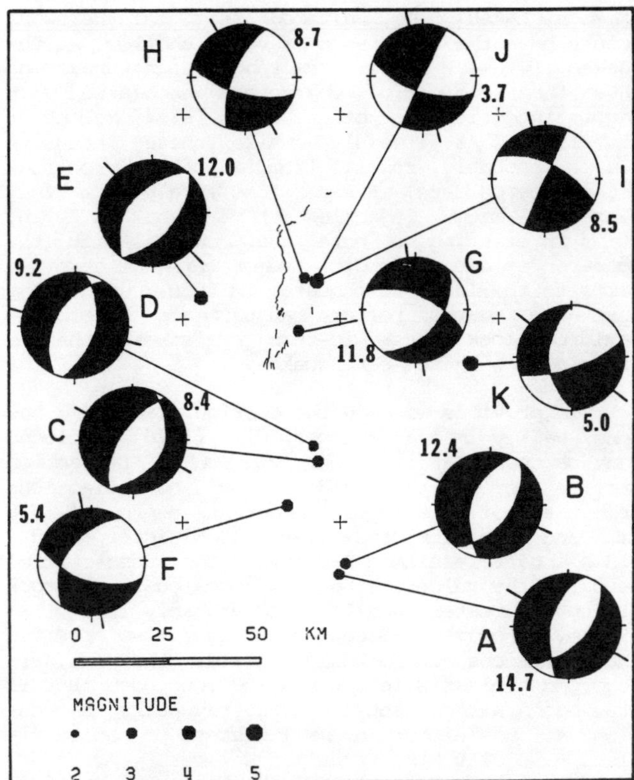

Figure 8. P-wave fault-plane solutions (lower hemisphere, equal-angle projection; compression quadrants shaded) for selected events in the Dixie Valley area. Orientation of T-axis (axis of minimum compressive stress) indicated by heavy line on each mechanism. Numbers indicate depth, km.

Mono-Excelsior Area. In the area east of Mono Lake (Figure 2) earthquakes occur in a roughly E-W zone through the Adobe Hills Tertiary volcanic center. Detailed inspection of earthquakes in this area indicates that the E-W zone is made up of a number of short NE-SW segments in an en-echelon arrangement. Gilbert et al. (1968) described the Adobe Hills center as the source of the most voluminous eruptions in the Mono basin during the last 4 m.y., and attributed the eruptions to zones of extension related to left-lateral motion on faults striking about N60°E. According to them, such motion would result in rotation of the blocks between the faults and produce open spaces where north-south faults within the range intersect the transverse faults. Focal mechanisms for this region (Figure 9) are consistent with the observation of predominently left-lateral slip on NE-striking faults by Gilbert et al.

Recent earthquake sequences in this region (Adel, Oregon, 1968; Denio, Nevada, 1973; Mammoth Lakes, California, 1975, 1976, 1978-1983; Mono Basin, California, 1974, 1976, 1978) show a distinct tendency for temporal clustering, which according to Richins (1974) may be characteristic of geothermally active areas in the Great Basin. Another indication of relatively high temperature at mid-crustal depth is the lack of seismic activity at depths greater than about 15 km.

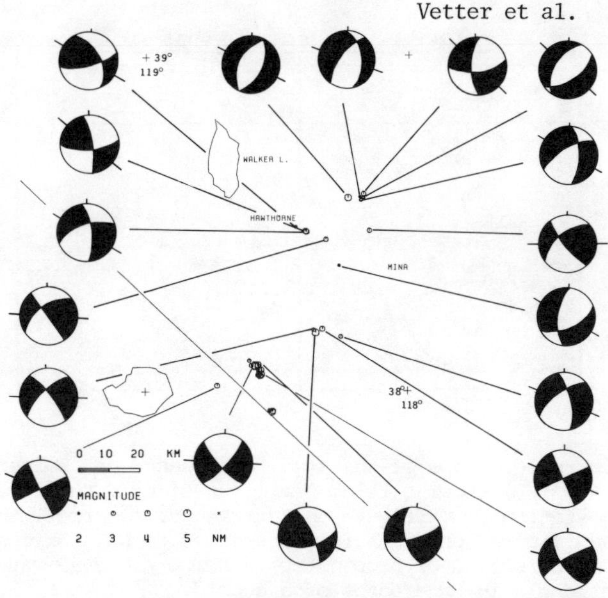

Figure 9. Focal mechanisms for the Excelsior Mountains-Luning area. Town of Luning is 10 km south of event 20 in upper right part of figure.

To investigate the possibility that earthquakes in the Adobe Hills area might be related to a zone of partial melting in the crust, VanWormer and Ryall (1980) analyzed teleseismic P residuals for 22 stations in the area north and east of Mono Lake. These residuals, taken relative to a station (Tonopah) outside the area of interest and corrected for elevation, are shown on Figure 10. For teleseismic sources to the southeast (azimuth 119° to 155°) Figure 10a shows an area of relatively late arrivals east of Mono Lake, elongated to the northeast around the southern edge of the Excelsior Mountains and the eastern side of the Garfield Hills. For sources to the northwest (azimuth 303° to 316°) Figure 10b shows a very similar feature. The amplitude of this traveltime anomaly is about 0.5 second, which is higher than the value of 0.3 second found by Steeples and Iyer (1976) for Long Valley caldera. The size of this this anomaly and the agreement in its location for waves propagating in opposite directions indicates that it is due to shallow structure. However, the plateau between the -0.1 and -0.2 second contours in the upper left part of Figure 10a, and the plateau between the 0.1 and 0.2 second contours in the lower right part of Figure 10b suggest an upper-mantle component of the anomalous zone in this area. Taken together with the volcanic history of the area, the P residuals on Figure 10 would be consistent with a region of partial melting in the shallow crust below the Adobe Hills (centered about 25 km east of Mono Lake), extending to the northeast along a zone of mapped faults and connected to an upper-mantle source below the Excelsior Mountains.

A similar result was obtained by Iyer and Evans (1983) for the Mono and Inyo Craters. Their interpretation of the traveltime anomaly was in terms of a low-velocity body extending from shallow depth to at least 25 km, under most of the volcanic chain.

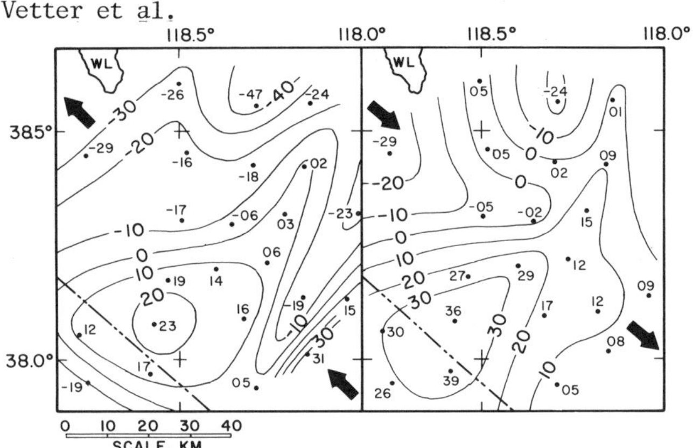

Figure 10. Maps of the area southeast of Walder Lake (WL) showing teleseismic P residuals relative to station TNP. Figure on left is for sources to SE; figure on right is for sources to NW. Arrows show direction of propagation. Numbers are P-wave residuals in hundredths of a second.

Mammoth Lakes Area. Over the last five years, starting in the fall of 1978, a major earthquake swarm has been in progress in the boundary zone between the Sierra Nevada and the Great Basin, near Mammoth Lakes, California. So far this swarm has produced four earthquakes with ML 6 or greater and uplift and spasmodic tremor in Long Valley caldera have been associated with probable magma injection. An Earthquake Hazard Watch was issued by the US Geological Survey for this area in May 1980, and a Volcano Hazard Notice was issued in May 1982. Evidence that led to the prediction of the larger shocks of this sequence is described by Ryall and Ryall (1981b).

Figure 11 shows more than 2,000 earthquakes of the Mammoth Lakes sequence for the period 1978-1982. The epicentral zone for shallow earthquakes of this sequence is irregular in shape, extending more than 30 km in a WNW-ESE direction, about 30 km from north to south, and 7 km into Long Valley caldera. While some lineups of events can be seen, spatial correlations of epicenters with either mapped faults or linear features on Landsat imagery are generally lacking. Instead, the epicentral distribution appears to reflect intense brecciation of the shallow crust. Earthquakes deeper than about 10 km, however, are located in a more restricted zone, 5 km wide and about 25 km from north to south. During this sequence no earthquakes have been recorded in the area between the caldera and Mono Lake.

Six weeks after the ML 6+ shocks in 1980, earthquakes in one small area just east of the town of Mammoth Lakes began to occur as intensive swarms, with a typical swarm lasting 1-2 hours, producing hundreds of microearthquakes and having the appearance of spasmodic tremor (Ryall and Ryall, 1981a; 1983). In volcanic regions spasmodic tremor is considered to represent intensive cracking within the volcanic system, due either to the injection of a tongue of magma or to gas released under high pressure from the magma chamber. In the Mammoth Lakes case, although some of the swarm events have the appearance of very shallow earthquakes (weak P onset, amorphous signature) none have the long-period character associated with magma injection in other areas (e.g., Malone et al., 1983). As a result, while these intensive swarms probably result from magmatic processes within the caldera, it seems less likely that they represent magma injection at very shallow depth. It is interesting to note that swarms with the appearance of spasmodic tremor were not observed prior to the large earthquakes in 1980, suggesting that the swarms reflect adjustments within the cauldron block, caused in turn by stress changes associated with the earthquakes.

In previous work on the configuration of the Long Valley magma chamber, Hill (1976) concluded that secondary arrivals observed on a refraction profile across the caldera could be reflections from the roof of a magma chamber at depth of 7-8 km, and Steeples and Iyer (1976) interpreted a 0.3-second teleseismic P-delay in the west-central part of the caldera as due to anomalously hot rock at depths greater than 7 km and probably less than 25 km. However, Steeples and Iyer also observed that stations inside and outside the caldera recorded teleseismic S-waves and concluded that if true magma was present along the ray paths in question it was either in small pockets or had sufficient viscosity to transmit S-waves. At a US Department of Energy (1980) workshop the participants concluded that low geothermal gradients measured in boreholes in Long Valley caldera suggested a temperature of about 600° C at 15 km depth, and appeared to preclude the existence of magma at depths accessible by drilling.

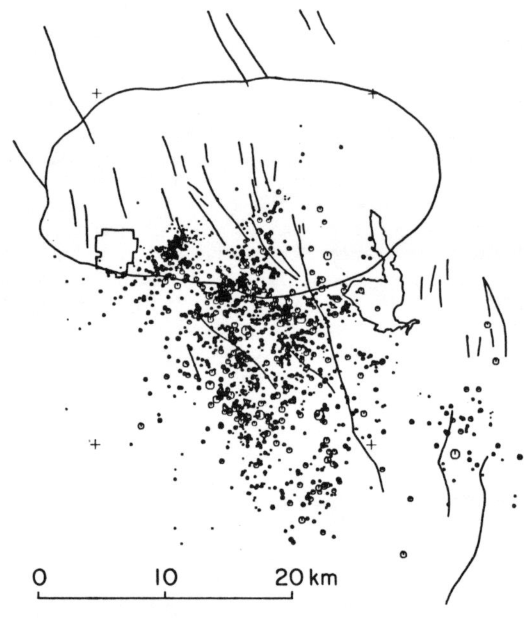

Figure 11. Mammoth Lakes earthquakes, 1978-1982. Heavy lines show mapped faults and caldera boundary. Town of Mammoth Lakes is shown in SW part of caldera, Lake Crowley is SE of caldera.

During analysis of the Mammoth Lakes earth-quakes Ryall and Ryall (1981) noticed that shallow earthquakes around the southwest boundary of the caldera were characterized by lack of S-waves at regional seismic stations to the north, and P-waves for the same station-event pairs were deficient in frequencies greater than about 2-3 Hz. They con-cluded that these observations were consistent with Hill's (1976) interpretation of a magma chamber at depth of 7-8 km. This study was extended by Sanders and Ryall (1983) based on detailed analysis of more than 200 well-located events of the Mammoth Lakes sequence on recordings of about 30 regional seismic stations. From a comparison of ray paths for signals with anomalous and normal character, they concluded that one or possibly two regions in the upper 13 km of the caldera were responsible for the observed S-wave and high-frequency filtering effects. These regions are shown in map view and cross section on Figures 12 and 13. The analysis suggests that the large magma body in the south-central part of the caldera is relatively massive between depths of about 7 to 13 km. For depths of 5-7 km path effects are less pronounced, suggesting the presence of smaller magma bodies (dikes, sills) alternating with solid rock. From 4.5 to 5 km the attenuating material seems even more diffuse and for depths shallower than 4.5 km paths through the caldera are associated with normal signals. A second magma body in the northwest part of the cal-dera is less well-defined, but coincides with the reflection tentatively identified by Hill (1976) as evidence for a shallow magma chamber.

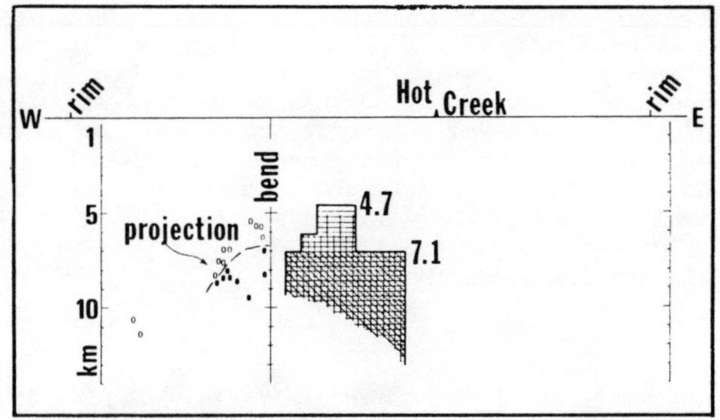

Figure 13. Cross-section along profile W-E on Fig-ure 12, showing approximate outline of magma body in central part of Long Valley caldera. Dashed line -- boundary between normal signals (open cir-cles) and anomalous signals (closed circles) for possible magma body in NW part of caldera. Note that section is bent.

Similar to the pattern observed by Ryall and Vetter (1982) for the Dixie Valley-Fairview Peak zone, P-wave fault-plane solutions for earthquakes in the Mammoth Lakes area show consistent strike-slip mechanisms for depths less than 9 km (Figure 14) and primarily oblique or normal faulting for greater depths (Figure 15). Orientation of the axis of minimum compressive stress (T-axis, Figure 16) for all of the solutions is consistent with crustal extension perpendicular to the NNW-trending Sierra Nevada frontal fault system -- with deeper events resulting from normal or oblique movement on faults striking NNW and dipping east, and shallow events reflecting conjugate right- and left-lateral shear on nearly vertical fractures striking, respectively WNW and NNE (upper right part of Fig-ure 16). Hill (1977) proposed a model in which conjugate shear failures of this type accompany the formation of magma-filled dikes, to explain the predominance of strike-slip mechanisms in volcanic regions prone to earthquake swarms. Comparison of his model with focal mechanisms for the Mammoth Lakes sequence (bottom of Figure 16) suggests that events with strike-slip mechanisms may be associ-ated with the formation of clusters of vertical fissures or dikes at depths less than 9 km.

Taken together, these observations support the suggestion of Lachenbruch and Sass (1978) that lithospheric extension in the Basin and Range pro-vince results in a combination of normal faulting and magmatic intrusion of the brittle crust. According to those authors bimodal volcanic centers like Long Valley caldera exist "because they are at places where the lithosphere is pulling apart rapidly, drawing up basalt from below to fill the void." In the case under consideration a major earthquake swarm in the Mammoth Lakes area appears to have started with a complex pattern of strike-slip faulting, possibly associated with the forma-tion of northwest-trending dikes, at shallow depth in a broad area south of Long Valley caldera. This activity reached a crescendo in the spring and sum-mer of 1980, with uplift of the resurgent dome (Savage and Clark, 1982) and the occurrence of several strong, complex ruptures along the caldera

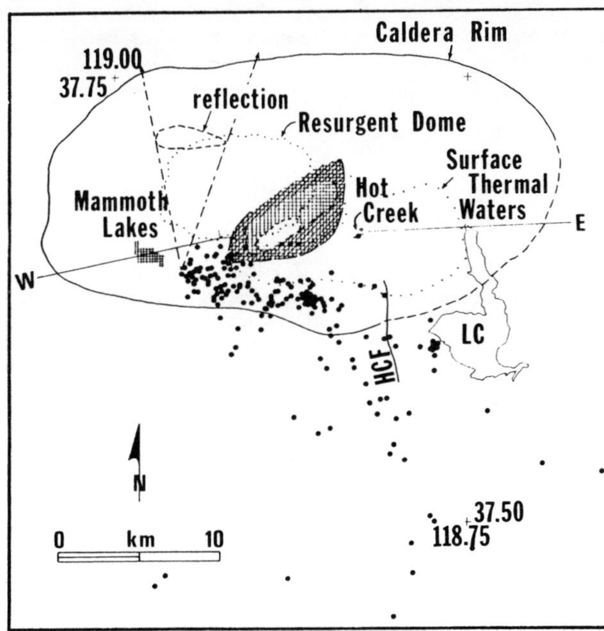

Figure 12. Map showing the location of probable magma bodies in Long Valley caldera. Dots -- epi-centers of events used in the analysis; shaded area in central part of caldera -- main magma body; shading patterns correspond to depth ranges shown on Figure 13; dashed lines -- zone in which a second, smaller body may be present in NW part of caldera; dotted lines -- outlines of resurgent dome and hot spring area.

and in the crustal block south of it. Following this maximum activity brecciation and possibly associated intrusion of the shallow crust spread rapidly to the south, north and west, with occasional bursts of spasmodic tremor marking an area of rapid crack formation just east of the town of Mammoth Lakes, due either to the injection of a small tongue of magma in the southwest part of the caldera or to gas expelled under high pressure from the shallow magma chamber.

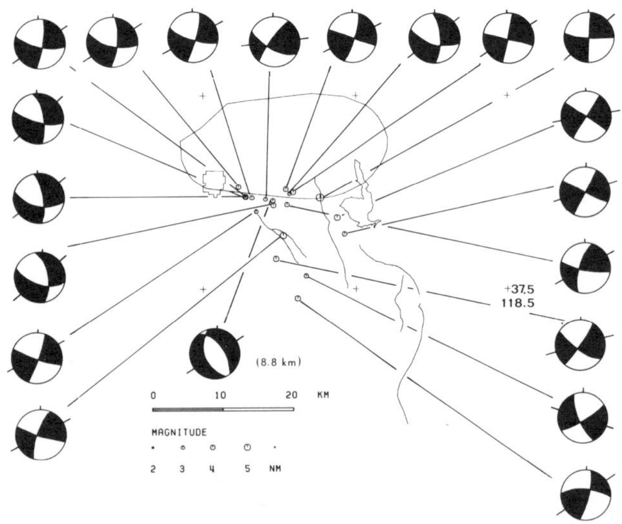

Figure 14. Focal mechanisms for Mammoth Lakes earthquakes shallower than 9 km.

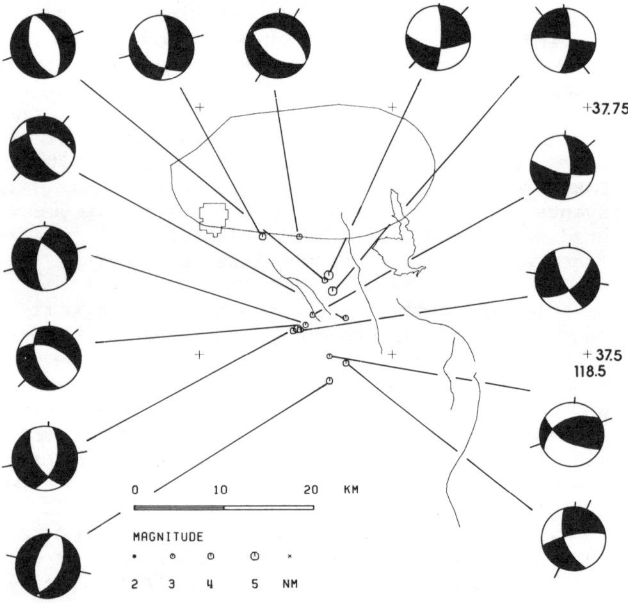

Figure 15. Focal mechanisms for Mammoth Lakes earthquakes deeper than 9 km.

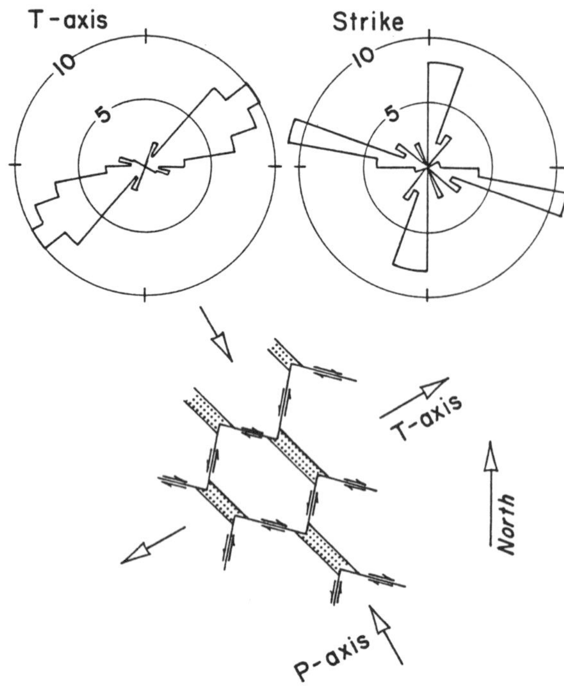

Figure 16. Model to explain strike-slip faulting for shallow Mammoth Lakes earthquakes (adapted from Hill, 1977). Top left -- rose diagram showing orientation of T-axes; top right -- rose diagram showing strike of the two planes of the fault-plane solution; bottom -- model with NW-striking vertical fissures or dikes together with conjugate right- and left-lateral shears on planes shown.

CHANGE OF FOCAL MECHANISM WITH DEPTH AND RELATED STRESS PATTERN

Analysis of P-wave first-motion for more than 150 earthquakes in all active areas covered by our network shows strong indications for a consistent pattern of stress change with depth (Vetter and Ryall, 1983). For the entire region studied, earthquakes in the uppermost crust (depth to about 9 km) are characterized by strike- or oblique-slip mechanisms; purely strike-slip events are strongly restricted to the upper 5-6 km. Below 9 km we find mostly oblique- or normal-slip events. There are some deeper strike-slip earthquakes, but no normal-faulting events were found for depths less than about 8.5 km. Similar to the change shown by Figures 14 and 15 for the Mammoth Lakes area the deeper events have a significant component of normal slip for most of the areas studied.

This change in mechanism with depth is well illustrated by Figure 17, which shows fault-plane solutions for two Mammoth Lakes earthquakes that occurred nine minutes apart in time and had almost identical epicenters but different depths. The first, with depth 8.2 km, had a strike-slip mechanism and the second, at depth 14.2 km, was oblique with a strong normal-slip component.

An exception to this pattern is observed for the Mono Basin-Excelsior Mountains, where primarily strike-slip earthquakes are found, even though a third of the events have depth in the 9-12 km range. This is illustrated by Figure 9, where the only mechanisms with a strong component of normal slip are located north of the Excelsior Mountains.

The observed change in mechanism with depth over much of the western Great Basin can be explained by increasing overburden pressure with depth (Vetter and Ryall, 1983). At shallow depth the predominance of strike-slip mechanisms requires that the greatest principal stress (S_1) is horizontal, while the occurrence of normal faulting at mid-crustal depths requires that S_1 be vertical. We assume that the vertical stress is the overburden pressure, $S_V = \rho\, g\, z$, where the mean crustal density $\rho \cong 2.7$ gm/cm.

The observation of changing focal mechanism with depth can be used to estimate the values of the principal stresses as a function of depth in the region, shown on Figure 18. On the figure, P_0 is the pore pressure, assumed to be hydrostatic. The least principal stress $S_3 = S_h$ is horizontal at all depths considered, since S_3 represents the axis of extension.

The mean depth of all the strike-slip earthquakes found in our study was about 5 km; for oblique-slip events with about equal strike- and normal-slip components the mean depth is 10.6 km; and for normal faulting it is about 13 km. From this observation we conclude that at about 10.6 km the lines S and S must cross (point 3 on Figure 18). The occurrence of both strike- and oblique-slip earthquakes deeper than about 7 km, and oblique- and normal-slip events deeper than about 9 km indicates that below 7 km the difference between the intermediate and greatest principal stresses is not great, and that local stress variations may determine which type of mechanism occurs. Zoback and Zoback (1980a, b) proposed previously that the intermediate and maximum principal stresses were about equal in the western Great Basin.

Measurements of maximum shear stress $\tau_{max} = (S_1 - S_3)/2$ in the upper few kilometers of the crust have been interpreted by Haimson (1977), McGarr and Gay (1978), McGarr (1980) and Zoback et al. (1977). These papers indicated that shear stress increases approximately linearly with depth, and is lower in soft rock than in hard rock. For the latter the stress change with depth follows a regression line

$$\tau_{max} \approx 5.67 + 6.37\, z \; (MPa), \tag{1}$$

while a regression line through all the data points measured in an extensional regime is

$$\tau_{max} \approx 1 + 6.7\, z \; (MPa) \tag{1a}$$

(both curves from McGarr, 1980).

The value of $(S_1 - S_3)$ from equation 1 is shown on Figure 18, extrapolated to mid-crustal depth. The difference between equations (1) and (1a) is greatest at shallow depth where we find strike-slip earthquakes, but it is small at greater depth where oblique- and normal-slip events predominate.

From equation (1) with $S_1 = S_V$ (normal faulting regime) we estimate S_3 at 10.6 and 20 km depth to be about 135 and 260 MPa, respectively (points 4 and 5 on Figure 18). At shallower depth (for a strike-slip regime with $S_H > S_V > S_h$) we can estimate S_1 and S_3 only for the special case where

$$S_V = S_2 \approx \frac{S_3 + S_1}{2} \tag{2}$$

For a depth of 3 km this relationship gives stress values of 105 and 55 MPa (points 6 and 7 on Figure 18).

For the normal faulting regime at depth $\gtrsim 10$ km we also use Byerlee's (1977) equation

$$S_3 = \frac{(S_1 - P_0)}{[(\mu^2 + 1)^{1/2} + \mu]^2} + P_0 \tag{3}$$

with P_0 = pore pressure and μ = the coefficient of friction. μ is experimentally determined by Byerlee to be about 0.85 for normal stress less than 200 MPa, representing the upper 6-8 km of the crust, and about 0.6 for normal stress in the range 200-2000 MPa, representing the deeper crust. P is assumed to be hydrostatic.

If we assume, then, that the horizontal stresses increase linearly with depth we obtain from equation (3) that the change in S_h with depth is about 13.4 MPa/km ($\mu = 0.85$), or 15 MPa/km ($\mu = 0.6$), in both cases for the minimum horizontal stress.

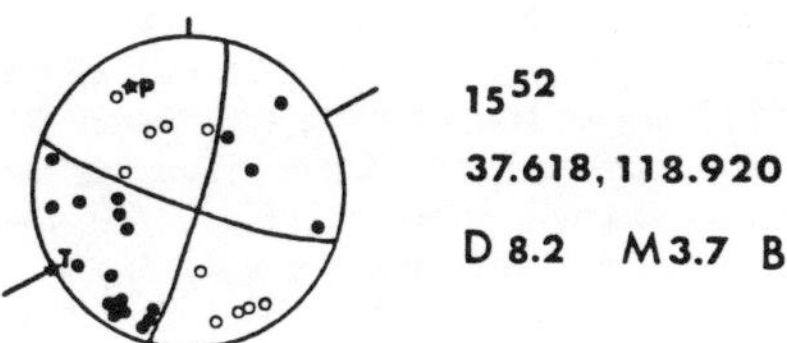

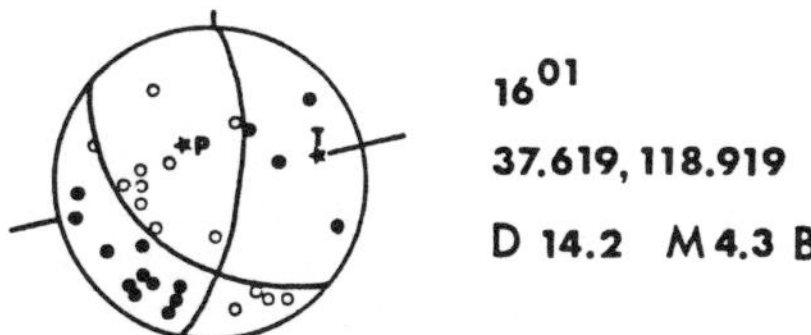

Figure 17. Focal mechanisms for two Mammoth Lakes earthquakes on 9 August 1981. D — depth, km; M — magnitude ML; B — quality.

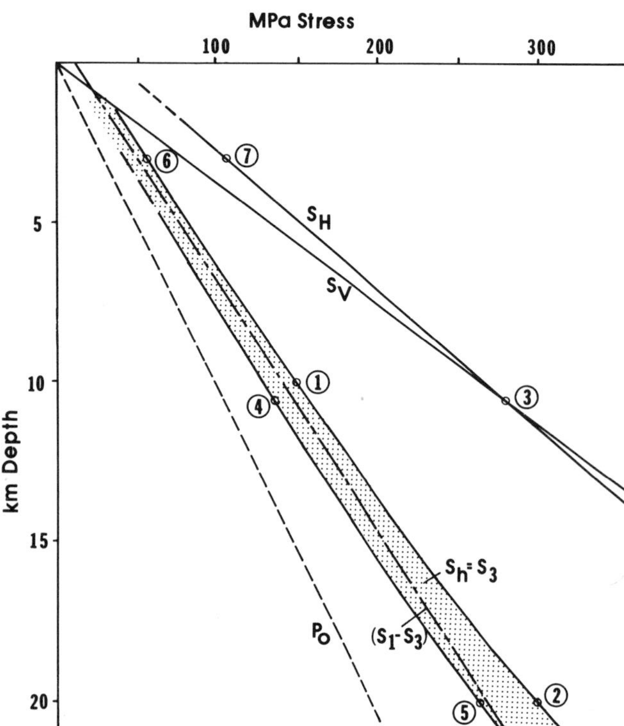

Figure 18. Estimates of the principal stresses as a function of depth. Symbols explained in the text.

For the maximum horizontal stress S_H also increasing linearly with depth we can write (McGarr, personal communication, 1983)

$$S_H \approx A + (\rho\, g - \delta)\, z \qquad (4)$$

where A and o are constants. The constant A can be determined at the surface, where for strike-slip faulting $\tau_m = (S - S)/2$. According to equations (1a) and (1) we calculate τ_m at depth 0 km as 1 and 5.7 MPa, respectively. These give values for A of 2 and 11.4 MPa, respectively. At depth 10.6 km where the S_H and S_V curves cross, $S_V = S_H$ and we obtain a value of 0.19 MPa/km for the constant δ with A = 2 MPa; with A = 11.4 MPa we find $\delta = 1.1$ MPa/km.

ACKNOWLEDGEMENT

Art McGarr and M. L. Zoback corrected errors in the original manuscript and made helpful suggestions regarding our presentation of the data. Figure 18 is based in part on a sketch by M. D. Zoback during the October 1982 Chapman Conference in Utah; his sketch was derived from our preliminary conclusions on systematic changes in mechanism with depth and possible implications for stress in the lithosphere. This research was partly supported by the U. S. Geological Survey under contract 14-08-0001-21248, and partly by the U. S. Department of Energy under contract DE-AS08-82ER12082. The research was also partly supported by the Advanced Research Projects Agency of the Department of Defense and was monitored by the Air Force Office of Scientific Research under contract F49620-83-C-0012.

REFERENCES

Byerlee, J., 1977, Friction of rocks, US Geol. Survey, Open-File Rpt., Experimental Studies of Rock Friction with Application to Earthquake Prediction, 55-77.

Gilbert, C. M. Christensen, M. N., Al-Rawi, Y. T. and Jajoie, K. R., 1968). Structural and volcanic history of Mono Basin, California-Nevada, Geol. Soc. Am. Memoir 116, 275-329.

Haimson, B. C., 1977, Crustal stress in the continental United States as derived from hydrofracturing tests, in The Earth's Crust, Geophys. Monogr. Ser., 20, ed. J. G. Heacock, 576-592.

Hill, D. P., 1976, Structure of Long Valley caldera, California, from a seismic refraction experiment, Jour. Geophys. Res., 81, 745-754.

Hill, D. P., 1977, A model for earthquake swarms, Jour. Geophys. Res., 82, 1347-1352.

Iyer, H. M. and Evans, J. R., 1983, Teleseismic traveltime anomaly at Mono and Inyo Craters, California: a dike-form low-velocity body? Earthquake Notes 54, 92 (abstract).

Lachenbruch, A. H., and Sass, J. H., 1978, Models of an extending lithosphere and heat flow in the Basin and Range province, Geol. Soc. Am. Mem., 152, 209-250.

Malone, S. D., Boyko, C. and Weaver, C. S., 1983, The seismic precursors to the 1981-1982 eruptions of Mount St. Helens, Washington, Science, in press.

McGarr, A., 1980, Some constraints on levels of shear stress in the crust from observations and theory, Jour. Geophys. Res., 85, 6231-6238.

McGarr, A., and Gay, N. C., 1978, State of stress in the earth's crust, Ann. Rev. Earth Planet. Sci., 6, 405-436.

Meissner, R., and Strehlau, J., 1982, Limits of stresses in continental crusts and their relation to the depth-frequency distribution of shallow earthquakes, Tectonics, 1, 73-89.

Mogi, K., 1969, Some features of recent seismic activity in and near Japan (2), activity before and after large earthquakes, Bull. Earthquake Res. Inst., Tokyo Univ. 47, 395-417.

Richins, W. D., 1974, Earthquake swarm near Denio, Nevada, February to April 1973, Univ. of Nevada M. S. Thesis, 57pp.

Romney, C., 1957, The Dixie Valley-Fairview Peak, Nevada, earthquakes of December 16, 1954: seismic waves, Bull. Seismol. Soc. Am., 47, 301-320.

Ryall, F. and Ryall, A., 1981a, Attenuation of P and S waves in a magma chamber in Long Valley caldera, California, Geophys. Res. Ltrs., 8, 557-560.

Ryall, A. and Ryall, F., 1981b, Spatial-temporal variations in seismicity preceding the May, 1980, Mammoth Lakes, California, earthquakes, Bull. Seism. Soc. Am., 71 (3, 747-760.

Ryall, A. and Vetter, U. R., 1982, Seismicity Related to Geothermal Development in Dixie Valley, Nevada , Final Rept. on U.S.D.O.E. Contract DE-AC08-79NV10054, Univ. of Nevada, Reno, 102 pp.

Ryall, A. and Ryall, F., 1983, Spasmodic tremor and possible magma injection in Long Valley caldera, eastern California, Science, 219, 1432-1433.

Sanders, C. O. and Ryall, F., 1983, Geometry of magma bodies beneath Long Valley, California, determined from anomalous earthquake signals, Geophys. Res. Letters, 10, 690-692.

Savage, J. C. and Clark, M. M., 1982, Magmatic resurgence in Long Valley caldera, California: Possible cause of the 1980 Mammoth Lakes earthquakes, Science, 217, 531-533.

Sibson, R. H., 1974, Frictional constraints on thrust, wrench and normal faults, Nature, 249, 542-544.

Sibson, R. H., 1982, Fault zone models, heat flow, and the depth distribution of earthquakes in the continental crust of the United States, Bull. Seismol. Soc. Am., 72, 151-163, 1982.

Slemmons, D. B., 1957, Geological effects of the Dixie Valley-Fairview Peak, Nevada, earthquakes of December 16, 1954, Bull. Seismol. Soc. Am., 47, 353-375.

Slemmons, D. B., 1967, Pliocene and Quaternary crustal movements of the Basin and Range province, USA, Osaka City Univ., Jour. of Geoscience 10, 91-103.

Steeples, D. W. and Iyer, H. M., 1976, Low-velocity zone under Long Valley as determined from teleseismic events, J. Geophys. Res., 81, 849-860.

U. S. Dept. of Energy, 1980, Comparative Assessment of Five Potential Sites for Hydrothermal-Magma Systems: Summary, USDOE/TIC-11303, 51 pp.

U. S. Geological Survey, 1983, Potential hazards from future volcanic eruptions in the Long Valley-Mono Lake area, east-central California and southwest Nevada - a preliminary assessment, U. S. Geol. Survey Circular 877, 10 pp.

VanWormer, J. D., and Ryall, A. S., 1980, Sierra Nevada-Great Basin boundary zone: Earthquake hazard related to structure, active tectonic processes, and anomalous patterns of earthquake occurence, Bull. Seismol. Soc. Am., 70, 1557-1572.

Vetter, U. R. and A. S. Ryall, 1983, Systematic change of focal mechanism with depth in the western Great Basin, Jour. Geophys. Res., in press.

Wallace, R. E., 1978, Patterns of faulting and seismic gaps in the Great Basin province, U. S. Geol. Survey, Open-File Rept. 78-943, 857-868.

Wright, L., 1976, Late Cenozoic fault patterns and stress fields in the Great Basin and westward displacement of the Sierra Nevada block, Geology 4, 489-494.

Zoback, M. D., Healy, J. H., and Roller, J. C., 1977, Preliminary stress measurements in central California using the hydraulic fracturing technique, Pure Appl. Geophys., 115, 135-152.

Zoback, M. L., and Zoback, M. D., 1980a, Faulting patterns in north-central Nevada and strength of the crust, J. Geophys. Res., 85, 275-284.

Zoback, M. L. and M. D. Zoback, 1980b, State of stress in the conterminous United States, Jour. Geophys. Res., 85, 6113-6156.

RESISTIVITY STRUCTURE OF THE NORTHERN BASIN AND RANGE

Philip E. Wannamaker

Earth Science Laboratory/University of Utah Research Institute
420 Chipeta Way, Suite 120
Salt Lake City, Utah   84108

ABSTRACT

Trustworthy models of northern Basin and Range resistivity have been lacking mainly due to an inadequate regard for the effects of upper crustal three-dimensional (3D) inhomogeneities upon the surface electromagnetic measurements. While geomagnetic deep sounding (GDS) is relatively insensitive to upper crustal complexity, the ability of GDS to resolve structure of interest at greater depths is limited.   The role of low-resistivity 3D structures near the surface has been documented most thoroughly for the magnetotelluric (MT) technique, although controlled-source electromagnetics (CSEM) using either grounded or ungrounded transmitters also is seriously complicated by inhomogeneities. The natural field methods of MT and GDS are preferred to CSEM for deep resistivity exploration due to the plane-wave nature of the source as well as to the availability of data at very low frequencies and of 2D and 3D modeling algorithms for treating upper crustal structure.

Magnetotelluric measurements in S.W. Utah have detected a low-resistivity layer from 35 to 65 km depth in the upper mantle that is proposed to reflect an accumulation of basaltic melt in peridotite.   GDS experiments and various tectonic indicators suggest that this melt is controlled by adiabatic upwelling and fusion along the eastern margin of the northern Basin and Range. Similar GDS anomalies and extensional processes appear active in the western margin of this province, implying a similar resistivity structure. This would leave the northern Basin and Range interior of central Nevada as a comparatively quiescent region with an upper mantle seismic low-velocity zone whose melt interconnection, and commensurate low resistivity, remain fairly intact.

INTRODUCTION

The gains in knowledge of earth processes by investigations of electrical resistivity structure traditionally have been modest compared to those by other geoscientific methods.   A major reason for this is that the electromagnetic (EM) measurements at the surface, from which one infers sub-surface resistivity, in general are contaminated by complexity in the uppermost crust to a degree that is more extreme than for other geophysical techniques.

Through the course of this paper, I will examine the pitfalls in the interpretation of EM measurements in tectonically active environments, review previous surveys of resistivity structure in the northern Basin and Range, and propose a model for the Late Cenozoic evolution of deep resistivity structure and physiochemical conditions throughout the northern Basin and Range.

RESISTIVITY SURVEYS IN THE NORTHERN BASIN AND RANGE

Several key surveys of northern Basin and Range resistivity structure will be studied in this section (see Figure 1). Methods utilizing

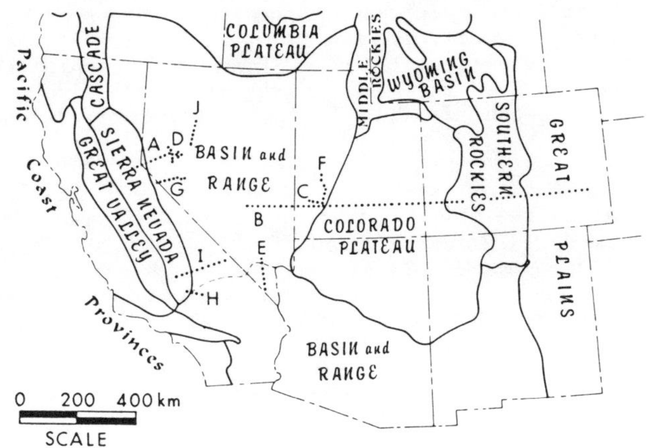

Figure 1.   Physiographic provinces of the south western United States, modified from Stewart (1978). Dashed line separates Great Basin from the southern Basin and Range (Eaton, 1982). Locations are plotted for resistivity surveys discussed in text: A, Schmucker (1970); B, Porath (1971); C, MT survey at the Roosevelt Hot Springs, this paper; D, Stanley et al., 1976; E, Keller, 1971; F, Petrick et al., 1981; G, Lienert and Bennett, 1977; H, Lienert, 1979; I, Towle, 1980; J, Wilt et al., 1982.

either natural or artificial sources of EM fields have been applied. Principles of the techniques will not be reviewed here; the reader is referred to the literature cited for this purpose. I will, however, emphasize the effects on EM observations of the upper crustal three-dimensional (3D) heterogeneity so prevalent in the northern Basin and Range.

Geomagnetic Deep Sounding

Geomagnetic deep sounding (GDS), which makes use of natural magnetic field temporal variations, is less affected by upper crustal complexity than are the other EM methods I cover. The analysis of Wannamaker et al. (1983) can be extended easily to show that the secondary magnetic field transfer functions of GDS (Schmucker, 1970) are band-limited in frequency, with relatively small-scale, shallow structure like graben sediments responding over a frequency range which is higher than that of the deep regional resistivity structure of interest approximately as the square of the geometric scale factor distinguishing the two classes of structure. However, details of the separation depend on the layered host for the structures (Wannamaker et al., 1983). This separation of responses in frequency enables discrimination against surface inhomogeneity. Nevertheless, the response of the deep structure one seeks may itself be complicated; use of relatively simple 1D or 2D interpretation algorithms likely will yield models that are only qualitatively correct.

Pioneering measurements by Schmucker (1970) established geomagnetic anomalies indicating conductive upwellings beneath the Rio Grande Rift and the northwestern Basin and Range. Concerning the latter area, Schmucker places the conductive mass at a depth of 40 km in the upper mantle (Figure 2).

Much of what we know about the deep electrical resistivity of the northern Basin and Range and the Colorado Plateau is the result of geomagnetic temporal variations observed by Reitzel et al. (1970). Three components of magnetic field were recorded at periods from 20 to 200 minutes along four E-W profiles about 150 km apart, with station spacings averaging about 120 km. Porath et al. (1970) and Porath and Gough (1971) showed that anomalies in vertical and E-W horizontal fields were of internal origin (see Figure 3), with the external fields varying smoothly over the recording array. This behavior of the external fields severely limits the ability of GDS to resolve horizontal resistivity layering (Schmucker, 1970).

From these estimates of normalized anomalous fields, Porath (1971) presented a preferred, two-dimensional resistivity model, also shown in Figure 3, with a 2 Ω-m layer of varying thickness from the northern Basin and Range to the Great Plains, and with superimposed conductive ridges beneath the Wasatch Fault Belt and the Southern Rocky Mountains. This model was correlated with

seismic LVZ estimates. Gough (1983) has reaffirmed the model of Porath (1971) and proposed that the conductive ridges in the upper mantle beneath western Utah and the southern Rocky Mountains reflect zones of partial melting. However, the intrinsic values of depth and resistivity are not well resolved (ibid.), especially given the assumption of purely 2D geometry. The secondary field amplitudes diminish to the south, and so presumably does the anomalous resistivity structure, but the features in Figure 3 appear to continue into northwesternmost Arizona.

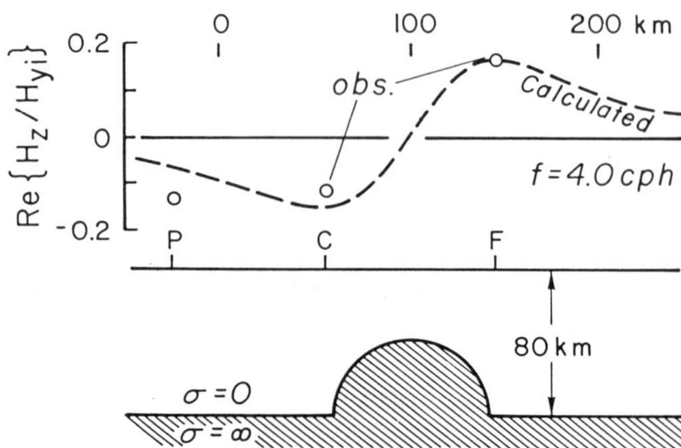

Figure 2. Observed and computed, normalized, in-phase, vertical magnetic field variation along WSW-ENE profile in western most Nevada. Also shown is simple, semi-cylindrical conductivity hump in the upper mantle which fits the observations. Urban centers located as abbreviations P (Pacific, CA), C (Carson City) and F (Fallon). Frequency is 4 cycles/hr. Adapted from Schmucker (1970).

Magnetotelluric Measurements

The magnetotelluric (MT) method, which makes use of both electric and magnetic natural field temporal variations, has been widely applied in the exploration of geothermal systems, sedimentary basins and the deep crust and upper mantle (Swift, 1967; Word et al., 1971; Vozoff, 1972; Larsen, 1975; Jupp and Vozoff, 1976; Stanley et al., 1977). While recent advances in instrumentation and data processing (Gamble et al., 1979; Weinstock and Overton, 1981; Stodt, 1983) enable accurate measurements of tensor MT responses, models of deep resistivity derived from MT responses commonly are suspect due to an unsatisfactory treatment of upper crustal lateral inhomogeneities (Wannamaker et al., 1983).

The Trouble with Upper Crustal Structure. - Serious problems with interpretation of MT soundings near electrically conductive graben sediments were first documented by Swift (1967) in his Arizona and New Mexico studies. A detailed

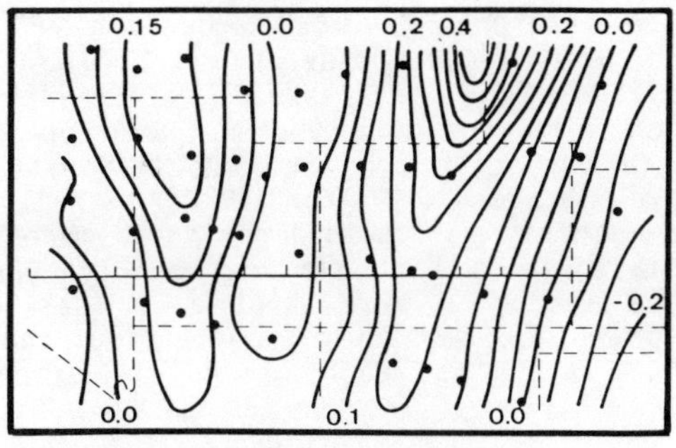

In-Phase Y$_{ia}$ / Y$_n$

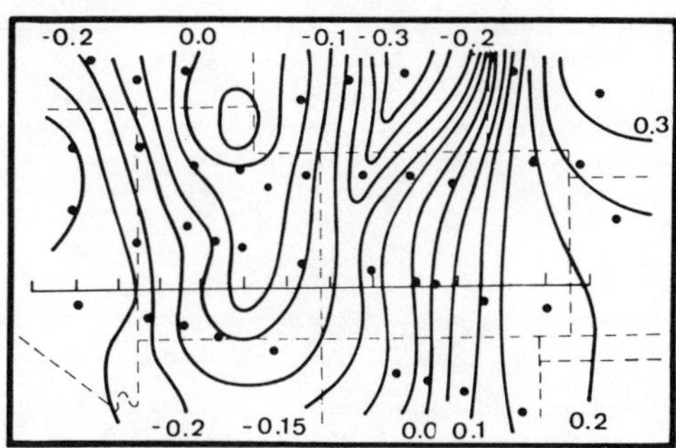

In-Phase Z$_{ia}$ / Y$_n$

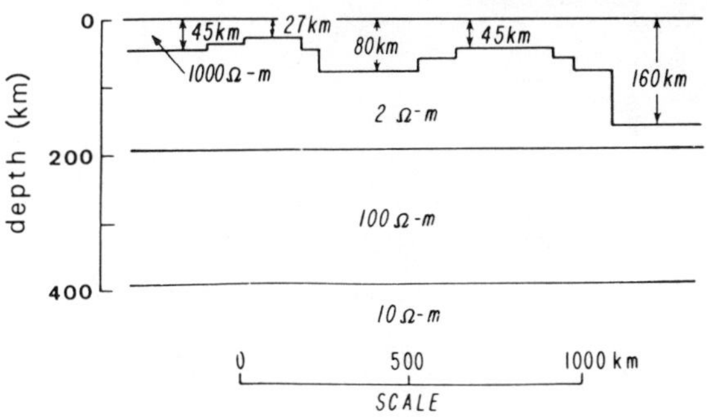

Figure 3. Normalized, anomalous, in-phase, horizontal E-W (Y_{ia}/Y_n) and vertical (Z_{ia}/Z_n) magnetic variation fields at a period of 60 minutes in the south western United States (Porath et al., 1970) along with the two-dimensional resistivity cross-section preferred by Porath (1971b) and Gough (1983) to explain the observations. The E-W profile for which the cross-section is defined is drawn on the maps of observed fields.

examination of the MT responses associated with graben alluvial fill has been performed by Wannamaker et al. (1983), who concluded that such responses are fundamentally three-dimensional in nature and that indiscriminate use of 1D or 2D modeling algorithms generally will incur serious errors.

Wannamaker et al. (1983) demonstrated in particular that two-dimensional transverse electric (TE) modeling algorithms are inappropriate for interpreting apparent resistivities, impedance phases or tipper in the northern Basin and Range identified as TE (Word et al., 1971; Vozoff, 1972). The 2D TE mode, consisting of the horizontal E-field parallel to strike and the orthogonal H-fields, involves no boundary charges and hence no current-gathering, whereas in the northern Basin and Range, all modes of MT responses due to upper crustal inhomogeneities are dominated by current-gathering.

If MT sounding data identified as TE in the presence of a 3D conductive inhomogeneity are inverted assuming a 1D model of deep resistivity structure, as I perceive has been overwhelmingly the case in the literature, then any such model will suffer systematically a compression of layer thicknesses and a downward bias in layer resistivities relative to the true 1D regional profile enclosing the inhomogeneity. This is because the TE mode apparent resistivity everywhere over and to the exterior of a conductive 3D body will be depressed throughout all frequencies relative to the apparent resistivity of the layered host containing the structure (Ting and Hohmann, 1981; Wannamaker et al., 1983).

Fortunately, accurate resistivity cross-sections through horst-graben morphology in the Great Basin can be obtained by selective application of a 2D transverse magnetic (TM) modeling algorithm (Wannamaker et al., 1983). The 2D TM mode consists of the H-field parallel to strike and the orthogonal E-fields. Boundary charges are included in both 3D and 2D TM formulations, allowing a proper treatment of current-gathering effects upon apparent resistivity and impedance phase. In defining data for transverse magnetic modeling of northern Basin and Range upper crustal structure, I recommend a uniform coordinate system based on tipper-strike (Vozoff, 1972). As with GDS transfer functions, tipper responses are band-limited in frequency (Wannamaker et al., 1983), so that one may choose an optimal frequency range for defining tipper-strike to minimize the contributions of secondary inhomogeneities much smaller than the conductive graben sediments for which one is primarily concerned in compensating.

MT Measurements at the Roosevelt Hot Springs.-
A thorough accounting of upper crustal heterogeneity in the course of resolving deep resistivity has been performed on a collection of tensor MT data at the Roosevelt Hot Springs thermal area in S.W. Utah (Ward et al., 1978; Ross et al., 1982; see Figure 4). The complexly 3D nature of the MT responses directly over the thermal anomaly area prevented a rigorous quantification of any resistivity structure associated with an economic hot brine reservoir or a mid-crustal magmatic heat source for the geothermal system (ibid.). However, soundings along line B-B' in Figure 4 at distance from the thermal anomaly area, carefully modeled to remove the effects of upper crustal lateral inhomogeneities,

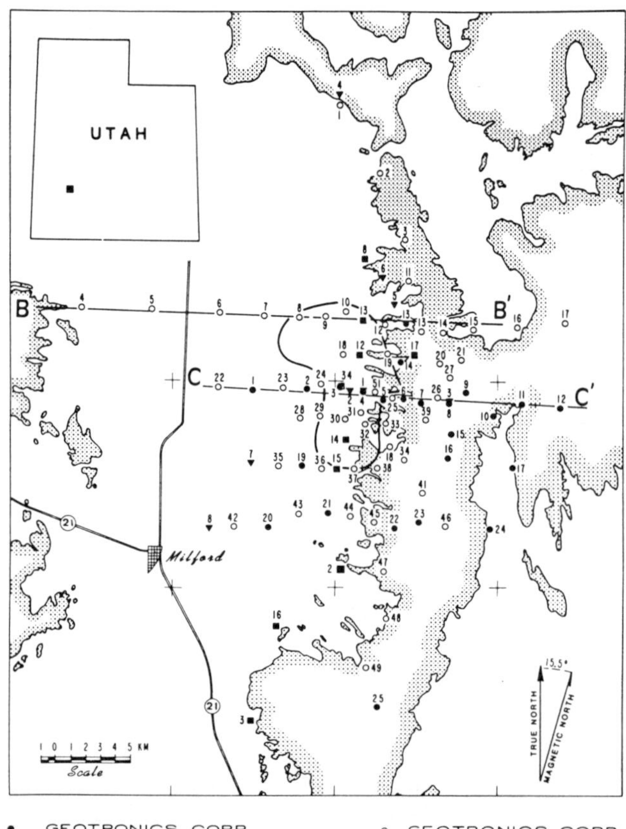

- GEOTRONICS CORP.
  (1976)

o GEOTRONICS CORP.
  (1978)

■ UNIV. OF TEXAS
  AT AUSTIN (1977)

▼ UNIV. OF CALIFORNIA
  AT BERKELEY (1979)

MT./AMT. STATIONS
ROOSEVELT HOT SPRINGS KGRA, UTAH

Figure 4.   Magnetotelluric site location map for
the Roosevelt Hot Springs thermal
area. Rock outcrop bounding the
Milford Valley sediments is lightly
stippled while the solid curve is the
400 mWm^{-2} thermal contour (Ward et al.,
1978). Station numbers referred to in
text are prefixed according to the year
in which they were occupied.

have yielded a regional resistivity profile for
S.W. Utah to depths of about 100 km.

The observed quantities ρ_{yx} and ϕ_{yx} of line
B-B', identified as TM, using a uniform coordinate
direction as recommended previously, have been
assembled in Figure 5 into pseudosections (Vozoff,
1972). The measured results for this line exhibit
a virtually classic horst-graben response for the
TM mode and bear a close resemblance to the
computations of Wannamaker et al. (1983). Note
especially the strong lateral gradients in ρ_{yx} and
ϕ_{yx} between stations 78-8 and 78-10, which are due
to the abrupt rangefront faulting bounding the
graben sediments on the east side.

To interpret these MT observations, I have
relied upon trial-and-error matching of results
calculated from an assumed resistivity cross-
section with the observed MT data of Figure 5
using a versatile and accurate 2D finite element
forward program (Rijo, 1977; Stodt, 1978).
Nonuniqueness in multidimensional modeling was
alleviated by a high station density in each MT
profile and by a wide spectrum of data at each
station. In particular concerning the latter
point, data at each sounding extended to
sufficiently high frequencies so that both principal apparent resistivities ρ_{xy} and ρ_{yx}, as well
as impedance phases ϕ_{xy} and ϕ_{yx}, have converged to
common values above some frequency. For this high
frequency range, the earth will likely be effectively one-dimensional, allowing one to constrain
the near-surface model resistivities close to
their true values. Such isotropic behavior occurs
for frequencies above 1 Hz for the valley stations
and above 30 Hz for the mountain sites.

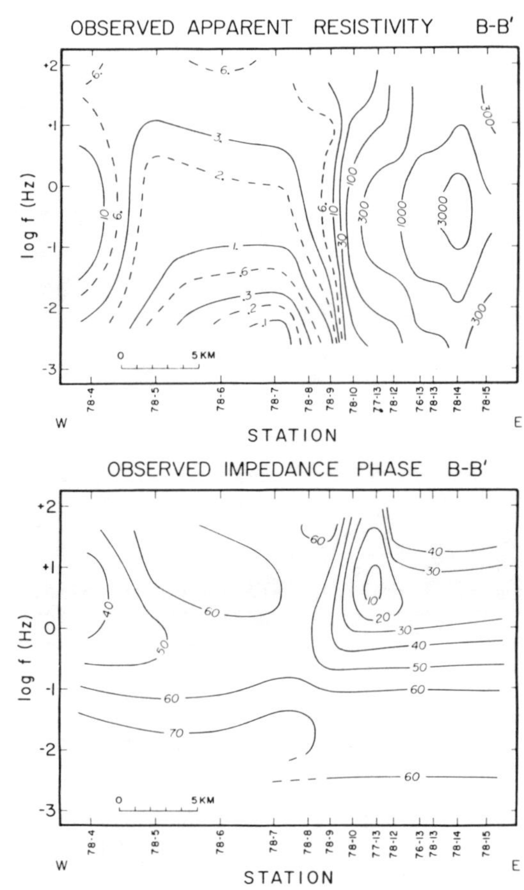

Figure 5.   Observed pseudosections of ρ_{yx} and ϕ_{yx},
identified as transverse magnetic,
along line B-B' of Figure 4. Contours
of ρ_{yx} in Ω-m and of ϕ_{yx} in degrees.

The computed pseudosections of ρ_{yx} and ϕ_{yx} for line B-B' appear in Figure 6. Note that the computed and observed pseudosections are virtually identical; the observations have been fit to within data scatter almost everywhere. Examples of data scatter typical of sites over the Mineral Mountains will be forthcoming.

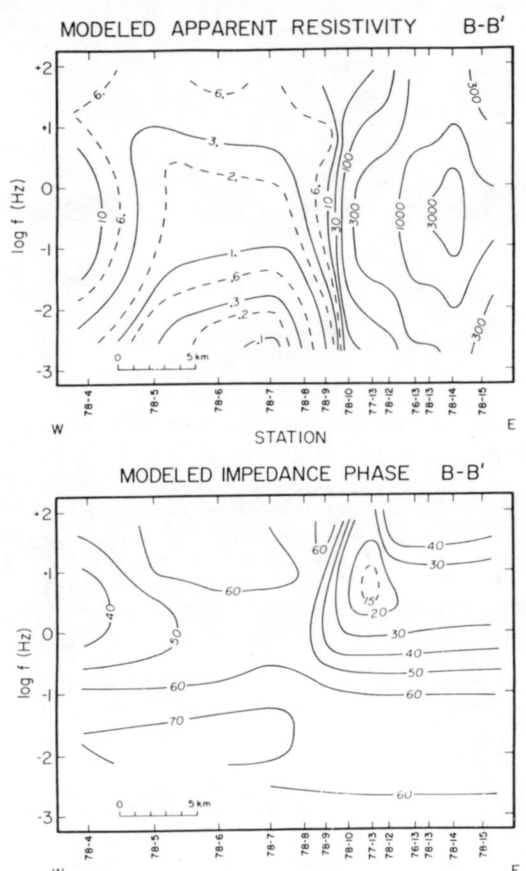

Figure 6. Best-fit pseudosections of ρ_{yx} and ϕ_{yx} obtained from transverse magnetic, finite element simulation of the observations of Figure 5. Contours of ρ_{yx} in Ω-m and of ϕ_{yx} in degrees.

The model of deep resistivity structure produced by application of the 2D TM finite element algorithm to the observations can be divided into two major parts: first, a shallow portion detailing just the extensive upper crustal lateral inhomogeneities in the upper 2 km; and second, a deep, purely layered portion extending to about 100 km which is determined by the physiochemical conditions of the crust and upper mantle below 2 km.

The upper 3 km of the finite element cross-section we have derived is shown in Figure 7. Within the Milford Valley, units within 140 m of the surface having resistivities from 3.5 to 18 Ω-m correspond to surficial clays, sands and gravels which are partially or completely water-saturated. Looking a bit deeper here, resistivities as low nearly as 1 Ω-m are caused by Pleistocene Lake Bonneville clays (Hintze, 1973, 1980), which we show residing to a maximum depth near 700 m below station 78-7. Deeper still exists a poorly resolved but substantial thickness (> 1 km) of modest resistivity material (25 Ω-m) related to pre-Bonneville alluvium and volcanics (Rowley et al., 1979). Note as well the especially steep dip of the eastern boundary faulting of the valley alluvium, inferred from the abrupt lateral gradients in ρ_{yx} and ϕ_{yx} between stations 78-9 and 78-10 in Figure 5 (cf. Wernicke and Burchfiel, 1982). The initial guess in the modeling process of this upper crustal resistivity section was guided by the refraction seismic model of Gertson and Smith (1979) and the gravity observations of Carter and Cook (1978), but the final section in Figure 7 shows important differences.

Over the Mineral Mountains, our model in Figure 7 shows resistivities of hundreds of ohm-m increasing to 3000 Ω-m over the depth interval of 2 to 3 km. This represents a decrease in porosity of the well-indurated basement rocks to well under 1%, probably due for the greater part to a decrease in fracture porosity below 2 to 3 km depth.

The 1400 Ω-m medium in Figure 7 begins the deep layered resistivity profile introduced previously and shown in its entirety in Figure 8. The best-fit model shows a maximum resistivity of 3000 Ω-m over the depth interval of 3 to 11 km. Subsequently, resistivity falls to 200 Ω-m by 35 km depth, whereupon an abrupt drop to 20 Ω-m is encountered. This low-resistivity material defines a deep, thick layer bottomed at 65 km depth by a sudden increase to a 200 Ω-m basal half-space. The fact that this deeper part of the model is purely layered is not due to poor resolution on the part of the measurements of Figure 5; the data quality for all soundings in line B-B' is good to excellent and yet the data are fit within scatter by imposing a purely 1D section for depths greater than 2 km. Furthermore, through application of a 2D TM program, the data of line C-C' in Figure 4 also are fit using the preferred deep profile of Figure 8 (Ross et al., 1982).

In Figure 9 are plotted both modes of apparent resistivity and impedance phase for station 76-13 along with several computed response curves. First, consider the solid curves passing through the observed data points of ρ_{yx} and ϕ_{yx}. These curves display the calculated response of the best-fit finite element model of Figures 7 and 8 at station 76-13, a response which seems very agreeable with the trends of the observed data points. In addition, curves of long dashes lying intermediate to the principal apparent resistivity and impedance phase observations are shown in Figure 9. The response of the best-fit one-dimensional deep resistivity profile of Figure 8 in the absence of all upper crustal lateral

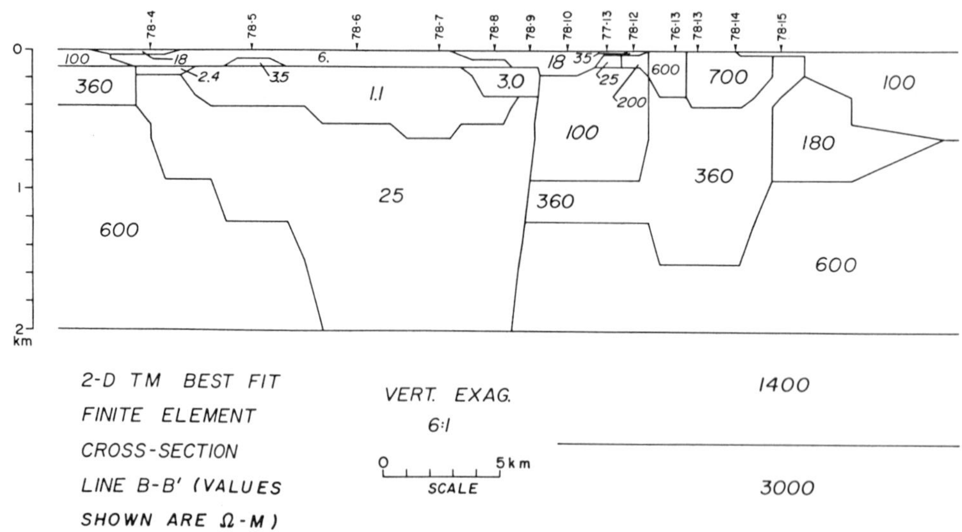

Figure 7. Best-fit, 2D TM finite element resistivity cross-section fitting the observations of line B-B' in Figure 4.

structure is represented by these latter curves. The difference between the observations and the purely 1D response illustrates again the importance of the Milford Valley in determining the MT signatures in this area. The inclusion of lateral inhomogeneities in models of deep resistivity structure is of paramount importance in tectonically disturbed regions like the Great Basin, and supercedes other considerations of the appropriateness of resistivity models vis-a-vis continuous vs. layered one-dimensional structures (Larsen, 1981; Parker and Whaler, 1981).

I turn now to resolution of detail in the deep resistivity profile of Figure 9. It is noted at the outset that, while lateral inhomogeneities, i.e., the Milford Valley, have induced anisotropy to a high degree over the Mineral Mountains, both tensor and 1D signatures in Figure 9 bear a certain mutual resemblance. In particular, the factor of ten drop in model resistivity at 35 km depth is responsible for the particularly steep gradients in the apparent resistivities and the values of impedance phases about 70° around 0.03 Hz. Also, without the abrupt rise in model resistivity at 65 km depth, the apparent resistivity data would not flatten out nor would the impedance phase data fall as strongly or rapidly as they do in Figure 9 at the lowest frequencies, especially below 0.003 Hz.

At frequencies less than 0.01 Hz in Figure 9, both ϕ_{xy} and ϕ_{yx} have converged to common values while ρ_{xy} and ρ_{yx} have shapes that are essentially identical except that they are offset by a factor of ten. Eventually, a convergence of this sort occurs as frequency falls for all stations in the Roosevelt Hot Springs area. This frequency dependence of the tensor sounding curves

particularly illustrates that a one-dimensional, regional resistivity profile containing local lateral variations in structure, chiefly the Milford Valley sediments, is a viable model for the resistivity makeup here in S.W. Utah.

To strengthen confidence in the best-fit deep resistivity profile, I consider three alternates in Figure 8. The first is a profile whose log resistivity decreases linearly with depth (plotted with dots and labelled the "Steady Decrease" model in Figure 8). I arrive at a second layered sequence drawn with alternating dots and dashes in Figure 8 and called the "Shallow" regional model by shrinking the depths and resistivities of individual layers by 30% such that layer contrasts and conductivity thickness products remain constant (Madden, 1971). The last parameter resolution examination deals solely with the deep low-resistivity layer from 35 to 65 km. I maintain that this 20 Ω-m medium is not a "thin" layer (Madden, 1971), and to prove so it is replaced by a unit of equivalent conductivity-thickness product of 10 Ω-m resistivity and 15 km thickness.

The TM mode computed response curves at site 76-13 for the valley cross-section contained in each of the alternate model regional profiles appear in Figure 9. The computed apparent resistivity or impedance phase or both for each of the alternate models do not fit the observed data points over much of the frequency range, causing them to be dismissed as candidates for deep resistivity structure in this area. Recall that all soundings on profiles B-B' and C-C' are explained using the same regional profile; several other soundings in the Mineral Mountains are comparable in quality to site 76-13 and show

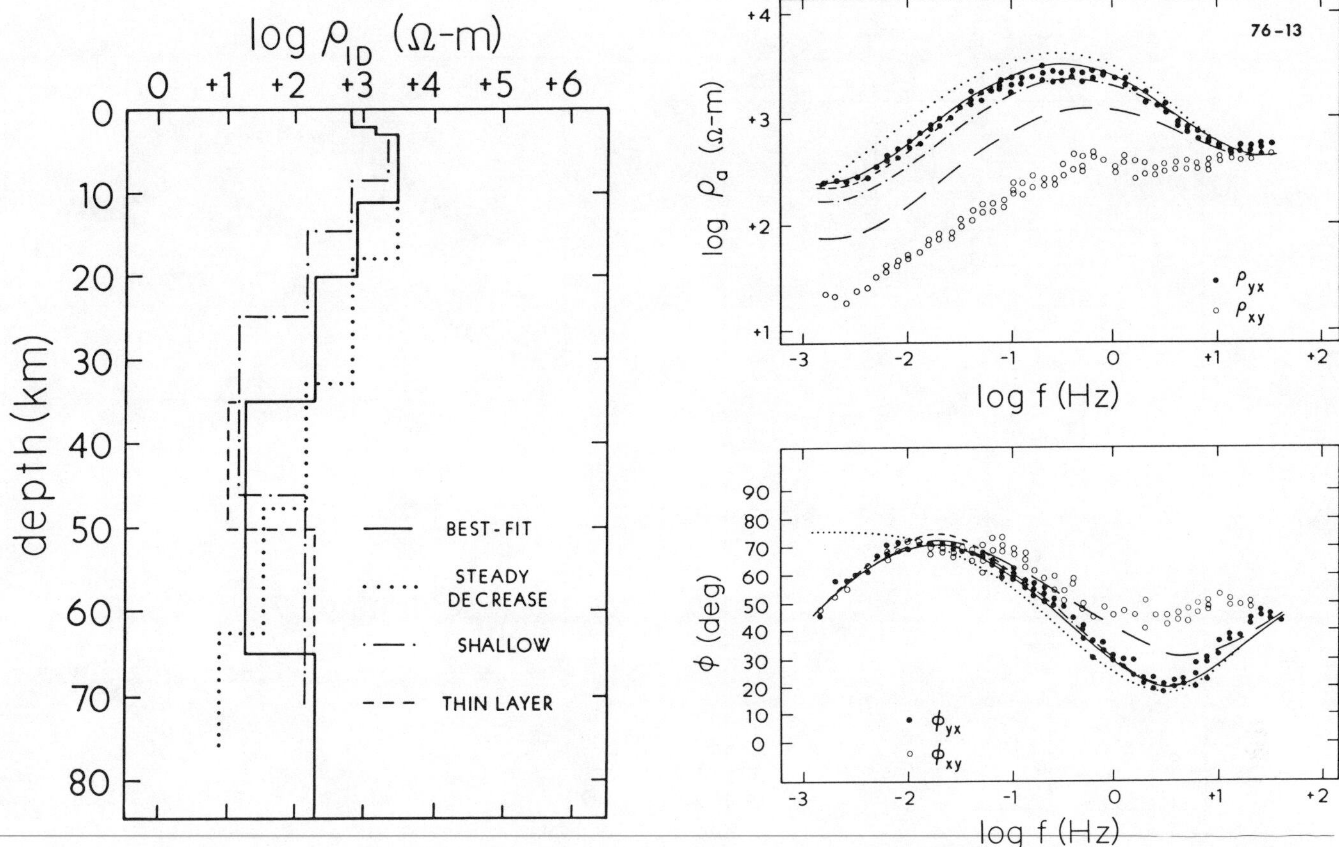

Figure 8. One-dimensional, deep resistivity profile in this area of S.W. Utah defining the lower portion of our 2D finite element model. Alternate resistivity profiles used in parameter resolution tests are drawn with dots and dashes.

Figure 9. Observed data ρ_{xy}, ρ_{yx}, ϕ_{xy} and ϕ_{yx} for sounding 76-13 of line B-B'. Solid curve represents goodness-of-fit of computed to observed results using the best-fit resistivity cross-sections of Figures 7 and 8. Curves of dots and shorter dashes relate to correspondingly drawn alternate resistivity profiles in Figure 8 used in parameter resolution tests. Curves of long dashes appearing between the two modes of data points represent the 1D forward computation using the best-fit regional profile in Figure 8.

equally well the inadequacy of the alternate regional profiles. I presume a maximum depth of sensitivity of the data to subsurface structure of about 120 km, as imposition of low-resistivity material here causes computed phases at 0.002 Hz which exeed those of the best-fit model by about four degrees and which cannot be made to agree with the observations by increasing the resistivity of the 200 Ω-m material.

Before closing this particular study, there is one more demonstration of the hazards of 1D interpretation of MT data in regions of extension like the northern Basin and Range. In Figure 10 are reproduced the best-fit deep profile of Figure 8 as well as model profiles computed through direct 1D inversion of the tensor MT observations of Figure 9 (I used the routine of Petrick et al., 1977). The display of these observations with the computed 1D responses of the aforesaid model profiles in Figure 11 shows the goodness of fit possible by a 1D inversion algorithm to data taken in 3D environments. The deep resistivity models derived thereby are unacceptable, however; their

departure from the preferred deep profile exceeds bounds that were already rejected in the rigorous parameter resolution studies of Figures 8 and 9. Moreover, due to local lateral inhomogeneity, computed layered models will vary drastically from site to site within the mountains, underscoring the unreliability of one-dimensional inversion in this environment.

Matters are even worse for 1D inversion of MT soundings taken within northern Basin and Range graben fill. Figure 12 shows layered earths computed through 1D inversion of apparent resistivities ρ_{xy} and ρ_{yx} and impedance phases ϕ_{xy} and ϕ_{yx}, identified as TE and TM (Word et al., 1971; Vozoff, 1972), for site 78-6 of Figure 4. Solid curves through the data in Figure 13 show

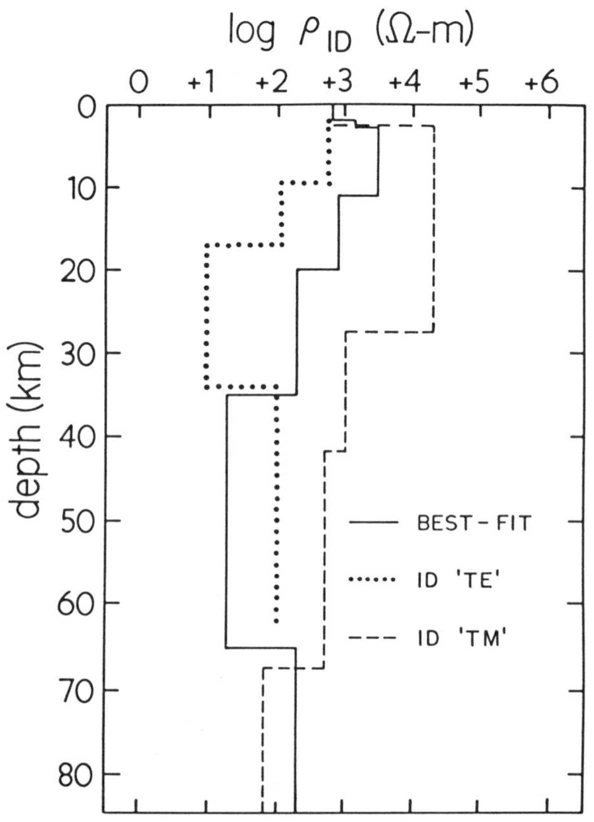

Figure 10. Best-fit regional resistivity profile for the Roosevelt Hot Springs thermal area compared to layered models obtained through 1D inversion of ρ_{xy} and ϕ_{xy} (dotted lines) and ρ_{yx} and ϕ_{yx} (dashed lines).

the goodness-of-fit obtained by the inversion, with the nominally very conductive basal half-spaces resolved by the observations below 0.2 Hz. However, given the best-fit model of Figure 8 derived through a rigorous accounting for the effects of the valley sediments, the layered models at site 78-6 are obviously unrealistic. Note that the data of Figure 13 bear a close resemblance to the theoretical computations of Wannamaker et al. (1983) over their model valley. I believe that 1D models of resistivity given by Stanley et al. (1976) for the Stillwater-Soda Lakes district, which were computed from soundings taken within deep conductive alluvium of the Carson Sink area, are similarly affected. The case for low resistivities at depths as little as 5 km in that region remains to be substantiated.

Physical State at Depth in S.W. Utah. -   In Figure 14, the best-fit regional profile for the Roosevelt Hot Springs is compared to laboratory experiments on the electrical properties of rocks. A crustal thickness slightly less than 30 km is assumed for this area (Keller et al., 1979; Allmendinger et al., 1983). Petrological studies and seismic velocities point to a mafic

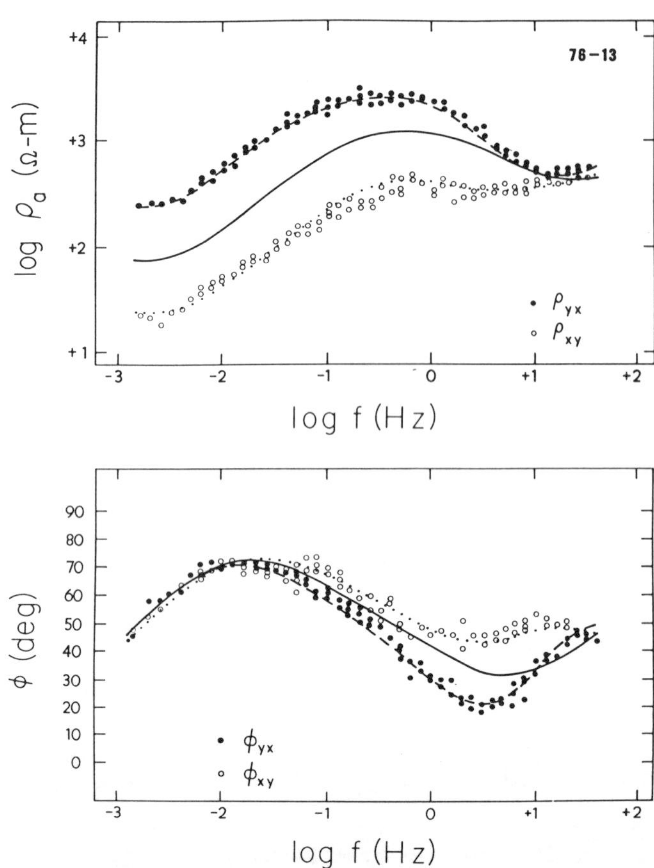

Figure 11. Observed data points of ρ_{xy}, ρ_{yx}, ϕ_{xy} and ϕ_{yx} for sounding 76-13 showing the goodness of fit that can be obtained to 3D data using a 1D inverse routine. However, the resultant layered models were dismissed as unrealistic. Solid curve appearing between data points is the 1D forward response of the best-fit regional profile.

metaigneous composition near gabbro for most of the lower half of the crustal section, with relatively felsic rocks above (Smith, 1978; Padovani et al., 1982; Nash, 1983). A peridotite composition is taken for the upper mantle rocks (Wyllie, 1979).

Adopting a conductive geotherm for a regional heat flow of 2.4 HFU (1 HFU = 41.6 mWm^{-2}) (Lachenbruch and Sass, 1978; Chapman et al., 1981), the lower bound on aqueous electrolytic conduction through rock pores of Brace (1971) and the mean dry gabbro semiconductivity results of Kariya and Shankland (1983) are combined to yield the curve of dots in Figure 14. Over the depth interval 15 to 30 km, the regional profile appears somewhat less resistive than what is predicted by dry solid-state measurements. If a gabbroic

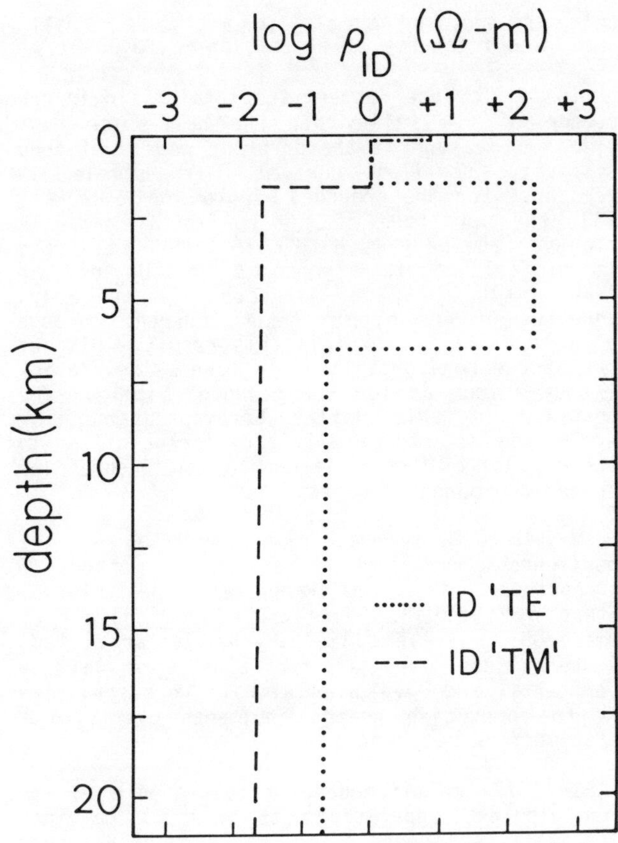

Figure 12. Layered resistivity models obtained through 1D inversion of ρ_{xy} and ϕ_{xy} (dotted lines) and ρ_{yx} and ϕ_{yx} (dashed lines) for sounding 78-6 of Figure 4. Note scale change from Figure 10.

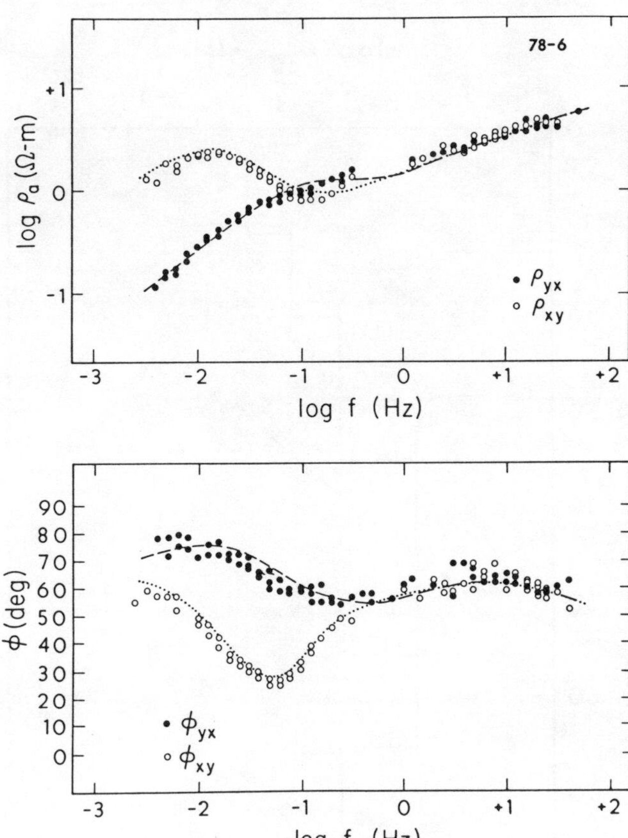

Figure 13. Observed data points of ρ_{xy}, ρ_{yx}, ϕ_{xy} and ϕ_{yx} for sounding 78-6. Curves of dots and dashes show responses of layered earth models of Figure 12.

composition is relevant then free water in this interval is improbable (Burnham, 1979a, 1979b). However, a reduction of solid-state resistivity through active deformation and non-zero partial pressure of water appears possible (Winkler, 1979; Duba and Heard, 1980; Kirby, 1983; Spear and Silverstone, 1983).

The curve of dashes in Figure 14 pertains to upper mantle rocks and results from the solid-state measurements on olivine by Duba et al. (1974) and the conductive geotherm. Of greatest importance, the shape of the solid-state curve is very different from the regional profile. In an earth where temperature increases steadily with depth, such as is the case with our conductive geotherm or even with any of the temperature profiles of Lachenbruch and Sass (1978) which incorporate convective components, the Arrhenius solid-state temperature dependence for a rock results in bulk resistivity which decreases monotonically with depth. During the discussion of parameter resolution, a deep resistivity profile that decreased steadily with depth was rejected as a candidate to explain the low-frequency MT observations in S.W. Utah; the order-of-magnitude rise of resistivity in the upper mantle in our

best-fit deep model is a strict requirement.

I therefore look to partial melting in the upper mantle as the cause of the 20 Ω-m deep layer. Again imposing the conductive geotherm, the curve of dots and dashes in Figure 14 arises using a dry peridotite melting curve (Mysen and Kushiro, 1977; Wyllie, 1979; Best et al., 1980), the measurements of alkali-olivine basalt conductivity of Rai and Manghnani (1978) and the melt phase interconnection model of Waff and Bulau (1979). The large liquid fractions implied by this melting curve are of course unrealistic and result from use of a conductive temperature profile. Given that melt conductivities are strongly temperature dependent, and that the interconnection model is strictly an abstraction, I simply conclude that the melt fraction in the deep, low-resistivity layer is rather small, less than 5%. In reality, the dry peridotite solidus is advocated as a representative pressure-temperature trajectory for the depth interval of the 20 Ω-m layer (Mysen and Kushiro, 1977; Wyllie, 1979; Best et al., 1980). It is argued shortly that this trajectory may be valid to much greater depths in the region.

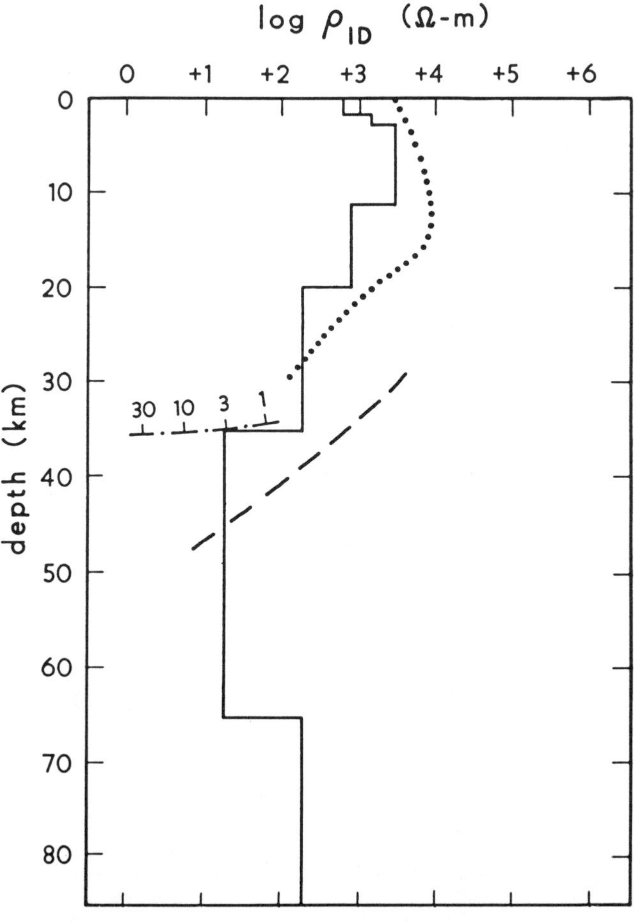

Figure 14. Best-fit, crust and upper mantle resistivity profile for southwestern Utah compared to physical model incorporating aqueous electrolytic conduction in rock pores, solid-state semiconduction in minerals and ionic conduction in partial melts. Definition of curves explained in text.

Controlled-Source EM Techniques

In controlled-source electromagnetics (CSEM), EM fields are impressed by a transmitter, either fixed-site or roving, and monitored as a function of position and/or frequency at a receiver (Grant and West, 1965; Ward, 1967). The current transmitter can be a grounded bipole, an ungrounded loop, or a very long, essentially ungrounded line source.

Effects of Upper Crustal Structure. - Ward (1983) has investigated theoretically the behavior of EM fields about typical northern Basin and Range alluvial fill for grounded bipoles and an ungrounded loop. The basin model he has chosen has a resistivity and dimensions similar to that of Wannamaker et al. (1983), except that Ward's

model was enclosed simply in a 400 Ω-m half-space. The frequency of excitation was 0.03 Hz.

The difference between the total field in the presence of the valley and the half-space host field is a measure of the error in models of deep resistivity profiles derived from simple 1D inversion. For the grounded bipole source of Ward (1983) the amplitude of the horizontal magnetic field over the body rose to more than 3 times the primary field. It asymptotes to 60% of the primary field at large distances from the body. Channeling of the bipole return current through the valley fill can explain this result. Distortions of the horizontal H-field were as severe for the square loop as for the grounded bipole. The importance of this latter behavior cannot be overemphasized; total and primary fields differ to extremely large distances and to low frequencies even for ungrounded sources.

Models of Deep Resistivity from CSEM. - Experiments using galvanic resistivity methods in the northern Basin and Range and the Colorado Plateau by the U.S. Geological Survey have been summarized by Keller (1971). As Keller stated, the techniques used at the time were largely experimental and developmental, and were sensitive only to conductive graben sediments overlying a resistive rock basement.

In 1977, multifrequency dipole-dipole (3 and 6 km dipoles, separations to n = 6) galvanic resistivity measurements were made in southwestern Utah by the Department of Geology and Geophysics, University of Utah, in an attempt to describe any resistivity structure associated with the Pioche-Marysvale trend (Rowley et al., 1979; Petrick et al., 1981). The data, from a profile extending 33 km northward from the Roosevelt Hot Springs, were dominated to an unexpectedly severe degree by upper crustal, 3D lateral inhomogeneities, especially graben sedimentary fill. These inhomogeneities lead to a lack of fit of layered inverse calculations to observed data, as well as dubious 1D and 2D earth models (W.R. Petrick and A.C. Tripp, unpub.). A 3D inversion of these results (Petrick et al., 1981) presents a concentration of low-resistivity structure within 2 km of the surface, evidence again of variable basin fill dominating the observations.

Towle (1980), utilizing an interstate power transmission line as a long grounded bipole, has studied deep resistivity in eastern California with direct current magnetic field variations. A waiting period of 10 minutes was necessary to allow transients in the current to decay to negligible levels. Towle's data is consistent with a shallow return current concentration lying east of the Sierran Front in fractured rocks and sediments in the uppermost crust of the Basin and Range.

Assuming the same interstate power system as Towle to be an essentially ungrounded AC line source of current, Lienert and Bennett (1977) and Lienert (1979) advanced 1D models of resistivity in eastern California and western Nevada to depths of 40 to 50 km. Included in their interpretation was 10-15 km thick layer of low resistivity (10-30 Ω-m) whose depth to top varied from 20 to 30 km (Figure 15). However, in light of Towle's experience, it appears that grounding terms were very important over the frequency range employed in the survey areas of Lienert and Bennett (1977) and Lienert (1979) (6-42 cycles/hr). Moreover, these authors did not evaluate quantitatively the effects of the 3D sedimentary basins common in their areas. It was shown by Ward (1983) that these structures can cause widespread distortions of EM fields to low frequencies even for ungrounded sources. Given these points, I do not believe their results demonstrated with certainty that a resistivity minimum exists in the lower crust.

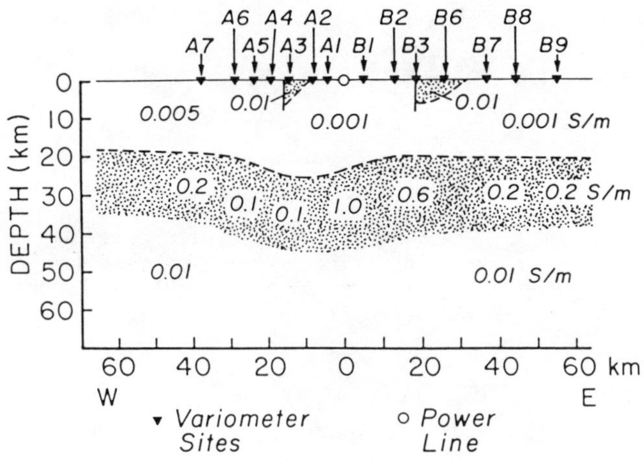

Figure 15. Cross-section of model conductivity structure beneath the Walker Lake, Nevada, district. Profile location also shown on Figure 1. Modified from Lienert and Bennett (1977).

Wilt et al. (1982) have described low-frequency CSEM soundings in the Buena Vista valley of N.W. Nevada. Through 1D inversion of the data, these authors advocate low resistivity (2-7 Ω-m) at depths of 4-7 km in basement rocks beneath the valley. A 1D inversion of a nearby MT sounding in the valley also showed relatively low resistivities (∿ 25 Ω-m) at rather shallow depths (∿ 11 km). However, Wilt et al. admit that preliminary 2D modeling suggests that 1D inversion of their CSEM data in the presence of the valley underestimated depth to conductive basement. Given the 3D model simulations of valley fill by Ward (1983) and Wannamaker et al. (1983) and the experience at the Roosevelt Hot Springs discussed previously, I take the approximate agreement between the 1D MT and 1D CSEM models at Buena Vista valley as further evidence that graben sediments seriously disrupt the CSEM observations. Again, I believe the case for widespread low resistivities at shallow depths in basement rocks of the Great Basin remains quite uncertain.

TECTONIC SIGNIFICANCE OF NORTHERN BASIN AND RANGE RESISTIVITY

The interpretation of deep resistivity at the Roosevelt Hot Springs has provided a point sampling of physiochemical conditions at depth in S.W. Utah. However, comprehension of these conditions will require consideration of the tectonic evolution of the northeastern Basin and Range during Late Cenozoic time. The model of resistivity and its implications finally are extended to the northern Basin and Range as a whole.

Evolution of Resistivity in the Northeastern Basin and Range

It is noted at the outset that the depth interval of interconnected basaltic melt corresponding to the 20 Ω-m layer in Figure 14 is significantly shallower than the seismic low-velocity zones of either the northern Basin and Range interior or the Colorado Plateau (Priestley and Brune, 1978; Thompson and Zoback, 1979), which are also interpreted to reflect partial melting in the upper mantle (Wyllie, 1979). These apparently disparate geological environments can be reconciled, with help from geomagnetic deep sounding anomalies and concepts of regional tectonic setting.

Diapirism and Melting. - Of paramount importance in fitting the deep physical state inferred in S.W. Utah to those of neighbouring regions is the well-documented, progressive concentration of extensional tectonism throughout the northeastern Basin and Range during the last 8-9 m.y. (e.g. Christiansen and McKee, 1978; Smith, 1978; Stewart, 1978; Rowley et al., 1979; Thompson and Zoback, 1979). This concentration has resulted in crustal thinning, low upper mantle velocities, bimodal volcanism and heat flow which are anomalous compared to the northern Basin and Range interior of central Nevada (Figures 16-19).

While much lithospheric extension throughout the northern Basin and Range prior to Mid-Miocene time may have been due to intrusion of calc-alkaline magmas derived ultimately from melting in the upper mantle associated with Farallon plate subduction (Zoback et al., 1981), such intrusion alone probably does not thin the crust (Gastil, 1979; Eaton, 1982). The Late Miocene to present extension defining today's horst graben morphology, however, is broadly distributed and indicates stretching and upwelling of both the crust and the upper mantle. Crustal thicknesses determined seismically hence can be used as a guide to average extension rates since the Late Miocene (see Figure 16).

If a value of 44 km for the thickness of the Colorado Plateau crust (Keller et al., 1979) can be used as a guide, then crustal thicknesses imply an average rate of extension of nearly 2% per m.y. for the northern Basin and Range interior of east-central Nevada in the last 10-12 m.y. However,

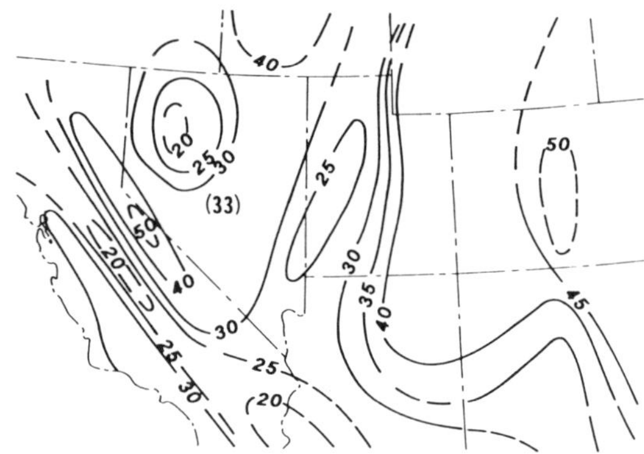

Figure 16.   Crustal thicknesses in the south western United States.   Contour values in kilometers.   From Smith (1978), Keller et al. (1979) and Priestly et al. (1980).

about 5% per m.y. is implied for much of the northeastern Basin and Range during the last 8-9 m.y.   If so, then a 1D estimate of upward velocity v of material at depth (related linearly to extension rate s and depth z through $v = sz$; Lachenbruch and Sass, 1978) ranges from about 1 $\frac{1}{2}$ km/m.y. at 35 km through 5 km/m.y. at 100 km to over 7 km/m.y. at 160 km.

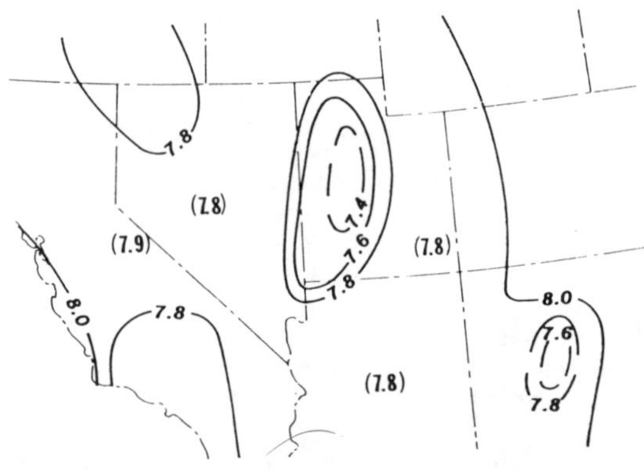

Figure 17.   Uppermost mantle compressional (P_n) velocities.   Contour values in km/s. From Smith (1978) and Keller et al. (1979).

While contributing a great deal to surface heat flow and temperatures within the lithosphere (Lachenbruch and Sass, 1978), extension rates tell something about the thermal state at greater depths as well.   Consider for simplicity a spherical diapir in the upper mantle of radius 50 km, a representative E-W length scale for the region of enhanced tectonism defining the northeastern Basin and Range.   For the thermal state of a rising diapir to be largely adiabatic (Oxburgh, 1980), its ascent time must be less than its thermal conduction time constant (Spera, 1980).   If the hypothetical spherical diapir has been rising with a velocity $v = 5$ km/m.y. from a

depth of 100 km or so during the last 8 m.y., then a distance of roughly 40 km would have been covered.   In covering this distance, the diapir ascent rate merely must exceed about 0.40 km/m.y. (Spera, 1980), much less than v, for the aforesaid thermal inequality to be met.   One may conclude that large volumes rising here in the upper mantle tend to retain much of their sensible heat during transport.

Figure 18.   Distribution of volcanics less than 5 m.y. old.   From Christiansen and McKee (1978), Ward et al. (1978) and Luedke and Smith (1978a, 1978b, 1981).

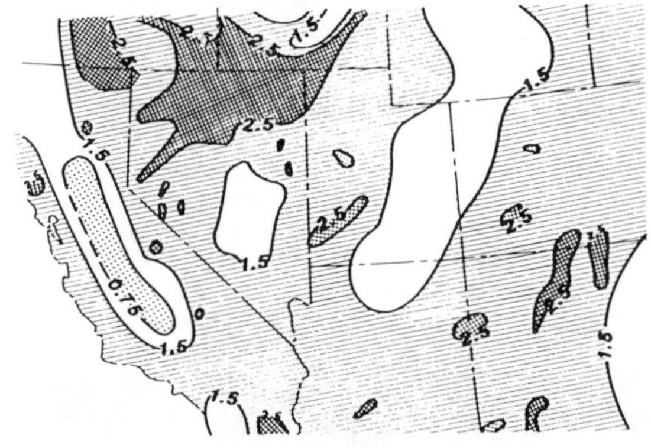

Figure 19.   Surface heat flow contours in heat flow units.   From Sass et al. (1981) and Bodell and Chapman (1982).

As a mass at depth rises adiabatically, its temperature trajectory may intersect the peridotite solidus so that melting will proceed (Yoder, 1976).   Upon crossing the solidus, the temperature within the ascending diapir is buffered by the melting behavior of peridotite (Oxburgh, 1980).   In particular, the great size of the latent heat of fusion relative to the heat capacity of peridotite along with the nearly invariant nature of basalt production serve to keep the temperature gradient within the diapir well within 1 °C/km of the solidus during adiabatic ascent (Yoder, 1976; Mysen and Kushiro,

1977; Oxburgh, 1980; Turcotte, 1982). The degree of melting is inevitably subdued through heat loss to the earth's surface from the top of the diapir (Spera, 1980).

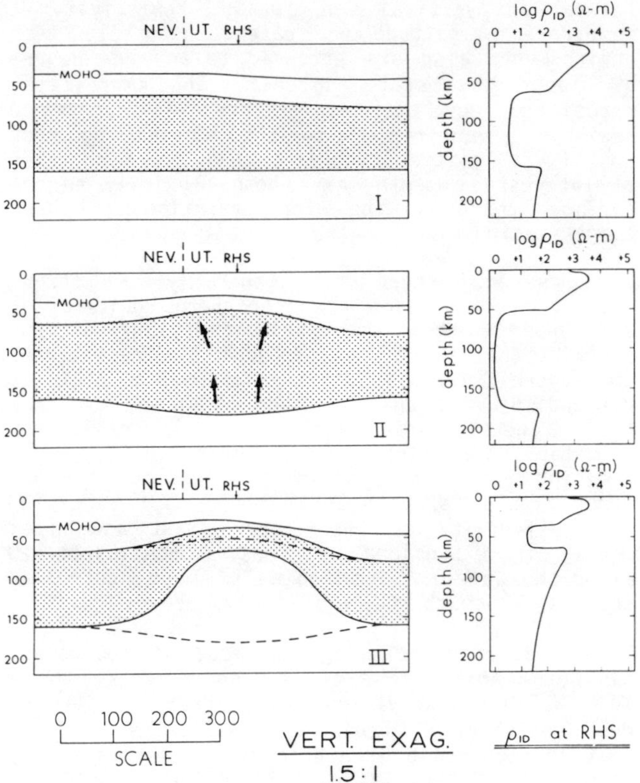

Figure 20. Schematic, highly simplified, three-stage model of evolution of resistivity structure in the eastern Great Basin. The region of continuous interconnection of melt in the upper mantle is defined by stippling and the predicted deep resistivity profile beneath the eastern Great Basin at each stage is drawn at the right side of diagram.

With this brief discussion of diapirism, I advance a schematic and highly simplified model for the Late Cenozoic evolution of deep resistivity in the northeastern Basin and Range. The model is depicted in Figure 20 in three stages: I, by about 10 m.y. ago, extensional activity responsible for the present horst-graben morphology had spread throughout the northern Basin and Range. Attendant upwelling of mantle peridotite had resulted in widely distributed, fundamentally basaltic volcanism, presumably involving the mechanism of decompression just described, along with a layer of partial melt of low melt fraction (∿ 5%) which did not differ greatly from that defining the present seismic low-velocity zone of central Nevada; II, by 8 m.y. ago, the well-documented concentration of extension in the northeastern Basin and Range was

underway. Accelerated diapiric uprise of mantle material leads to a further amplification of the thickness and degree of melting of the LVZ. The extent of the amplification is difficult to gauge, as it depends on the degree to which the geotherm just above and just below the former LVZ was superadiabatic (Oxburgh, 1980); III, the melt fraction in the magnified zone of fusion can build up only to a point, after which much of it drains buoyantly upward. This convection of basalts, which in effect are superheated with respect to the environment into which they rise, advances the front of melting to 35 km depth and helps establish the realm of interconnected melt corresponding to the 20 Ω-m layer of Figure 14 by warming the host peridotite during passage.

The model of present distribution of partial melting in the upper mantle of the northern Basin and Range and western Colorado Plateau in Figure 20 is similar to that of Thompson and Zoback (1979). The latter authors utilize the resistivity model of Porath (1971), however, to suggest that interconnected melt exists to a depth of about 160 km beneath the northeastern Basin and Range. Nevertheless, the magnetotelluric data preclude such an interconnection below about 65 km and the resolution by geomagnetic deep sounding of such details is doubtful (Gough, 1983). Given the chemistry of certain basaltic lavas in the northeastern Basin and Range (Wyllie, 1979; Best et al., 1980) and our model of diapiric upwelling, some melting events are probable at depths well below 65 km - they just don't result in a broad zone of interconnected liquid. On the other hand, the uniqueness of the model of Thompson and Zoback beneath the northeastern Basin and Range must be questioned.

I suggest that the zone of interconnected melt below S.W. Utah lies where it is because the very much reduced rigidity of the crystalline peridotite matrix at greater depths (Darot and Gueguen, 1981) promotes both the formation of vertical fissures under tensile stress (Spera, 1980) and the ductile collapse of the crystalline matrix as the melt leaves to enter the fissures (Stolper et al., 1981; Turcotte, 1982). In addition, any melt percolating up from great depth along intergranular passageways (Turcotte, 1982) would be impeded from rising beyond the zone of our low resistivity layer by the increasingly stiff rheology of peridotite.

Thermal State at Depth in the Eastern Great Basin. - Previously, I assigned a temperature trajectory through the deep low-resistivity layer that was coincident with the dry peridotite solidus (nominally that of Wyllie, 1979). Based upon compositions of basalts in S.W. Utah, such a thermal profile seems appropriate to depths near 100 km (op. cit.). In fact, if the abstraction of adiabatic diapiric uprise in the northeastern Basin and Range is favored, a dry solidus trajectory may be definitive to a depth exceeding 160 km.

In Figure 14, a conductive geotherm based

upon a regional heat flow of 2.4 HFU for S.W. Utah predicted fusion of dry peridotite at essentially the same depth as the top of the model low-resistivity layer. However, with the high extension rates proposed for the northeastern Basin and Range, the foregoing coincidence should not be (Lachenbruch and Sass, 1978). As a simple example, purely solid-state stretching at a rate of 5% per m.y. is consistent with a depth of melting of nearly 40 km, but a surface heat flux reaching 3 HFU is to be expected (ibid.).

One would conclude on this basis that the crust and upper mantle of northeastern Basin and Range has not reached a steady thermal state. The onset of extension is fundamentally an increase of heat flux into the lithosphere, and adjustment of surface heat flow to this form of thermal disturbance may take longer than the 8 m.y. inferred as the interval of concentrated extension in the northeastern Basin and Range (Lachenbruch and Sass, 1978; Rowley et al., 1979). By coincidence, a conductive geotherm based on present surface heat flow and a non-equilibrium convective geotherm considering the concentration of extension in the northeastern Basin and Range do not differ greatly.

Application of Interpretation to Other Areas

Attenuated crustal thicknesses in northwestern Nevada (Figure 16a), the pronounced extension near Yerrington interpreted by Proffett (1977) and the concentration of rifting, heat flow and volcanism in western Nevada and southeastern California (Figure 16b-d) indicate that the western margin of the northern Basin and Range experiences an enhanced degree of extension relative to the northern Basin and Range interior. The similarity in tectonic styles between the eastern and western margins of the northern Basin and Range suggests that a deep electrical structure corresponding to Stage III of Figure 17 is likeliest for the northwestern Basin and Range. The deep resistivity investigation by Schmucker (1970) permits this conclusion, but does not uniquely favor it given the limited resolution.

Another tectonic environment which probably fits into the framework of Figure 17 is the Rio Grande Rift and nearby regions. Of particular interest is the continuation of this rift zone northward into Colorado. As described by Williams (1982), extensional deformation and volcanism diminishes in this direction although some Late Tertiary mafic volcanism persists into Wyoming. In light of the modest degree of extension, I propose that the continuation of the Rio Grande Rift into Colorado may be a good place to search for deep electrical structure representing Stage II in Figure 17. The northward increase in geomagnetic deep sounding anomalies in this area (Figure 3) is furthermore suggestive of this notion, but magnetotelluric profiling with a proper multidimensional interpretation is needed to assess accurately the deep resistivity section here.

CONCLUSIONS

In active extensional environments, significant temperature perturbations may exist, possibly with interconnected melt phases, creating lateral and vertical contrasts in resistivity of an order-of-magnitude or greater. Unluckily, such environments also are attended to a large degree by upper crustal lateral inhomogeneities, especially graben sedimentary fill, whose resistivity contrast with the surrounding rock host is at least as high as that of the structures of interest at depth, and whose proximity to the surface obscures the interpretation of deep targets using electromagnetic measurements.

However, there is generally a strong preferred orientation of lithospheric deformation and upper mantle processes, and thus of the distribution of lateral resistivity inhomogeneities, in rift environments that is perpendicular to the direction of spreading. For the magnetotelluric method, a 2D transverse magnetic mode algorithm is most suitable for removing the distortion of MT soundings by such upper crustal structures and hence for assessing the resistivity of the deeper crust and upper mantle in extensional regimes. Equivalent 2D algorithms are not available for most controlled sources, although they are sorely needed.

An appropriately rigorous modeling technology has been applied to a collection of MT soundings in S.W. Utah and yielded an accurate profile of deep resistivity to depths exceeding 100 km. A feature required by the data is a low-resistivity (nominally 20 Ω-m) layer residing from about 35 to 65 km in the upper mantle beneath the Roosevelt Hot Springs. No low-resistivity layer in the middle to lower crust has been detected, however, nor do I think the case has yet been made on the basis of EM observations for such a layer to be widespread in the northern Basin and Range (cf. Eaton, 1982; Jiracek et al., 1983).

The present-day deep resistivity structure of southwestern Utah derives from the evolution of the northeastern Basin and Range through Late Cenozoic time. The low-resistivity layer interpreted in the upper mantle here is a manifestation of diapiric uprise and melting events occurring over an interval that extends to much greater depths than does the layer itself, perhaps in excess of 160 km. Both the resistivity structure and the timing and amplitude of extensional activity suggest that temperatures in the deep crust and uppermost mantle here exceed those which would be inferred from surface heat flow. Similar processes appear to be occurring in the northwestern Basin and Range, leaving central Nevada as a fairly quiescent region whose low-resistivity layering in the upper mantle, ironically, may be relatively pronounced. Unfortunately, deep resistivity surveys are lacking in the northern Basin and Range interior.

ACKNOWLEDGEMENTS

Thoughtful reviews and criticisms of this work by Drs. J.R. Bowman, D.S. Chapman, G.W. Hohmann, W.P. Nash, D.L. Nielson, W.T. Parry, W.R. Sill, C. M. Swift, Jr., and S.H. Ward are gratefully recognized. Technical assistance came from Sandra Bromley, Doris Cullen and Joan Pingree.

This work was supported by NSF contract EAR 8116602 and DOE/DGE contract DE-AC07-80ID12079.

REFERENCES

Allmendinger, R. W., Sharp, J. W., Von Tish, D., Serpa, L., Brown, L., Kaufman, S., Oliver, J., and Smith, R. B., 1983, Cenozoic and Mesozoic structure of the eastern Basin and Range province, Utah, from COCORP seismic-reflection data: Geology, 11, p. 532-536.

Best, M. G., McKee, E. H., and Damon, P. E., 1980, Space-time-composition patterns of Late Cenozoic mafic volcanism, southwestern Utah and adjoining areas: Am. J. Sci., 280, p. 1035-1050.

Bodell, J. M., and Chapman, D. S., 1982, Heat flow in the north-central Colorado Plateau: J. Geop. Res., 87(B4), p. 2869-2884.

Brace, W. F., 1971, Resistivity of saturated crustal rocks to 40 km based on laboratory measurements, In The Structure and Physical Properties of the Earth's Crust, ed. by J. G. Heacock, AGU Mono. 14, p. 243-256.

Burnham, C. W., 1979a, The importance of volatile constituents, In The Evolution of the Igneous Rocks: Fiftieth Anniversary Perspectives, ed. by H. S. Yoder, Jr., Princeton Univ. Press, p. 439-482.

________, 1979b, Magmas and hydrothermal fluids, In Geochemistry of Hydrothermal Ore Deposits, ed. by H. L. Barnes, John Wiley and Sons, New York, p. 71-136.

Carter, J. A., and Cook, K. L., 1978, Regional gravity and aeromagnetic surveys of the Mineral Mountains and vicinity, Millard and Beaver Counties, Utah: ESL Report 77-11, 178 p.

Chapman, D. S., Clement, M. D., and Mase, C. W., 1981, Thermal regime of the Escalante Desert, Utah, with an analysis of the Newcastle geothermal system: J. Geop. Res. 86(B12), p. 11735-11746.

Christiansen, R. L., and McKee, E. H., 1978, Late Cenozoic volcanic and tectonic evolution of the Great Basin and Columbia intermontane region, In Cenozoic Tectonics and Regional Geophysics of the Western Cordillera, ed. by R. B. Smith and G. P. Eaton, GSA Mem. 152, p. 283-311.

Darot, M., and Gueguen, Y., 1981, High-temperature creep of forsterite single crystals: J. Geop. Res., 86(B7), p. 6219-6234.

Duba, A. G., Heard, H. C., and Schock, R. N., 1974, Electrical conductivity of olivine at high pressure and under controlled oxygen fugacity: J. Geop. Res., 79(11), p. 1667-1673.

Duba, A. G., and Heard, H. C., 1980, Effect of hydration on the electrical conductivity of olivine: EOS Transactions, 61(17), p. 404.

Eaton, G. P., 1982, The Basin and Range province: origin and tectonic significance: Ann. Rev. Earth Plan. Sci., 10, p. 409-440.

Gamble, T. D., Goubau, W. M., and Clarke, J., 1979, Magnetotellurics with a remote reference: Geophysics, 44(1), p. 53-68.

Gastil, R. G., 1979, A conceptual hypothesis for the relation of differing tectonic terranes to plutonic emplacement: Geology, 7, p. 542-544.

Gertson, R. C., and Smith, R. B., 1979, Interpretation of a seismic refraction profile across the Roosevelt Hot Springs, Utah and vicinity: ESL Report DOE/ID/78-1701.a.3, 116 p.

Gough, D. I., 1983, Electromagnetic geophysics and global tectonics: J. Geop. Res., 88(B4), p. 3367-3378.

Grant, F. S., and West, G. F., 1965, Interpretation Theory in Applied Geophysics: McGraw-Hill Book Company, Toronto, 584 p.

Hintze, L. F., 1973, Geologic history of Utah: Brigham Young Univ. Geology Studies, 20, pt. 3, 181 p.

________, 1980, Geologic map of Utah: Utah Geological and Mineralogical Survey, Salt Lake City.

Jiracek, G. R., Mitchell, P. S., and Gustafson, E. P., 1983, Magnetotelluric results opposing magma origin of crustal conductors in the Rio Grande rift: Tectonophysics, 94, p. 299-326.

Jupp, D. L., and Vozoff, K., 1976, Discussion on "The magnetotelluric method in the exploration of sedimentary basins" by K. Vozoff: Geophysics, 41(2), p. 325-328.

Kariya, K. A., and Shankland, T. J., 1983, Interpretation of electrical conductivity of the lower crust: Geophysics, 48(1), p. 52-61.

Keller, G. V., 1971, Electrical studies of the crust and upper mantle, In the Structure and

Physical Properties of the Earth's Crust, ed. by J. G. Heacock, AGU Mono. 14, p. 107-125.

Keller, G. R., Braile, L. W., and Morgan, P., 1979, Crustal structure, geophysical models and contemporary tectonism of the Colorado Plateau: Tectonophysics, 61, p. 131-147.

Kirby, S. H., 1983, Rheology of the lithosphere: Reviews of Geophysics and Space Physics, 21(6), p. 1458-1487.

Lachenbruch, A. H., and Sass, J. H., 1978, Models of an extending lithosphere and heat flow in the Basin and Range province, In Cenozoic Tectonics and Regional Geophysics of the Western Cordillera, ed. by R. B. Smith and G. P. Eaton, GSA Mem. 152, p. 209-250.

Larsen, J. C., 1975, Low frequency (0.1-6.0 cpd) electromagnetic study of the deep mantle electrical conductivity beneath the Hawaiian Islands: Geop. J. Roy. Ast. Soc., 43, p. 17-46.

________, 1981, A new technique for layered earth magnetotelluric inversion: Geophysics, 46(9), p. 1247-1257.

Lienert, B. R., 1979, Crustal electrical conductivities along the eastern flank of the Sierra Nevadas: Geophysics, 44(11), p. 1830-1845.

Lienert, B. R., and Bennett, D. J., 1977, High electrical conductivities in the lower crust of the northwestern Basin and Range: an application of inverse theory to a controlled-source deep-magnetic-sounding experiment, In The Earth's Crust, ed. by J. G. Heacock, AGU Mono. 20, p. 531-552.

Luedke, R. G., and Smith, R. L., 1978a, Map showing distribution, composition, and age of Late Cenozoic volcanic centers in Arizona and New Mexico, U.S.G.S. Miscellaneous Investigations Series, Map 1-1091-A.

__________, 1978b, Map showing distribution, composition, and age of Late Cenozoic volcanic centers in Colorado, Utah, and southwestern Wyoming, U.S.G.S. Miscellaneous Investigations Series, Map 1-1091-B.

__________, 1981, Map showing distribution, composition, and age of Late Cenozoic volcanic centers in California and Nevada, U.S.G.S. Miscellaneous Investigation Series, Map 1-1091-C.

Madden, T. R., 1971, The resolving power of geoelectric measurements for delineating resistive zones within the crust, In The Structure and Physical Properties of the Earth's Crust, ed. by J. G. Heacock, AGU Mono. 14, p. 95-105.

Mysen, B. O., and Kushiro, I., 1977, Compositional variations of coexisting phases with degree of melting of peridotite in the upper mantle: Am. Min., 62, p. 843-865.

Nash, W. P., 1983, Silicic volcanism along the eastern margin of the Basin and Range province: Utah and Idaho: GSA Abstracts with Programs, p. 402.

Oxburgh, E. R., 1980, Heat flow and magma genesis, In Physics of Magmatic Processes, ed. by R. B. Hargraves, Princeton Univ. Press, p. 161-200.

Padovani, E. R., Hall, J., and Simmons, G., 1982, Constraints on crustal hydration below the Colorado Plateau from V_p measurements on crustal xenoliths: Tectonophysics, 84, p. 313-328.

Parker, R. L., and Whaler, K. A., 1981, Numerical methods for establishing solutions to the inverse problem of electromagnetic induction: J. Geop. Res., 86(B10), p. 9574-9584.

Petrick, W. R., Pelton, W. H., and Ward, S. H., 1977, Ridge regression inversion applied to crustal resistivity sounding data from South Africa: Geophysics, 42(5), p. 995-1005.

Petrick, W. R., Sill, W. R., and Ward, S. H., 1981, Three-dimensional resistivity inversion using alpha centers: Geophysics, 46(8), p. 1148-1162.

Porath, H., 1971, Magnetic variation anomalies and seismic low-velocity zone in the western United States: J. Geop. Res., 76(11), p. 2643-2648.

Porath, H., and Gough, D. I., 1971, Mantle conductive structures in the western United States from magnetometer array studies: Geop. J. Roy. Ast. Soc., 22, p. 261-275.

Porath, H., Oldenburg, D. W., and Gough, D. I., 1970, Separation of magnetic variation fields and conductive structures in the western United States: Geop. J. Roy. Ast. Soc., 19, p. 237-260.

Priestley, K., and Brune, J. N., 1978, Surface waves and the structure of the Great Basin of Nevada and western Utah: J. Geop. Res., 83 (B5), p. 2265-2272.

Priestley, K., Orcutt, J. A., and Brune, J. N., 1980, Higher-mode surface waves and structure of the Great Basin of Nevada and western Utah: J. Geop. Res., 85(B12), p. 7166-7174.

Proffett, J. M., Jr., 1977, Cenozoic geology of the Yerrington district, Nevada, and implications for the nature and origin of Basin and Range faulting: GSA Bull., 88, p. 247-266.

Rai, C. S., and Manghnani, M. H., 1978, Electrical conductivity of basalts to 1550°C, In Proceedings of Chapman Conference on Partial Melting in the Earth's Upper Mantle, ed. by H. J. B. Dick, Oreg. Dept. Geol. Min. Ind., Bull. 96, p. 219-232.

Reitzel, J. S., Gough, D. I., Porath, H., and Anderson, C. W., III, 1970, Geomagnetic deep soundings and upper mantle structure in the western United States: Geop. J. Roy. Ast. Soc., 19, p. 213-235.

Rijo, L., 1977, Modeling of electric and electromagnetic data: Ph.D. Thesis, Dept. of Geology and Geophysics, Univ. of Utah.

Ross, H. P., Nielson, D. L., and Moore, J. N., 1982, Roosevelt Hot Springs geothermal system, Utah - case study: Am. Assoc. Petr. Geol. Bull., 66(7), p. 879-902.

Rowley, P. D., Steven, T. A., Anderson, J. J., and Cunningham, C. G., 1979, Cenozoic stratigraphic and structural framework of southwestern Utah: U.S.G.S Professional Paper 1149, 22 p.

Sass, J. H., Blackwell, D. D., Chapman, D. S., Costain, J. K., Decker, E. R., Lawver, L. A., and Swanberg, C. A., 1981, Heat flow from the crust of the United States, in Physical Properties of Rocks and Minerals, ed. by Y. S. Touloukian and C. Y. Ho, McGraw-Hill/CINDAS Data Series on Material Properties, II-2, New York, p. 503-548.

Schmucker, U., 1970, Anomalies of geomagnetic variations in the southwestern United States, Bull. Scripps Inst. Oceanography, 13, 165 p.

Smith, R. B., 1978, Seismicity, crustal structure and intraplate tectonics of the interior of the western Cordillera, In Cenozoic Tectonics and Regional Geophysics of the Western Cordillera, ed. by R. B. Smith and G. P. Eaton, GSA Mem. 152, p. 111-144.

Spear, F. S., and Silverstone, J., 1983, Water exsolution from quartz: implications for the generation of retrograde metamorphic fluids: Geology, 11(2), p. 82-85.

Spera, F. J., 1980, Aspects of magma transport, In Physics of Magmatic Processes, ed. by R. B. Hargraves, Princeton Univ. Press, p. 265-324.

Stanley, W. D., Wahl, R. R., and Rosenbraum, J. G., 1976, A magnetotelluric study of the Stillwater-Soda Lakes, Nevada, geothermal area: U.S.G.S. Open-File Report 76-80, 38 p.

Stanley, W. D., Boehl, J. E., Bostick, F. X., Jr., and Smith, H. W., 1977, Geothermal significance of magnetotelluric soundings in the Snake River Plain - Yellowstone region: J. Geop. Res., 82(17), p. 2501-2514.

Stewart, J. H., 1978, Basin and Range structure in western North America: a review, In Cenozoic Tectonics and Regional Geophysics of the Western Cordillera, ed. by R. B. Smith and G. P. Eaton, GSA Mem. 152, p. 1-31.

__________, 1980, Geology of Nevada: Nevada Bureau of Mines and Geology Special Pub. 4, 136 p.

Stodt, J. A., 1978, Documentation of a finite element program for solution of geophysical problems governed by the inhomogeneous 2-D scalar Helmholtz equation: NSF Program Listing and Documentation, ESL, 66 p.

__________, 1983, Bias removal for conventional magnetotelluric data: ESL report, in press, Salt Lake City.

Stolper, E., Walker, D., Hager, B. H., and Hays, J. F., 1981, Melt segregation from partially molten source regions: the importance of melt density and source region size: J. Geop. Res., 86(B7), p. 6261-6272.

Swift, C. M., 1967, A magnetotelluric investigation of an electrical conductivity anomaly in the southwestern United States: Ph.D. thesis, Massachusetts Institute of Technology, 211 p.

Thompson, G. A., and Zoback, M. L., 1979, Regional geophysics of the Colorado Plateau: Tectonophysics, 61, p. 149-181.

Ting, S. C., and Hohmann, G. W., 1981, Integral equation modeling of three-dimensional magnetotelluric response: Geophysics, 46(2), p. 182-197.

Towle, J. N., 1980, Observations of a direct current concentration on the eastern Sierran front: evidence for shallow crustal conductors on the eastern Sierran front and beneath the Coso Range: J. Geop. Res., 85(B5), p. 2484-2490.

Turcotte, D. L., 1982, Magma migration: Ann. Rev. Earth Plan. Sci., 10, p. 397-408.

Vozoff, K., 1972, The magnetotelluric method in the exploration of sedimentary basins: Geophysics, 37(1), p. 98-141.

Waff, H. S., and Bulau, J. R., 1979, Equilibrium fluid distribution in an ultramafic partial melt under hydrostatic stress conditions: J. Geop. Res., 84(B11), p. 6109-6114.

Wannamaker, P. E., Hohmann, G. W., and Ward, S. H., 1983, Magnetotelluric responses of three-dimensional bodies in layered earths: submitted to Geophysics.

Ward, S. H., 1967, Electromagnetic theory for geophysical application, In Mining Geophysics, II, Society of Exploration Geophysics, Tulsa, p. 10-196.

Ward, S. H., 1983, Controlled source electrical methods for deep exploration: Geop. Surv., in press.

Ward, S. H., Parry, W. T., Nash, W. P., Sill, W. R., Cook, K. L., Smith, R. B., Chapman, D. S., Brown, F. H., Whelan, J. A., and Bowman, J. R., 1978, A summary of the geology, geochemistry and geophysics of the Roosevelt Hot Springs thermal area, Utah: Geophysics, 43(7), p. 1515-1542.

Weinstock, H., and Overton, W. C., Jr., 1981, SQUID applications in geophysics: SEG, Tulsa, 208 p.

Wernicke, B., and Burchfiel, B.C., 1982, Modes of extensional tectonics: J. Struc. Geol., 4(2), p. 105-115.

Williams, L. A. J., 1982, Physical aspects of magmatism in continental rifts, In Continental and Oceanic Rifts, ed. by G. Palmasson, AGU Geodynamics Series, 8, p. 193-222.

Wilt, M. J., Goldstein, N. E., Haught, J. R., and Morrison, H. F., 1982, Deep electromagnetic soundings in central Nevada: Expanded abstract of paper presented at the 52nd annual meeting of the SEG, Dallas, TX.

Winkler, H. C. F., 1979, Petrogenesis of metamorphic rocks: Fifth ed., Springer-Verlag, New York, 348 p.

Word, D. R., Smith, H. W., and Bostick, F. X., Jr., 1971, Crustal investigations by the magnetotelluric impedance method, In The Structure and Physical Properties of the Earth's Crust, ed. by J. G. Heacock, AGU Mono. 14, p. 145-167.

Wyllie, P. J., 1979, Petrogenesis and the physics of the earth, In The Evolution of the Igenous Rocks: Fiftieth Anniversary Perspectives, ed. by H. S. Yoder, Jr., Princeton Univ. Press, p. 483-520.

Yoder, H. S., Jr., 1976, Generation of basaltic magma: National Academy of Sciences, Washington, 265 p.

Zoback, M. L., Anderson, R. E., and Thompson, G. A., 1981, Cenozoic evolution of the state of stress and style of tectonism of the Basin and Range province of the western United States: Phil. Trans. Roy. Soc. Lond., A 300, p. 407-434.

Style of Basin-Range Faulting as Inferred from Seismic Reflection Data in
the Great Basin, Nevada and Utah

Mary Lou Zoback
U. S. Geological Survey, 345 Middlefield Road, Mail Stop 77
Menlo Park, California 94025

R. Ernest Anderson
U. S. Geological Survey, Denver Federal Center, Mail Stop 966
Denver, Colorado 80225

ABSTRACT

A review of seismic reflection data from the
Great Basin indicates three general modes of
modern basin formation: (A) relatively simple
asymmetric sags bounded by one or more major
steep ($\sim 60°$) planar normal faults, (B) tilted
ramps associated with moderately to deeply
penetrating listric normal faults, and (C)
assemblages of complexly deformed subbasins
associated with both listric and planar normal
faults that sole into a gently dipping
detachment surface. Faults of each mode are
known to be active, as evidenced by historical,
Holocene, or latest Pleistocene surface ruptures.
Reflection data from north-central Nevada
indicate that many of the high temperature hot
string systems in this area are situated along
normal fault zones bounding basins of mode A.
The concentration of geothermal activity in both
north-central and northwestern Nevada coincides
generally with a region of higher than normal
heat flow for the province and a relatively high
rate and density of Quaternary faulting and
seismicity. These facts, together with the
observation that most hot springs occur along
range-front fault zones, have led to the popular
interpretation that these hot-spring systems
probably are largely the result of meteoric
water circulating along the relatively steep
fault zones to moderate depths (3-5 km) in the
warm crust of the area.

In contrast, reflection data from central
Utah and from the Raft River geothermal area in
southwestern Idaho indicate that some of the
geothermal systems in these regions are
associated with basins and faulting of mode C,
important components of which are low-angle
($5°$-$35°$ dip) detachment faults at depths of less
than 4 km. The transport of both geothermal
water and magma across these detachment faults
however, implies structural continuity between
the upper and lower plates. Quaternary faulting
is also prevalent in the central Utah geothermal
systems; however, in contrast to central Nevada,
many of the Utah geothermal areas (excluding the
Raft River area) are related to young volcanic
features.

INTRODUCTION

Geologic studies, particularly in the last
ten years, have emphasized a long period of
extensional tectonism during the Cenozoic
throughout the Basin and Range province of the
western United States. Basin-range faulting,
the deformation responsible for the distinctive
modern physiography of much of the Great Basin
portion of the province, probably represents a
unique late-stage event in this long history of
extension. Major normal fault zones, commonly
50-80 km in length, bound the north-northwest to
north-northeast trending modern range blocks. ·
Their orientation, geometry, spacing, and depth
of penetration are the chief factors that
control the structural fabric of the region.
Gravity and seismic reflection data suggest that
the bedrock structure of the intervening
sediment-filled basins may be graben-like
(bounded on both sides by major normal fault
zones) or half-graben-like (bounded on only one
margin by a major "master" fault zone) or some
variation between these two end-members.

In this paper we review the current status
of our understanding of the geometry, mechanics,
and history of major basin-range normal fault
zones as these faults are known to exert a
major control on geothermal systems. Topics
covered include fault geometry, depth of
penetration, proximity to magmatic systems or
regional heat flow anomalies, and recency of
movement. A description and discussion of the
full range of the Cenozoic extensional tectonic
styles as well as the timing of different events
in the Great Basin region can be found elsewhere
in this volume (Stewart, 1983).

MODES OF BASIN-RANGE FAULTING

Various modes of formation of the modern
basins and ranges in the Great Basin have been
proposed and reviewed extensively in the
literature (cf., Stewart, 1971, 1978; Anderson
and others, 1983). While they serve as useful
end members representing failure in the brittle
regime, we now know that basin-range structure
is complex and probably includes a continuum of

variation between the end member modes. Rather than describe the individual modes of formation in detail, we have chosen to emphasize the geologic and geophysical characteristics which distinguish each from the others. The primary variants in all modes are fault geometry and depth of penetration of faulting. Another important variant is the mechanics of lower crustal/upper mantle accomodation for the shallow extension and the nature of the coupling between that accomodation and the brittle upper crustal extension.

In a horst-graben model (Figure 1a) basins typically 10-20 km wide are bounded by approximately planar normal faults dipping about 60° basinward. A localized zone of extension must exist beneath the basin at depths of 8-17 km coinciding to the depth interval in which the planar graben faults intersect. Deep accomodation for extension may be by plastic flow, or by penetrative fracturing and subjacent penetrative shearing down to a detachment near the base of the seismogenic zone. Some localized extension beneath the basin may be accomodated by dike intrusion (Thompson, 1966). A major drawback of the horst and graben model is that it is difficult to accomodate the regional tilt patterns of both individual range blocks and sets of ranges (Stewart, 1971, 1978, and 1980), which is suggestive of a faulting style incorporating block tilting.

In its extreme, a tilted block mode with planar faults (Figure 1b) can result in domino-style extension with basins being created solely by the offset and down-tilting of the blocks adjacent to the master normal fault zone (e.g. Morton and Black, 1975). It is unlikely, however, that domino style extension will develop under conditions of widely spaced major normal fault zones such as in the Great Basin, where spacing of the fault zones (25-35 km), is approximately twice the depth penetration of faulting as determined by the depth of the seismogenic zone (10-15 km). The large ratio of fault spacing to fault depth requires enormous areas of cross-sectional accomodation between blocks in the tilted block/planar fault mode (Gross and Hillemeyer, 1982). To get around this space problem Stewart (1978, 1980) has suggested a model in which the large-scale tilted block structure actually is the buoyant response of blocks bounded by dipping faults and floating on a plastically extending substratum. Eaton (1982) objected to this model on the basis of the rheological properties of the crust in the Basin and Range province.

Probably a more realistic variation of the tilted block mode involves master curved faults that decrease in dip with depth (listric faults) (Figure 1c). Stratal downbending (reverse drag) and/or antithetic normal faulting toward such faults accentuate basin development adjacent to the fault zone. Rotation due to slip on listric faults can also help explain the observed tilts of the major blocks. In either the planar or listric fault variation of the tilted block model, basin development is influenced both by the spacing of the master fault zones as well as

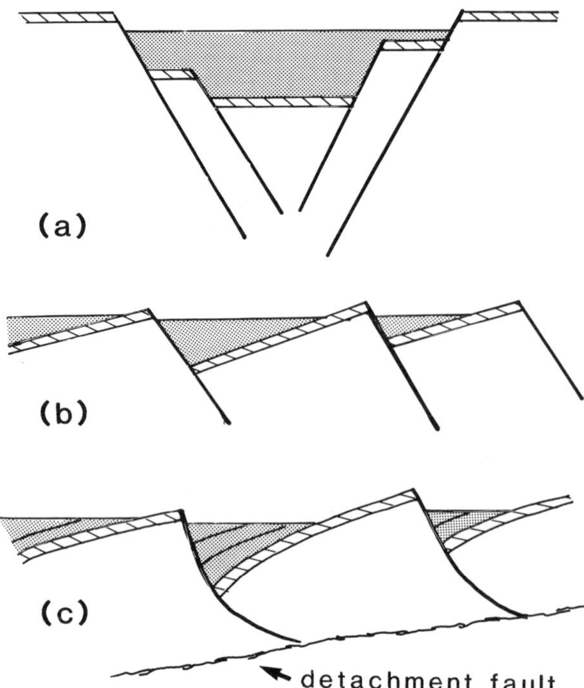

Figure 1. Proposed models of basin-range faulting. (a) Horst and graben; (b) tilted block/planar faults; (c) tilted block/listric faults.

the depth penetration of faulting. Even without details of fault curvature, it can be argued geometrically, that basin width must be roughly proportional to the depth of penetration of the master normal fault zone. Clay models of extensional tectonism suggest that basin width is approximately twice the thickness of the faulted layer (e.g., Cloos, 1968, Figure 16, p. 429). The curvature of the fault surface and hence the long-enduring controversy regarding listric versus planar normal faulting proves to be of little importance, the critical criterion being the depth penetration of faulting.

Various models have been proposed to explain the relationship between upper crustal extension by faulting and lower crustal extension. Most models involve decoupling of the faulted part of the crust from the underlying crust on some sort of subhorizontal to gently dipping detachment fault or shear zone. Types of accomodation suggested for the crust beneath the gently dipping features include: none, for the special case where the feature penetrates the entire crust and possibly lithosphere (Wernicke, 1981); stretching and attenuation by steady state creep (Stewart, 1971, 1978; Eaton, 1982); and faulting and intrusive dilation (Cape and others, 1983; Miller and others, 1983). In all three cases lower crustal extension need not be directly beneath the loci of upper crustal extension; the largest horizontal offset between the two is associated with low-angle faults or shear zones that penetrate the entire crust and possibly lithosphere.

SEISMIC REFLECTION DATA

Publicly available and previously published seismic reflection profiles in the Great Basin provide a data base that is adequate for reviewing the large scale features related to basin-range faulting, particularly the basin shape and the continuity and patterns of dip of basin-fill strata. Anderson and others (1983) noted several limitations in using the available reflection data to directly study the subsurface fault geometry and the mode of deep accomodation of extension. First, the existing data are largely restricted to the upper few kilometers of crust (generally < 5 km). Second, it is not generally possible to accurately define the subsurface fault geometry with seismic reflection data for several reasons: (1) steeply dipping features (common surface dips of normal faults range between 40°-70°) generally are not recognizable as discrete reflectors with standard reflection processing even though sharp velocity contrasts may exist across them; (2) reflections from intrabasin strata adjacent to range-bounding faults (where they are most needed to resolve the detailed geometry of subsurface structures) are generally weak or absent probably because of the common presence there of coarse clastic wedges that are poorly sorted, poorly stratified, or chaotic; (3) reflections from pre-basin strata are generally of poor to very poor quality (largely due to extensive and complex pre-existing structure) and cannot be matched across major faults to obtain constraints on the geometry or magnitude of fault offset; (4) many of the existing profiles terminate at or short of the trace of the range-bounding faults. Low-angle detachment faults in the Great Basin, however, have proven to be prominent reflectors, due in part to their shallow dip (generally less than 15°) and also, to apparent high impedance contrasts associated with these fault zones (McDonald, 1976; Allmendinger and others, 1983).

To help get around these limitations, Anderson and others (1983) established several criteria based on features generally easily observed on the reflection data to distinguish between the various modes of faulting and basin development discussed earlier. These criteria include: overall basin geometry (symmetric or asymmetric), dip of basin-fill strata adjacent to major basin-bounding normal fault zones, and the pattern of basin sedimentation at basin margins. Anderson and others suggested three general modes of basin formation based on their analysis of available seismic reflection data: (A) relatively simple, commonly asymmetric, sags bounded on one or both sides by deeply penetrating relatively steep normal fault zones (either planar or gently listric); (B) tilted ramps associated with moderately to deeply penetrating listric normal faults; and (C) assemblages of complexly deformed subbasins associated with both listric and planar normal faults that sole into a gently dipping detachment surface.

Anderson and others (1983) discussed in detail all available seismic reflection data for the Great Basin as of the end of 1981. In this paper we summarize these data by giving examples of each of these three modes of basin formation. Some recently acquired seismic reflection data are also discussed. Figure 2 gives the location and source of all current publicly available seismic reflection data in the Great Basin together with the interpreted style of faulting keyed to the letter designations in the previous paragraph.

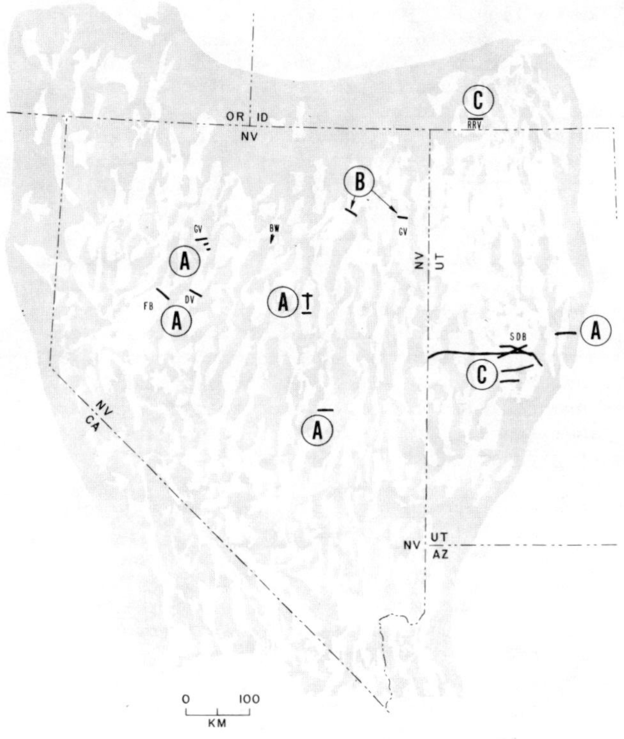

Figure 2. Available seismic reflection profiles in the Great Basin (heavy lines). Interpreted mode of basin structure indicated by letters (see text and Figure 14): (A) symmetric or assymmetric sag, (B) tilted ramp, (C) complexly deformed sub-basins. Place names indicated by small letters: FB-Fallon Basin, DV-Dixie Valley, GV (western Nevada) - Grass Valley, BW-Beowawe, GV (eastern Nevada)-Goshute Valley, SDB-Sevier Desert Basin, and RRV-Raft River Valley. References for Nevada seismic data are given in Anderson and others (1983) with the exception of Grass Valley, Nevada which is described in the text. Data in Utah are from COCORP (long E-W profile, Allmendinger and others, 1983), McDonald (1976), and Zoback (1982). Data in Raft River Valley, southwestern Idaho are from Covington (1983) and Ackermann (written communication, 1983).

Sag-like Basin Structure

The sag-like basin structure which is considered related to deeply-penetrating, approximately planar master normal faults is typified by the Fallon Basin area (Figure 2). Figure 3a shows an unmigrated, 22 km long, seismic reflection profile across the main part of the basin from Hastings (1979). Across most of the basin the sedimentary sequences (primarily lacustrine) are well-bedded and lack internal structural disruption as indicated by numerous continuous reflectors 10 km or more in lateral extent and a few that are more than 15 km in length. Many of these reflectors show a reversal of dip as they are traced from northwest to southeast and ultimately die out beneath the southeastern part of the basin (see migrated line drawing--Figure 3b). A broad asymmetric sag controlled by a steep planar normal fault is suggested as the predominant basin structure. Available data indicate that the Stillwater Range block, which is only 5-10 km wide and separates the Fallon basin on the west from Dixie Valley on the east, is bounded by structural basins that are approximate mirror images of one another (Anderson and others, 1983). Reflection data for Dixie Valley were described briefly by Anderson and others (1983) and are discussed in detail by Okaya and Thompson (1983). The latter authors combined analysis of the seismic reflection data with studies of the 1954 Fairview Peak (M = 7.2) earthquake to conclude that the major normal fault zone bounding Dixie Valley on the west is approximately planar and extends to a depth of about 15 km. Though the earthquake and reflection seismic data from the Dixie Valley area are consistent with a model that includes the "master" normal fault; it is instructive to consider an alternative model for formation of the asymmetric sag structure.

Figure 3c is a cross-sectional sketch of a clay model of extension done by R.K. Hose (written communication, 1983). The experimental design was similar to that described by Cloos (1968, p. 425) and was such that the faults terminate downward against overlapping tin sheets. The line of overlap between the two sheets remained fixed during the experiment and coincided with the furthest right hand faulting in the clay. Extensional traction was applied to the clay layer by sliding the underlying tin sheet to the left. If the top of the clay model corresponds to the base of a basin we see that it has the form of an asymmetric sag underlain by highly fractured material. Faults with major throw are not seen. Instead, the sag is produced by a combination of stratal attenuation and offset on many small faults. The stratal attenuation is probably accomplished by displacement on abundant unidentified microfractures and/or by grain-grain ajustments. Analogous attenuation of sub-basin rocks of the scale of Fallon basin (Fig. 3a) could be associated with displacements on a class of small fractures and faults with displacements ranging from a few centimeters to a few tens of meters. Such structures would be undetected by seismic reflection profiling even if good subbasin reflectors were present. Clearly, they could be extremely important to fracture, porosity, and premeability and hence to the potential for sub-basin water circulation. The absence of a master fault would not preclude strain accumulation sufficient to produce a large earthquake and associated extensive surface faulting. Some of the faults seen in the clay model, especially the sag-marginal ones, extend from the surface to the tin sheets. The overall width of the sag is nearly twice the depth to the tin sheets. If extrapolated to Fallon Basin or Dixie Valley these faults, though not of great displacement, would have a depth pentration of 9-10 km. The stress drop associated with a reasonable displacement of 3 m on such a fault could produce a large earthquake.

Tilted-Ramp Basin Structure

Anderson and others (1983) discuss two examples of tilted ramp basin structure from eastern Nevada. One of these, a profile published by Effimoff and Pinezich (1981) across Goshute Valley, extends from near the bedrock-alluvium contact on the east side of the Pequop Range eastward to a point several kilometers west of the base of the Toano Range (Figure 4a, for location see Figure 2). A line drawing of the data is shown in Figure 4b. In its eastern part the profile crosses a broad sediment covered pediment which is bounded on the west by a buried fault zone that appears to be the major structure controlling basin development. The overall gentle easterly tilt of the basin fill wedge as well as the dip of reflectors into the fault zone led Anderson and others to interpret this fault structure as a moderately deeply penetrating listric fault zone. This interpretation is at odds with that of Effimoff and Pinezich (1981) who suggest a sharply listric fault zone flattening only a few kilometers beneath the valley which is genetically related to a pre-existing thrust fault in the Toana Range.

The bedrock structure beneath the Goshute Valley may be very similar to the sketch of the clay model (model done by Cloos, 1968; sketch by Stewart, 1971) shown in Figure 4c. In this model the tilted ramp structure is largely the result of rotation related to slip on the master listric fault. Note however, that a distributed zone of antithetic faulting adjacent to the master fault zone tends to reduce the overall amount of tilting in the basin floor.

Another significant feature of the clay model shown in Figure 4c is the development of a cross-structure within the basin which is related to irregularities in the geometry of the master fault zone. This cross-fault is characterized by oblique slip (both horizontal and down dip components of motion). Possibly

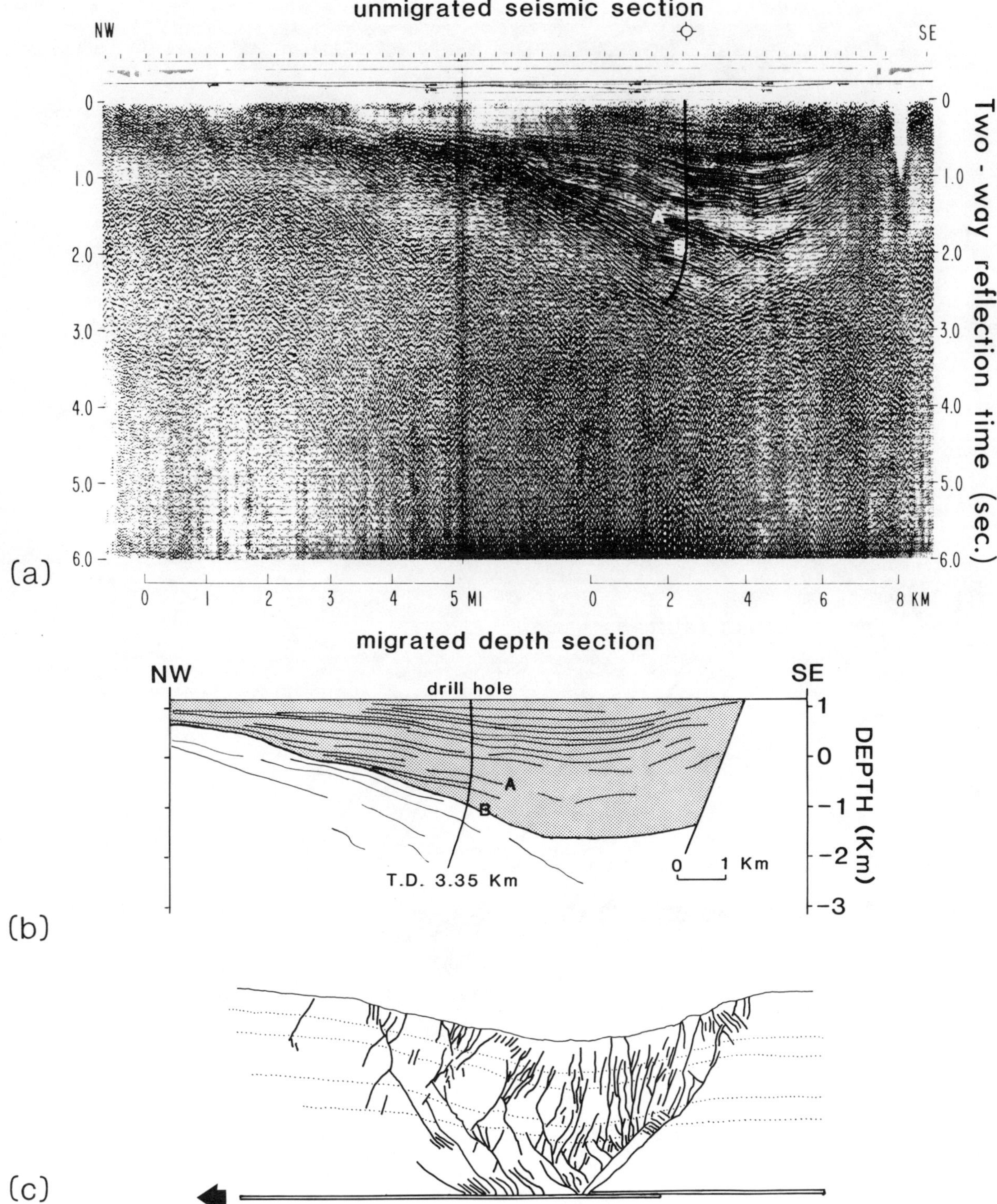

Figure 3. Sag-like basin structure. (a) Unmigrated seismic section across Fallon Basin, central Nevada (Hastings, 1979 and Anderson and others, 1983), (b) Line drawing of migrated depth section (Hastings, 1979 and Anderson and others, 1983), (c) Sketch of clay model of possible bedrock structure and faulting beneath the basin (clay model done by R. K. Hose, written communication, 1983; sketch by Stewart, 1971).

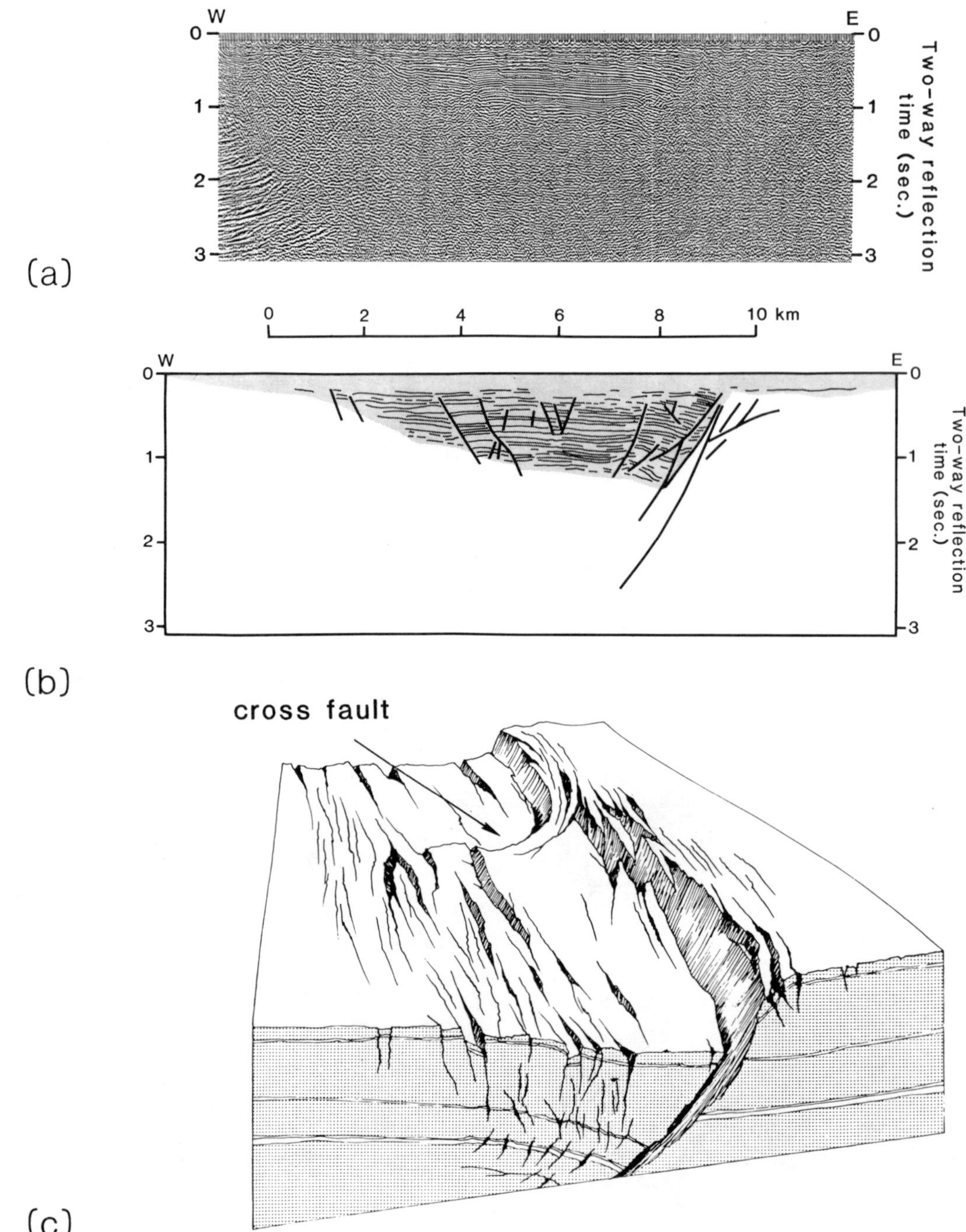

Figure 4. Basin above a tilted ramp structure. (a) Migrated seismic profile across Goshute Valley (Effimoff and Pinezich, 1981). (b) Line drawing of seismic profile with fault interpretation (Anderson and others, 1983). (c) Sketch (Stewart, 1971) of clay model by Cloos (1968) of possible bedrock structure and faulting beneath the basin.

analogous cross-strike zones of bedrock offset have been identified beneath numerous basins within the Great Basin by both seismic reflection (Effimoff and Pinezich, 1981) and gravity (c.f., Zoback, 1983) data.

Complexly deformed sub-basins above a shallow detachment fault

This style of basin development is typified in the shallow seismic reflection data in the Sevier Desert basin reported by McDonald (1976) and discussed in detail in Anderson and others (1983). The upper plate rocks in this area have been extended by numerous planar and listric normal faults. Major fault zones identified on the seismic reflection profiles have an average spacing of about 5 km. Dip components of sediments within the basin show a complex pattern of combined reverse and normal drag adjacent to these fault zones. The detachment fault itself, the Sevier Desert detachment, is apparent as a strong, relatively continuous reflector from the near surface to depths of 12-15 km (Allmendinger and others, 1983).

Another example of basin development above a shallow detachment fault comes from geological and geophysical studies and seismic reflection data from the Raft River geothermal area. Interpretation of a detailed seismic refraction study in the area indicate that there are no major normal fault offsets in the bedrock beneath the valley; however, faults with 50 m or less of vertical displacement could have been missed because of scatter in the data (Ackermann, 1979). A detachment fault has been interpreted to lie at the base of the Tertiary basin sediments along their contact with a thin sectin of Precambrian and lower Paleozoic metasedimentary rocks which mantle the crystalline basement (Mabey and others, 1978; Covington, 1980 and 1983). This interpretation is supported by an abrupt increase in fracturing near the base of the basin fill section that can readily be observed in drill core and also by low-angle normal faults in core sections obtained less than 200 m above the inferred detachment fault (Guth and others, 1981). This detachment fault is located at depths of less than 2 km beneath the valley in the vicinity of the geothermal area. Further eastward beneath the valley (downdip) the detachment fault has been interpreted as separating fault blocks of metamorphosed to nonmetamorphosed Paleozoic and Proterozoic rocks from the underlying basement complex at depths of 3 km or greater (Covington, 1983). Figure 5a shows a seismic reflection

Line 4 migrated

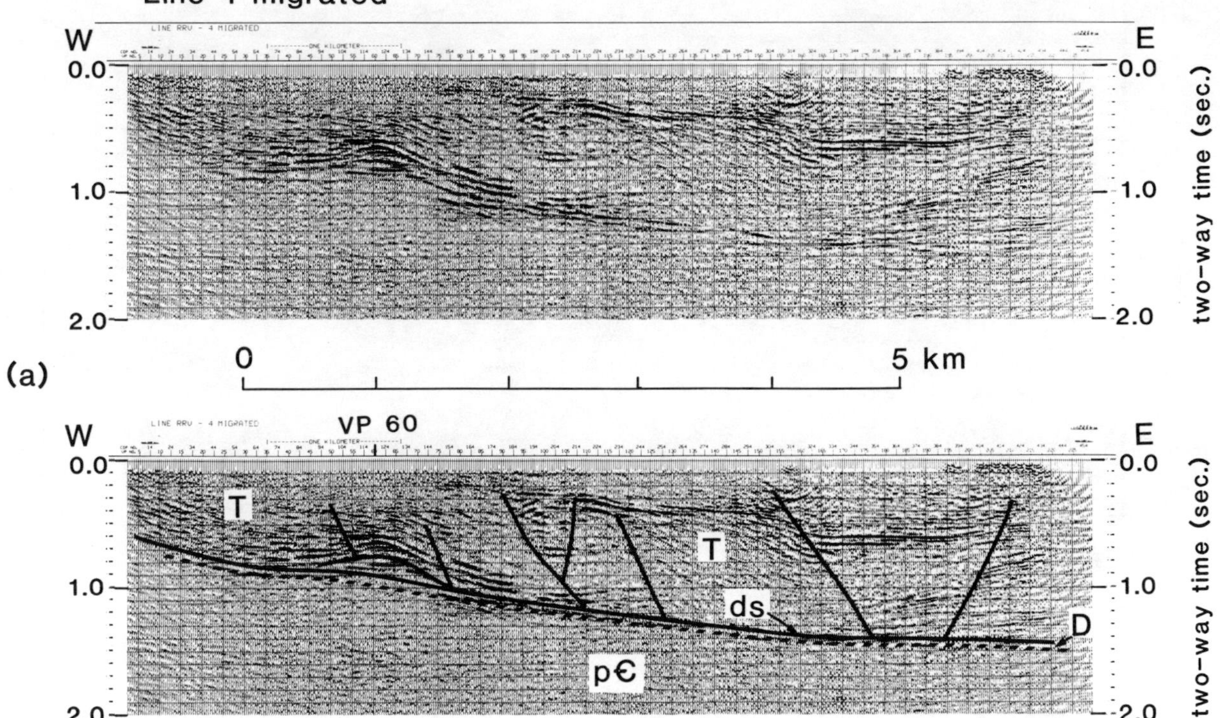

Figure 5. (a). Migrated seismic profile across a portion of Raft River Valley (H. D. Ackermann, 1983, written communication and Covington (1983). (b) Interpretation of Raft River profile showing major faults, modified from Covington (1983). Heavy dashed line indicates top of Precambrian crystaline basement (p Є), based on an integrated interpretatin of seismic reflection and refraction data (Ackermann, written communication, 1983). T, Tertiary basin sediments; ds, detachment surface; D, ductily deformed lower Paleozoic and Proterozoic schists and quartzites. Small antiform near VP 60 may represent a bedrock klippe above detachment fault.

profile across part of the valley in the vicinity of the geothermal area, an interpretation of the profile somewhat modified from that of Covington (1983) is given on Figure 5b. The inferred top of crystalline basement on Figure 5b comes from an integrated analysis of the seismic refraction and reflection data (H. D. Ackermann, 1983, written communication). Unlike the Sevier Desert detachment, the inferred Raft River detachment is not easily recognized as a prominent, nearly continuous reflector. However, as indicated on Figure 5b, several major normal faults within the basin fill do not appear to offset the top of crystalline basement or, by inference, the detachment fault in the metasedimentary rocks lying above the crystalline basement.

Of special significance in both the Raft River area and the Sevier Desert basin region is the general lack of evidence on the seismic profiles of breakage of the relatively shallow (< 4 km deep) detachment fault by the normal faults above it. However, as discussed more fully in the section on geothermal implications, there is ample evidence that both minor faulting and fluid motion must occur across the detachment surface in both areas.

Case Study: Grass Valley, Nevada

While the broad generalization and categorization of basin structure and fault geometry outlined above is appealing, detailed analysis of several closely spaced lines within a single basin commonly reveals many complexities. As an example, four seismic profiles in Grass Valley, Nevada, collected as part of the

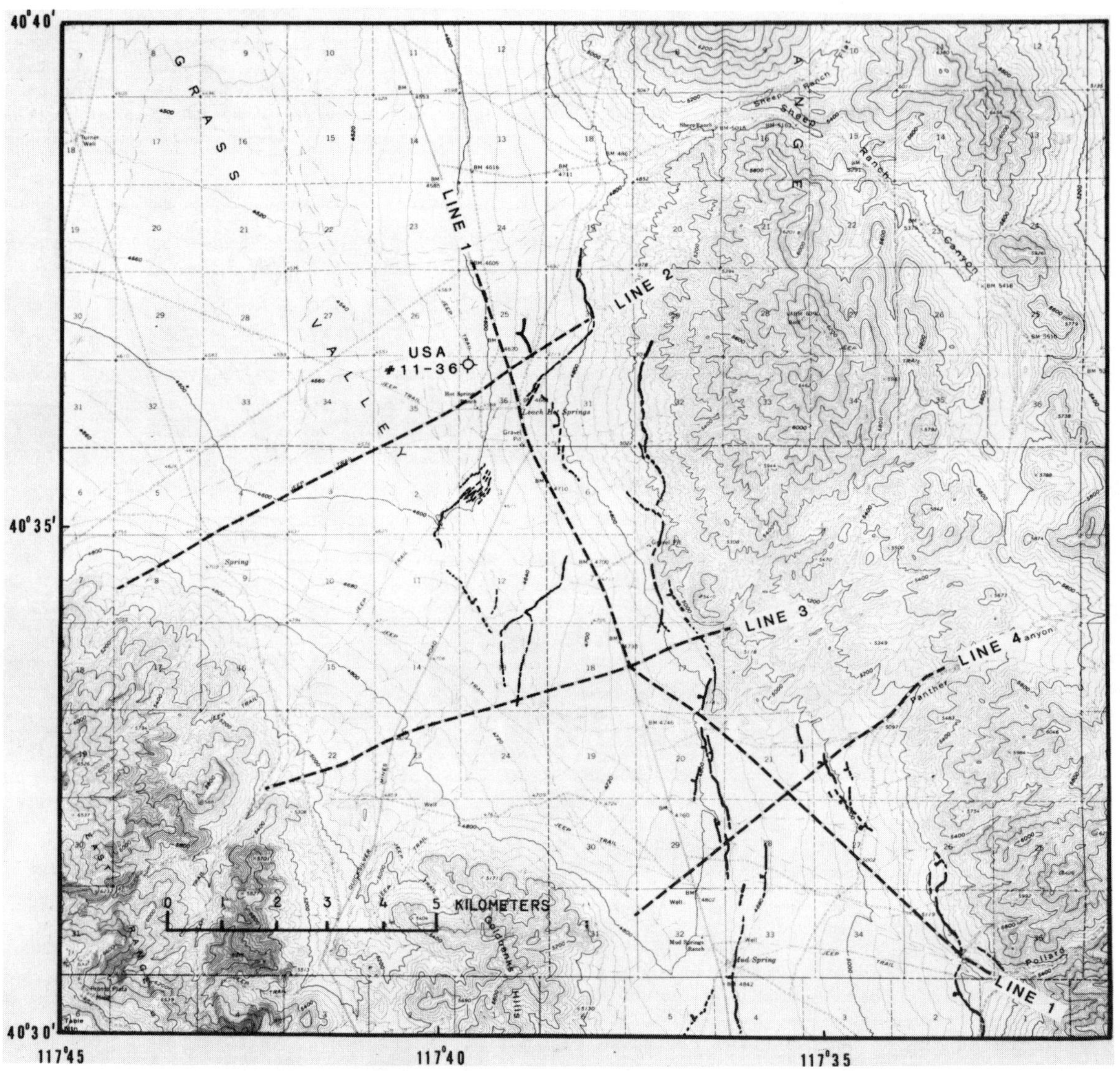

Figure 6. Map of Grass Valley, Nevada, showing Quaternary faults (heavy lines) from Wallace (1980b), locations of seismic reflection profiles (dashed lines), and well USA # 11-36.

Department of Energy/Division of Geothermal Energy (DOE/DGE) industry-coupled geothermal program, are discussed. Grass Valley lies between the Sonoma Range on the east and the East Range to the west (see Figure 2 for location). The valley is bounded on its east side by a major north-northwest-trending normal fault zone containing numerous Quaternary scarps including the faulting related to the 1915 M = 7.8 Pleasant Valley earthquake (Wallace, 1980a). Leach Hot Springs is located within this fault zone, near the intersection of a northeast-and northwest-trending fault.

Three cross-valley vibroseis seismic reflection lines were run in Grass Valley along with one axial line, subparallel to the range front fault zone. The locations of the Quaternary fault scarps and the seismic lines are shown on Figure 6. The data were collected and processed by Sunoco Energy Development Company. A standard processing sequence was performed including deconvolution, velocity analysis, statics, stacking and filtering. Wave equation migration was also performed.

The overall basin structure of Grass Valley, as interpreted from the seismic reflection data, is asymmetric with a major normal fault zone on the east side and only minor faulting and/or downwarping on the western side. This overall asymmetry is consistent with the observed eastward tilt of the flanking range in this region (Stewart, 1980).

The asymmetric synclinal sag structure of the basin is most clearly demonstrated on Line 2 (Figure 7a) which begins at the base of a scarp in Quaternary-Tertiary gravels and fanglomerates along the base of the Sonoma Range, and extends westward across the valley and ends near an exposure of Cretaceous quartz monzonite in the East Range. Maximum basin fill thickness along Line 2 of about 1.7 km was determined from a combined analysis of gravity (Goldstein and Paulsson, 1978), well (USA #11-36, northwest corner section 36, T32N, R38E; Wilde and Koenig, 1980) and the seismic data. An interpretation of Line 2 is given in Figure 7b. A major west-dipping fault zone with an offset of ~ 1.2 km is interpreted to separate the basin from an alluvial covered pediment along the east margin of the valley. A smaller east-dipping fault zone occurs within the western half of the basin. The basin fill section is divided into two units based in part on an observed disconformity and also on seismic refraction data in the vicinity of the profile (Majer, 1978). A zone of relatively dense, probable alluvial fan material within the basin fill adjacent to the main fault zone was required to match the gravity data (Figure 7c). Other geometries shown in the gravity model in Figure 7c were interpreted from the reflection data. Also note on Figure 7b the highlighted series of subhorizontal reflectors in the bedrock beneath the basin between about 1.25 and 2.4 seconds depth. These are discussed below.

Line 3 begins in Paleozoic and Triassic sedimentary rocks, 0.5 km east of the main range fault zone. The final stack for Line 3 is shown in Figure 8a, the migrated version is shown on

Figure 8b. Maximum fill thickness interpreted near VP 151 is about 1.0 km. The overall basin structure observed on Line 3 appears to be that of a tilted ramp; a prominent band of westward dipping reflectors in bedrock beneath the western edge of the basin (between about 1.25 to 2.4 seconds on the migrated section-Figure 8b) could be interpreted as a zone of a shallowly dipping normal faults or possibly the shallow portions of a series of listric faults. Apparent dip on these reflectors is between 24°-28°W. A tilted ramp basin structure would be consistent with a listric fault that flattened at such shallow depths (about 3 to 6 km). However, close inspection of the unmigrated section for Line 3 (Figure 8a) reveals an overlap of reflectors of opposing dip, a so called bow-tie structure, in the basin fill suggestive of a synclinal or sag structure bounded by a relatively steep, planar fault. On the unmigrated section the beds adjacent to the main fault zone are seen to dip basinward, not into the fault. Furthermore, this band of bedrock reflectors beneath the western edge of the basin occurs in approximately the same depth interval as subhorizontal reflectors (mentioned above) beneath the basin on Line 2, suggesting that they may correspond to a bedrock structure unrelated to the range-front fault. It is possible that the westerly-dipping reflectors beneath the basin on Line 3 may represent a faulted extension of near horizontal reflectors seen at about 0.70-0.75 seconds east of the main fault zone. These reflectors located east of the main fault zone as well as a number of shallower reflectors can be observed on the axial line, Line 1.

Line 4 crosses the narrowest part of Grass Valley, and begins in bedrock in the draw separating the Sonoma Range to the north from the Tobin Range further south (Figure 6). Line 4 lacks the characteristic well-defined reflections from the basin fill section (Figure 9). The gravity data (Goldstein and Paulsson, 1979) indicate that a transverse bedrock high, approximately coincident with Line 4, separates the Grass Valley basin from the Pleasant Valley basin to the south. Thus, Line 4 may cross a part of the valley where the basin fill is a thin veneer resting, at shallow depths, on complexly faulted bedrock. The only major reflector seen on Line 4 is a shallow west-dipping feature on the east end of the line. This surface is located east of most of the mapped Quaternary fault scarps transected by the line. The apparent dip of this reflector is between 17°-20°. It is located approximately 0.25-0.5 km below the fan surface. An updip projection of this surface suggests that it may coincide with the buried extension of the range and may possibly represent a buried pediment on bedrock. This basinward dipping surface is located near a left stepping en-echelon offset in the range front. However, the main zone of Quarternary faulting is located about 1.4 km basinward of the western, downdip truncation of the reflectors. Some minor Quarternary scarps do occur near the truncation of the reflectors.

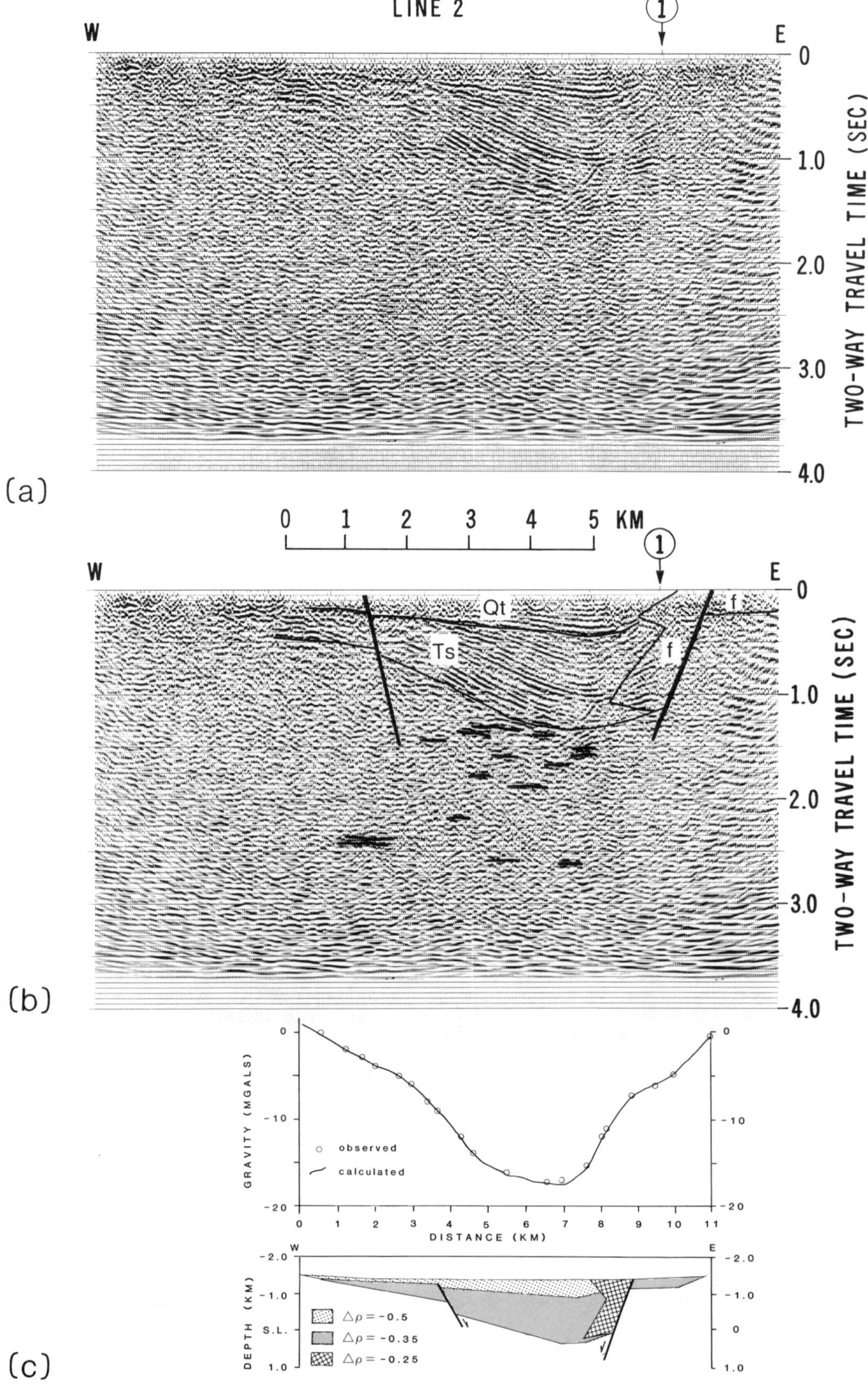

LINE 2
1
W
E
0
1.0
2.0
3.0
4.0
TWO-WAY TRAVEL TIME (SEC)
(a)
0 1 2 3 4 5 KM
1
W
E
Qt
Ts
f
f
0
1.0
2.0
3.0
4.0
TWO-WAY TRAVEL TIME (SEC)
(b)
0
-10
-20
GRAVITY (MGALS)
0
-10
-20
observed
calculated
0 1 2 3 4 5 6 7 8 9 10 11
DISTANCE (KM)
W
E
-2.0
-1.0
S.L.
0
1.0
DEPTH (KM)
-2.0
-1.0
0
1.0
△ρ = -0.5
△ρ = -0.35
△ρ = -0.25
(c)

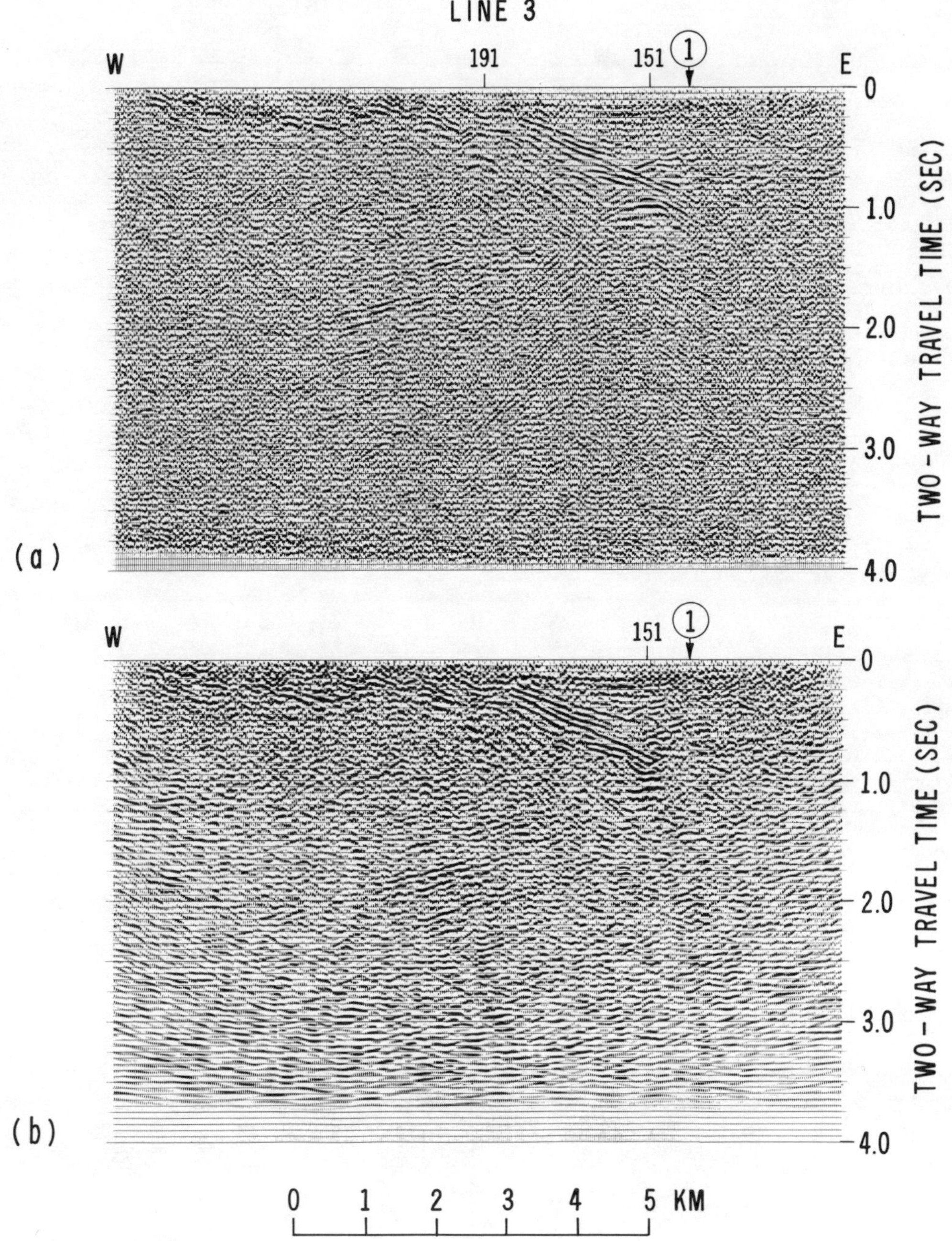

Figure 8. (a) Final stack for Line 3, Grass Valley. Circled number on top of section indicates location of cross line. (b) Migrated time section for Line 3.

Figure 7. (a) Migrated time section of Line 2, Grass Valley. Circled number on top of section indicates location of cross line. (b) Interpretation of Line 2. Qt, young basin fill; Ts, older basin fill, probably includes interstratified volcanic rocks; f, fan and pediment material. See text for further discussion. (c) Gravity model for Line 2, density contrasts (in g/cm^3) are indicated . Bodies just west of main fault zone are thought to represent fan material, somewhat denser than the rest of the basin fill section.

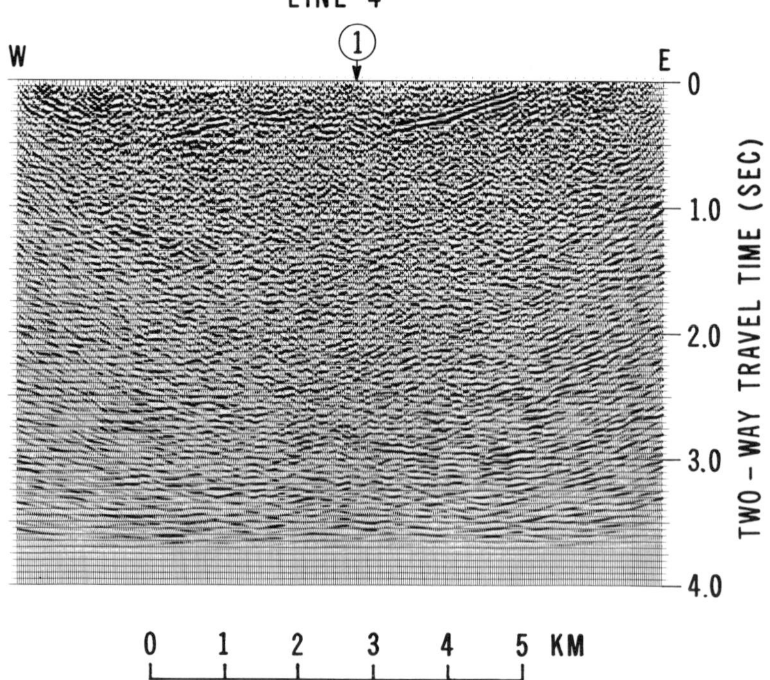

Figure 9. Migrated time section for Line 4, Grass Valley. Circled number on top of section indicates location of cross line.

Line 1 begins 2.75 km northwest of Leach Hot Springs and runs parallel to the axis of the valley (Figure 6). The reflection pattern observed on Line 1 is quite complex (Figure 10). Reflectors are difficult to interpret even in the vicinity of the cross valley tie lines. At the northernmost end of the line a section of basin fill sediments appear to have been crossed obliquely, however, for the most part the line runs along the steep gravity gradient associated with the major westward-dipping range-bounding fault zone. The high-amplitude sub-horizontal reflectors located in the upper 1.0 second, particularly in the southern half of the section, may represent fault zone reflections parallel to strike. A rather steep northward-dipping reflection observed near the intersection with Line 2 may represent a reflection off the northeast-striking, northwest-dipping fault that terminates near Leach Hot Springs.

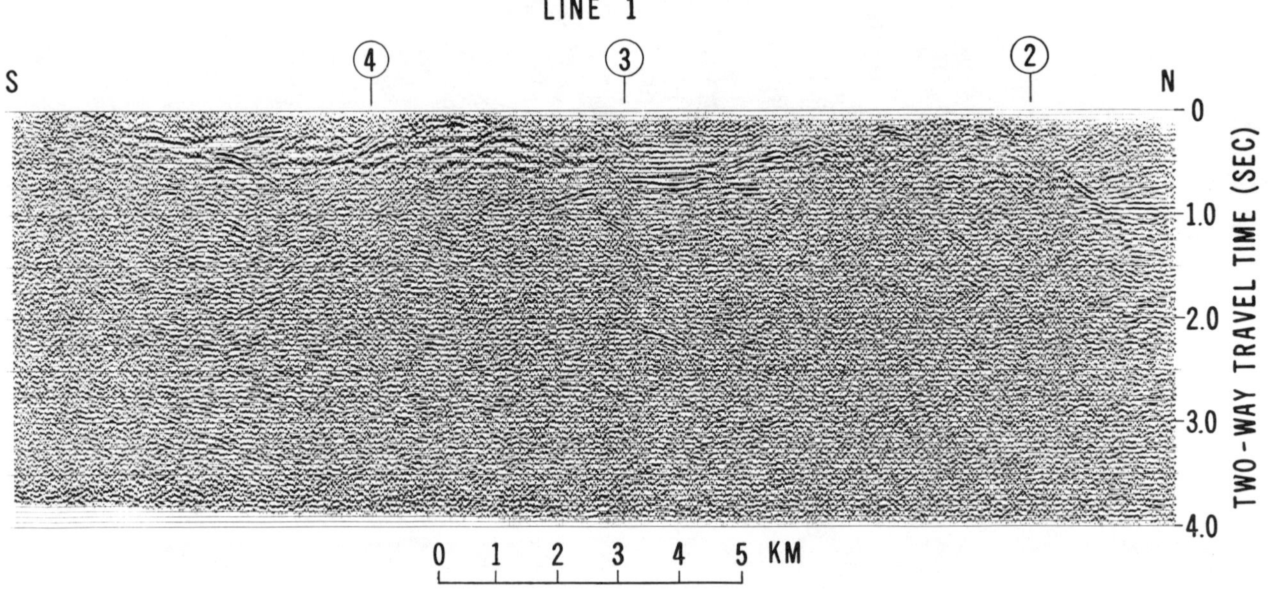

Figure 10. Migrated time section for Line 1, Grass Valley. Circled number on top of section indicates location of cross line.

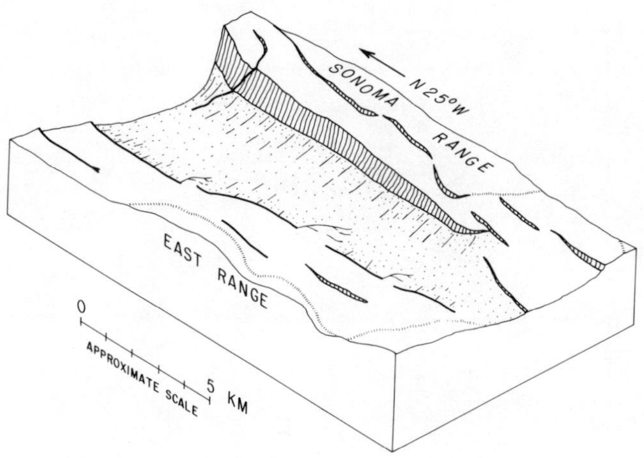

Figure 11. Generalized block diagram of bedrock structure beneath part of Grass Valley inferred from gravity and seismic reflection data.

In summary, the detailed seismic reflection coverage in Grass Valley indicate complex internal structure (Figure 11). The overall structure in the deeper parts of the valley (seen most clearly on Line 2) appears to be that of an asymmetrical sag. The major fault zone lies on the east side of the basin although there does appear to be minor faulting on the west side of the basin. Gently westward dipping sub-basin reflectors on Line 3, if viewed independent of the other lines, could be interpreted as reflections off a series of low angle faults. Based on similar, but horizontal, reflections seen in the bedrock beneath the basin on Line 2, these reflectors are thought to originate in bedrock structures unrelated to the main range front fault. A relatively steep, planar west-dipping fault zone decreasing in displacement to the south toward Line 4 is the favored interpretation of the master fault bounding the basin.

ADDITIONAL GEOPHYSICAL CONSTRAINTS ON THE SYTLE OF BASIN-RANGE FAULTING

Seismicity studies provide information on currently active brittle deformation in the upper crust. Several regional characteristics of extensional deformation in the Great Basin have emerged as a result of such studies: (1) brittle deformation is largely limited to the upper 15 km of crust; a summary of earthquake focal depths by Eaton (1980) indicated that about 80% of the events occur at depths of 10 km or less, (2) focal mechanisms indicate that slip is occurring on relatively high angle planes ($\geq 30°$; c.f., Arabasz 1981; Vetter and Ryall, 1983 and Zoback, 1983), slip on the low-angle fault planes such as the Sevier Desert detachment may be largely aseismic (Smith, 1981), (3) in both regional and detailed local earthquake studies there is

generally a very poor correlation between earthquake hypocenters and the likely downward projection of mapped surface faults (c.f., Arabasz and others, 1980 and 1981). This third point, together with studies of fault scarps formed in the M = 7.8, 1915 Pleasant Valley earthquake, lead Wallace (1979, 1980b), to suggest that upper crustal extension may in large part be controlled by localized deep zones of extension not necessarily related to the surface pattern of Basin and Range blocks. Some of the microseismicity may also be related to minor intrablock deformation.

Smith (1983) recently presented seismic reflection profiles across several segments of the Wasatch fault zone. The data were interpreted as indicating that some of the main basin-bounding faults along the Wasatch fault zone are sharply listric, becoming horizontal at depths of only 3-4 km. Earthquake focal depth studies in a 100 km wide zone centered along the Wasatch fault zone indicate that most events occur at depths of less than 10 km; however, 63% of the events occurred at depths greater than 4 km below the surface (Arabasz and others, 1980). Most of the events however, cannot be placed directly on the Wasatch fault zone. Many occur near the intersection of the Wasatch fault and transverse structural zones that often delimit distinct structural basins along the Wasatch fault (Zoback, 1983). An analysis of nodal plane dips for the larger of these events indicate slip on high-angle planes ($\geq 30°$, with mean and median values of 49°-54°; Zoback, 1983).

Cape and others (1983) proposed a model to explain the disparity between earthquake focal depth studies and the shallow fault structure determined from seismic reflection data. Their model (Figure 12) involves two tiers of faulting within the upper crust, above and below a major detachment fault at a depth of 4-5 km, interpreted on the COCORP Rio Grande rift profile. This detachment fault was inferred, on the basis of seismic reflection and gravity data, to occur near the top of the Precambrian crystalline basement. This tiered fault model has faulting and intrusion in the layer beneath the detachment and may explain the broad focal depth distribution (generally 0-15 km) in areas where shallow detachments have been recognized.

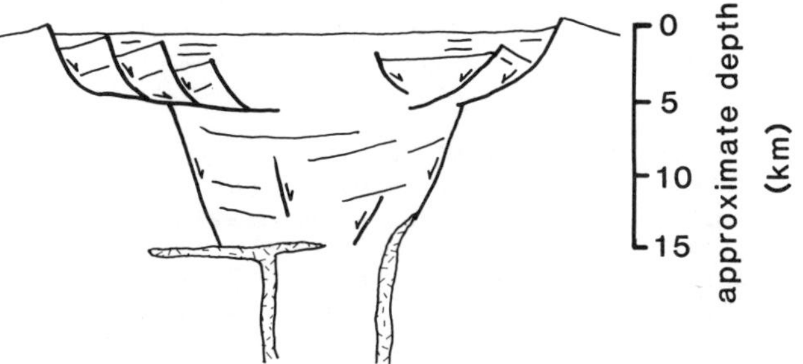

Figure 12. Proposed crustal model for extension in the Rio Grande Rift. Note deep intrusion and normal faulting above and below a shallow (< 5 km) detachment fault (from Cape and others, 1983).

GEOTHERMAL IMPLICATIONS

The seismic reflection data indicate the presence in some areas of the northern Basin and Range Province of predominantly planar and listric master basin-bounding faults that penetrate to depths of 8 to greater than 12 km in the upper crust. The commonly observed asymmetry in overall basin cross-sectional geometry as well as the regional pattern of tilted range blocks (Stewart, 1980) indicates that many basins are bounded only on one side by a deeply-penetrating master fault system. These master faults are clearly a critical component of geothermal convective systems in the Great Basin region as evidenced by the common occurrence of most modern geothermal systems along range-bounding fault zones. The basin margin bounded by the master fault zone can generally be predicted from the overall tilt of the surrounding ranges.

Cross-fault structures beneath the basins (as shown in Figure 4b) may play an important role in localizing a geothermal system along a master fault zone. Geothermal activity at Beowawe, north-central Nevada (see Figure 2 for location), may be largely controlled by cross-fault structures (Zoback, 1979).

A comparison of Quaternary fault scarps, surface heat flow, and the distribution of known geothermal systems with reservoir temperatures greater than 90°C (Figure 13) in the Great Basin indicates a direct correlation, attesting to the critical role that faulting plays in the development and maintenance of the geothermal systems particularly in regions of above average heat flow. Although the fault map does not represent uniform province-wide coverage, the general distribution pattern of faulting is probably correct. High levels of faulting occur in western Nevada and along a N-S belt in central Utah. These zones of dense Quaternary faulting compare well with the distribution of modern seismicity in the province (c.f., Smith 1978). In fact, seismicity and geologically-determined moment rates for these areas also compare quite well (Smith and Bauer, 1983). Thus, the observation that major geothermal areas tend to occur in zones of Quaternary faulting, (which are also seismically active today) suggest that the fault/fracture premeability, so important to the northern Basin and Range geothermal systems, is probably maintained by seismicity which counteracts the geochemical self-sealing nature of these systems.

The occurrence of geothermal systems with reservoir temperatures greater than 90°C is densest in western Nevada, especially for the hottest systems (reservoir temperatures >150°C). Significantly, many of the hottest geothermal systems occur within the regions of highest heat flow, particularly the Battle Mountain high region of northern Nevada (Sass and others, 1971; Sass and others, 1981) where heat flow values are approximately double the continental mean. The available seismic reflection data in this region suggest sag-like basins which we have interpreted as being bounded by relatively steep, deeply

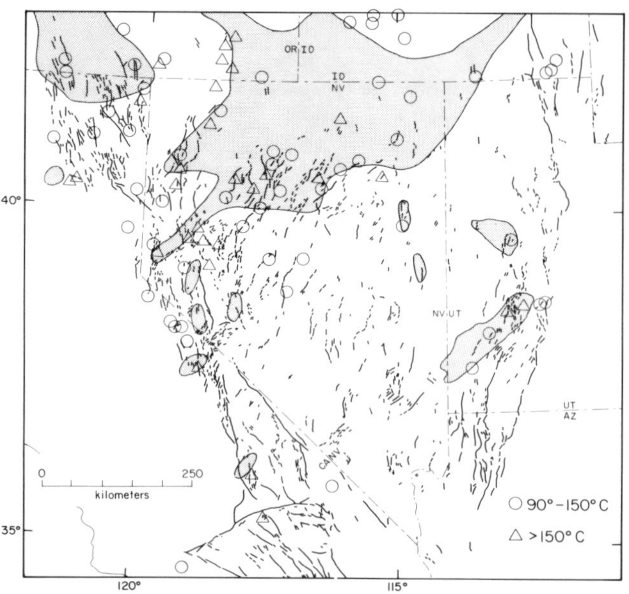

Figure 13. Geothermal systems with reservoir temperatures > 90°C (Muffler, 1979) superimposed on a map of Quaternary faulting in the Great Basin (Nakata and others, 1983) and regions of high heat flow (shaded regions heat flow $\geq$ 2.5 HFU, 104 mW/M^2, Sass and others, 1981). Circles indicate systems with reservoir temperatures between 90°-150°C; triangles indicate systems with reservoir temperatures greater than 150°C.

penetrating master fault zones. As mentioned above, the high level of seismicity and Quaternary faulting in this area are probably a major contributing factor in the development and maintenance of fracture permeability allowing the deep circulation of meteoric water in these systems. There remains a serious problem however with invoking conductive heating of deeply circulating meteoric water as the primary heat source for many of the long-lived Great Basin hot-water geothermal systems. It seems unlikely that fracture porosity could be great enough to supply enough heated surface area to maintain geothermal systems for several hundred thousand years. Zoback (1979) estimated an age of 200,000 years for the Beowawe, north-central Nevada geothermal system (See Figure 2 for location).

The geothermal systems in the Black Rock Desert region of central Utah (Figure 13) are located in or near zones of late Quaternary basalt and rhyolite volcanism and Quaternary faulting. Strontium isotopes indicate that the basalts have a mantle source (Hoover, 1974). The zone of Quaternary faulting and volcanism in the Black Rock Desert lie along a buried major normal fault zone that can be inferred from gravity data (Figure 14). This localized zone of tectonism occurs above the Sevier Desert detachment fault in an area where the detachment is at a depth of less than 3 km (McDonald, 1976), and must represent a major break in the otherwise largely continuous detachment fault. The magmatic,

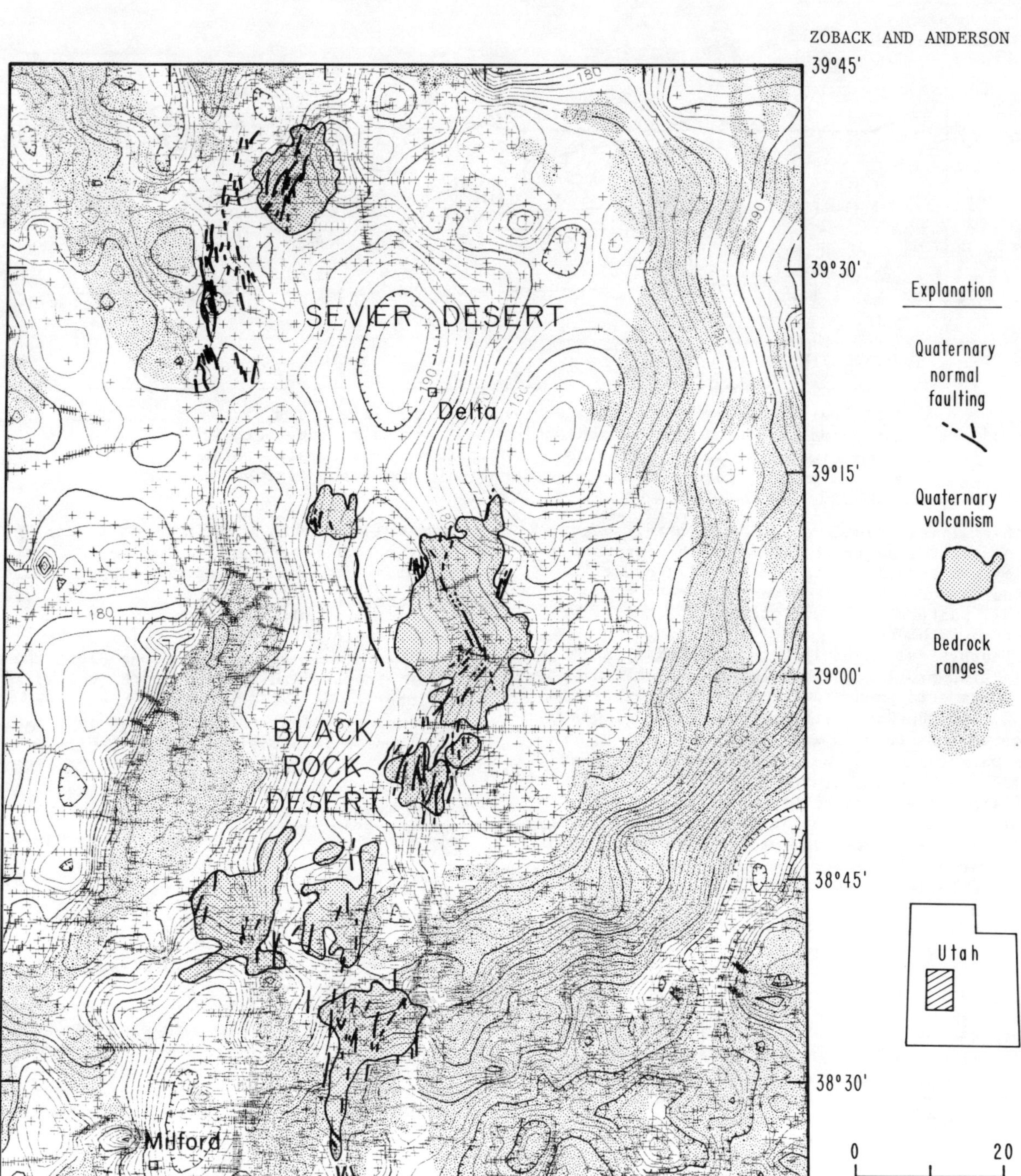

Figure 14. Complete Bouguer gravity map of the Black Rock Desert area of central Utah. The pronounced north-south gravity low near the center of the map indicates the presence of an intrabasin graben. The gradients on either side of the low define the graben-bounding fault zones. Sites of primary basaltic volcanism are patterned; Quaternary faults are approximately located by heavy lines, and gravity stations are indicated by crosses. Sources of gravity data are U. S. Geological Survey gravity files and Serpa and Cook (1980).

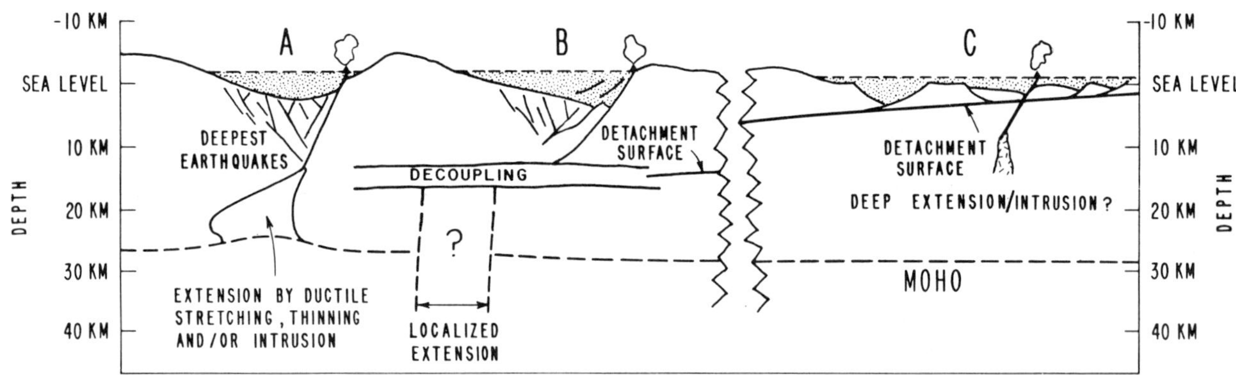

Figure 15. Diagramatic cross section showing three modes of basin formation and normal faulting that may exist in the Great Basin together with possible lower crustal styles of extension. Major structures related to geothermal systems are indicated. Stipple pattern indicates basin fill sediments. Modified from Anderson and others (1983).

geothermal, and faulting activity in this area are no doubt genetically linked. They probably indicate a localized zone of lower(?) crust extension which is largely unrelated to the overall pattern of extension tied to the Sevier Desert detachment.

Geophysical investigations (primarily seismic reflection profiling) have indicated shallow low-angle detachment faults in other geothermal areas. In the Roosevelt Hot Springs area, Utah (located less than 50 km southwest of the Black Rock Desert region) the geothermal reservoir is formed by a complex system of faults and joints in Precambrian and Tertiary crystalline rocks. A major component of this fracture system is a very shallow low-angle detachment fault (at a depth of less than one km, Bruhn and others, 1982). The ultimate source of heat for this system is believed to be an igneous intrusion related to young (less than about 0.6 myBP) rhyolitic volcanism (Ward and others, 1978; Robinson and Iyer, 1981).

Drilling has verified that the geothermal reservoir in the Raft River Valley is a highly fractured zone near the base of the Cenozoic basin fill (Mabey and others, 1978). Mabey and others (1978) speculated that the geothermal waters encountered in the reservoir were conductively heated by circulation to depths of 3 to 6 km. As the reservoir is located in the basin fill section, above the inferred detachment fault at the the base of basin fill at a depth less than 2 km, it appears that there must be major vertical flow of geothermal water across the detachment fault implying some sort of structural break across the detachment.

The style of deep accomodation for brittle, upper crust extension also has geothermal implications. Thermo-mechanical modelling of Lachenbruch and Sass (1978) demonstrated that the observed heat flux and extension rates in the Great Basin are consistent with either stretching or intrusive dilation of the lower crust. Localized intrusion beneath basins to accomodate extension produced by intersecting, approximately planar graben faults could provide a method of

elevating geotherms to structural levels where they can be reached by circulating groundwater. Obviously, intrusion along the deeper portions of major basin-bounding faults, if common, could result in a significant geothermal potential at shallow depth throughout the region. The relatively small volumes of young (less than 6 my BP; cf. Stewart and Carlson, 1976) volcanism in the interior of the Great Basin have been cited as evidence against the widespread applicability of intrusion into basin-bounding faults. Anderson and others (1983) presented additional arguments suggesting that while intrusion beneath basins may exist locally, it probably isn't a province-wide phenomena.

SUMMARY

On the basis of available seismic reflection data, three main modes of normal faulting and basin development can be defined for the Great Basin region. Figure 15 summarizes basin structure and fault style for each mode along with salient features possibly related to geothermal potential. Lower crustal styles of extension are also shown. Faults in basins formed by each mode are known to be active as evidenced by historic, Holocene or latest Pleistocene surface rupture (Anderson and others, 1983).

Potentially the most important geothermal resource type in the Great Basin province is a convective system in which cold meteoric water decends along faults and fractures in an area of high thermal gradient, is heated, and convects upward through other faults and fractures. From the viewpoint of resource potential, it is critical to identify the master fault zones, which penetrate deeply and control circulation. The distributed zones of synthetic and antithetic minor faulting shown in modes A and B probably only modify or complicate the picture. The close association in western Nevada of belts of major geothermal systems, high heat flow, dense Quaternary faulting, and seismicity indicate the critical role that active faulting and fracturing play in the development and maintenance of convective geothermal systems. Quaternary faulting is also prevalent in the central Utah geothermal systems; however, in contrast to western Nevada, most of the Utah geothermal areas are related to late Quaternary volcanic features.

Low-angle normal faults (detachment faults) and fractures related to these faults are important structural features of many of the Utah geothermal areas. Both geothermal water and magma transport across these shallowly- dipping detachments imply some structural continuity between the upper and lower plates. The detachment faults do not appear to act as any sort of seal for the geothermal systems and must be broken in places. The possible role of detachment faults in providing significant lateral permeability remains to be investigated.

ACKNOWLEDGEMENTS

Discussions with J. H. Stewart on the nature of basin-range faulting (as well as his review of this manuscript) were instrumental in formulating some of the ideas presented here and are greatly appreciated, as are the numerous figures which he provided. Special thanks to D. A. Okaya and H. M. Iyer for helpful reviews of earlier versions of the manuscript and to H. A. Ackermann for providing the seismic reflection data from Raft River Valley, Idaho.

Large size paper copies of the Grass Valley seismic reflection data are available (for a minor reproduction charge) through the University of Utah Research Institute, 420 Chipeta Way, Salt Lake City, Utah, file number NV/LCH/AMN-6.

REFERENCES

Ackermann, H. D., 1979, Seismic refraction study of the Raft River geothermal area, Idaho: Geophysics, v. 44, p. 216-225.

Allmendinger, R. W., Sharp, J. W., Von Tish, D., Serpa, L., Brown, L., Kaufman, S., Oliver, J., and Smith, R. B., 1983, Cenozoic and Mesozoic structure of the eastern Basin and Range, COCORP seismic reflection data: Geology, (in press).

Anderson, R. E., Zoback, M. L., and Thompson, G., 1983, Implications of selected subsurface data on the structural form and evolution of some basins in the northern Basin and Range province: Geol. Soc. America Bull., (in press).

Arabasz, W. J., 1981, Seismicity and listric faulting in central and southwest Utah: EOS, (Trans. Amer. Geophys. Union), v. 62, p. 960-961.

Arabasz, W. J., Richins, W. D., and Langer, C. J., 1981, The Pocatello Valley (Idaho-Utah border) earthquake sequence of March to April 1975: Seis. Soc. America Bull., v. 71, p. 803-826.

Arabasz, W. J., Smith, R. B., and Richins, W. D., 1980, Earthquake studies along the Wasatch front, Utah: Network monitoring, seismicity, and seismic hazards: Seis. Soc. America Bull., v. 70, p. 1479-1500.

Bruhn, R. L., Yusas, M. R., and Huertas, F., 1982, Mechanics of low-angle normal faulting: an example from Roosevelt Hot Springs geothermal area, Utah: Tectonophysics, v. 86, p. 343-361.

Cape, C. D., McGeary, S., Bracken, R. E., Gagnon, L. D., and Thompson, G. A., 1983, Cenozoic normal faulting and the shallow structure of the Rio Grande Rift near Socorro, New Mexico: Geol. Soc. America Bull., v. 94, p. 3-14.

Cloos, E., 1968, Experimental analysis of Gulf Coast fracture patterns: Amer. Assoc. of Petrol. Geol. Bull., p. 420-444.

Covington, H. R., 1980, Subsurface geology of the Raft River geothermal area, Idaho: Geothermal Resources Council, Trans., v. 4, p. 113-115.

Covington, H. R., 1983, Structural evolution of the Raft River Basin, Idaho: Geol. Soc. America Memoir 157, p. 229-237.

Eaton, G. P., 1980, Geophysical and geological characteristics of the crust in the Basin and Range province: in Burchfiel, B. C., Oliver, J. E., and Silver, L. T., eds., Continental Tectonics: National Research Council, Washington, D.C., p. 96-110.

Eaton, G. P., 1982, The Basin and Range province, origin and tectonic significance: Ann. Rev. Earth Planet. Sci., v. 10, p. 409-410.

Effimoff, I., and Pinezich, A. R., 1981, Tertiary structural development of selected valleys based on seismic data, Basin and Range Province, northeastern Nevada: in Vine, F. J., and Smith, A. G., eds., Extensional tectonics associated with convergent plate boundaries: Phil. Trans. Roy. Soc. London, A, v. 300, p. 217-230.

Goldstein, N. E. and Paulson, B., 1978, Interpretation of gravity surveys in Grass and Buena Vista Valleys, Nevada: Geothermics, v. 7, no. 1, p. 29-50.

Gross, W. W., and Hillemeyer, F. L., Geometric analysis of upper plate fault patterns in the Whipple-Buckskin detachment terrane, California and Arizona: in Frost, E. G., and Martin, D. L., eds., Mesozoic-Cenozoic tectonic evolution of the Colorado River region, California, Arizona, and Nevada: Cordilleran Publishers, San Diego, CA, p. 256-266.

Guth, L. R., Bruhn, R. L., and Beck, S. L., 1981, Fault and joint geometry at Raft River geothermal area, Idaho: Tech. report, DOE/DGE contract DOE-AC07-80ID12079, Univ. of Utah, 19 p.

Hastings, D. D., 1979, Results of exploratory drilling northern Fallon Basin, western Nevada: in Newman, G. W., and Goode, H. D., eds., 1979 Basin and Range symposium: Rocky Mountain Assoc. of Geologists and Utah Geol. Assoc., p. 515-522.

Hoover, J. D., 1974, Periodic Quarternary volcanism in the Black Rock Desert, Utah: Brigham Young Univ. Geol. Studies, v. 21, p. 3-72.

Lachenbruch, A. H., and Sass, J. H., 1978, Models of an extending lithosphere and heat flow in the Basin and Range Province: Geol. Soc.America Memoir 152, p. 209-250.

Mabey, D. R., Hoover, D. B., O'Donnell, J. E., and Wilson, C. W., 1978, Reconnaissance geophysical studies of the geothermal system in southern Raft River Valley, Idaho: Geopysics, v. 43, p. 1470-1484.

Majer, E. L., 1978, Seismological investigations in geothermal regions; Lawrence Berkeley Laboratory Report LBL-7054, 225 p.

McDonald, R. E., 1976, Tertiary tectonics and sedimentary rocks along the transition: Basin and Range Province to Plateau and Thrust Belt Province, Utah: in Hill, J. G., ed., RMAG Symp. Geol. Cordilleran Hingeline: Rocky Mountain Assoc. Geol., p. 281-317.

Miller, E. L., Gans, P. B., and Garing, J. E., 1983, The Snake Range Decollement: an exhumed mid-Tertiary ductile-brittle transition: Tectonics, v. 2, p. 239-263.

Morton, W. H., and Black, R., 1975, Crustal attenuation in Afar: in Pilger, A., and Rossler, A., eds., Afar Depression of Ethiopia: Deutsche Forschungsgemeinschaft, Stuttgart, p. 55-65.

Muffler, L. J. P., ed., 1979, Geothermal energy in the western United States, Map 1: in Assessment of geothermal resources of the United States--1978: U. S. Geological Survey Circular 790, 163 p.

Nakata, J. K., Wentworth, C. M., Machette, M. N., 1982, Quaternary fault map of the Basin and Range and Rio Grande Provinces, western United States: U. S. Geological Survey Open-File Report 82-579.

Okaya, D. A., and Thompson, G. A., 1983, Geometry of Cenozoic extensional faulting, Dixie Valley, Nevada: Tectonics, in press.

Robinson, R., and Iyer, H. M., 1981, Delineation of a low velocity body under the Roosevelt Hot Springs geothermal area, Utah, using teleseismic P-wave data: Geophysics, v. 46, p. 1456-1466.

Sass, J. H., Lachenbruch, A. H., Munroe, R. J., Greene, G. W., and Moses, T. H., Jr., 1971, Heat flow in the western United States: Jour. of Geophys. Res., v. 76, p. 6376-6413.

Sass, J. H., Blackwell, D. D., Chapman, D. S., Costain, J. K., Decker, E. R., Lawver, L. A., and Swanberg, C. A., 1981, Heat Flow from the crust of the United States, in Touloukian, Y. S., Judd, W. R., and Roy, R. F., eds., Physical Properties of Rocks and Minerals: McGraw-Hill Book Co., p. 503-548.

Serpa, L. F., and Cook, K. L., 1980, Detailed gravity and areomagnetic surveys in the Black Rock desert area, Utah: Dept. of Energy, Topical Report IDO-78-1701.a.5.3, DOE/ET28392-39, 210 p.

Smith, R. B., 1978, Seismicity, crustal structure, and intraplate tectonics of the interior of the western Cordillera: in Smith, R. B., and Eaton, G. P., eds., Cenozoic tectonics and regional geophysics of the western Cordillera: Geol. Soc. America Memoir 152, p. 111-144.

Smith, R. B., 1981, Listric faults and earthquakes, what evidence?: EOS, (Trans. Amer. Geophys. Union), v. 62, p. 961.

Smith, R. B., 1983, Cenozoic tectonics of the eastern Basin-Range: Inferences on the origin and mechanism from seismic reflection and earthquake data: Geol. Soc. America, Abs. with Programs, v. 15, no. 5, p. 287.

Smith, R. B., and Bauer, M. S., 1983, Elastic strain release in The Basin-Range and southern California: Inferences from seismicity and geologically derived moment rates: Earthquake Notes, v. 54, no. 1, p. 45.

Stewart, J. H., 1971, Basin and Range structure: a system of horsts and grabens produced by deep-seated extension: Geol. Soc. America Bull., v. 92, p. 1019-1044.

Stewart, J. H., 1978, Basin and Range structure in western North America--A review: in Cenozoic tectonics and regional geophysics of the western Cordillera: Geol. Soc. America Memoir 152, p. 1-31.

Stewart, J. H., 1980, Regional tilt patterns of late Cenozoic basin-range fault blocks, western United States: Geol. Soc. America Bulletin, v. 91, p. 460-464.

Stewart, J. H., 1983, Cenozoic structure and tectonics of the northern Basin and Range Province, California, Nevada, and Utah: Geothermal Resources Council Trans. (this volume).

Stewart, J. H., and Carlson, J. E., 1976, Cenozoic rocks of Nevada: Nevada Bur. Mines and Geology, Map 52 Vetter, U. R., and Ryall, A. S., 1983, Systematic change of focal mechanism with depth in western Great Basin: Jour. Geophys. Res., (in press).

Wallace, R. E., 1979, Strain pattern represented by scarps formed during the earthquakes of October 2, 1915, Pleasant Valley, Nevada: Tectonophysics, v. 52, p. 599.

Wallace, R. E., 1980a, Map of fault scarps formed during earthquake of October 2, 1915, Pleasant Valley, Nevada and other young fault scarps; U. S. Geological Survey Open File Rept. of 80-608.

Wallace, R. E., 1980b, Listric faulting and seismicity, northern Nevada: Earthquake Notes, v. 50, no. 4, p. 67-68.

Ward, S. H., Parry, W. T., Nash, W. P., Sill, W.
R., Cook, K. L., Smith, R. B., Chapman, D.
S., Brown, F. H., Whelon, J. A., and Bowman,
J. R., 1978, A summary of the geology,
geochemistry, and geophysics of the Roosevelt
Hot Springs thermal area, Utah: Geophysics,
v. 43, p. 1515-1542.

Wernicke, B., 1981, Low-angle normal faults in
the Basin and Range province: nappe
tectonics in an extending origin: Nature, v.
291, p. 645-648.

Wernicke, B., and Burchfiel, B. C., 1982, Modes
of extensional tectonics: Jour. Struct.
Geology, v. 4, p. 105-115.

Zoback, M. L., 1979, A geologic and geophysical
investigation of the Beowawe geothermal area,
north-central Nevada: Stanford Univ. Publ.
Geol. Sci., v. 16, 76 p.

Zoback, M. L., 1982, Interpretation of a seismic
reflection profile across the Wasatch fault
zone, central Utah (abs.): EOS, Trans.
American Geophys. Union, v. 63, p. 1033.

Zoback, M. L., 1983, Structure and Cenozoic
tectonism along the Wasatch fault zone,
Utah: GSA Memoir 157, p. 3-27.

LIST OF AUTHORS